WELDING HANDBOOK

Eighth Edition

Volume 3

MATERIALS AND APPLICATIONS – PART 1

The Four Volumes of the Welding Handbook, Eighth Edition

1) WELDING TECHNOLOGY

2) WELDING PROCESSES

3) MATERIALS AND APPLICATIONS – PART 1

4) MATERIALS AND APPLICATIONS – PART 2

WELDING HANDBOOK

Eighth Edition

Volume 3

MATERIALS AND APPLICATIONS PART 1

William R. Oates
Editor

AMERICAN WELDING SOCIETY
550 N.W. LEJEUNE ROAD
MIAMI, FL 33126

Library of Congress Number: 96-083169
International Standard Book Number: 0-87171-470-1

American Welding Society, 550 N.W. LeJeune Road, Miami, FL 33126

THE WELDING HANDBOOK is a collective effort of many volunteer technical specialists to provide information to assist with the design and application of welding and allied processes.

Reasonable care is taken in the compilation and publication of the Welding Handbook to insure authenticity of the contents. However, no representation is made as to the accuracy or reliability of this information, and an independent, substantiating investigation should be undertaken by any user.

The information contained in the Welding Handbook shall not be construed as a grant of any right of manufacture, sale, use, or reproduction in connection with any method, process, apparatus, product, composition, or system, which is covered by patent, copyright, or trademark. Also, it shall not be construed as a defense against any liability for such infringement. Whether or not use of any information in the Handbook would result in an infringement of any patent, copyright, or trademark is a determination to be made by the user.

Printed in the United States of America

CONTENTS

WELDING HANDBOOK COMMITTEE

May 31, 1992

C. W. Case, Chairman	Inco Alloys International
D. R. Amos, 1st Vice Chairman	Westinghouse Turbine Plant
B. R. Somers, 2nd Vice Chairman	Consultant
W. R. Oates, Secretary	American Welding Society
J. R. Condra	E. I. duPont de Nemours and Company
J. G. Feldstein	Teledyne McKay
J. M. Gerken	Consultant
K. F. Graff	Edison Welding Institute
L. Heckendorn	Consultant
J. C. Papritan	Ohio State University
P. I. Temple	Detroit Edison
M. J. Tomsic	Plastronic, Incorporated
R. M. Walkosak	Westinghouse Electric Corporation

WELDING HANDBOOK COMMITTEE

May 31, 1993

B. R. Somers, Chairman	Lehigh University
W. R. Oates, Secretary	American Welding Society
H. R. Castner	Edison Welding Institute
J. G. Feldstein	Foster Wheeler Energy Corporation
J. M. Gerken	Consultant
L. Heckendorn	Consultant
S. D. Kiser	Inco Alloys International
J. C. Papritan	Ohio State University
P. I. Temple	Detroit Edison
M. J. Tomsic	Plastronic, Incorporated
R. M. Walkosak	Westinghouse Electric Corporation

WELDING HANDBOOK COMMITTEE

May 31, 1994

B. R. Somers, Chairman	Lehigh University
P. I. Temple, 1st Vice Chairman	Detroit Edison
R. M. Walkosak, 2nd Vice Chairman	Westinghouse Electric Corporation
W. R. Oates, Secretary	American Welding Society
H. R. Castner	Edison Welding Institute
J. G. Feldstein	Foster Wheeler Energy Corporation
J. M. Gerken	Consultant
L. Heckendorn	Intech R&D, USA
S. D. Kiser	Inco Alloys International
J. C. Papritan	Ohio State University
M. J. Tomsic	Plastronic, Incorporated

WELDING HANDBOOK COMMITTEE

May 31, 1995

B. R. Somers, Chairman	Lehigh University
P. I. Temple, 1st Vice Chairman	Detroit Edison
R. M. Walkosak, 2nd Vice Chairman	Westinghouse Electric Corporation
W. R. Oates, Secretary	American Welding Society
B. J. Bastian	Benmar Associates
H. R. Castner	Edison Welding Institute
D. J. Corrigall	Miller Electric Company
J. G. Feldstein	Foster Wheeler Energy Corporation
J. M. Gerken	Consultant
L. Heckendorn	Intech R&D, USA
S. D. Kiser	Inco Alloys International
J. C. Papritan	Ohio State University
D. V. Rypien	American Bureau of Shipping
R. P. Schuster	Harnischfeger Industries, Inc.
M. J. Tomsic	Plastronic, Incorporated
C. L. Tsai	Ohio State University

PREFACE

This is Volume Three of the Eighth Edition of the *Welding Handbook*. The title of this third volume, Materials and Applications—Part 1, suggests correctly that the Eighth Edition will have a fourth volume, Materials and Applications—Part 2. Volume Three begins the process of updating and substantially expanding the material previously presented in Volume Four of the Seventh Edition and Volume 5 of the Sixth Edition.

Originally, the Welding Handbook Committee had planned to publish the Materials and Applications title as a single publication. However, as the Committee began to compile the text, it became apparent that the material was too voluminous to be contained in a single text. Therefore, it was necessary to separate Materials and Applications into two volumes. Unfortunately, this decision was made after the first two volumes of the Eighth Edition had been published, and so the front matter in Volumes One and Two indicate Volume Three to be the Materials and Applications material in its entirety.

Volume Three, Materials and Applications—Part 1, covers many nonferrous metals, plastics, composites, ceramics, and new specialized topics: maintenance and repair welding and underwater welding and cutting. Volume Four, Materials and Applications—Part 2, will contain information mostly on ferrous materials, but it also will cover titanium, clad and dissimilar metals, tube-to-tubesheet welding, and the reactive, refractory, and precious metals and alloys. Volume Four, Materials and Applications—Part 2, is scheduled to be published two years following the publication of Volume Three, Materials and Applications—Part 1.

This volume represents a considerable expansion of the information on these topics contained in previous editions. As with the first two volumes of this Eighth Edition, practical application data and color illustrations have been added.

This volume, like the others, was a voluntary effort by the Welding Handbook Committee and the Chapter Committees. The Chapter Committee Members and the Welding Handbook Committee Member responsible for each chapter are recognized on the title page of that chapter. Other individuals also contributed in a variety of ways, particularly in chapter reviews. All participants contributed generously of their time and talent, and the American Welding Society expresses its sincere appreciation to them and to their employers for supporting this work.

The Welding Handbook Committee expresses its appreciation to the AWS staff members who assisted with this volume, especially the Technical Division staff engineers for technical assistance, Deborah Givens for editorial assistance, and Doreen Kubish for editorial and production assistance with text, graphics, and layout.

The Welding Handbook Committee welcomes your comments on the *Welding Handbook* as well as your potential interest in contributing to future volumes. Communications may be addressed to the Editor, Welding Handbook, American Welding Society, 550 N. W. LeJeune Road, Miami, Florida, 33126.

B. R. Somers, Chairman
Welding Handbook Committee
1992-1996

W. R. Oates, Editor
Welding Handbook

CHAPTER 1

ALUMINUM AND ALUMINUM ALLOYS

PREPARED BY A COMMITTEE CONSISTING OF:

W. H. Kielhorn, Chairman
LeTourneau University

F. Armao
Aluminum Company of America

H. A. Chambers
TRW Nelson Stud Welding Division

P. Dent
Grumman Aircraft Systems

P. B. Dickerson
Consultant

R. Donnelly
Vertex Communications

S. E. Gingrich
Grove Worldwide

D. R. Hill
Alumatech

L. S. Kramer
Martin Marietta

J. Murphy
Consultant

J. R. Pickens
Martin Marietta

J. Schuster
Omni Technologies Corporation

J. Stokes
Consultant

A. Szabo
Martin Marietta

WELDING HANDBOOK COMMITTEE MEMBER:
P. I. Temple
Detroit Edison Company

ALUMINUM AND ALUMINUM ALLOYS

INTRODUCTION

ALUMINUM IS JOINED with most of the known joining processes. It is readily joined by welding, brazing, soldering, adhesive bonding, or mechanical fastening. In many instances, aluminum is joined by the conventional equipment and techniques used with other metals. Occasionally, specialized equipment or techniques, or both, are required. The alloy, joint configuration, strength requirement, appearance, and cost are factors that dictate the choice of joining process. Each process for joining aluminum has certain advantages and limitations.

GENERAL CHARACTERISTICS OF ALUMINUM

ALUMINUM IS LIGHT in weight, yet some of its alloys have strengths exceeding mild steel. It retains good ductility at subzero temperatures, has high resistance to corrosion, and is not toxic. Aluminum has good electrical and thermal conductivity as well as high reflectivity to both heat and light. It is nonsparking and nonmagnetic.

Aluminum is easy to fabricate. It can be cast, rolled, stamped, drawn, spun, stretched, or roll-formed. It can also be hammered, forged, or extruded into a wide variety of shapes. Machining ease and speed are important factors in using aluminum parts. Aluminum can be given a wide variety of mechanical, electromechanical, chemical, or paint finishes.

Pure aluminum melts at 1220 °F (660 °C). Aluminum alloys have an approximate melting range of from 900 to 1220 °F (482 to 660 °C), depending upon the alloy. There is no color change in aluminum when heated to the welding or brazing temperature range. This necessitates a welder assuming a position where the melting of the base and filler metals under the arc or flame can be witnessed.

High thermal conductivity (as compared to steel) necessitates a high rate of heat input for fusion welding. Thick sections may require preheating. When resistance spot welding, aluminum's high thermal and electrical conductivities require higher current, shorter weld time, and more precise control of the welding variables than when welding steel.

Since aluminum is nonmagnetic and no arc blow is experienced when welding with direct current, aluminum is being used for backing and fixtures in welding.

Aluminum and its alloys rapidly develop a tenacious, refractory oxide film when exposed to air. This natural oxide film can be removed using either a protective atmosphere or fluxes suitable during arc welding, brazing, or soldering. Exposure to the elevated temperatures of thermal treatments or exposure to moist environments cause the aluminum oxide film to markedly thicken, necessitating removal prior to joining.

Heat-treated aluminum can have an exceptionally thick aluminum oxide film that causes poor wetting and flow of filler metal.[1] Aluminum oxide melts at about 3700 °F (2037 °C) and should be removed by chemical or mechanical means prior to welding. The composition and melting range of an aluminum alloy are the primary considerations for selection of the joining process.

Anodic electrolytic treatments applied to aluminum result in forming thick, dense oxide coatings that must be removed prior to fusion welding, brazing or soldering.[2] The anodic coatings can resist 400 volts or more, so welding arcs cannot be initiated. This oxide must be removed not only from the joint but also at the location of the work lead when arc welding.

The properties and performances of aluminum parts are influenced by microstructural changes that occur during any elevated-temperature joining process. The strength, fatigue life, ductility, and formability originally in the parts are changed depending upon the amount of annealing, overaging, and formation of cast structure occurring during the joining process. The results of these changes are presented in subsequent sections of this chapter devoted to specific joining processes.

ALUMINUM FORMS

PURE ALUMINUM IS readily alloyed with many other metals to produce a wide range of physical and mechanical properties. Table 1.1 lists the major alloying elements in the wrought aluminum alloys. The means by which the alloying elements strengthen aluminum are used as the basis to classify aluminum alloys into two categories: nonheat-treatable and heat-treatable. Wrought aluminum alloys have similar joining characteristics regardless of product form. Cast aluminum alloys are produced in the form of sand, permanent mold, and die castings. Substantially, the same welding, brazing, and soldering practices are used on both cast and wrought metal.

Conventional die castings are not recommended where welded or brazed construction is required. They can be adhesively bonded and, to a limited degree, soldered. New developments in vacuum die casting have improved the quality of castings to the point where some may be satisfactorily welded.

---

1. The term *film* is used to describe the naturally occurring oxide film on aluminum due to the reaction with oxygen in air.
2. The term *coating* is used to describe a deliberate chemical buildup of aluminum oxide on aluminum surfaces that results in a very thick, dense oxide layer on the aluminum surface.

Table 1.1
Designations for Wrought Alloy Groups

A system of four-digit numerical designations is used to identify wrought aluminum alloys. The first digit indicates the alloy group as follows:

Aluminum, 99.0% and greater	1XXX*
Major Alloying Element:	
Copper	2XXX
Manganese	3XXX
Silicon	4XXX
Magnesium	5XXX
Magnesium and Silicon	6XXX
Zinc	7XXX
Other elements	8XXX
Unused series	9XXX

* For 1XXX series, the last two digits indicate the minimum aluminum purity (e.g., 1060 is 99.60% Al minimum). The second digit in all groups indicates consecutive modifications of an original alloy, such as 5154, 5254, 5454, and 5654 alloys.

To increase corrosion resistance, some alloys are clad with high-purity aluminum or a special aluminum alloy. The cladding, usually from 2 1/2 to 15 percent of the total thickness on one or both sides, not only protects the composite due to its inherent corrosion resistance, but generally exerts a galvanic effect which further protects the core material. Special composites also are produced for brazing, soldering, and finishing purposes.

WROUGHT ALLOYS

Nonheat-Treatable Aluminum Alloys

THE INITIAL STRENGTH of the nonheat-treatable aluminum alloys depends primarily upon the hardening effect of alloying elements such as silicon, iron, manganese, and magnesium. These elements affect increases in strength either as dispersed phases or by solid-solution strengthening. The nonheat-treatable alloys are mainly found in the 1XXX, 3XXX, 4XXX, 5XXX alloy series (Table 1.2) depending upon their major alloying elements.

Iron and silicon are the major impurities in commercially pure aluminum, but they add strength to these 1XXX series alloys. Silicon is the major element in many welding and brazing filler alloys. Magnesium is the most effective solution-strengthening element in the nonheat-treatable alloys. Aluminum-magnesium alloys of the 5XXX series have relatively high strength in the annealed condition. The strength of all of the nonheat-treatable alloys may be improved by strain hardening (sometimes called *work hardening* or *cold working*.)

Table 1.2
Compositions and Applications of Nonheat-Treatable Wrought Alloys

Aluminum Association Designation	Nominal Composition (% Alloying Element)				Typical Applications
	Cu	Mn	Mg	Cr	
1060	99.60% minimum aluminum				Chemical process equipment, tanks, piping.
1100	99.00% minimum aluminum				Architectural and decorative applications, furniture, deep drawn parts, spun hollow ware.
1350	99.50% minimum aluminum				Electrical conductor wire, bus and cable.
3003	0.12	1.2	—	—	General purpose applications where slightly higher strength than 1100 is required. Process and food handling equipment, chemical and petroleum drums and tanks.
3004	—	1.2	1.0	—	Sheet metal requiring higher strength than 3003.
5005	—	—	0.8	—	Electrical conductor and architectural applications.
5050	—	—	1.4	—	Similar to 3003 and 5005 but stronger. Has excellent finishing qualities.
5052, 5652	—	—	2.5	—	Sheet metal applications requiring higher strength than 5050. Formable and good corrosion resistance. Storage tanks, boats, appliances. Alloy 5652 has closer control of impurities for H_2O_2 service.
5083	—	0.7	4.4	0.15	Marine components, tanks, unfired pressure vessels, cryogenics structures, railroad cars, drilling rigs.
5086	—	0.45	4.0	0.15	Marine components, tanks, tankers, truck frames.
5154, 5254	—	—	3.5	0.25	Unfired pressure vessels, tankers. Alloy 5254 has closer control of impurities for H_2O_2 service.
5454	—	0.8	2.7	0.12	Structural applications and tanks for sustained high-temperature service.
5456	—	0.8	5.1	0.12	Structures, tanks, unfired pressure vessels, marine components.

To remove the effects of strain hardening and improve ductility, the nonheat-treatable alloys may be annealed by heating to a uniform temperature in the range of 650 to 775 °F (343 to 410 °C). The exact annealing schedule will depend upon the alloy and its temper. Although the rate of cooling from the annealing temperature is unimportant, fixturing may be required to prevent distortion or warping. Basic temper designations applicable to the nonheat-treatable alloys are indicated in Table 1.3.

When fusion welded, the nonheat-treatable alloys lose the effects of strain hardening in a narrow heat-affected zone (HAZ) adjacent to the weld; the strength in the HAZ will approach that of the annealed condition. Table 1.4 contains information on the relative weldability and properties of common nonheat-treatable alloys in various tempers.

Heat-Treatable Aluminum Alloys

THE INITIAL STRENGTH of aluminum alloys in this group depends upon the alloy composition, just as in the nonheat-treatable alloys. Since elements such as copper, magnesium, zinc, and silicon, either singularly or in various combinations, show a marked increase in solid solubility in aluminum with increasing temperature, it is possible to subject them to thermal treatments that will impart pronounced strengthening. Heat-treatable aluminum alloys develop their properties by solution heat treating and quenching, followed by either natural or artificial aging. Cold working may add additional strength. The heat-treatable alloys may also be annealed to attain maximum ductility. This treatment involves holding at an elevated temperature and controlled cooling to achieve maximum softening. Basic

Table 1.3
Basic Temper Designations Applicable to the Nonheat-Treatable Aluminum Alloys

Designation*	Description	Application
-0	Annealed, recrystallized	Applies to wrought products which are annealed to obtain the lowest strength temper, and to cast products which are annealed to improve ductility and dimensional stability.
-F	As fabricated	Applies to products of shaping processes in which no special control over thermal conditions or strain hardening is employed. For wrought products, there are no mechanical property limits.
-H1	Strain hardened only	Applies to products which are strain hardened to obtain the desired strength without supplementary thermal treatment. The number following this designation indicates the degree of strain hardening.
-H2	Strain hardened and then partially annealed	Applies to products which are strain hardened more than the desired final amount and then reduced in strength to the desired level by partially annealing. For alloys that age soften at room temperature, the H2 tempers have the same minimum ultimate tensile strength as the corresponding H3 tempers. For other alloys, the H2 tempers have the same minimum ultimate tensile strength as the corresponding H1 tempers and slightly higher elongation. The number following this designation indicates the degree of strain hardening remaining after the product has been partially annealed.
-H3	Strain hardened and then stabilized	Applies to products which are strain hardened and whose mechanical properties are stabilized by a low-temperature thermal treatment which results in slightly lowered tensile strength and improved ductility. This designation is applicable only to those alloys which, unless stabilized, gradually age soften at room temperature. The number following this designation indicates the degree of strain hardening before the stabilization treatment.

* The digit following the designation H1, H2, and H3 indicates the degree of strain hardening. Numeral 8 has been assigned to indicate tempers having an ultimate tensile strength equivalent to that achieved by a cold reduction of approximately 75% following a full anneal. Tempers between 0 (annealed) and 8 are designated by numerals 1 through 7. Material having an ultimate tensile strength about midway between that of the 0 temper and that of the 8 temper is designated by the numeral 4; about midway between the 0 and 4 tempers by the numeral 2; and about midway between the 4 and 8 tempers by the numeral 6. Numeral 9 designates tempers whose minimum tensile strength exceeds that of the 8 temper by 2.0 ksi (137.9 MPa) or more. For two-digit H tempers whose second digit is odd, the standard limits for ultimate tensile strength are exactly midway between those of the adjacent two-digit H tempers whose second digits are even. The third digit, when used, indicates a variation of a two-digit temper. It is used when the degree of control of temper or the mechanical properties are different from but close to those for the two-digit H temper designation to which it is added, or when some other characteristic is significantly affected.

temper designations applicable to the heat-treatable alloys are indicated in Table 1.5.

The heat-treatable alloys are found primarily in the 2XXX, 6XXX, and 7XXX alloy series, although some 4XXX series alloys are heat treatable depending on the combination of elements. Some of the widely used heat-treatable alloys are listed in Table 1.6 with their nominal composition and general application. Tables 1.7A , 1.7B, and 1.7C show the relative joining characteristics as well as the properties of common heat-treatable wrought alloys in their usual tempers.

Table 1.4A
Weldability and Properties of Nonheat-Treatable Wrought Aluminum Alloys
WELDABILITY[a,b]

Aluminum Alloy	Oxyfuel Gas	Arc with Flux	Arc with Inert Gas	Resistance	Pressure	Brazing	Soldering with Flux
1060	A	A	A	B	A	A	A
1100	A	A	A	A	A	A	A
1350	A	A	A	B	A	A	A
3003	A	A	A	A	A	A	A
3004	B	A	A	A	B	B	B
5005	A	A	A	A	A	B	B
5050	A	A	A	A	A	B	B
5052, 5652	A	A	A	A	B	C	C
5083	C	C	A	A	C	X	X
5086	C	C	A	A	B	X	X
5154, 5254	B	B	A	A	B	X	X
5454	B	B	A	A	B	X	X
5456	C	C	A	A	C	X	X

a. Weldability ratings are based on the most weldable temper:
   A. Readily weldable.
   B. Weldable in most applications; may require special technique or preliminary trials to establish welding procedures, performance, or both.
   C. Limited weldability.
   X. Particular joining method is not recommended.

b. All alloys can be adhesive bonded, ultrasonically welded, or mechanically fastened.

Table 1.4B
Weldability and Properties of Nonheat-Treatable Wrought Aluminum Alloys
PHYSICAL PROPERTIES

Aluminum Alloy	Temper	Density		Approximate Melting Range		Thermal Conductivity [a]		Electrical
		lb/in.^{3}	kg/m^3	°F	°C	BTU/(h•ft•°F)	W/(m•K)	Conductivity (% IACS)[b]
1060	-0	0.0975	2699	1195-1215	646-657	135	234	62
	-H18					133	231	61
1100	-0	0.098	2713	1190-1215	643-657	128	222	59
	-H18					126	218	57
1350	-0	0.0975	2699	1195-1215	646-657	135	234	62
	-H19					135	234	62
3003	-0	0.099	2740	1190-1210	643-654	112	193	50
	-H14					92	158	41
	-H18					89	154	40
3004	-0	0.098	2713	1165-1210	629-654	94	163	42
	-H34					94	163	42
	-H38					94	163	42
5005	-0	0.098	1713	1170-1210	632-654	116	200	52
	-H34					116	200	52
	-H38					116	200	52
5050	-0	0.097	2685	1155-1205	624-652	112	193	50
	-H34					112	193	50
	-H38					112	193	50
5052, 5652	-0	0.097	2685	1125-1200	607-649	80	138	35
	-H34					80	138	35
	-H38					80	138	35
5083	-0	0.096	2657	1065-1180	574-638	68	117	29
5086	-0	0.096	2657	1085-1185	585-641	72	125	31
5154, 5254	-0	0.096	2657	1100-1190	593-643	72	125	32
	-H12					72	125	32
	-H34					72	125	32
	-H38					72	125	32
5454	-0	0.097	2685	1115-1195	602-646	77	134	34
	-H32					77	134	34
	-H34					77	134	34
5456	-0	0.096	2657	1055-1180	568-638	68	117	29

a. Thermal conductivity at 77 °F (25 °C).

b. Percentage of International Annealed Copper Standard (IACS) value for Volume Electrical Conductivity, which equals 100 percent at 68 °F (20 °C).

Table 1.4C
Weldability and Properties of Nonheat-Treatable Wrought Aluminum Alloys
TYPICAL MECHANICAL PROPERTIES

Aluminum Alloy	Temper	Ultimate Tensile Strength		Yield Strength (0.2% Offset)		Elongation % in 2 in. (50.8 mm)		Shear Strength		Fatigue Strength[a]		Brinnell Hardness[b]
						1/16 in. (1.6 mm) Sheet	1/2 in. (12.7 mm) Round					
		ksi	MPa	ksi	MPa			ksi	MPa	ksi	MPa	(500 kg load)
1060	-0	10	69	4	28	43	—	7	48	3	21	19
	-H18	19	131	18	124	6	—	11	76	7	45	35
1100	-0	13	90	5	34	35	45	9	62	5	34	23
	-H18	24	166	22	152	5	15	13	90	9	62	44
1350	-0	12	83	4	28	—	23	8	55	—	—	—
	-H19	27	186	24	166	—	1.5	15	103	7	48	—
3003	-0	16	110	6	41	30	40	11	76	7	48	28
	-H14	22	152	21	145	8	16	14	96	9	62	40
	-H18	29	200	27	186	4	10	16	110	10	69	55
3004	-0	26	179	10	69	20	25	16	110	14	96	46
	-H34	35	241	29	200	9	12	18	124	15	103	63
	-H38	41	283	36	248	5	6	21	145	16	110	77
5005	-0	18	124	6	41	25	—	11	76	—	—	28
	-H34	23	159	20	138	8	—	14	96	—	—	41
	-H38	29	200	27	186	5	—	16	110	—	—	51
5050	-0	21	145	8	55	24	—	15	103	12	83	36
	-H34	28	193	24	166	8	—	18	124	13	90	53
	-H38	32	221	29	200	6	—	20	138	14	96	63
5052, 5652	-0	28	193	13	90	25	30	18	124	16	110	47
	-H34	38	262	31	214	10	14	21	145	18	124	68
	-H38	42	290	37	255	7	8	24	166	20	138	77
5083	-0	42	290	21	145	—	22	25	172	—	—	—
	-H116	46	317	33	228	—	16	—	—	23	159	—
	-H321	46	317	33	228	—	16	—	—	23	159	—
5086	-0	38	262	17	117	22	—	23	159	—	—	—
	-H116	42	290	30	207	12	—	—	—	—	—	—
	-H34	47	324	37	255	10	—	27	186	—	—	—
5154, 5254	-0	35	241	17	117	27	—	22	152	17	117	58
	-H112	35	241	17	117	25	—	—	—	17	117	63
	-H34	42	290	33	228	13	—	24	166	19	131	73
	-H38	48	331	39	269	10	—	28	193	21	145	80
5454	-0	36	248	17	117	22	—	23	159	—	—	62
	-H32	40	276	30	207	10	—	24	166	—	—	73
	-H34	44	303	35	241	10	—	26	179	—	—	81
5456	-0	45	310	23	159	—	24	—	—	—	—	—
	-H112	45	310	24	166	—	22	—	—	—	—	—
	-H116	51	352	37	255	—	16	30	207	—	—	90

a. Fatigue strength for round specimens and 500 million cycles.

b. 10-mm ball used.

Table 1.5
Basic Temper Designations Applicable to the Heat-Treatable Aluminum Alloys

Designation*	Description	Application
-0	Annealed	Applies to wrought products which are annealed to obtain the lowest strength temper, and to cast products which are annealed to improve ductility and dimensional stability. The 0 may be followed by a digit other than zero.
-F	As fabricated	Applies to products of shaping processes in which no special control over thermal conditions or strain hardening is employed. For wrought products, there are no mechanical property limits.
-W	Solution heat treated	An unstable temper applicable only to alloys which spontaneously age at room temperature after solution heat treatment. This designation is specific only when the period of natural aging is indicated, for example: W 1/2 h.
-T1		Cooled from an elevated-temperature shaping process and naturally aged to a substantially stable condition. Applies to products which are not cold worked after cooling from an elevated-temperature shaping process, or in which the effect of cold work in flattening or straightening may not be recognized in mechanical property limits.
-T2		Cooled from an elevated-temperature shaping process, cold worked, and naturally aged to a substantially stable condition. Applies to products which are cold worked to improve strength after cooling from an elevated-temperature shaping process, or in which the effect of cold work in flattening or straightening is recognized in mechanical property limits.
-T3		Solution heat treated, cold worked, and naturally aged to a substantially stable condition. Applies to products which are cold worked to improve strength after solution heat treatment, or in which the effect of cold work in flattening or straightening is recognized in mechanical property limits.
-T4		Solution heat treated and naturally aged to a substantially stable condition. Applies to products which are not cold worked after solution heat treatment, or in which the effect of cold work in flattening or straightening may not be recognized in mechanical property limits.
-T5		Cooled from an elevated-temperature shaping process and then artificially aged. Applies to products which are not cold worked after cooling from an elevated-temperature shaping process, or in which the effect of cold work in flattening or straightening may not be recognized in mechanical property limits.
-T6		Solution heat treated and stabilized. Applies to products which are not cold worked after solution heat treatment, or in which the effect of cold work in flattening or straightening may not be recognized in mechanical property limits.
-T7		Solution heat treated and stabilized. Applies to products which are stabilized after solution heat treatment to carry them beyond the point of maximum strength to provide control of some special characteristic.
-T8		Solution heat treated, cold worked, and then artificially aged. Applies to products which are cold worked to improve strength, or in which the effect of cold work in flattening or straightening is recognized in mechanical property limits.
-T9		Solution heat treated, artificially aged, and then cold worked. Applies to products which are cold worked to improve strength.
-T10		Cooled from an elevated-temperature shaping process, cold worked, and then artificially aged. Applies to products which are cold worked to improve strength, or in which the effect of cold work in flattening or straightening is recognized in mechanical property limits.

* Additional digits, the first of which shall not be zero, may be added to designation T1 through T10 to indicate a variation in treatment which significantly alters the characteristics of the product.

Table 1.6
Compositions and Application of Heat-Treatable Wrought Aluminum Alloys

Base Alloy	Nominal Composition (% Alloying Element)							Typical Applications
	Cu	Si	Mn	Mg	Zn	Ni	Cr	
2014	4.4	0.8	0.8	0.50	—	—	—	Structures, structural and hydraulic fittings, hardware, and heavy-duty forgings for aircraft or automotive uses.
2017	4.0	0.50	0.7	0.6	—	—	—	Same as 2014; screw machine parts.
2024	4.4	—	0.6	1.5	—	—	—	Structural, aircraft sheet construction, truck wheels; often clad for strength with good corrosion resistance
2036	2.6	—	0.25	0.45	—	—	—	Automotive body sheet.
2090[a]	2.7	—	—	—	—	—	—	Structural; high strength and damage tolerant aerospace applications.
2218	4.0	—	—	1.5	—	2.0	—	Forging alloy; engine cylinder heads, pistons, parts requiring good strength and hardness at elevated temperature.
2219[b]	6.3	—	0.30	—	—	—	—	Structural; high-temperature strength; aerospace tanks; good weldability.
2519[c]	5.8	—	0.30	0.17	0.06	—	—	Structural; high-strength armor.
2618[d]	2.3	0.18	—	1.6	—	1.0	—	Same as 2218.
6005	—	0.8	—	0.50	—	—	—	Structural and architectural.
6009	0.40	0.8	0.50	0.6	0.25	—	0.10	Automotive body sheet.
6010	0.40	1.0	0.50	0.8	0.25	—	0.10	Automotive body sheet.
6013	0.9	0.25	0.35	0.95	—	—	—	General structural applications, improved strength over 6061.
6061	0.25	0.6	—	1.0	—	—	0.20	Structural, architectural, automotive, railway, and marine applications; pipe and pipe fittings; good formability, weldability, corrosion resistance, strength.
6063	—	0.40	—	0.7	—	—	—	Pipe, railings, hardware, architectural applications.
6070	—	1.4	—	0.8	—	—	—	Structural applications; piping.
6101	0.50	—	—	0.6	—	—	—	Electrical conductors.
6262[e]	0.28	—	—	1.0	—	—	0.09	Screw machine products, fittings.
6351	—	1.0	0.6	0.6	—	—	—	Same as 6061.
6951	—	0.35	—	0.6	—	—	—	Brazing sheet core alloy.
7004[f]	—	—	—	1.5	4.2	—	—	Truck trailer, railcar extruded shapes.
7005[g]	—	—	0.45	1.4	4.5	—	0.13	Truck trailer, railcar extruded shapes.
7039	—	—	0.30	2.8	4.0	—	0.20	Armor plate; military bridges.
7075	1.6	—	—	2.5	5.6	—	0.23	High-strength aircraft and other applications; cladding gives good corrosion resistance.
7079	0.6	—	—	3.3	4.3	—	0.20	Strongest aluminum alloy where section thickness exceeds 3 in. (76.2 mm), large and massive parts for aircraft and allied construction.
7178	2.0	—	—	2.8	6.8	—	0.23	Aircraft construction; slightly higher strength than 7075.

a. Also 2.2 Li and 0.12 Zr
b. Also 0.06 Ti, 0.10 V and 0.18 Zr
c. Also 0.06 Ti 0.17 Zr and 0.10 Va
d. Also 1.1 Fe and 0.07 Ti

e. Also 0.6 Pb and 0.6 Bi
f. Also 0.15 Zr
g. Also 0.15 Zr and 0.035 Ti

Table 1.7A
Weldability and Properties of Heat-Treatable Wrought Aluminum Alloys
WELDABILITY[a,b]

Aluminum Alloy	Oxyfuel Gas	Arc with Flux	Arc with Inert Gas	Resistance	Pressure	Brazing	Soldering with Flux
2014	X	C	C	B	C	X	C
2017	X	C	C	B	C	X	C
2024	X	C	C	B	C	X	C
2036	X	C	B	B	C	X	C
2090	X	X	B	B	C	X	C
2218	X	C	C	B	C	X	C
2219	X	C	A	B	C	X	C
2519	X	C	B	B	C	X	C
2618	X	C	C	B	C	X	C
6005	A	A	A	A	B	A	B
6009	C	C	B	B	B	X	C
6010	C	C	B	B	B	X	C
6013	C	C	B	A	B	X	C
6061	A	A	A	A	B	A	B
6063	A	A	A	A	B	A	B
6070	C	C	B	B	B	X	C
6101	A	A	A	A	A	A	A
6262	C	C	B	A	B	B	B
6351	A	A	A	A	B	A	B
6951	A	A	A	A	A	A	A
7004	X	X	A	A	B	B	B
7005	X	X	A	A	B	B	B
7039	X	X	A	A	B	C	B
7075	X	X	C	B	C	X	C
7079	X	X	C	B	C	X	C
7178	X	X	C	B	C	X	C

a. Weldability ratings are based on the most weldable temper:
   A. Readily weldable.
   B. Weldable in most applications; may require special technique or preliminary trials to establish welding procedures, performance, or both.
   C. Limited weldability.
   X. Particular joining method is not recommended.

b. All alloys can be adhesive bonded, ultrasonically welded, or mechanically fastened.

Table 1.7B
Weldability and Properties of Heat-Treatable Wrought Aluminum Alloys
PHYSICAL PROPERTIES

Aluminum Alloy	Temper	Density lb/in.³	Density kg/m³	Approximate Melting Range °F	Approximate Melting Range °C	Thermal Conductivity [a] BTU/(ft·h·°F)	Thermal Conductivity [a] W/(m·K)	Electrical Conductivity (% IACS)[b]
2014	-0	0.101	2796	945-1180	507-638	112	193	50
	-T4					77.5	134	34
	-T6					89	154	40
2017	-0	0.101	2796	955-1185	513-641	112	193	50
	-T4					77.5	134	34
2024	-0	0.101	2796	935-1180	502-638	112	193	50
	-T3					70	121	30
	-T4					70	121	30
	-T361					70	121	30
2036	-T4	0.100	2768	1030-1200	554-649	92	159	41
2090	-T8	0.093	2574	1042-1091	561-589	51	88	17
2218	-T72	0.101	2796	945-1180	507-638	89	154	40
2219	-0	0.103	2851	1010-1190	543-643	99	172	44
	-T31					65	112	28
	-T62					70	121	30
	-T81					70	121	30
	-T87					70	121	30
2519	-T87	0.102	2823	1030-1190	554-643	76.7	133	33
2618	-T61	0.100	2768	1020-1180	549-638	86.7	150	39
6005	-T1	0.097	2685	1125-1210	607-654	104	180	47
	-T5					109	189	49
6009	-T4	0.098	2713	1040-1200	560-649	96.7	167	44
6010	-T4	0.098	2713	1040-1200	560-649	86.7	150	39
6013	-T4	0.098	2713	1075-1200	579-649	86.7	150	38
	-T6					94	163	42
6061	-0	0.098	2713	1080-1205	582-652	104	180	47
	-T4					89	154	40
	-T6					96.7	167	43
6063	-0	0.097	2685	1140-1210	616-655	126	218	58
	-T1					112	193	50
	-T5					121	209	55
	-T6					116	200	53
6070	-T6	0.098	2713	1050-1200	566-649	99	172	44
6101	-H111	0.097	2685	1150-1210	621-654	—	—	—
	-T6					126	218	57
6262	-T9	0.098	2713	1080-1205	582-652	99	172	44
6351	-T5	0.098	2713	1105-1205	596-652	102	176	46
	-T6					102	176	
6951	-0	0.098	2713	1040-1210	616-654	114	198	56
	-F6					114	198	52
7004	-T5	0.100	2768	—	—	—	—	—
	-T6					114	198	52
7005	-T53	0.100	2768	1125-1195	607-646	—	—	—
7039	-T64	0.099	2740	1070-1180	577-638	89	154	34
7075	-0	0.101	2796	890-1175	477-635	99	172	45
	-T6					75	130	33
	-T73					87	151	39
7079	-0	0.099	2740	900-1180	482-638			
	-T6					72.5	125	32
7178	-0	0.102	2823	890-1165	477-629	—		
	-T6					—	—	31

a. Thermal conductivity at 77 °F (25 °C).

b. Percentage of International Annealed Copper Standard (IACS) value for Volume Electrical Conductivity, which equals 100 percent at 68°F (20 °C).

Table 1.7C[a]
Weldability and Properties of Heat-Treatable Wrought Aluminum Alloys
TYPICAL MECHANICAL PROPERTIES

Aluminum Alloy	Temper	Ultimate Tensile Strength ksi	Ultimate Tensile Strength MPa	Yield Strength (0.2% Offset) ksi	Yield Strength (0.2% Offset) MPa	Elongation % in 2 in. (50.8 mm) 1/16 in. (1.6 mm) Sheet	Elongation % in 2 in. (50.8 mm) 1/2 in. (12.7 mm) Round	Shear Strength ksi	Shear Strength MPa	Fatigue Strength[b] ksi	Fatigue Strength[b] MPa	Brinnell Hardness[c] (500 kg load)
2014	-0	27	186	14	96	—	18	18	124	13	89.6	45
	-T4	62	428	42	290	—	20	38	262	20	138	105
	-T6	70	483	60	414	—	13	42	290	18	124	135
2017	-0	26	180	10	69	—	22	18	124	13	89.6	45
	-T4	62	428	40	276	—	22	38	262	18	124	105
2024	-0	27	186	11	76	20	22	18	124	13	89.6	45
	-T3	70	483	50	345	18	—	41	283	20	138	120
	-T4	68	469	47	324	20	19	41	283	20	138	120
	-T361	72	496	57	393	13	—	42	290	18	124	120
2036	-T4	49	338	28	193	24	—	—	—	18	124	—
2090	-T8	78	538	72	496	7.5	6	—	—	—	—	—
2218	-T72	48	331	37	255	—	11	30	207	—	—	95
2219	-0	25	172	11	76	18	—	—	—	—	—	—
	-T31	52	358	36	248	17	—	—	—	—	—	—
	-T62	60	414	42	290	10	—	—	—	15	103	—
	-T81	66	455	51	352	10	—	—	—	15	103	—
	-T87	69	476	57	393	10	—	—	—	15	103	—
2519	-T87	72	496	64	441	10	—	44	303	28	189	132
2618	-T61	64	441	54	372	—	10	38	262	18	124	115
6005	-T1	28	193	18	124	—	18	—	—	—	—	—
	-T5	44	303	39	269	—	12	26	179	—	—	—
6009	-T4	33	228	18	124	25	—	22	152	17	117	—
6010	-T4	42	290	25	172	24	—	28	193	18	124	—
6013	-T4	43	296	23	159	22	—	—	—	—	—	—
	-T6	59	407	54	372	9	—	34	234	—	—	—
6061	-0	18	124	8	55	25	30	12	83	9	62.1	30
	-T4	35	241	21	145	22	25	24	166	14	96.5	65
	-T6	45	310	40	276	12	17	30	207	14	96.5	95
6063	-0	13	90	7	48	—	—	10	69	8	55.2	25
	-T1	22	152	13	90	20	—	14	96	9	62.1	42
	-T5	27	186	21	145	12	—	17	117	10	68.9	60
	-T6	35		31		12	—	22	—	10	68.9	73
6070	-T6	55	379	51	352	10	—	34	234	14	96.5	—
6101	-H111	14	96	11	76	—	—	—	—	—	—	—
	-T6	32	221	28	193	15	—	20	138	—	—	71
6262	-T9	58	400	55	379	—	10	35	241	13	89.6	120
6351	-T5	45	310	41	283	—	12	27	186	—	—	—
	-T6	48	331	45	310	—	11	27	186	—	—	—

a. Continued on next page.

b. Fatigue strength for round specimens and 500 million cycles.

c. 10-mm ball used.

Table 1.7C (continued)
Weldability and Properties of Heat-Treatable Wrought Aluminum Alloys
TYPICAL MECHANICAL PROPERTIES

| Aluminum Alloy | Temper | Ultimate Tensile Strength | | Yield Strength (0.2% Offset) | | Elongation % in 2 in. (50.8 mm) | | Shear Strength | | Fatigue Strength[b] | | Brinnell Hardness |
		ksi	MPa	ksi	MPa	1/16 in. (1.6 mm) Sheet	1/2 in. (12.7 mm) Round	ksi	MPa	ksi	MPa	(500 kg load)[c]
6951	-0	16	110	6	41.4	30	—	11	76	—	—	28
	-T6	39	269	33	228	13	—	26	179	—	—	82
7004	-T5	57	393	48	331	—	—	34	234	—	—	—
7005	-T53	53	365	47	324	—	12	30	207	—	—	—
7039	-T64	65	448	55	379	13	10	38	262	—	—	120
7075	-0	33	228	15	103	17	16	22	152	—	—	60
	-T6	83	572	73	503	11	11	48	331	23	159	150
	-T73	73	503	63	434	—	13	44	303	22	152	—
7079	-0	33	228	15	103	17	16	—	—	—	—	—
	-T6	78	538	68	469	—	14	45	310	23	159	145
7178	-0	33	228	15	103	15	16	22	152	—	—	60
	-T6	88	607	78	538	10	11	52	358	22	152	160

b. Fatigue strength for round specimens and 500 million cycles.

c. 10-mm ball used.

CAST ALLOYS

THE ALLOY DESIGNATIONS for cast aluminum alloys are described in Table 1.8.

Cast alloys are either nonheat-treatable or heat-treatable based upon their composition, as described for the wrought alloys. The cast alloys also may be classified according to the casting method for which the alloy is suitable, such as sand casting, permanent mold casting, or die casting. Tables 1.9 through 1.12 give the characteristics of nonheat-treatable and heat-treatable casting alloys. The temper designation system for castings is the same as that for wrought aluminum products.

Table 1.8
Designations for Casting Alloy Groups

A system of four-digit numerical designations is used to identify aluminum and aluminum alloys in the form of castings or foundry ingot. The first digit indicates the alloy group as follows:

Aluminum, 99.00% and greater	1XX.X
Major alloying element	
Copper	2XX.X
Silicon, with added Copper, Magnesium, or both	3XX.X
Silicon	4XX.X
Magnesium	5XX.X
Zinc	7XX.X
Tin	8XX.X
Other Element	9XX.X
Unused series	6XX.X

For the 1XX.X series, the second two digits identify the minimum aluminum percentage. For all castings, the last digit, which is separated from the others by a decimal point, indicates the product form. Castings are indicated by XXX.0, while XXX.1 and XXX.2 indicate ingot types. A modification of the original alloy or impurity limits is indicated by a serial letter before the numerical designation. The serial letters are assigned in alphabetical sequence starting with "A". "X" is reserved for experimental alloys.

Table 1.9
Compositions, Types, and Applications of Nonheat-Treatable Cast Aluminum Alloys

Base Alloy	Nominal Composition (% Alloying Element)				Suitable Types of Castings			Typical Applications
	Cu	Si	Mg	Zn	Sand	Permanent Mold	Die	
208.0	4.0	3.0	—	—	X	X	—	General-purpose alloy; manifolds, valve housings, and applications requiring pressure tightness.
213.0	7.0	2.0	—	—	X	X	—	—
238.0	10.0	4.0	0.25	—	—	X	—	High as-cast hardness. Sole plates for electric hand irons.
360.0	—	9.5	0.5	—	—	—	X	General-purpose die castings, cover plates, and instrument cases. Excellent casting characteristics.
380.0	3.5	8.5	—	—	—	—	X	General purpose. Good casting characteristics and mechanical properties.
413.0	—	12.0	—	—	—	—	X	General-purpose die casting alloy for large, intricate parts with thin sections, as typewriter frames, instrument cases, etc. Excellent casting characteristics; very good corrosion resistance.
443.0	0.6 max	5.25	—	—	X	X	—	General-purpose alloy, cooking utensils, pipe fittings,
A443.0	0.3 max	5.25	—	—	X	X	—	architectural and marine applications. Excellent castability
B443.0	0.15 max	5.25	—	—	X	X	—	and pressure tightness.
A444.0	—	7.0		—		X	—	Structural applications (AASHTO)
511.0	—	0.5	4.0	—	X	—	—	Anodically treated architectural parts and ornamental hardware. Takes uniform anodic finish.
512.0	—	1.8	4.0	—	X	—	—	Cooking utensils and pipe fittings.
513.0	—	—	4.0	1.8		X	—	Cooking utensils and ornamental hardware; takes uniform anodic finish.
514.0	—	—	4.0	—	X	—	—	Chemical process fittings, special food-handling equipment, and marine hardware. Excellent corrosion resistance.
518.0	—	—	8.0	—		—	X	Marine fittings and hardware. High strength, ductility, and resistance to corrosion.
535.0[a]	—	—	6.9	—	X	—	—	High welded strength and ductility.
710.0	0.5	—	0.7	6.5	X	—	—	General-purpose sand castings for subsequent brazing. Good machinability.
711.0[b]	0.5	—	0.35	6.5		X	—	Torque converter blades and brazed parts. Good machinability.
712.0[c]	—	—	0.6	5.8	X	—	—	Same as 710.0 above, good corrosion resistance.

a. Also 0.18 Mn, 18 Ti, and 0.005 Be

b. Also 1.0 Fe

c. Also 0.5 Cr

Table 1.10
Compositions, Types, and Applications of Heat-Treatable Cast Aluminum Alloys

Base Alloy	Nominal Composition (% Alloying Element)				Suitable Types of Castings (X)		Typical Applications
	Cu	Si	Mg	Ni	Sand	Permanent Mold	
A201.0[a]	4.5	—	0.25	—	X	—	—
222.0	10.0		0.25		X	X	Bushings, bearing caps, meter parts, and air-cooled cylinder heads. Retains strength at elevated temperatures.
240.0[b]	8.0	—	6.0	0.50	X	—	—
242.0	4.0	—	1.5	2.0	X	X	Heavy-duty pistons and air-cooled cylinder heads. Good strength at elevated temperatures.
A242.0[c]	4.1	—	1.5	2.0	X	—	Heavy-duty pistons and air-cooled cylinder heads. Good strength at elevated temperatures.
295.0	4.5	1.1		—	X	—	Machinery and aircraft structural members, crankcases, and wheels.
319.0	3.5	6.0		—	X	X	General purpose, engine parts, automobile cylinder heads, piano plates.
332.0[d]	3.0	9.5	1.0	—	—	X	Automotive pistons. Good properties at operating temperatures.
333.0	3.5	9.0	0.30	—	—	X	Engine parts, gas meter housing, regulator parts, and general purpose.
336.0	1.0	12.0	1.0	2.5	—	X	Heavy-duty diesel pistons. Good strength at elevated temperatures.
354.0	1.8	9.0	0.50	—	—	X	Aircraft, missile, and other applications requiring premium-strength castings.
355.0	1.3	5.0	0.50	—	X	X	General-purpose castings, crankcases, accessory housings, and aircraft fittings.
C355.0	1.3	5.0	0.50	—	X	X	Aircraft, missile, and other structural applications requiring high strength.
356.0	—	7.0	0.35	—	X	X	General-purpose castings, transmission cases, truck-axle housings and wheels, cylinder blocks, railway tank-car fittings, marine hardware, bridge railing parts, architectural uses.
A356.0	—	7.0	0.35	—	X	X	Aircraft, missile, and other structural applications and aircraft fittings.
A357.0[e]	—	7.0	0.55	—	X	X	Aircraft, missile, and other structural applications requiring high strength.
359.0	—	9.0	0.6	—	X	X	Aircraft, missile, and other structural applications requiring high strength.
520.0	—	—	10.0	—	X	—	Sand castings requiring strength and shock resistance, such as aircraft structural members. Excellent corrosion resistance. Not recommended for use over 250 °F (121 °C).

a. Also 0.7 Ag, 0.30 Mn, and 0.25 Ti

b. Also 0.5 Mn

c. Also 0.20 Cr

d. Also 1.0 Zn

e. Also 0.05 Be

Table 1.11A
Weldability and Properties of Nonheat-Treatable Cast Aluminum Alloys
WELDABILITY[a,b]

Aluminum Alloy	Oxyfuel Gas	Arc with Flux	Arc with Inert Gas	Resistance	Pressure	Brazing	Soldering with Flux
			Sand Castings				
208.0	C	C	B	B	X	X	C
213.0	C	C	B	B	X	X	C
430.0	A	A	A	A	X	C	C
443.0	A	A	A	A	X	C	C
A443.0	A	A	A	A	X	C	C
B443.0	A	A	A	A	X	C	C
511.0	X	X	A	A	X	C	C
512.0	X	X	B	B	X	C	C
514.0	X	X	A	A	X	C	C
535.0	X	X	A	A	X	X	C
710.0	C	C	B	B	X	A	B
712.0	C	C	A	B	X	A	B
			Permanent Mold Castings				
208.0	C	C	B	B	X	X	C
213.0	C	C	B	B	X	X	C
238.0	C	C	B	A	X	X	C
443.0	A	A	A	A	X	C	C
B443.0	A	A	A	A	X	C	C
A444.0	B	B	A	A	X	C	C
513.0	X	X	A	A	X	C	C
711.0	C	C	A	A	X	A	C
			Die Castings				
360.0	C	X	C	B	X	X	X
380.0	C	X	C	B	X	X	X
413.0	C	C	C	B	X	X	X
518.0	X	X	C	B	X	X	X

a. Weldability ratings are based on the most weldable temper:
   A. Readily weldable.
   B. Weldable in most applications; may require special technique or preliminary trials to establish welding procedures, performance, or both.
   C. Limited weldability.
   X. Particular joining method is not recommended.

b. All alloys can be adhesive bonded, ultrasonically welded, or mechanically fastened.

FILLER METAL SELECTION

THE END USE of the weldment and desired performance are important considerations in selecting an aluminum alloy filler metal. Many base-metal alloys and alloy combinations can be joined using any one of several filler metals, but only one may be the optimum for a specific application. Standard wrought aluminum welding filler metals are listed in Table 1.13.

The primary factors commonly considered when selecting an aluminum alloy filler metal are

(1) Freedom from cracks
(2) Tensile or shear strength of the weld metal
(3) Weld ductility
(4) Service temperature
(5) Corrosion resistance
(6) Color match after anodizing

When fillet welding thick assemblies, substantial cost savings may be realized by using a higher strength filler metal that permits a reduction in weld passes. When making foundry repairs of castings, a homogeneous

Table 1.11B
Weldability and Properties of Nonheat-Treatable Cast Aluminum Alloys
PHYSICAL PROPERTIES

Aluminum Alloy[c]	Density		Approximate Melting Range		Thermal Conductivity[a,c]		Electrical Conductivity (% IACS)[b,c]
	lb/in.^{3}	kg/m^3	°F	°C	BTU/(ft•h•°F)	W/(m•K)	
Sand Castings							
208.0	0.101	2796	970-1170	521-632	70	121	31
213.0	0.106	2934	965-1160	518-627	70	121	30
430.0	0.097	2685	1070-1170	577-632	84	146	37
A443.0	0.097	2685	1070-1170	577-632	84	146	37
B443.0	0.097	2685	1070-1170	577-632	84	146	37
511.0	0.096	2658	1090-1180	588-638	82	141	36
512.0	0.096	2658	1090-1170	588-632	84	146	38
514.0	0.096	2658	1110-1180	599-638	79	137	35
535.0	0.091	2519	1020-1170	549-632	58	100	23
710.0	0.102	2823	1110-1200	599-649	79	137	35
712.0	0.102	2823	1110-1180	599-638	92	158	40
Permanent Mold Castings							
208.0	0.101	2796	970-1170	521-632	70	121	31
213.0	0.106	2934	965-1160	518-627	70	121	30
238.0	0.107	2962	950-1110	510-599	60	104	25
443.0	0.097	2685	1070-1170	577-632	84	146	37
B443.0	0.097	2685	1070-1170	577-632	84	146	37
A444.0	0.097	2685	1070-1170	577-632	92	158	41
513.0	0.097	2685	1080-1180	583-638	77	133	34
711.0	0.103	2851	1110-1190	599-644	92	158	40
Die Castings							
360.0	0.097	2685	1060-1090	515-588	84	146	37
380.0	0.099	2740	970-1090	521-588	62	108	27
413.0	0.096	2658	1070-1090	577-588	89	154	39
518.0	0.091	2519	1000-1150	538-621	58	100	24

a. Thermal conductivity at 77 °F (25 °C).

b. Percentage of International Annealed Copper Standard (IACS) value for Volume Electrical Conductivity, which equals 100 percent at 68°F (20 °C).

c. All casting alloys are in the "F" temper.

structure is often desired for the weld, so the filler metal should have the same composition as the casting alloy.

Cracking

IN GENERAL THE nonheat-treatable aluminum alloys can be welded with a filler metal of the same basic composition as the base alloy. The heat-treatable alloys are somewhat more complex metallurgically, and are more sensitive to "hot short" cracking during the weld cooling cycle. Generally, a dissimilar filler metal having a lower melting temperature and similar or lower strength than the base metal is used for the heat-treatable alloys, e.g., alloy 4043 [1070 °F (577 °C) solidus] or alloy 4145 [970 °F (510 °C) solidus]. By

allowing the low-melting constituents of the base metal adjacent to the weld to solidify before the weld metal, stresses are minimized in the base metal during cooling, and intergranular cracking tendencies are minimized.

The relative sensitivity to cracking based upon weld metal composition of four aluminum alloy systems (Al-Si 4XXX series; Al-Mg 5XXX series, Al-Cu 2XXX series and Al-Mg$_2$Si 6XXX series) is shown in Figure 1.1. These curves show that the high silicon content and the high magnesium content aluminum alloys are easy to weld due to a low sensitivity to cracking.

Heat-treatable alloy 2219 (6.3% Cu) is easily welded using its companion alloy 2319 filler metal. The 6XXX series alloys are very sensitive to cracking if the weld metal composition remains close to the base metal

Table 1.11C
Weldability and Properties of Nonheat-Treatable Cast Aluminum Alloys
TYPICAL MECHANICAL PROPERTIES

Aluminum Alloy	Temper	Ultimate Tensile Strength		Yield Strength (0.2% Offset)		Elongation % in 2 in. (50.8 mm), 1/2 in. (12.7 mm) Diameter Round	Shear Strength		Fatigue Strength[a]		Brinell Hardness (500 kg load)[b]
		ksi	MPa	ksi	MPa		ksi	MPa	ksi	MPa	
						Sand Castings					
208.0	F	21	145	14	96	2.5	17	117	11	76	55
213.0	F	24	166	15	103	1.5	20	138	9	62	70
430.0	F	19	131	8	55	8.0	14	96	8	55	40
A443.0	F	19	131	8	55	8.0	14	96	8	55	40
B443.0	F	19	131	8	55	8.0	14	96	8	55	40
511.0	F	21	145	12	83	3.0	17	117	8	55	50
512.0	F	20	138	13	90	2.0	17	117	8.5[c]	59	50
514.0	F	25	172	12	83	9.0	20	138	7	48	50
535.0	F	36	248	18	124	9.0	—	—	—	—	65
710.0[c]	F	35	241	25	172	5.0	26	179	8	55	75
712.0[c]	F	35	241	25	172	5.0	26	179	9	62	75
						Permanent Mold Castings					
208.0	F	28	193	16	110	2.0	22	152	13	90	70
213.0	F	30	207	24	166	1.5	24	166	9.5	66	85
238.0	F	30	207	24	166	1.5	24	166	—	—	100
443.0	F	23	159	9	62	10.0	16	110	8	55	45
B443.0	F	23	159	9	62	10.0	16	110	8	55	45
A444.0	F	35	241	18	124	8.0	—	—	11	76	70
513.0	F	27	186	16	110	7.0	22	152	10	69	60
711.0[c]	F	35	241	18	124	8.0	—	—	11	76	70
						Die Castings					
360.0	F	47	324	25	172	3.0	30	307	19	131	75
380.0	F	48	331	24	166	3.0	31	214	21	145	80
413.0	F	43	296	21	145	2.5	28	193	19	131	80
518.0	F	45	310	27	186	8.0	29	200	20	138	80

a. Fatigue strength for round specimens and 500 million cycles.

b. 10-mm ball used.

c. Tests made approximately 30 days after casting.

composition, as illustrated in the square groove joint of Figure 1.2. These can be welded easily if beveled to permit an excess of filler metal admixture with the base metal. For alloy 6061, the weld metal should possess at least 50% 4043 filler metal or 70% alloy 5356 filler metal. Fillet welds permit this admixture by the filler metal naturally, provided the base metal is not excessively melted.

Alloy 4145 filler metal provides the least cracking susceptibility with the 2XXX series alloys, such as 2014 and 2618, as well as the Al-Cu and Al-Si-Cu-type casting alloys. The 7XXX series alloys exhibit a wide range of crack sensitivity related to their copper content. The alloys having a low copper content (7004, 7005, and 7039) are welded with 5356, 5183 or 5556 alloy filler metals. Aluminum alloys with higher copper content, such as alloy 7075 or alloy 7178, are not acceptable for arc welding.

Filler metals with a high silicon content (4XXX series) should not be used to weld high magnesium content 5XXX series alloys; excessive magnesium-silicide eutectics developed in the weld structure decrease ductility and increase crack sensitivity. Mixing the high-magnesium content and high-copper content alloys results in high sensitivity to weld cracking and low weld ductility.

Strength

IN MANY CASES several filler metals are available that meet the minimum as-welded mechanical properties.

Table 1.12A
Weldability and Properties of Heat-Treatable Cast Aluminum Alloys
WELDABILITY[a,b]

Aluminum Alloy	Oxyfuel Gas	Arc with Flux	Arc with Inert Gas	Resistance	Pressure	Brazing	Soldering with Flux
			Sand Castings				
A201.0	C	C	B	B	X	X	C
222.0	X	C	B	B	X	X	X
240.0	X	X	C	B	X	X	X
242.0	X	X	C	B	X	X	X
A242.0	X	X	C	B	X	X	X
295.0	C	C	B	B	X	X	X
319.0	C	C	B	B	X	X	X
355.0	B	B	B	B	X	X	X
C355.0	B	B	B	B	X	X	X
356.0	A	A	A	A	X	C	C
A356.0	A	A	A	A	X	C	C
A357.0	B	B	A	A	X	C	C
359.0	B	B	A	A	X	C	C
520.0	X	X	B	C	X	X	X
			Permanent Mold Castings				
222.0	X	C	B	B	X	X	X
242.0	X	X	C	B	X	X	X
319.0	C	C	B	B	X	X	X
332.0	X	X	B	B	X	X	X
333.0	X	X	B	B	X	X	X
336.0	C	C	B	B	X	X	X
354.0	C	C	B	B	X	X	X
355.0	B	B	B	B	X	X	X
C355.0	B	B	B	B	X	X	X
356.0	A	A	A	A	X	C	C
A356.0	A	A	A	A	X	C	C
A357.0	B	B	A	A	X	C	C
359.0	B	B	A	A	X	C	C

a. Weldability ratings are based on the most weldable temper:
   A.   Readily weldable.
   B.   Weldable in most applications; may require special technique or preliminary trials to establish welding procedures and performance.
   C.   Limited weldability.
   X.   Particular joining method is not recommended.

b. All alloys can be adhesive bonded, ultrasonically welded, or mechanically fastened.

Table 1.12B
Weldability and Properties of Heat-Treatable Cast Aluminum Alloys
PHYSICAL PROPERTIES

Aluminum Alloy	Temper	Density lb/in.^{3}	Density kg/m^3	Approximate Melting Range °F	Approximate Melting Range °C	Thermal Conductivity[a] BTU/(ft·hr·°F)	Thermal Conductivity[a] W/(m·K)	Electrical Conductivity (% IACS)[b]
				Sand Castings				
A201.0	-T6	0.101	2796	1060-1200	571-649	840	1452	30
222.0	-T61	0.107	2962	970-1160	521-627	890	1539	33
240.0	-F	0.100	2768	960-1120	516-604	660	1141	23
242.0	-T77	0.102	2823	980-1180	527-638	1040	1798	38
	-T571	—	—	—	—	920	1591	34
A242.0	-T75	0.102	2823	980-1180	527-638	—	—	—
295.0	-T4	0.102	2823	970-1190	521-643	950	1642	35
	-T62	—	—	—	—	980	1694	35
319.0	-T5	0.101	2796	970-1120	521-604	—	—	—
	-T6	—	—	—	—	—	—	—
355.0	-T51	0.098	2713	1020-1150	549-621	1160	2006	43
	-T6	—	—	—	—	980	1694	36
	-T7	—	—	—	—	1130	1954	42
C355.0	-T6	0.098	2713	1020-1150	549-621	1010	1746	39
356.0	-T51	0.097	2685	1040-1140	560-616	1160	2006	43
	-T6	—	—	—	—	1040	1798	39
	-T7	—	—	—	—	1070	1850	40
A356.0	-T6	0.098	2713	1040-1130	560-610	1040	1798	40
A357.0	-T6	0.098	2713	1030-1130	554-610	1100	1902	40
359.0	-T6	0.097	2713	1050-1110	566-599	950	1642	35
520.0	-T4	0.093	2574	840-1110	449-599	600	1037	21
				Permanent Mold Castings				
222.0	-T65	0.107	2962	970-1160	521-627	900	1556	33
242.0	-T571	0.102	2823	980-1180	527-638	920	1591	34
	-T61	—	—	—	—	920	1591	33
319.0	-T6	0.101	2796	970-1120	527-604	—	—	—
332.0	-T5	0.100	2768	970-1180	527-638	720	1245	26
333.0	-T5	0.100	2768	970-1090	527-643	840	1452	29
	-T6	—	—	—	—	810	1400	29
	-T7	—	—	—	—	980	1694	35
336.0	-T551	0.098	2713	1000-1060	538-627	810	1400	29
	-T65	—	—	—	—	—	—	—
354.0	-T62	0.098	2713	1000-1110	538-599	870	1504	32
355.0	-T6	0.098	2713	1020-1150	549-621	1040	1798	39
C355.0	-T61	0.098	2713	1020-1150	549-621	1010	1746	39
356.0	-T6	0.097	2685	1040-1140	560-616	1040	1798	41
	-T7	—	—	—	—	1070	1850	40
A356.0	-T61	0.098	2713	1040-1130	560-610	1040	1798	40
A357.0	-T6	0.098	2713	1030-1130	554-610	1100	1902	40
359.0	-T62	0.097	2685	1050-1110	566-599	950	1642	35

a. Thermal conductivity at 77 °F (25 °C)

b. Percentage of International Annealed Copper Standard (IACS) value for Volume Electrical Conductivity, which equals 100 percent at 68°F (20 °C)

Table 1.12C
Weldability and Properties of Heat-Treatable Cast Aluminum Alloys
TYPICAL MECHANICAL PROPERTIES

Aluminum Alloy	Temper	Ultimate Tensile Strength		Yield Strength (0.2% Offset)		Elongation % in 2 in. (50.8 mm) / 1/2 in. (12.7 mm) Diameter Round	Shear Strength		Fatigue Strength[a]		Brinnell Hardness[b] (500 kg load)
		ksi	MPa	ksi	MPa		ksi	MPa	ksi	MPa	
Sand Castings											
A201.0	-T6	65	448	55	379	8.0	42	290	—	—	130
222.0	-T61	41	283	40	276	<0.5	32	221	8.5	59	115
240.0	-F	34	234	28	193	1.0	—	—	—	—	90
242.0	-T77	30	207	23	159	2.0	24	166	10.5	72	75
	-7571	32	221	30	207	0.5	26	179	11	76	85
A242.0	-T75	31	214	—	—	2.0	—	—	—	—	—
295.0	-T4	32	221	16	110	8.5	26	179	7	48	60
	-T62	41	283	32	221	2.0	33	228	8	55	90
319.0	-T5	30	207	26	179	1.5	24	166	11	76	80
	-T6	36	248	24	166	2.0	29	200	11	76	80
355.0	-T51	28	193	23	159	1.5	22	152	8	55	65
	-T6	35	241	25	172	3.0	28	193	9	62	80
	-T7	38	262	26	179	0.5	28	193	10	69	85
C355.0	-T6	39	269	29	200	5.0	—	—	—	—	85
356.0	-T51	25	172	20	138	2.0	20	138	8	55	60
	-T6	33	228	24	166	3.5	26	179	8.5	59	70
	-T7	34	234	30	207	2.0	24	166	9	62	75
A356.0	-T6	40	276	30	207	6.0	—	—	—	—	75
A357.0	-T6	46	317	36	248	3.0	40	276	12	83	85
359.0	-T6	—	—	—	—	—	—	—	—	—	—
520.0	-T4	48	331	26	179	16.0	34	234	8	55	75
Permanent Mold Castings											
222.0	-T65	48	331	36	248	<0.5	26	179	9	62	140
242.0	-T571	40	276	34	234	1.0	30	207	10.5	72	105
	-T61	47	324	42	290	0.5	35	241	9.5	66	110
319.0	-T6	40	276	27	186	3.0	—	—	—	—	95
332.0	-T5	36	248.	28	193	1.0	—	—	—	—	105
333.0	-T5	34	234	25	172	1.0	27	186	12	83	100
	-T6	42	290	30	207	1.5	33	228	15	103	105
	-T7	37	255	28	193	2.0	28	193	12	83	90
336.0	-T551	36	248	28	193	0.5	28	193	13.5	93	105
	-T65	47	324	43	296	0.5	36	248	—	—	125
354.0	-T62	57	393	46	317	3.0	40	276	—	—	110
355.0	-T6	42	290	27	186	4.0	34	234	10	69	90
C355.0	-T61	46	317	34	234	6.0	32	221	14	96	100
356.0	-T6	38	262	27	186	5.0	30	207	13	80	80
	-T7	32	221	24	234	6.0	25	172	11	76	70
A356.0	-T61	41	283	30	207	10.0	28	193	13	90	90
A357.0	-T6	52	358	45	310	5.0	35	241	15	103	100
359.0	-T62	50	345	42	290	5.5	—	—	16	110	—

a. Fatigue strength for round specimens and 500 million cycles.

b. 10-mm ball used.

Table 1.13
Chemical Composition of Wrought Aluminum Filler Metals

| Filler Alloy | Elements, wt. %[a] | | | | | | | | Other Elements | | Al |
	Si	Fe	Cu	Mn	Mg	Cr	Zn	Ti	Each	Total	
1100	Note b	Note b	0.05-0.20	0.05	—	—	0.10	—	0.05[c]	0.15	99.0 min.
1188	0.06	0.06	0.005	0.01	0.01	—	0.03	0.01	0.01[c]	—	99.88 min.
2319	0.20	0.03	5.8-6.8	0.20-0.40	0.02	—	0.10	0.10-0.20	0.05[c]	0.15	Remainder
4009[d]	4.5-5.5	0.20	1.0-1.5	0.10	0.45-0.6	—	0.10	0.10-0.20	0.05[c]	0.15	Remainder
4010[e]	6.5-7.5	0.20	0.20	0.10	0.30-0.45	—	0.10	0.20	0.05[c]	0.15	Remainder
4011[f]	6.5-7.5	0.20	0.20	0.10	0.45-0.7	—	0.10	0.04-0.20	0.05[f]	0.15	Remainder
4043	4.5-6.0	0.8	0.30	0.05	0.05	—	0.10	0.20	0.05[c]	0.15	Remainder
4047	11.0-13.0	0.8	0.30	0.15	0.10	—	0.20	—	0.05[c]	0.15	Remainder
4145	9.3-10.7	0.8	3.3-4.7	0.15	0.15	0.15	0.20	—	0.05[c]	0.15	Remainder
4643	3.6-4.6	0.8	0.10	0.05	0.10-0.30	—	0.10	0.15	0.05[c]	0.15	Remainder
5183	0.40	0.40	0.10	0.50-1.0	4.3-5.2	0.05-0.25	0.25	0.15	0.05[c]	0.15	Remainder
5356	0.25	0.40	0.10	0.05-0.20	4.5-5.5	0.05-0.20	0.10	0.06-0.20	0.05[c]	0.15	Remainder
5554	0.25	0.40	0.10	0.50-1.0	2.4-3.0	0.05-0.20	0.25	0.05-0.20	0.05[c]	0.15	Remainder
5556	0.25	0.40	0.10	0.50-1.0	4.7-5.5	0.05-0.20	0.25	0.05-0.20	0.05[c]	0.15	Remainder
5654	Note g	Note g	0.05	0.01	3.1-3.9	0.15-0.35	0.20	0.05-0.15	0.05[c]	0.15	Remainder

a. Single values are maximum, except where otherwise specified.

b. Silicon plus iron shall not exceed 0.95 percent.

c. Beryllium shall not exceed 0.0008 percent.

d. Same composition as C355.0 cast alloy.

e. Same composition as A356.0 cast alloy.

f. Beryllium content is 0.04 to 0.07 percent. Same composition as A357.0 cast alloy. Used for GTAW rod only.

g. Silicon plus iron shall not exceed 0.45 percent.

Typical all-weld-metal tensile strengths and minimum shear strengths of several filler metals are listed in Table 1.14. The diffusion of alloying elements from base metal may increase the as-welded mechanical properties.

When heat-treatable alloy weldments are to be postweld heat treated, the filler metal selection is more limited. When welding alloy 2219 and alloy 2014, the heat-treatable filler metal alloy 2319 will provide the highest strength. In most cases, the filler metal is not a heat-treatable composition or is only mildly responsive to strengthening by thermal treatments. For example, when welding less that 1/2 in. (12.7 mm) thick alloy 6061-T6 with alloy 4043 filler metal, magnesium in alloy 6061 migrates into weld metal to produce magnesium silicide in sufficient quantity to respond to heat treatment. In thicker groove welds, the wider bevels can prevent diffusion of magnesium to the center of the weld, and little or no response results from postweld heat treatments. Filler metal alloy 4643 contains sufficient magnesium so that a weld in 3 in. (76.2 mm) thick alloy 6061 has attained a transverse ultimate tensile strength of 45 ksi (440 MPa) when postweld heat treated and aged to meet the original alloy 6061-T6 properties.

The strength of fillet welds is highly dependent upon the filler metal composition and minimum shear strength values, which are listed in Table 1.14. Filler metal alloys 5356, 5183, and 5556 provide high shear strength for structural fillet welds.

The 1XXX and 5XXX alloy series filler metals produce very ductile welds and are preferred when the weldment is subjected to forming or spinning operations or postweld straightening operations.

Elevated and Cryogenic Temperature Service

FILLER METALS CONTAINING in excess of 3% Mg nominal composition (alloys 5183, 5356, 5556, and

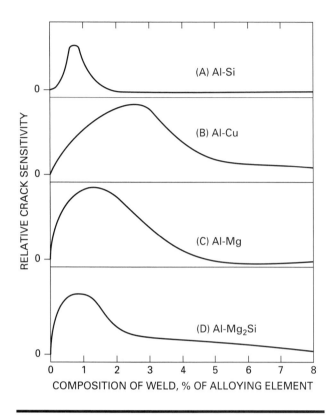

Figure 1.1—Hot Cracking Sensitivity of Welds in Aluminum Alloyed with (A) Silicon, (B) Copper, (C) Magnesium, and (D) Magnesium Silicide

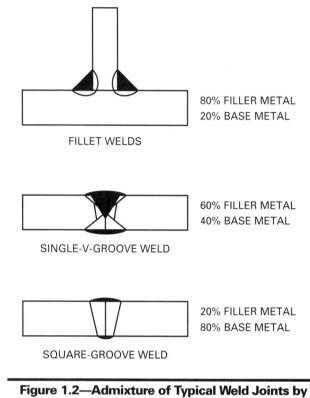

Figure 1.2—Admixture of Typical Weld Joints by Base Metal and Filler Metal

5654) are not suitable for applications where temperatures are sustained above 150 °F (66 °C), because they can be sensitized to stress-corrosion cracking. This would include the lengthy aging treatments used in postweld thermal treatments.

Filler metal alloy 5554 and all other filler metals listed in Table 1.13 are suitable for sustained elevated temperature service. All aluminum filler metals are suitable for cryogenic temperature applications.

Corrosion Resistance

ASSEMBLIES, VESSELS, DRUMS, and tankers for use in certain corrosive environments or with certain chemicals may require special filler metals. These alloys may be of higher purity, such as alloy 1188 filler metal for alloy 1060 chemical drums, or may have closer composition limits on some alloying elements. A good example is the tight control over copper and manganese impurities in alloy 5254 plate and alloy 5654 filler metal for hydrogen peroxide service.

Aluminum-magnesium filler metals are highly resistant to general corrosion when used with base alloys having similar magnesium content. However, the 5XXX alloy series filler metals can be anodic to the 1XXX, 3XXX, and 6XXX alloy series base metals with which they might be used. In immersed service, the weld metal will pit and corrode, to protect the base metal, at varying rates based upon the difference in electrical potential of weld metal and base metal. Thus an aluminum-silicon filler metal, such as alloy 4043 or alloy 4047, would be preferred for improved corrosion resistance over alloy 5356 filler metal when welding alloy 6061 base metal for an immersed-service application.

Color Match

COLOR MATCH BETWEEN the weld metal and base metal is often desired for ornamental or architectural applications that are given chemical or electrochemical finishes. The final color is highly dependent upon the composition of the filler alloy and how closely it matches specific elements in the base alloy. The two elements of primary interest are silicon and chromium.

**Table 1.14
Typical Aluminum Filler Metal Properties
(As-Welded Condition)**

Filler Alloy	Minimum Shear Strength		All-Weld-Metal Ultimate Tensile Strength	
	ksi	MPa	ksi	MPa
1100	7.5	52	13.5	93
2319	16	110	37.5	258
4043	11.5	79	29	200
5183	18.5	128	41	283
5356	17	117	38	262
5554	17	117	33	230
5556	20	138	42	290
5654	12	83	32	221

Silicon in an alloy will create a gray-to-black color with increasing percentages of silicon. Thus, welds made with Al-Si filler metal will exhibit a sharp color contrast with all base alloys except those clad with an Al-Si alloy or with the Al-Si casting alloys. Chromium causes an alloy to develop a yellow or gold shading when anodically treated, so a 5XXX alloy series filler metal with a similar chromium content as the base metal would be the preferred filler metal. Copper and manganese in aluminum alloys have slight darkening effects that need to be considered.

Alloy 1188 filler metal will produce a good color match in welds of the 1XXX alloy series, as well as with alloys 3003, 5005, and 5050. Alloy 5356 filler metal is a good choice for welding the 5XXX and 6XXX series alloys when a color match is needed.

Selection of Filler Metal

THE SELECTION OF the correct filler metal greatly influences the service life of an aluminum weldment. A guide to the selection of filler metals for general purpose welding of various aluminum alloy combinations, including castings, is presented in Table 1.15. In the repair of castings, the filler metal is usually chosen to match the composition of the casting, and often it is cast from the same lot in the foundry.

Storage and Use of Aluminum Filler Metal

A MAJOR STEP toward producing good aluminum welds is the use of quality filler metal of the correct size and alloy. It should be free of gas and nonmetallic impurities, with a clean, smooth surface free of moisture, lubricant, or other contaminants. Care must be taken during storage and use to prevent contamination that would cause poor welds.

The quality of the filler metal is particularly important with the gas metal arc welding process. In this consumable-electrode process, relatively small-diameter filler wire is fed through the welding gun at a high rate of speed. To feed properly, the electrode must be uniform in diameter; of a suitable temper; free from slivers, scratches, inclusion kinks, waves or sharp bends; and spooled so that it is free to unwind without restriction. Proper pitch and cast also are important to prevent wandering of the wire as it emerges from the contact tube. If the electrode surface is not clean, the high rate of wire feed can carry a relatively large amount of foreign material into the weld pool, resulting in porosity or poor quality welds.

To avoid contamination, filler-metal supplies must be kept covered and stored in a dry place at a relatively uniform temperature. Electrode spools temporarily left unused on the welding machine, as between work shifts, should be covered with a clean cloth or plastic bag if the feed unit does not have its own cover. If a spool of wire will not be used overnight, it should be returned to its carton and tightly sealed, unless it is in a spool enclosure that provides a dry or protective atmosphere. Original electrode containers should not be opened until the contents are to be used. The 5XXX series electrodes are most likely to develop a hydrated oxide and, when not in use, should be stored in cabinets maintaining a relatively low humidity (less than 35% RH).

Information on selection, manufacture, packaging, and testing of bare aluminum filler materials is covered in the latest editon of ANSI/AWS A5.10, *Specification for Bare Aluminum and Aluminum Alloy Welding Electrodes and Rods.*

SURFACE PREPARATION GUIDELINES

QUALITY ARC WELDING of aluminum depends upon clean and dry metal with a thin oxide film. The basic cause of porosity in aluminum welds is hydrogen; any moisture or hydrocarbons in the arc area will dissociate to provide hydrogen. Hydrated oxides and "water stain" on material that has been improperly stored can be particularly troublesome. The moisture penetrates and causes inward oxide growth. This hydrated-oxide layer possesses chemically combined water to cause porosity. Also, the thickened oxide is difficult to remove by the arc action; it melts at 3700 °F (2038 °C), which is three times the melting point of the aluminum alloy, and prevents proper fusion.

When material is received, it should be properly stored. This will minimize future cleaning operations. Storing aluminum outside, or in buildings without adequate climate controls, results in moisture condensation on the metal, permitting moisture to be drawn in between layers of material by capillary action. If the

Table 1.15
Guide to the Selection of Filler Metal for General Purpose Welding[a,b,c]

Base Metal	201.0, 206.0, 224.0	319.0, 333.0, 354.0, 355.0, C355.0	356.0, A356.0, 357.0, A357.0, 413.0, 443.0, A444.0	511.0, 512.0, 513.0, 514.0, 535.0	7004, 7005, 7039, 701.0, 712.0	6009, 6010, 6070	6005, 6061, 6063, 6101, 6151, 6201, 6351, 6951	5456	5454
1060, 1070, 1080, 1350	ER4145	ER4145	ER4043[d,e]	ER5356[e,f,g]	ER5356[e,f,g]	ER4045[d,e]	ER4043[e]	ER5356[g]	ER4043[e,g]
1100, 3003, Alc. 3003	ER4145	ER4145	ER4043[d,e]	ER5356[e,f,g]	ER5356[e,f,g]	ER4043[d,e]	ER4043[e]	ER5356[g]	ER4043[e,g]
2014, 2036	ER4145[h]	ER4145[h]	ER4145	—	—	ER4145	ER4145	—	—
2219	ER2319[d]	ER4145[h]	ER4145[e,f]	ER4043[e]	ER4043[e]	ER4043[d,e]	ER4043[d,e]	—	ER4043[e]
3004, Alc. 3004	—	ER4043[e]	ER4043[e]	ER5356[i]	ER5356[i]	ER4043[e]	ER4043[e,i]	ER5356[g]	ER5356[i]
5005, 5050	—	ER4043[e]	ER4043[e]	ER5356[i]	ER5356[i]	ER4043[e]	ER4043[e,i]	ER5356[g]	ER5356[i]
5052, 5652[l]	—	ER4043[e]	ER4043[e,i]	ER5356[i]	ER5356[i]	ER4043[e]	ER5356[f,i]	ER5356[i]	ER5356[i]
5083	—	—	ER5356[e,f,g]	ER5356[g]	ER5183[g]	—	ER5356[g]	ER5183[g]	ER5356[g]
5086	—	—	ER5356[e,f,g]	ER5356[g]	ER5356[g]	—	ER5356[g]	ER5356[g]	ER5356[g]
5154, 5254[l]	—	—	ER4043[e,i]	ER5356[g]	ER5356[i]	—	ER5356[i]	ER5356[i]	ER5356[i]
5454	—	ER4043[e]	ER4043[e,i]	ER5356[i]	ER5356[i]	ER4043[e]	ER5356[f,i]	ER5356[i]	ER5554[h,i]
5456	—	—	ER5356[e,f,g]	ER5356[g]	ER5556[g]	—	ER5356[g]	ER5556[g]	—
6005, 6061, 6063, 6101, 6151, 6201, 6351, 6951	ER4145	ER4145[e,f]	ER4043[e,i,j]	ER5356[i]	ER5356[e,f,i]	ER4043[d,e,j]	ER4043[e,i,j]	—	—
6009, 6010, 6070	ER4145	ER4145[e,f]	ER4043[d,e,j]	ER4043[e]	ER4043[e]	ER4043[e,i,j]	—	—	—
7004, 7005, 7039, 710.0, 712.0	—	ER4043[e]	ER4043[e,i]	ER5356[i]	ER5356[g]	—	—	—	—
511.0, 512.0, 513.0, 514.0, 535.0	—	—	ER4043[e,i]	ER5356[i]	—	—	—	—	—
356.0, A356.0, 357.0, A357.0, 413.0 443.0, A444.0	ER4145	ER4145[e,f]	ER4043[e,k]	—	—	—	—	—	—
319.0, 333.0, 354.0, 355.0, C355.0	ER4145[h]	ER4145[e,f,k]	—	—	—	—	—	—	—
201.0, 206.0, 224.0	ER2319[d,k]	—	—	—	—	—	—	—	—

a. Service conditions such as immersion in fresh or salt water, exposure to specific chemicals, or a sustained high temperature [over 150°F (66 °C)] may limit the choice of filler metals. Filler metals ER5183, ER5356, ER5556, and ER5654 are not recommended for sustained elevated-temperature service.

b. Recommendations in this table apply to gas shielded arc welding processes. For oxyfuel gas welding, only ER1188, ER1100, ER4043, ER4047, and ER4145 filler metals are ordinarily used.

c. Where no filler metal is listed, the base metal combination is not recommended for welding.

d. ER4145 may be used for some applications.

e. ER4047 may be used for some applications.

f. ER4043 may be used for some applications.

g. ER5183, ER5356, or ER5556 may be used.

h. - m. See table footnotes on next page.

Table 1.15 (continued)
Guide to the Selection of Filler Metal for General Purpose Welding

Base Metal	5154, 5254^l	5086	5083	5052, 5652^l	5005, 5050	3004, Alc. 3004	2219	2014, 2036	1100, 3003, Alc. 3003	1060, 1070, 1080, 1350
1060,1070, 1080,1350	ER5356^{e,f,g}	ER5356^g	ER5356^g	ER4043^{e,g}	ER1100^{e,f}	ER4043^{e,g}	ER4145^{e,f}	ER4145	ER1100^{e,f}	ER1188^{e,f,k,m}
1100, 3003, Alc. 3003	ER5356^{e,f,g}	ER5356^g	ER5356^g	ER4043^{e,g}	ER1100^{e,f}	ER4043^{e,g}	ER4145^{e,f}	ER4145	ER1100^{e,f}	—
2014, 2036	—	—	—	—	ER4145	ER4145	ER4145^h	ER4145^h	—	—
2219	ER4043^e	—	—	ER4043^{e,g}	ER4043^{d,e}	ER4043^{d,e}	ER2319^d	—	—	—
3004, Alc. 3004	ER5356^i	ER5356^g	ER5356^g	ER5356^{e,f,i}	ER5356^{f,i}	ER5356^{f,i}	—	—	—	—
5005, 5050	ER5356^i	ER5356^g	ER5356^g	ER5356^{e,f,g}	ER5356^{f,i}	—	—	—	—	—
5052, 5652^l	ER5356^i	ER5356^g	ER5356^g	ER5654^{f,i,l}	—	—	—	—	—	—
5083	ER5356^g	ER5356^g	ER5183^g	—	—	—	—	—	—	—
5086	ER5356^g	ER5356^g	—	—	—	—	—	—	—	—
5154, 5254^l	ER5654^{i,l}	—	—	—	—	—	—	—	—	—
	—	—	—	—	—	—	—	—	—	—
	—	—	—	—	—	—	—	—	—	—
	—	—	—	—	—	—	—	—	—	—
	—	—	—	—	—	—	—	—	—	—
	—	—	—	—	—	—	—	—	—	—
	—	—	—	—	—	—	—	—	—	—
	—	—	—	—	—	—	—	—	—	—
	—	—	—	—	—	—	—	—	—	—
	—	—	—	—	—	—	—	—	—	—

a. - g.   See table footnotes on preceding page.

h.   ER2319 may be used for some applications. It can supply high strength when the weldment is postweld solution heat -treated and aged.

i.   ER5183, ER5356, ER5554, ER5556, and ER5654 may be used. In some cases, they provide: (1) improved color match after anodizing treatment, (2) highest weld ductility, and (3) higher weld strength. ER5554 is suitable for sustained elevated-temperature service.

j.   ER4643 will provide high strength in 1/2 in. (12 .7 mm) and thicker groove welds in 6XXX alloys when postweld solution heat treated and aged.

k.   Filler metal with the same analysis as the base metal is sometimes used. Filler alloys ER4009 or R4009, ER4010 or R4010, and R4011 meet the chemical composition limits of R-C355.0, R-A356.0 and R-A357.0 alloys, respectively.

l.   Base metal alloys 5254 and 5652 are useful for hydrogen peroxide service. ER5654 filler metal is used for welding both alloys for low-temperature service [150 °F (66 °C) and below].

m.   ER1100 may be used for some applications.

product is interleaved with paper, a "wicking" action can occur to further contribute to the water stains and hydrated-oxide conditions. Good storage practice is to position the sheets vertically and far enough apart to permit moisture to run off and to allow air circulation to dry the surfaces.

Prior to welding, the first operation should be to remove all grease, oil, dirt, paint, or other surface contaminants that can generate hydrogen gas or interfere with weld fusion. Degreasing may be done by wiping, spraying, or dipping in a solvent or by steam cleaning. The cleaning and welding areas must be well ventilated. If chlorinated solvents are used, the degreasing should be done at a location remote from the welding area. Highly toxic phosgene-type gases can result from the dissociation of the vapors of chlorinated hydrocarbons (such as triclorethylene and other chlorinated hydrocarbons) by arc radiation. Petroleum-base solvents, those solvents that are sufficiently volatile to leave little residue, are non-toxic when used in the welding area.

The low flash points of petroleum-base solvents require special storage and handling.

The naturally formed thin aluminum oxide film is removed by the arc of the inert gas shielded process or by the fluxes used in by other joining methods. Metal having a thick oxide film or coating resulting from thermal treatments, chemical or electrochemical treatments, or poor storage conditions should have the thick oxide coating removed. Oxide removal is accomplished with caustic soda, acids, or proprietary solutions. Some of these are listed in Table 1.16. Special attention should be given to the rinsing and drying cycles to avoid residual chemicals or production of a hydrated oxide. If the metal surface is rough from machining, sawing, forming, or improper handling, the welding surface may possess folds on the surface which would entrap oxide or lubricants. Chemical etching is desirable to open and remove any metal folds or burrs.

Mechanical oxide removal, although not as consistent as chemical means, is usually satisfactory if performed properly. These methods include wire brushing, scraping, or filing. Grinding or sanding with wheels or discs can be done with proper materials, although it is easy to imbed abrasive particles in the aluminum surface that may result in unacceptable inclusions in the weld. Contamination of the surface with a binder compound may create poor fusion or porosity in the weld. Wire brushing is by far the most widely used method. Stainless steel brushes with 0.010 to 0.015 in. (0.254 to 0.381 mm) diameter bristles give good scratching action. The brush must be kept clean of all contaminants, and it should be used with a light pressure to avoid burnishing the aluminum surface and entrapping oxide particles. Hand brushes are as effective as power brushes but at a greater expenditure of labor.

It should be noted that chemical cleaning and brushing of joint components is done prior to assembly for welding. Fitted joints may retain solvents or contaminants that result in weld discontinuities.

SAFE PRACTICES

THIS CHAPTER CONCENTRATES on the unique aspects of welding processes and procedures as they pertain to the joining of aluminum and its alloys.[1] Since most welding processes use enough heat to produce molten filler and base metal to accomplish the joining, the following conditions warrant safety precautions in the welding of aluminum alloys:

(1) High levels of fumes are produced when using the 5XXX (magnesium-bearing) aluminum filler metals.

(2) The use of argon-based shielding gas blends results in the production of ozone, especially with 4XXX filler metals.

The use of filtering masks or airline respirators will be required if it is determined that personnel are being exposed to excessive pollutants.

Caution must also be observed in the reaction between aluminum and certain solvents and cleaners. Consult information provided by manufacturers for the necessary safe practices in the use of their products.

---

1. For a thorough and more general coverage of welding processes, including safety recommendations, refer to the chapter in the *Welding Handbook*, Vol. 2, 8th Ed., which describes the desired process. Safe practices also are included in the *Welding Handbook*, Vol. 1, 8th Ed., Chapter 16, and in ANSI/ASC Z49.1, *Safety in Welding, Cutting and Allied Processes*.

Table 1.16
Chemical Treatments for Oxide Removal Prior to Welding or Brazing Aluminum*

Type of Solution	Concentration	Temperature	Type of Container	Procedure	Purpose
1. Sodium Hydroxide (Caustic Soda) followed by	1. NaOH 1.76 oz (50 grams) .26 per gal (1 L) of water	1. 140-160 °F (60-71 °C)	1. Mild steel	1. Immerse for 10 to 60 seconds. Rinse in cold water.	Removes thick oxide for all welding and brazing procedures. Active etchant.
2. Nitric Acid	2. Equal parts of HNO_3 (68%) and water	2. Room	2. Type 347 stainless steel	2. Immerse for 30 seconds. Rinse in cold water. Rinse in hot water and dry.	
Sulfuric-Chromic	H_2SO_4 - 1 gal (3.79 L) CrO_3 - 45 oz (1.28 kg) Water - 9 gal (34.1 L)	160-180 °F (60-82 °C)	Antimonal lead lined steel tank	Dip for 2 to 3 minutes. Rinse in cold water. Rinse in hot water and dry.	For removal of heat-treatment and annealing films and stains and for stripping oxide coatings.
Phosphoric-Chromic	H_3PO_4 (75%)- 3.5 gal (13.3 L) CrO_3 - 1.75 lb (79.4 grams) Water - 100 gal (379 L)	200 °F (93 °C)	Type 347 stainless steel	Dip for 5 to 10 minutes. Rinse in cold water. Rinse in hot water and dry.	For removal of heat-treatment and annealing films and stains and for stripping oxide coatings.
Sulfuric Acid	H_2SO_4 - 5.81 oz (165 grams) Water - 0.26 gal (1 L)	165 °F (73 °C)	Polypropylene-lined steel tank	Immerse for 5 to 10 minutes. Rinse in cold water. Rinse in hot water and dry.	Oxide removal. Mild etchant.
Ferrous Sulfate	$Fe_2SO_4 \, H_2O$ 10% by volume	80 °F (26.7 °C)	Polypropylene	Immerse for 5 to 10 minutes. Rinse in cold water. Rinse in hot water and dry.	Oxide removal.

* All chemicals used as etchants are potentially dangerous. All persons using any of these etchants should be thoroughly familiar with all of the chemicals involved and the proper procedures for handling and mixing these chemicals. Always add the chemicals to water. Never add the water to the chemicals. Safety glasses must be worn at all times when using chemical etchants. Avoid contact with the skin. When working with unfamiliar chemicals or cleaning products, always review the Material Safety Data Sheets (MSDS) for these products. These are available from the suppliers., and must be available to all employees for guidance on the proper handling of the materials.

ARC WELDING

JOINT GEOMETRY

THE RECOMMENDED JOINT geometries for arc welding aluminum are similar to those for steel. Aluminum joints have smaller root openings and larger groove angles than are generally used for steel since aluminum is more fluid and welding gun nozzles are larger. Typical joint geometries for arc welding aluminum are shown in Figure 1.3.

A special joint geometry, shown in Figure 1.4, is recommended for gas tungsten arc welding (GTAW) or gas metal arc welding (GMAW) when only one side of the joint is accessible and a smooth root surface is required. The effectiveness of this design for complete joint penetration is dependent upon the surface tension of the weld metal. It can be used with section thicknesses over 0.12 in. (3.0 mm) and in all welding positions. The abutting sections are designed so that complete joint penetration is possible with the first welding pass. It should be noted that this design has a large groove area and requires a relatively large amount of filler metal to fill the joint. Distortion may be greater than with conventional joint designs and its principal application is for circumferential joints in aluminum pipe.

V-groove joint designs are adequate for butt joints that are accessible from both sides. As a rule, a 60 degree groove angle is the minimum practical size for section thickness greater than 0.12 in. (3.0 mm). Thick sections may require even larger groove angles, such as 75 or 90 degrees, depending upon the welding process.

With thick plates, U-grooves are preferred to V-grooves to minimize the amount of deposited metal and to permit torch access to the root. Special joint geometries, shown in Figure 1.5, may be warranted to minimize porosity caused by entrapment of hydrogen when welding horizontally.

GAS TUNGSTEN ARC WELDING

Process Characteristics

IN GAS TUNGSTEN arc welding (GTAW), the base metal is melted by an ac or dc arc between the base metal and a virtually nonconsumable electrode of tungsten (or tungsten alloy) in a GTAW torch.[2] Filler metal is not melted directly by the arc, but by the molten base metal. If the arc contacts the filler, the electrode becomes contaminated by attraction forces between the arc and filler metal. The electrode is shielded by an inert gas that flows through the nozzle in the GTAW torch. The shielding gas, which prevents any reaction between the weld metal and air, is usually argon, helium, or a mixture of the two. The gas also protects unfused base metal adjacent to the weld metal. The weld metal may be composed of base metal alone or a mixture of base and filler metal. Filler metal may or may not be added.

Gas Tungsten Arc Welding Equipment

THE USUAL POWER source used in GTAW is a constant-current type (drooping volt/amp characteristic). In these power sources, the slope of the volt-ampere curve is relatively steep so that a change in the arc voltage (arc length) will not create a major change in arc current. Such a power source may be a dc-motor-driven generator, a rectifier, or a transformer.[3]

Welding Current and Polarity

ALUMINUM CAN BE gas tungsten arc welded using conventional ac or square-wave ac or dc with the electrode either negative (DCEN) or positive (DCEP). DCEP is rarely used due to the limited current-carrying ability of the tungsten electrode without overheating it.

Alternating Current

THE OXIDE REMOVAL action will take place only during the portion of the current cycle when the electrode is positive. To ensure the initiation of this half cycle, the power source must either have a high enough open-circuit voltage [125 volts root-mean-square (rms) for argon and 150 volts rms for helium] or a high voltage at high frequency must be impressed on the arc gap at the time that the current passes through zero and the electrode becomes positive. If the arc must be initiated without touching the electrode to the work, the high-frequency voltage must be used. A stable arc (one with steady uninterrupted current flow in each direction) is characterized by the absence of a snapping or cracking sound, a smooth flow of filler metal into the molten pool of weld metal, easy arc starting, and elimination of tungsten inclusions in the weld. The magnitude of the current when the electrode is negative will be greater than when the electrode is positive unless the

---

2. For a more detailed discussion of gas tungsten arc welding refer to Chapter 3, *Welding Handbook*, Vol. 2, 8th Ed.

3. For more details on GTAW power sources, see Chapters 1 and 3, *Welding Handbook*, Vol. 2, 8th Ed.

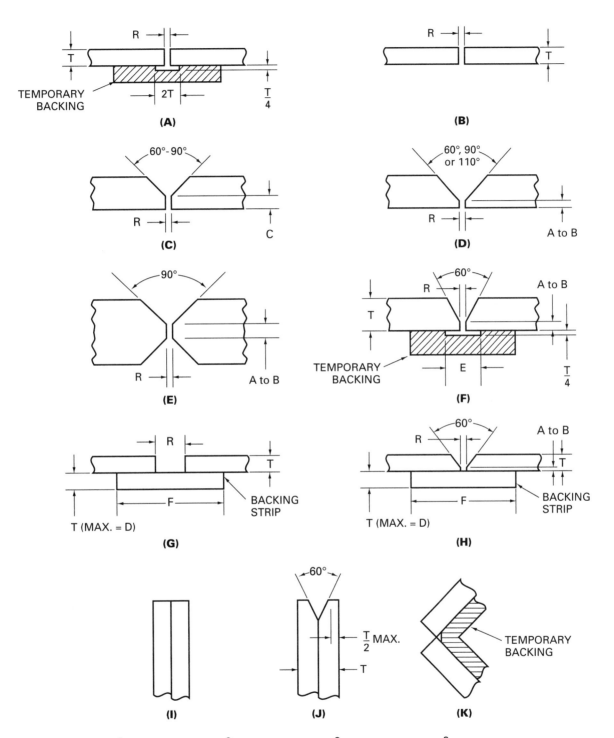

Note: A = $^1/_{16}$ in. (1.6 mm), B = $^3/_{32}$ in. (2.4 mm), C = $^3/_{16}$ in. (4.8 mm), D = $^3/_8$ in. (9.5 mm), E = $^1/_2$ in. (12.7 mm), F = $1^1/_2$ in. (38.1 mm), R = root opening, T = thickness.

Figure 1.3—Typical Joint Geometries for Arc Welding of Aluminum

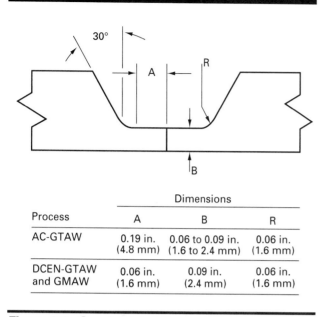

Process	Dimensions		
	A	B	R
AC-GTAW	0.19 in. (4.8 mm)	0.06 to 0.09 in. (1.6 to 2.4 mm)	0.06 in. (1.6 mm)
DCEN-GTAW and GMAW	0.06 in. (1.6 mm)	0.09 in. (2.4 mm)	0.06 in. (1.6 mm)

Figure 1.4—Special Joint Geometry for Arc Welding from One Side Only for Complete Joint Penetration

transformer is provided with electrical circuitry to prevent the imbalance.

Alternating current is used with a shielding gas of argon, or a mixture of argon and helium with 50 percent or more argon. The aluminum oxide on the surface is removed by arc action. For mixtures with high percentages of helium, very short arc lengths are needed for arc stability, the oxide removal action of the arc is lessened, and preweld oxide removal is usually required for good fusion. Helium and helium-rich gas mixtures are seldom used in ac welding but are most common for dc-GTAW. Pure tungsten or zirconia tungsten electrodes which form spherical tips are recommended for ac-GTAW. Excessive spitting of thoriated tungsten occurs when using GTAW with ac. Thoriated tungsten is used with dc-GTAW.

Tables 1.17, 1.18, and 1.19 give typical procedures for manual GTAW with ac power. These tables are intended to serve only as guides to establish welding procedures for any specific application.

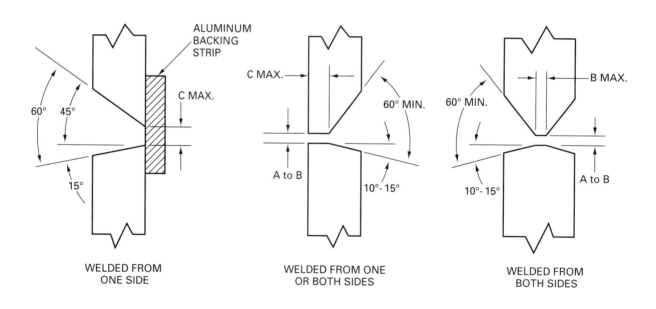

Note: A = 0 in. (0 mm), B = 1/16 in. (1.6 mm), C = 1/8 in. (3.2 mm).

Figure 1.5—Special Weld Joint Geometries for Arc Welding Aluminum in the Horizontal Position

Table 1.17
Typical Procedures for Manual Gas Tungsten Arc Welding of Butt Joints in Aluminum with AC and Argon Shielding

Section Thickness		Welding Position[a]	Joint Geometry[b]	Root Opening		Number of Weld Passes	Filler Rod Diameter		Electrode Diameter		Welding Current, A	Travel Speed	
in.	mm			in.	mm		in.	mm	in.	mm		in./min	mm/s
0.062	1.6	F,V,H	B	0-0.062	0-1.6	1	0.062,0.094	1.6,2.4	0.062	1.6	60-80	8-10	3.4-4.2
		O	B	0-0.062	0-1.6	1	0.062,0.094	1.6,2.4	0.062	1.6	60-75	8-10	3.4-4.2
0.094	2.4	F	B	0-0.094	0-2.4	1	0.125	3.2	0.094	2.4	95-115	8-10	3.4-4.2
		V,H	B	0-0.094	0-2.4	1	0.094,0.125	2.4, 3.2	0.094	2.4	85-110	8-10	3.4-4.2
		O	B	0-0.094	0-2.4	1	0.094,0.125	2.4, 3.2	0.094	2.4	90-110	8-10	3.4-4.2
0.125	3.2	F	B	0-0.125	0-3.2	1-2	0.125,0.156	3.2, 4.0	0.094	2.4	125-150	10-12	4.2-5.1
		V,H	B	0-0.094	0-2.4	1-2	0.125	3.2	0.094	2.4	110-140	10-12	4.2-5.1
		O	B	0-0.094	0-2.4	1-2	0.125,0.156	3.2, 4.0	0.094	2.4	115-140	10-12	4.2-5.1
0.188	4.8	F	D-60°	0-0.125	0-3.2	2	0.156,0.188	4.0, 4.8	0.125	3.2	170-190	10-12	4.2-5.1
		V	D-60°	0-0.094	0-2.4	2	0.156	4.0	0.125	3.2	160-175	10-12	4.2-5.1
		H	D-90°	0-0.094	0-2.4	2	0.156	4.0	0.125	3.2	155-170	10-12	4.2-5.1
		O	D-110°	0-0.094	0-2.4	2	0.156	4.0	0.125	3.2	165-180	10-12	4.2-5.1
0.25	6.4	F	D-60°	0-0.125	0-3.2	2	0.188	4.8	0.156	4.0	220-275	8-10	3.4-4.2
		V	D-60°	0-0.094	0-2.4	2	0.188	4.8	0.156	4.0	200-240	8-10	3.4-4.2
		H	D-90°	0-0.094	0-2.4	2-3	0.156,0.188	4.0, 4.8	0.156	4.0	190-225	8-10	3.4-4.2
		O	D-110°	0-0.094	0-2.4	2	0.188	4.8	0.156	4.0	210-250	8-10	3.4-4.2
0.375[c]	9.5	F	D-60°	0-0.125	0-3.2	2	0.188,0.25	4.8, 6.4	0.188	4.8	315-375	8-10	3.4-4.2
		F	E	0-0.094	0-2.4	2	0.188,0.25	4.8, 6.4	0.188	4.8	340-380	8-10	3.4-4.2
		V	D-60°	0-0.094	0-2.4	3	0.188	4.8	0.188	4.8	260-300	8-10	3.4-4.2
		V,H,O	E	0-0.094	0-2.4	2	0.188	4.8	0.188	4.8	240-300	8-10	3.4-4.2
		H	D-90°	0-0.094	0-2.4	3	0.188	4.8	0.188	4.8	240-300	8-10	3.4-4.2
		O	D-110°	0-0.094	0-2.4	3	0.188	4.8	0.188	4.8	260-300	8-10	3.4-4.2

a. F – flat; H – horizontal; V – vertical; O – overhead.

b. See Figure 1.3. Angle dimension is the appropriate groove angle.

c. May be preheated. Caution: Preheat above 250 °F (121 °C) may significantly affect the as-welded strength of heat-treatable aluminum alloys.

Welding Technique

FOR MANUAL GTAW of aluminum, the torch is held in one hand and the filler rod (if used) in the other. To reduce the possible occurrence of tungsten inclusions in weld starts, the arc may be initiated on a starting block in the joint or away from the base material surface. The arc is then broken and restarted on the warm tungsten in the aluminum weld joint. The arc is held at the starting point until the metal melts and a weld pool is established. Establishment and maintenance of a suitable weld pool is important, and welding must not proceed ahead of the weld pool.

If filler metal is required, it may be added to the front or leading edge of the pool but to one side of the centerline. Both hands are moved in unison, with a slight backward and forward motion along the joint. The tungsten electrode should not touch the filler rod, and the end of the filler rod should not be withdrawn from the argon shield.

A short arc length must be maintained to obtain sufficient penetration and to avoid undercutting, excessive width of the weld bead, and consequent loss of control of penetration and weld contour. One rule is to use an arc length approximately equal to the diameter of the tungsten electrode. The arc and weld pool must be seen by the welder, and this is more difficult with shorter arcs. The gas nozzle should be as small as possible while still providing adequate shielding of the weld pool. The torch should be vertical to the centerline of the joint, at a forehand angle of about 75 to 85 degrees from the plane of the work. When welding unequal sections, the arc should be directed toward the heavier section.

Table 1.18
Typical Procedures for Manual Gas Tungsten Arc Welding of Fillet Welds in Aluminum with AC and Argon Shielding

Section Thickness		Welding Position[a]	No. of Passes	Filler Rod Diameter		Electrode Diameter		Welding Current, A	Travel Speed	
in.	mm			in.	mm	in.	mm		in./min	mm/s
0.062	1.6	F,H,V	1	0.062, 0.094	1.6, 2.4	0.062	1.6	70-110	8-10	3.4-4.2
		O	1	0.062, 0.094	1.6, 2.4	0.062	1.6	65-90	8-10	3.4-4.2
0.094	24	F	1	0.094, 0.125	2.4-3.2	0.094	2.4	110-145	8-10	3.4-4.2
		H,V	1	0.094	2.4	0.094	2.4	90-125	8-10	3.4-4.2
		O	1	0.094	2.4	0.094	2.4	110-135	8-10	3.4-4.2
0.125	3.2	F	1	0.125	3.2	0.094	2.4	135-175	10-12	4.2-5.1
		H,V	1	0.125	3.2	0.094	2.4	115-145	8-10	3.4-4.2
		O	1	0.125	3.2	0.094	2.4	125-155	8-10	3.4-4.2
0.188	4.8	F	1	0.156	4.0	0.125	3.2	190-245	8-10	3.4-4.2
		H,V	1	0.156	4.0	0.125	3.2	175-210	8-10	3.4-4.2
		O	1	0.156	4.0	0.125	3.2	185-225	8-10	3.4-4.2
0.25	6.4	F	1	0.188	4.8	0.156	4.0	240-295	8-10	3.4-4.2
		H,V	1	0.188	4.8	0.156	4.0	220-265	8-10	3.4-4.2
		O	1	0.188	4.8	0.156	4.0	230-275	8-10	3.4-4.2
0.375[b]	9.5	F	2	0.188	4.8	0.188	4.8	325-375	8-10	3.4-4.2
		V	2	0.188	4.8	0.188	4.8	280-315	8-10	3.4-4.2
		H	3	0.188	4.8	0.188	4.8	270-300	8-10	3.4-4.2
		O	3	0.188	4.8	0.188	4.8	290-335	8-10	3.4-4.2

a. F – flat; H – horizontal; V – vertical; O – overhead.

b. May be preheated. Caution: Preheats above 250°F (121 °C) may significantly affect the as-welded strength of heat-treatable aluminum alloys.

Table 1.19
Typical Procedures for Manual Gas Tungsten Arc Welding of Edge and Corner Joints in Aluminum with AC and Argon Shielding

Section Thickness		Joint Geometry[a]	No. of Weld Passes	Filler Rod Diameter		Electrode Diameter		Welding Current,[b,c] A	Travel Speed[c]	
in.	mm			in.	mm	in.	mm		in./min	mm/s
0.062	1.6	I,K	1	0.062,0.094	1.6, 2.4	0.062	1.6	60-85	10-16	4.2-6.8
0.094	2.4	I,K	1	0.125	3.2	0.062	1.6	90-120	10-16	4.2-6.8
0.125	3.2	I,K	1	0.125,0.156	3.2, 4.0	0.094	2.4	115-150	10-16	4.2-6.8
0.188	4.8	J,K	1	0.156	4.0	0.125	3.2	160-220	10-16	4.2-6.8
0.250	6.4	J,K	2	0.188	4.8	0.125	3.2	200-250	8-12	3.4-5.1

a. See Figure 1.3.

b. Use current in low end of range for welding in the horizontal and vertical positions.

c. Higher welding current and travel speed may be used for corner joints if temporary backing is employed.

Welding speed and frequency of adding filler metal are governed by the skill of the welder. When using the correct current, travel speed is higher with less heat dissipation into the part. This promotes progressive solidification and better weld bead control. When the arc is broken, shrinkage cracks may occur in the weld crater resulting in a defective weld. This defect can be prevented by gradually lengthening the arc while adding filler metal to the crater, quickly breaking and restriking the arc several times while adding more filler metal into the crater, or by using foot control to reduce the current while adding filler at the end of a weld. Crater filling devices may be used if properly adjusted and timed.

When the electrode has been contaminated with aluminum, it must be replaced or cleaned. Minor contamination can be burned off by increasing the current while holding the arc on a piece of scrap metal. Severe contamination can be removed with a grinding wheel or by breaking off the contaminated portion of the electrode and reforming the correct electrode contour on a piece of scrap aluminum.

Tacking before welding is helpful in controlling distortion. Tack welds should be of ample size and strength and should be chipped out or tapered at the ends before welding over them.

The joint designs in Figures 1.3, 1.4, and 1.5 and in the accompanying Tables 1.17, 1.18, and 1.19 are applicable to GTAW with minor exceptions. Inexperienced welders who cannot maintain a very short arc may require a wider edge preparation, included angle, or joint spacing. Edge and corner welds in 3003 and the 1XXX series alloys with a tight fit are rapidly made without addition of filler metal and have good appearance. Such joints are not recommended for welding alloys 6061, 6063, 3004, 5052, 7005, 7039, or similar alloys that tend to be hot-short without the addition of filler metal.

Direct Current Electrode Negative

DIRECT CURRENT ELECTRODE negative (DCEN) was once considered unsuitable for welding aluminum because of the absence of any arc cleaning action. Currently this process, using 100% helium and thoriated tungsten electrodes, has proven advantageous for many automatic welding operations, especially in the welding of heavy sections. Since there is less tendency to heat the electrode, small electrodes can be used for a given welding current. This contributes to keeping the weld bead narrow. The use of DCEN provides a greater heat input rate than can be obtained with alternating current. Greater heat density is developed in the weld pool, which results in a weld that is deeper and narrower.

The greater heat input produces rapid melting of the base metal and excellent penetration. It is not necessary to preheat thick sections before welding. Edge preparation can be eliminated and the groove reduced in size so that less filler metal is required. The heating rate is rapid with DCEN; the weld pool is formed immediately resulting in less distortion of the base metal.

The surface appearance of DCEN welds differs from that of alternating current welds, and welders accustomed to alternating current expect to see clean, bright metal on the weld surface during welding. Using DCEN, the welded surface is dulled by an oxide film that is easily removed by light wire brushing. The surface oxide does not indicate lack of fusion, porosity, or inclusions in the weld. There is no arc cleaning action using DCEN and thorough preweld cleaning of the base metal is necessary. This normally involves degreasing, chemical cleaning, wire brushing, plus scraping or filing of the joint area.

GTAW with DCEN has distinct advantages compared to ac power, particularly with machine welding where a consistent, short arc length can be easily maintained. The deep penetration possible with helium shielding is particularly useful for welding thick sections. With thin sections, DCEN permits much higher travel speed than ac.

Argon shielding may be used with DCEN, but penetration will be less than with helium. Arc length control will not be so critical, and may be beneficial, when manually welding thin sections.

Joint Geometry. A square-groove joint geometry may be satisfactory for some section thicknesses and alloy types suitable for autogenous welding that would usually require a V-groove joint with ac power. A square-grooved weld in 0.75 in. (19 mm) thick aluminum plate is shown in Figure 1.6. The weld was made with two passes, one from each side, without filler metal. Helium shielding and DCEN power were used.

When a V-groove joint is necessary, the root face may have a greater width and the groove angle may be less than those needed with an ac arc. Examples of joint designs for thick sections are shown in Figure 1.7.

Manual GTAW. The technique for manual GTAW with DCEN power is somewhat different from that used with ac power. Superimposed high frequency should be used to initiate the arc, and it should shut off automatically when the arc is established. Generally, it is not necessary to hesitate to form a molten weld pool because melting by the arc is rapid. However, a delay in travel may be necessary to achieve desired penetration at the start of the weld. A foot-operated control is recommended for adjusting the welding current as needed.

The welding torch should be moved steadily forward, and the filler rod should be fed evenly into the leading edge of the molten weld pool. The crater should be

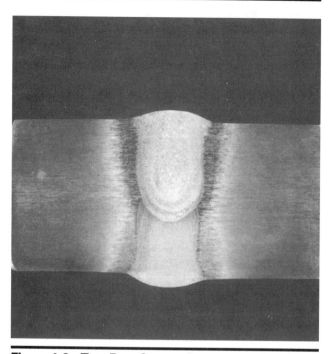

Figure 1.6—Two-Pass Square-Groove Weld in 0.75 in. (19 mm) Thick Aluminum Plate Welded from Both Sides without Filler Metal with DCEN Power (x2)

completely filled to eliminate cracking. Typical procedures for manual GTAW with DCEN power are given in Tables 1.20 and 1.21. These are only guides and have not been substantiated by appropriate tests to a particular standard.

Machine GTAW-DCEN. Precision equipment is available for mechanized welding with DCEN power. It may provide upslope and downslope of welding current as well as automatic control of shielding gas flow, wire feed, and travel. In addition, a programmer may be included for automation of the welding torch movement, welding current, and arc voltage.

For welding square-groove butt joints, a very short arc is used, and the tip of the tungsten electrode may be positioned at or below the base metal surface. The position of the tungsten electrode tip affects weld-bead penetration and width as well as undercutting and weld-metal turbulence. The optimum position of the electrode tip will likely depend upon the base-metal composition, thickness, and the welding procedure. In most cases, weld backing is not needed when machine welding square-groove butt joints from one side.

Typical machine welding procedures that may be used as guides in establishing suitable procedures for a

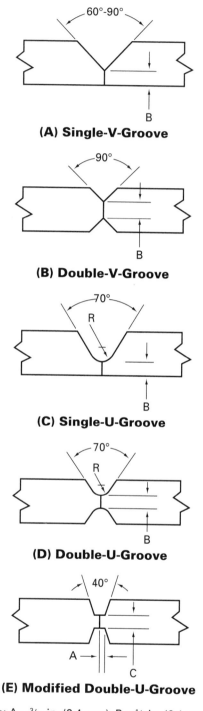

(A) Single-V-Groove

(B) Double-V-Groove

(C) Single-U-Groove

(D) Double-U-Groove

(E) Modified Double-U-Groove

Note: A = $^3/_{32}$ in. (2.4 mm), B = $^1/_4$ in. (6.4 mm), C = $^3/_8$ in. (9.5 mm), R = $^1/_8$ in. (3.2 mm).

Figure 1.7—Typical Joint Designs for Gas Tungsten Arc Welding Thick Aluminum Sections with DCEN Power

Table 1.20
Typical Procedures for Manual Gas Tungsten Arc Welding Butt Joints in Aluminum
with DCEN and Helium Shielding in Flat Position

| Section Thickness | | Joint Geometry* | No. of Weld Passes | Filler Rod Diameter | | Electrode Diameter | | Arc Voltage, V | Welding Current, A | Travel Speed | |
in.	mm			in.	mm	in.	mm			in./min	mm/s
0.032	0.8	A	1	0.045	1.1	0.040	1.0	21	20	17	7.2
0.040	1.0	A	1	0.062	1.6	0.040	1.0	20	26	16	6.8
0.062	1.6	A	1	0.062	1.6	0.040	1.0	20	44	20	8.5
0.094	2.4	A	1	0.094	2.4	0.062	1.6	17	80	11	4.7
0.125	3.2	A	1	0.125	3.2	0.062	1.6	15	118	16	6.8
0.25	6.4	A	1	0.156	4.0	0.125	3.2	14	250	7	3.0
0.50	12.7	B	2	0.156	4.0	0.125	3.2	14	310	5.5	2.3
0.75	19.0	C	2	0.156	4.0	0.125	3.2	17	300	4	1.7
1.00	25.4	C	5	0.25	6.4	0.125	3.2	19	360	1.5	0.6

* A – square-groove; B – single-V-groove, per Figure 1.7(A); C – double-V-groove, per Figure 1.7(B)

particular application are given in Tables 1.22 through 1.25.

Direct Current Electrode Positive

WELDING WITH DIRECT current electrode positive (DCEP) provides good surface cleaning action and permits welding of thin aluminum sections with sufficient current to maintain a stable arc. Argon shielding should be used because helium or argon-helium mixtures would contribute to electrode overheating. The weld bead tends to be wide and penetration is shallow.

Edge or square-groove joint geometries with or without filler metal may be applicable. Typical conditions

for manual welding with DCEP power are given in Table 1.26.

Square-Wave Alternating Current

SQUARE-WAVE ALTERNATING current (SWAC) differs from conventional ac balanced wave power with respect to the current wave form. The SWAC power source is designed to produce dc power and rapidly shift the polarity to produce a square alternating wave form. The relative percentage of electrode negative time within one cycle of current is adjustable within limits.

This type of power combines the advantage of surface cleaning associated with conventional ac power

Table 1.21
Typical Procedures for Manual Gas Tungsten Arc Welding of Fillet Welds in Aluminum
with DCEN and Helium Shielding

| Section Thickness | | Welding Position* | Fillet Size | | Filler Rod Diameter | | Electrode Diameter | | Arc Voltage, V | Welding Current, A | Travel Speed | |
in.	mm		in.	mm	in.	mm	in.	mm			in./min	mm/s
0.094	2.4	H	0.125	3.2	0.094	2.4	0.094	2.4	14	130	21	8.9
0.125	3.2	H	0.125	3.2	0.094	2.4	0.094	2.4	14	180	18	7.6
0.25	6.4	H	0.188	4.8	0.156	4.0	0.125	3.2	14	255	15	6.3
0.25	6.4	V	0.188	4.8	0.156	4.0	0.125	3.2	14	230	10	4.2
0.375	9.6	H	0.188	4.8	0.156	4.0	0.125	3.2	14	335	14	5.9
0.375	9.6	H	0.312	7.9	0.250	6.4	0.125	3.2	14	290	7	3.0
0.50	12.7	H,V	0.312	7.9	0.250	6.4	0.125	3.2	16	315	6-7	2.5-3.0

* H – horizontal; V – vertical

Table 1.22
Typical Procedures for Machine Gas Tungsten Arc Welding of Square-Groove
Butt Joints in Aluminum with DCEN and Helium Shielding[a]

Section Thickness		Electrode Diameter		Filler Wire Feed Rate[b]		Arc Voltage, V	Welding Current, A	Travel Speed	
in.	mm	in.	mm	in./min	mm/s			in./min	mm/s
0.040	1.0	0.040	1.0	60	25.4	14	65	54	22.9
0.094	2.4	0.094	2.4	75	31.7	13	180	54	22.9
0.125	3.2	0.094	2.4	55	23.3	11	240	40	16.9
0.25	6.4	0.125	3.2	40	16.9	11	350	15	6.3
0.375	9.6	0.156	4.0	30	12.7	11	430	8	3.4

a. Single-pass weld in flat position.

b. Filler wire of 0.062 in. (1.6 mm) diameter.

and deep penetration obtainable with DCEN power. However, one advantage is gained with some sacrifice in the other. If longer electrode-positive time is needed for acceptable cleaning, penetration will decrease with a specific welding current.

The square-wave shape enhances arc re-ignition during polarity reversal. In early power supplies, the superimposed high frequency was needed only to start the arc and was not needed continuously during welding to stabilize the arc. Present equipment with a lower open-circuit voltage (OCV) requires the high frequency to be superimposed continuously, even when using SWAC.

Welding techniques similar to those for conventional ac welding, such as the electrode tip shape, are suitable

Table 1.23
Typical Procedures for Machine Gas Tungsten Arc Welding of Square-Groove Butt Joints
in 2219 Aluminum Alloy with DCEN and Helium Shielding[a,b]

Section Thickness		Welding Position[c]	Electrode				Filler Wire Feed Rate[e]		Welding Current, A	Travel Speed	
			Diameter		Position[d]						
in.	mm		in.	mm	in.	mm	in./min	mm/s		in./min	mm/s
0.25	6.4	F	0.125	3.2	0.10	2.5	36	15.2	145	8	3.4
		H							135	10	4.2
0.38	9.7	F	0.125	3.2	0.10	2.5	32	13.5	220	8	3.4
		H							180	10	4.2
0.50	12.7	H,V	0.125	3.2	0.10	2.5	10	4.2	250	8	3.4
0.63	16.0	H,V	0.125	3.2	0.10	2.5	5-7	2.1-3.0	300	7	3.0
0.75	19.0	H,V	0.125	3.2	0.10	2.5	5-7	2.1-3.0	340	6	2.5
0.88	22.4	H,V	0.156	4.0	0.12	3.0	4-6	1.7-2.5	385	5	2.1
1.00	25.4	H,V	0.188	4.8	0.15	3.8	3-5	1.3-2-1	425	4	1.7

a. Weld with two passes, one from each side.

b. Arc voltage is adjusted at 11.5 to 12.5 V.

c. F – flat; H – horizontal; V – vertical.

d. Distance of the electrode tip below the base metal surface.

e. ER 2319 filler metal, 0.062 in. (1.6 mm) diameter.

Table 1.24
Typical Procedures for Machine Gas Tungsten Arc Welding of Square-Groove Butt Joints in 7039 Aluminum with DCEN and Helium Shielding

Section Thickness		Welding Position[a]	Electrode Diameter		Welding Current, A	Arc Voltage, V	Travel Speed	
in.	mm		in	mm			in./min	mm/s
0.25	6.4	F	0.094	2.4	200	10	20	8.5
		H	0.125	3.2	280	10	20	8.5
		V-up	0.125	3.2	260	9.5	20	8.5
		V-down	0.125	3.2	285	9	26	11.0
0.38	9.7	F	0.125	3.2	300	12	16	6.8
		H	0.125	3.2	380	9	12	5.1
		V-up	0.125	3.2	300	11	16	6.8
		V-down	0.156	4.0	410	11	18	7.6
0.50	12.7	F	0.250	6.4	350	18	13	5.5
		H	0.156	4.0	400	9.5	9	3.8
		V-up	0.156	4.0	400	10	13	5.5
		V-down	0.156	4.0	400	10	10	4.2
0.75	19.0	F	0.125	3.2	375	22	6	2.5
		H	0.188	4.8	500	9	5	2.1
		V-up	0.156	4.0	470	11	5	2.1
		V-down	0.188	4.8	550	11	10	4.2

a. No root opening. Welded with two passes, one from each side, without filler metal.

b. F – flat; H – horizontal; V – vertical with direction of torch travel up or down.

with SWAC welding. Argon shielding is preferred, but argon-helium mixtures can be used. With 100% helium, the required arc voltage may exceed the OCV of present SWAC welders. If this is the case, the arc cannot be initiated with 100% helium.

Joint Geometry. Square-wave ac welding offers some savings in weld joint preparation over conventional ac balanced-wave GTAW. Smaller V-grooves, U-grooves, and a thicker root face can be used, and the greater depth-to-width weld ratio is conducive to less weldment distortion, more favorable welding residual stress distribution, and less use of filler wire. With some slight modification, the same joint geometry can be used as in GTAW-DCEN.

Welding Technique. The welding technique is very similar to conventional ac welding, and it is necessary to have either superimposed high frequency or high open-circuit voltage for arc stabilization. Pure or zirconia-tungsten electrodes, which "ball" on the end, are used as in conventional ac-GTAW. Argon should be used to start the arc and can be used as the shielding gas, but helium will give deeper penetration. A combination of the two gases may be used.

Electrodes

THE CHOICE OF tungsten electrode depends upon the type of welding current selected for the application. With conventional ac, better arc action is obtained when the electrode has a hemispherical-shaped tip, as shown in Figure 1.8. AWS classifications EWP (pure tungsten) and EWZr (tungsten-zirconia) electrodes retain this tip shape well.[4] Fine tungsten inclusions result when using thoria-tungsten electrodes with ac power.

Classifications EWTh-1 or EWTh-2 (tungsten-thoria) electrodes are suitable with direct current only. Both classifications have higher electron emissivity, better current-carrying capacity, and longer life than do EWP electrodes. Consequently, arc starting is easier and the arc is more stable.

Pure or zirconiated tungsten electrodes are satisfactory for ac welding. The zirconiated electrode is less prone to electrode contamination by base or filler metal and has a slightly higher current rating. Thoriated

---

4. ANSI/AWS A5.12, *Specification for Tungsten and Tungsten Alloy Electrodes for Arc Welding and Cutting.*

Table 1.25
Typical Procedures for Machine Gas Tungsten Arc Welding of Single-V-Groove Joints in 7039 Aluminum with DCEN and Helium Shielding

Section Thickness		Joint Design*	Welding Position	Pass No.	Welding Current, A	Arc Voltage, V	Travel Speed		Filler Metal Diameter		Filler Metal Feed Rate	
in.	mm						in./min	mm/s	in.	mm	in./min	mm/s
0.25	6.4	1	Vertical-up	1	175	10	12	5.1	0.063	1.6	70	29.6
				2	185	9	9	3.8	None	None	None	None
0.38	9.7	2	Vertical-up	1,2,3	215-220	11-11.5	12	5.1	0.063	1.6	51-56	21.6-23.7
				4	215-220	10	8	3.4	None	None	None	None
0.50	12.7	2	Vertical-up	1	250-270	11-13	10	4.2	0.063	1.6	60-69	25.4-29.2
				2,3	250-270	11-13	7.5	3.2	0.063	1.6	60-69	25.4-29.2
				4	250-270	11.5	8.5	3.6	None	None	None	None
0.63	16.0	3	Vertical-up	1	320	12	12	5.1	0.093	2.4	32	13.5
				2,3,4	320	12	12	5.1	0.093	2.4	28-44	11.8-18.6
				5	320	11	12	5.1	None	None	None	None
0.75	19.0	3	Vertical-up	1	380	10.5	8	3.4	0.093	2.4	39	16.5
				2,3	370-380	12	8	3.4	0.093	2.4	39	16.5
				4	370	11.5	8	3.4	0.093	2.4	33	14.0
				5	380	10	6		None	None	None	None
0.25	6.4	1	Horizontal	1	225	9.5	12	5.1	0.093	2.4	25	10.6
				2	225	9	9	3.8	None	None	None	None
0.38	9.7	2	Horizontal	1,2	300	11.5	10	4.2	0.063	1.6	68	28.8
				3	300	11	6.5	2.7	0.063	1.6	56	24.5
				4	310	10	6.5	2.7	None	None	None	None
0.50	12.7	2	Horizontal	1	310	11	10	4.2	0.063	1.6	68	28.8
				2,	310	10	8.5-9	3.6-3.8	0.063	1.6	56-68	23.7-28.8
				4	320	9	6	2.5	None	None	None	None
0.63	16.0	3	Horizontal	1	330	11	9	3.8	0.093	2.4	30	12.7
				2.3	315-325	12-12.5	8-9	3.4-3.8	0.093	2.4	40-44	16.9-18.6
				4	310	13	8	3.4	0.093	2.4	34	14.4
				5	330	11	6	2.5	None	None	None	None
0.75	19.0	3	Horizontal	1	350	10.5	8	3.4	0.093	2.4	39	16.5
				2,3	350	12.5	8	3.4	0.093	2.4	39	16.5
				4	350	12.5	8	3.4	0.093	2.4	33	14.0
				5	360	10	6	2.5	None	None	None	None

* The following joint designs are referred to by number in the indicated Joint Design column.

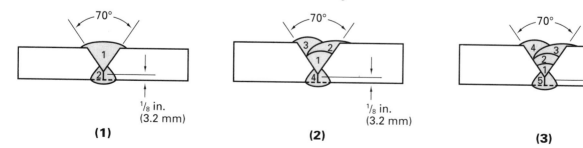

(1)   (2)   (3)

Table 1.26
Typical Conditions for Manual Gas Tungsten Arc Welding of Aluminum with DCEP and Argon Shielding*

Section Thickness		Electrode Diameter		Welding Current,	Filler Rod Diameter	
in.	mm	in.	mm	A	in.	mm
0.010	0.25	0.094	2.4	15-25	None	None
0.020	0.51	0.125, 0.156	3.2, 4.0	40-55	0.020	.51
0.030	0.76	0.188	4.8	50-65	0.020-0.047	.51
0.040	1.0	0.188	4.8	60-80	0.047	1.2
0.050	1.3	0.188	4.8	70-90	0.047-0.062	1.2-1.6

* An edge or square-groove weld should be used.

tungsten electrodes are not generally recommended for welding with ac.

Figure 1.9 shows current ranges for standard tungsten electrodes with both DCEN and ac, while Figure 1.10 pertains to thoriated tungsten electrodes.

Shielding Gases For GTAW

ARGON IS THE most commonly used shielding gas, particularly for manual welding, but helium is used in special cases. Arc voltage characteristics with argon permit greater arc length variations with minimal effect on arc power than does helium. Argon also provides better arc starting characteristics and improved cleaning action with alternating current.

Helium is used primarily for machine welding with DCEN power. It permits welding at higher travel speeds or with greater penetration than does argon.

Helium-argon mixtures are sometimes used to take advantage of the higher heat input rate with helium while maintaining the favorable arc characteristics of argon. A mixture of 25% helium, 75% argon will permit higher travel speeds with ac power. Cleaning action is still acceptable. If helium is over 25%, rectification of the ac may be experienced with low open-circuit voltage equipment. A mixture of 90% helium, 10% argon

will provide better arc starting characteristics with dc power than does pure helium.

GAS METAL ARC WELDING

Equipment

A WIDE CHOICE of equipment is available for gas metal arc welding (GMAW).[5] Selection of the best type for a particular job in a particular shop depends on many factors. Some of these are welding position, size of weldment, amount of welding, and desired production rate. A wise choice can be made only after all of the factors have been considered in terms of the types available.

The three main equipment elements for GMAW are the electrode feeding mechanism, the gun, and the power source. When making spot welds, these are augmented by a controller to regulate the time of current application.

Electrode Drive Systems. In semiautomatic welding, there are three drive systems for delivering the electrode from the spool to the arc. These systems are designated push, pull, and push-pull. In the push systems, the drive rolls are located near the spool and the wire electrode is pushed to the gun through a conduit that is about 10 to 12 ft (3.0 to 3.6 m) long. In the pull system, the rolls are located at the gun and the electrode is pulled through the conduit. The pull distance through a conduit also is limited to 10 to 12 ft (3.0 to 3.6 m) for aluminum electrodes. In the push-pull system, the electrode is fed through the conduit by two or more sets of rolls, one near the spool and one in the gun. This permits conduit lengths of 25 ft (6.6 m) or more. Some systems include conduit modules with a drive at one end of

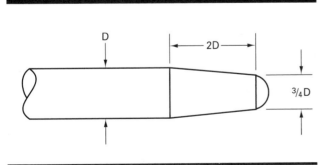

Figure 1.8—Approximate Tungsten Electrode Tip Shape for Arc Welding Aluminum with AC Power

5. For a more detailed discussion of gas metal arc welding refer to Chapter 4, *Welding Handbook*, Vol. 2, 8th Ed.

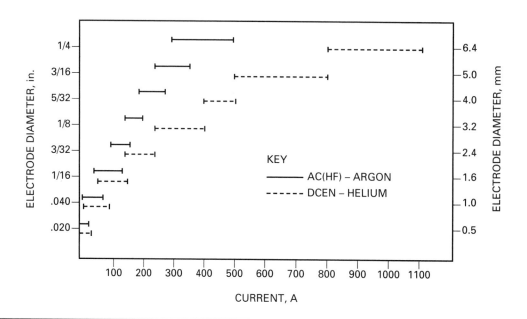

Figure 1.9—Current Ranges for Pure Tungsten Electrodes, AC vs. DCEN

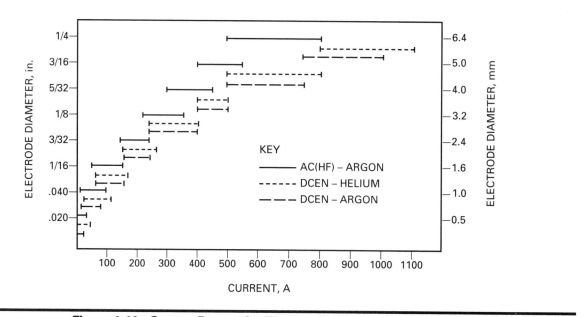

Figure 1.10—Current Ranges for Thoriated Tungsten Electrodes

each. Several modules can be connected in series to feed over long distances.

Systems selection is based largely upon the diameter and tensile strength of the electrode and the distance between the electrode supply and the work. The push system is limited to high-strength electrodes, 0.047 in. (1.2 mm) diameter and larger, that have adequate stiffness to overcome the drag in the conduit and welding gun.

Pull-type feed systems are common with mechanized welding heads and with some semiautomatic welding guns. These systems are used with 3/64 in. (1.2 mm) and smaller electrode sizes of any aluminum alloy. In automatic or machine welding, the electrode drive unit is normally located just above the welding gun. The push-pull systems are applicable to most electrodes and available sizes.

The electrode wire drive motor must have ample power to feed the wire at uniform speed. Wire passages should be free of sharp bends or discontinuities and be perfectly aligned. Radiused (not V-grooved) drive rolls on top and bottom are preferred over serrated rolls because they do not mark the wire nor load the electrode conduit with fine aluminum chips shaved from the electrode or wire by the drive rolls. The grooves must be perfectly aligned, and there should be a minimum of friction or eccentricity. Drive-roll pressure is critical, and a means for accurate adjustment must be provided. Drive rolls should be located as close to the arc as practical, thus affording a minimum electrode wire column to the tip of the contact tube to assure good transfer of current at the end of the tube. Wire straighteners are sometimes used to reduce the cast in the electrode wire as it comes off the spool and to keep the electrode and arc directed properly.

Bent contact tubes restrict feeding of low-strength aluminum electrodes. Contact tubes of short lengths [0.75 to 1 in. (19 to 25 mm)] limit the points of current commutation to the aluminum electrode, whose oxide has insulating characteristics. High current surges resulting from a fast electrode run-in with constant-voltage power cause arcing inside the contact tube and result in frequent burn-backs when attempting arc starts. Straight contact tubes of sufficient length [4 to 6 in. (102 to 152 mm)] are preferred so that the electrode has numerous contacts with the inside diameter of the contact tube to minimize arcing. A slow run-in feed for arc starting substantially reduces the amplitude of current surges, thus minimizing the arcing. The longer contact tubes also can straighten the lower-strength aluminum alloy electrodes to stabilize the position of the electrode as it emerges from the contact tube. When spool, drive rolls, and electrode holder are suitably oriented and aligned, wire should feed uniformly without wandering.

If a constant-current power source is used, a slow electrode run-in will be required for machine GMAW. If the wire strikes the plate at full welding speed, a constant current power supply will not be able to initiate an arc. The short-circuit current is not high enough. This is overcome by using a slow run-in, which advances the electrode to the work at a slow speed, switching to full welding speed when the arc has been established. A slow run-in also is desirable with a constant-arc-voltage power source to minimize arcing in

the contact tube, reduce weld buildup at starts, and provide improved weld fusion at starts. The reduced buildup at starts also reduces defects when overlapping the start in circumferential welds.

When semiautomatic gas metal arc welding with a constant-current (drooping volt/amp characteristic) power supply, a touch-start feature is preferred; however, a slow electrode run-in can be used.

Guns. Semiautomatic welding guns may be either pistol-shaped or possess straight bodies with slightly curved nozzles. Both water-cooled and air-cooled guns are available.

For mechanized and automatic welding, the guns are almost always straight and of circular cross section. Gas shielding should be concentric around the electrode.

The choice of gun design depends upon a number of factors. The most important factor is the ease with which the gun can be manipulated for a particular joint design and welding position.

Power Sources. Welding power sources are designated as having one of two types of electrical characteristics, variable voltage (constant current) or constant voltage (constant potential), which refer to the change in output voltage as the welding current varies. Both may be used satisfactorily if current is of relatively constant magnitude. The most widely used is the constant-voltage machine since it also is used with ferrous metals. The constant-current, variable-voltage power source is preferred for most uniform heat input and highest quality welds, particularly when the gun is oscillated or manipulated. Under stable welding conditions, no difference can be discerned. Differences become evident when the welding conditions must be adjusted or when transient events occur such as arc ignition or changes in gun-to-work distance.

Shielding Gas for GMAW

THE SHIELDING GAS serves a secondary function in addition to shielding the molten metal. By adjusting the composition of the shielding gas (which is either argon, helium, or a mixture of the two), some control is provided over the distribution of heat to the weld. This, in turn, influences the shape of the weld-metal cross section and the speed of welding. Adjustment of the cross section by modification of the shielding-gas composition can control penetration without changing electrode melting rate to facilitate evolution of gas from the weld metal. However, use of helium only as a shielding gas usually results in an unstable arc and is seldom used for that reason.

Typical cross sections of bead welds that were made using argon, helium, and argon-helium mixtures are shown in Figure 1.11. The relatively narrow and deep

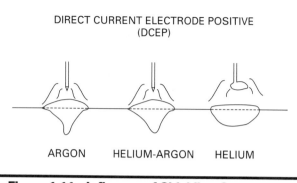

DIRECT CURRENT ELECTRODE POSITIVE
(DCEP)

ARGON   HELIUM-ARGON   HELIUM

**Figure 1.11—Influence of Shielding Gas on Weld
Profile**

argon-shielded weld may favor gas entrapment. The deep and narrow projection at the center of the argon-shielded weld can be used to gain penetration at a given level of current density; however, when making a joint with one weld on each side, the weld placement must be carefully controlled to ensure overlap of the two passes at the center. For a specific arc length, the addition of helium will increase the arc volts by 2 or 3 volts as compared to argon alone. Since the weld profile broadens and deepens to a maximum at 75% He-25% Ar, this mixture has been a common choice for achieving sound welds when welding heavy plate from both sides. For most welding, argon is the preferred gas.

Shielding gas is supplied from cylinders or a bulk supply via pipelines. Cylinders are used if the rate of consumption is low, the working area is large, or the location of weldments varies greatly. Bulk supplies are favored in high-volume production where the welding is done at a number of fixed stations. The purity of the shielding gas is of utmost importance. Only gases having a dew point of -76 °F (-60 °C) or better should be used, and care must be taken to prevent contamination. Dust, dirt, and moisture can accumulate in the cylinder fitting, which should be carefully cleaned and blown out before use. Plastic hose is recommended. All hose connections and other fittings must be pressure-tight since entrance of air or escape of shielding gas will affect the weld.

Argon is the most commonly used shielding gas for GMAW semiautomatic welding in the spray transfer mode. It provides excellent arc stability, bead shape, and penetration. This gas may be used in all welding positions. Helium is suitable for mechanized and automatic welding with high currents in the flat position. Helium-argon mixtures are preferred for semiautomatic welding with 5XXX aluminum alloy electrodes and are sometimes used instead of helium to take advantage of the arc stability provided by argon for mechanized and automatic welding. The preferred mixtures contain

from 50 to 75% helium. With a given arc length, arc voltage increases with increasing helium content.

Metal Transfer

WITH DCEP POWER, the filler metal will be transferred across the arc as a stream of fine, superheated droplets when the welding current and arc voltage are above certain threshold values. These values will depend upon the electrode alloy, size, and feed rate. This mode of transfer is known as *spray transfer.* It is the one normally used for GMAW of aluminum. The spray transfer mode may be continuous or intermittent. Intermittent transfer is called *pulsed spray welding.*

Pulsed spray welding may be used for welding in all positions. Metal transfer takes place during the periods of high welding current, but ceases during the intervening periods of low current. However, the cleaning action is continuous at both current levels. This pulsing action reduces the overall heat input to the base metal for good control of the molten weld pool and penetration. The lower heat input makes it easier to gas metal arc weld thin aluminum sections. Pulsing also allows the use of larger electrodes than what is practical with steady current, especially on thin metal.

The amount of spatter during welding may be more severe with electrodes that contain low-vapor-pressure elements. The aluminum-magnesium alloy electrodes (ER5XXX series) commonly cause the most spatter. The vapor pressure of the magnesium tends to cause disintegration of the droplets as they separate from the electrode tip. This produces small spatter balls that are often thrown clear of the arc.

When the arc voltage (arc length) is decreased to below a certain value for a specific electrode and amperage, the size of the droplets will increase and the form of the arc will change. Electrode melting rate and penetration will increase. This has a distinct advantage when welding thick sections.

When the arc voltage is in the spray transfer range and the welding current is decreased to below some threshold value, metal transfer will change from spray to globular type. This latter type is not suitable for aluminum welding because fusion with the base metal will be incomplete. Conversely, when the arc voltage (length) is decreased significantly with adequate current, short circuiting will occur. This type of transfer also is not recommended for aluminum for the same reason.

Welding Procedures

TYPICAL GMAW PROCEDURES using small and large diameter electrodes are given in Tables 1.27 through 1.30. Small diameter electrodes can be used for semiautomatic welding in all positions; large diameter

Table 1.27
Typical Procedures for Gas Metal Arc Welding of Groove
Welds in Aluminum Alloys with Argon Shielding

Section Thickness		Welding Position[a]	Joint Geometry[b]	Root Opening		No. of Weld Passes	Electrode Diameter		Welding Current (DCEP), A	Arc Voltage, V	Shielding Gas Flow Rate[c]		Travel Speed	
in.	mm			in.	mm		in.	mm			ft³/h	L/min	in./min	mm/s
0.06	1.6	F	A	0	0	1	0.030	0.8	70-110	15-20	25	12	25-45	10.5-19.0
		F	G	0.09	2.4									
0.09	2.4	F	A	0	0	1	.030-.047	0.8-1.2	90-150	18-22	30	14	25-45	10.5-19.0
		F,V,H,O	G	0.12	3.2	1	0.030	0.8	110-130	18-23	30	14	23-30	9.7-12.7
0.12	3.2	F,V,H	A	0.09	2.4	1	.030-.047	0.8-1.2	120-150	20-24	30	14	24-30	10.2-12.7
		F,V,H,O	G	0.19	4.6	1	.030-.047	0.8-1.2	110-135	19-23	30	14	18-28	7.6-11.8
0.19	4.8	F,V,H	B	0.06	1.6	2	.030-.047	0.8-1.2	130-175	22-26	35	16	24-30	10.3-12.7
		F,V,H	F	0.06	1.6	1	.047	1.2	140-180	23-27	35	16	24-30	10.3-12.7
		O	F	0.06	1.6	2	.047	1.2	140-175	23-27	60	28	24-30	10.3-12.7
		F,V	H	0.09-0.19	2.4-4.8	2	.047-.062	1.2-1.6	140-185	23-27	35	16	24-30	10.3-12.7
		H,O	H	0.19	4.8	3	.047	1.2	130-175	23-27	60	28	25-35	10.5-14.8
0.25	6.4	F	B	0.09	2.4	2	.047-.062	1.2-1.6	175-200	24-28	40	19	24-30	10.3-12.7
		F	F	0.09	24	2	.047-.062	1.2-1.6	185-225	24-29	40	19	24-30	10.3-12.7
		V,H	F	0.09	2.4	3F,1R	.047	1.2	165-190	25-29	45	21	25-35	10.5-14.8
		O	F	0.09	2.4	3F,1R	.047-.062	1.2-1.6	180-200	25-29	60	28	25-35	10.5-14.8
		F,V	H	0.12-0.25	3.3-6.4	2-3	.047-.062	1.2-1.6	175-225	25-29	40	19	24-30	10.3-12.7
		O,H	H	0.25	6.4	4-6	.047-.062	1.2-1.6	170-200	25-29	60	28	25-40	10.5-16.9
0.38	9.6	F	C-90°	0.09	2.4	1F,1R	.062	1.6	225-290	26-29	50	24	20-30	8.5-12.7
		F	F	0.09	2.4	2F,1R	.062	1.6	210-275	26-29	50	24	24-35	10.3-14.8
		V,H	F	0.09	2.4	3F,1R	.062	1.6	190-220	26-29	55	26	24-30	10.3-12.7
		O	F	0.09	2.4	5F,1R	.062	1.6	200-250	26-29	80	38	25-40	10.5-16.9
		F,V	H	0.25-0.38	6.4-9.6	4	.062	1.6	210-290	26-29	50	24	24-30	10.3-12.7
		O,H	H	0.38	9.6	8-10	.062	1.6	190-260	26-29	80	38	25-40	10.5-16.9
0.75	19.0	F	C-60°	0.09	2.3	3F,1R	.062-.094	1.6-2.4	340-400	26-31	60	28	14-20	5.9-8.5
		F	F	0.12	3.2	4F,1R	.094	2.4	325-375	26-31	60	28	16-20	6.8-8.5
		V,H,O	F	0.06	1.6	8F,1R	.062	1.6	240-300	26-31	80	38	24-30	10.3-12.7
		F	E	0.06	1.6	3F,3R	.062	1.6	270-330	26-31	60	28	16-24	6.8-10.3
		V,H,O	E	0.06	1.6	6F,6R	.062	1.6	230-280	26-31	80	38	16-24	6.8-10.3

a.   F – flat; V – vertical; H – horizontal; O – overhead

b.   Refer to Figure 1.3

c.   Nozzle ID = 5/8 to 3/4 in. (16 to 19 mm)

[1/8 in. (3.2 mm) or longer] electrodes can be used only in the flat position with automatic welding.

Constant-voltage power and constant-speed electrode drives are normally used with small diameter electrodes less than 3/64 in. (1.2 mm). The electrode feed rate is adjusted to obtain the desired welding current for good fusion and penetration. Arc voltage is adjusted to give a spray transfer mode of filler metal. A constant-current power source and constant-speed electrode drive should be used with 3/64 in. (1.2 mm) and larger diameter electrodes. The welding current is set at the desired value, and the arc voltage is set by adjusting the wire feed speed to the desired arc length. The arc length (voltage) is critical with respect to good fusion with the groove faces. If the voltage is too low, short circuiting will take place between the electrode and the weld pool.

Table 1.28
Typical Procedures for Gas Metal Arc Welding of Fillet Welds in Aluminum Alloys with Argon Shielding

Section Thickness		Welding Position[a]	No. of Passes	Electrode Diameter		Welding Current (DCEP), A	Arc Voltage, V	Shielding Gas Flow Rate[b]		Travel Speed	
in.	mm			in.	mm			ft³/h	L/min	in./min	mm/s
0.094	2.4	F,V,H,O	1	0.030	0.8	100-130	18-22	30	14	24-30	10-13
0.125	3.2	F	1	0.030-0.047	0.8-1.2	125-150	20-24	30	14	24-30	10-13
		V,H	1	0.030	0.8	110-130	19-23	30	14	24-30	10-13
		O	1	0.030-0.047	0.8-1.2	115-140	20-24	40	19	24-30	10-13
0.19	4.8	F	1	0.047	1.2	180-210	22-26	30	14	24-30	10-13
		V,H	1	0.030-0.047	0.8-1.2	130-175	21-25	35	16	24-30	10-13
		O	1	0.030-0.047	0.8-1.2	130-190	22-26	45	21	24-30	10-13
0.25	6.4	F	1	0.047-0.062	1.2-1.6	170-240	24-28	40	19	24-30	10-13
		V,H	1	0.047	1.2	170-210	23-27	45	21	24-30	10-13
		O	1	0.047-0.062	1.2-1.6	190-220	24-28	60	28	24-30	10-13
0.38	9.6	F	1	0.062	1.6	240-300	26-29	50	24	18-25	8-11
		H,V	3	0.062	1.6	190-240	24-27	60	28	24-30	10-13
		O	3	0.062	1.6	200-240	25-28	65	31	24-30	10-13
0.75[c]	19.0	F	4	0.094	2.4	360-380	26-30	60	28	18-25	8-11
		H,V	4-6	0.062	1.6	260-310	25-29	70	33	24-30	10-13
		O	10	0.062	1.6	275-310	25-29	85	40	24-30	10-13

a.   F – flat; V – vertical; H – horizontal; O – overhead.

b.   Nozzle ID = 5/8 to 3/4 in. (15.9 to 19 mm)

c.   For thicknesses of 0.75 in. (19 mm) and larger, double-bevel joint with a 50 degree minimum groove angle and 0.09 to 0.13 in. (2.3 mm to 3.3 mm) root face is sometimes used.

Welding procedures for pulsed spray welding depend upon the design of the power source. The welding variables may include part or all of the following:

(1)   Electrode diameter
(2)   Wire feed speed
(3)   Shielding gas
(4)   Pulse rate
(5)   Pulse peak voltage
(6)   Average welding current
(7)   Background current
(8)   Travel speed

The recommendations of the welding power source manufacturer should be followed when developing procedures for pulse spray welding of aluminum.

Automatic GMAW

THE GMAW PROCESS is easily adapted to automatic welding. Because of the fixed torch angle and nozzle-to-work distance, the weld receives adequate gas coverage, and the automatic torch travel allows for higher welding speeds than in manual welding. Longitudinal or circumferential welds can be made without intermediate stops and starts, which virtually eliminates the weld termination crater-crack problem. With automatic welding, higher current can be used than with the manual method (the upper range is limited only by arc stability).

Welding current above 360 A is not satisfactory with constant-voltage power. Welding currents of up to 500 A can be used with drooping volt-amp characteristic power supplies and argon shielding. Considerably higher welding current, even greater than 750 A, is practical with helium or argon-helium mixtures. The

Table 1.29
Typical Conditions for Flat Position GMAW Groove Welds in Aluminum
Alloys with Large Diameter Electrodes

Section Thickness		Joint Geometry				Electrode Diameter		Shielding Gas	Weld Pass[2]	Arc Voltage, V	Welding Current (DCEP), A[3]	Travel Speed	
				F									
in.	mm	Type[1]	α, Degrees	in.	mm	in.	mm					in./min	mm/s
0.75	19.0	A	90	0.25	6.3	0.156	4.0	Ar	1 2	28	450 500	16	6.8
1.00	25.4	A	90	0.13	3.3	0.188	4.8	Ar	1,2	26.5	500	12	5.1
1.25	31.8	A	70	0.18	4.6	0.188	4.8	Ar	1,2	26.5	550	10	4.2
1.25	31.8	B	45	0.25	6.3	0.156	4.0	Ar	1 2 Back	25 27 26	500	10 10 12	4.2 4.2 5.1
1.50	38.1	A	70	0.18	4.6	0.188	4.8	Ar	1 2 3,4	26 27 29	550 575 600	10	4.2
1.50	38.1	A	70	0.18	4.6	0.219	5.56	Ar	1 2	27 27.5	650 675	8	3.4
1.75	44.5	A	70	0.13	3.3	0.219	5.56	Ar	1,2 3,4	26 27	650 600	10	4.2
1.75	44.5	B	45	0.25	6.3	0.188	4.8	Ar	1,2 3,4 Back	28 30 30	600 550 550	10 14 10	4.2 5.9 4.2
2.00	50.8	A	70	0.2	4.6	0.188	4.8	He	1,2,3,4	32	550	10	4.2
2.00	50.8	B	45	0.3	6.3	0.188	4.8	Ar	1,2 3-7 Back	28 26 28	600 500 550	10 14 10	4.2 5.9 4.2
3.00	76.2	A	70	0.18	4.6	0.219	5.56	Ar-25%He	1,2 3,4 5,6 7-10	25 23 26 27	650 500 650 625	9 10 9 9	3.8 4.2 3.8 3.8
3.00	76.2	C	30	0.50	12.7	0.219	5.56	He	1,2 3-6	29 31	650	10	4.2

1. The following joint types are referred to by letter in the indicated column under joint geometry:

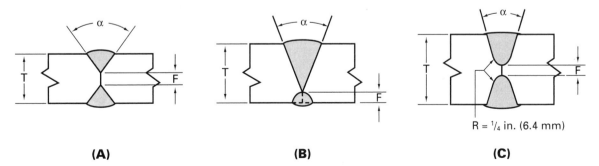

(A)   **(B)**   **(C)**

2. All passes are welded in the flat position, odd numbers from one side and even numbers from other side with joint designs (A) and (C). Joint is back gouged prior to depositing the back weld.

3. Constant-current dc power source and constant-speed electrode drive unit.

Table 1.30
Typical Procedures for Gas Metal Arc Welding of Fillet Welds in Aluminum Alloys with Large Diameter Electrodes and Argon Shielding

Fillet Size		Electrode Diameter					Travel Speed	
in.	mm	in.	mm	Weld Pass[a]	Welding Current, A[b]	Arc Voltage, V	in./min	mm/s
0.5	12.7	0.156	4.0	1	525	22	12	5.1
0.5	12.7	0.188	4.8	1	550	25	12	5.1
0.63	16	0.156	4.0	1	525	22	10	4.2
0.75	19	0.156	4.0	1	600	25	10	4.2
0.75	19	0.188	4.8	1	625	27	8	3.4
1	25.4	0.156	4.0	1 2,3	600 555	25 24	12 10	5.1 4.2
1	25.4	0.188	4.8	1 2,3	625 550	27 28	8 12	3.4 5.1
1.25	31.8	0.156	4.0	1,2,3	600	25	10	4.2
1.25	31.8	0.188	4.8	1 2,3	625 600	27 28	8 10	3.4 4.2

a. Welded in the flat position with one or three passes, using stringer beads.

b. Constant-current power source and constant-speed electrode wire drive unit.

high current allows welds to be made in fewer weld passes with little or no edge preparation.

Single-pass welds in butt joints are made in aluminum up to a 0.5 in. (12.7 mm) in thickness without edge preparation. Material as thick as 1.5 in. (38 mm) has been welded with one weld pass on each side, without edge preparation, using helium as a shielding gas. Recent developments in high-current welding with large-diameter electrodes and double-U-grooves have resulted in one-pass welds from each side in plate up to 3 in. (76.2 mm) thick. In heavy material, it may be necessary to bevel the plate edges in order to reduce the height of weld reinforcement.

The greater control over the welding variables possible with automatic welding allows material as thin as 0.02 in. (0.5 mm) to be welded, provided that adequate fixturing is used. The higher travel speed possible when automatic welding light-gauge material results in better weld appearance, minimal distortion, and lower welding costs. In general, automatic welding results in fewer weld passes, lower heat inputs, less edge preparation, and reduced labor costs.

Major areas that require attention in automatic welding are those affecting uniformity of the electrode wire drive speed, power pickup in the contact tube, and maintenance of a constant welding current. A

high-quality electrode with uniform surface resistance is necessary to provide this uniform current pickup. For thicknesses less than 1/8 in. (3.2 mm), automatic arc voltage control may be used in the conventional manner by governing the feed speed, or using a constant-voltage power source with stringer-bead techniques. A slow run-in wire feed speed is used to avoid poor fusion at starts. Proper operation with either variable-voltage or constant-voltage power requires uniform and dependable electrode feeding.

The need for proper fixturing in automatic welding cannot be overstated. Joint edges must be properly aligned and joint spacing maintained over the length of the weld. This can be accomplished by the use of tack welds, but some form of fixturing that will maintain the required conditions is strongly recommended.

Adequate shielding gas coverage is vitally important, and the torch should be capable of providing adequate coverage at all currents and travel speeds. The automatic welding carriage must have sufficient power and mass to give stability to the equipment.

For vertical and lateral adjustments, the torch must be rigidly mounted so that the torch angle and work angle can be accurately set and maintained. The torch carriage and track should be rigid enough to prevent vibrations from being transmitted to the welding head, which would produce erratic results.

Resolution of the actual welding variables is influenced by many factors and has to be done with the actual application in mind. Fabricators should become familiar with the available data before attempting to plan an automatic welding operation.

Gas Metal Arc Spot Welding

LAP JOINTS IN aluminum sheet can be spot welded by GMAW. Basically, the arc melts through the top sheet of less than 0.125 in. (3.2 mm) and penetrates into the bottom sheet. This action produces a round nugget of weld metal that joins the sheets, similar to resistance spot welding. The filler metal forms a convex reinforcement on the face of the nugget. Penetration into the bottom sheet may be partial or complete. For partial penetration, the bottom sheet should be thicker than the top sheet.

Equipment for spot welding is similar to that for manual GMAW. A "pull" or "push-pull" drive system and an insulated shielding gas nozzle to bear against the top sheet should be used. A "push" drive system produces inconsistent results and should not be used. A series of spot welds is more consistent when the gun system exerts a uniform force against the top sheet during each weld.

A control unit is needed to time the durations of welding current, electrode feed, and shielding gas flow. Welding current and wire-feed timers should have a range of 0 to 2 seconds with an accuracy of about 0.015 seconds.

A slow run-in electrode feed is used until arc initiation. At the weld termination, the welding current should flow briefly after the wire feed stops; a predetermined length of wire is melted, thus preventing the tip of the electrode from freezing in the weld metal.

Argon, helium, or mixtures of the two may be used for shielding. Argon is usually used, but helium is preferred for welding thin sheet since it produces a nugget with a larger cone angle than does argon. A rough weld surface and more spatter are disadvantages of GMAW spot welding with helium shielding.

Typical conditions for spot welding lap joints between two aluminum sheets with a 0.047 in. (1.19 mm) diameter electrode are given in Table 1.31. These conditions may vary somewhat with the compositions of the base metal and electrode, surface conditions, fit-up, shielding gas, and equipment design. Welding conditions for each application should be established by appropriate destructive tests.

Penetration into the back sheet depends upon the arc voltage, welding current, and welding time for a particular electrode size and shielding gas. Increasing any of these variables will increase penetration. Obtaining good penetration with thick sheets requires high

Table 1.31
Typical Settings for Gas Metal Arc Spot Welding with Constant-Voltage Power Supplies
For Various Aluminum Sheet Thicknesses

Sheet Thickness				Partial Penetration Welds				Complete Penetration Welds			
Top		Bottom		Open-Circuit	Electrode Feed		Weld	Open-Circuit	Electrode Feed*		Weld
in.	mm	in.	mm	Voltage, V	in./min	mm/s	Time, s	Voltage, V	in./min	mm/s	Time, s
0.020	0.51	0.020	0.51	—	—	—	—	27	250	106	0.3
0.020	0.51	0.030	0.76	—	—	—	—	28	300	127	0.3
0.030	0.76	0.030	0.76	25.5	285	120	0.3	28	330	140	0.3
0.030	0.76	0.050	1.27	25.5	330	140	0.3	31	430	182	0.3
0.030	0.76	0.064	1.63	30	360	152	0.3	31	450	190	0.3
0.050	1.27	0.050	1.27	31	385	163	0.4	32	450	190	0.4
0.050	1.27	0.064	1.63	32	400	169	0.4	32	500	212	0.4
0.064	1.63	0.064	1.63	32	420	178	0.4	32	550	233	0.5
0.064	1.63	0.125	3.17	32.5	650	275	0.5	34.5	675	286	0.5
0.064	1.63	0.187	4.75	35	700	296	0.5	39	700	296	0.5
0.064	1.63	0.250	6.35	39	775	328	0.5	41	800	338	0.5
0.125	3.2	0.125	3.17	39.5	800	338	0.5	41	850	360	0.6
0.125	3.2	0.187	4.75	41	850	360	0.75	41	900	381	0.75
0.125	3.2	0.250	6.35	41	900	381	1.0	—	—	—	—

* Electrode is 0.047 in. (1.2 mm) diameter ER5554. The welding current in amperes is about 0.5 times the electrode feed rate in in./min.

electrode feed rates that provide high current with constant-voltage power sources. If a small adjustment in penetration is desired, only the wire feed speed (amperage) needs to be changed.

A welding time of approximately 0.5 second is usually satisfactory for making arc spot welds. Times less than 0.25 second may result in nonuniform welds because the arc starting time is appreciable compared to the total welding time. Long welding times may help to reduce porosity. Short welding times may be best for vertical and overhead welding, or when a flat nugget is desired. With short weld times, arc length must be adjusted by the open-circuit-voltage setting on constant-voltage welding power sources.

Fit-up between parts can cause variations in gas metal arc spot welds. If the welds are inconsistent despite careful control of the variables, better methods of fixturing are needed. Good contact between components is necessary for heat transfer into the bottom member.

SHIELDED METAL ARC WELDING

SHIELDED METAL ARC welding (SMAW) of aluminum is primarily used in small shops for noncritical applications and repair work. It can be done with simple, low-cost equipment that is readily available and portable.[6] Welding speed is slower with SMAW than that achieved with gas metal arc welding. SMAW is not recommended for joining aluminum when good welding quality and performance are required. One of the gas-shielded arc welding processes should be used for making high-quality welds.

The electrode covering contains an active flux that combines with aluminum oxide to form a slag. This slag is a potential source of corrosion and must be completely removed after each weld pass. The covering on aluminum electrodes is prone to deterioration with time and exposure to moisture, and they must be stored in a dry atmosphere. Otherwise, before use the electrodes should be oven-dried in accordance with the manufacturer's recommendations. Typically this entails drying in an oven at 150 to 200 °F (66 to 93 °C). Welding is done with direct current electrode positive (DCEP) power. The following are some important factors to be considered in welding aluminum with SMAW:

(1)   Moisture content of the electrode covering
(2)   Cleanliness of the electrode and base metal
(3)   Preheating of the base metal
(4)   Proper slag removal between passes and after welding

The minimum recommended base metal thickness for SMAW of aluminum is 0.12 in. (3.0 mm). For thicknesses less than 0.25 in. (6.35 mm), no edge preparation other than a relatively smooth, square cut is required. Sections over 0.25 in. (6.35 mm) thickness should be beveled to provide a 60 to 90 degree single-V-groove. On very thick material, double-V- or U-grooves may be used. Depending upon base-metal thickness, root-face width should be between 0.06 and 0.25 in. (1.52 and 6.35 mm). A root opening of 0.03 to 0.06 in. (0.76 to 1.52 mm) is desirable.

For welding applications involving plate or complicated welds, it is desirable to apply preheat. Preheating is nearly always necessary on thick sections to obtain good penetration. Preheating also prevents porosity at the start of the weld due to rapid cooling and reduces distortion. Preheating may be done using an oxyfuel gas torch or a furnace.

Single-pass welds should be made whenever possible. If multiple-pass welds are needed, thorough removal of slag between passes is essential for optimum results. After the completion of welding, the bead and work should be thoroughly cleaned to remove all traces of slag. Most of the slag can be removed by mechanical means, such as a rotary wire brush or a slag hammer, and the residual traces by steam cleaning or a hot water rinse. In the test for complete slag removal, weld area is swabbed with a solution of 5% silver nitrate; foaming will occur if slag is present. Any remaining can be removed by soaking the weld in hot 5% nitric acid or warm 10% sulfuric acid solution for a short time, followed by a thorough rinse in hot water.

One difficulty encountered by interruption of the arc is the formation of a fused slag coating over the end of the electrode. It must be removed to restrike the arc.

Table 1.32 suggests approximate procedures for SMAW of aluminum.

PLASMA ARC WELDING

PLASMA ARC WELDING (PAW) is similar to gas tungsten arc welding (GTAW) except that the arc is constricted in size by a water-cooled nozzle.[7] Arc constriction increases the energy density, directional stability, and focusing effect of the arc plasma. Welding with the plasma arc is done with two techniques: melt-in and keyhole. Welding with the melt-in technique is similar to GTAW. The keyhole technique involves welding with a small hole through the molten weld pool at the leading edge. This technique is normally used when welding relatively thick sections.

PAW is normally done with direct current electrode negative (DCEN) and there is no arc cleaning action as

---

6. For a more detailed discussion of shielded metal arc welding refer to Chapter 2, *Welding Handbook*, Vol. 2, 8th Ed.

7. For a more detailed discussion of plasma arc welding refer to Chapter 10, *Welding Handbook*, Vol. 2, 8th Ed.

Table 1.32
Suggested Procedure for Shielded Metal Arc Welding of Aluminum

| Thickness | | Electrode or Filler Diameter | | Approximate DCEP Current, A | No. of Passes | | Filler Metal Consumption | | | | | |
| | | | | | | | Butt | | Lap | | Fillet | |
in.	mm	in.	mm		Butt	Lap and Fillet	lb/100 ft	kg/100 m	lb/100 ft	kg/100 m	lb/100 ft	kg/100 m
0.081	2.0	0.13	3.23	60	1	1	4.7	0.70	5.3	0.79	6.3	0.93
0.102	2.6	0.13	3.2	70	1	1	5.0	0.74	5.7	0.85	6.3	0.93
0.125	3.2	0.13	3.2	80	1	1	5.7	0.85	6.3	0.93	6.3	0.93
0.156	4.0	0.13	3.2	100	1	1	6.3	0.93	6.5	0.97	6.5	0.97
0.188	4.8	0.16	4.0	125	1	1	8.7	1.3	9.0	1.34	9.0	1.34
0.250	6.4	0.19	4.8	160	1	1	12	1.8	12	1.8	12	1.8
0.375	9.5	0.19[a] 0.25[b]	4.8[a] 6.4[b]	200	2	3	25	3.7	29	4.3	29	4.3
0.50	12.7	0.19[a] 0.25[b]	4.8[a] 6.4[b]	300	3	3	35	5.2	35	5.2	35	5.2
1.00	25.4	0.31	7.9	450	3	3	130	19.3	150	22.3	150	22.3
2.00	50.8	0.31-0.38	7.9-9.5	550	8	8	400	60	450	67	450	67

a. For laps and fillets

b. For butts

with GTAW. Surface cleaning prior to welding is necessary. Aluminum can be plasma arc welded also with conventional or square-wave ac or DCEP to take advantage of arc cleaning. Power sources for PAW are similar to those for GTAW.

Deeper penetration and faster welding speeds are advantages of PAW over GTAW. Using the melt-in technique, these conditions tend to increase the presence of porosity in aluminum welds because gases have less time to escape to the surface of the molten weld pool. Preweld cleaning of the base metal, clean filler wire, and adequate inert gas shielding of the weld are needed to minimize weld porosity. However, the oxides at the abutting edges are dispersed, and porosity-free welds are produced using the key-hole technique with a sophisticated variable polarity system. Inert gas shielding is desirable on both sides of the weld with this method, especially with 5XXX series and Al-Li alloys.

CRATER FILLING

ALUMINUM ALLOY WELD metal has a tendency to crater crack when fusion welding is stopped abruptly.

This problem can be avoided when the proper technique is used to fill the crater.

With manual welding, the forward travel should be stopped and the crater filled before it solidifies. When using gas tungsten arc or plasma arc welding with remote current control, the welding current should be decreased gradually, while adding filler, until the molten weld pool freezes. Without remote current control, the arc should be moved rapidly back on the weld bead as the welding torch speed is accelerated to reduce the size of the molten pool substantially prior to breaking the arc. This is often referred to as *tailing-out* the weld.

With semiautomatic gas metal arc or shielded metal arc welding, arc travel should be reversed and accelerated to tail-out the weld over a short length before the arc is broken. With automatic welding, the wire feed speed, the welding current, arc voltage, and the orifice gas with plasma arc welding should be programmed to fill the crater when travel stops. These variables are gradually decreased as the crater fills. Removable aluminum run-off tabs can be used to terminate the welds with any of the processes.

STUD WELDING

THE DISCUSSION OF stud welding in this chapter is limited to the requirements unique to stud welding aluminum alloys. Processes used are arc stud welding and capacitor-discharge stud welding.[7]

ARC STUD WELDING

FOR ARC STUD welding (SW) of aluminum, an inert shielding gas is used (with or without a ferrule), and the gun is equipped with a controlled-plunge dampening device. A special gas-adaptor-foot ferrule holder is used to contain the high-purity inert shielding gas during the weld cycle. Argon shielding is usually used, but helium may be useful with large studs to take advantage of using direct current electrode positive power and the higher arc energy. A typical equipment connection is shown in Figure 1.12. Equipment is now available in which the power source and gun-timing device are integrated into one unit.

A special process technique in aluminum SW, referred to as *gas arc*, uses an inert gas for shielding the arc and molten metal from the atmosphere. A ferrule is not used. This technique is suitable for aluminum stud welding and is limited to production applications

7. For a more detailed discussion of stud welding refer to Chapter 9, *Welding Handbook*, Vol. 2, 8th Ed.

because a fixed setup must be maintained and the welding variables must be in a narrow range. A typical example of this type of aluminum arc stud welding is the fastening of tapped studs to aluminum utensils for securing handles and legs. Figure 1.13 illustrates a variety of aluminum studs for SW.

CAPACITOR-DISCHARGE STUD WELDING

THE EQUIPMENT FOR capacitor-discharge stud welding consists of a stud gun and a combination power-control unit with associated interconnecting cables.

Capacitor-discharge units use ac power to charge banks of capacitors that provide welding power. Each capacitor-discharge unit is an integrated power-control system similar to that used with aluminum arc stud welding. There are three capacitor-discharge welding methods: drawn arc, initial contact, and initial gap. Regardless of the capacitor-discharge method used, no ceramic arc shield (ferrule) is used. The drawn-arc capacitor-discharge method requires the use of inert gas to shield the weld area. Of the capacitor-discharge methods, the initial-gap method is superior for welding aluminum. Since weld time is short, the length and size of weld cable is important. The ground path for current is critical and the ground area must be thoroughly cleaned. Two ground cables and clamps are required for even current distribution. When the initial-gap method is used with a manual stud gun, an additional return path must be milled beneath one leg of the gun to complete the control circuit. Portable equipment setup is illustrated in Figure 1.14. Commonly used aluminum studs for capacitor-discharge stud welding are shown in Figure 1.15.

CAPABILITIES AND LIMITATIONS

WHEN STUD WELDING precipitation-hardened aluminum alloys, a short weld cycle minimizes over-aging and softening of the adjacent base metal. Metallurgical compatibility between stud material and the base metal must be considered.

Small studs can be welded to thin sections by the capacitor-discharge method. Studs have been welded to sheet as thin as 0.020 in. (0.5 mm) without melt-through. Because the depth of melting is very shallow, capacitor-discharge welds can be made without damage to a prefinished opposite side. No subsequent cleaning or finishing is required. Capacitor-discharge power permits the welding of dissimilar aluminum alloys and aluminum to die-cast zinc.

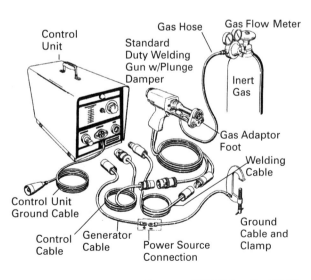

Courtesy of TRW Nelson Stud Welding

Figure 1.12—Basic Equipment Setup for Arc Stud Welding of Aluminum

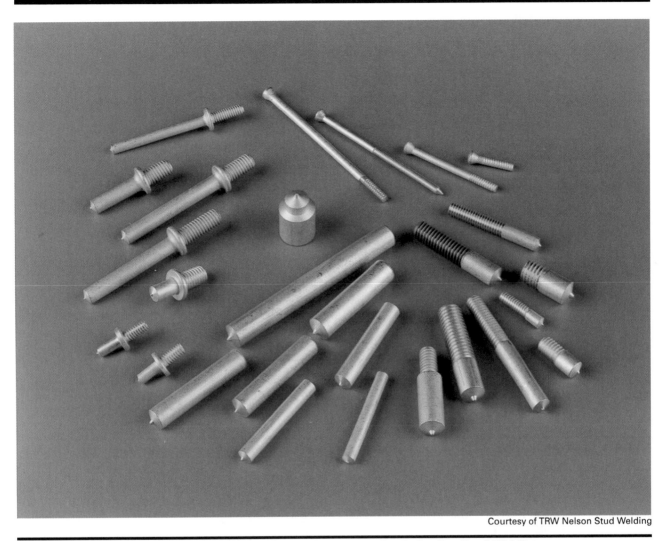

Figure 1.13—Commonly Used Aluminum Studs for Arc Stud Welding

STUD AND BASE METAL DETAILS

AN ALUMINUM STUD differs from a steel stud in that no flux is used on the weld end. A cylindrical or cone-shaped projection is used on the base of the aluminum stud. The projection serves to initiate the long arc used for aluminum stud welding. Figure 1.16 shows a typical fastener and weld-end projection for arc stud welding (SW). Aluminum arc studs range in weld-base diameters from 3/16 to 1/2 in. (5 to 13 mm). Studs less than 1/2 in. (13 mm) are most common.The stud-base diameter welded by the capacitor-discharge process is 1/16 to 5/16 in. (2 to 8 mm), although 3/8 in. (9.5 mm) may be welded under carefully limited circumstances.

Aluminum studs are commonly made of aluminum magnesium alloys, including alloys 5183, 5356, and 5556, which have a typical tensile strength of 40 ksi (276 MPa). These alloys have high strength and good ductility. They are metallurgically compatible with most aluminum alloys used in fabrication.

Base metal of the 1100, 3000, and 5000 series alloys, 2219, and low-copper-content 7000 series alloys are considered excellent for stud welding. The 4000 and 6000 series alloys are considered fair. The 2000 and 7000 series alloys are generally poor. Tables 1.33 and 1.34 summarize the weldability of aluminum base metal and stud materials. Table 1.35 identifies common pairings of aluminum base and stud materials used in applications, and it footnotes some base alloys with good weldability.

The typical aluminum arc stud weld macrosection in Figure 1.17 shows that molten weld metal is pushed to

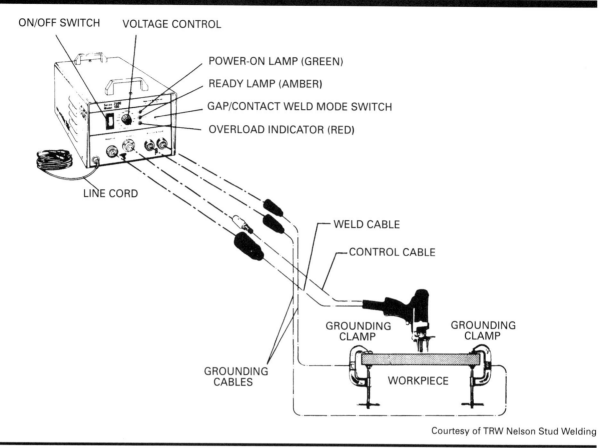

ON/OFF SWITCH   VOLTAGE CONTROL

POWER-ON LAMP (GREEN)
READY LAMP (AMBER)
GAP/CONTACT WELD MODE SWITCH
OVERLOAD INDICATOR (RED)

LINE CORD

WELD CABLE
CONTROL CABLE

GROUNDING
CLAMP   GROUNDING
                 CLAMP

GROUNDING
CABLES   WORKPIECE

Courtesy of TRW Nelson Stud Welding

Figure 1.14—Portable Equipment Setup for Initial Gap or Initial Contact Capacitor-Discharge Stud Welding of Aluminum

the perimeter of the stud to form a flash. The amount of weld metal (cast structure) in the joint is minimal. Aluminum arc welded studs typically reduce in length 1/8 in. (3 mm) during welding. For studs welded by the capacitor-discharge process, the volume of metal that is melted by all methods is almost negligible. Aluminum capacitor-discharge stud lengths are reduced between 0.008 to 0.015 in. (0.20 to 0.38 mm). Because of the short welding cycle, the heat-affected zones common to arc welding are present, but they are small.[8]

Table 1.36 indicates the normal time and current settings as well as shielding gas flow rates for arc stud welded aluminum fasteners. Whenever possible, flanged studs are preferred when using capacitor-discharge methods for the following two reasons:

(1) Larger weld area. For a given stud size, a flanged stud has a larger cross-sectional area. This increases the weld strength and the strength of the finished assembly.

(2) Economy. Most flanged studs are easier to produce and therefore less expensive than nonflanged studs.

Part of the material from the length reduction of arc stud welds appears as flash in the form of a fillet around the stud base. This flash should not be confused with a conventional fillet weld because it is formed in a different manner. When properly formed and contained, the flash indicates complete fusion over the full cross section of the stud base. The dimensions of the flash are closely controlled by the design of the ferrule. The diameter of the flash increases the effective diameter of the stud, and this must be considered when designing mating parts. Figure 1.18 shows methods for accommodating the stud flash. Table 1.37 shows the clearance for round studs. Flash size varies with size and shape of the stud material and ferrule clearance. Test welds should be made to check the necessary clearances. The stud

8. Chapter 4, "Welding Metallurgy," *Welding Handbook,* Vol. 1, 8th Ed., contains helpful information on welding metallurgy that is applicable to arc stud welding of various materials.

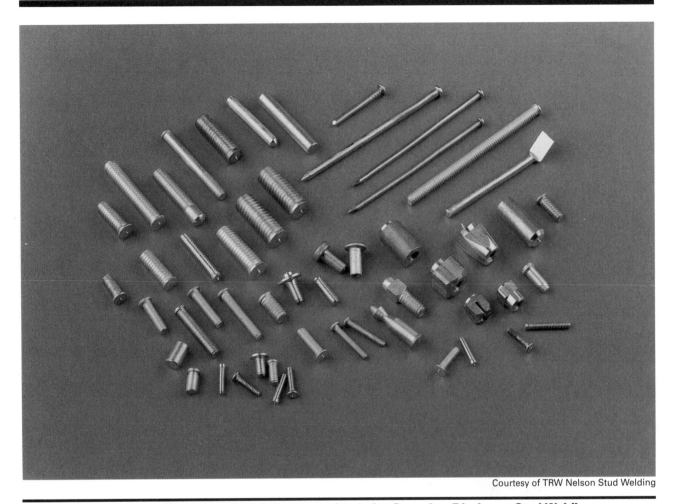

Courtesy of TRW Nelson Stud Welding

Figure 1.15—Commonly Used Aluminum Studs for Capacitor-Discharge Stud Welding

weld flash may not be fused along its vertical and horizontal legs, but this lack of fusion is not considered detrimental to the joint quality.

There are some types of applications for which the capabilities of arc stud welding and capacitor-discharge processes overlap, but the selection between the two processes is well-defined. A process selection chart is shown in Table 1.38. The more-difficult selection is in the method that should be used, e.g., contact, gap, or drawn arc. The main criteria used in selection should be the fastener size, base-metal thickness, and need to minimize reverse-side marking.

PREPARATION OF BASE METAL FOR STUD WELDING

CLEANING OF THE base metal to assure good welding results may be necessary. Mill-finish aluminum usually requires no cleaning but should be free of dirt, oil, paint, and similar contaminants. Polished aluminum is usually taped, which guarantees the surface will be clean enough to weld, and etching to remove the aluminum oxide film is usually unnecessary. Heavy coatings of aluminum oxide should be removed by brushing with a stainless steel wire brush prior to welding.

Anodizing creates a high surface resistance and should not be performed until the studs have been welded (the studs also benefit from anodizing). When anodizing cannot be avoided prior to welding, the surface must be milled at locations where the studs are to be welded. On anodized sheet aluminum, milling may reduce the sheet thickness to the point where previously-absent reverse-side marking or melt-through may occur. This may be prevented by using a slightly smaller stud or heavier gauge aluminum sheet metal. In addition to milling the anodized surface for good stud-to-base metal contact, it is necessary to provide a

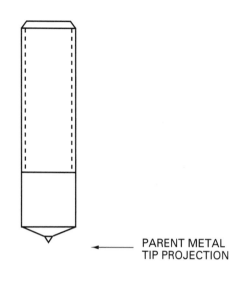

← PARENT METAL
TIP PROJECTION

Figure 1.16—Full Diameter Weld Base Aluminum Stud for Arc Stud Welding (Note Tip Projection)

low-resistance path for weld current through a ground clamp. For the initial-gap method of capacitor-discharge welding, an additional return path must be milled beneath one leg of the gun to complete the control circuit.

STRENGTH OF ALUMINUM STUDS

THE AVERAGE TENSILE strength of the aluminum alloys commonly used for arc stud welding is 42 ksi (290 MPa) with an average yield strength of 30 ksi (207 MPa). Table 1.39 shows the strengths of welded threaded studs that may be expected using these figures.

Table 1.33
Weldability (Arc Stud Welding) of Aluminum Alloy Base Plate Material

Base Plate Alloy Series	Weldability	Strength
1000	Excellent	High
2000	Poor	Low
3000	Excellent	High
4000	Good	Low
5000	Excellent	High
6000	Good	High
7000	Poor	Low

Table 1.34
Weldability (Arc Stud Welding) of Aluminum Alloys Used as Stud Material

Stud Alloy	Weldability	Strength
1100	Fair	Low
2319	Excellent	High
5356	Excellent	High
5183	Excellent	High
5556	Excellent	High
4043	Excellent	Medium

Table 1.35
Aluminum Stud Alloys Commonly Used with Selected Aluminum Base Alloys

Base Alloys	Stud Alloy Used
1000	1100
2219[a]	2319
3000	1100, 4043
4000	4043
5000	5256, 5183, 5556
6000	4043, 5000
7000[b]	5000 type

a. Good weldability

b. Good weldability (7004, 7005, 7039)

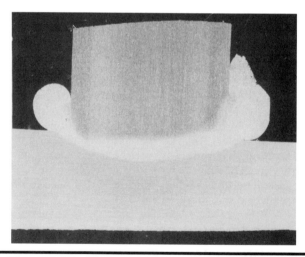

Figure 1.17—Macrostructure of a 3/8 in. (9.5 mm) Diameter Type 5356 Aluminum Alloy Arc Stud Welded to a 1/4 in. (6.4 mm) Type 5053 Aluminum Alloy Plate (Reduced to 75%)

Table 1.36
Typical Conditions for Arc Stud Welding of Aluminum Alloys

Stud Weld Base Diameter		Weld Time, Seconds	Welding Current, A[a]	Shielding Gas Flow[b]	
in.	mm			ft³/h	L/min
1/4	6.4	0.33	250	15	7.1
5/16	7.9	0.50	325	15	7.1
3/8	9.5	0.67	400	20	9.4
7/16	11.1	0.83	430	20	9.4
1/2	12.7	0.92	475	20	9.4

a. The currents shown are actual welding current and do not necessarily correspond to power source dial settings.

b. Shielding gas – 99.95% pure argon.

The torque figures are based on a torque coefficient factor of 0.20. The section on inspection and testing has a discussion on the significance of torque coefficient factors. For a particular application, ultimate torque and tensile strengths of the studs should be established and acceptable proof load ranges specified.

For design purposes, the smallest cross-sectional area of the stud should be used for load determination, and adequate safety factors should be considered.

To develop full stud strength using arc stud welding (SW), the aluminum plate (base metal) thickness should be a minimum of about half the weld-base diameter. A minimum plate thickness is required for each stud size to permit arc stud welding without melt-through or excessive distortion, as shown in Table 1.40.

Typical strengths of aluminum capacitor-discharge welded studs are given in Table 1.41.

QUALITY CONTROL AND INSPECTION

ANSI/AWS D1.2, *Structural Welding Code – Aluminum*, Section 7, outlines code requirements for qualifying, testing, and inspecting aluminum studs welded by either arc stud welding or capacitor-discharge welding processes. The procedure qualification is identical for both. Failed studs may be repaired or replaced in accordance with the ANSI/AWS D1.2 code requirements.

The appearance of arc stud welded aluminum studs is radically different from that of studs welded with the capacitor-discharge process. The amount of weld metal expelled during the arc welding process is greater than that expelled by capacitor-discharge welding and is usually formed, even with gas arc welding (where no ferrule is used), into a flash (fillet) around the periphery of the stud base.

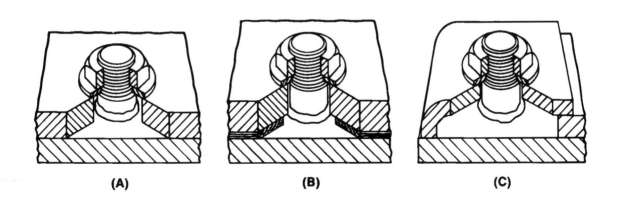

| (A) | (B) | (C) |

Figure 1.18—Methods of Accommodating Flash

Table 1.37
Weld Fillet Clearances for Aluminum Arc Stud Welds

Stud Base Diameter		Counterbore				Countersink	
		A*		B*		C*	
in.	mm	in.	mm	in.	mm	in.	mm
3/16	4.8	0.390	9.9	0.156	4.0	0.156	4.0
1/4	6.4	0.469	11.9	0.136	4.0	0.156	4.0
5/16	7.9	0.531	13.5	0.187	4.7	0.187	4.7
3/8	9.5	0.656	16.7	0.218	5.6	0.187	4.7
7/16	11.1	0.750	19.1	0.250	6.4	0.250	6.4
1/2	12.7	0.844	21.4	0.281	7.1	0.250	6.4

* A, B, and C dimensions are measured as shown in these diagrams:

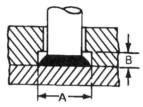 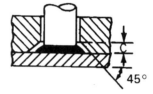

For aluminum arc stud welds, Figure 1.19(A) illustrates a typical satisfactory flash (fillet) structure. Examples of unsatisfactory visual appearances are illustrated in Figure 1.19(B) through (F).

The weld flash around the stud base is inspected for consistency and uniformity, and lack of flash may indicate a faulty weld. To pass visual inspection, the studs should have a well-formed and fully wetted circumferential flash over at least 75 percent of the periphery. In contrast to arc stud welding, capacitor-discharge welded studs do not exhibit a flash (fillet) of sufficient formation to easily determine visual deficiencies. There should be some flash and full wetting at the weld joint. Figure 1.20 shows in macrosection how the expelled weld metal forms a low-profile flash and wets the stud periphery.

Figure 1.21 shows visual examples of satisfactory and unsatisfactory capacitor-discharge welded aluminum studs.

SAFETY PRECAUTIONS

PERSONNEL OPERATING STUD welding equipment should be provided with face and skin protection to guard against burns from spatter produced during welds. Eye protection in the form of safety glasses with side shields, or a face shield with a No. 3 filter lens, should be worn to protect against arc radiation.

When inert gas is used to shield aluminum stud welding, proper procedures in handling, storing, refilling, and using the gas storage cylinders should be observed. Capacitor-discharge stud welding, particularly initial-gap or initial-contact methods, are characterized by a sharp noise when the arc disintegrates the stud tip. Continual exposure to this elevated noise level may be damaging to hearing, and the use of hearing protection is advised.

Table 1.38
Aluminum Stud Welding Process Selection Chart[1]

Factors to be Considered	Arc Stud Welding	Capacitor-Discharge Stud Welding		
		Initial Gap	**Initial Contact**	**Drawn Arc**
Stud Diameter				
1/16 to 1/8 in. (1.6 to 3.2 mm)	D	A	B	A
1/8 to 1/4 in. (3.2 to 6.4 mm)	C	A	B	A
1/4 to 5/16 in. (6.4 to 7.9 mm)	A	A	B	B
5/16 to 3/8 in. (7.9 to 9.5 mm)	A	C	C	C
3/8 to 1/2 in. (9.5 to 12.7 mm)	A	D	D	D
Stud Material				
1100	D	A	B	B
4043	A	A	A	A
5183	B	A	B	B
5356	B	A	B	B
5556	B	A	B	B
6061	B	A	B	B
7000[2]	D	C	D	D
Base Material				
1000 Series	B	A	B	B
2000 Series[3]	D	C	D	C
3000 Series	B	A	B	B
4000 Series	C	B	C	B
5000 Series	B	A	B	B
6000 Series	B	A	B	B
7000 Series[4]	D	C	D	D
Base Metal Thickness				
Under 0.015 in. (0.4 mm)	D	A	B	B
0.015 to 0.062 in (0.4 to 1.6 mm)	C	A	A	A
0.062 to 0.125 in. (1.6 to 3.2 mm)	B	A	A	A
over 0.125 in. (3.2 mm)	A	A	A	A
Strength Criteria				
Heat effect on exposed surfaces	B	A	B	A
Weld fillet clearance	B	A	A	A
Strength of stud governs	A	A	B	A
Strength of base metal governs	A	A	B	A

1. Applicability of factors in chart are coded as follows:
   A – Applicable without special procedures, equipment, etc.
   B – Applicable with special techniques or on specific applications which justify preliminary trials or
       testing to develop welding procedure and technique.
   C – Limited application.
   D – Not recommended.

2. Consult the stud or equipment manufacturer; 7000 Series aluminum requires specific conditions and special techniques.

3. 2219 is easily welded with 2319 stud.

4. 7004, 7005, 7039 are very weldable.

Table 1.39
Typical Strengths of Aluminum Arc Welded Studs[a]

Diameter/ Stud Thread	META[b]		Yield Tensile Load[c]		Yield Torque Load[c]		Ultimate Tensile Load		Ultimate Torque Load		Ultimate Shear Load[d]	
	in.²	mm²	lb	kg	in.·lb	N·m	lb	kg	in.·lb	N·m	lb	kg
10-24 UNC	0.017	11.0	510	231	19	2	714	324	27	3	428	194
10-32 UNF	0.020	12.9	600	272	22	2.5	840	381	32	4	504	229
1/4-20 UNC	0.032	19.4	960	435	48	5	1344	610	67	7.6	806	366
1/4-28 UNF	0.036	23.2	1080	490	54	6	1512	686	76	8.6	907	411
5/16-18 UNC	0.052	33.5	1560	708	97	11	2184	991	136	15	1310	594
5/16-24 UNF	0.058	37.4	1740	789	108	12	2436	1105	152	17	1462	663
3/8-16 UNC	0.078	50.3	2340	1061	175	20	3276	1486	246	28	1966	892
3/8-24 UNF	0.088	56.8	2640	1197	198	22	3696	1676	277	31	2217	1005
7/16-14 UNC	0.106	68.4	3180	1442	278	31	4452	2019	389	44	2671	1211
7/16-20 UNF	0.118	76.1	3540	1606	310	35	4956	2248	434	49	2973	1349
1/2-13 UNC	0.142	91.6	4260	1932	426	48	5964	2705	596	67	3578	1623
1/2-20 UNF	0.160	103.2	4800	2177	480	54	6720	3048	672	76	4032	1829

a. Mechanical properties are based on 42 ksi (290 MPa) ultimate strength and 30 ksi (207 MPa) yield strength.

b. Mean effective thread area (META) is based on a mean diameter between the minor and pitch diameters, providing a closer correlation with actual yield and tensile strengths.

c. Typically, studs should be used at no more than 60% of the yield load figures.

d. Ultimate shear load is calculated at 60% of the ultimate tensile load.

Table 1.40
Recommended Minimum Aluminum Plate Thicknesses for Arc Stud Welding

Stud Base Diameter		Aluminum			
		Without Backup		With Backup*	
in.	mm	in.	mm	in.	mm
0.19	4.8	0.13	3.3	0.13	3.3
0.25	6.4	0.13	3.3	0.13	3.3
0.31	7.9	0.19	4.8	0.13	3.3
0.38	9.5	0.19	4.8	0.19	4.8
0.44	11.1	0.25	6.4	0.19	4.8
0.50	12.7	0.25	6.4	0.25	6.4

*   A metal backup to prevent melt-through of the plate

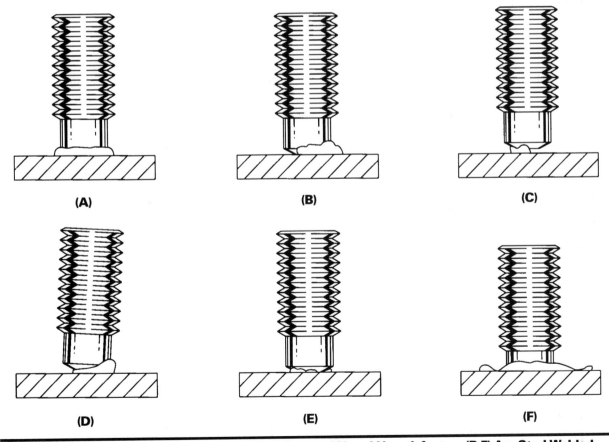

(A)   (B)   (C)

(D)   (E)   (F)

Figure 1.19—Visual Appearances of Satisfactory (A) and Unsatisfactory (B-F) Arc Stud Welded Aluminum Studs

Table 1.41
Typical Strengths of Aluminum Capacitor-Discharge Welded Studs

Stud Thread Diameter	META[a]		Yield Tensile Load[b]		Yield Torque Load[b]		Ultimate Tensile Load		Ultimate Torque Load		Ultimate Shear Load[c]	
	in.²	mm²	lb	kg	in.·lb	N·m	lb	kg	in.·lb	N·m	lb	kg
Aluminum Alloy 1100[d]												
4-40	0.006	3.9	120	54	2.7	0.3	126	57	2.8	0.3	76	34
6-32	0.009	5.8	180	82	5	0.6	189	86	5.2	0.6	113	52
8-32	0.014	9.0	280	127	9	1.0	294	133	9.6	1.1	176	80
10-24	0.017	11.0	340	154	13	1.5	375	170	14	1.6	225	102
10-32	0.020	12.9	400	181	15	1.7	420	191	16	1.8	252	115
1/4-20	0.032	19.4	640	290	32	3.6	672	305	34	3.8	403	183
1/4-28	0.036	23.2	720	327	37	4	756	343	38	4.3	454	206
5/16-18	0.052	33.5	1040	472	65	7	1092	495	68	7.7	655	297
5/16-24	0.058	37.4	1160	526	73	8	1218	552	76	8.6	731	331
Aluminum Alloys 5183, 5356, 6061[e]												
4-40	0.006	3.8	180	82	4	0.5	252	114	5.6	0.6	151	68
6-32	0.009	5.8	270	122	7.4	0.8	378	171	10	1.1	227	103
8-32	0.014	9.0	420	191	13.7	1.6	588	267	19	2.1	353	160
10-24	0.017	11.0	510	231	19	2	714	324	27	3	428	194
10-32	0.020	12.9	600	272	22	3	840	381	32	4	504	229
1/4-20	0.032	19.4	960	435	48	5	1344	610	67	7	806	366
1/4-28	0.036	23.2	1080	490	54	6	1512	686	76	8	907	411
5/16-18	0.052	33.5	1560	708	97	11	2184	991	136	15	1310	594
5/16-20	0.058	37.4	1740	789	108	12	2436	1105	152	17	1462	663

a. Mean effective thread area (META) is based on a mean diameter between the minor and pitch diameters, providing a closer correlation with actual yield and tensile strengths.

b. Typically, studs should be used at no more than 60% of the yield load figures.

c. Ultimate shear load is calculated at 60% of the ultimate tensile load.

d. Mechanical properties of 1100 alloy are based on 21 ksi (145 MPa) ultimate strength and 20 ksi (138 MPa) yield strength.

e. Mechanical properties of 5183, 5356, and 6061 alloys are based on 42 ksi (290 MPa) ultimate strength and 30 ksi (207 MPa) yield strength.

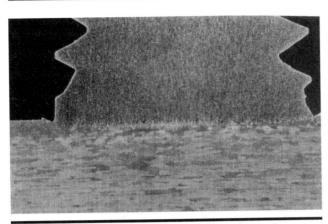

Figure 1.20—Macrostructure of a Capacitor-Discharge Stud Welded 6061-T6 Aluminum Stud, 3/8 in. (9.5 mm) Diameter, to 1/8 in. (3.2 mm) Aluminum Sheet of the Same Alloy (10x, Reduced to 75%)

As with any welding process, the work area should be kept clean of any paper, wood, rags, or other combustible materials that can be ignited by flying sparks. Before repairs to equipment are attempted, electrical power should be turned off and electric switch boxes locked. Capacitors used in capacitor-discharge equipment should be completely drained of electrical charge before attempting repairs. Most capacitor-charged systems have an interlock device that drains the capacitors upon removal of the equipment cover and does not permit welding with the cover removed. Users should follow manufacturer's instructions completely for equipment installation and repair. Other safety details are outlined in ANSI/AWS C5.4, *Recommended Practices for Stud Welding.*

(A) Good Weld   (B) Weld Power Too High   (C) Weld Power Too Low

Figure 1.21—Visual Appearances of Satisfactory and Unsatisfactory Capacitor-Discharge Welded Aluminum Studs

ELECTRON BEAM WELDING

ELECTRON BEAM WELDING (EBW) is generally applicable to edge, butt, T, corner, and lap joints.[7] Welding can be done in a chamber under high vacuum at a pressure of 10^{-6} to 10^{-3} torr (1.22×10^{-4} to 0.133 Pa), at medium vacuum from 10^{-3} to 25 torr (0.133 to 333 Pa), or at atmospheric pressure using helium shielding. In the latter case, the electron beam is generated in a vacuum and exited to atmospheric pressure through a series of ports.

Most aluminum alloys can be electron beam welded, but cracking may be experienced with some of the heat-treatable alloys, such as 2XXX, 6XXX, and 7XXX. The addition of filler metal may prevent weld cracking. Wire feeders have been developed for use in EBW chambers. Figure 1.22 shows a cross section from an alloy 6061-T6 weld made using a wire feeder. Also, filler in the form of preplaced shims has been used effectively.

7. For a more detailed discussion of electron beam welding refer to Chapter 21, *Welding Handbook*, Vol. 2, 8th Ed.

JOINT GEOMETRY

SQUARE-GROOVE JOINTS are normally used to make complete joint penetration EBW welds. Figures 1.23 and Figure 1.24 show cross sections of this type of weld. A concave bead can occur in some instances of this type of weld which can be corrected by a second weld pass with or without addition of filler metal. Root spiking and variation in penetration can occur when joining thick sections, using a single pass, with partial joint penetration EB welds. These problems are minimized by using multiple-pass autogenous or wire-fed electron beam welding.

Figure 1.25 illustrates the multiple-pass autogenous method that uses the base metal for filler metal, supplementing it with a boss that is dimensioned to suit the particular weld. The weld boss is removed by machining subsequent to welding when a smooth surface is required.

Details of this and other joint geometries are shown in Figure 1.26. Joints premachined for dimensional

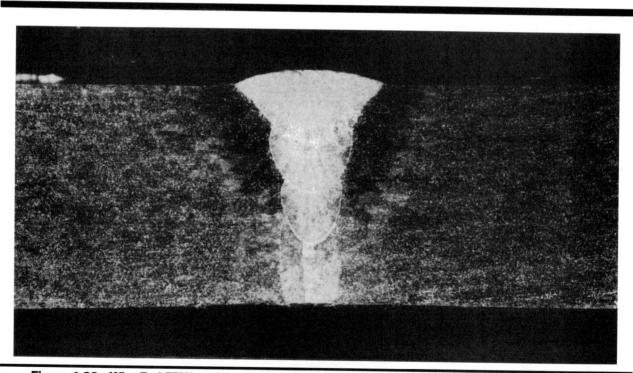

Figure 1.22—Wire-Fed EBW on Aluminum Alloy 6061-T6 [0.213 in. (5.4 mm) Thick] with ER 4043 Filler Metal (10x)

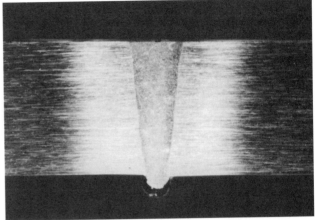

Figure 1.24—Single-Pass EBW on Aluminum Alloy 7049-T73 [0.16 in. (4.1 mm) Thick] (10x) (Reduced to 67%)

Figure 1.23—Single-Pass EBW on Aluminum Alloy 5083 (Both Sides Machined After Welding) [0.760 in. (19.3 mm) Thick] (7x)

(A) Root Bead

(B) Multiple-Pass

Figure 1.25—Autogenous EBW on 5083 Aluminum [0.31 in. (7.9 mm) Thick] (7x) (Reduced to 62%)

compensation, illustrated in Figure 1.26, can correct weld distortion. The joints shown in Figure 1.26 for multiple-pass welding were developed for specific thickness applications but may be modified for other thicknesses. For example, details of the groove for joint (C) may be used instead of the rectangular groove in joint (D) for filling a thicker joint where filler metal is used. Figure 1.27 shows a macrosection from a multiple-pass weld with an autogenous root bead and three wire-fed

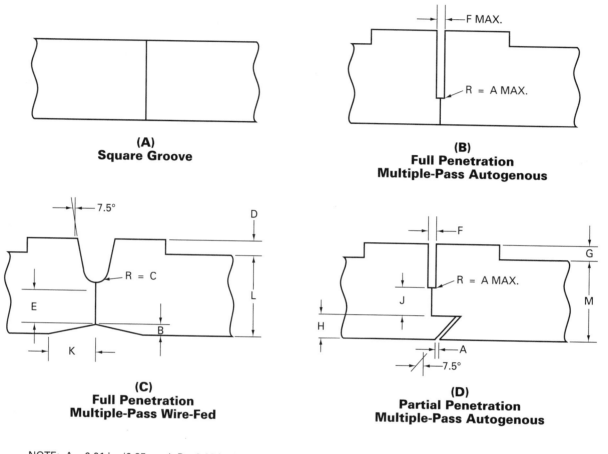

(A)
Square Groove

(B)
Full Penetration
Multiple-Pass Autogenous

(C)
Full Penetration
Multiple-Pass Wire-Fed

(D)
Partial Penetration
Multiple-Pass Autogenous

NOTE: A = 0.01 in. (0.25 mm), B = 0.02 in. (0.51 mm), C = 0.025 in. (0.64 mm), D = 0.030 in. (0.76 mm),
E = 0.040 in. (1.02 mm), F = 0.042 in. (1.07 mm), G = 0.060 in. (1.52 mm), H = 0.075 in. (1.91 mm),
J = 0.100 in. (2.54 mm), K = 0.125 in. (3.18 mm), L = 0.183 in. (4.65 mm), M = 0.310 in. (7.87 mm),
R = corner radius.

Figure 1.26—Joint Geometries for EBW of Aluminum

fill beads in 0.31 in. (7.9 mm) thick alloy 5083. The addition of filler metal was not necessary to prevent cracking in this alloy but was needed to fill the joint.

EQUIPMENT AND WELDING CONDITIONS

ALUMINUM ALLOYS CAN be welded with either low or high voltage EBW equipment. EBW machines are available from 60 to 175 kV and up to 100 kw of power. The choice of equipment and procedures will depend upon the alloy, material thickness, joint design, and service requirements. Typical conditions for single-pass EBW of several alloys and thicknesses are given in Table 1.42.

Tables 1.43, 1.44, and 1.45 show conditions developed for multiple-pass welding alloys 6061-T6 and 5083-0 with the addition of filler metal and for multiple-pass autogenous welding alloy 5083-0.

Computer-controlled EBW machines are particularly useful for both multiple-pass wire-fed filler metal and multiple-pass autogenous welding.

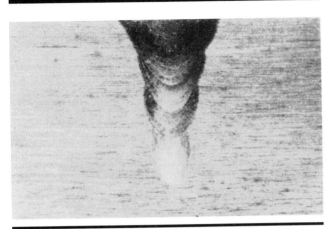

Figure 1.27—Autogenous EB Root Bead and Three Cold-Wire-Fed EB Fill Beads with 5356 Filler [0.31 in. (7.9 mm) Thick], Weld Boss Removed (8x) (Reduced to 67%)

DEPLETION OF ELEMENTS AND PROPERTIES

FIGURE 1.28 ILLUSTRATES that magnesium can be depleted from electron beam welds when alloy 5083 is EB welded in a vacuum environment. These welds were made using the joint shown in Figure 1.26 (D) except that a bevel groove was used to accommodate the wire-fed filler metal. The loss of magnesium from the autogenous root bead was between 0.6 to 1.1 weight percent. Lower losses of magnesium experienced in the second and third autogenous beads resulted in a total loss of 1.1 weight percent. There is an increased loss of magnesium from the third weld pass of the 5% magnesium cold-wire-fed weld. The loss of other elements, such as zinc, can be minimized.

The loss of alloying elements generally does not result in a significant reduction in yield strength, but this must be determined for each application. EB welds in non-heat-treatable aluminum alloys (1XXX, 3XXX, and 5XXX) have properties equal to or better than those of gas tungsten arc welds and yield strength joint efficiency of approximately 100 percent. Typical properties of EB welds in some aluminum alloys are shown in Table 1.46.

Table 1.42
Typical Conditions for EBW of Aluminum Alloys

Thickness[a]		Alloy	Welding Atmosphere[b]	Beam Voltage, kV	Beam Current, mA	Travel Speed		Energy Input	
in.	mm					in./min	mm/s	kJ/in.	kJ/m
0.050	1.27	6061	HV	18	33	100	42.3	0.36	14.17
0.050	1.27	2024	HV	27	21	71	30.0	0.48	18.90
0.120	3.05	2014	HV	29	54	75	31.7	1.3	51.18
0.125	3.17	6061	HV	26	52	80	33.8	1.00	39.37
0.125	3.17	7075	HV	25	80	90	38.1	1.3	51.18
0.38	9.65	2219	Helium	175	40	55	23.3	7.6	299.21
0.50	12.70	2219	HV	30	200	95	40	3.8	149.61
0.63	16.00	6061	HV	30	275	75	31.7	6.6	259.84
0.75	19.05	2219	HV	145	38	50	21.1	6.6	259.84
1.00	25.40	5086	HV	35	222	30	12.7	15	590.55
2.00	50.80	5086	HV	30	500	36	15.2	25	984.25
2.38	60.45	2219	HV	30	1000	43	18.2	42	1653.54
6.00	152.40	5083	HV	58	525	10	4.2	182	7165.34

a. Square-groove butt joint

b. HV = high vacuum [10^{-5} torr (1.33×10^{-3} Pa)]

Table 1.43
Welding Conditions for Multiple-Pass EBW of Aluminum: Welding with Cold-Wire-Fed Filler Metal of 0.213 in. (5.4 mm) Thick Alloy 6061-T6. Joint Geometry (C)[a]

Weld Pass[b]	Travel Speed		Beam Focus, mA[c]	Wire Feed Rate[d]	
	in./mm	mm/s		in./min	mm/s
1	18	7.6	410	18	7.6
2	18	7.6	420	30	12.7
3	12	5.1	430	30	12.7
4	12	5.1	430	30	12.7

a. Refer to Figure 1.26 for joint geometry.

b. A hairpin filament gun was used. Voltage was constant at 75 kV. Beam current was constant at 9 mA.

c. Sharp focus at the root of the joint was 385 mA at a beam current of 2 mA.

d. Alloy 4043 wire was 0.030 in. (0.76 mm) diameter.

Table 1.44
Welding Conditions for Multiple-Pass EBW of Aluminum: Multiple-Pass, Autogenous Welding of Alloy 5083. Joint Geometry (D)[a]

Weld Pass[b]	Beam Current, mA	Weld Speed		Beam Focus (mA)[c]	Beam Oscillation[d]	
		in./min	mm/s		in.	mm
1	9.3	40	16.9	495	None	None
2	17.0	40	16.9	520	0.080	0.03
3	17.0	40	16.9	520	0.080	0.03
4	17.0	40	16.9	540	0.080	0.03

a. Refer to Figure 1.26 for joint geometry.

b. A ribbon filament gun was used. Voltage was constant at 120 kV.

c. Sharp focus at the root of the joint was 470 mA at a beam current value of 2 mA

d. The electron beam was oscillated transverse to the weld joint (x-direction) at 1 kHz to the dimension shown.

Table 1.45
Welding Conditions for Multiple-Pass EBW of Aluminum: Autogenous Root Weld and Cold-Wire-Fed Welding of Alloy 5083. Joint Geometry (D) with 7.5° Bevel Angle[a]

Weld Pass[b]	Beam Current, mA	Weld Speed		Beam Focus mA[c]	Beam Oscillation[d]		Wire Feed Rate[e]	
		in./min	mm/s		in.	mm	in./min	mm/s
1	11	40	16.9	500	(0.100) Y	(2.54)Y	None	None
2,3,4	13	7.04	3.0	510	(0.080) Y	(2.03)Y	45.0	19.0

a. Refer to Figure 1.26 for joint geometry.

b. A ribbon filament gun was used. Voltage was constant at 120 kV.

c. Sharp focus at the root of the joint was 470 mA at a beam current value of 2 mA.

d. The electron beam was oscillated longitudinal (Y-direction) or transverse (X-direction) at 1kHz to the dimension shown.

e. Aluminum alloy 5356 wire was 0.035 in. (0.9 mm) in diameter.

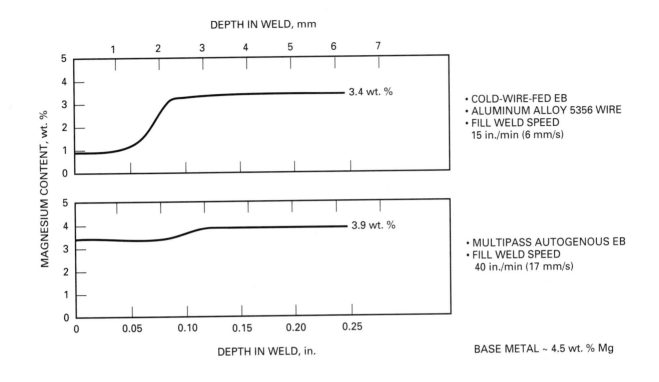

Figure 1.28—Weight Percent Magnesium Content in EB Welds in Aluminum Alloy 5083

Table 1.46
Properties of Aluminum EB Welds Compared to Base Metal Properties

Base Metal	Specimen Identity[a,b]	Average Tensile Strength		Average Yield Strength, 0.2% Offset		Average Tensile Elongation	
		ksi	MPa	ksi	MPa	% in 2 in. (50.8 mm)	% in 1 in. (25.4 mm)
Nonheat-treatable Alloys							
5083-0	BM	42.0	290	21.0	145	22.0	—
	AW	37.8	261	21.6	149	9.6	—
	AW[c]	36.5	252	21.0	145	11.3	—
5456-H321[d]	BM	46.0	317	33.0	228	12.0	—
	AW	45.0	310	38.4	265	4.0	—
Heat-treatable Alloys							
2219-T87[e]	BM	69.0	476	55.0	379	10.0	—
	AW	46.0	317	33.0	228	3.0	—
6061-T6	BM[f]	45.0	310.	40.0	276	18.0	—
	AW	34.5	238	28.9	199	10.0	—
	AW[g]	30.4	210	22.3	154	—	7.0
7039-T64	BM	60.6	418	51.6	356	—	10.0
	AW	43.2	298	37.2	256	—	3.8
	WHT	44.8	309	44.2	305	—	1.8
7039-0	BM	33.1	228	16.4	113	—	17.8
	AW	32.9	227	16.9	116	—	14.7
7075-T6	BM	76.0	524	67.0	462	—	11.0
	AW	50.5	348	43.3	299	—	2.0
	WHT	70.1	483	60.4	416	—	6.0

a. Welds are autogenous except where addition of wire is indicated.

b. BM (base-metal specimen), AW (as-welded specimen), WHT (welded in heat-treated condition then reheat-treated specimen).

c. ER 5356 filler wire.

d. From Hamilton Standard Electron Beam Welding Data Manual, Data Sheet No. 4.1.12 and 4.1.20.

e. From Brennecke, M. W., "Electron Beam Welded Heavy Gage Alloy 2219," *Welding Journal*, 44(1) 27S-29S.

f. Base metal properties from *Metals Handbook*, Properties and Selection of Metals, Vol. 1, 8th Edition.

g. ER 4043 filler metal.

LASER BEAM WELDING

LASER BEAM WELDING (LBW) is performed using the heat generated by a high-energy-density photon (light) beam to produce melting.[8] Because the laser beam can be focused very sharply, laser welds can penetrate comparatively thick joints while producing a very narrow weld, and correspondingly narrow heat-affected zone (HAZ), as compared to arc welding processes. Because a smaller volume of metal is melted, less heat energy goes into the workpiece.

In welding aluminum alloys, the reduction of total heat input can have a beneficial effect. Almost all industrial aluminum alloys are strengthened either by precipitation hardening or strain hardening. The temperatures reached in the HAZ during arc welding are sufficient to result in local over-aging of precipitation-hardened alloys or local annealing of strain-hardened alloys. The effect of this over-aging or annealing is a degradation of HAZ mechanical properties as opposed to base-metal properties. This degradation can be substantial. In arc-welded, high-strength 2XXX series alloys, reduction of mechanical properties of up to 50 percent or more is not uncommon.

The much narrower weld and HAZ exhibited by laser beam welds as compared to arc welds results in much less metal volume becoming over-aged or annealed. This in turn results in laser beam welds generally exhibiting higher yield and tensile strengths in transverse tension testing than arc welds of equal thicknesses. There is also a disadvantage to a very narrow HAZ; the mechanical property mismatch (i.e., the weak HAZ compared to the stronger base material) is highly localized. In applications that involve plastic deformation, the deformation strains are localized in the HAZ. As a consequence, transverse tension tests of laser beam welds typically exhibit low elongation, not because of lack of ductility in the weld or HAZ, but because all of the deformation occurs in the HAZ. This can be a

problem when forming components that have been laser beam welded; failure occurs in the HAZ. Strain localization can also have an adverse effect on fatigue and impact properties of laser beam welds.

The major difficulty in LBW of aluminum alloys is that aluminum does not couple well with the 1.06 μm or 10.5 μm wavelength light emitted, respectively, by Nd:YAG and CO lasers. In other words, the laser beam energy tends to be reflected rather than being absorbed by the aluminum, which does not contribute to the energy required to melt the metal. On polished aluminum surfaces, as much as 90 percent of the laser energy is reflected. Additionally, once a weld pool and keyhole are established, the reflectivity goes down dramatically, resulting in a power density that is too high.

Early LBW control systems were unable to accommodate this change in reflectivity. Control systems have been developed which can vary the energy input to compensate for the reflectivity change when the weld keyhole is established. Further development of control systems is needed.

Another way to reduce the reflectivity of the aluminum is to modify the surface by mechanical or chemical roughening, deposition of various absorptive paint-on coatings, or anodizing and dyeing the aluminum surface. These methods have been tried with varying degrees of success.

These difficulties make aluminum more difficult to laser beam weld than other common structural materials. Despite these difficulties, the aerospace industry is successfully welding 2XXX and 6XXX series alloys in many applications. The aluminum-lithium alloys 2090 and 2091 have been successfully laser beam welded in laboratory development programs. The automotive industry has reported preliminary success in LBW thin gauge [0.03 to 0.08 in. (0.76 to 2.0 mm)] 2XXX and 5XXX alloy series sheet. Although the general application of LBW to aluminum alloys requires further equipment and process developments, the use of LBW for aluminum alloys is increasing at a moderate rate.

---

8. For a more detailed discussion of laser beam welding refer to Chapter 22, *Welding Handbook*, Vol. 2, 8th Ed.

RESISTANCE WELDING

WELDABILITY

SOME ALUMINUM ALLOYS are easier to resistance weld than others. The relative weldabilities of various wrought and casting alloys were given previously in Tables 1.4A, 1.7A, 1.11A, and 1.12A. In general, the casting alloys considered weldable by other processes can also be resistance welded. Permanent mold and sand casting alloys can be successfully spot welded, but die castings are considered difficult to join by this method. Casting alloys may be spot welded to themselves, to other casting alloys, and to wrought alloys.

The temper of an aluminum alloy affects its weldability. Aluminum alloys in the annealed condition are more difficult to resistance weld consistently than are the alloys in a work-hardened or solution heat-treated condition due to deeper indentations, distortion, and increased tip pickup. Electrode life and weld consistency are improved when welding the harder tempers.

JOINT DESIGN

THE BEST JOINT properties for spot welds are obtained when welding material of equal thickness in the range of 0.028 to 0.125 in. (0.71 to 3.2 mm). Acceptable spot welds can be made between material of unequal thicknesses with thickness ratios ranging up to 3 to 1. As the thickness ratio increases, the welding conditions become more critical; closely controlled welding conditions are required to ensure acceptable weld quality. While it is practical to join three parts by spot welding, the spot-welding schedule should be adequately verified by procedure qualification testing before the joint design is used in production.

Joints in aluminum require greater edge distances and joint overlaps than those used for steel. Suggested design dimensions are given in Table 1.47. When a flange is used on one or both components, it should provide the required overlap and be flat. Sometimes, spring-back will permit only the edge of the flange to contact the other component. If the faying surfaces cannot be brought together with light hand pressure, the flange should be reworked.

Table 1.47 also contains suggested spot-welding spacings for varying material thicknesses. Closer spacings than those indicated will require adjustment of the spot-welding schedule to account for increased shunting of current through previous welds.

Spot welds normally should be used to carry shear loads, not tensile or peel loads. When tension or combined loadings are applied, special tests should be conducted to determine the expected strength of the joint under service conditions. The strength of the spot welds

Table 1.47
Minimum Design Dimensions for Spot-Welded Joints in Aluminum Sheet

Sheet Thickness[a]		Nugget Diameter		Weld Spacing[b]		Edge Distance[c]		Joint Overlap	
in.	mm	in.	mm	in.	mm	in.	mm	in.	mm
0.032	0.81	0.14	3.6	0.50	12.7	0.25	6.3	0.50	12.7
0.040	1.02	0.16	4.1	0.50	12.7	0.25	6.3	0.56	14.2
0.050	1.27	0.18	4.6	0.63	16.0	0.31	7.9	0.63	16.0
0.063	1.60	0.20	5.1	0.63	16.0	0.38	9.7	0.75	19.0
0.071	1.80	0.21	5.3	0.75	19.0	0.38	9.7	0.81	20.6
0.080	2.03	0.23	5.8	0.75	19.0	0.38	9.7	0.88	22.4
0.090	2.29	0.24	6.1	0.88	22.4	0.44	11.2	0.94	23.9
0.100	2.54	0.25	6.3	1.00	25.4	0.44	11.2	1.00	25.9
0.125	31.7	0.28	7.1	1.25	31.8	0.50	12.7	1.13	28.7

a. Data for other thicknesses may be obtained by interpolation, or the data for the next smaller thickness may be used.

b. Distance between centers of adjacent spot welds.

c. Distance from the center of a spot weld to the edge of the sheet or flange.

in tension may vary from 20 to 90 percent of the shear strength.

The shear strength of individual spot welds will vary considerably with alloy composition, section thickness, welding schedule, weld spacing, edge distance, and overlap. Joint strength will depend upon the average individual spot strength, the number of welds, and the positioning of the welds in the joint.

The minimum strength per weld and the minimum average strength requirements of a military specification are given in Table 1.48. Somewhat lower strengths may be used for industrial applications where a lesser degree of quality is needed. The strengths listed in the table are based upon spot welds having the given minimum nugget diameters. Weld nuggets of smaller diameter than those shown in the table are not recommended. Nugget diameter and penetration are directly affected by changes in the welding schedule and any changes in the electrode face geometry such as are caused by wear.

SURFACE PREPARATION

WELDS OF UNIFORM strength and good appearance depend upon a consistently low surface resistance. The surface condition of as-received material may be satisfactory for many commercial spot and seam welding applications. On the other hand, applications such as in aircraft and other special equipment require very consistent welds of high quality, particularly when failure of the weldment would result in loss of the equipment. Proper cleaning of parts and monitoring of the cleaning operation are important in maintaining uniform weld quality.

For most applications, some cleaning, such as degreasing, is necessary before resistance welding. If a thick oxide is present, a non-etching deoxidizer is used. For critical applications, the cleanliness of the components must be monitored, and time lapse between cleaning and welding must not exceed a specified period. A two- to six-minute immersion in a room-temperature nitric-hydrofluoric solution consisting of the following has produced excellent results:

(**1**) Nitric acid [technical grade (68% HNO_3)] — 15 oz/gal (120 g/l)
(**2**) Hydrofluoric acid (48% HF) — 0.15 oz/gal (2 g/l)
(**3**) Wetting agent — 0.14 oz/gal (2 g/l)

Caution: All acids used in cleaning solutions are potentially dangerous! Personnel using them should be thoroughly familiar with all chemicals involved and adequate safety equipment should be used.

The immersion of the work is followed by:

(**1**) 30 seconds cold running-water rinse
(**2**) 10 seconds rinse in hot water, 140 to 160 °F (60 to 71 °C)
(**3**) Dry in warm air blast

SURFACE CONTACT RESISTANCE

MEASUREMENT OF SURFACE contact resistance is an effective method of monitoring a cleaning operation. Figure 1.29 is a diagram of a surface-resistance measuring device. Two cleaned coupons are overlapped and placed between two 3 in. (76.2 mm) radius-faced spot-welding electrodes. A standardized current and electrode force are applied; a current of 50 mA and a force of 600 lb (2.67 kN) are frequently used. The voltage drop between the two coupons is measured with a milli-volt meter or a Kelvin bridge. The resistance between the two coupons is then calculated as follows from the current and voltage readings:

$$R = E/I \qquad\qquad (1.1)$$

where
R = resistance in ohms
E = electromotive force in volts
I = current in amperes

It is important that all tests are carried out under identical conditions because the results are sensitive to small changes in procedure. Average surface resistance is usually obtained from at least five readings on each set of coupons. The coupons must not move as the electrode force is applied. Movement may break the oxide coating and cause false readings. With this test, the contact resistance between properly cleaned aluminum sheets will range from 10^{-5} to about 10^{-4} ohms. That of uncleaned stock may range up to 10^{-2} ohms or higher.

RESISTANCE SPOT WELDING

SPOT WELDING IS a practical joining method for fabricating aluminum sheet structures. It may be useful with all wrought alloys as well as many permanent mold and sand casting alloys.

The procedures and equipment for spot welding of aluminum are similar to those used for steels. However, the higher thermal and electrical conductivities of aluminum alloys require some variations in the equipment and the welding schedules. For example, the current

Table 1.48
Minimum Tension-Shear Strength for Resistance Spot Welds in Aluminum Alloys (MIL-W-6858D)

Thickness of Thinner Sheet, in. (mm)	Minimum Nugget Diameter, in. (mm)	Base Metal Tensile Strength							
		Below 19.5 ksi (134 MPa)		19.5 - 35 ksi (134-241 MPa)		35 - 56 ksi (241-386 MPa)		56 ksi (386 MPa) and above	
		Tension-Shear Strength per Weld							
		Min. lb (N)	Min. Avg.* lb (N)	Min. lb (N)	Min. Avg.* lb (N)	Min. lb (N)	Min. Avg.* lb (N)	Min. lb (N)	Min. Avg.* lb (N)
0.016 (0.4)	0.085 (2.2)	50 (222)	65 (289)	70 (311)	90 (400)	100 (445)	125 (556)	110 (489)	140 (623)
0.020 (0.5)	0.10 (2.5)	80 (336)	100 (445)	100 (445)	125 (556)	135 (601)	170 (756)	140 (623)	175 (778)
0.025 (0.6)	0.12 (3.0)	110 (489)	140 (623)	145 (645)	185 (823)	175 (778)	200 (890)	185 (823)	235 (1045)
0.032 (0.8)	0.14 (3.6)	165 (734)	210 (934)	210 (934)	265 (1179)	235 (1045)	295 (1312)	260 (1157)	325 (1446)
0.040 (1.0)	0.16 (4.1)	225 (1001)	285 (1268)	300 (1334)	375 (1668)	310 (1379)	390 (1735)	345 (1535)	435 (1935)
0.050 (1.3)	0.18 (4.6)	295 (1312)	370 (1646)	400 (1779)	500 (2224)	430 (1913)	540 (2402)	465 (2068)	585 (2602)
0.063 (1.6)	0.20 (5.1)	395 (1757)	495 (2202)	570 (2535)	715 (3180)	610 (2713)	765 (3403)	670 (2980)	840 (3737)
0.071 (1.8)	0.21 (5.3)	450 (2002)	565 (2513)	645 (2869)	810 (3603)	720 (3203)	900 (4003)	825 (3670)	1035 (4604)
0.080 (2.0)	0.23 (5.8)	525 (2335)	660 (2936)	765 (3403)	960 (4270)	855 (3803)	1070 (4760)	1025 (4559)	1285 (5716)
0.090 (2.3)	0.24 (6.1)	595 (2647)	745 (3314)	870 (3870)	1090 (4849)	1000 (4448)	1250 (5560)	1255 (5583)	1570 (6984)
0.100 (2.5)	0.25 (6.4)	675 (3003)	845 (3759)	940 (4181)	1175 (5227)	1170 (5204)	1465 (6517)	1490 (6628)	1865 (8296)
0.125 (3.2)	0.28 (7.1)	785 (3492)	985 (4381)	1050 (4671)	1315 (5849)	1625 (7228)	2035 (9052)	2120 (9430)	2650 (11 788)
0.140 (3.6)	0.30 (7.6)	— —	— —	— —	— —	1920 (8541)	2400 (10 676)	2525 (11 232)	3160 (14 056)
0.160 (4.1)	0.32 (8.1)	— —	— —	— —	— —	2440 (10 854)	3050 (13 567)	3120 (13 878)	3900 (17 348)
0.180 (4.6)	0.34 (8.6)	— —	— —	— —	— —	3000 (13 345)	3750 (16681)	3725 (16 570)	4660 (20 729)
0.190 (4.8)	0.35 (8.9)	— —	— —	— —	— —	3240 (14 412)	4050 (18 015)	4035 (17 949)	5045 (22 441)
0.250 (6.4)	— —	— —	— —	— —	— —	6400 (28 469)	8000 (35 586)	7350 (32 694)	9200 (40 924)

* Average of three or more tension-shear tests

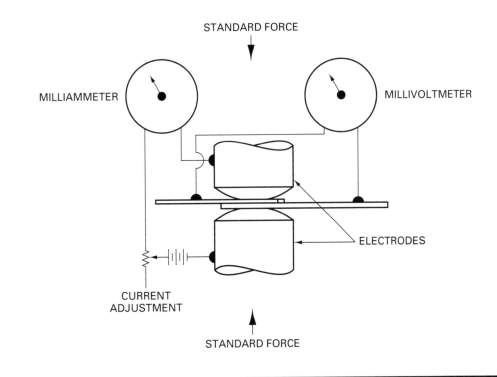

Figure 1.29—Arrangement for Measuring Surface Contact Resistance for Monitoring Cleanness

must be two to three times that required for a comparable joint between steel sections.

Spot Weld Equipment

ALUMINUM CAN BE welded with both ac and dc power.[9] High welding current is required because of the high electrical conductivity of aluminum. Consequently, the primary power demand is higher than that required when spot welding an equivalent thickness of steel. For the best quality, machines that produce continuous or pulsed dc power are preferred. Acceptable welds for some applications can be made with single-phase ac equipment. In any case, the welding machine should be equipped with a low-inertia force system to provide fast electrode follow-up as the weld nugget is formed. In addition, a forging force system may be necessary to consistently produce crack-free welds, particularly in the 2XXX, 6XXX, and 7XXX series heat-treatable aluminum alloys.

Spot Weld Electrodes

STANDARD ELECTRODES WITH radius-contoured faces are used on both sides of the joint for most welding operations. When surface marking is undesirable, a flat-faced electrode may be used on one side. Heat balance in the joint must be considered when this is done. Resistance welding electrode materials are classified by the Resistance Welders Manufacturing Association (RWMA).[10]

RWMA Group A, Class 1, copper alloy electrodes are recommended for spot welding aluminum. A Class 1 alloy has the highest electrical conductivity of the Group A alloys. A Class 2 alloy is sometimes used when the higher hardness is needed to maintain contour with high electrode forces.

A copper-aluminum alloy tends to form on the face of each electrode during use. The condition is commonly called *electrode pickup*, and the alloy that forms is brittle with relatively low electrical conductivity. As this alloy coating builds up, the contact resistance

---

9. For a more detailed discussion of resistance welding equipment refer to Chapter 19, *Welding Handbook*, Vol. 2, 8th Ed.

10. See ANSI/RWMA Bulletin No. 16, "Resistance Welding Equipment Standards." Resistance Welders Manufacturers Association, Philadelphia.

increases and the electrode pickup tends to stick to the aluminum surface. Electrode pickup mars the surface, produces an unpleasant appearance, and may also pull particles of the Cu-Al alloy from the electrode face.

This action increases the surface roughness and the rate of electrode deterioration.

Excessive electrode pickup is generally the result of improper surface preparation of the parts prior to welding, insufficient electrode force, or excessive welding current. The coating on the electrode faces may be removed by dressing periodically with an appropriately shaped tool covered with a fine abrasive cloth. Care must be taken to maintain the original face contour. A file should not be used to dress the electrode.

Spot Welding Schedules

SUGGESTED SCHEDULES FOR spot welding with three types of machines are given in Tables 1.49, 1.50 and 1.51. These tables can be used as guides for establishing production welding schedules.

Roll Spot And Seam Welding

ROLL SPOT AND seam welding are very similar to spot welding except that copper alloy wheel electrodes are used. Seam welding consists of a series of overlapping weld nuggets that form a gas- or liquid-tight joint.

Roll spot welding consists of a series of uniformly spaced individual spot welds. The same equipment that is used for seam welding may be used for roll spot welding by adjusting the interval between weld heat cycles (cool time or off time).

Welding may be done while the wheel electrodes and the work are in motion or while they are momentarily stopped. Surface appearance and weld quality will be better when the electrodes and work are stationary. The welding force is maintained on the nugget during solidification and cooling. With moving electrodes, the weld nugget moves from between these electrodes before the nugget is adequately cooled.

Roll Spot and Seam Welding Equipment

EQUIPMENT USED FOR roll spot or seam welding should have features similar to those of spot welding machines. Somewhat higher welding currents and electrode forces may be required for roll spot and seam welding because shunting of current through the previous nugget may be greater than with spot welding. Excessive travel speed can contribute to aluminum pick-up on the wheel electrodes. This can be corrected by increasing the time between welds and decreasing the travel speed to give the desired number of welds per unit length.

Table 1.49
Suggested Schedules for Spot Welding of Aluminum Alloys with Single-Phase AC Machines

Thickness*		Electrode Face Radii, Top-Bottom		Net Electrode Force		Approximate Welding Current, kA	Welding Time, Cycles (60 Hz)
in.	mm	in.	mm	lb	N		
0.032	0.81	2-2 or 2-Flat	51-51 or 51- flat	500	2224	26	7
0.040	1.02	3-3 or 3-Flat	76-76 or 76- flat	600	2669	31	8
0.050	1.27	3-3 3-Flat	76-76 or 76- flat	680	3025	33	8
0.062	1.57	3-3 or 3-Flat	76-76 or 76-flat	750	3336	36	10
0.070	1.78	4-4	102-102	800	3559	38	10
0.081	2.06	4-4	102-102	880	3914	42	10
0.090	2.29	6-6	152-152	950	4226	46	12
0.100	2.54	6-6	152-152	1050	4671	56	15
0.110	2.79	6-6	152-152	1150	5115	64	15
0.125	3.17	6-6	152-152	1300	5783	76	15

* Thickness of one sheet of a two-sheet combination

Table 1.50
Suggested Schedules for Spot Welding of Aluminum Alloys with
Three-Phase Frequency Converter Machines

Sheet Thickness[a]		Electrode Face Radii[b]		Electrode Force				Current, kA[c]		Time, Cycles (60Hz)	
				Weld		Forge					
in.	mm	in.	mm	lb	N	lb	N	Weld	Postheat	Weld	Postheat
0.025	0.64	3	76	500	2224	1500	6672	34	8.5	1	3
0.032	0.81	4	102	700	3114	1800	8007	36	9.0	1	4
0.040	1.02	4	102	800	3559	2000	8896	42	12.6	1	4
0.050	1.27	4	102	900	4003	2300	10 231	46	13.8	1	5
0.063	1.60	6	152	1300	5783	3000	13 345	54	18.9	2	5
0.071	1.80	6	152	1600	7117	3600	16 014	61	21.4	2	6
0.080	2.03	6	152	2000	8896	4300	19 127	65	22.8	3	6
0.090	2.29	6	152	2400	10 676	5300	23 575	75	30.0	3	8
0.100	2.54	8	203	2800	12 455	6800	30 248	85	34.0	3	8
0.125	3.17	8	203	4000	17 793	9000	40 034	100	45.0	4	10

a. Thickness of thinnest sheet of a two-sheet combination.

b. Top and bottom electrode face radii are identical.

c. Suitable for alloys 2014-T3, 4, and 6; 2024T-3 and 4; 7075-T6. Somewhat lower current may be used for softer alloys, such as 5052, 6009, and 6010.

Table 1.51
Suggested Schedules for Spot Welding of Aluminum Alloys with
Three-Phase Rectifier Machines

Sheet Thickness[a]		Electrode Face Radii[b]		Electrode Force				Current, kA[c]		Time, Cycles (60Hz)	
				Weld		Forge					
in.	mm	in.	mm	lb	N	lb	N	Weld	Postheat	Weld	Postheat
0.032	0.81	3	76	670	2980	1540	6850	28	0	2	0
0.040	1.02	3	76	730	3247	1800	8007	32	0	3	0
0.050	1.27	8	203	900	4003	2250	10 008	37	30	4	4
0.063	1.60	8	203	1100	4893	2900	12 900	43	36	5	5
0.071	1.80	8	203	1190	5293	3240	14 412	48	38	6	7
0.080	2.03	8	203	1460	6494	3800	16 903	52	42	7	9
0.090	2.29	8	203	1700	7562	4300	19 127	56	45	8	11
0.100	2.54	8	203	1900	8452	5000	22 241	61	49	9	14
0.125	3.17	8	203	2500	11 121	6500	28 913	69	54	10	22

a. Thickness of one sheet of a two-sheet combination.

b. Top and bottom electrode face radii are identical.

c. Suitable for 2014-T3, 4, and 6; 2024T-3 and 4; 7075-T6. Somewhat lower current may be used for softer alloys, such as 5052, 6009, and 6010.

Roll Spot and Seam Welding Electrodes

RADIUS-FACED WHEEL electrodes are normally used. Face radii generally range from 1 to 10 in. (25.4 to 254 mm). Normally, the face radius should be about the same as the wheel radius to approach a spherical radius in contact with the workpieces. The faces should be cleaned after each 3 to 5 revolutions of continuous welding, and after each 10 to 20 revolutions of roll spot welding. An appropriate cutting tool may be used. The electrodes also may be cleaned continuously with a medium to fine grade abrasive material bearing against each electrode face under 5 to 10 lb (22.4 to 44.8 N) of force.

Roll Spot and Seam Welding Schedules

TYPICAL SETTINGS FOR seam welding alloy 5052-H34 aluminum sheet with single-phase ac seam welding machines are given in Table 1.52. This data may be used as a guide when developing welding schedules for other alloys or tempers. Quality control for roll spot and seam welding is the same as for spot welding.

Weld Quality

THE QUALITY OF spot and seam welds in aluminum alloys is more sensitive to process variations than are similar welds is steels.[11] This is related to the high resistivity of aluminum oxides and the high electrical and thermal conductivities of the metal. The size of the weld nugget is very sensitive to the heat energy developed by the resistance of the workpieces to the welding current. The energy must be produced rapidly to overcome losses to the surrounding base metal and the electrodes.

The contact resistance between the faying surfaces and between the electrodes and the workpieces are a significant part of the total resistance in the circuit. Significant variations in these contact resistances can cause large changes in the welding current density. Since the condition of the aluminum surfaces affects the contact resistances, uniform cleanness is essential for consistent weld quality.

The contact resistance between the electrodes and the workpieces increases with electrode pickup on the electrode faces. As this contact resistance increases, so does electrode heating and wear. As the contact area increases, welding current density decreases and results in reduced nugget size and penetration. Weld strength decreases at the same time. Electrode wear requires constant attention.

Other important factors that affect weld quality are surface appearance, internal discontinuities, sheet separations, metal expulsion, weld strength, and weld ductility. Uniform weld quality can be obtained only by the use of proper equipment and trained operators and adherence to qualified welding procedures and schedules. These should be developed and maintained during production by a regular program of quality control.

---

11. For a more detailed discussion of weld schedules and weld quality refer to Chapter 17, *Welding Handbook*, Vol. 2, 8th Ed.

Table 1.52
Typical Schedules for Gas-Tight Seam Welds in 5052-H34 Aluminum Alloy
with Single-Phase AC Machines

Sheet Thickness[a]		Welds		On + Off Time, Cycles (60 Hz)[b]	Travel Speed[c]		On Times, Cycles (60 Hz)		Electrode Force		Welding Current,	Approximate Weld Width[d]	
in.	mm	per in.	per m		ft/min	mm/s	Min.	Max	lb	N	kA	in.	mm
0.025	0.64	18	709	5 1/2	3.0	15.2	1	1 1/2	600	2669	26.0	0.11	2.8
0.032	0.81	16	630	5 1/2	3.4	17.2	1	1 1/2	690	3069	29.0	0.13	3.3
0.040	1.02	14	551	7 1/2	2.9	14.7	1 1/2	2 1/2	760	3381	32.0	0.14	3.6
0.050	1.27	12	472	9 1/2	2.6	13.2	1 1/2	3	860	3825	36.0	0.16	4.1
0.063	1.60	10	394	11 1/2	2.6	13.2	2	3 1/2	960	4270	38.5	0.19	4.8
0.080	2.03	9	354	15 1/2	2.1	10.6	3	5	1090	4849	41.0	0.22	5.6
0.100	2.54	8	315	20 1/2	1.8	9.1	4	6 1/2	1230	5471	43.0	0.26	6.6
0.125	3.17	7	276	28 1/2	1.5	7.6	5 1/2	9 1/2	1350	6005	45.0	0.32	8.1

a. Thinner of a two-sheet combination.

b. Use next higher full cycle setting if timer is not equipped for synchronous initiation.

c. Should be adjusted to give the desired number of spots per inch.

d. Welding force, welding current, or both should be adjusted to produce the desired weld width. Use lower force for soft alloys or tempers and higher force for hard alloys or tempers.

Factors that tend to produce cracks or porosity in welds are excessive heating of the nugget, a high cooling rate, and improper application of the electrode force. Spot welds in some high-strength alloys, such as alloy 2024 and alloy 7075, are subject to cracking if the welding current is too high or the electrode force is too low. Cooling rate can be controlled by application of current downslope or a postheat cycle. With dual-force machines, proper adjustment of the forge delay time may also prevent cracking.

Quality Control

QUALITY OF ALUMINUM spot and seam welds depends upon the welding schedule, electrode condition, and surface preparation. All three must be controlled to maintain acceptable weld quality.

Quality criteria for resistance welds should be established for the intended application. A range of quality standards, based upon service-reliability requirements, is suggested when service conditions for resistance welds vary in a particular product. Quality criteria for military hardware and aircraft are established by military specifications.

FLASH WELDING

ALL ALUMINUM ALLOYS may be joined by the flash welding process.[12] This process is particularly adapted to making butt or miter joints of two parts of similar cross sections. Flash welding can be used to join aluminum to copper.

Good mechanical properties are obtained in flash welded joints, and joint efficiencies of at least 80 percent are readily obtained. Strength is generally higher when the alloy is in a hard temper. Heat treatment after flash welding may provide a slight increase in joint efficiency.

Equipment used to flash weld aluminum is similar to that for welding steel except that more rapid platen acceleration and upset force application and higher welding currents are necessary. Electrodes may be fabricated from tool steels to prevent sticking of the aluminum to copper surfaces and also to provide a sharp edge to shear off the flash at the conclusion of the upset. Bend or tensile tests, or both, are used to determine joint strength and to assess weld quality.

HIGH-FREQUENCY RESISTANCE WELDING

RESISTANCE WELDING WITH high-frequency current is used primarily for high-speed production of tubing. In this application, squeeze rolls forge the edges of the sheet together after they are heated to welding temperature with high-frequency current. Tubing with wall thicknesses of 0.03 to 0.125 in. (0.76 to 3.2 mm) can be welded at high travel speeds.

HIGH-FREQUENCY WELDING EQUIPMENT

HIGH-FREQUENCY GENERATORS ARE electrical devices and require all usual safety precautions in handling and repairing such equipment. Voltages are in the range from 400 to 20 000 V and may be of either low or high frequency. Proper care and safety precautions should be taken while working on high-frequency generators and their control systems. Units must be equipped with safety interlocks on access doors and automatic safety grounding devices to prevent operation of the equipment when access doors are open. The equipment should not be operated with panels or high-voltage covers removed or with interlocks and grounding devices blocked.

The output high-frequency primary leads should be encased in metal ducting and should not be operated in the open. Induction coils and contact systems should always be properly grounded for operator protection. High-frequency currents are more difficult to ground than low-frequency currents, and grounding lines must be kept short and direct to minimize inductive impedance. The magnetic field from the output system must not induce heat in adjacent metallic sections and cause fires or burns.

Injuries from high-frequency power, especially at the upper range of welding frequencies, tend to produce severe local surface-tissue damage. However, they are not likely to be fatal because current flow is shallow.

SAFETY

THE MAIN HAZARDS that may arise with resistance welding processes and equipment are

(1) Electric shock from contact with high voltage terminals or components

(2) Ejection of small particles of molten metal from the weld

(3) Crushing of some part of the body between the electrodes or other moving components of the machine

For information on the electrical hazard, see the preceding section, "High-Frequency Welding Equipment." For a detailed discussion of safety information on resistance welding equipment, refer to Chapter 19, *Welding Handbook*, Volume 2, 8th Edition.

---

12. For a more detailed discussion of flash welding refer to Chapter 18, *Welding Handbook*, Vol. 2, 8th Ed.

SOLID-STATE WELDING

COLD WELDING

COLD WELDING (CW) is performed without the addition of heat.[13] An external pressure is applied to the two pieces to be welded, resulting in a substantial amount of plastic deformation. A fundamental requisite for CW is that at least one of the metals is highly ductile and does not markedly work harden. Both butt and lap joints can be cold welded.

The weld flash (metal expelled from the joint during welding) in cold-welded butt joints must be removed mechanically (i.e., by grinding or machining). The plastic deformation occurring during cold welding breaks up aluminum oxides on the surface, and the oxides are expelled from the butt joint during cold welding. Preweld cleaning is not as critical for CW butt joints as for lap joints, but precleaning of lap joints for cold welding is critical. The preferred method is to degrease and wire brush the faying surfaces.

Since there is no heat-affected zone, the weld in the butt joint is as strong or nearly as strong as the base material. Many aluminum alloys that cannot be arc welded because of their crack susceptibility can be successfully cold welded. For example, butt joints in alloys 2024 and 7075 have been successfully cold welded, but lap joints have not.

Butt and miter joints can be made in most aluminum alloy wire, rod, tubing and simple extruded shapes, and lap joints can be welded in sheet material. Butt joints in soft annealed alloys will require a total upset distance of approximately 1.5 times the material thickness. Higher strength alloys require greater upset distance, approximately 4 to 5 times the material thickness, to obtain an acceptable weld. Welds in lap joints require a thickness reduction of about 70 percent at the weld location and are only practical in the low-strength 1XXX and 3XXX series alloys. These welds provide good shear strength, but do not perform well when subjected to a bending or peeling type of loading.

ULTRASONIC WELDING

ULTRASONIC WELDING IS accomplished by the local application of high-frequency, low-amplitude vibratory motion to the workpieces while they are held together under a low clamping pressure.[14] The process is used to join foil and sheet gauges of aluminum alloys as well as to join thin wires to sheet or foil. The top-sheet thickness limit in lap-joint ultrasonic welding of aluminum alloys is usually 0.060 in. (1.5 mm), though spot welds have been made in thicknesses up to 0.125 in. (3.2 mm). The bottom sheet can be up to 1 in. (25.4 mm) thick. The types of possible ultrasonic welds are spot, roll spot, seam, ring, and linear (line) welds.

All aluminum alloys can be ultrasonically welded, but the degree of weldability varies with the alloy and temper. Aluminum alloys can also be joined to other metals with this process.

Ultrasonic welding can be done without extensive surface preparation, with minimum deformation, and with low compressive loads. Ultrasonic welds look much like resistance spot or seam welds, but they often are characterized by a localized roughened surface. Reduction in thickness at the weld in a lap joint will be about 5 percent as compared to 70 percent in the cold weld.

For relatively low-strength aluminum alloys, ultrasonic spot welds exhibit approximately the same strength as resistance spot or seam welds. With the higher strength alloys, the strength of ultrasonic welds can exceed the strength of resistance welds. The primary reasons are that ultrasonic welding produces no heat-affected zones in the base metal, and the size of the weld is generally greater.

Ultrasonic welding generally requires less surface preparation than does resistance welding. Degreasing of the aluminum is normally advisable. To obtain uniform welds, heat-treated alloys and alloys containing high percentages of magnesium should be cleaned of surface oxides before welding.

EXPLOSION WELDING

EXPLOSION WELDING USES energy from the detonation of an explosive to produce a solid-state weld.[15] The force produced by the detonation of the explosive drives the two components together to create a high-strength weld with minimum diffusion and deformation at the interface. Explosion welding is limited to lap joints and to cladding of parts with a second metal having special properties.

A common application is to clad carbon steel, stainless steel, copper, or titanium alloys with aluminum. Explosion welded bi-metallic sections are primarily used as transition segments. Then conventional welding processes are used to weld similar metal on each side of the segment. The net effect is to join aluminum to

---

13. For a more detailed discussion of cold welding refer to Chapter 29, *Welding Handbook*, Vol. 2, 8th Ed.

14. For a more detailed discussion of ultrasonic welding refer to Chapter 25, *Welding Handbook*, Vol. 2, 8th Ed.

15. For a more detailed discussion of explosion welding refer to Chapter 24, *Welding Handbook*, Vol. 2, 8th Ed.

another desired alloy, such as steel. Surface preparation for explosive welding of aluminum is similar to that for other welding processes. The faying surfaces should be cleaned shortly before welding. The normal surface oxides are broken up and dispersed during welding.

DIFFUSION WELDING

IN DIFFUSION WELDING, the principal method of joint formation is migration of atoms from each of the pieces being welded into the other piece.[16] The weld is produced by the application of high temperatures and high pressures for long times. There is no melting or macroscopic deformation. A thin, solid foil of filler metal may be inserted between the faying surfaces to help activate the diffusion process.

In diffusion welding aluminum alloys, some means must be provided to prevent, disrupt, or dissolve the surface oxides. Weld strength and ease of bonding are achieved with a thin intermediate layer of another metal such as silver, copper, or gold-copper alloy. A wide range of temperatures, times, and pressures may be used. The welding operation must be performed under vacuum or in an inert gas atmosphere.

FRICTION WELDING

FRICTION WELDING IS a solid-state welding process that produces coalescence of materials under compressive force by rotating or moving the pieces to be joined relative to each other to produce heat and plastically displacing material from the faying surfaces.[17] The plastically displaced material, commonly called flash, may be removed in a later operation to produce a smooth surface.

Most friction welding is based on rotary motion and is best applied to joining circular parts (i.e., rod, bar, wire, tube, and pipe). Recent developments in linear friction welding, in which a linear oscillating motion is used, have shown that this technique can be successfully applied to aluminum alloys so that parts without circular symmetry can be joined.

Almost all aluminum alloys can be friction welded, including the 7XXX alloys that cannot be arc welded because of crack susceptibility. The metal, softened by the frictional heating, is expelled from the joint, and joint strength approaches that of the base material, even for high-strength heat-treatable aluminum alloys. The aluminum oxide present on the faying surfaces is broken up and expelled from the joint, and preweld cleaning is not as critical as for other welding processes.

While not all material combinations are possible, aluminum alloys can be readily friction welded to many other materials. Two of the more common weld combinations are aluminum to copper alloy, used in the electrical industry, and aluminum to stainless steel, used as transition couplings in piping systems and pressure vessels.

---

16. For a more detailed discussion of diffusion welding refer to Chapter 26, *Welding Handbook*, Vol. 2, 8th Ed.

17. For a more detailed discussion of friction welding refer to Chapter 23, *Welding Handbook*, Vol. 2, 8th Ed.

OXYFUEL GAS WELDING

ALUMINUM CAN BE welded by the oxyfuel gas welding process.[18] However, it should only be used for noncritical or repair applications when suitable inert-gas-shielded arc welding equipment is not available. Table 1.53 gives suggested procedures for welding aluminum with the oxyfuel gas flame. The advantage of the process is the simplicity, portability, and low cost of the equipment, but the disadvantages when compared to arc welding are more numerous. The disadvantages are:

(1) The aluminum oxide on the surface melts at a much higher temperature than the aluminum. Coupled with this is that aluminum does not change color when the melting temperature is reached. With the low rate of heat input from oxyfuel gas, the unskilled operator may melt a hole in the aluminum before the surface has reached the molten state; the unmelted oxide on the surface cannot support the molten metal, and it falls through the bottom of the section.

(2) An active welding flux is required to remove oxide and protect the molten aluminum from oxidation.

(3) Welding speeds are slower.

(4) Heat-affected zones are wider.

(5) Weld-metal solidification rates are slower, increasing the possibility of hot cracking.

(6) The gas flame offers no surface cleaning action.

(7) Distortion of the weldment is greater.

(8) The welding flux must be completely removed.

---

18. For a more detailed discussion of oxyfuel gas welding refer to Chapter 11, *Welding Handbook*, Vol. 2, 8th Ed.

Table 1.53
Suggested Procedures for Oxyfuel Gas Welding

Metal Thickness		Oxyhydrogen						Oxyacetylene					
		Diameter of Orifice in Tip		Oxygen Pressure		Hydrogen Pressure		Diameter of Orifice in Tip		Oxygen Pressure		Acetylene Pressure	
in.	mm	in.	mm	psi	kPa	psi	kPa	in.	mm	psi	kPa	psi	kPa
0.020	0.51	0.035	0.89	1	6.9	1	6.9	0.025	0.63	1	6.9	1	6.9
0.032	0.81	0.045	1.14	1	6.9	1	6.9	0.035	0.89	1	6.9	1	6.9
0.051	1.30	0.065	1.65	2	13.8	1	6.9	0.045	1.14	2	13.8	2	13.8
0.081	2.06	0.075	1.91	2	13.8	1	6.9	0.055	1.40	3	20.7	3	20.7
0.125	3.17	0.095	2.41	3	20.7	2	13.8	0.065	1.65	4	27.6	4	27.6
0.250	6.35	0.105	2.67	4	27.6	2	13.8	0.075	1.91	5	34.5	5	34.5
0.312	7.92	0.115	2.92	4	27.6	2	13.8	0.085	2.16	5	34.5	5	34.5
0.375	9.52	0.125	3.17	5	34.5	3	20.7	0.095	2.41	6	41.4	6	41.4
0.625	15.87	0.150	3.81	8	55.2	6	41.4	0.105	2.67	7	48.3	7	48.3

Standard oxyfuel gas welding torches are suitable for welding aluminum sections about 0.03 to 1 in. (0.76 to 25.4 mm) in thickness. Thicker sections are seldom welded because good fusion is difficult with the limited heat available from the flame.

FUEL GASES

ACETYLENE IS THE most commonly used fuel for oxyfuel gas welding of aluminum because of its high combustion intensity and flame temperature. A slightly reducing flame (excess acetylene) is used since it reduces the possibility of forming unwanted aluminum oxides. This produces a carbonaceous deposit which obscures vision of the weld pool and requires good skills by the welder in working the filler rod.

Hydrogen is a preferred fuel for welding aluminum. Hydrogen is used with a neutral flame, produces good visibility of the weld, and is easiest to use. A larger tip is used for oxyhydrogen welding than for oxyacetylene to compensate for the lower intensity.

WELDING FLUX

ALUMINUM WELDING FLUX is designed to remove the aluminum oxide surface film and exclude oxygen from the molten weld pool. It is generally furnished in powder form and mixed with water to form a thin free-flowing paste. The filler metal should be uniformly coated with flux either by dipping or painting, and the joint faces and adjacent surfaces should be coated with flux to prevent oxidation of these surfaces during welding.

Flux residues are corrosive to aluminum when moisture is present, and thorough cleaning after welding is of prime importance. Weldments in small parts or assemblies may be cleaned by immersion in an acid cleaning solution. Any one of the following solution choices may be used (these are not steps; choose only one):

(1)   10 percent sulfuric acid at room temperature for 20 to 30 minutes

(2)   5 percent sulfuric acid at 150 °F (66 °C) for 5 to 10 minutes

(3)   40 to 50 percent nitric acid at room temperature for 10 to 20 minutes

Acid cleaning should be followed by a hot water rinse and then a cold water rinse.

Caution: All acids used in cleaning solutions are potentially dangerous! Personnel using them should be thoroughly familiar with all chemicals involved and adequate safety equipment should be used.

Steam cleaning may be used to remove flux residue, particularly on parts that cannot be immersed. Brushing may be necessary to remove adhering flux particles.

JOINT DESIGNS

JOINT DESIGNS FOR oxyfuel gas welding are the same as those for gas tungsten arc welding, which are shown in Figures 1.3 and 1.4.

For sections over 0.18 in. (4.5 mm) thick, penetration is best achieved by beveling the edges to be joined. Single V-groove joints are sometimes used on plate up to about 0.50 in. (12.7 mm) thick. Permanent backings are not recommended for gas welding due to possible entrapment of welding flux and the probability of subsequent corrosion. Fillet-welded lap joints generally are not recommended for the same reason.

PREWELD CLEANERS

GREASE AND OIL should be removed from the welding surfaces with a safe solvent. Fluxes will perform better if thick oxide layers are removed from the surfaces prior to welding.

FILLER METALS

BARE FILLER RODS ER1100, ER4043, ER4047, and ER4145, may be used for oxyfuel gas welding.[19] Also, covered electrodes E1100, E3003, and E4043 can be used.[20] Although the welding of the 5XXX series alloys is not recommended because of poor wetting characteristics, thin sections possessing no more than 2.5% Mg may be welded with a single pass. The proper choice of filler metal depends upon the alloy being used and the end-use requirements.

The size of the filler metal rod is related to the thickness of the sections being welded. A large rod may melt

---

19. See ANSI/AWS A5.10, *Specification for Bare Aluminum and Aluminum Alloy Welding Electrodes and Rods.*
20. See ANSI/AWS A5.3, *Specification for Aluminum and Aluminum Alloy Electrodes for Shielded Metal Arc Welding.*

too slowly and tend to freeze the weld pool prematurely. A small rod tends to melt rapidly and makes addition of filler metal into the molten weld pool difficult.

PREHEATING

PREHEATING IS NECESSARY when the mass of base metal is so great that the heat is conducted away from the joint too fast to accomplish welding. Preheat will also improve control of the molten weld pool.

WELDING TECHNIQUE

INITIALLY, THE FLAME is moved in a circular motion to preheat both edges of the joint uniformly. The flame is then held where the weld is to begin until a small molten pool forms. The end of the filler rod is fed into the molten pool to deposit a drop of metal and then withdrawn. This is repeated as welding progresses using the forehand technique. The flame should be oscillated to melt both joint faces simultaneously. The inner flame cone should not touch the molten weld pool but be kept from 0.062 to 0.25 in. (1.6 to 6.4 mm) away. The crater should be filled before removing the flame.

WELDING ALUMINUM CASTINGS

CASTINGS ARE SOMETIMES welded to correct foundry defects, to repair castings damaged in service, or to assemble castings into weldments.

The nominal compositions, weldability, and physical and mechanical properties of various casting alloys were given previously in Tables 1.9 through 1.12.

Standard casting alloy filler metals are listed in Table 1.54. Sand and permanent-mold castings can be welded in a similar manner as wrought aluminum alloys, and weldability is based primarily on the chemical composition and melting range of the casting alloy. Die castings tend to be rather gassy due to the entrapment of die lubricants, and welds that penetrate their "skin" (surface layer) will be extremely porous. Vacuum die castings can possess very sound internal structures and have been welded satisfactorily.

When repairing newly made castings in the foundry, a filler metal of the same alloy is used to provide a homogeneous structure. If the same filler composition of the casting is not available, the foundries can cast their own filler metal in suitable size and chemical composition.

New castings are clean, and it is usually only necessary to remove any sand or other surface contaminants before repair welding. Internal defects, determined by radiography, need to be gouged out by chipping, manual routers, deburring tools, or similar means to permit weld penetration into sound metal. Gas tungsten arc welding is commonly used to repair new castings. To repair sections 3/16 in. (4.8 mm) and thinner, ac power is commonly used. Direct current electrode negative power is often preferred for thicker sections to minimize preheating requirements.

Repair welding of castings that have been in service requires different consideration. These castings have usually been exposed to oil, grease, or other contaminants and must be thoroughly cleaned before welding. Also, a filler metal of the same composition may not be available. In such circumstances, it is often acceptable to use another standard filler alloy from either Table 1.15 or 1.54.

Typical groove-weld tensile strengths for casting alloys are presented in Table 1.55. The heat-treatable

Table 1.54
Composition of Standard Filler Metals for Welding and Repair of Aluminum Castings, wt. %[a]

Filler Alloy	Si	Fe	Cu	Mn	Mg	Ni	Zn	Ti	Other Elements Each	Other Elements Total	Al
206.0	0.10	0.15	4.2-5.0	0.20-0.50	0.15-0.35	0.05	0.10	0.15-0.30	0.05	0.15	Remainder
C355.0	4.5-5.5	0.20	1.0-1.5	0.10	0.40-0.6	—	0.10	0.20	0.05	0.15	Remainder
4009[b]	4.5-5.5	0.20	1.0-1.5	0.10	0.45-0.6	—	0.10	0.20	0.05[b]	0.15	Remainder
A356.0	6.5-7.5	0.20	0.20	0.10	0.25-0.45	—	0.10	0.20	0.05	0.15	Remainder
4010[b]	6.5-7.5	0.20	0.20	0.10	0.30-0.45	—	0.10	0.20	0.05[b]	0.15	Remainder
357.0	6.5-7.5	0.15	0.05	0.03	0.45-0.6	—	0.05	0.20	0.05	0.15	Remainder
A357.0[c]	6.5-7.5	0.20	0.20	0.10	0.40-0.7	—	0.10	0.04-0.20	0.05[c]	0.15	Remainder
4011[c]	6.5-7.5	0.20	0.20	0.10	0.45-0.7	—	0.10	0.04-0.20	0.05[c]	0.15	Remainder

a. Single values are maximum, except when otherwise specified.

b. Beryllium shall not exceed 0.0008%.

c. Beryllium content shall be 0.04 to 0.07%.

casting alloys will exhibit a partial loss of mechanical properties from the heat of welding in the same manner as the wrought alloys. By selecting the proper filler metal to respond to subsequent heat treatment, these heat-treatable alloys can be postweld heat-treated to restore their original heat-treated properties.

When making an assembly by welding an aluminum casting to a wrought aluminum alloy, the strength of the weldment will be controlled by the lower strength heat-affected zone. If heat-treatable alloys are joined and postweld heat treatment is desired, the compatibility of the solution heat-treating and artificial aging practices is an important criteria in the selection of the cast and

Table 1.55
Typical As-Welded Tensile Strength of Gas Metal Arc Welds in Aluminum Castings

Base Alloy	Filler Alloy	Ultimate Tensile Strength ksi	Ultimate Tensile Strength MPa
208.0-F	4043	20	138
295.0-T6	2319	32.5	224
354.0 -T4,T61,-T62	4043, 4047	22.2	153
356.0-T6,T7,T71	4043	27	186
A356.0-T6, -T61	4043	27	186
443.0-F	4043	18	124
A444.0-F	4043	24.5	169
514.0-F	5654	25	170
520.0-T4	5356	25	170
535.0-F	5356,5556	37	196
710.0-F	5356	34	193

wrought alloys. The selection of filler metal for the cast-to-wrought alloy weldment is normally a compromise, and each case requires special consideration.

For highest strengths and greatest ductility, castings with a high silicon content should be welded with an Al-Si filler alloy, such as alloy 4043. Cast or wrought alloys having a high magnesium content should be welded with an Al-Mg filler metal, such as Alloy 5356. Mixing large amounts of magnesium and silicon in the weld metal will result in the formation of large quantities of brittle magnesium-silicide, which increases susceptibility to weld cracking and may affect corrosion resistance.

Welding a high-silicon-content casting alloy, such as 356.0, to a high-magnesium-content wrought alloy, such as alloy 5083, should be avoided. Whether a 4XXX series or a 5XXX series filler metal is selected, the magnesium-silicide problem will occur in one of the weld transition zones. Best overall performance is experienced when joining a 5XXX wrought alloy to a 5XX.0 casting alloy. The 3XX.0 and 4XX.0 cast alloys can be joined to the 1XXX, 2XXX, 3XXX, 4XXX, and 6XXX series wrought alloys with a 4XXX series filler metal.

When welding thick-to-thin sections, thermal strains may result in cracking or distortion. It may be necessary to preheat the casting for welding. The temperature used depends upon the casting shape, alloy, and prior heat treatment, but it generally is between 400 and 900 °F (205 and 482 °C). Up to 600 °F (316 °C), properties of the T6 temper will be affected, but little loss will occur for the T5 and T7 tempers. Over 600 °F (316 °C), all alloys become annealed, and postweld heat treatment is necessary to restore properties.

BRAZING

BRAZEABLE ALLOYS

MANY ALUMINUM ALLOYS can be brazed with commercially available filler metals.[21] The alloy may be a casting or a wrought product, heat-treatable or non-heat-treatable. Aluminum alloys are rated for brazeability in Tables 1.4A, 1.7A, 1.11A and 1.12A. All of the 1XXX and 3XXX series alloys are brazeable, as well as the 5XXX series alloys that contain less than 2% magnesium. The high magnesium alloys of the 5XXX series are difficult to braze because of poor wetting characteristics, melting points below available filler metals, and poor joint properties.

The 6XXX series are the easiest of the heat-treatable alloys to braze. The 2XXX series and most of the 7XXX series are not brazeable because their melting temperatures are below those of commercially available filler metals.

Brazeable casting alloys include 356.0, 357.0, 359.0, 443.0, and 712.0. High-quality castings are as easy to braze as their equivalent wrought alloys. Problems arise when the casting quality is low and the metal is porous. Die castings are difficult to braze.

---

21. For a more detailed discussion of brazing refer to Chapter 12, *Welding Handbook*, Vol. 2, 8th Ed.

FILLER METALS

COMMERCIAL BRAZING FILLER metals used for aluminum are described in Table 1.56. They are all based on the aluminum-silicon eutectic. Increasing the silicon decreases the liquidus temperature until the eutectic composition (12% silicon) is reached at 1070 °F (577 °C). The addition of 4% copper, as in BAlSi-3, lowers the solidus temperature so that the alloy can be used to braze Al-Si casting alloys.

Magnesium is added to Al-Si alloys for fluxless vacuum brazing applications. It is better for the furnace atmosphere, and it also modifies the oxide layer to permit wetting. Magnesium lowers the melting point during initial melting, but as magnesium is lost by vaporization, the solidus temperature increases.

Filler metals are produced in the form of wire, rod, foil, shim, powder, or paste. Powdered filler metal products normally contain brazing flux and binder for forming a paste. Premixed paste for torch brazing applications contain flux.

In some cases, filler metal is applied as cladding to one or both sides of an aluminum core sheet. This product is known as *brazing sheet* and can be formed and worked by conventional means. It is widely used for assemblies that are furnace or dip brazed. Table 1.57 describes common brazing sheet. Those sheets clad with

Table 1.56
Aluminum Brazing Filler Metals

Class[a]	Designation	Nominal Composition,%[b]		Melting Range		Brazing Range		Standard Forms[c]	Applicable Brazing Processes[d]
		Si	Mg	°F	°C	°F	°C		
BAlSi-2	4343	7.5	—	1070-1142	577-617	1110-1150	599-621	C	D,F
BAlSi-3[e]	4145	10	—	970-1085	521-585	1060-1120	571-604	R	D,F,T
BAlSi-4	4047	12	—	1070-1080	577-582	1080-1120	582-604	P,R,S	D,F,T
BAlSi-5	4045	10	—	1070-1110	577-599	1090-1120	588-604	C	D,F
BAlSi-7	4004	10	1.5	1038-1105	559-596	1090-1120	588-604	C	Vf
BAlSi-9	4147	12	0.3	1044-1080[f]	562-582	1080-1120	582-604	C	Vf
BAlSi-11[g]	4104	10	1.5	1038-1105[f]	559-596	1090-1120	588-604	C	Vf
—	4044	8.5	—	1070-1115[f]	577-602	1100-1135	593-613	C	D,F

a. See ANSI/AWS A5.8, *Specification for Filler Metals for Brazing and Braze Welding*

b. Remainder Al.

c. C – cladding on sheet; P – powder; R – rod or wire; S – sheet or foil.

d. D – dip; F – furnace; T – torch; Vf – vacuum furnace.

e. Also contains 4% Cu.

f. Melting range in air. Melting range in vacuum is different.

g. Also contains 0.1% Bi.

Table 1.57
Aluminum Brazing Sheet[a]

Designation No.	Side Clad	Core Alloy	Cladding Alloy
7	1	3003	BAlSi-7
8	2		
11	1	3003	BAlSi-2
12	2		
13	1	6951	BAlSi-7
14	2		
21	1	6951	BAlSi-2
22	2		
23	1	6951	BAlSi-5
24	2		
33	1	6951	4044
34	2		
44[b]	2	6951	4044
			7072
—	1 or 2	3003	BAlSi-9
—	1 or 2	3105	BAlSi-11[c]

a.   Not all designations are allowable as commercial products.

b.   One side is clad with 7072 alloy for corrosion resistance.

c.   Maximum brazing temperature is 1110 °F (593 °C).

BAlSi-7, 9, or 11 filler metal are designed for vacuum brazing applications.

FLUXES

CHEMICAL FLUXES ARE required for conventional aluminum brazing operations, except for vacuum brazing. Magnesium is used as an oxide modifier in vacuum or inert-gas brazing without salt fluxes.

Aluminum brazing fluxes are mixes of fluoride and chloride inorganic salts. They are supplied as dry powder. For torch and furnace brazing, the flux is mixed with water or alcohol to make a paste. The paste, or slurry, is brushed, sprayed, dipped, or flowed onto the entire joint area and the filler metal as well. Fluxes are hydroscopic and will absorb water from the atmosphere if left in unopened containers.

In dip brazing, the hot bath is molten flux. The bath remains stable and active for months or years with only moderate maintenance requirements.

A flux made of complex fluoride salts is available for torch and furnace brazing. The flux residuals are not water soluble and therefore are non-corrosive to aluminum.

FLUX RESIDUE REMOVAL

AFTER BRAZING, THE corrosive flux residue must be removed. The most effective treatment is soaking in hot agitated water 180 to 200 °F (82 to 93 °C). For some fluxes, a chemical cleaning may be required subsequent to hot-water soaking.

Mechanical cleaning, such as wire brushing or grinding, is not adequate. It breaks up the residue into fine particles that may become embedded in the aluminum surface. Scrubbing with a fiber brush under running hot water is an effective practice.

Immersion in hot water before the brazement has cooled is effective in removing a major portion of the flux. The assembly should be allowed to cool to prevent distortion by thermal shock. The hot-water soak will then remove virtually all of the flux residues.

Final traces of residues can be removed chemically. Acceptable solutions for this are given in Table 1.58. A number of commercial proprietary cleaners also are available for this purpose. All methods require a final thorough clean-water rinse to remove the cleaning agent.

A silver nitrate test is used to check for flux residues. The brazed assembly or joint is soaked in a minimum amount of distilled or deionized water. Heating the water increases the accuracy of the test. A small sample of the water is tested for chlorides by adding several drops of a 5% silver nitrate solution. A white precipitate indicates the presence of residue, and cleaning should be repeated. An alternate method is to place a few drops of distilled water on the joint area. After one minute, collect the water and drop it into a 5% silver nitrate solution. A white cloud indicates flux residues.

BRAZING PROCESSES

ALUMINUM IS BRAZED by the torch, dip, furnace, or vacuum processes. Vacuum brazing of aluminum is a fluxless process, while furnace brazing is usually done with conventional salt fluxes. Several specialized processes exist, but the majority of aluminum is brazed by the torch, dip, furnace, or vacuum methods.

DESIGN CONSIDERATIONS

MANY BRAZING APPLICATIONS use flux and filler metal. The preferred design for such brazing is a lap joint. The simple, square butt joint used in some welding applications does not function well for brazing because it does not accommodate the flow of flux and filler metal. Also, the required joint spacing is difficult

Table 1.58
Chemical Solutions for Removing Aluminum Brazing Flux Residue

Solution	Composition[a]	Bath Temperature	Procedure[b]
Nitric acid	1 gal (3.8 L) HNO$_3$, 7 gal (26.5 L) water	Ambient	Immerse 10 to 20 minutes, rinse in hot or cold water.[c]
Nitric-hydro-fluoric acid	1 gal (3.8 L) HNO$_3$ 0.06 gal (0.23 L) HF 10 gal (37.8 L) water	Ambient	Immerse for 5 minutes, rinse in cold and hot water.[d]
Hydrofluoric acid	0.3 gal (1.14 L) HF 10 gal (37.8 L) water	Ambient	Immerse for 2 to 5 minutes, rinse in cold water, immerse in nitric acid solution, rinse in hot or cold water.
Phosphoric acid-chromium trioxide	0.4 gal (1.5 L) 85% H$_3$PO$_4$ 1.8 lb (0.82 kg) CrO$_3$ 10 gal (37.8 L) water	180 °F (82 °C)	Immerse for 10 to 15 minutes, rinse in hot or cold water.[e]

a. Acids are concentrated technical grades.

b. In all cases, remove the major portion of flux residue with a hot-water rinse. To check for flux residue, place a few drops of distilled water on the area. After 1 minute, collect the water and drop it into a 5% silver nitrate solution. A white precipitate indicates the presence of residue, and cleaning should be repeated.

c. Chloride concentration should not exceed 0.68 oz/gal (5 g/L). Not recommended for sections less then 0.020 in. (0.5 mm) thick.

d. Chloride concentration should not exceed 0.40 oz/gal (3g/L). Not recommended for sections less than 0.020 in. (0.5 mm) thick.

e. Chloride concentration should not exceed 13.4 oz/gal (100 g/L). Suitable for thin sections.

to maintain. The lap joint spacing produces a capillary effect which draws the molten brazing filler metal into the space between the faying surfaces. The amount of lap to obtain full-strength joints is usually about 3T (where T equals the section thickness).

The strength of the joint varies inversely with the space between the faying surfaces (also called *joint thickness* or *clearance*). To obtain adequate strength and capillary action with torch, furnace, induction, or dip brazing, the joint thickness is usually 0.002 to 0.004 in. (0.05 to 0.10 mm) with a 3T lap. If the joint thickness exceeds 0.010 in. (0.25 mm), capillary action and joint strength are very low. If the amount of lap exceeds 0.25 in. (6.35 mm), a greater joint thickness may be required. As the brazing alloy flows through increased lap distances, there may be a pickup of alloys as mutual diffusion occurs between the base and filler metals. This may increase filler metal viscosity. In this case, the joint thickness is increased so that the joint can be completely filled even though the flowability of the filler metal has been somewhat decreased. The amount of joint thickness increase may have to be determined by trial.

In fluxless brazing, the mating parts must be in contact with each other throughout the brazing operation. This is accomplished by tightly fitting or press-fitting them together.

Figure 1.30 illustrates 16 joint designs that are typical of those suitable for the brazing of aluminum and aluminum alloys. The joint designs should facilitate easy assembly and inspection. When possible, the parts should be self-locating. Figure 1.31 shows designs which include self-locating features. Unvented pockets should be avoided where pressure could build when the brazing heat is applied. Such pressure could cause parts to become misaligned or could prevent a leak-tight joint from developing.

Fixturing devices to hold mating parts in the desired orientation can be used to achieve low-cost manufacturing for high-volume production. The device may include loading features that will decrease the joint thickness as the flux and brazing alloy are brought to the flowing temperature. Decreasing joint thickness increases joint strength and joint filling. Material selection for fixturing is important, especially when flux is used. Flux may inadvertently come in contact with the fixture and may react with the fixture materials. It is therefore important to design the fixture so that the flux cannot come in contact or use materials which will not react. Stainless steel or protectively coated carbon steels are commonly used. The design of the fixture must not cause misalignment of the brazement parts when heat is applied.

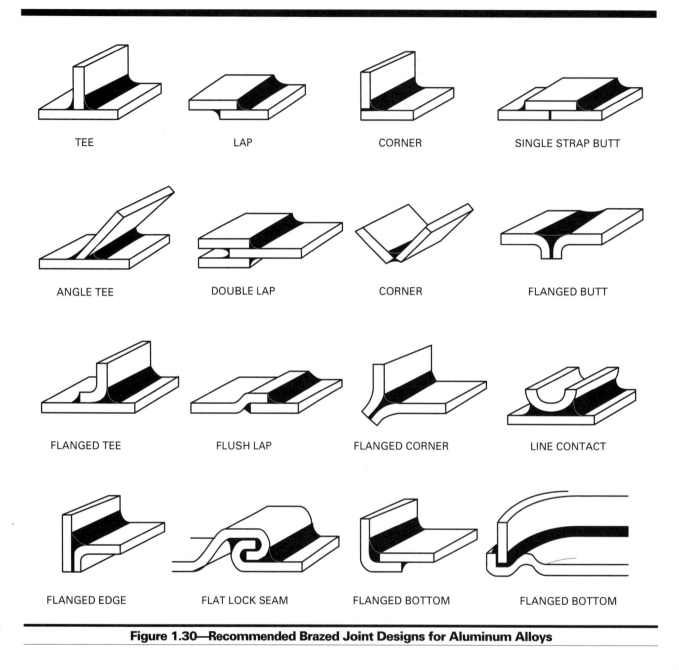

Figure 1.30—Recommended Brazed Joint Designs for Aluminum Alloys

SURFACE PREPARATION

AT THE VERY least, degreasing of the aluminum surface is recommended prior to brazing. This would be adequate for fluxless brazing and when flux is used for nonheat-treatable alloys, where contamination and oxide film formation are minor. A solvent vapor degreasing is recommended when cleaning thin sections such as parts used for heat exchangers.

Light etching is permissible with thin-gauge clad material. Care must be exercised not to remove an excessive amount of cladding.

Chemical cleaning is usually a necessary supplement to degreasing for heat-treatable alloys to remove the thicker oxide film. Metalworking operations, even hand hammering, can embed oxides into the surface. The flowability of filler metals in the form of wire or sheet

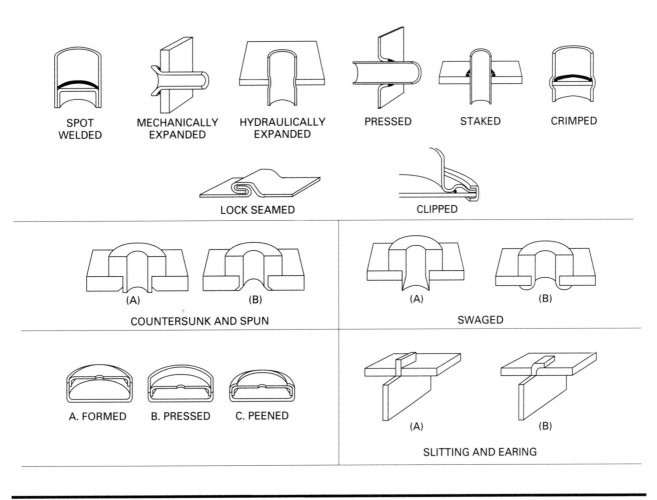

SPOT WELDED   MECHANICALLY EXPANDED   HYDRAULICALLY EXPANDED   PRESSED   STAKED   CRIMPED

LOCK SEAMED   CLIPPED

(A)   (B)
COUNTERSUNK AND SPUN

(A)   (B)
SWAGED

A. FORMED   B. PRESSED   C. PEENED

(A)   (B)
SLITTING AND EARING

Figure 1.31—Typical Self-Jigging Joints for Aluminum Alloys

may be enhanced by light etching, mechanical abrasion, or chemical cleaning before preplacement.

A caustic or nitric-hydrofluoric acid solution can be effective in chemically cleaning the heavy oxides from the surface. Caustic cleaning is capable of removing thick aluminum oxide layers. Several effective proprietary cleaning solutions are available. One of the following two general cleaning procedures could be used:

Procedure 1:

(**1**)   Degrease.
(**2**)   Dip for 60 seconds into 5% (weight) sodium hydroxide at 140 °F (60 °C).
(**3**)   Rinse with tap water at ambient temperature.
(**4**)   Dip for 10 seconds in 50% (volume) nitric acid at room temperature.
(**5**)   Rinse in hot or cold water and dry.

Procedure 2:

(**1**)   Degrease.
(**2**)   Dip for five minutes in a solution of 10% (volume) nitric acid, and 0.25% (volume) hydrofluoric acid at room temperature.
(**3**)   Rinse in hot or cold water and dry.

Caution: All acids used in cleaning solutions are potentially dangerous! Personnel using them should be thoroughly familiar with all chemicals involved and adequate safety equipment should be used.

The purity of the rinse water used may need to be considered. Tap water purity varies with geographic location. The demands of the brazing application, as well as the tap water purity, will dictate whether or not a filtering operation or special water may be necessary.

JOINT PROPERTIES

SINCE THE BRAZING temperatures are much closer to the melting temperature of aluminum than when brazing most other base metals, it is important to carefully control the brazing temperature [usually within 5 °F (within 3 °C) for furnace and dip brazing], as well as the time at that temperature.

If the upper limit of one or both of these conditions is exceeded, or an excessive amount of diffusion results, melting or incipient melting may occur at the grain boundaries, resulting in deleterious grain structure changes and decreased corrosion resistance. In any event, annealing will occur at brazing temperatures. When the brazement is a nonheat-treatable alloy, the annealed temper mechanical properties of the alloy will result. For heat-treatable aluminum alloys, strength and corrosion resistance can be improved by postbraze heat treatment and quenching from the brazing temperature.

A complex geometry of the brazement may preclude quenching from the brazing temperature due to dimensional changes which could cause the newly formed joints to fail. Complex brazements should be allowed to cool before performing a postbraze heat treatment.

When quenching from the brazing temperature is permissible, it can be accomplished by one of the following methods:

(1)   Spraying with water
(2)   Immersion in a tank of water, hot or cold
(3)   A cold air blast (the slowest of these quenching methods)

Solidification of the brazed joint must take place before quenching to prevent rupturing.

The corrosion resistance of brazements depends upon the grain structure after brazing and the thoroughness of flux residue removal. If the cleaning operation does not remove all flux residue, the combination of residual flux and moisture may result in corrosion of the joint.

SOLDERING

ALTHOUGH ALUMINUM AND many aluminum base alloys can be soldered by techniques similar to those used for other metals, problems can arise if insufficient thought is given to the application involved.[22] Reaction and abrasion soldering methods are more frequently used with aluminum than with other metals. Aluminum flux soldering requires special fluxes. Rosin fluxes are unsuitable for removing surface oxides for soldering aluminum.

SOLDERABLE ALUMINUM ALLOYS

THE MOST COMMONLY soldered aluminum alloys usually contain less than 1 percent magnesium or 5 percent silicon. Magnesium in aluminum alloys forms a tenacious oxide on the aluminum surface that results in poor solder flow and wetting, and rapid solder penetration in these alloys can result in loss of mechanical properties. Aluminum casting alloys, because of their composition, usually have poor solderability, and the surface condition of castings makes them more difficult to solder than wrought alloys. Cleaning and oxide

removal on castings is more difficult, and surface porosity can result in incomplete flux removal.

Residual stresses from quenching or cold working may interfere with forming a satisfactory solder joint. Stress accelerates penetration of solder along grain boundaries and causes cracking or loss of mechanical properties. Intergranular solder penetration can be minimized through stress relieving by heating, although this occurs naturally when soldering with high-temperature zinc solders.

Clad aluminum alloys often possess improved soldering characteristics over the bare alloys. The cladding of the alloys can improve flux and solder wetting properties and reduce diffusion of solder into the alloy. This is especially useful when low-temperature soldering the 2000 to 7000 series aluminum alloys. In addition to aluminum cladding on aluminum alloys, other metals such as copper, brass, nickel, zinc, or silver can be applied to the aluminum surface to facilitate soldering. Copper, in particular, can be electroplated or rolled onto the aluminum to permit low-temperature soldering with solders and fluxes normally used with copper. Copper-clad aluminum wire is used in many electrical applications.

---

22.   For a more detailed discussion of soldering refer to Chapter 13, *Welding Handbook*, Vol. 2, 8th Ed.

SOLDERS FOR ALUMINUM

SOLDERS FOR ALUMINUM can be classified into three groups according to their melting points, as shown in Table 1.59.

Low-Temperature Solders

LOW-TEMPERATURE SOLDERS have melting points in the range of 300 to 500 °F (149 to 260 °C). These solders contain lead, tin, or both, with small additions of zinc, cadmium, or bismuth. These additional elements increase corrosion resistance of soldered joints. A tin-zinc solder has increased corrosion resistance over lead-tin solders.

The low-temperature solders produce joints with the lowest corrosion resistance, but they are the easiest to use in soldering operations. The mechanical strength of low-temperature soldered joints approaches that of soldered joints in copper, which exhibit a shear strength of about 6 ksi (41 MPa).

Intermediate-Temperature Solders

SOLDERS IN THIS group have melting points in the range of 500 to 700 °F (260 to 371 °C). These solders contain tin or cadmium in combination with zinc. Additions of copper, lead, nickel, silver, and sometimes aluminum are made to improve various properties of the solders. Among the most commonly used intermediate-temperature solders, the 70% tin - 30% zinc and the 60% zinc - 40% cadmium are the most popular. Because of their higher zinc content, these solders generally wet aluminum readily and produce stronger and more corrosion-resistant joints than the low-temperature solders.

High-Temperature Solders

THE HIGH-TEMPERATURE solders have melting points in the range of 700 to 800 °F (371 to 430 °C). These solders are zinc based and contain up to 10% aluminum. Small amounts of other metals such as copper, cadmium, iron, and nickel are sometimes added to modify their melting and wetting characteristics. Of the aluminum solders, the high-zinc solders have the highest strength with shear strengths in excess of 15 ksi (103 MPa). These solders are usually the least expensive and exhibit the greatest corrosion resistance.

FLUXES

PROBABLY THE MOST widely used method of removing aluminum oxide films is to employ a flux. Commercially available fluxes are conveniently described as organic and reaction types.

Organic fluxes, as the name implies, contain various organic compounds. They are usually viscous materials ranging in color from pale yellow to brown. Care is required not to overheat them or they will carbonize and inhibit, rather than promote, soldering. These fluxes are normally used with low-temperature solders, and will deteriorate rapidly above 500 °F (260 °C). Most organic flux residues are mildly corrosive and should be removed from parts thinner than 0.005 in. (0.13 mm). Alcohols are effective in removing organic flux residues. Some modified fluxes contain zinc and ammonia compounds. The residues of these fluxes are significantly more corrosive than those of the unmodified fluxes and should be removed if possible. Flux residues are electrically conductive in the presence of moisture and should be removed from soldered electrical joints.

Table 1.59
General Characteristics of Aluminum Solders

Type	Melting Range	Common Constituents	Ease of Application	Wetting of Aluminum	Relative Strength	Relative Corrosion Resistance
Low-Temperature	300 to 500 °F (149 to 260 °C)	Tin or lead base plus zinc, cadmium, or both	Best	Poor to Fair	Low	Low
Intermediate-Temperature	500 to 700 °F (260 to 371 °C)	Zinc-cadmium or zinc-tin base	Moderate	Good to Excellent	Moderate	Moderate
High-Temperature	700 to 800 °F (371 to 430 °C)	Zinc base plus aluminum, copper, cadmium	Most Difficult	Good to Excellent	High	Good

The major component of reaction fluxes is zinc chloride. Compositions of these fluxes vary according to the application. Fluxes for furnace or automatic flame soldering, where the filler solder is preplaced, contain a high percentage of zinc chloride and are completely expended when the reaction temperature is reached.

For manual torch soldering, fluxes contain a high proportion of other halides in conjunction with the zinc chloride to provide an effective flux cover during manual addition of solder. Reaction fluxes, upon reaching a specific temperature, penetrate the aluminum oxide film, react with the underlying aluminum to deposit metallic zinc, and evolve gaseous aluminum chloride which appears as white smoke. The molten zinc formed when reaction fluxes are used is sometimes enough to produce soldered joints without additional solder. These fluxes are used at temperatures close to 700 °F (371 °C) because the zinc solders melt at this temperature. Reaction flux residues are highly corrosive and should be removed by thorough rinsing in hot or cold water.

Organic Flux Residue Removal

THE CHLORIDE-FREE ORGANIC fluxes are usually noncorrosive or at most only slightly corrosive, and the residues of most organic fluxes are nonhygroscopic. Flux residues are not normally removed from aluminum foil assemblies thicker than 0.005 in. (0.13 mm) and aluminum wire greater than 0.010 in. (0.25 mm) in diameter. Where flux removal is considered necessary (in high voltage applications or where insulating coatings are applied), this operation is best performed as soon after soldering as possible. Flux residues are most easily removed when they have not been overheated to their char point or "cooked" by unduly long exposure to soldering temperature.

Cold or hot water immersion is not recommended for removal of organic flux residues. Water may penetrate the solder into the aluminum-solder interface and promote electrochemical attack and rapid failure of the joint. Flux residue removal is best accomplished with organic solvents such as alcohol or chlorinated hydrocarbon solvents. The soldered joint may be dipped into the solvent, but more effective cleaning is accomplished if scrubbing the joint is performed, either with a fiber brush or by mechanical agitation.

Inorganic Flux Residue

RESIDUES REMAINING WHEN a reaction flux or chloride flux is used should be removed as quickly as possible after soldering. These residues are highly hygroscopic and strongly corrosive to aluminum. Both the chloride fluxes and reaction fluxes contain inorganic chlorides or other halides that are best removed with water. The assemblies, while still hot, may be immersed directly into hot or boiling water, or have a hot water spray directed onto the part. Flux removal is usually accomplished in a line using several tanks of wash water in a cascade system where the final rinse is clean, chloride-free water.

Flux residues containing high percentages of zinc chloride are difficult to remove with water alone, and chemical cleaning baths often are used to remove these residues. Typically, a cleaning procedure may consist of a hot-water wash followed by a soak in a dilute hot hydrochloric acid bath, another rinse, a soak in a dilute hot-alkaline bath, followed by a thorough rinse in hot water. The parts may be dried in a stream of hot air if water staining is to be avoided.

Check for Complete Flux Removal

THE SILVER NITRATE test is used to determine any remaining flux residues on a workpiece after it has been washed and dried. The silver nitrate test solution is made by adding 5 grams of silver nitrate to 100 milliliters of triple distilled water to which 3 or 4 drops of nitric acid have been added. The solution should test acid to litmus.

To test for residual chlorides, a few drops of distilled water are placed on an area where flux residue is suspected. After a few minutes, an eye dropper is used to transfer the test water to a sample of the test solution. If the test solution becomes cloudy, chlorides are present. This test is extremely sensitive and can detect airborne contamination and chlorides in tap water. Because of this, the test requires judgment on the part of the user. It is good practice to run a blank whenever this test is used. With careful techniques, chloride residues as low as 1 ppm can be detected.

The silver nitrate test is essentially a test for the presence of chlorides and should not be used with fluxes that do not contain chlorides. For these fluxes, a spectrophotometer can be used to detect the elements comprising organic flux residues.

SOLDERING PROCESSES

THE EXCELLENT THERMAL conductivity of aluminum, combined with the higher melting temperature of many of the solders used to join aluminum assemblies, normally requires that a large-capacity heat source be used to bring the joint area to proper temperature. Uniform, well-controlled heating is a necessity.

Torch Soldering

AIR-FUEL GAS TORCHES are commonly used to solder aluminum assemblies. The operation may be either manual or automatic. Abrasion or ultrasonic techniques may be used, or flux can be applied to the joint and the solder either preplaced or manually fed. The best torch soldering technique is to apply heat to both sides of the assembly until solder flow is initiated. The flame is then moved directly over the joint slightly behind the front of the solder flow. Since the flame does not come into direct contact with the flux, there is no premature flux reaction. A major application of torch soldering is in soldering "U" return bends on heat exchangers.

Soldering Irons

IT IS DIFFICULT to heat aluminum sheet thicker than 0.064 in. (1.6 mm) with a soldering iron even though the cross-sectional area of an assembly is small. Auxiliary heat sources such as ovens or hot plates often are used. Flux may be applied to the joint, and solder wire may be fed manually into the joint. The soldering iron is brought into direct contact with the joint immediately behind the front of solder flow. When an organic flux is used, the iron should never be in direct contact with the flux because charring of the flux will occur, and fluxing activity will be reduced or destroyed. Organic flux residues should be removed with alcohol.

Furnace Soldering

FURNACE SOLDERING CAN be used with all types of solder for aluminum. Distortion of parts due to differential thermal expansion is low. Solder is preplaced at the joints and flux is applied by spraying, brushing, or immersion. Temperature control is critical to assure that fluxing action and solder flow occur simultaneously, or else poor wetting is experienced.

Reaction Soldering

REACTION SOLDERING IS especially good for aluminum applications. It is particularly adaptable to heat exchanger fabrication as illustrated in Figures 1.32 and 1.33. Reaction fluxes contain a high percentage of zinc chloride. When the reaction temperature of 700 to 725 °F (371 to 385 °C) is reached, the zinc chloride reacts chemically with the aluminum to deposit zinc at the joints. This reaction evolves aluminum chloride fumes. Sufficient zinc to form line-contact joints is usually deposited from the flux alone. If additional filler is required, zinc particles can be mixed with the flux, or preplaced zinc or zinc-base solder can be used. Heating may be supplied by furnace, gas flame, resistance, or induction.

Reaction fluxes are usually very hygroscopic and must be handled and stored with care. Moisture absorption leads to formation of hydrated zinc chloride; after this stands, it forms oxychlorides that hamper satisfactory fluxing action and solder flow. For maximum performance of reaction fluxes, anhydrous vehicles such a n-propyl alcohol, n-butyl alcohol, or methyl-ethyl ketone should be used rather than water. When using reaction fluxes, adequate exhaust of the work areas is

Figure 1.32—Heat Exchanger Fabricated by Reaction Soldering, with Heat Applied by a Torch

Figure 1.33—Heat Exchanger (Reduced to 67%)

essential to exhaust the alcohol vehicle as well as the copious fumes of the flux reaction.

Dip Soldering

DIP SOLDERING IS well-suited to joining aluminum because the solder pot itself is an excellent large-capacity heat source. This method is ideal for joining assemblies at a high production rate and can employ the same techniques and production schedules normally used to solder other metals. The low-temperature solders and organic fluxes lend themselves more readily to this process, although the high-temperature solders and reaction fluxes are sometimes employed.

Resistance Soldering

RESISTANCE SOLDERING MAY be used to join aluminum to itself or to other metals. It is also suitable for spot or tack soldering.

In resistance soldering, the work to be soldered is connected between a ground and a movable electrode, or between two movable electrodes, to complete an electrical circuit. The heat is applied to the joint both by the electrical resistance of the metal being soldered and by conduction from the electrode, which is usually carbon.

For resistance soldering aluminum, flux is usually applied to the joint area by brushing, and the solder is either preplaced in the joint or fed manually. A metal or carbon electrode is then brought into contact with the joint area and maintained in position while current passes through the joint until solder flow occurs. Because the heat is generated in the aluminum itself, better temperature control can be maintained and there is less danger of damaging the flux by overheating. After soldering, the flux residues should be removed.

JOINT DESIGN

THE DESIGN USED for soldering aluminum assemblies are similar to those used for soldering other metals. The most commonly used designs are forms of simple lap, crimp, and T-type joints. A good joint design will provide the following:

(1) An ample area for solder contact
(2) An adequate clearance and a path for flux or solder, or both, to flow into the entire joint area
(3) A means for locating solder properly
(4) A means for flux escape or removal
(5) Minimum access or entry for corrosive attack
(6) A contour suitable for subsequent protective coating where desired

Joint clearance will vary with the specific soldering method, base alloy composition, solder composition, and flux employed. As a general guide, joint clearances ranging from 0.005 to 0.020 in. (0.12 to 0.50 mm) are required when chemical fluxes are used, and 0.002 to 0.010 in. (0.05 to 0.25 mm) when reaction fluxes are employed.

PREPARATION FOR SOLDERING

ALUMINUM SURFACES MUST be free of grease, dirt, and other foreign material before soldering. In most instances, solvent degreasing is sufficient for surface preparation. Wire brushing or chemical cleaning occasionally are required for heavily oxidized surfaces, especially surfaces of alloys containing higher levels of magnesium, silicon, or both, which have particularly tenacious oxides which may need abrasive or chemical cleaning.

OXIDE REMOVAL

TO OBTAIN SUCCESSFUL soldered joints with aluminum, the oxide layer on the aluminum surface must be removed. This can be accomplished during soldering by mechanical abrasion, ultrasonic dispersion, or fluxing. The latter was discussed in the section on fluxes, Page 91.

Abrasion

ABRADING THE ALUMINUM surface under a molten solder layer permits soldering. Oxide removal may be accomplished through the molten solder by brushing with fiberglass or stainless steel brushes or buffing with stainless steel wool. Alternatively, the solder rod itself, particularly when using zinc-based solders, can be used to break up the oxide and allow molten solder to contact and bond to the aluminum.

Ultrasonic Dispersion

ANOTHER MEANS OF removing the oxide film is to erode it from the aluminum using ultrasonic energy. When ultrasonic energy is introduced into molten solder, cavitation occurs, forming numerous voids within the molten metal. Collapse of these voids creates an abrasive effect that removes the aluminum oxide films and allows the solder to wet the aluminum. Ultrasonic soldering can be accomplished when an ultrasonically vibrating tip is brought into contact with molten solder on the part being soldered. A more effective approach is to dip the assembly to be soldered into a pot of molten solder which is agitated by ultrasonic energy.

PERFORMANCE OF SOLDERED ALUMINUM JOINTS

Mechanical Strength

THE MECHANICAL STRENGTH developed by soldered aluminum assemblies, although often of secondary importance, ranges approximately from a minimum equal to the strength of soft solders [0.6 ksi (4 MPa)] to a maximum greater than 40 ksi (276 MPa) shear stress. The intermediate- and high-temperature solders retain considerable strength at temperatures where soft solders melt [about 350 °F (177 °C)]. High-temperature solders are fully effective up to about 212 °F (100 °C) and can be exposed to temperatures up to 350 °F (177 °C) without loss of strength. However, creep can be experienced under conditions of stress at temperatures above 250 °F (121 °C).

Corrosion Resistance

ALUMINUM SOLDER JOINTS, like all solder joints, corrode when two or more parts of the joint are in contact with an electrolyte. The corrosion process is essentially electrochemical in nature, and the metals joined by the electrolyte form a galvanic couple. The anodic element (the most negative cell element) corrodes more rapidly, protecting the remaining elements until it is consumed. The rate of corrosion depends on the voltage difference between the elements, their distance apart, and the composition of the electrolyte. When soldered joints are properly cleaned, no residual salts remain. As long as no moisture is present, no corrosion will occur. In practice, these conditions are not attained, and some corrosion will occur.

If solder containing tin is used to solder aluminum, the intermetallic interface will contain a high tin content. This interface is highly negative to the other constituents of the joint. When exposed to an electrolyte, this highly negative interface will corrode rapidly, and the joint fails catastrophically. The solder separates from the aluminum as if cut by a knife. For maximum life, such a joint should be used in a dry atmosphere, or it should be coated to protect it from exposure to the elements.

When a joint is formed with zinc solder, which is anodic to aluminum, galvanic corrosion is spread over the face of the zinc. Since there is no highly anodic thin intermetallic interface present, the zinc solder will protect the aluminum, and the joint will remain intact until all the solder is consumed. Assemblies prepared with pure zinc or zinc-aluminum solders have withstood corrosive attack for many years and are considered satisfactory for most applications requiring long outdoor service. An outstanding example of a zinc-soldered assembly is the condenser coil of an automobile air conditioner.

ADHESIVE BONDING

MOST ALUMINUM ALLOYS are readily joined by adhesive bonding.[23] Depending on the application, adhesive bonding offers some unique advantages over other joining processes.

The primary advantage of adhesive bonding is the ability to distribute applied loads over a wide area, thereby reducing the working stress levels. Additionally, adhesive joints often can be made as strong as the base materials. There is no weakening by a heat-affected zone as with fusion welds. Joint strength can be increased merely by increasing the joint area (e.g., the overlap distance).

Another advantage to adhesive bonding is that aluminum can be joined to most other structural metals, plastics, and composites. Additionally, the adhesive forms an electrically insulating barrier between two metals being joined. This insulating barrier reduces the likelihood of galvanic corrosion which occurs when metals with different electric potentials are in direct contact with one another, as is the usual case when mechanical fasteners are used for joining. In addition, adhesive bonding also can be used to attenuate noise and structural vibration if the adhesive and joint design are properly selected.

23. For a more detailed discussion of Adhesive Bonding of Metals refer to Chapter 27, *Welding Handbook*, Vol. 2, 8th Ed.

ADHESIVE MATERIALS

MOST ADHESIVES IN use today for joining aluminum are organic polymeric materials, and as such their long-term mechanical performance is dependent on their chemical nature and the environment in which they are expected to perform. Fortunately, commercial adhesives are available in a wide range that spans a large process and performance spectrum. Most notable for joining aluminum are the epoxy adhesives, which can be formulated to cure either rapidly or slowly at room temperature, and whose high-temperature performance and moisture resistance can be enhanced by curing at elevated temperatures. When epoxy adhesives are used, the cleanliness and preparation of the surface to be bonded is of paramount importance.

Another class of adhesives gaining wide acceptance for bonding aluminum is anaerobic acrylics. This group of adhesives has found use in applications where only minimal surface preparation can be afforded, but high strength and durable joints are required. As a class, these adhesives can provide joint strengths equivalent to those obtained with epoxies, without the need to insure oil-free and pristine bonding surfaces. These adhesives are capable of dissolving thin films of mill oil into the bulk adhesive layer without adversely affecting joint strength or performance. In the case of aluminum, however, the weak natural oxide film found on nonanodized bonding surfaces will be the determining factor for maximizing the bond strength. Undermining corrosion at the metal-to-adhesive interface may occur in harsh environments when additional surface treatments and corrosion-inhibiting primers are not used. Nevertheless, anaerobic acrylics offer a viable alternative to epoxies for many metal bonding applications.

SURFACE PREPARATION

TO DEVELOP MAXIMUM adhesion between the epoxy resin and aluminum surface, complete wetting of the metal bonding surface by the adhesive must be obtained. Any oxide present on the surfaces to be bonded must first be removed by grinding or wire brushing. In addition to removing aluminum oxide from the bonding surface, mechanically abrading the bonding surface increases the surface area available for bonding and decreases the joint stress for a given load. The rough surface permits greater penetration and mechanical interlocking of the adhesive into the bonding surface, thereby increasing the overall joint strength.

Any mill oil present on the metal bonding surface must be removed by vapor degreasing with an organic solvent, such as trichloroethane, prior to the application of adhesive. If this is not practical or the bonding surface is small, wiping with a lint-free cloth saturated with a suitable solvent until all traces of dirt and oil contamination are gone from the bonding surface is acceptable. For some applications, this is all the surface preparation required prior to adhesive application and bonding.

When bonded metallic materials will be exposed to harsh environments or maximum bond strength is required, then additional surface treatments are needed. Corrosion-inhibiting primers, usually containing chromium compounds, are often used on aluminum surfaces to be bonded to prevent corrosion of the metal at the bond line. This is especially important for aluminum bonded joints that will experience corrosive environments such as salt. In these cases, it is of paramount importance to take precautions to prevent undermining corrosion at the bond line between the adhesive and metal. Ultimately this will control the long-term durability and bond strength of the joint.

In the case of aluminum, a weak oxide film is always present on surfaces exposed to air. This oxide film is often the weak link in adhesively bonded structures, with joint failure occurring in the oxide layer rather than the adhesive. To achieve maximum strength and long-term durability with aluminum bonded joints, the weak oxide film first must be removed and a new stronger oxide formed in its place. For critical applications, this dictates the use of mineral acid etching to remove the natural oxide film from the aluminum surface to be bonded, and either phosphoric or chromic acid anodizing to form a stronger, more durable oxide suitable for adhesive bonding. Mechanical removal of the natural oxide film in lieu of acid etching is not recommended, since the newly exposed aluminum is highly reactive and oxidizes almost immediately when in contact with oxygen, thereby reducing the effectiveness of the oxide removal procedure.

JOINING TO OTHER METALS

ALUMINUM CAN BE joined to most other metals either directly or indirectly by precoating the other metal or using bimetallic transition pieces. In addition to riveting, bolting, and adhesive bonding, which can join aluminum to non-metallics as well as other metals, the following processes have been used with specific dissimilar metals:

(1) Low-temperature soldering can be used to join aluminum to silver, bronze, copper, magnesium, nickel, lead, tin, titanium, zinc, precious metals, ceramics, cermet, and glass.

(2) High-temperature soldering with a high-zinc-content solder is used to join copper tubes into flared aluminum tubes at the ends of the circuits of aluminum air conditioning condensers for automotive and household usage.

(3) Ultrasonic dip soldering is used in making air-conditioning condenser copper-to-aluminum tubular connections, as well as pigtailed wire connections between aluminum and copper wire for electrical applications.

(4) Ultrasonic soldering irons have been used to precoat zinc or other lower-melting-temperature solders onto aluminum and other metal to facilitate the dissimilar metal joining.

(5) Brazing has been used to join aluminum to copper whose joint surfaces are precoated with a silver-base solder, and to steel that has been coated with aluminum. Aluminum coatings can be applied by dipping clean steel, with or without fluxing, into molten aluminum at 1275 to 1300 °F (690 to 705 °C).

(6) Diffusion bonding can be used to join aluminum to copper, nickel, stainless steel, zirconium, uranium, and a multitude of silver- and copper-plated materials.

(7) Flash welding is an excellent method for making aluminum-to-copper tubular, rod, and plate transition segments for refrigeration tubing and electrical connectors. The brittle Cu-Al intermetallic phase can be squeezed out of the joint during the upset operation. Arc welding can then complete the Al-Al and Cu-Cu assembly joints.

(8) Ultrasonic welding joins aluminum foil and sheet directly to silver, gold, beryllium, copper, iron, germanium, magnesium, manganese, nickel, palladium, platinum, silicon, tin, tantalum, titanium, tungsten, and zinc.

(9) Cold pressure welding (at room temperature) can be used to make lap welds and butt welds in wire, rod, and bar to join aluminum to copper primarily for electrical applications.

(10) Elevated-temperature pressure welding (forge welding) can join aluminum to copper, steel, stainless steel, and zinc. Bimetallic sheets and plates are produced by hot rolling for cookware, electrical connections, and welding transition assemblies.

(11) Explosion welding is used to join aluminum to copper, steel, and stainless steel for arc welding transition sheets, plates, tube, and ring sections.

(12) Friction or inertial welding can join aluminum to copper, bronze, brass, steel, stainless steel, magnesium, nickel, titanium, and zirconium.

(13) Gas metal arc plug welds have been made through holes in copper sheet, mild steel, stainless steel, and aluminized steel sheets to obtain a combination of electrical and mechanical connections in the form of a "fused rivet."

(14) Gas tungsten arc welding using low-amperage direct current electrode positive power can be used to join aluminum to aluminum-coated steel when the arc is directed onto the aluminum alloy and the molten weld pool flows over the coating without breaking the aluminum-steel bond.

BIMETALLIC TRANSITION INSERTS

BIMETALLIC TRANSITION INSERTS permit gas tungsten arc welding and gas metal arc welding of the aluminum side to another aluminum member, and the steel or copper side can be welded to its mating component. This procedure is used for refrigeration and air-conditioning tube connections, cryogenic tank and piping applications, joining aluminum tubes to carbon or stainless steel tube sheets in heat exchangers, as well as structural applications such as welding an aluminum deck house onto the steel deck of a ship, or a large diameter [120 ft (365 m)] aluminum sphere to a ship's steel structure to haul liquid natural gas.

CORROSION RESISTANCE

OXYFUEL GAS WELDING, brazing, and some soldering methods used for joining aluminum to other metals are carried out with fluxes containing a variety of chlorides. Residues from these fluxes are hygroscopic and upon absorption of moisture become active electrolytes to accelerate corrosion in the aluminum member. All flux residues must be removed from aluminum parts and assemblies after joining.

When aluminum is in direct contact with another metal, the presence of an electrolyte will set up a galvanic cell between the two metals and cause preferential attack. Aluminum is anodic to most common metals except for magnesium and zinc, so it will sacrifice itself to protect steel, copper, lead, and other metals that are cathodic to it on the galvanic scale. These joints should be painted, coated, wrapped, or protected by a convenient means to eliminate moisture or any electrolyte at the contact area.

ARC CUTTING

TWO ARC CUTTING processes used with aluminum are plasma arc cutting and air carbon arc cutting.[24]

PLASMA ARC CUTTING

PLASMA ARC CUTTING can be used to sever aluminum alloys. The metal is melted and blown away by a high-velocity gas jet to form a kerf. Cutting can be done in any position. Nitrogen or argon-hydrogen is commonly used as the plasma-producing gas, with carbon dioxide or nitrogen as the shielding gas. Both manual and machine cutting equipment are available.

Aluminum can be cut with a rather wide range of operating conditions. The quality of the cut will be related to these conditions and the equipment used. Typical conditions for machine cutting are given in Table 1.60. Section thicknesses of 0.125 to 6 in. (3.2 to 152 mm) can be cut with mechanized equipment. The maximum practical thickness for manual cutting is about 2 in. (51 mm).

A plasma arc can also be used to gouge aluminum to produce J- and U-groove joint designs. Special torch orifice designs are needed to give the proper shape to the groove.

The aluminum is melted during cutting, and the heat produces a heat-affected zone (HAZ) similar to fusion welding on each side of the cut. The metallurgical behavior of aluminum alloys during cutting is similar to that during welding.

No significant problems occur when cutting nonheat-treatable alloys. Shallow shrinkage cracks may develop in the cut surface with all heat-treatable alloys. Example of these cracks are shown in Figures 1.34 and 1.35.

The HAZ next to the cut surface in some high-strength, heat-treatable alloys, such as alloys 2014, 2024, and 7075, may display reduced corrosion resistance. Cutting does not contaminate the metal as do machining methods that require a cutting fluid or a lubricant. Any aluminum oxide formed on the cut edges should be removed prior to welding the nonheat-treatable alloys. Common methods of oxide removal may be used. Before welding the arc-cut surface of heat-treatable alloys, mechanical removal of the edge cracks, up to a depth of 1/8 in. (3.2 mm), is recommended.

Plasma arc cutting apparatus is generally sold as a package, including the torch and the power source. Since the fumes evolved with aluminum are voluminous, a water table and exhaust hood are recommended for operator safety. When cutting close to a

24. For a more detailed discussion of plasma arc and air carbon arc cutting refer to Chapter 15, *Welding Handbook*, Vol. 2, 8th Ed.

Table 1.60
Typical Conditions for Machine Plasma Arc Cutting of Aluminum Alloys

Thickness		Speed		Orifice Diameter*		Current (DCEN), A	Power, kW
in.	mm	in./min	mm/s	in.	mm		
1/4	6	300	127	1/8	3.2	300	60
1/2	13	200	86	1/8	3.2	250	50
1	25	90	38	5/32	4.0	400	80
2	51	20	9	5/32	4.0	400	80
3	76	15	6	3/16	4.8	450	90
4	102	12	5	3/16	4.8	450	90
6	152	8	3	1/4	6.4	750	170

* Plasma gas flow rates vary with orifice diameter and gas used from about 100 ft³/h (47 L/min) for a 1/8 in. (3.2 mm) orifice to about 250 ft³/h (118 L/min) for a 1/4 in. (6.4 mm) orifice. The gases used are nitrogen and 65% argon-35% hydrogen. The equipment manufacturer should be consulted for each application.

water surface, hydrogen is generated when molten aluminum reacts with water. A cross flow of air is required to avoid a buildup of an explosive atmosphere. When cutting under water, a perforated air line should be used to bubble air through the water to avoid hydrogen concentrations.

AIR CARBON ARC CUTTING

THIS PROCESS USES a carbon arc and an air blast to remove metal. It is more effective for gouging of grooves than for cutting. Grooves up to 1 in. (25.4 mm) deep can be made in a single pass, but depth increments of 0.25 in. (6.35 mm) provide better process control. The width of the groove is determined primarily by the size of the carbon electrode. The depth of the groove is affected by the torch angle and travel speed.

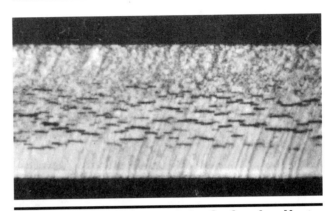

Figure 1.34—A Plasma Arc Cut Surface in a Heat-Treatable Aluminum Alloy Showing Cracking (x3)

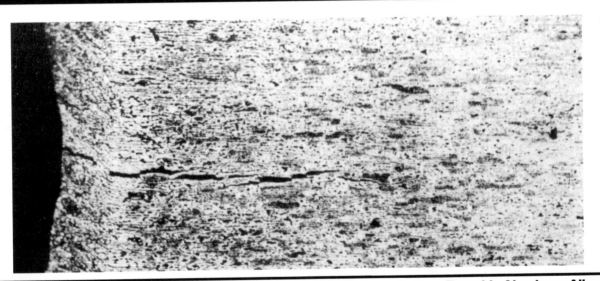

Figure 1.35—Transverse Section Through the Heat-Affected Zone of a Heat-Treatable Aluminum Alloy Showing an Intergranular Crack (x100)

The arc is operated with direct current electrode negative power. Operating conditions must be closely controlled to ensure that all molten metal is blown from the work surface. The arc length must be great enough to permit the air stream to pass under the tip of the electrode.

Carbon contamination of the cut aluminum surface commonly occurs. This method causes intergranular cracking of the cut edges in heat-treatable alloys, as described for plasma arc cutting.

PROPERTIES AND PERFORMANCE OF WELDMENTS

METALLURGICAL EFFECTS

THE PROPERTIES AND performance of an aluminum welded joint are influenced by many factors, including composition, form and temper of the base metals, the filler metal used, the welding process, rate of cooling, joint design, postweld mechanical or thermal treatments, and the service environment. Although the heat affect of welding in softening the base material adjacent to the weld is generally the controlling factor relative to the as-welded strength of an aluminum weldment, the weld metal composition and structure can also significantly influence the final strength, ductility, and toughness.

Weld Metal

THE PROPERTIES OF the deposited aluminum weld metal are influenced by the composition and rate of solidification. The solidification rate depends on the welding process and technique, plus all the factors affecting heat input and transfer away from the molten pool. A higher rate of solidification generally produces a finer microstructure with greater strength and less tendency for hot cracking.

The composition of the weld metal is dependent upon the base metal and filler metal chemical composition and the resultant admixture based upon the joint design, welding process, and procedure employed. When welding the nonheat-treatable aluminum alloys, the chemical composition of the filler metal is generally similar to that of the base metal. The heat-treatable aluminum alloys possess a much wider melting range and are more sensitive to hot cracking. A dissimilar filler metal with a lower solidus temperature than the base metal is generally employed so that the heat-treatable base metal is allowed to completely solidify and develop some strength along the fusion zone before the weld solidification shrinkage stresses are applied. Many of the filler metals used are not heat-treatable and depend upon base-metal admixture of the weld to form a weld-metal chemical composition responsive to postweld heat treatments. These welds do not exhibit as high ductility and toughness characteristics in the weld metal as obtained in nonheat-treatable aluminum weldments. Also, depending upon the filler metal used, the as-welded strength may be considerably lower.

The molten weld metal exhibits a high solubility for hydrogen. Hydrogen can be introduced into the weld metal from residual hydrocarbons or hydrated oxides on the surfaces of the base and filler metals as well as from faulty welding equipment or improper gas shielding that permits moist atmospheric contamination.

During solidification of the weld, hydrogen precipitates out of the solidifying molten aluminum and becomes entrapped in the solid weld metal as porosity. Thus, when using welding procedures that result in rapid solidification of the weld metal, it becomes very important to eliminate sources of hydrogen from the weld area. Particularly, bare filler wire must be properly cleaned and stored in a dry environment until needed for production.

Heat-Affected Zone

THE EFFECTS OF welding upon aluminum base alloys vary with the distance from the weld and may be divided roughly into areas that reflect the temperature attained by the metal. The length of time at a specific temperature can be significant for the heat-treatable alloys. The width of the heat-affected zone (HAZ) in all alloys, as well as the extent of metallurgical changes in the heat-treatable alloys, is dependent upon the rate of heat input and heat dissipation. These are influenced by the welding process, thickness or geometry of the part, speed of welding, preheat and interpass temperatures, and types of backing or fixturing. Figure 1.36 illustrates the effect of three heat-input conditions upon the properties in the heat-affected zone and width of the HAZ in heat-treatable alloy 6061-T6.

With the nonheat-treatable alloys, recrystallization of cold-worked metal and some grain growth are likely to occur in the HAZ. The strength of the HAZ will be similar to that of the annealed material, since that portion of the zone subjected to temperatures above the annealing temperature will be instantaneously annealed (see Table 1.61). Time at temperature and cooling rate are unimportant for these alloys. A typical metallurgical structure across the weld of a nonheat-treatable alloy is illustrated in Figure 1.37. Very little change in microstructure can be noted in the base metal. Its behavior in service, when stressed parallel to the weld, will depend somewhat on the ratio of the width of the HAZ relative to that of the unaffected base metal; when a member is stressed transverse to the weld, the mechanical properties obtained are independent of the welding process and technique employed in making the weld.

The heat-treatable alloys contain alloying elements that exhibit a marked change in solubility with temperature change. The high strengths of these alloys are due to the controlled solution heat treatment and precipitation hardening (aging) of some of the microconstituents. The heat-treatable alloys are normally welded in the aged condition, in which they have a controlled amount of hardening microconstituents precipitated from solid

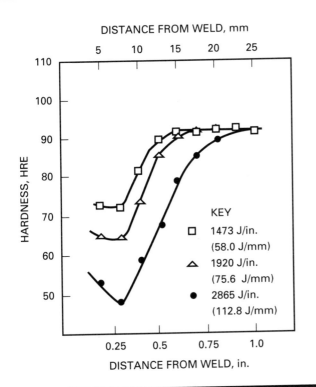

DISTANCE FROM WELD, mm

KEY

□ 1473 J/in.
   (58.0 J/mm)

△ 1920 J/in.
   (75.6 J/mm)

● 2865 J/in.
   (112.8 J/mm)

DISTANCE FROM WELD, in.

Figure 1.36—Rockwell E Scale Hardness Profiles of the Heat-Affected Zone of Gas Tungsten Arc Welds on 6061-T6 using Various Heat Inputs

solution. The heat of welding causes re-solution of the hardening microconstituents in the HAZ, followed by an uncontrolled precipitation of the microconstituents in the HAZ upon cooling. This overaging in the HAZ lowers the base-metal strength adjacent to the weld.

The response to welding is much more complex than observed with the nonheat-treatable alloys, because the effect depends upon the peak temperature and time at temperature to which the metal is exposed (see Figure 1.36) Thus, variations in the metallurgical structure will vary with distance from the fusion line, as illustrated in Figures 1.38 and 1.39. Several hours at the annealing temperature are required to completely soften these alloys. The annealing temperature is exceeded in the HAZ for a short time period. The welding process, technique, preheat and interpass temperatures, and the rate of cooling greatly influence the degree of micro-structural changes, which influences the sensitivity to hot cracking and the amount of softening that occurs.

Immediately adjacent to the weld metal, the base metal is heated to a sufficiently high temperature to rapidly redissolve any precipitates. These solid-solution zones will possess an intermediate strength and be quite ductile. Next to this will be a zone which has been subjected to temperatures in excess of the precipitation-age-hardening temperature but below the solution-heat-treatment temperature. This zone will exhibit varying degrees of overaging and softening depending upon the temperature and time held at the overaging temperature. The portion of overaged (or partially annealed)

Table 1.61
Mechanical Properties of Gas Shielded Arc Welded Butt Joints in Nonheat-Treatable Aluminum Alloys

Base Alloy	Filler Alloy	Average Ultimate Tensile Strength		Minimum Ultimate Tensile Strength		Minimum Tensile Yield Strength*		Tensile Elongation, % in 2 in. (50.8 mm)	Free Bend Elongation, %
		ksi	MPa	ksi	MPa	ksi	MPa		
1060	1188	10	69	8	55	2.5	17	29	63
1100	1100	13	90	11	76	4.5	31	29	54
1350	1188	10	69	8	55	2.5	17	29	63
3003	1100	16	110	14	97	7	48	24	58
5005	5356	16	110	14	97	7	48	15	32
5050	5356	23	158	18	124	8	55	18	36
5052	5356	28	193	25	172	13	90	19	39
5083	5183	43	296	40	276	24	165	16	34
5086	5356	39	269	35	241	17	117	17	38
5154	5654	33	228	30	207	15	103	17	39
5454	5554	35	241	31	214	16	110	17	40
5456	5556	46	317	42	290	26	179	14	28

* 0.2% offset in a 10 in. (254 mm) gauge length across a butt weld

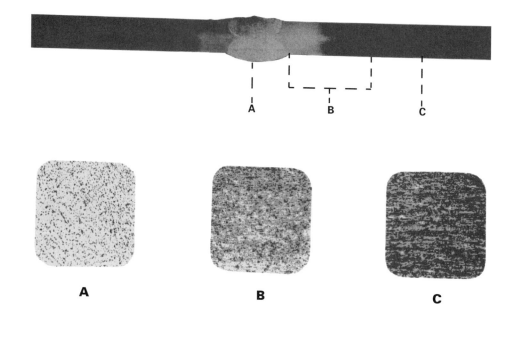

Figure 1.37—Cross Sections of Gas Metal Arc Welds Showing (A) Weld Metal, (B) Heat-Affected Zone, and (C) Unaffected Zone in 3/8 in. (9.5 mm) Thick 5456-H321 Plate (x100; top plate, 2x)

material closest to the solid-solution zone will generally possess the lowest strength. Postweld aging treatments have little effect upon the strength of this zone. The entire weldment must be solution heat-treated and aged to reproduce the previous properties of the base material. It is best to weld the heat-treated alloys with fast methods that rapidly dissipate the heat in order to minimize the degree of overaging and reduction of strength in the HAZ.

Complete heat treatment (solution treating followed by age hardening) of weldments may be practical in some cases. Where this is not the case, only postweld aging can be used to improve the strength of the solid-solution zones, but this method should not be expected to strengthen the already overaged material. If the base metal is welded in the solution heat-treated (-T4) condition, postweld aging can most effectively be employed to improve as-welded strengths. This procedure is most effective for low-heat-input welding processes or techniques that avoid excessive precipitation aging. Typical examples of this strength improvement are evident in Figure 1.39 and Tables 1.62 and 1.63

TENSILE STRENGTH AND DUCTILITY

Nonheat-Treatable Alloys

THE HEAT OF welding causes the nonheat-treatable alloys to lose the effects of strain (work) hardening in the heat-affected zone (HAZ). This zone adjacent to the weld reaches the annealed condition characterized by lowered strength and increased ductility. Since the metal is annealed by the initial welding operation, repeated welding during a repair operation does not further reduce the strength except as influenced by a possible wider HAZ. For this reason, the minimum annealed tensile strength of the base metal is generally considered as the minimum strength of butt welds in the nonheat-treatable alloys.

Mechanical property data for groove welds in commonly welded nonheat-treatable alloys is given in Table 1.61. It may be noted that no temper designations are given for the nonheat-treatable alloys in the table. The weld properties of these alloys are much less affected by the base-metal temper than the heat-treatable alloys.

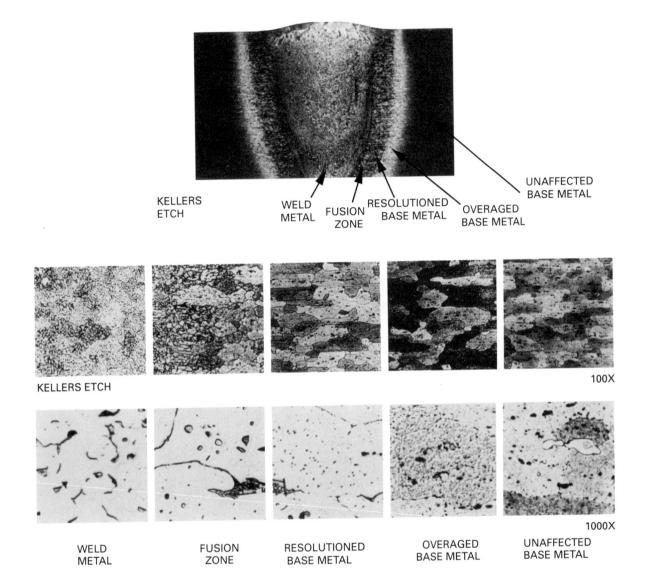

Note: Single-pass gas tungsten arc weld (DCEN) in 1/2 in. (12.7 mm) thick 2219-T87 aluminum alloy plate (as-welded).

Figure 1.38—The Five Microstructural Zones Characteristic of the Heat-Treatable Alloys

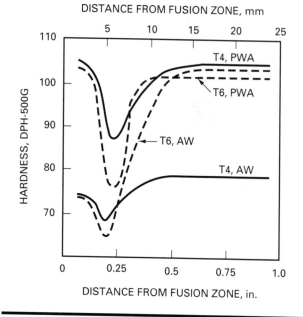

Figure 1.39—Hardness Profiles of the Heat-Affected Zone for 6061-T4 and T6 Starting Material in the As-Welded (AW) and Postweld-Aged (PWA) Conditions

The same is generally true for the effect of metal thickness.

Excellent ductility is exhibited by welds in the nonheat-treatable alloys. These welds are capable of developing extensive deformation prior to failure. The higher strength 5XXX series alloys are particularly favorable because of the closer matching of strength and ductility in the various zones across the weld joint. The aluminum-magnesium-manganese alloys (such as alloys 5083, 5086, and 5456) have found wide application in welded construction due to their high annealed strength and good ductility.

Heat-Treatable Alloys

THE STRENGTHS OF the heat-treatable alloys are also decreased by the heat of welding, but the substantial time at temperature required to anneal these alloys is not attained during welding. As a result, only partially annealed properties are observed in the HAZ in the as-welded condition. These properties vary considerably with chemical composition, heat input, and cooling rate, making it difficult to establish minimum mechanical property values for design. Performance data is best established for a specific alloy, temper, thickness, and joining process at the time of qualifying the welding procedure.

Typical properties of groove welds in some of the heat-treatable alloys are given in Table 1.62. The mechanical property values given are average and cover a wide range of material thicknesses, welded by either the gas tungsten arc or gas metal arc welding process, using many different joint designs and welding procedures. The heat-treatable alloy weldments generally exhibit lower weld ductility in the as-welded condition than do similarly as-welded nonheat-treatable alloys. The major exception involves the weldable 7000 series alloys, which naturally age quite rapidly after welding to provide increased strength with good ductility.

The section thickness can have a significant effect upon the as-welded properties of the heat-treatable alloys. The thinner gauges can be welded with less total heat input and dissipate the heat more rapidly than the thicker gauges. The reduced time at temperature results in higher as-welded strengths in the thinner heat-treatable alloys.

Preheating can markedly reduce the strength of welds in the heat-treatable alloys, particularly when applied at temperatures in excess of 250 °F (121 °C), and corrosion resistance may be impaired. Preheating heat-treatable alloys is rarely recommended except when the welds are subject to postweld heat treatment.

Repair welding may slightly lower weld joint strength when compared to the original weld strength. Because of the microstructural changes that take place during the original welding and the greater restraint generally associated with repairs, hot-cracking tendencies can be greater in the HAZ or in previously deposited weld metal when repair welding the heat-treatable alloys.

Abnormal microstructure in aluminum, as in other materials, may sometimes lead to difficulties in welding. Areas of segregation, such as stringers of low-melting-temperature constituents, can lead to porosity and cracking. Grain size and orientation may have a marked effect on the weldability and resultant weld performance. As illustrated in Figure 1.40, the adverse grain size and orientation in the alloy 2014-T6 forging contributed to integranular cracking of the low-melting-temperature grain-boundary constituents.

Postweld Heat Treatment

THE HEAT-TREATABLE ALLOYS may be reheat-treated after welding to restore the base metal in the HAZ to nearly its original strength. The joint will then fail in the weld metal except where the weld bead reinforcement is left intact. In this case, failure normally occurs in the fusion zone at the edge of the weld. The strength obtained in the weld metal after postweld heat treatment will depend on the filler metal used. Where filler metal of other than the base-metal composition is used, strength will depend upon the admixture of the filler metal with base metal. To obtain the highest

Table 1.62
Typical Mechanical Properties of Gas Shielded Arc Welded Butt Joints in Heat-Treated Aluminum Alloys

Base Alloy and Temper	Filler Alloy	As-Welded						Postweld Heat-Treated and Aged					
		Tensile Strength		Yield Strength[a]		% Elongation Tensile in 2 in. (50.8 mm)	Free Bend	Tensile Strength		Yield Strength[a]		% Elongation Tensile in 2 in. (50.8 mm)	Free Bend
		ksi	MPa	ksi	MPa			ksi	MPa	ksi	MPa		
2014-T6	4043	34	234	28	193	4	9	50	345	—	—	2	5
2014-T6	2319	35	241	28	193	5	—	60	414	46	317	5	—
2036-T4	4043	37	255	25	172	5	—	—	—	—	—	—	—
2219-T81, T87	2319	35	241	26	179	3	15	55	379	38	262	7	5
2219-T31, T37	2319	35	241	26	179	3	15	40[b]	276[b]	33[b]	227[b]	2[b]	12[b]
2519-T87	2319	37	255	33	227	4	15	56	386	—	—	5.5	—
2519-T37	2319	—	—	—	—	—	—	41[b]	283[b]	40[b]	276[b]	5.0[b]	—
6009-T4	4043	32	221	20	138	9	—	—	—	—	—	—	—
6010-T-4	4043	34	234	21	145	10	—	—	—	—	—	—	—
6061-T6	4043	27	186	18	124	8	16	44[c]	303[c]	40[c]	276[c]	5[c]	11
6061-T6	5356	30	207	19	131	11	25	—	—	—	—	—	—
6061-T4	4043	27	186	18	124	8	16	35[b]	241[b]	24[b]	165[b]	3[b]	—
6063-T6	4043	20	138	12	83	8	16	30	207	—	—	13	11
6063-T6	5356	20	138	12	83	12	25	—	—	—	—	—	—
7004-T5	5356	40	276	24	165	8	38	—	—	—	—	—	—
7005-T53	5556	44	303	25	172	10	33	50	345	33	227	4	25
7039-T63	5183	47	324	32	221	10	34	—	—	—	—	—	—

a. 0.2% offset in 2 in. (50.8 mm) gauge length.

b. Postweld aged only.

c. For thickness greater than 0.5 in. (12.7 mm), 4643 filler is required.

strength, it is essential that the weld metal respond to the postweld heat-treatment (PWHT) used. Although PWHT increases the tensile strength, some loss in weld ductility occurs.

In very thick material, the mechanical properties of a PWHT weldment may be lower than the base metal if the filler metal used is not heat-treatable and the weld must rely on admixture resulting in sufficient alloy pickup from the base alloy to achieve a heat-treatable composition. This effect is shown in line F of Table 1.63 for a single-V-groove weld in 3 in. (76 mm) thick alloy 6061 plate. When the joint design provides insufficient admixture and the weldment is to be PWHT, consider-

ation should be given to the use of an Al-Si-Mg filler metal, (e.g., alloy 4643) in place of Al-Si filler metal (e.g., alloy 4043). Line G of Table 1.63 describes properties for alloy 4643 filler welds in thick alloy 6061 base metal.

In cases where complete PWHT of a weldment is not practical, parts can be welded in the solution heat-treated condition and then artificially aged after welding. A substantial increase in properties over the normal as-welded strengths can sometimes be obtained in this manner when high welding rates are employed. For example, if alloy 6061 is welded in the T4 temper and then aged to T6, the strength of the joint may approach

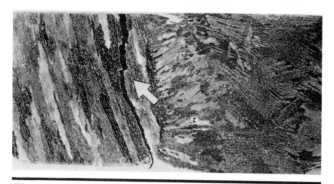

Figure 1.40—Fracture (Arrow) in a 2014-T6 Plate-to-Forging Weld. Grain Size and Orientation in the Forging (left) Contributed to the Failure

40 ksi (275 MPa), which is a great improvement over the 27 ksi (186 MPa) as-welded strength (Line C, Table 1.63). The mechanical properties rarely attain those of a fully PWHT weldment (with solution heat treatment followed by aging). An additional benefit that results from welding heat-treatable alloys in the solution heat-treated T4 condition is that hot cracking is minimized due to the more uniform microstructure with the micro-constituents being held in solid solution.

The low-copper-content 7XXX series heat-treatable alloys, such as 7004, 7005, and 7039, combine good weldability with high as-welded strength. These alloys are less sensitive to the rate of quench from their solution heat-treating temperature and are solution heat-treated at lower temperatures compared to the other heat-treatable alloy types. These alloys will naturally age quite rapidly at room temperature (in 2 to 4 weeks) to provide high tensile and yield strength with a high joint efficiency. An artificial aging treatment after welding can be used in place of natural aging to improve the yield strength further. A full postweld solution heat treatment followed by precipitation hardening (aging) produces the highest strength.

SHEAR STRENGTH

FILLET WELDS ARE designed on the basis of shear on a plane area through the weld. This area is the product of the effective throat and the length of the weld. The composition of the weld metal is nearly the same as the filler metal because admixture with the base metal is low. The shear strength of fillet welds with various filler alloys is given in Table 1.64. Longitudinal loading develops the lowest strength and is used as the basis for minimum design criteria values. Highest as-welded shear strengths are obtained with the high magnesium content 5XXX series filler metals.

IMPACT STRENGTH

ALUMINUM WELDMENTS HOLD up quite well under impact loadings, particularly the nonheat-treatable alloys. Aluminum and aluminum alloys do not exhibit a brittle transition range at low temperatures (as do many ferrous materials) but maintain their ductility and resistance to shock loading at extremely low temperatures. Their tensile and yield strengths actually improve as temperature decreases.

A considerable amount of impact testing of ferrous materials and weldments at room and subzero temperature has been done with the Charpy V-Notch or Izod tests, and some data has also been obtained on aluminum. In some aluminum alloys and tempers, the specimens do not fracture but merely bend, giving high impact values but invalidating the test. In general, Charpy V-Notch and Izod test values obtained for aluminum should only be used for comparison purposes.

FATIGUE STRENGTH

THE FATIGUE STRENGTH of welded structures follows the same general rules that apply to other kinds of fabricated assemblies. Fatigue strength is governed by the peak stresses at points of stress concentration rather than by nominal stresses. Anything that can be done to reduce the peak stresses by eliminating stress raisers will tend to increase the life of the assembly under repeated loads.

Curves representing the average fatigue strengths for transverse welded butt joints in four aluminum alloys are shown in Figure 1.41. For the shorter cyclic lives, the fatigue strengths reflect differences in static strength. However, for the very large numbers of cycles, the difference between the alloys is small. The fatigue strength of a groove weld may be significantly increased by such means as removing weld bead reinforcements or peening the weldment. If such procedures are not practical, it is desirable that the reinforcement blend into the base plate smoothly to avoid any abrupt changes in thickness. For welding processes that produce relatively smooth weld beads, there is little or no increase in strength with further smoothening. The benefit of smooth weld beads can be nullified by excessive spatter during welding. Adhering spatter creates severe stress raisers in the base metal adjacent to the weld.

Investigations have been made on the effect of stress concentrations. For long life, the allowable stress for plate joined by transverse fillet welds is less than half that of plate joined with transverse groove welds. Welds in aluminum alloys usually perform very well under repeated load conditions. Tests have shown that properly designed and fabricated welded joints will generally perform as well under repeated load conditions as riveted joints designed for the same static loading.

Table 1.63

Effect of Welding Conditions on Strength of Groove Joints in Alloy 6061 with Filler Metal ER 4043[a]

Key[b]	Base Alloy and Temper	Thickness in.	Thickness mm	Welding Process and Conditions	As Welded Tensile Strength ksi	MPa	Yield Strength[c] ksi	MPa	% Elong. in 2 in. (50.8 mm) Gauge	Aged After Welding Tensile Strength ksi	MPa	Yield Strength[c] ksi	MPa	% Elong. in 2 in. (50.8 mm) Gauge	Solution Heat-Treated and Aged After Tensile Strength ksi	MPa	Yield Strength[c] ksi	MPa	% Elong. in 2in. (50.8 mm) Gauge
A	6061-T4	1/32	0.8	AC-GTA 95.9 in./min (40.6 mm/s)	33	227	21	145	6	—	—	—	—	—	—	—	—	—	—
B	6061-T6	1/32	0.8	AC-GTA 95.9 in./min (40.6 mm/s)	33	227	26	179	2	—	—	—	—	—	—	—	—	—	—
C	6061-T4	1/8	3.2	SP-DC GTA 20.1 in./min (8.5 mm/s)	34	234	21	145	8	41	283	26	179	3	44	303	40	276	5
D	6061-T6	1/8	3.2	SP-DC GTA 34.9 in./min (14.8 mm/s) Single pass	36	248	24	165	6	—	—	—	—	—	44	303	40	276	5
E	6061-T6	1/4	6.4	Auto-GMA One pass each side 39.9 in./min (16.9 mm/s)	37	255	20	138	6	—	—	—	—	—	43	296	40	276	5
F	6061-T4	3	76	Auto-GMA	25	172	13	90	10	—	—	—	—	—	34	234	—	—	6
G	6061-T6[c]	3	76	Auto-GMA Multipass V-Groove	27	186	14	97	13	—	—	—	—	—	45	310	40	276	4

a. 4643 filler metal was used in line G.

b. Letters in this column are used only as line references.

c. 0.2% offset in 2 in. (50.8 mm) gauge.

Table 1.64
Minimum Shear Strength of Fillet Welds

| Filler Alloy | Shear Strength | | | |
| | Longitudinal | | Transverse | |
	ksi	MPa	ksi	MPa
1100	7.5	52	7.5	52
2319[a]	16	110	16	110
2319[b]	22	152	29	200
4043	11.5	79	15	103
4643	13.5	93	20	138
5183	18.5	127	28	193
5356	17	117	26	179
5554	15	103	23	158
5556	20	138	30	207
5654	12	83	18	124

a. Naturally or artificially postweld aged

b. Solution heat-treated and artificially aged after welding

EFFECT OF TEMPERATURE

THE MINIMUM TENSILE strength of aluminum welds at other than room temperature is listed in Table 1.65. The performance of welds follows closely that of the annealed base metal in the nonheat-treatable alloys. At temperatures as low as -320 °F (-196 °C) the strength of aluminum increases without loss of ductility; it is therefore a particularly useful metal for low temperature or cryogenic applications.

As the temperature rises above room temperature, aluminum alloys lose strength. The 2XXX series alloys exhibit highest strength at elevated temperatures. Alloys in the aluminum-magnesium group with 3.5 percent or more magnesium content are not recommended for sustained temperatures exceeding 150 °F (66 °C) because of a susceptibility to stress corrosion.

FRACTURE CHARACTERISTICS

THE FRACTURE CHARACTERISTICS of weldments may be described in terms of their resistance to rapid crack propagation at elastic stresses or their ability to deform plastically in the presence of stress raisers, avoiding the low-energy initiation and propagation of cracks. Resistance to rapid crack propagation may be measured in terms of tear resistance.

A fracture toughness test is useful for this purpose but only for relatively low ductility material. Most aluminum alloy weldments are too ductile for this test to be of any significance. The ability of weldments to deform plastically and redistribute load to adjacent regions may be expressed in terms of notch toughness. The relationship of the tensile strength of notched specimens to the joint yield strength provides alternate but generally coarser measures of these characteristics.

Regardless of the criterion used to measure fracture characteristics, and depending on the filler metal used, aluminum alloy weldments are usually at least as tough as the base aluminum alloy. In nonheat-treatable alloys, where welding anneals a narrow zone on each side of the weld, the weld has the same high resistance to rapid crack propagation as annealed material, and much more than the cold-worked base metal. The elongation of specimens taken across a weld in cold-worked base metal may be low, suggesting that this is not the case. This low value is actually the result of nonuniform strength of regions within the gauge length and the resultant strain concentration at the weld.

The same situation exists for heat-treated alloys in the as-welded condition, where tear resistance or notch toughness is much greater than that of the heat-treated base metal. Heat treatment or aging after welding increases the strength of the weldment and brings its fracture characteristics more in line with those of the base material, particularly when base and filler metal alloys are of essentially the same chemical composition. Where composition of a heat-treatable filler alloy differs from the base metal, such as alloy 4043 filler metal and alloy 6061 base metal, the weldment may be less tough than the base metal after PWHT of the welded assembly. A relative merit rating of aluminum alloy weldments based upon notch toughness and tear resistance is shown in Figure 1.42

CORROSION RESISTANCE

MANY ALUMINUM ALLOYS can be welded without reducing their resistance to corrosion. In general, the welding method does not affect corrosion resistance unless the residual flux residue is not completely removed after using either the oxyfuel gas or shielded metal arc welding processes.

The excellent corrosion resistance of the nonheat-treatable alloys is not changed appreciably by welding. Combinations of these alloys have good resistance to corrosion. For installations involving prolonged elevated-temperature service [above 150 °F (66 °C)], however, limitations are placed on the amount of cold work that is permissible for some of the 5XXX series alloys,

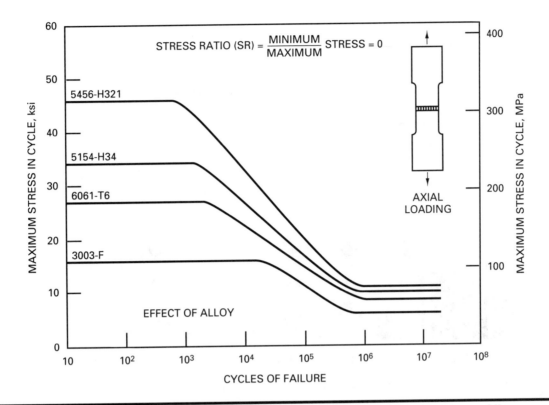

Figure 1.41—Average Fatigue Strength for Groove Welds in Four Aluminum Alloys

Table 1.65
Ultimate Tensile Strength at Various Temperatures for GMAW and GTAW
Welded Groove Joints in Aluminum Alloys

Alloy and Temper	Filler Alloy	Ultimate Tensile Strength											
		-300 °F (-184 °C)		-200 °F (-129 °C)		-100 °F (-73 °C)		100 °F (38 °C)		300 °F (149 °C)[a]		500 °F (260 °C)[a]	
		ksi	MPa	ksi	MPa	ksi	MPa	ksi	MPa	ksi	MPa	ksi	MPa
2219-T37[b]	2319	48.5	334	40.0	276	36.0	248	35.0	241	31.0	214	19.0	131
2219[c]	2319	64.5	445	59.5	410	55.0	379	50.0	345	38.0	262	22.0	152
3003	1100	27.5	190	21.5	148	17.5	121	14.0	97	9.5	66	5.0	34
5052	5356	38.0	262	31.0	214	26.5	183	25.0	172	21.0	145	10.5	72
5083[a]	5183	54.5	376	46.0	317	40.5	279	40.0	276	—	—	—	—
5086[a]	5356	48.0	331	40.5	279	35.5	245	35.0	241	—	—	—	—
5454	5554	44.0	303	37.0	255	32.0	221	31.0	214	26.0	179	15.0	103
5456	5556	56.0	386	47.5	327	42.5	293	42.0	290	—	—	—	—
6061-T6[b]	4043	34.5	238	30.0	207	26.5	183	24.0	165	20.0	138	6.0	41
6061[c,d]	4043[d]	55.0	379	49.5	341	46.0	317	42.0	290	31.5	217	7.0	48

a. Alloys not listed at 300 and 500 °F (149 and 260 °C) are not recommended for use at sustained operating temperatures of over 150 °F (66 °C).

b. As-welded.

c. Postweld solution heat-treated and aged.

d. 4643 filler alloy for 3/4 in. (19 mm).

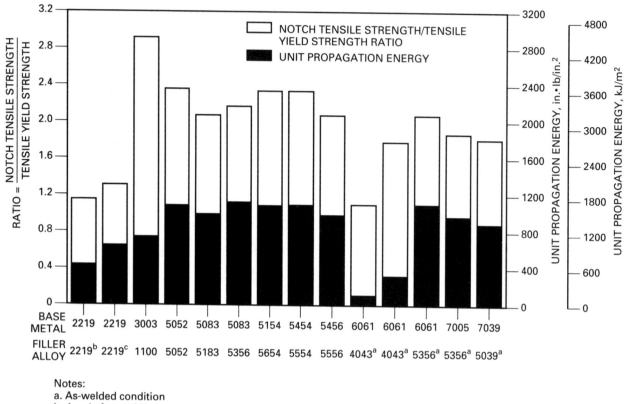

Notes:
a. As-welded condition
b. Aged after welding
c. Heat treated and aged after welding

Figure 1.42—Relative Tear Resistance Ratings of Aluminum Alloy Weldments. Tear Resistance (Unit Propagation Energy) is at Right and Notch Toughness (Ratio of Notch Tensile Strength to Tensile Yield Strength) is on Left.

particularly those with higher magnesium content, which may show susceptibility to stress corrosion.

The aluminum-magnesium-silicon alloys, such as 6061 and 6063, have high resistance to corrosion in both the unwelded and welded conditions and are not noticeably affected by such factors as temper, operating temperature, nature and magnitude of stress, and environment.

The 2XXX and 7XXX series heat-treatable alloys containing substantial amounts of copper and zinc may have their resistance to corrosion lowered by the heat of welding. Grain-boundary precipitation in the HAZ of these alloys creates an electrical potential different from the remainder of the weldment and, in the presence of an electrolyte, selective corrosion at the grain boundaries may take place. In the presence of stress, this corrosion can proceed rapidly.

Postweld heat treatment provides a more homogeneous structure and improves the corrosion resistance of heat-treatable alloys. If the assembly cannot be solution heat-treated and artificially aged after welding, better resistance to corrosion is observed if the original material is welded in the T6 temper rather than in the T4 temper.

The corrosion resistance of welded assemblies fabricated of clad material is superior to that of welded assemblies in nonclad material. Paint protection is recommended when welded joints of the bare 2000 or 7000 series alloys are employed in outdoor environments.

APPLICATIONS

COMMUNICATIONS SATELLITE DISH ANTENNAS

MANY COMMUNICATIONS SATELLITE dish antennas are made from aluminum. They range in diameter from 8 to 100 ft (2.5 to 30 m). The antenna shown in Figure 1.43 has a 20-ft (6-m) diameter. The dish material is 5052-0 condition, which has the ability to maintain the desired curvature dimensional accuracy under various service conditions. If designed for extremely high winds, the surface may be perforated, but the signal efficiency is less than with the solid sheet surface.

A slotted "Z" section as shown in Figure 1.44 is used to provide structural stability to the dish surface. The section is slotted to provide ease of conformance to the desired curvature of the dish. Then structural stability is attained as it is adhesively bonded to the carefully formed dish sheet material by a specially developed acrylic adhesive. Figure 1.45 shows a portion of the completed assembly with the excess adhesive clearly visible where it oozed out around the edges of the slotted "Z" section. The adhesive serves not only as the

Courtesy of Vertex Communications

Figure 1.44— Portion of Slotted "Z" Section Used for Ease of Curvature Conformance and Structural Stiffness

Courtesy of Vertex Communications

Figure 1.43—Six-Meter Communications Satellite Dish Antenna

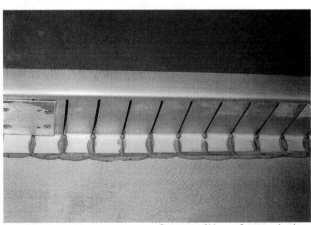

Courtesy of Vertex Communications

Figure 1.45— Slotted "Z" Section Adhesively Joined to Dish Panel (Note Excess Adhesive Material Between "Z" and Dish Material)

bonding agent but also as shim or spacer material to float out irregularities so as not to alter the desired curvature of the dish material.

Adhesive bonding avoids the shrinkage stress associated with welding, but the curing process still results in minor ripples that can be detected if precision examination is made of the surface.

Other components for the satellite antennas, such as those shown in Figures 1.46 and 1.47, are made from 5052 or 6061 aluminum and are joined by welding using gas tungsten arc welding (GTAW) and gas metal arc welding (GMAW). ER4043 filler metal is used for either base metal. The shielding gas is argon with a flow rate of 25 to 40 ft^3/h (12 to 19 L/min) through nozzle sizes of 3/8 to 5/8 in. (9.5 to 15.8 mm).

ALUMINUM DUMP TRAILER

IN THE EARLY 1960s, attempts were made to replace steel in dump trailers with aluminum alloys because aluminum weighs about 30 percent less than steel. This should have resulted in greater payloads and productivity. Initially, the design was based on successful steel

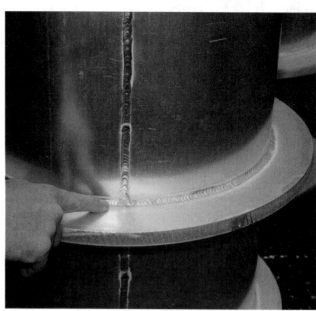

Courtesy of Vertex Communications

Figure 1.47—Satellite Feed Component Welded by GMAW with ER4043 Filler Metal

Courtesy of Vertex Communications

Figure 1.46—Satellite Hub Welded by GMAW with ER4043 Filler Metal

fabrications, simply substituting an aluminum alloy. Failures resulted because the welding procedures reduced the heat-affected zone yield strength by 40 percent. As aluminum section thicknesses were increased to eliminate failures, the weight savings was eliminated.

In the early 1990s, the aluminum end-dump trailer shown in Figure 1.48 was designed. A round profile for the bed was the best geometric configuration to produce the required strength. This design eliminated the need for ribs and cross-members that require circumferential welds, which would weaken the trailer in the area of highest stress.

Specially designed rolls form the three 5454-H34 sheets into 40 ft (12.2 m) lengths which are joined longitudinally to 6061-T6 extrusions. Most of the welding is by mechanized gas metal arc welding using ER5556 filler metal and argon shielding. This design concept is illustrated in Figure 1.49. The sheets are placed in grooves in the extrusions, forming tongue-in-groove type offset lap joints.

This design required 70 total pieces as compared to 350 in previous designs and reduced the length of welds required by 70 percent. The trailer weight also was reduced by about 2500 pounds, and the aerodynamics

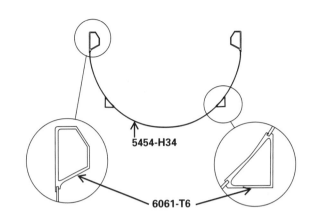

Figure 1.49—Cross Section of Aluminum Trailer Bed of Dump Truck

of this design decreased the fuel consumption rate by about 10 percent. These latter two features increased a trailer's profit potential by close to $10,000 annually.

Courtesy of Alumatec

Figure 1.48—Aluminum Dump Trailer made from 5454-H34 Sheet, 6061-T6 Extrusions, and Gas Metal Arc Welded with 5556 Filler Metal

SPACE SHUTTLE EXTERNAL TANK

LITHIUM IS A very effective addition to aluminum alloys because lithium reduces alloy density and increases stiffness. Aluminum lithium alloys often contain copper or magnesium, or both, as the primary alloying elements for increased strength. Most commercial aluminum-lithium alloys also contain zirconium, because of its ability as a grain refiner and a recrystallization inhibitor. With proper selection of composition, aluminum-lithium alloys can be very strong.

Most aluminum-lithium alloys were developed for aircraft applications where welding is not a preferred joining technique. An exception to this is Al-5Mg-2Li-0.14Zr (wt. %) alloy 1420 which was developed in the 1960s in the Soviet Union for welded applications.[25,26] In the early to middle 1980s, the potential to reduce the weight of welded launch systems by using aluminum-lithium alloys was recognized.[27] Studies were initiated to assess the weldability of the recently developed Al-Cu-Li-(Mg)-Zr alloys, such as 2090, 2091, and 8090. Other studies were designed specifically to develop aluminum-lithium alloys with weldability as a first-tier design property.[28,29] These latter alloys were trademarked Weldalite® alloys and include registered variants 2094 and 2195, as well as others.[30] Their compositions are shown in Table 1.66.

Weldability of Aluminum-Lithium Alloys

THE WELDABILITY OF aluminum-lithium alloys has been reviewed on several occasions.[31,32] Weldability for aluminum alloys is generally defined as resistance to hot cracking. Aluminum-lithium alloys can be weldable, but hot cracking has been reported for some alloys using conventional fillers. A particular problem in welding Al-Li alloys is their propensity for weld-zone

Table 1.66
Typical Compositions of Selected Aluminum-Lithium Alloys within Their Registered Ranges (wt. %, balance Al)

Alloy	Cu	Li	Mg	Ag	Zr
1420*	—	2.0	5.0	—	0.14
2090	2.7	2.2	—	—	0.12
2091	2.0	1.9	1.3	—	0.12
2094	4.5	1.25	0.4	0.4	0.14
2195	4.0	1.0	0.4	0.4	0.14

* Variants of 1420 exist also with chromium, manganese, or both.

porosity, which results from the reactivity of lithium. Nevertheless, low-porosity weldments can be fabricated by use of proper prewelding preparation, which includes chemically or mechanically milling the surfaces to be joined, coupled with the use of an inert gas cover during welding.

Welding Weldalite® Alloys

BY CONSIDERING THE known hot-cracking susceptibility of binary and ternary model alloys as a starting point for alloy design, Al-Cu-Li Weldalite® alloys were specifically designed to be weldable. Moreover, the first such alloy was designed to be above the solid solubility limit where hot-cracking resistance is often very good because of the ability of the solute-rich eutectic liquid to heal weld cracks by a backfilling mechanism during the final stages of solidification. Consequently, initial heats contained 6.3% copper, the level in conventional alloy 2219, whose excellent weldability is in part responsible for its use as the main structural alloy on the space shuttle external tank.

Welding trials as a function of copper content showed that the alloy Al-XCu-1.3Li-0.4Ag-0.4Mg-0.14Zr could be readily welded with no discernible hot cracking at copper levels down to 4.5 percent and possibly lower. Because all of the alloy compositions in this study were above the solid solubility limit, decreasing the copper level did not reduce strength in this composition range. As expected, fracture toughness and ductility increased with decreasing copper level, so the nominal 4.5% level was selected and the alloy was registered as 2094.

As the properties of commercial lots of 2094 plate became available, trade studies revealed that greater weight savings could be achieved at higher toughness levels even with a slight decrease in strength. This gave rise to alloy 2195 with the nominal composition

25. Fridlyander, I. N. British Patent 1172736, 1967.
26. Pickens, J. R., Langan, T. J., and Barta, E. "The weldability of Al-5Mg-2Li-0.1Zr alloy 01420," *Aluminum Lithium III*, Eds. Baker, C., Gregson, P. J., Harris, S. J., and Peel, C. J., 137-147. London: The Institute of Metals, 1986.
27. Pickens, J. R. *Journal of Materials Science*, 20: 4247-58, 1985.
28. Pickens, J. R., Heubaum, F. H., Langan, T. J., and Kramer, L. S. "Al-(4.5-6.3)Cu-1.3Li-0.4Ag-0.4Mg-0.14Zr alloy Weldalite™ 049." *Aluminum Lithium Alloys*, Eds. Sanders, T. H. and Starke, E. A., Jr., 1397-1414. Birmingham, U.K.: Materials and Components Pub. Ltd.
29. Pickens, J. R., Heubaum, F. H., and Langan, T. J., Proceedings of Symposium on Air Frame Materials, Eds. Holt, R.T. and Lee, S., 3-26. Ottawa, Canada: 30th Annual CIM Conference on Met., August 21, 1991.
30. Weldalite is a registered trademark of Martin Marietta Corp.
31. Pickens, J. R. *Journal of Materials Science*, 20: 4247-58, 1985.
32. Pickens, J. R. "Recent developments in the weldability of lithium containing aluminum alloys," *Journal of Materials Science*, 25(7): 3035-47, 1990.

Al-4.0Cu-1.0Li-0.4Ag-0.4Mg-0.14Zr. One concern with this modification was that hot cracking might appear when the large subcomponents of launch systems tanks were "fit up" to be welded. Commercially available filler alloy 2319, with a high copper level of 6.3 percent, was selected to ensure hot-cracking resistance via parent-filler mixing.

Two major obstacles to obtaining sound, x-ray-clear welds in any Al-Li alloy are the fairly thick surface oxide layer, which must be removed prior to welding, and the high reactivity of lithium with oxygen and hydrogen, necessitating additional inert gas shielding requirements. Experiments with variable polarity plasma arc welding (VPPAW) confirmed that defects are consistently encountered in welds unless 0.010 to 0.015 in. (0.254 to 0.396 mm) of material is mechanically or chemically removed from weld lands and adjoining surfaces before welding. Once the initial mill-produced oxide layer is removed, however, it does not reform to the same extent, even if the material is left exposed to shop environments for extended periods of time.

A standard "backside" purge chamber was used in the laboratory to shield the root sides of weld joints. Initially, pure helium was used as the purge gas; however, the addition of argon proved to be quite effective in enhancing the gas coverage of the entire weld joint. Oxygen levels were monitored inside the purge chamber, and a maximum allowable level of 3% oxygen was established as the requirement for x-ray-clear welds. Sealing the chamber to minimize leak paths allowed oxygen levels in the vicinity of the weld joint to reach 0.2 to 0.5% with consistently good results.

Backside shielding of a production VPPA weld joint posed an interesting challenge to welding engineers. Although the purge chamber was effective in laboratory conditions, sealing off the entire backside of a 27.5-ft (8.4-m) diameter external tank assembly was impractical and, due to numerous leaks in the weld tooling, the desired oxygen levels could not be achieved. A small, traveling gas cup was designed to move in a synchronized manner with the welding torch and direct the inert gas mixture locally to the area of the keyhole weld.

Because of the existing external tank production tool configurations, the cup was made to be only 0.75 in. wide x 9 in. long x 6 in. deep (19 x 228 x 152 mm), so that it would fit between the massive backing chill bars in the tools. Through carefully designed experiments, an optimum location of the cup with respect to the keyhole was established, both laterally and longitudinally, and the argon-helium mixture was tailored to the specific application.

At this point, the only Weldalite® alloy plate that was immediately available was alloy 2094. After welding several full-length "confidence panels" of alloy 2094 in external tank production tools (specifically, the tools where dome gore panels and the associated frame chord are welded into a quarter-dome assembly), the first full size Al-Li launch vehicle assembly was successfully VPPA welded in late 1991. The backside shield cup, although somewhat distorted by the heat from the plasma plume during keyhole welding, performed as well as expected. Over 500 in. (12.7 m) of weld were completed with no discernible porosity. Room temperature as-welded tensile strengths from 2094 weldments were typically 50 ksi (345 MPa). The weld schedule is detailed in Table 1.67, and a photo of the completed 2094 quarter-dome assembly is shown in Figure 1.50.

A study was performed to compare the as-welded weldment properties of 2195/2319 with those of 2219/2319 using VPPAW. Long transverse tensile properties and weldment toughness, as measured by fracture strength of surface-cracked toughness panels, were measured for each alloy weldment as a function of temperature. Surface cracks were introduced at the weld centerline or the fusion line on selected specimens.

Neither 2195 nor 2219 displayed any hot-cracking problems. The weldment tensile strength of 2195/2319 was significantly greater than that of 2219/2319 at both ambient and cryogenic temperatures (see Figure 1.51). The mean 70 °F (21 °C) tensile strength of the 2195 VPPA weldments was 56.6 ksi (390 MPa), significantly higher than that of 2219 and higher than that of any known aluminum alloy VPPA weldments at that time. The weldment toughness of 2195/2319 is also superior

Table 1.67
VPPAW Schedule for Welding 0.200 in. (5.16 mm) Thick 2094-T87 Quarter-Dome Components Using 0.062 in. (1.59 mm) 2319 Weld Wire

Pass #	Welding Current, A	Arc Voltage, V	Travel Speed in./min	Travel Speed mm/s	Wire Speed in./min	Wire Speed mm/s	Plasma Gas ft³/h	Plasma Gas L/min	Torch Gas ft³/h	Torch Gas L/min	Polarity Ratio, N:P	Backing Gas ft³/h	Backing Gas L/min
1	105	22	8.0	3.4	40	16.9	6.0 Ar	2.83 Ar	70 He	33.0 He	19:4	65 He, 21 Ar	31 He, 10 Ar
2	95	21	8.0	3.4	20	8.5	2.0 Ar	0.94 Ar	70 He	33.0 He	19:4	—	—

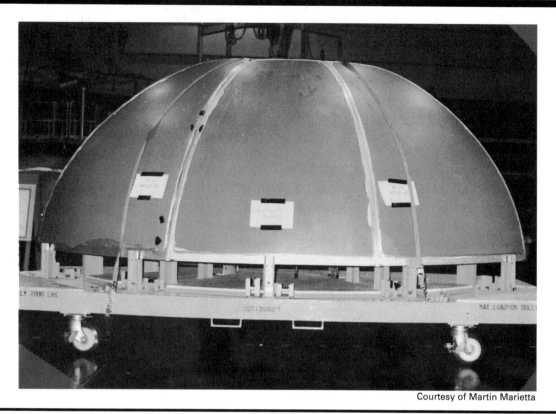

Courtesy of Martin Marietta

Figure 1.50—Full-Scale, Quarter-Dome Assembly of the Space Shuttle External Tank Fabricated Using Weldalite® Alloy 2094

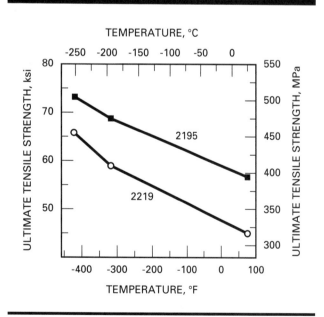

Figure 1.51—Weldment Tensile Strength vs. Temperature for Weldalite® Alloy 2195 Compared with 2219; Each Welded Using 2319 Weld Wire

to that of 2219/2319 at both ambient and cryogenic temperatures (see Figure 1.52). Thus, Weldalite® alloy 2195 can produce VPPA weldments that are both stronger and tougher than those of the current leading launch system alloy, 2219.

This successful demonstration of welding in existing launch vehicle production conditions paves the way for successful implementation of Weldalite® alloys in the next generation of high-performance launch vehicles, with expected weight savings of up to 20 percent over vehicles made using conventional alloy 2219.

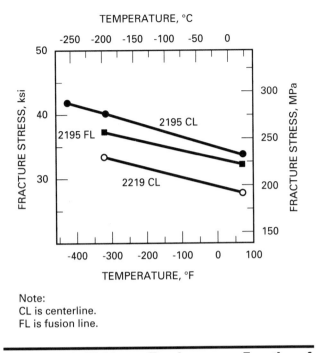

Note:
CL is centerline.
FL is fusion line.

Figure 1.52—Weldment Toughness as a Function of Temperature for Weldalite® Alloy 2195 and Conventional Alloy 2219; Each Welded Using 2319 Weld Wire

SUPPLEMENTARY READING LIST

Aluminum Association, Inc. *Aluminum standards and data*, Publication No. 1. Washington, D.C.: Aluminum Association, 1993.

American Society for Metals. "Fabrication and finishing." *Aluminum*, Vol. II, 383-574. Metals Park, Ohio: American Society for Metals, 1967.

ASM International, "Selection and weldability of aluminum alloys." *ASM Handbook*, Vol. 6. 528-559. Metals Park, Ohio: ASM International, 1993.

—————. "Practice consideration for arc welding aluminum alloys." *ASM Handbook*, Vol. 6. 722-739. Metals Park, Ohio: ASM International, 1993.

—————. "Brazing aluminum alloys." *ASM Handbook*, Vol. 6. 937-940. Metals Park, Ohio: ASM International, 1993.

American Society for Testing and Materials. "Standard methods of tension testing wrought and cast aluminum and aluminum-alloy and magnesium-alloy products," ASTM B-557. *ASTM Annual Book of Standards*, Volume 02.02. Philadelphia: American Society for Testing and Materials, 1989.

American National Standards Institute. *Safety in Welding and Cutting*, ANSI/ASC Z49.1. Miami: American Welding Society, 1988.

American Welding Society. *Recommended practices for stud welding*, ANSI/AWS C5.4. Miami: American Welding Society, 1993.

—————. *Structural welding code–aluminum*, ANSI/AWS D1.2. Miami: American Welding Society, 1990.

—————. *Recommended practices for gas shielded arc welding of aluminum and aluminum alloy pipe*, ANSI/AWS D10.7. Miami: American Welding Society, 1986.

American Society of Civil Engineers. "Suggested specifications for structures of aluminum alloy 6063-T5 and 6063-T6," Report to the Task Committee on Light-weight Alloys. *Journal of the Structural Division, Proceedings of the American Society of Civil Engineers* 88 (ST6): 47. New York: American Society of Civil Engineers, December 1962.

Andrew, R. C., and Waring, J. "Effect of porosity on transverse weld fatigue behavior." *Welding Journal* 53(2): 85s-90s, 1974.

Baeslack, W. A., Fayer, G., Ream, S., and Jackson, C. E. "Quality control in arc stud welding." *Welding Journal* 54(11): 789-798, 1975.

Barhorst, S. "The cathodic etching technique for automated aluminum tube welding." *Welding Journal* 64(5): 28-31, 1985.

Brennecke, M. W. "Electron beam welded heavy gage aluminum alloy 2219." *Welding Journal* 44(1): 27s-39s, 1965.

Collins, F. R. "Porosity in aluminum alloy welds." *Welding Journal* 37(6), 589-593, 1958.

Dalziel, Charles, F. "Effects of electric current on man." *ASEE Journal*, 43(10): 4436-4536, 1964.

Dowd, J. D. "Weld cracking of aluminum alloys." *Welding Journal* 31(10): 448s-456s, 1952.

Dudas, J. H., and Collins, F. R. "Preventing weld cracks in high-strength aluminum alloys." *Welding Journal* 45(6): 241s-249s, 1966.

Hill, H. N. "Residual welding stresses in aluminum alloys." *Metal Progress*, 80(2): 92-96, August 1961.

Howden, D. G. "An up-to-date look at porosity formation in aluminum weldments." *Welding Journal* 50(2): 112-114, 1971.

Krumpen, R. P., Jr., and Jordan, C. R. "Reduced fillet weld sizes for naval ships." *Welding Journal* 63(4): 34-41, 1984.

Marsh, C. "Strength of aluminum fillet welds." *Welding Journal* 64(12): 335s-338s, 1985.

————. "Strength of aluminum T-joint fillet welds." *Welding Journal* 67(8): 171s-176s, 1988.

Murphy, J. L., Huber, R. A., and Lever, W. E. "Joint preparation for electron beam welding thin aluminum alloy 5083." *Welding Journal* 69(4): 125s-132s, 1990.

Murphy, J. L., Mustaleski, T. M., Jr., and Watson, L. C. "Multipass autogenous electron beam welding." *Welding Journal* 67(9): 187s-195s, 1988.

Murphy, J. L., and Turner, P. W. "Wire feeder and positioner for narrow groove electron beam welding." *Welding Journal* 55(3): 181-190, 1976.

Nelson, F. G., Kaufman, J. G., and Wanderer, E. T. "Aluminum and notch toughness of groove welds in wrought and cast aluminum alloys at cryogenic temperatures." *Advances in Cryogenic Engineering*, Vol. 14, 1969.

Nelson, F. G., and Rolf, R. L. "Shear strengths of aluminum fillet welds." *Welding Journal* 45(2): 82s - 84s, 1966.

Nordmark, G. E. "Peening increases fatigue strength of welded aluminum." *Metal Progress* 84(5): 101-103, 1963.

Pease, C. C., Preston, F. J., and Taranto J. "Stud welding on 5083 aluminum and 9% nickel steel for cryogenic use." *Welding Journal* 52(4): 232-237, 1973.

Pense, A. W., and Stout, R. D. "Influence of weld defects on the mechanical properties of aluminum weldments." Bulletin 152. New York: Welding Research Council, July 1970.

Rolf, R. L. "Welded aluminum cylinders under external pressure." Presented at the ASCE Natural Water Resources and Ocean Engineering Convention, San Diego, Calif.; April 1976.

Sanders, W. W., Jr., and Day, R. H. "Fatigue behavior of aluminum alloy weldments." Bulletin No. 286. New York: Welding Research Council, August 1983.

Saperstein, Z. P., et al. "Porosity in aluminum welds." *Welding Journal* 43(10): 443s-453s, 1964.

Sharp, M. L., Rolf, R. L., Nordmark, G. E., and Clark, J. W. "Tests of fillet welds in aluminum." *Welding Journal* 61(4): 117s-124s, 1982.

Shore, R. J., and McCauley, R. B. "Effects of porosity on high strength aluminum 7039," *Welding Journal* 49(7): 311s-321s, 1970.

Shoup, T. E. "Stud welding." Bulletin 214. New York: Welding Research Council, April 1976.

Singleton, R. C. "The growth of stud welding." *Welding Engineer*, 257-263, July 1963.

TRW Nelson Stud Welding Division. *Design and application handbook for light gauge metalworking.* TRW Nelson Stud Welding Division, Publication CD-92, 1992.

PERFORMANCE OF WELDS

Aluminum Association. *Specifications for Aluminum Structures.* Publication No. 30. Washington, D.C.: Aluminum Association, 1986.

Burk, J. D., and Lawrence, F. V., Jr. "Effects of lack-of-penetration and lack-of-fusion on the fatigue properties of 5083 aluminum alloy welds." Bulletin 234. New York: Welding Research Council, January 1978.

Kaufman, J. G., and Stickley, G. W. "Notch toughness of aluminum sheet and welded joints at room and subzero temperatures." *Cryogenic Technology,* July/August 1967.

Lawrence, F. V., Jr., and Munse, W. H. "Effects of porosity on the tensile properties of 5083 and 6061 aluminum alloy weldments." Bulletin 181. New York: Welding Research Council, February 1973.

Lawrence, F. V., Jr., Munse, W. H., and Burke, J. D. "Effects of porosity on the fatigue properties of 5083 aluminum alloy weldments," Bulletin 206. New York: Welding Research Council, June 1975.

McCarthy, W. A., Jr., Lamba, H., and Lawrence, F. V., Jr. "Effects of porosity on the fracture toughness of 5083, 5456 and 6061 aluminum alloy weldments." Bulletin 261. New York: Welding Research Council, September 1980.

Nelson, F. G., Kaufman, J. G., and Holt, M. "Fracture characteristics of welds in aluminum alloys." *Welding Journal* 45(7): 321s-329s, 1966.

Sharp, M. L. *Behavior and design of aluminum structures*. New York: McGraw-Hill, 1993.

BRAZING

Aluminum Association. *Aluminum brazing handbook*. Publication No. 21. Washington, D.C.: Aluminum Association, 1990.

Dickerson, P. B. "Working with aluminum? Here's what you can do with dip brazing." *Metals Progress* 87(5): 80-85, 1965.

——————. "How to dip braze aluminum assemblies." *Metals Progress* 87(6): 73-78, 1965.

Patrick, E. P. "Vacuum brazing of aluminum." *Welding Journal* 54(3): 159-163, 1975.

Swaney, O. W., Trace, D. E., and Winterbottom, W. L. "Brazing aluminum automotive heat exchangers in vacuum: process and materials." *Welding Journal* 65(5): 49-57, 1986.

SOLDERING

Aluminum Association. *Aluminum soldering handbook*. Publication No. 22. Washington, D.C.: Aluminum Association, 1985.

PLASMA-GMAW

"Plasma-mig welding proves fast and versatile." *Metals and Materials*, 46-48, March 1978.

Schevers, A. A. "Plasma-mig welding of aluminum." *Welding and Metal Fabrication* 44(1): 17-20, 1976.

Swart, J. "Plasma-mig improves mechanized welding." *Metal Construction*, 140-142, March 1982.

ALUMINUM-LITHIUM ALLOYS

Cross, E., Ph.D. thesis. Colorado School of Mines, 1986.

Cross, C. E., Olson, D. L., Edwards, G. R., and Capes, J. F. *Aluminum lithium alloys II*, Eds. Sanders, T. H., Jr., and Starke, E. A., Jr., 675-682. Warrendale, Pa.: The Materials Society – American Institute of Mining, Metallurgical, and Petroleum Engineers, 1983.

Edwards, M. R., and Stoneham, V. E. *Journal de Physics Coll.* 3C [Suppl. au(a)], C3-293, 1987.

Fridlyander, I. N. British Patent 1172736, 1967.

Gayle, F. W., Heubaum, F. H., and Pickens, J. R. "Structure and properties during aging of an ultra-high strength Al-Cu-Li-Ag-Mg alloy." *Scripta Metallurgica et Materialia* 24: 79-84, 1990.

Gittos, M. F. "Gas shielded arc welding of Al-Li alloy 8090." Report 7944.01/87/556.2. Abingdon, Cambridge, U.K.: The Welding Institute, May 1987.

Kramer, L. S., Heubaum, F. H., and Pickens, J. R. "The weldability of high strength Al-Cu-Li alloys." *Aluminum lithium alloys*, Eds. Sanders, T. H., and Starke, E. A., Jr., 1415-24. Birmingham, U.K.: Materials and Components Engineering Publications, Ltd., 1989.

Langan, T. J., and Pickens, J. R. "Identification of strengthening phases in the Al-Cu-Li alloy Weldlite™ 049." *Aluminum lithium alloys*, Eds. V. Sanders, T. H., and Starke, E. A., Jr., 691-700. Birmingham, U.K.: Materials and Components Engineering Publications Ltd., 1989.

Martukanitz, R. P., Natalie, C. A., and Knoefel, J. O. *Journal of Metals* 39:38, 1987.

Mironenko, V. N., Evstifeev, V. S., and Korshunkova, S. A. *Welding Prod.* 24(12): 44, 1974.

Moore, K. M., Langan, T. J., Heubaum, F. H., and Pickens, J. R. "Effects of Cu content on the corrosion and stress corrosion behavior of Al-Cu-Li Weldalite™ alloys." *Aluminum lithium alloys*, Eds. Sanders, T. H., and Starke, E. A., Jr., 1281-1291. Birmingham, U.K.: Materials and Components Engineering Publications, Ltd., 1989.

Pickens, J. R., Langan, T. J., and Barta, E. "The weldability of Al-5Mg-2Li-0.1Zr alloy 01420." *Aluminum Lithium III*, Eds. Baker, C., Gregson, P. J., Harris, S. J., and Peel, C., J. 137-147. London: The Institute of Metals, 1986.

Pickens, J. R. *Journal of Materials Science* 20: 4247-4258, 1985.

——————. "Recent developments in the weldability of lithium containing aluminum alloys." *Journal of Materials Science* 25(7): 3035-3047, 1990.

Pickens, J. R., Heubaum, F. H., Langan, T. J., and Kramer, L. S. "Al-(4.5-6.3)Cu-1.3Li-0.4Ag-0.4Mg-0.14Zr alloy Weldalite™ 049." *Aluminum Lithium Alloys*, Eds. Sanders, T. H. and Starke, E. A., Jr., 1397-1414. Birmingham, U.K.: Materials and Components Engineering Pub. Ltd., 1989.

Pickens, J. R., Heubaum, F. H., and Kramer, L. S. "Ultra high strength, forgeable Al-Cu-Li-Ag-Mg alloy." *Scripta Metallurgica et Materialia* 24: 457-465, 1990.

Pickens, J. R., Heubaum, F. H., and Langan, T. J. *Proceedings of Symposium on Air Frame Materials*, Eds. Holt, R. T. and Lee, S., 3-26. Ottowa, Canada: 30th Annual CIM Conference on Met., August 21, 1991.

Pumphrey, W. F., and Lyons, J. V. *Journal of The Institute for Metals* 74: 439-455, 1947-1948.

MAGNESIUM AND MAGNESIUM ALLOYS

PREPARED BY A COMMITTEE CONSISTING OF:

A. T. D'Annessa, Chairman
Consultant

P. B. Dickerson
Consultant

E. Willner
*Technical Consultant,
Lockheed Missiles & Space
Company*

WELDING HANDBOOK COMMITTEE MEMBER
J. M. Gerken
Consultant

CHAPTER 2

MAGNESIUM AND MAGNESIUM ALLOYS

INTRODUCTION

MAGNESIUM ALLOYS ARE used in a wide variety of applications where light weight is important. Structural applications include industrial, materials-handling, commercial, and aerospace equipment. In industrial machinery, such as textile and printing machines, magnesium alloys are used for parts that operate at high speeds and must be lightweight to minimize inertial forces. Materials-handling equipment examples are dockboards, grain shovels, and gravity conveyors. Commercial applications include luggage and ladders. Good strength and stiffness at both room and elevated temperatures, combined with light weight, make magnesium alloys useful for some aerospace applications.

Unlike most other metals and alloy systems, the application of welded magnesium alloy structures has been declining. This trend is the result of increased use of mechanical assembly along with declining use of wrought material in aerospace applications. Advances in casting and forging technologies producing near net-shape parts has also reduced the need for welding to produce complex shapes. Also, new applications for thorium-bearing alloys will diminish significantly due to environmental emphases to eliminate radioactive substances.

GENERAL PROPERTIES

Chemical Properties

MAGNESIUM AND ITS alloys have a hexagonal close-packed crystal structure. The amount of deformation they can sustain at room temperature is limited when compared to aluminum alloys. However, their formability increases rapidly with temperature, and the metals can be severely worked between 400 and 600 °F (204 and 316 °C). Forming and straightening operations and weld peening are generally done at elevated temperatures.

When heated in air, magnesium oxidizes rapidly forming an oxide that inhibits wetting and flow during welding, brazing, or soldering. For this reason, a protective shield of inert gas or flux must be used during exposure to elevated temperatures to prevent oxidation.

The oxide layer formed on magnesium surfaces recrystallizes at high temperatures and becomes flaky. It tends to break up more readily during welding than does the oxide layer on aluminum. Magnesium oxide (MgO) is highly refractory and insoluble in both liquid and solid magnesium. Magnesium nitride (MgN_2) is relatively unstable and readily decomposes in the presence of moisture.

Under normal operating conditions, the corrosion resistance of many magnesium alloys in nonindustrial atmospheres is better than that of ordinary iron and equal to some aluminum alloys. Formation of a gray oxide film on the surface is usually the extent of attack. For maximum corrosion resistance, chemical surface treatments, paint finishes, and plating can be used. Galvanic corrosion can be serious when magnesium is in direct contact with other metals in the presence of an electrolyte. This can be avoided by the following:

(**1**) Proper design
(**2**) Careful selection of metals in contact with the magnesium
(**3**) Insulation from dissimilar metals

Like other metals, some magnesium alloys are susceptible to stress-corrosion cracking if residual stresses from welding or fabrication are not reduced to a safe level by heat treatment.

Physical Properties

MAGNESIUM IS WELL known for its extreme lightness, machinability, weldability, and the high strength-to-weight ratio of its alloys. It has a density of about 0.06 lb/in.^{3} (1660 kg/m^3). On an equal-volume basis, it weighs about one-fourth as much as steel and two-thirds as much as aluminum.

Pure magnesium melts at 1200 °F (649 °C), about the same temperature as aluminum. However, magnesium boils at about 2025 °F (1107 °C), which is low compared to other structural metals. The average coefficient of thermal expansion for magnesium alloys at temperatures from 65 to 750 °F (18 to 399 °C) is about 15 x 10^{-6} per °F (27 x 10^{-6} per °C), which is about the same as that of aluminum and twice that of steel. The thermal conductivity of magnesium is about 89 Btu/(ft•h•°F) [154 W/(m•K)], and for magnesium alloys ranges from 40 to 64 Btu/(ft•h•°F) [70 to 110 W/(m•K)]. Magnesium's electrical resistivity is about 1.7 times that of aluminum.

Magnesium requires a relatively low heat input for melting because of its comparatively low latent heat of fusion and specific heat per unit volume. On an equal volume basis, the total heat of fusion is approximately two-thirds that for aluminum and one-fifth that for steel. The high coefficient of thermal expansion tends to cause considerable distortion during welding. In this respect, the fixturing required for the welding of magnesium is very similar to that needed for aluminum. However, it must be more substantial than the fixturing for the welding of steel.

Mechanical Properties

MAGNESIUM HAS A modulus of elasticity of about 6500 ksi (44 800 MPa) compared to 10 000 ksi (68 950 MPa) for aluminum and 30 000 ksi (206 800 MPa) for steel. This means that magnesium exhibits greater displacement than aluminum and steel under similar loads.

Cast magnesium has a yield strength of about 3 ksi (21 MPa) and a tensile strength of about 13 ksi (90 MPa). Wrought magnesium products have tensile strengths in the range of 24 to 32 ksi (165 to 221 MPa). Their yield strengths in compression will be lower than those in tension. The reason for this is that it is easier for deformation to occur within a magnesium grain under compression. Alloying significantly increases the mechanical properties of magnesium.

Magnesium has low ductility when compared to aluminum. Tensile elongations range from 2 to 15 percent at room temperature. However, ductility increases rapidly at elevated temperatures.

Magnesium and its alloys are notch-sensitive, particularly in fatigue, because of the low ductility. Tensile properties increase and ductility decreases with decreasing testing temperature.

ALLOYS

ALLOY SYSTEMS

MOST MAGNESIUM ALLOYS are ternary types. They may be considered in four groups based on the major alloying element: aluminum, zinc, thorium, or rare earths.[1] There also are two binary systems employing manganese and zirconium.

---

1. A group of 15 similar metals with atomic numbers 57 through 71.

Magnesium alloys may also be grouped according to service temperature. The magnesium-aluminum and magnesium-zinc alloy groups are suitable only for room-temperature service. Their tensile and creep properties degrade rapidly when the service temperature is above about 300 °F (149 °C). The magnesium-thorium and magnesium-rare earth alloys are designed for elevated-temperature service. They have good tensile and creep properties up to 700 °F (371 °C).

DESIGNATION METHOD

MAGNESIUM ALLOYS ARE designated by a combination letter-number system composed of four parts. Part 1 indicates the two principal alloying elements by code letters arranged in order of decreasing percentage. The code letters are listed in Table 2.1.

Part 2 indicates the percentages of the two principal alloying elements in the same order as the code letters. The percentages are rounded to the nearest whole number. Part 3 is an assigned letter to distinguish different alloys with the same percentages of the two principal alloying elements. Part 4 indicates the condition of temper of the product. It consists of a letter and number similar to those used for aluminum, as shown in Table 2.2. They are separated from Part 3 by a hyphen.

An example is alloy AZ63A-T6. The AZ indicates that aluminum and zinc are the two principal alloying elements. The 63 indicates that the alloy contains nominally 6 percent aluminum and 3 percent zinc. The following A shows that this was the first standardized alloy of this composition. The fourth part, T6, shows that the product has been solution heat-treated and artificially aged.

COMMERCIAL ALLOYS

MAGNESIUM ALLOYS ARE produced in the form of castings and wrought products including forgings, sheet, plate, and extrusions. A majority of the alloys produced in these forms can be welded. Commercial magnesium alloys designed for either room-temperature or elevated-temperature service are listed in Tables 2.3 and 2.4, respectively, with their nominal compositions.

Wrought Alloys

WELDED CONSTRUCTION FOR room-temperature service frequently is designed with AZ31B alloy. It offers a good combination of strength, ductility, toughness, formability, and weldability in all wrought product forms. The alloy is strengthened by work hardening. AZ80A and ZK60A alloys can be artificially aged to

Table 2.1
Code Letters for Magnesium Alloy Designation System

Letter	Alloying Element
A	Aluminum
E	Rare earths
H	Thorium
K	Zirconium
M	Manganese
Q	Silver
Z	Zinc

Table 2.2
Temper Designations for Magnesium Alloys

F	As fabricated
O	Annealed, recrystallized (wrought products only)
H	Strain-hardened
T	Thermally treated to produce stable tempers other than F, O, or H
W	Solution heat-treated (unstable temper)

Subdivisions of H

H1, plus one or more digits	Strain-hardened only
H2, plus one or more digits	Strain-hardened and then partially annealed
H3, plus one or more digits	Strain-hardened and then stabilized

Subdivisions of T

T1	Cooled and naturally aged.
T2	Annealed (cast products only)
T3	Solution heat-treated and then cold-worked
T4	Solution heat-treated
T5	Cooled and artificially aged
T6	Solution heat-treated and artificially aged
T7	Solution heat-treated and stabilized
T8	Solution heat-treated, cold-worked, and artificially aged
T9	Solution heat-treated, artificially aged, and cold-worked
T10	Cooled, artificially aged, and cold-worked

develop good strength properties for room temperature applications.

AZ10A, M1A, and ZK21A alloy weldments are not sensitive to stress-corrosion cracking. Therefore, postweld stress relieving is not required for weldments made of these alloys. The alloys are strengthened by work hardening for room-temperature service.

HK31A, HM21A, and HM31A alloys are designed for elevated-temperature service. They are strengthened by a combination of work hardening followed by artificial aging.

Table 2.3
Commercial Magnesium Alloys for
Room-Temperature Service

ASTM Designation	Nominal Composition, % (Remainder Mg)					
	Al	Zn	Mn	RE*	Zr	Th
Sheet and Plate						
AZ31B	3.0	1.0	0.5	—	—	—
M1A	—	—	1.5	—	—	—
Extruded Shapes and Structural Sections						
AZ10A	1.2	0.4	0.5	—	—	—
AZ31B	3.0	1.0	0.5	—	—	—
AZ61A	6.5	1.0	0.2	—	—	—
AZ80A	8.5	0.5	0.2	—	—	—
M1A	—	—	1.5	—	—	—
ZK21A	—	2.3	—	—	0.6	—
ZK60A	—	5.5	—	—	0.6	—
Sand, Permanent Mold or Investment Castings						
AM100A	10.0	—	0.2	—	—	—
AZ63A	6.0	3.0	0.2	—	—	—
AZ81A	7.6	0.7	0.2	—	—	—
AZ91C	8.7	0.7	0.2	—	—	—
AZ92A	9.0	2.0	0.2	—	—	—
K1A	—	—	—	—	0.6	—
ZE41A	—	4.2	—	1.2	0.7	—
ZH62A	—	5.7	—	—	0.7	1.8
ZK51A	—	4.6	—	—	0.7	—
ZK61A	—	6.0	—	—	0.8	—

* As mischmetal (approximately 52% Ce, 26% La, 19% Nd, 3% Pr)

Table 2.4
Commercial Alloys for Elevated-Temperature
Service

ASTM Designation	Nominal Composition, % (Remainder Mg)					
	Th	Zn	Zr	RE*	Mn	Ag
Sheet and Plate						
HK31A	3.0	—	0.7	—	—	—
HM21A	2.0	—	—	—	0.5	—
Extruded Shapes and Structural Sections						
HM31A	3.0	—	—	—	1.5	—
Sand, Permanent Mold or Investment Castings						
EK41A	—	—	0.6	4.0	—	—
EZ33A	—	2.6	0.6	3.2	—	—
HK31A	3.2	—	0.7	—	—	—
HZ32A	3.2	2.1	0.7	—	—	—
QH21A	1.1	—	0.6	1.2	—	2.5

* As mischmetal (approximately 52% Ce, 26% La, 19% Nd, 3% Pr)

Cast Alloys

THE MOST WIDELY used casting alloys for room-temperature service are AZ91C and AZ92A. These alloys are more crack-sensitive than the wrought Mg-Al-Zn alloys with lower aluminum content. Consequently, they require preheating prior to fusion welding.

EZ33A alloy has good strength stability for elevated-temperature service and excellent pressure tightness. HK31A and HZ32A alloys are designed to operate at higher temperatures than is EZ33A. QH21A alloy has excellent strength properties up to 500 °F (260 °C). All of these alloys require heat treatment to develop optimum properties. They have good welding characteristics.

Mechanical Properties

TYPICAL STRENGTH PROPERTIES for magnesium alloys at room temperature are given in Table 2.5. For castings, the compressive and tensile yield strengths are about the same. For wrought products, however, the yield strength in compression is often lower than in tension.

Tensile and creep properties of representative magnesium alloys at three elevated temperatures are given in Table 2.6. The alloys containing thorium (HK, HM, and HZ) have greater resistance to creep at 400 and 600 °F (204 and 316 °C) than do the Mg-Al-Zn alloys.

MAJOR ALLOYING ELEMENTS

WITH MOST MAGNESIUM alloy systems, the solidification range increases as the alloy addition increases. This contributes to a greater tendency for cracking during welding. At the same time, the melting temperature as well as the thermal and electrical conductivities decrease. Consequently, less heat input is required for fusion welding as the alloy content increases.

Aluminum and zinc show decreasing solubility in solid magnesium with decreasing temperature. These elements will form compounds with magnesium. Consequently, alloys containing sufficient amounts of these elements can be strengthened by a precipitation-hardening heat treatment. Other alloying elements also behave similarly in ternary alloy systems.

Aluminum

WHEN ADDED TO magnesium, aluminum gives the most favorable results of the major alloying elements. It increases both strength and hardness. Alloys containing more than about 6% aluminum are heat-treatable. The aluminum content of an alloy has no adverse effect on weldability. Weldments of alloys containing more than about 1.5% aluminum require a postweld stress-relief heat treatment to prevent susceptibility to stress-corrosion cracking.

Table 2.5
Room-Temperature Mechanical Properties of Magnesium Alloys

ASTM Designation	Tensile Strength		Tensile Yield Strength*		Compressive Yield Strength*		Elongation, in 2 in. (51 mm), %
	ksi	MPa	ksi	MPa	ksi	MPa	
Sheet and Plate							
AZ31B-0	37	255	22	152	16	110	21
AZ31B-H24	42	290	32	221	26	179	15
HK31A-H24	33	228	30	207	22	152	9
HM21A-T8	34	234	25	172	19	131	10
M1A-0	34	234	19	131	—	—	18
M1A-H24	39	269	29	200	—	—	10
Extruded Shapes and Structural Sections							
AZ10A-F	35	241	22	152	11	76	10
AZ31B-F	38	262	29	200	14	97	15
AZ61A-F	45	310	33	228	19	131	16
AZ80A-F	49	338	36	248	22	152	11
AZ80A-T5	44	303	38	262	27	186	8
HM31A-T5	44	303	38	262	27	186	8
M1A-F	37	255	26	179	12	83	11
ZK21A-F	42	290	33	228	25	172	10
ZK60A-F	49	338	37	255	28	193	14
ZK60A-T5	52	358	44	303	36	248	11
Sand, Permanent Mold, or Investment Castings							
AM100A-T6	40	276	22	152	22	152	1
AZ63A-F	29	200	14	97	—	—	6
AZ63A-T4	40	276	13	90	—	—	12
AZ63A-T6	40	276	19	131	19	131	5
AZ81A-T4	40	276	12	83	12	83	15
AZ91C-F	24	165	14	97	—	—	2
AZ91C-T4	40	276	12	83	—	—	14
AZ91C-T6	40	276	21	145	21	145	5
AZ92A-F	24	165	14	97	—	—	2
AZ92A-T4	40	276	14	97	—	—	9
AZ92A-T6	40	276	21	145	21	145	2
EK41A-T5	25	172	13	90	—	—	3
EZ33A-T5	23	159	15	103	15	103	3
HK31A-T6	32	221	15	103	15	103	8
HZ32A-T5	27	186	14	97	14	97	4
K1A-F	25	172	7	48	—	—	19
QH21A-T6	40	276	30	207	—	—	4
ZE41A-T5	30	207	20	138	20	138	4
ZH62A-T5	35	241	25	172	25	172	4
ZK51A-T5	30	206	24	165	24	165	4
ZK61A-T6	45	310	28	193	28	193	10

* 0.2% offset yield strength

Beryllium

THE TENDENCY FOR magnesium alloys to burn during melting and casting is reduced by adding beryllium up to about 0.001%. Beryllium is added to magnesium filler metals to reduce oxidation and the danger of ignition at elevated temperatures during joining operations.

Manganese

THIS ELEMENT HAS little effect upon tensile strength, but increases yield strength slightly. Its most important function is to improve the saltwater corrosion resistance of magnesium-aluminum and magnesium-aluminumzinc alloys. Magnesium-manganese alloys have relatively high melting temperatures and thermal conductivities. Therefore, they require somewhat more welding heat input than do the magnesium-aluminum-zinc alloys. Joint efficiency is low in magnesium-manganese alloys because of grain growth in the heat-affected zone.

Table 2.6
Elevated-Temperature Properties of Some Representative Magnesium Alloys

Alloy	300 °F (148 °C)						400 °F (204 °C)						600 °F (316 °C)					
	Tensile Strength		Tensile Yield Strength		Creep Strength*		Tensile Strength		Tensile Yield Strength		Creep Strength[a]		Tensile Strength		Tensile Yield Strength		Creep Strength[a]	
	ksi	MPa	ksi	MPa	ksi	MPa	ksi	MPa	ksi	MPa	ksi	MPa	ksi	MPa	ksi	MPa	ksi	MPa
Sheet and Plate Alloys																		
AZ31B-H24	22	152	13	90	1.0	6.9	13	90	8	55	—	—	6	41	2	14	—	—
HK31A-H24	26	179	24	165	—	—	24	165	21	145	6.0	41.4	12	83	7	48	—	—
HM21A-T8	23	159	21	145	—	—	19	131	18	124	11.4	78.6	15	103	13	90	5.0	34.5
Extrusion Alloys																		
AZ31B-F	25	172	15	103	3.0	20.7	15	103	9	62	—	—	6	41	2	14	—	—
AZ80A-T5	35	241	23	159	3.5	24.1	22	152	15	103	—	—	9	62	3	21	—	—
HM31A-T5	28	193	25	172	—	—	24	165	21	145	10.9	75.2	18	124	15	103	7.6	52.4
ZK60A-T5	25	172	22	152	1.0	6.9	15	103	12	83	—	—	—	—	—	—	—	—
Casting Alloys																		
AZ92A-T6	28	193	17	117	3.8	26.2	17	117	12	83	—	—	8	55	5	34	—	—
AZ63A-T6	24	165	15	103	4.1	28.3	18	124	12	83	—	—	8	55	6	41	—	—
EZ33A-T5	22	152	14	97	—	—	21	145	12	83	8.0	55.2	12	83	8	55	1.2	8.3
HK31A-T6	27	186	15	103	—	—	24	165	14	97	9.5	65.5	20	138	12	83	2.9	20.0
HZ32A-T5	22	152	12	83	—	—	17	117	10	69	7.8	53.8	12	83	8	55	3.0	20.7
QH21A-T6	33	228	29	200	—	—	30	207	27	186	12	82.7	14	97	13	90	—	—

* Creep strength based on 0.2% total extension in 100 h.

Rare Earths

RARE EARTHS ARE added either as mischmetal or didymium. These additions are beneficial in reducing weld cracking and porosity in castings because they narrow the freezing range of the alloy.

Silver

SILVER IMPROVES THE mechanical properties of magnesium alloys. The QH21A alloy in the T6 condition has the highest room-temperature strength of the commercial magnesium casting alloys. This alloy has good weldability.

Thorium

THORIUM ADDITIONS GREATLY increase the strengths of magnesium alloys at temperatures up to 700 °F (371 °C). The most common alloys contain 2 to 3% thorium in combination with zinc, zirconium, or manganese. Thorium improves the weldability of alloys containing zinc.

Zinc

ZINC IS OFTEN used in combination with aluminum to improve the room temperature strength of magnesium. It increases hot shortness when added in amounts over 1% to magnesium alloys containing 7 to 10% aluminum. In amounts greater than 2% in these alloys, zinc is likely to cause weld cracking.

Zinc also is used in combination with zirconium, thorium, or rare earths to produce precipitation-hardenable magnesium alloys with good strength properties.

Zirconium

ZIRCONIUM IS A powerful grain-refining agent in magnesium alloys. It is added to alloys containing zinc, thorium, rare earths, or combinations of these. Zirconium is believed to confer a slightly beneficial effect on the weldability of Mg-Zn alloys by increasing the solidus temperature.

HEAT TREATMENT

MAGNESIUM ALLOYS ARE heat-treated to improve mechanical properties. The type of heat treatment depends upon the alloy composition, product form, and service requirements.

A solution heat treatment improves strength, toughness, and impact resistance. A precipitation heat treatment following solution heat treatment increases the yield strength and hardness at some sacrifice in toughness. A precipitation heat treatment alone will simultaneously increase the tensile properties and stress relieve as-cast components.

Combinations of solution heat treating, strain hardening, and precipitation hardening are often used with wrought products. Use of intermediate strain hardening with thermal heat treatment produces some of the highest strength properties.

WELDABILITY

THE RELATIVE WELDABILITY of magnesium alloys by gas shielded arc and resistance spot welding processes are shown in Table 2.7. Castings are not normally resistance welded. The Mg-Al-Zn alloys and alloys that contain rare earths or thorium as the major alloying element have the best weldability. Alloys with zinc as the major alloying element are more difficult to weld. They have a rather wide melting range, which makes them sensitive to hot cracking. With proper joint design and welding conditions, joint efficiencies will range from 60 to 100 percent, depending upon the alloy and temper.

Most wrought alloys can be readily resistance spot welded. Due to short weld cycles and heat transfer characteristics, fusion zones are fine-grained and heat-affected zones experience only slight degradation from grain coarsening.

Table 2.7
Relative Weldability of Magnesium Alloys

Alloy	Gas Shielded Arc Welding	Resistance Spot Welding
	Wrought Alloys	
AZ10A	Excellent	Excellent
AZ31B, AZ31C	Excellent	Excellent
AZ61A	Good	Excellent
AZ80A	Good	Excellent
HK31A	Excellent	Excellent
HM21A	Excellent	Good
HM31A	Excellent	Good
M1A	Excellent	Good
ZK21A	Good	Excellent
ZK60A	Poor	Excellent
	Cast Alloys	
AM100A	Good	—
AZ63A	Fair	—
AZ81A	Good	—
AZ91C	Good	—
AZ92A	Fair	—
EK41A	Good	—
EZ33A	Excellent	—
HK31A	Good	—
HZ32A	Good	—
K1A	Excellent	—
QH21A	Good	—
ZE41A	Good	—
ZH62A	Poor	—
ZK51A	Poor	—
ZK61A	Poor	—

SURFACE PREPARATION

AS WITH OTHER metals, the cleanliness of magnesium alloy components and filler metals is important for obtaining sound joints of acceptable quality. Any surface contamination will inhibit wetting and fusion. Magnesium alloys are supplied with an oil coating, an acid-pickled surface, or a chromate conversion coating for protection during shipping and storage. The surfaces and edges to be joined must be cleaned just before joining to remove the surface protection as well as any dirt or oxide present.[2] Chemical cleaners commonly used for magnesium alloys are given in Table 2.8.

Oil, grease, and wax are best removed by either washing with organic solvents or vapor degreasing in a chlorinated hydrocarbon solvent.[3] Subsequent cleaning in alkaline or emulsion-type cleaners is recommended to be sure that the surfaces are absolutely free of oil or

2. For additional information on cleaning refer to "Surface Cleaning, Finishing, and Coating" *Metals Handbook*, Vol. 5, 9th Ed. Metals Park, Ohio: American Society for Metals, 1982.
3. Vapors of these solvents are toxic. For additional information, refer to ANSI/ASC Z49.1, *Safety in Welding, Cutting and Allied Processes*. Miami: American Welding Society (latest edition).

Table 2.8
Chemical Cleaning of Magnesium Alloys

Type of Treatment	Composition of Solution		Method of Application	Uses
Alkaline cleaner	Sodium carbonate Sodium hydroxide Water to make Temperature Solution pH	3 oz (85 gm) 2 oz (57 gm) 1 gal (3.8 L) 190-212 °F (87-100 °C) 11 or greater	3 to 10 min immersion followed by cold water rinse and air dry.	Used to remove oil and grease films, as well as old chrome pickle or dichromate coatings.
Bright pickle	Chromic acid Ferric nitrate Potassium fluoride Water to make Temperature	1.5 lb (680 gm) 5.3 oz (150 gm) 0.5 oz (14.2 gm) 1 gal. (3.8 L) 60-100 °F (16-38 °C)	0.25 to 3 min immersion followed by cold and hot water rinse and air dry.	Used after degreasing to prepare surfaces for welding and brazing. Gives bright clean surfaces; resistant to tarnish.
Spot weld cleaners	*No. 1 Bath* Conc. sulfuric acid Water to make Temperature *No. 2 Bath* Chromic acid Conc. sulfuric acid Water to make Temperature *No. 3 Bath* Chromic acid Water to make Temperature	 1.3 fl oz (38 mL) 1 gal (3.8 L) 70-90 °F (21-32 °C) 1.5 lb (680 gm) 0.07 fl. oz (2.1 mL) 1 gal (3.8 L) 70-90 °F (21-32 °C) 0.33 oz (9.4 gm) 1 gal (3.8 L) 70-90 °F (21-32 °C)	Immerse 0.25-1 min in No. 1 bath. Rinse in cold water. Follow by immersing in either No. 2 or No. 3 bath. For No. 2 bath, immerse 3 min and follow by cold water rinse and air dry. For No. 3 bath, immerse 0.5 min followed by cold water rinse and air dry.	Used after degreasing to remove oxide layer and prepare surface for spot welding. Gives low consistent surface resistance.
Flux remover clean	Sodium dichromate Water to make Temperature	0.5 lb (23 gm) 1 gal (3.8 L) 180-212 °F (82-100 °C)	2 hour immersion in boiling bath, followed by cold and hot water rinse and air dry.	Used after hot water cleaning and chrome pickling to remove or inhibit any flux particles remaining from welding or brazing.
Chrome pickle MIL-M-3171 Type 1 AMS 2475	Sodium dichromate Conc. nitric acid Water to make Temperature	1.5 lb (680 gm) 24 fl. oz (710 mL) 1 gal (3.8 L) 70-90 °F (21-32 °C)	0.5 to 2 min immersion, hold in air 5 s, followed by cold and hot water rinse and air or forced dry, max 250 °F (121°C). When brushed on, allow 1 min before rinse.	Used as paint base and for surface protection. Second step in flux removal. Applied with brush for touch-up of welds and treatment of large structures.

grease. Alkaline (caustic) cleaner will also remove previously applied chemical surface treatments. Cleaners of this type that are suitable for steel are generally satisfactory for magnesium. Cleaning may be by either the immersion or the electrolytic method. Thorough water rinsing, preferably by spray, is necessary after alkaline cleaning to avoid degradation of subsequent acid chemical baths used to treat the parts.

After all oil or other organic material has been removed, the part or joint is ready for chemical or mechanical cleaning. A bright pickle will produce suitable clean surfaces for welding. A final mechanical cleaning is preferred for most critical production applications to ensure uniform surface cleanliness. Stainless steel wool or a wire brush is recommended. Wire brushing should not gouge the surface. For resistance spot welding, chemical cleaning is preferred to provide a uniformly low surface resistance (see Table 2.8).

If neutralization is desired after chemical cleaning and prior to rinsing, a water solution of 6.5 oz/gal (49 gm/L) of sodium metasilicate, operating at 180 °F (82 °C), may be used. After cleaning, special care must be taken to protect the components from contamination during all subsequent handling operations.

Any oxide film or smut deposited on the surface of weldments may be removed by wire brushing or by chemical treatment in a water solution of 16 oz/gal (120 gm/L) of tetrasodium pyrophosphate and 12 oz/gal (90 gm/L) of sodium metaborate operating at 180 °F (82 °C).

ARC WELDING

APPLICABLE PROCESSES

THE GAS TUNGSTEN arc and gas metal arc welding processes are commonly used for joining magnesium alloy components. Inert gas shielding is required with these processes to avoid excessive oxidation and entrapment of oxide in the weld metal. Processes that use a flux covering do not provide adequate oxidation protection for the molten weld pool and the adjacent base metal.

JOINT DESIGN

JOINT DESIGNS SUITABLE for gas shielded arc welding are shown in Figure 2.1. The thickness limitations for welding these joint designs are given in Table 2.9. Because of the high deposition rate of the gas metal arc welding process, a root opening, a beveled joint, or both, should be used to provide space for the deposited metal. Increasing the travel speed to maintain a conventional bead size is not acceptable because undercutting, incomplete fusion, or inadequate penetration may result.

A backing strip is employed when welding sheet metal components to help control joint penetration, root surface contour, and heat removal. Magnesium, aluminum, copper, mild steel, or stainless steel is employed as a backing material. When a temporary backing strip is used, the root side of the joint should be shielded with inert gas to prevent oxidation of the root surface. The gas is supplied through holes in the backing strip. In those instances where a backing strip cannot be used because of space limitations, a chemical flux of the type used in oxyfuel gas welding is sometimes painted on the root side of the joint to smooth the root surface and to control joint penetration. Chemical fluxes must be completely removed after welding to avoid corrosion problems.

FILLER METALS

THE WELDABILITY OF most magnesium alloys is good when the proper filler metal is employed. A filler metal with a lower melting point and a wider freezing range than the base metal will provide good weldability and minimize weld cracking. The recommended filler metals for various magnesium alloys are given in Table 2.10.

ER AZ61A or ER AZ92A (Mg-Al-Zn) filler metal may be used to weld alloys of similar composition and also ZK21A (Mg-Zn-Zr) alloy. ER AZ61A filler metal is preferred for welding wrought products of those alloys because of its low cracking tendencies. On the other hand, ER AZ92A filler metal shows less crack sensitivity for welding the cast Mg-Al-Zn and AM100A (Mg-Al) alloys. The deposited metal will respond to the precipitation heat treatments applied to the repaired casting. ER AZ101A filler metal may also be used to weld those casting alloys.

ER EZ33A (RE-Zn-Zr) filler metal is used to weld wrought and cast alloys designed for high-temperature service, either to themselves or each other. The welded joints will have good mechanical properties at elevated temperatures.

ER AZ92A filler metal is recommended for welding the room-temperature service wrought and cast alloys together or to one of the wrought or cast elevated-temperature service alloys. It will minimize weld cracking tendencies. ER EZ33A filler metal should not be

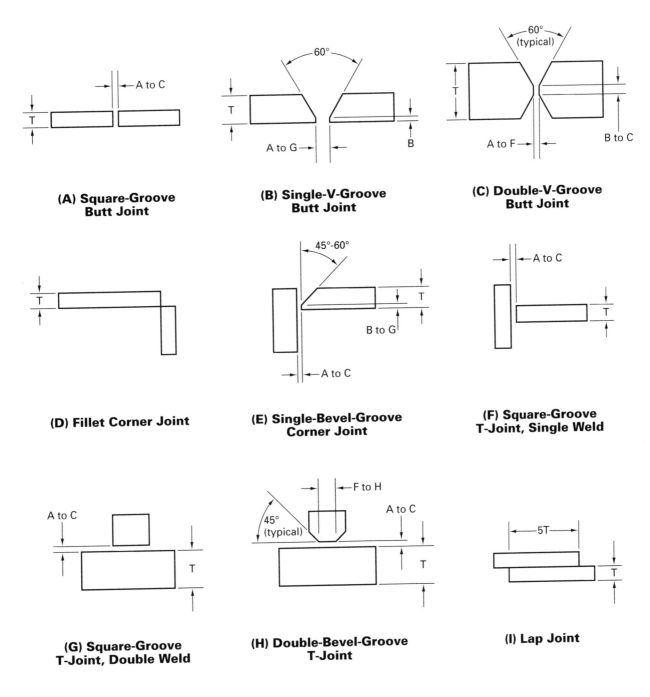

(A) Square-Groove Butt Joint

(B) Single-V-Groove Butt Joint

(C) Double-V-Groove Butt Joint

(D) Fillet Corner Joint

(E) Single-Bevel-Groove Corner Joint

(F) Square-Groove T-Joint, Single Weld

(G) Square-Groove T-Joint, Double Weld

(H) Double-Bevel-Groove T-Joint

(I) Lap Joint

Note:
A = 0 in. (0 mm), B = $^1/_{16}$ in. (1.6 mm), C = $^3/_{32}$ in. (2.4 mm),
F = $^1/_8$ in. (3.2 mm), G = $^3/_{16}$ in. (4.8 mm), H = $^1/_4$ in. (6.4 mm), T = thickness.

Figure 2.1—Typical Arc Welded Joint Designs for Magnesium Alloys

Table 2.9
Thickness Limitations for Arc Welded Joints in Magnesium Alloys[a]

Joint Design[b]	Gas Tungsten Arc Welding with						Gas Metal Arc Welding with				Remarks
	ac		dcen		dcep		Short Circuiting Transfer		Spray Transfer		
	t (min)	t (max)	t (min)	t (max)	t (min)	t (max)	t (min)	t (max)	t (min)	t (max)	
A	0.025 in. (0.64 mm)	1/4 in. (6.4 mm)	0.025 in. (0.64 mm)	1/2 in. (12.7 mm)	0.025 in. (0.64 mm)	3/16 in. (4.8 mm)	0.025 in. (0.64 mm)	3/16 in. (4.8 mm)	3/16 in. (4.8 mm)	3/8 in. (9.5 mm)	Single-pass, complete penetration weld. Used on lighter material thicknesses.
B	1/4 in. (6.4 mm)	3/8 in. (9.5 mm)	1/4 in. (6.4 mm)	3/8 in. (9.5 mm)	3/16 in. (4.8 mm)	3/8 in. (9.5 mm)	Not recommended		1/4 in. (6.4 mm)	1/2 in. (12.7 mm)	Multipass complete penetration weld. Used on thick material. On material thicker than suggested maximum, use the double-V-groove weld, butt joint C, to minimize distortion.
C	3/8 in. (9.5 mm)	Note (c)	3/8 in. (9.5 mm)	Note (c)	3/8 in. (9.5 mm)	Note (c)	Not recommended		1/2 in. (12.7 mm)	Note (c)	Multipass complete penetration weld. Used on thick materials. Minimizes distortion by equalizing shrinkage stress on both sides of joint
D	0.040 (1.0 mm)	1/4 in. (6.4 mm)	0.040 (1.0 mm)	1/4 in. (6.4 mm)	0.040 (1.0 mm)	1/4 in. (6.4 mm)	1/16 (1.6 mm)	3/16 in. (4.8 mm)	3/16 in. (4.8 mm)	1/2 in. (12.7 mm)	Single-pass complete penetration weld. For material thicker than suggested maximum, use single-bevel-groove corner joint E. It requires less welding, especially if a square corner is required.
E	3/16 in. (4.8 mm)	Note (c)	3/16 in. (4.8 mm)	Note (c)	3/16 in. (4.8 mm)	Note (c)	Not recommended		1/4 in. (6.4 mm)	Note (c)	Single or multipass complete penetration weld. Used on thick materials to minimize welding. Produces square joint corners.
F	0.025 in. (0.64 mm)	1/4 in. (6.4 mm)	0.025 in. (0.64 mm)	1/2 in. (12.7 mm)	0.025 in. (0.64 mm)	5/32 (4.0 mm)	1/16 (1.6 mm)	5/32 (4.0 mm)	5/32 (4.0 mm)	3/8 in. (9.5 mm)	Single-welded T-joint. Suggested thickness limits based on 40% joint penetration.
G	1/16 (1.6 mm)	3/16 in. (4.8 mm)	1/16 (1.6 mm)	3/8 in. (9.5 mm)	1/8 (3.2 mm)	1/8 (3.2 mm)	1/16 (1.6 mm)	3/32 (2.4 mm)	5/32 (4.0 mm)	3/4 (19 mm)	Double-welded T-joint. Suggested thickness limits based on 100% joint penetration.
H	3/16 in. (4.8 mm)	Note (c)	3/8 in. (9.5 mm)	Note (c)	1/8 (3.2 mm)	Note (c)	Not recommended		3/8 in. (9.5 mm)	Note (c)	Double-welded T-joint. Used on thick material requiring 100% joint penetration.
J	0.040 (1.0 mm)	Note (c)	0.040 (1.0 mm)	Note (c)	0.025 in. (0.64 mm)	Note (c)	0.040 (1.0 mm)	5/32 (4.0 mm)	5/32 (4.0 mm)	Note (c)	Single- or double-welded lap joint. Strength depends upon size of fillet welds. Maximum strength in tension on double-welded joints is obtained when lap equals five times thickness of thinner member.

a. Based on good welding practices and the ability to gas tungsten arc weld with 300 A of ac or DCEN power or with 125 A of DCEP power, as well as gas metal arc weld with 400 A of DCEP power.

b. Refer to Figure 2.1 for the appropriate joint design.

c. Thickest material in commercial use may be welded this way.

Table 2.10
Recommended Filler Metals for Arc Welding Magnesium Alloys

Alloys	ER AZ61A	ER AZ92A	ER EZ33A	ER AZ101A	Base Metal
		Recommended Filler Metal*			
		Wrought Alloys			
AZ10A	X	X			
AZ31B	X	X			
AZ61A	X	X			
AZ80A	X	X			
ZK21A	X	X			
HK31A			X		
HM21A			X		
HM31A			X		
M1A					X
		Cast Alloys			
AM100A	X			X	X
AZ63A	X			X	X
AZ81A	X			X	X
AZ91C	X			X	X
AZ92A	X			X	X
EK41A			X		X
EZ33A			X		X
HK31A			X		X
HZ32A			X		X
K1A			X		X
QH21A			X		X
ZE41A			X		X
ZH62A			X		X
ZK51A			X		X
ZK61A			X		X

* Refer to ANSI/AWS A5.19, *Specification for Magnesium Alloy Welding Electrodes and Rods*, for additional information.

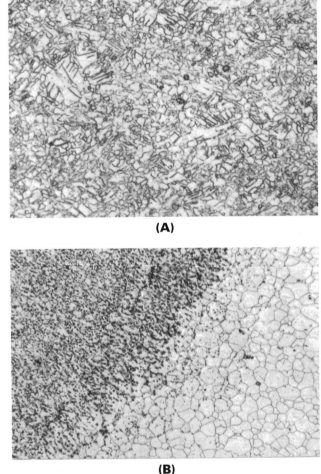

(A)

(B)

Figure 2.2—(A) Microstructure of 1/8 in. (3.18 mm) Thick AZ31B-H24 Magnesium Sheet (x250), (B) Microstructure of Weld Interface between the AZ31B Heat-Affected Zone and AZ61A Weld Metal (x100) [Base Metal to Right, Weld to Left]

used for welding aluminum-bearing magnesium alloys because of severe weld cracking problems.

Casting repairs should be made with a filler metal of the same composition as the base metal when good color match, minimum galvanic effects, or good response to heat treatment is required. For these and other unusual service requirements, the material supplier should be consulted for additional information.

Typical base metal and weld interface microstructures of AZ31B-H24, HK31A-H24, HM21A-T8, and HM31A-T5 alloys are shown in Figures 2.2 through 2.5. Three of these alloys, excepting HM21A, which had been recrystallized prior to welding, showed a significant amount of recrystallization and grain growth in the heat-affected zones.

Radiographs of welds in alloys containing rare earths and thorium will often show segregation along the edges of the weld metal. This segregation is caused by incipient melting in the base metal. A white line will show along the weld interface because of the x-ray absorption characteristics of the rare earths and thorium segregated there. This type of microstructure is shown in Figure 2.6.

PREHEATING

THE NEED TO preheat the components prior to welding is largely determined by the product form, section thickness, and the degree of restraint on the joint. Thick sections may not require preheating unless the joint restraint is high. Thin sections and highly restrained joints require preheat to avoid weld cracking. This is particularly true of alloys high in zinc.

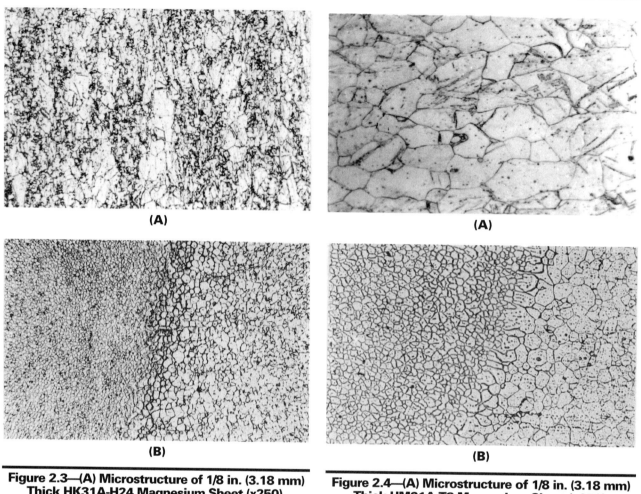

Figure 2.3—(A) Microstructure of 1/8 in. (3.18 mm) Thick HK31A-H24 Magnesium Sheet (x250), (B) Microstructure of Weld Interface Between the HK31A Heat-Affected Zone and EZ33A Weld Metal (x100) [Base Metal to Right, Weld to Left]

Figure 2.4—(A) Microstructure of 1/8 in. (3.18 mm) Thick HM21A-T8 Magnesium Sheet (x250), (B) Microstructure of Weld Interface Between the HM21A Heat-Affected Zone and EZ33A Weld Metal (x100) [Base Metal to Right, Weld to Left]

Recommended preheat temperature ranges for cast magnesium alloys are given in Table 2.11. The maximum preheat temperature should not exceed the solution heat-treating temperature for the alloy. Otherwise, the mechanical properties of the weldment may be altered significantly.

The method of preheating will depend upon the component size. Furnace heating is preferred, but large components may have to be preheated locally. The welding fixture also may have to be heated when joining thin sections to maintain an acceptable interpass temperature.

An air circulating furnace with a temperature control of ±10 °F (±5.6 °C) is recommended for preheating of castings. The furnace temperature should not cycle above the maximum temperature indicated in Table

2.11. A temperature-limit control set at 10 °F above (5.6 °C above) the maximum acceptable temperature should be provided to override the automatic controls. Solution heat-treated castings, or solution heat-treated and aged castings, can be charged into a furnace operating at the preheat temperature without damage. They should remain in the furnace until they are uniformly heated throughout.

Welding should proceed immediately after the castings are removed from the furnace. It should be discontinued if the temperature of the castings drops below the acceptable minimum preheat. In that case, the castings should be reheated in the furnace before proceeding with the welding operation.

Castings can be cooled in still, ambient air after welding without danger of cracking. However, castings of intricate

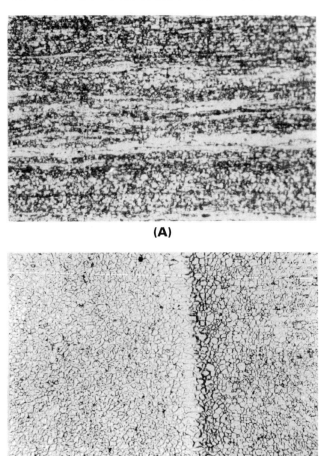

(A)

(B)

Figure 2.5—(A) Microstructure of 1/8 in. (3.18 mm) Thick HM31A-T5 Magnesium Extrusion (x500), (B) Microstructure of Weld Interface Between the HM31A Heat-Affected Zone and EZ33A Weld Metal (x100) [Base Metal to Right, Weld to Left]

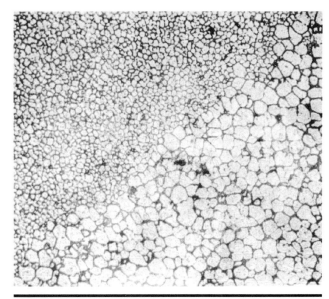

Figure 2.6—Microstructure of the Weld Interface in an EZ33A Magnesium Alloy Casting Showing Pools of a Eutectic (x100) [Base Metal is on the Right]

design should be cooled more slowly to room temperature to avoid distortion from nonuniform cooling.

GAS TUNGSTEN ARC WELDING

GAS TUNGSTEN ARC welding is used for joining magnesium components and repair of magnesium castings. It is well suited for welding thin sections. Control of heat input and the molten weld pool is better than with gas metal arc welding.

Welding Current

MAGNESIUM ALLOYS ARE welded by this process using techniques and equipment similar to those used for aluminum. They may be welded with alternating current or direct current. Alternating current is preferred because of the good arc-cleaning action. Conventional ac power of 60 Hz with arc stabilization or square-wave alternating current may be used. With square-wave alternating current, the electrode positive and negative periods are adjustable within limits. This type of power can provide adequate cleaning action as well as good joint penetration and arc stability. A section through a weld made with 60 Hz ac power is shown in Figure 2.7(A).

Direct current power with the electrode positive (DCEP) provides an arc with excellent cleaning action. However, it can only be used to weld thin sections because the welding current is limited by heating of the tungsten electrode. Joint penetration tends to be wide and shallow, as shown in Figure 2.7(B). Welds in relatively thick sections are typified by low welding speeds, wide bead faces, and wide heat-affected zones with large grain size.

Direct current electrode negative (DCEN) power is not commonly used for welding magnesium alloys because of the absence of arc-cleaning action. However, this type of power is sometimes used for mechanized welding of square-groove butt joints in sections up to 0.25 in. (6.4 mm) thickness. Careful preweld cleaning and good fit-up are needed to produce sound welds. DCEN power with helium shielding can produce narrow, deep joint penetration, as shown in Figure 2.7(C).

Table 2.11
Recommended Weld Preheat and Postweld Heat Treatments for Cast Magnesium Alloys

Alloy	Metal Temper Before Welding	Desired Temper After Welding	Weld Preheat*	Postweld Heat Treatment*
AZ63A	T4	T4	Heavy and unrestrained sections; none or local. Thin and restrained sections: 350-720 °F (177-382 °C)	1.5 h at 730 °F (388 °C)
	T4 or T6	T6	Same as for T4 above	1.5 h at 730 °F (388 °C) plus 5 h at 425 °F (219 °C)
	T5	T5	Heavy and unrestrained sections: none or local. Thin and restrained sections: None to 500 °F (260 °C) [1.5 h max at 500 °F (260 °C)]	5 h at 425 °F (219 °C)
AZ81A	T4	T4	Heavy and unrestrained sections: none or local. Thin and restrained sections: 350-750 °F (177-399 °C)	0.5 h at 780 °F (416 °C)
AZ91C	T4	T4	Same as for AZ81A-T4 above	0.5 h at 780 °F (416 °C)
	T4 or T6	T6	Same as for AZ81A-T4 above	1.5 h at 780 °F (416 °C) plus either 4 h at 420 °F (216 °C) or 16 h at 335 °F (168 °C)
AZ92A	T4	T4	Same as for AZ81A-T4 above	0.5 h at 770 °F (410 °C)
	T4 or T6	T6	Same as for AZ81A-T4 above	0.5 h at 770 °F (410 °C) plus either 4 h at 500 °F (260 °C) or 5 h at 425 °F (219 °C)
AM100A	T6	T6	Same as for AZ81A-T4 above	0.5 h at 780 °F (416 °C) plus 5 h at 425 °F (219 °C)
EK41A	T4 or T6	T6	None to 500 °F (260 °C) [1.5 h max at 500 °F (260 °C)]	16 h at 400 °F (204 °C)
	T5	T5	None to 500 °F (260 °C) [1.5 h max at 500 °F (260 °C)]	16 h at 400 °F (204 °C)
EZ33A	F or T5	T5	None to 500 °F (260 °C) [1.5 h max at 500 °F (260 °C)]	5 h at 420 °F (216 °C); or 2 h at 625 °F (329 °C) plus 5 h at 420 °F (216 °C)
HK31A	T4 or T6	T6	None to 500 °F (260 °C)	16 h at 400 °F (204 °C); or 1h at 600 °F (316 °C) plus 16 h at 400 °F (204 °C)
HZ32A	F or T5	T5	None to 500 °F (260 °C)	16 h at 600 °F (316 °C)
K1A	F	F	None	None
ZE41A	F or T5	T5	None to 600 °F (316 °C)	2 h at 625 °F (329 °C); or 2 h at 625 °F (329 °C) plus 16 h at 350 °F (177 °C)
ZH62A	F or T5	T5	None to 600 °F (316 °C)	16 h at 480 °F (249 °C); or 2 h at 625 °F (329 °C) plus 16 h at 350 °F (177 °C)
ZK51A	F or T5	T5	None to 600 °F (316 °C)	16 h at 350 °F (177 °C); or 2 h at 625 °F (329 °C) plus 16 h at 350 °F (177 °C)
ZK61A	F or T5	T5	None to 600 °F (316 °C)	48 h at 300 °F (149 °C)
	T4 or T6	T6	None to 600 °F (316 °C)	2 to 5 h at 930 °F (499 °C) plus 48 h at 265 °F (129 °C)

* Temperatures shown are maximum allowable; furnace controls should be set so temperature does not cycle above maximum.

(A) AC Power

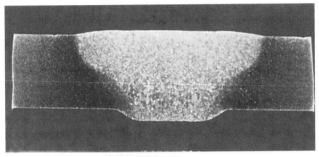

(B) DCEP Power

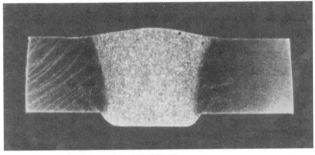

(C) DCEN Power

Figure 2.7—Cross Sections through Gas Tungsten Arc Welds Made with Different Types of Power in 3/16 in. (4.8 mm) AZ31B Magnesium Alloy

Shielding Gases

ARGON, HELIUM, AND mixtures of these gases can be used. The factors governing the selection of the shielding gas for magnesium alloys are the same as those for other metals, particularly aluminum.

Electrodes

PURE TUNGSTEN (EWP), tungsten-thoria (EWTh-1 or -2), and tungsten-zirconia (EWZr) electrodes can be used with magnesium alloys. The selection depends primarily upon the type of welding power and the welding amperage to be used. The tungsten-thoria electrodes should be restricted to use with dc power.

Welding Conditions

TYPICAL CONDITIONS FOR manual gas tungsten arc welding of butt joints in magnesium alloys are given in Table 2.12. Conditions for automatic gas tungsten arc welding butt joints in two thicknesses of AZ31B magnesium alloy are shown in Table 2.13. The welding machine should produce balanced ac power for good operating characteristics. These data may be used as guides for establishing joint welding procedures for a specific application.

Welding of Castings

GAS TUNGSTEN ARC welding is recommended for magnesium alloy castings. Welding is limited to the repair of defects in clean metal including broken sections, sand or blow holes, cracks, and cold shuts. Repair welding is not recommended in areas containing gross porosity or inclusions of oxide or flux. Castings that have been organically impregnated for pressure tightness or that may contain oil in pores should not be welded. Many castings are parts of aircraft structures that are heat treated to meet strength requirements. These castings must be heat treated again if they are welded.

Factors that need to be considered when welding castings include the type of alloy, previous thermal treatment, size and intricacies of sections, and degree of restraint. Alloy composition can be identified by designation markings on the castings. If not, a chemical or spectrographic analysis should be made.

Castings can be welded in the as-cast, solution heat-treated, or solution heat-treated and aged condition. However, the welding of some alloys in the as-cast condition is not recommended because of the greater risk of cracking and the possibility of grain growth in the weld zone during the long solution heat treating times required. The heat-treated condition of the casting before welding may influence the preheat temperature selection.

Castings should be stripped of paint and degreased before welding. Conversion coatings should be removed from around defective areas with stainless steel wool or a wire brush. A rotary deburring tool is recommended for removing defects and preparing the area for welding. Broken pieces should be clamped in position for welding. The appropriate joint preparation should be determined from Table 2.12. Where large holes or defective areas are to be filled with weld metal, a backing strip can be used to prevent excessive melt-through.

The casting should be preheated if the section to be repaired is relatively thick or highly restrained by surrounding structure. Welding of broken pieces should

Table 2.12
Typical Conditions for Manual Gas Tungsten Arc Welding of Magnesium Alloys

Thickness		Joint Design[1]	No. of Passes	Welding Current (ac), A	Pure Tungsten (EWP) Electrode Diameter		Argon Flow[2]		Welding Rod Diameter	
in.	mm				in.	mm	ft³/h	L/min	in.	mm
0.040	1.0	A	1	35	1/16	1.6	12	5.7	3/32	2.4
0.063	1.6	A	1	50	3/32	2.4	12	5.7	3/32	2.4
0.080	2.0	A	1	75	3/32	2.4	12	5.7	3/32	2.4
0.100	2.5	A	1	100	3/32	2.4	12	5.7	3/32	2.4
0.125	3.2	A	1	125	3/32	2.4	12	5.7	1/8	3.2
0.190	4.8	A	1	160	1/8	3.2	15	7.1	1/8	3.2
0.250	6.4	B	2	175	5/32	4.0	20	9.4	1/8	3.2
0.375	9.5	B	3	175	5/32	4.0	20	9.4	5/32	4.0
0.375	9.5	C	2	200	3/16	4.8	20	9.4	1/8	3.2
0.500	12.7	B	3	175	5/32	4.0	20	9.4	5/32	4.0
0.500	12.7	C	2	250	3/16	4.8	20	9.4	1/8	3.2

1. A - Square-groove butt joint, 0 root opening
   B - Single-V-groove butt joint, 1/16 in. (1.6 mm) root face, 0 root opening, 60° minimum included V-bevel
   C - Double-V-groove butt joint, 3/32 in. (2.4 mm) root face, 0 root opening, 60° minimum included V-bevel

2. Helium shielding will reduce the welding current about 20 to 30 A. Thorium-bearing alloys will require about 20% higher current.

Table 2.13
Conditions for Automatic Gas Tungsten Arc Welding of Square-Groove Butt Joints in AZ31B Magnesium Alloy

Type of Power	Welding Speed		Welding[a] Current, A	Filler Metal Feed Rate[b]		Electrode Diameter[c]		Arc Length	
	in./min	mm/s		in./min	mm/s	in.	mm	in.	mm
			Thickness = 0.063 in. (1.6 mm)						
AC (balanced wave)	12	5.1	55	35	14.8	3/32	2.4	0.025	0.6
	24	10.2	60	50	21.2	3/32	2.4	0.025	0.6
	36	15.2	70	54	22.9	1/8	3.2	0.025	0.6
	45	19.1	95	96	40.6	1/8	3.2	0.025	0.6
	70	29.6	170	160	67.7	3/16	4.8	0.025	0.6
	80[d]	33.9	195	190	80.4	3/16	4.8	0.025	0.6
	95[d]	40.2	200	203	85.9	3/16	4.8	0.025	0.6
			Thickness = 0.190 in. (4.8 mm)						
DCEN	48	20.3	75	80	33.9	1/8	3.2	0.025	0.6
DCEP	80[d]	33.9	120	184	77.9	1/4	6.4	0.020	0.5
AC (bal. wave)	34[d]	14.4	300	159	67.3	1/4	6.4	0.020	0.5
DCEN	20[d]	8.5	170	70	29.6	1/8	3.2	0.030	0.8
DCEP	7[d]	3.0	120	10[e]	4.2	1/4	6.4	0.020	0.5

a. With helium shielding

b. AZ61A or AZ92A filler metal of 1/16 in. (1.6 mm) diameter except where noted (see Note e)

c. Pure or zirconia-tungsten for ac; thoria-tungsten for dc power

d. Maximum speed for arc stability and freedom from undercutting

e. Filler metal diameter is 3/32 in. (2.4 mm).

commence at the center of the joint and progress toward the ends. Medium-size weld beads are preferred. Low welding current may cause cold laps, oxide contamination, or porous welds. High welding current may cause weld cracking or incipient melting in the heat-affected zone.

The filling of holes is the most critical type of repair from the standpoint of cracking. The arc should be struck at the bottom of the hole and welding should progress upward. The arc should not be held too long in one area to avoid the possibility of weld cracking or incipient melting in the heat-affected zone. The arc should be extinguished by gradually reducing the welding current to zero with appropriate current controls. This will permit the molten weld pool to solidify slowly and avoid crater cracking.

GAS METAL ARC WELDING

THE FUNDAMENTAL PRINCIPLES for gas metal arc welding (GMAW) of magnesium alloys are the same as for other metals. Welding can be done with this process at speeds that are two to three times faster than those with gas tungsten arc welding. Higher welding speeds reduce the heat input which, in turn, results in less distortion and some improvement in the tensile yield strength of the joint. The higher filler-metal deposition rates reduce welding time and fabrication costs.

Shielding Gases

ARGON SHIELDING IS used for GMAW. Occasionally, mixtures of argon and helium are used to aid filler metal flow and alter the arc characteristics for deeper joint penetration. Pure helium is undesirable for shielding because it raises the current required for spray arc transfer and increases weld spatter.

Metal Transfer

TYPICAL MELTING RATES for standard sizes of magnesium alloy electrodes using DCEP power are given in Figure 2.8. This figure shows the relationship between electrode feed rate and welding current for each size. It also illustrates the respective operating ranges for the three types of metal transfer used for GMAW. These are the short circuiting, pulsed spray, and spray transfer modes. The pulsed-spray operating region lies between the spray and short circuiting transfer regions. Without pulsing, the welding amperages between the short circuiting and spray transfer ranges would produce highly unstable globular transfer, which is not suitable for welding magnesium alloys.

Like short circuiting transfer, spray transfer is only stable over a limited welding current range. Excessive welding current causes arc turbulence, which must be avoided. The approximate arc voltage ranges corresponding to each type of metal transfer are 13 to 16 V for short circuiting transfer, 17 to 25 V for pulsed spray transfer, and 24 to 30 V for spray transfer.

Equipment

THE EQUIPMENT USED for GMAW of magnesium alloys is similar to that used for other nonferrous alloys. An appropriate power source is used to produce the desired method of filler metal transfer. A constant current power source may be preferred for spray transfer at the lower end of the recommended current range for the applicable electrode size. It will minimize weld spatter. A wire feeder with a "touch-start" or "slow run-in" feature is normally used with the constant current power. The power source for pulsed spray welding must be designed to produce two current levels. Spray transfer takes place during the periods of high current and ceases between them when the current is low.

Welding Conditions

TYPICAL CONDITIONS FOR gas metal arc welding various thicknesses of magnesium alloys are given in Table 2.14. These may be used as a guide when establishing welding conditions for a specific application. The short circuiting transfer mode is used for thin sections, and the spray transfer mode for thick sections. Pulsed spray transfer is recommended for the intermediate thicknesses because there is less heat input than with continuous spray transfer.

Recommended electrode sizes for welding various thicknesses of magnesium alloys are given in Table 2.15. With both spray and pulsed spray transfer, the lowest welding cost is achieved with the largest applicable electrode. With short circuiting transfer, only one or two electrode sizes can be used to produce welds with good fusion and joint penetration.

Spot Welding

ARC SPOT WELDING can be used to join magnesium sheet and extrusions in a variety of thicknesses. Welding schedules for suitable thickness combinations of AZ31B alloy are given in Table 2.16. These may be used as guides with other alloys.

Commercially available gas metal arc spot welding equipment is suitable for magnesium alloys. A constant potential power source, DCEP power, and argon shielding are recommended. The strength of gas metal arc spot welds may meet or exceed the strength of

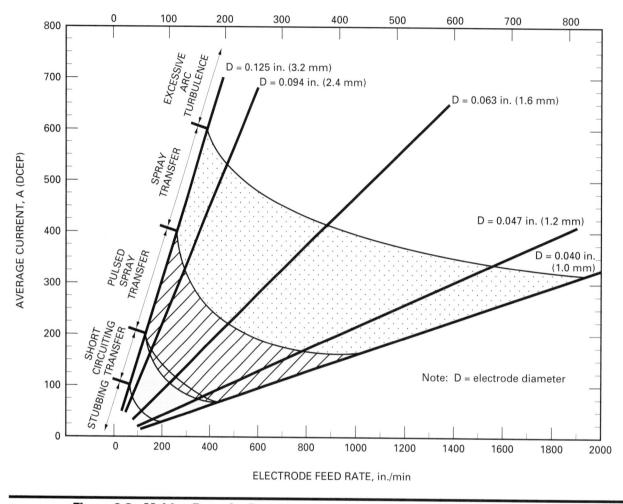

Figure 2.8—Melting Rates for Bare Magnesium Alloy Electrodes with Argon Shielding

resistance spot welds. Postweld stress relief of gas metal arc spot welds is required on all alloys sensitive to stress corrosion cracking.

STRESS RELIEVING

HIGH RESIDUAL STRESSES from welding or forming will promote stress-corrosion cracking in magnesium alloys that contain more than about 1.5% aluminum. Thermal treatments are used with these alloys to reduce residual stresses to safe levels to avoid this problem. Other magnesium alloys do not appear to be sensitive to this type of cracking.

Stress-corrosion cracking in welded structures usually occurs in the area adjacent to the weld bead. It is almost always a transcrystalline type of crack. Cracking may be delayed somewhat by painting. However, this will not ensure crack-free service for long periods and should not be substituted for stress relieving of the weldment.

Stress relieving can be accomplished either in a furnace or with a torch. Furnace stress relieving is preferred. The time and temperature necessary to stress relieve weldments of the various alloys and product forms are shown in Table 2.17. When a furnace is used, a fixture should be used to support the weldment during heating to prevent distortion and correct any warpage. The temperature of large weldments should be monitored with thermocouples to make certain that all

Table 2.14
Typical Conditions for Gas Metal Arc Welding of Magnesium Alloys[a]

Thickness		Joint[b] Design	No. of Weld Passes	Electrode						Voltage		Welding[c] Current (DCEP), A
				Diameter			Feed Rate					
in.	mm			in.	mm	in./min	mm/s			Pulse	Arc	
Short Circuiting Transfer												
0.025	0.6	A	1	0.040	1.0	140	59.3			—	13	25
0.040	1.0	A	1	0.040	1.0	230	97.4			—	14	40
0.063	1.6	A	1	0.063	1.6	185	78.3			—	14	70
0.090	2.3	A	1	0.063	1.6	245	104			—	16	95
0.125	3.2	B	1	0.094	2.4	135	57.2			—	14	115
0.160	4.1	B	1	0.094	2.4	165	69.9			—	15	135
0.190	4.8	B	1	0.094	2.4	205	86.8			—	15	175
Pulsed Spray Transfer												
0.063	1.6	A	1	0.040	1.0	360	152			55	21	50
0.125	3.2	A	1	0.063	1.6	280	119			55	24	110
0.190	4.8	A	1	0.063	1.6	475	201			52	25	175
0.250	6.4	C	1	0.094	2.4	290	123			55	29	210
Spray Transfer												
0.250	6.4	C	1	0.063	1.6	530	224			—	27	240
0.375	9.5	C	1	0.094	2.4	285-310	121-131			—	24-30	320-350
0.500	12.7	C	2	0.094	2.4	320-360	135-152			—	24-30	360-400
0.625	15.9	D	2	0.094	2.4	330-370	140-157			—	24-30	370-420
1.000	25.4	D	4	0.094	2.4	330-370	140-157			—	24-30	370-420

a. Argon shielding flow rate is 40 to 60 ft³/h (18.9 to 28.3 L/min) for short circuiting and pulsed spray; 50-80 ft³/h (23.6 to 37.8 L/min) for spray transfer. Arc travel speed is 24 to 36 in./min (258 to 387 mm/s). These conditions may also be used for fillet welds in thicknesses of 0.25 to 1.0 in. (6.4 to 25.4 mm).

b. A - Square groove, no root opening
   B - Square groove, 0.09 in. (2.3 mm) root opening
   C - Single-V-groove, 0.06 in. (1.5 mm) root opening, 60° included-V
   D - Double-V-groove, 0.13 in. (3.3 mm) root opening, 60° included-V

c. Average amperage with pulsed spray transfer

sections reach the proper temperature. In torch stress relieving, a temperature-indicating device should be used to avoid overheating.

POSTWELD HEAT TREATMENT

WELDED CASTINGS ARE heat treated to obtain desired properties. The appropriate postweld heat treatment depends upon the temper of the casting before welding and the desired temper after welding, as shown in Table 2.11. Because of the fine grain size and extensive dispersion of the precipitates in the weld zone, aluminum-bearing castings in the T4 or T6 condition may be solution heat treated for relatively short heating times after welding. In the case of AM100A, AZ81A, AZ91C, and AZ92A alloy castings, the solution heat-treating time must not exceed 30 minutes at

temperature to avoid excessive grain growth in the weld zone. A protective atmosphere must be used when the solution treating temperature is above 750 °F (399 °C) to prevent oxidation and active burning of the weldment.

The postweld heat treatments specified for the various alloys will produce the best weldment properties and also stress relieve castings to prevent cracking. If a postweld solution or temper heat treatment is not required, aluminum-bearing castings should be stress relieved.

WELD PROPERTIES

TYPICAL TENSILE STRENGTH properties at room and elevated temperatures are given in Tables 2.18 and 2.19, respectively, for gas tungsten arc welds in

Table 2.15
Recommended Electrode Sizes for Gas Metal Arc Welding of Magnesium Alloys

Electrode Diameter		Applicable Base Metal Thickness Range					
		Short Circuiting Transfer		Pulsed Spray Transfer*		Spray Transfer*	
in.	mm	in.	mm	in.	mm	in.	mm
0.040	1.0	0.03-0.06	0.8-1.5	0.06-0.09	1.5-2.3	0.16-0.25	4.1-6.4
0.045	1.1	0.04-0.07	1.0-1.8	0.07-0.12	1.8-3.0	0.19-0.25	4.8-6.4
0.063	1.6	0.06-0.09	1.5-2.3	0.10-0.25	2.5-6.4	0.20-.030	5.1-7.6
0.094	2.4	0.09-0.19	2.3-4.8	0.20-0.31	5.1-7.9	≥ 0.30	≥ 7.6

* Pulsed spray and spray transfer thickness schedules should provide good welding characteristics at minimum filler metal cost.

Table 2.16
Typical Conditions for Gas Metal Arc Spot Welding of AZ31B Magnesium Alloy Sheet

Sheet Thickness				Electrode Diameter[a]		Welding Current (DCEP), A	Arc Voltage, V	Weld Time, Cycles[b]	Shear Strength	
Front		Back								
in.	mm	in.	mm	in.	mm				lb/Spot	N/Spot
0.040	1.0	0.090	2.3	0.040	1.0	175	22-26	30-70	40-1085	178-4826
						200	24-26	25-50	760-1190	3405-5295
		0.125	3.2	0.040	1.0	175	22-26	40-100	310-1370	1379-6138
						200	22-28	20-90	460-1710	2046-7606
		0.190	4.8	0.040	1.0	175	22-26	40-100	50-1265	222-5627
						200	22-28	20-100	360-1725	1601-7672
0.063	1.6	0.063	1.6	0.040	1.0	200	26	50-60	348-889	1548-3954
		0.090	2.3	0.040	1.0	200	22-26	20-80	250-710	1112-3158
						225	24-26	25-45	360-850	1601-3781
		0.125	3.2	0.040	1.0	200	22-26	30-100	515-1165	2291-5182
						225	22-28	20-100	230-1340	1023-5960
		0.250	6.4	0.040	1.0	200	22-26	30-100	250-880	1112-3914
						225	22-28	20-100	320-2000	1423-8896
0.090	2.3	0.090	2.3	0.040	1.0	250	24-28	50-100	513-1129	2282-5022
				0.063	1.6	275	25-28	40-90	314-1078	1397-4795
		0.125	3.2	0.063	1.6	275	22-26	30-100	520-1060	2313-4715
						300	24-28	25-80	230-960	1023-4271
		0.190	4.8	0.063	1.6	275	22-26	30-100	290-1700	1290-7562
						300	22-28	20-100	200-1700	890-7562
0.125	3.2	0.125	3.2	0.094	2.4	325	24-27	40-100	583-1675	2593-7450
						350	24-25	40-100	680-1337	3024-5947
		0.156	4.0	0.094	2.4	350	22-26	30-100	530-1875	2357-8340
						375	24-26	30-80	640-1600	2847-7116
		0.190	4.8	0.094	2.4	350	22-24	30-100	190-1900	845-8451
						375	24-28	30-100	270-1640	1200-7295
0.156	4.0	0.156	4.0	0.094	2.4	375	24-26	80-150	517-1437	2300-6392
0.190	4.8	0.190	4.8	0.094	2.4	375	24-26	80-150	782-1323	3479-5885
						400	24-26	60-110	853-1194	3794-5311

a. AZ61A electrode

b. 60 Hz

Table 2.17
Recommended Stress-Relieving Heat Treatments for Magnesium Alloys

	Castings				Sheet				Extrusions		
	Temperature		Time,		Temperature		Time,		Temperature		Time,
Alloy	°F	°C	min	Alloy	°F	°C	min	Alloy	°F	°C	min
AM100A	500	260	60	AZ31B-O	500	260	15	AZ10A-F	500	260	15
AZ63A	500	260	60	AZ31B-H24	300	149	60	AZ31B-F	500	260	15
AZ81A	500	260	60	M1A-O	500	260	15	AZ61A-F	500	260	15
AZ91C	500	260	60	M1A-H24	400	204	60	AZ80A-F	500	260	15
AZ92A	500	260	60	HK31A-H24	600	316	30	AZ80A-T5	400	204	60
—	—	—	—	HM21A-T81	750	399	30	HM31A-T5	800	427	60
—	—	—	—	—	—	—	—	M1A-F	500	260	15
—	—	—	—	—	—	—	—	ZK21A-F	400	204	60
—	—	—	—	—	—	—	—	ZK60A-F	500	260	15
—	—	—	—	—	—	—	—	ZK60A-T5	300	149	60

Table 2.18
Typical Tensile Properties of Gas Tungsten Arc Welds in Magnesium Alloys at Room Temperature

Alloy and Temper	Filler Metal	Tensile Strength		Yield Strength*		Elongation in 2 in. (51 mm), %	Joint Efficiency, %
		ksi	MPa	ksi	MPa		
			Sheet				
AZ31B-O	AZ61A, AZ92A	35-36	241-248	17-19	117-131	10-11	95-97
AZ31B-H24	AZ61A, AZ92A	36-37	248-255	19-22	131-152	5	86-88
HK31A-H24	EZ33A	31-32	214-221	20-22	138-152	2-4	82-84
HM21A-T8	EZ33A	28-31	193-214	19-20	131-138	2-4	80-89
ZE41A-T5	ZE41A	30	207	20	138	4	100
ZH62A-T5	ZH62A	38	262	25	172	5	95
			Extrusions				
AZ10A-F	AZ61A, AZ92A	32-33	221-228	15-18	103-124	6-9	91-94
AZ31B-F	AZ61A, AZ92A	36-37	248-255	19-22	131-152	5-7	95-97
AZ61A-F	AZ61A, AZ92A	38-40	262-276	21-24	145-165	6-7	84-89
AZ80A-F	AZ61A	36-40	248-276	22-26	152-179	3-5	74-82
AZ80A-T5	AZ61A	34-40	234-276	24-28	165-193	2	62-73
HM31A-T5	EZ33A	28-31	193-214	19-24	131-165	1-2	64-70
ZK21A-F	AZ61A, AZ92A	32-34	221-234	17	117	4-5	76-81
			Castings				
AZ63A-T6	AZ92A, AZ101A	31	214	—	—	2	77
AZ81A-T4	AZ101A	34	234	13	90	8	85
AZ91C-T6	AZ101A	35	241	16	110	2	87
AZ92A-T6	AZ92A	35	241	21	145	2	87
EZ33A-T5	EZ33A	21	145	16	110	2	100
HK31A-T6	HK31A	29	200	16	110	9	94
HZ32A-T5	HZ32A	29	200	17	117	5	97
K1A-F	EZ33A	23	159	8	55	10	100

* 0.2% offset in a 2 in. (51 mm) gauge length

Table 2.19
Typical Elevated-Temperature Tensile Properties of Gas Tungsten Arc Welds in Magnesium Alloys[a]

Alloy[b]	Filler Metal	Test Temperature		Tensile Strength		Yield Strength[d]		Elongation in 2 in. (51 mm), %	Joint Efficiency, %
		°F	°C	ksi	MPa	ksi	MPa		
					Sheet				
HK31A-H24	EZ33A	400	204	21	145	13	90	18	88
		600	316	13	90	10	69	24	100
HM21A-T8	EZ33A	400	204	18	124	12	83	16	100
		600	316	14	97	10	69	14	93-100
					Extrusions				
HM31A-T5	EZ33A	400	204	21	144	12	83	22	87
		600	316	13	90	9	62	27	72
					Castings				
EZ33A-T5	EZ33A	400	204	19	131	11	76	13	90
		600	316	11	76	7	48	50	92
HK31A-T6	HK31A[c]	400	204	18	124	11	76	33	82
		600	316	15	103	9	62	25	79
HZ32A-T5	HZ32A[c]	400	204	19	131	13	90	33	100
		600	316	12	83	10	69	26	92

a. Weld reinforcement removed.

b. Alloys designed for elevated-temperature service.

c. EZ33A filler metal will give equivalent joint strengths.

d. 0.2% offset in a 2 in. (51 mm) gauge length.

wrought and cast magnesium alloys. Properties of joints made by gas metal arc welding are similar to or slightly higher than these strengths because of the reduced heat input. Table 2.20 gives the tensile properties of magnesium alloy weld metals produced by several filler metal and base metal combinations. The strengths of welds in most magnesium alloys are near those of the base metals. This may be shown by comparing the tensile strength data in Tables 2.18 and 2.20 for welded joints with similar data for the base metals in Table 2.5.

Table 2.20
Tensile Properties of Magnesium Alloy Weld Metals Produced From Various Filler Metal and Base Metal Combinations

Filler Metal	Base Metal	Ultimate Tensile Strength		Tensile Yield Strength*		Elongation in 2 in. (51 mm), %
		ksi	MPa	ksi	MPa	
AZ61A	AZ31B	34.3	236	14.5	100	10.0
AZ92A	AZ31B	36.8	254	18.9	130	8.0
EZ33A	HK31A	32.0	221	17.8	123	9.0
EZ33A	HM31A	26.8	185	19.8	137	3.5
EZ33A	HM21A	30.0	207	21.2	146	6.3
HK31A	HM31A	26.0	179	13.8	95	10.5
HK31A	HM21A	27.0	186	13.8	95	13.3

* 0.2% offset in a 2 in. (51 mm) gauge length

When the base metal is in the strain-hardened condition, recrystallization and some grain growth will take place in the heat-affected zone during welding. The heat-affected zone will then be weaker than the base metal, and sometimes have lower strength than the weld metal. The latter is due to the fine grain size of the weld metal. The grain sizes of AZ61A, AZ92A, and EZ33A weld metals are shown in Figure 2.9.

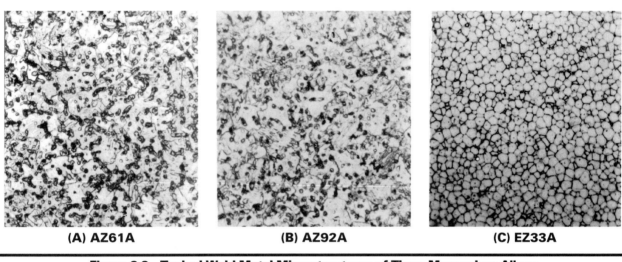

(A) AZ61A   **(B) AZ92A**   **(C) EZ33A**

**Figure 2.9—Typical Weld Metal Microstructures of Three Magnesium Alloys
[x250, Phospho-Picral Etch]**

RESISTANCE WELDING

SPOT WELDING

MAGNESIUM ALLOY SHEET and extrusions can be joined by resistance spot welding in thicknesses ranging from about 0.02 to 0.13 in. (0.5 to 3.3 mm). Alloys recommended for spot welding are M1A, AZ31B, AZ61A, HK31A, HM21A, HM31A, and ZK60A. Spot welding is used for low-stress applications where vibration is low or nonexistent. Magnesium alloys are spot welded using procedures similar to those for aluminum alloys.

Preweld Cleaning

CAREFUL PREWELD CLEANING is essential for the production of spot welds of consistent size and soundness. A uniform electrical surface resistance of about 50 microhms or less is necessary to obtain consistency. Chemical cleaning procedures for spot welding are given in Table 2.8. Chemically cleaned parts will maintain a low, consistent surface resistance for about 100 hours when stored in a clean, dry environment. However, the time between cleaning and welding for critical applications should be limited to 24 hours. Mechanically cleaned surfaces will develop progressively higher, inconsistent surface resistance after 8 to 10 hours. For best results, mechanically cleaned material should be spot welded within this time.

Equipment

BECAUSE OF THE relatively high thermal and electric conductivities of magnesium alloys, high welding currents and short weld times are required for spot welding. Spot welding machines designed for aluminum alloys are suitable for magnesium. As with aluminum, very rapid electrode follow-up is required to maintain pressure on the weld nugget as the metal softens and deforms rapidly. For this reason, low-inertia welding machines should be used. A dual force system is not required for spot welding magnesium alloys. However, dual electrode force is sometimes used to reduce internal discontinuities by applying a higher forging force on the nugget during solidification. Timing of this forging force application is important for it to be beneficial.

Electrodes

SPOT WELDING ELECTRODES for magnesium alloys should be made of RWMA Group A, Class 1 or Class 2 alloy. The faces of the electrodes must be kept clean and smooth to minimize the contact resistance between the electrode and the adjacent part. Cleaning should be done with an electrode dressing tool with the proper face contour covered with a very fine polishing cloth of 280-grit abrasive coarseness.

Electrode life between cleanings is limited by the transfer of copper to the adjacent part and subsequent sticking. The number of welds that can be produced between cleanings depends upon the electrode alloy and cooling efficiency, the method of base metal cleaning, the magnesium alloy composition, and the welding conditions. Table 2.21 shows the relative effectiveness of mechanical and chemical surface preparations on electrode life for some magnesium alloys. Chemical cleaning will give better electrode life than cleaning by wire brushing. The proper cleaning solution must be used for the magnesium alloy to be welded. In any case, the longest electrode life will be obtained when the welding conditions produce a weld nugget no larger than that necessary to meet design strength requirements.

Copper pickup on the spot weld surfaces increases the corrosion susceptibility of magnesium. Therefore, the copper should be completely removed from the surfaces by a suitable mechanical cleaning method. The presence of copper on spot welds can be determined by applying 10% acetic acid solution. A dark spot will form if copper is present on the surface.

Joint Design

THE JOINT DESIGNS for spot welding magnesium alloys are much the same as those for aluminum alloys.

Minimum recommended spot spacing and edge distance for the location of spot welds are given in Table 2.22.

Where two unequal thicknesses are to be spot welded, the thickness ratio should not exceed 2.5 to 1. With three thicknesses, the thickness variation should not exceed about 25 percent, and the thickest section should be in the center.

Welding Schedules

THE FOLLOWING ARE important factors that must be considered when developing a welding schedule:

(1) Dimensions, properties, and characteristics of the alloys to be welded
(2) Type of welding equipment to be employed
(3) Joint design

A welding schedule can be established for any particular combination of these. Typical schedules for spot welding magnesium alloys with four types of equipment are given in Tables 2.23 through 2.26. These data are intended only as guides in establishing schedules for specific applications.

The welding and postheat currents are approximate values. The magnitude of the welding current is adjusted by transformer taps, phase shift heat control, or both. To obtain the required current, simply start with a low value of weld heat and a corresponding percentage of postheat. The current is gradually increased until the desired shear strength, nugget diameter, and penetration are obtained. In some cases, it may be necessary to readjust the weld time to achieve the desired properties.

With single-phase ac equipment, welding current may be determined by primary or secondary measurement methods. The nugget diameter and the minimum indicated shear strength given in Table 2.24 should be

---

Table 2.21
Effect of Surface Preparation on Spot Welding Electrode Life with Magnesium Alloys

| | | No. of Spot Welds[b] | | |
| | | | Spot Weld Cleaner[c] | |
Alloy	Electrode Classification[a]	Wire Brushing	No. 2	No. 3
AZ31B	Class 1	30	over 200	over 550
AZ31B	Class 2	15	50	270
HK31A	Class 2	40	400	195
HM21A	Class 2	5	80	5

a. RWMA Group A

b. Between electrode cleanings

c. Refer to Table 2.8 for solution compositions.

Table 2.22
Suggested Spot Spacing and Edge Distance for Spot Welds in Magnesium Alloys

Thickness*		Spot Spacing				Edge Distance			
		Minimum		Nominal		Minimum		Nominal	
in.	mm	in.	mm	in.	mm	in.	mm	in.	mm
0.020	0.5	0.25	6.4	0.50	12.7	0.22	5.6	0.31	7.9
0.025	0.6	0.25	6.4	0.50	12.7	0.22	5.6	0.31	7.9
0.032	0.8	0.31	7.9	0.62	15.7	0.25	6.4	0.36	9.1
0.040	1.0	0.38	9.7	0.75	19.0	0.28	7.1	0.38	9.7
0.050	1.3	0.44	11.2	0.80	20.3	0.31	7.9	0.41	10.4
0.063	1.6	0.50	12.7	1.00	25.4	0.38	9.7	0.48	12.2
0.080	2.0	0.63	16.0	1.25	31.7	0.44	11.2	0.54	13.7
0.100	2.5	0.88	22.3	1.50	38.1	0.47	11.9	0.56	14.2
0.125	3.2	0.94	23.9	1.75	44.5	0.56	14.2	0.67	17.0

* Thinner section if thicknesses are unequal.

obtained when the measured welding current is within 5 percent of the listed value.

When dual electrode force is used, timing of application of the forging force is very important. If the forging force is applied too late, the temperature of the nugget will be too low for this higher force to consolidate the nugget. If the forging force is applied too soon, the nugget size may be too small or the electrode indentation excessive. Insufficient electrode force may cause weld metal expulsion, internal discontinuities in the nugget, surface burning, or excessive electrode sticking. Excessive electrode force is evidenced by deep electrode indentation, large sheet separation and distortion, or unsymmetrical weld nuggets.

Weld nugget diameter and penetration can be determined by sectioning through the center of the nugget. The exposed edge is polished, and then etched with a 10% acetic or tartaric acid solution. Penetration should be uniform in equal sheet thicknesses. If not, subsequent welds may require the use of a smaller electrode radius against the side with the lesser penetration. It also may be necessary to clean the electrodes more frequently or the part surfaces more thoroughly.

When spot welding dissimilar alloys, differences in thermal and electrical conductivities can be compensated for by using an electrode with a smaller radius in contact with the alloy that requires the higher heat input. For example, to center the weld nugget in a joint between equal thicknesses of M1A and AZ31B sheets, a smaller radius face should be used against the M1A alloy.

Joint Sealing

SPOT WELDED ASSEMBLIES can be given either a chrome pickle or a dichromate treatment, followed by painting and finishing as desired. Where sealed joints are required or the weldment is to be exposed to a corrosive atmosphere, a suitable sealing compound should be placed between the faying surfaces of the joint before welding. Several proprietary compounds are available for this purpose. Sealers should not be so viscous as to prevent metal-to-metal contact when the electrode force is applied. Welding should be done soon after applying the compounds, and frequent tests should be made to monitor weld quality.

Joint Strength

TYPICAL SHEAR STRENGTHS for spot welds in several thicknesses of three magnesium alloys are given in Table 2.27. Although higher shear strengths are readily obtainable, these values represent the average strengths for welds of maximum soundness and consistency.

SEAM WELDING

SEAM WELDS CAN be made in magnesium alloys under conditions similar to those required for spot welding. Shear strengths of about 750 to 1500 lb/in. of seam (130 to 265 N/mm of seam) can be obtained in M1A alloy in thicknesses of 0.040 to 0.128 in. (1.0 to 3.3 mm). Strengths of seam welds in AZ31B alloy sheet material are approximately 50 percent higher.

FLASH WELDING

FLASH WELDING EQUIPMENT and techniques similar to those used for aluminum alloys can be used for magnesium alloys. High current densities and extremely rapid flashing and upsetting rates are required. Upsetting current should continue for about 5 to 10 cycles (60 Hz) after upset. Special shielding atmospheres are not necessary.

Table 2.23
Schedules for Spot Welding Magnesium Alloys with Three-Phase Frequency Converter Machines

| Thickness[a] | | Electrode[b] | | | | Electrode Force | | | | Forge Delay Time, cycles[c] | Weld Heat or Pulse Time, cycles[c] | No. of Pulses | Postheat Time, cycles[c] | Approximate Current, A | | Nugget Diameter | | Min. Average Shear Strength | |
| | | Diameter | | Face Radius | | Weld | | Forge | | | | | | | | | | | |
in.	mm	in.	mm	in.	mm	lb	kg	lb	kg					Weld	Postheat	in.	mm	lb	N
													AZ31B Alloy						
0.020	0.5	1/2	12.7	3	76	800	360	—	—	—	1	2	—	25 400	—	0.19	4.8	195	865
0.025	0.6	1/2	12.7	3	76	800	360	—	—	—	1	1	2	20 200	4000	0.14	3.6	200	890
0.032	0.8	1/2	12.7	3	76	1000	454	—	—	—	1	2	—	26 400	—	0.20	5.1	330	1470
0.040	1.0	5/8	15.9	3	76	1200	545	—	—	—	1	2	5	28 300	10 300	0.21	5.3	425	1890
0.050	1.3	5/8	15.9	4	102	1400	635	3500	1590	2	2	1	—	29 000	—	0.19	4.8	435	1935
0.050	1.3	5/8	15.9	4	102	1600	725	—	—	—	2	1	—	31 000	—	0.19	4.8	440	1955
0.063	1.6	5/8	15.9	4	102	1750	795	—	—	—	3	1	—	35 200	—	0.22	5.6	580	2580
0.063	1.6	7/8	22.2	4	102	1200	545	3900	1770	3	3	1	—	43 600	—	0.25	6.4	690	3070
0.063	1.6	5/8	15.9	4	102	1200	545	1920	870	3	3	1	6	43 600	24 800	0.29	7.4	800	3560
0.090	2.3	7/8	22.2	4	102	2000	905	4300	1950	2	3	1	5	42 700	15 000	0.26	6.6	910	4050
0.125	3.2	7/8	22.2	6	152	4500	2040	—	—	—	5	6	—	66 900	—	0.46	11.7	2095	9320
													HK31A Alloy						
0.040	1.0	1/2	12.7	3	76	1000	454	—	—	—	1	1	—	19 600	—	0.17	4.3	310	1380
0.050	1.3	5/8	15.9	4	102	1400	635	—	—	—	2	2	—	31 600	—	0.23	5.8	530	2355
0.063	1.6	3/4	19.1	4	102	2400	1090	—	—	—	3	1	—	39 400	—	0.25	6.4	660	2935
0.080	2.0	3/4	19.1	4	102	3400	1540	—	—	—	4	1	—	50 500	—	0.29	7.4	890	3960
0.125	3.2	7/8	22.2	6	152	5000	2270	—	—	—	5	6	—	65 900	—	0.33	8.4	1300	5780
0.125	3.2	3/4	19.1	6	152	2400	1090	3200	1450	2	5	6	—	50 900	—	0.37	9.4	1380	6140
													HM21A Alloy						
0.040	1.0	1/2	12.7	3	76	800	360	—	—	—	1	2	—	21 600	—	0.18	4.6	355	1580
0.050	1.3	5/8	15.9	4	76	1200	545	—	—	—	2	2	—	30 700	—	0.21	5.3	470	2090
0.063	1.6	5/8	15.9	4	102	1600	725	—	—	—	3	2	—	40 600	—	0.23	5.8	560	2490
0.071	1.8	5/8	15.9	4	102	2200	995	—	—	—	4	2	—	47 400	—	0.29	7.4	770	3425
0.090	2.3	3/4	19.1	4	102	3000	1360	—	—	—	4	2	—	53 200	—	0.26	6.6	950	4225
0.125	3.2	7/8	22.2	6	152	3800	1725	—	—	—	5	2	—	66 700	—	0.32	8.1	1180	5250
0.125	3.2	7/8	22.2	6	152	2000	905	3600	1630	5	5	6	—	56 500	—	0.37	9.4	1405	6250

a. Two equal thicknesses

b. Spherical radius-faced electrodes on both sides

c. Cycles of 60 Hz

Table 2.24
Schedules for Spot Welding Magnesium Alloys with Single-Phase AC Machines

Thickness[a]		Electrode[b]				Electrode Force		Weld Time, cycles[c]	Approx. Welding Current, A	Nugget Diameter		Min. Average Shear Strength	
		Diameter		Face Radius									
in.	mm	in.	mm	in.	mm	lb	kg			in.	mm	lb	N
						AZ31B Alloy							
0.016	0.4	3/8	9.5	2	51	300	135	2	16 000	0.10	2.5	140	620
0.020	0.5	3/8	9.5	3	76	350	160	3	18 000	0.14	3.6	175	780
0.025	0.6	3/8	9.5	3	76	400	180	3	22 000	0.16	4.1	215	955
0.032	0.8	3/8	9.5	3	76	450	205	4	24 000	0.18	4.6	270	1200
0.040	1.0	1/2	12.7	3	76	500	225	5	26 000	0.20	5.1	345	1535
0.050	1.3	1/2	12.7	4	102	550	250	5	29 000	0.23	5.8	430	1915
0.063	1.6	1/2	12.7	4	102	600	270	6	31 000	0.27	6.9	545	2425
0.071	1.8	1/2	12.7	4	102	650	295	7	32 000	0.29	7.4	610	2715
0.080	2.0	1/2	12.7	4	102	700	315	8	33 000	0.31	7.9	690	3070
0.090	2.3	1/2	12.7	4	102	750	340	9	34 000	0.32	8.1	770	3425
0.100	2.5	1/2	12.7	6	152	800	360	10	36 000	0.34	8.6	865	3850
0.125	3.2	1/2	12.7	6	152	1000	455	12	42 000	0.38	9.7	1080	4805
						M1A Alloy							
0.016	0.4	3/8	9.5	2	51	300	135	3	17 000	0.08	2.0	70	310
0.020	0.5	3/8	9.5	3	76	300	135	3	20 000	0.12	3.0	95	425
0.025	0.6	3/8	9.5	3	76	350	160	4	24 000	0.14	3.6	130	580
0.032	0.8	3/8	9.5	3	76	400	180	5	26 000	0.16	4.1	175	780
0.040	1.0	3/8	9.5	3	76	450	205	6	28 000	0.18	4.6	225	1000
0.050	1.3	1/2	12.7	4	102	500	225	7	30 000	0.21	5.3	295	1310
0.060	1.5	1/2	12.7	4	102	550	250	8	32 000	0.24	6.1	385	1710
0.071	1.8	1/2	12.7	4	102	600	270	9	33 000	0.26	6.6	430	1915
0.080	2.0	1/2	12.7	4	102	650	295	10	35 000	0.28	7.1	495	2200
0.090	2.3	1/2	12.7	4	102	700	315	11	36 000	0.29	7.4	560	2490
0.100	2.5	1/2	12.7	6	152	750	340	12	38 000	0.31	7.9	680	3025
0.125	3.2	1/2	12.7	6	152	950	430	14	45 000	0.35	8.9	800	3560

a. Two equal thicknesses

b. Spherical radius-faced electrodes on both sides

c. Cycles of 60 Hz

Table 2.25[a]
Schedules for Spot Welding AZ31B Magnesium Alloy with Capacitor-Discharge Stored-Energy Machines[b]

| Thickness[c] | | Electrode[d] | | | | Electrode Force | | Charging Voltage, | Spot Diameter | | Min. Average Shear Strength | |
| | | Diameter | | Tip Radius | | | | | | | | |
in.	mm	in.	mm	in.	mm	lb	kg	kV	in.	mm	lb	N
0.020	0.5	1/2	12.7	3	76	650	295	1.4	0.14	3.6	145	645
0.040	1.0	1/2	12.7	3	76	725	330	2.2	0.20	5.1	336	1495
0.051	1.3	5/8	15.9	4	102	835	380	2.2	0.23	5.8	435	1935
0.064	1.6	5/8	15.9	4	102	1080	490	2.2	0.27	6.9	560	2490
0.100	2.5	3/4	19.1	6	152	1800	490	2.2	0.34	8.6	985	4380
0.125	3.2	7/8	22.2	6	152	2275	1030	2.2	0.38	9.6	1208	5375

a. Courtesy of The Dow Chemical Company

b. Transformer turns ratio of 480:1

c. Two equal thicknesses

d. Spherical radius-faced electrodes on both sides

Table 2.26
Schedules for Spot Welding AZ31B Magnesium Alloy with DC Rectifier Machines

| Thickness[a] | | Electrode[b] | | | | Electrode Force | | | | Forge Delay Time, cycles[c] | Weld Time, cycles[c] | Post-heat Time, cycles[c] | Approximate Current, A | | Nugget Diameter | | Min. Average Shear Strength | |
| | | Diameter | | Face Radius | | Weld | | Forge | | | | | Weld | Postheat | | | | |
in.	mm	in.	mm	in.	mm	lb	kg	lb	kg						in.	mm	lb	N
0.020	0.5	5/8	15.9	3	76	300	135	600	270	0.6	1	1	21 000	14 700	0.14	3.5	145	645
0.032	0.8	5/8	15.9	3	76	400	180	880	400	1.0	2	1	24 000	16 900	0.18	4.6	245	1090
0.040	1.0	5/8	15.9	3	76	480	215	1000	455	1.2	2	2	26 000	18 000	0.20	5.1	336	1495
0.051	1.3	5/8	15.9	3	76	580	265	1270	575	1.5	3	2	28 500	20 000	0.22	5.6	435	1935
0.064	1.6	5/8	15.9	4	102	700	315	1540	700	1.8	3	2	29 300	20 500	0.27	6.9	560	2490
0.081	2.1	7/8	22.2	4	102	860	390	1890	855	2.4	3	3	35 750	25 000	0.31	7.9	740	3290
0.093	2.4	7/8	22.2	6	152	970	440	2150	975	3.9	4	4	38 750	27 100	0.32	8.1	855	3805
0.102	2.6	7/8	22.2	6	152	1050	475	2320	1050	4.5	7	4	41 300	28 800	0.34	8.6	985	4380
0.125	3.2	7/8	22.2	6	152	1270	575	2780	1260	7.7	10	6	48 000	33 400	0.38	9.6	1208	5375

a. Two equal thicknesses

b. Spherical radius-faced electrodes on both sides

c. Cycles of 60 Hz

Table 2.27
Typical Shear Strengths of Single Spot Welds in Wrought Magnesium Alloys

| Thickness | | Average Spot Diameter | | Spot Shear Strength | | | | | |
| | | | | AZ31B | | HK31A | | HM21A | |
in.	mm	in.	mm	lb	N	lb	N	lb	N
0.020	0.5	0.14	3.5	220	980	—	—	—	—
0.025	0.6	0.16	4.1	270	1200	—	—	—	—
0.032	0.8	0.18	4.6	330	1465	300	1335	—	—
0.040	1.0	0.20	5.1	410	1825	375	1670	360	1600
0.050	1.3	0.23	5.8	530	2355	550	2445	—	—
0.063	1.6	0.27	6.9	750	3335	720	3200	660	2935
0.080	2.0	0.31	7.9	890	3960	—	—	—	—
0.100	2.5	0.34	8.6	1180	5250	—	—	—	—
0.125	3.2	0.38	9.7	1530	6805	1490	6625	1220	5425

Flash welds in AZ31B, AZ61A, and HM31A magnesium alloys have typical tensile strengths of 36, 42, and 38 ksi (248, 290, and 262 MPa), respectively, with elongations of about 4 to 8 percent. The typical microstructures of various zones in a flash welded joint in an HM31A-T5 magnesium alloy extrusion are shown in Figure 2.10.

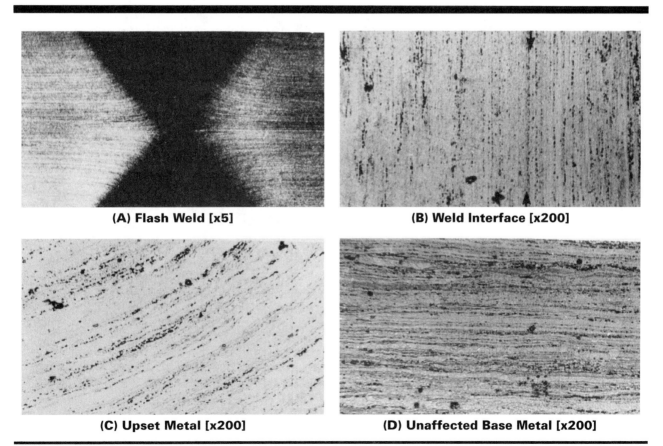

(A) Flash Weld [x5]

(B) Weld Interface [x200]

(C) Upset Metal [x200]

(D) Unaffected Base Metal [x200]

Figure 2.10—Flash Weld and Microstructures in HM31A-T5 Magnesium Alloy Rod

OXYFUEL GAS WELDING

OXYFUEL GAS WELDING should only be used for emergency field repair work when suitable arc welding equipment is not available. Its use is restricted almost exclusively to simple groove welds where residual flux can be effectively removed. The repair welds should be considered only temporary until they can be replaced with arc welds or a new part can be put in service.

FUEL GASES

THE FUEL GASES most commonly used are acetylene or a mixture of about 80% hydrogen and 20% methane. The latter fuel gas is well-suited for welding sheets up to 0.064 in. (1.6 mm) thick because of its soft flame. For welding thicker gauges, acetylene is desirable because of its higher heat of combustion. The oxyacetylene flame may cause slight pitting of the weld surface, but it is rarely serious enough to impair the strength of the weld.

FLUXES

FLUXES SPECIFICALLY RECOMMENDED for oxyfuel gas welding of magnesium should be used. These fluxes are prepared by mixing them with water or alcohol to form a heavy slurry or paste. They should be used soon after mixing. Prior to fluxing, the area to be welded should be cleaned to remove any dirt, oil, grease, oxide, or conversion coating.

One flux composition suitable for welding with various fuel gases is a mixture of 53% KCl, 29% $CaCl_2$, 12% NaCl, and 6% NaF, by weight. Another mixture suitable only for oxyacetylene welding consists of 45% KCl, 26% NaCl, 23% LiCl, and 6% NaF. The sodium compounds in these welding fluxes will give an intense yellow color to the flame. Welders should use suitable eye protection and ventilation when using these fluxes.

WELDING TECHNIQUE

A LIBERAL COATING of flux should be applied to both sides of the joint and to the welding rod. If needed, the joint should be tack welded at 1 to 3 in. (25 to 76 mm) intervals depending upon the metal thickness. All tack welds and overlapping weld beads should be remelted to float out any flux inclusions. Parts of relatively large mass should be preheated.

All traces of flux must be removed from the weldment in hot water. Then, the weldment is given a chrome pickle, followed by immersion for 2 hours in boiling flux remover (see Table 2.8).

OTHER WELDING PROCESSES

ELECTRON BEAM WELDING

IN GENERAL, MAGNESIUM alloys that can be arc welded can also be electron beam welded. The same preweld and postweld operations apply to both processes.

Close control of electron beam operating variables is required to prevent overheating and porosity at the root of the weld. The high vapor pressures in vacuum of magnesium and zinc in alloys contribute to this problem. It is very difficult to produce sound welds in magnesium alloys containing more than 1% zinc. Beam manipulation may be helpful in overcoming porosity.

A photomicrograph of an electron beam weld in 0.25 in. (6.4 mm) thick HM31A-T5 magnesium alloy is shown in Figure 2.11.

STUD WELDING

THE GAS SHIELDED arc stud welding process used for aluminum is also applicable to magnesium alloys. A ceramic ferrule is not needed. Helium shielding and DCEP power are used. The stud welding gun should be equipped with controlled plunge to avoid excessive spatter and undercutting of the base metal.

Typical conditions for welding 0.25 and 0.5 in. (6.4 and 12.7 mm) diameter AZ31B magnesium alloy studs to plate and the average breaking loads of the welded studs are given in Table 2.28. Figure 2.12 shows a cross section through a stud weld. The soundness of magnesium stud welds is, in general, very similar to that of aluminum stud welds.

Figure 2.11—Electron Beam Weld in 1/4 in. (6.4 mm) Thick HM31A-T5 Magnesium Alloy Extrusion (x10)

Figure 2.12—Typical Arc Stud Weld in AZ31B Magnesium Alloy

Table 2.28
Typical Welding Conditions and Breaking Loads for AZ31B-F Magnesium Alloy Studs
Joined to 0.25 in. (6.4 mm) AZ31B-O Alloy Plate

Stud Diameter		Welding[a] Current, A	Weld Time,[b] cycles	Lift		Plunge		Avg. Breaking Load	
in	mm			in.	mm	in.	mm	lb	kN
1/4	6.4	125	45	1/8	3.2	1/4	6.4	1530	6.80
1/2	12.7	375	40	1/8	3.2	3/16	4.8	4100	18.25

a. DCEP power and helium shielding

b. 60 Hz

BRAZING

BRAZING TECHNIQUES FOR magnesium alloys are similar to those used for aluminum alloys. However, the brazing of magnesium is not widely practiced. Furnace, torch, and dip brazing can be employed, but furnace and torch brazing experience is limited to M1A magnesium alloy.

FILLER METAL

ONLY ONE MAGNESIUM brazing filler metal is covered by specifications. It is BMg-1 filler metal having a nominal composition of 92% magnesium, 9% aluminum, and 2% zinc. Although it is similar to AZ92A magnesium alloy, the filler metal contains a small amount of beryllium to prevent excessive oxidation while it is molten. The brazing temperature range for this filler metal is 1120 to 1160 °F (604 to 626 °C). It is suitable for brazing only AZ10A, K1A, and M1A magnesium alloys. These alloys will be annealed when exposed to brazing temperature.

PREBRAZE CLEANING

AS WITH OTHER metals, all parts to be brazed should be thoroughly clean and free of burrs. All dirt, oil, or grease should be removed by vapor or solvent degreasing. Surface films, such as chromates or oxides, should be removed by mechanical or chemical cleaning. Abrasive cloth or steel wool is satisfactory for mechanical cleaning. Chemical cleaning should consist of immersion in hot alkaline cleaner and then in a suitable chemical cleaner (see Table 2.8).

FLUXES

FLUXES USED FOR brazing magnesium alloys are chloride-based, similar to those used for oxyfuel gas welding. The composition and melting point of two suitable brazing fluxes are given in Table 2.29.

BRAZING PROCEDURES

Furnace Brazing

ELECTRIC OR GAS furnaces with automatic temperature controls capable of holding the temperature within ±5 °F (±2.8 °C) should be used for brazing. A special atmosphere is not required. Sulfur dioxide (SO_2) or products of combustion in gas-fired furnaces will inhibit brazing filler metal flow and must be avoided.

Table 2.29
Composition and Melting Point of Magnesium Brazing Fluxes

Applicable Brazing Processes	Flux Composition, %		Approximate Melting Point
Torch	KCl	45	1000 °F (538 °C)
	NaCl	26	
	LiCl	23	
	NaF	6	
Torch, dip, furnace	KCl	42.5	730 °F (388 °C)
	NaCl	10	
	LiCl	37	
	NaF	10	
	AlF$_3$-3NaF	0.5	

Parts to be brazed should be assembled with the filler metal preplaced in or around the joint. Joint clearances of 0.004 to 0.010 in. (0.10 to 0.25 mm) should be used for good capillary flow of the brazing filler metal. Best results are obtained when dry powdered flux is sprinkled along the joint. Flux pastes made with water or alcohol will retard the flow of brazing filler metal. Flux pastes made with benzol, toluene, or chlorbenzol may be used, but they are more difficult to apply because the pastes are not smooth. Flux pastes should be dried by heating the assembly at 350 to 400 °F (175 to 205 °C) for 5 to 15 minutes in drying ovens or circulating air furnaces. Flame drying is not recommended because improper oxyacetylene flame adjustment may cause a heavy soot deposit.

Brazing time will depend upon the metal thickness at the joint and the amount of fixturing necessary to position the parts. The time should be the minimum necessary to obtain complete filler metal flow with minimum diffusion between the filler and base metals. One to two minutes at the brazing temperature is sufficient.

Torch Brazing

TORCH BRAZING IS done with a neutral oxyfuel gas or air-fuel gas flame. Natural gas is well suited for torch brazing because of its relatively low flame temperature. The brazing filler metal can be placed on the joint and fluxed before heating, or it may be face-fed. Flux pastes can be made with either water or alcohol. However, pastes made with alcohol give better results. Heat should be applied to the joint until the filler metal melts and flows in the joint. Overheating of the base metal must be avoided.

Dip Brazing

DIP BRAZING IS accomplished by immersing the assembly into a molten brazing flux held at brazing temperature. The flux serves the dual functions of both heating and fluxing. Temperature control should be accurate to within ±5 °F (±2.8 °C) of the desired brazing temperature. Joint clearance should be from 0.004 to 0.010 in. (0.10 to 0.25 mm).

After preplacing the filler metal, the parts should be assembled in a brazing fixture, preferably of stainless steel to resist the corrosive action of the flux. The fixtured assembly is preheated in a furnace to between 850 and 900 °F (454 and 482 °C). This is done to minimize distortion and the time in the flux bath. Immersion time in the flux bath should be relatively short because the

parts are heated rapidly by the molten flux. For example, 1/16 in. (1.6 mm) thick sheet can be heated in 30 to 45 seconds. Large assemblies with fixturing may require immersion for 1 to 3 minutes.

POSTBRAZE CLEANING

COMPLETE REMOVAL OF all traces of flux from the brazement is required to avoid subsequent corrosion. Brazed parts should be rinsed thoroughly in flowing hot water to remove the flux from the surface of the part.

A stiff-bristled brush may be used to scrub the surface and speed up flux removal. The brazement is then given a one to two minute immersion in chrome-pickle, followed by 2 hours in boiling flux remover cleaner. The compositions of these solutions are given in Table 2.8. The corrosion resistance of brazed joints depends primarily upon complete flux removal.

SOLDERING

BARE MAGNESIUM ALLOYS can be soldered only by the abrasion and ultrasonic methods. These methods can dislodge the oxide film on the surfaces to be soldered. No suitable flux is available to remove this film and permit the solder to wet the surfaces.

Conventional heating methods, including soldering irons and gas torches, may be used. Soldering is not recommended if the joint will be required to withstand moderately high stress. Soldered joints are low in strength and ductility. They also are unsatisfactory for service in the presence of an electrolyte. The marked difference in solution potential between a magnesium alloy and a solder can lead to severe galvanic attack. A suitable protective coating should be applied to soldered joints for good serviceability.

SOLDERS

THE SOLDERS LISTED in Table 2.30 are used for magnesium. Lead-containing solders, such as the 50% tin-50% lead alloy, can be used, but severe galvanic attack may take place in the presence of moisture. The tin-zinc solders have lower melting points and better wetting characteristics than the tin-zinc-cadmium solders, but they may form joints of low ductility. The high-cadmium solders produce the strongest and most ductile joints.

SURFACE PREPARATION

BARE MAGNESIUM SURFACES to be joined should be degreased with a suitable solvent and then mechanically cleaned immediately before soldering. A clean stainless steel wire brush, stainless steel wool, or aluminum oxide abrasive cloth is a suitable cleaning tool.

Electroplated coatings on magnesium offer an excellent soldering base. A zinc-immersion (zincate) coating, the first step in plating magnesium, followed by a 0.0001 to 0.0002 in. (0.0025 to 0.0051 mm) thick copper plate over the zinc coating provides a solderable surface. Tin or silver plating also may be used for this purpose. Soldering of electroplated surfaces is carried out using the procedures used for the deposited metal. Fusing a tin coating improves its protective value by flowing the deposited tin and sealing the pores. This technique consists of electroplating a 0.0003 to 0.0005 in. (0.0076 to 0.0127 mm) tin coating over the copper electroplate. The part is then immersed in a hot oil bath to flow the tin coating and close the pores. The process is being used on a large number of magnesium electronic parts to permit easy soldering.

JOINT TYPES

FLUXLESS SOLDERING OF bare magnesium alloys is limited to fillet joints and to the filling of surface defects

Table 2.30
Solders for Magnesium

Composition, %	Temperature				Use
	Solidus		Liquidus		
	°F	°C	°F	°C	
60 Cd-30 Zn-10 Sn	315	157	550	288	Low temperature - below 300 °F (149 °C)
90 Cd-10 Zn	509	265	570	299	High temperature - above 300 °F (149 °C)
72 Sn-28 Cd	350	177	470	243	Medium temperature - below 300 °F (149 °C)
91 Sn-9 Zn	390	199	390	199	High temperature - above 300 °F (149 °C)
60 Sn-40 Zn	390	199	645	341	High temperature - above 300 °F (149 °C)
70 Sn-30 Zn	390	199	592	311	Precoating solder
50 Sn-50 Pb	361	183	421	216	Filler solder on precoated surfaces
80 Sn-20 Zn	390	199	518	270	Precoating solder
40 Sn-33 Cd-27 Zn	—	—	—	—	Filler solder

in noncritical areas of wrought and cast products prior to painting. Conventional solder joints can be used with solderable electroplated surfaces.

PROCEDURES

BARE MAGNESIUM SURFACES must be precoated with a solder having good wetting characteristics. Solder coating with the friction method is done by rubbing the solder stick, soldering iron, or other tool on the magnesium under the molten solder to break up the oxide film. The ultrasonic method of precoating utilizes a hot soldering bit vibrating at ultrasonic frequencies. When in contact with the molten solder on the magnesium, the vibration causes an abrasive effect known as *cavitation erosion*. This action dislodges the surface oxides and permits wetting. After the surfaces are precoated with solder, the joint can be soldered using a soldering iron, torch, or hot plate.

PLASMA ARC CUTTING

MAGNESIUM ALLOYS CAN be cut with a plasma arc cutting torch. An argon-hydrogen mixture is used for the orifice and shielding gases. A mixture of 80% argon and 20% hydrogen is recommended for manual operation. A mixture of 65% argon and 35% hydrogen is recommended for automatic cutting. Typical conditions for automatic cutting are given in Table 2.31. An exhaust system is needed because of the evolution of large amounts of fumes.

Table 2.31
Typical Conditions for Automatic Plasma Arc Cutting of Magnesium Alloy Plates[a]

Thickness		Plasma		Cutting Speed		Shielding Gas Flow[b]		Remarks
in.	mm	Current (DCEP), A	Voltage, V	in./min	mm/s	ft³/h	L/min	
1/4	6.4	200	60	225	95	70	33	Minimum fume
1/4	6.4	400	75	150	64	70	33	Squarest cut
1/4	6.4	400	80	300	127	70	33	Maximum speed
1/2	12.7	240	60	150	64	70	33	Minimum fume
1/2	12.7	420	75	100	42	100	47	Squarest cut
1/2	12.7	460	70	300	127	100	47	Maximum speed
1	25.4	300	75	60	25	70	33	Minimum fume
1	25.4	450	80	60	25	100	47	Squarest cut
1	25.4	660	80	75	32	100	47	Maximum speed
2	50.8	350	100	25	10.6	100	47	Minimum fume
2	50.8	520	100	25	10.6	100	47	Squarest cut
2	50.8	600	90	50	21	100	47	Maximum speed
4	101.6	500	210	12	5	200	94	Squarest cut
6	152.4	750	225	12	5	300	142	Squarest cut

a. AZ31B magnesium alloy

b. 65% argon-35% hydrogen

SAFE PRACTICES

GENERAL SAFETY ISSUES are covered in Chapter 16, "Safe Practices," in Volume 1 of the *Welding Handbook*, 8th Edition.

Welding fumes from commercial magnesium alloys, except those containing thorium, are not harmful when the amount of fumes remains below the welding fume limit of 5 mg/m³. Welders should avoid inhalation of fumes from the thorium-containing alloys because of the presence of alpha radiation in the airborne particles. However, the concentration of thorium in the fumes is sufficiently low so that good ventilation or local exhaust systems will provide adequate protection. This radiation concern, however, is primarily responsible for the decline in use of the thorium containing alloys. No external radiation hazard is involved in the handling of the thorium-containing alloys.

The possibility of ignition when welding magnesium alloys in thicknesses greater than 0.01 in. (0.25 mm) is extremely remote. Magnesium alloy product forms will not ignite in air until they are at fusion temperature. Then, sustained burning will occur only if the ignition temperature is maintained. Inert gas shielding during welding prevents ignition of the molten weld pool.

Magnesium fires occur with accumulations of grinding dust or machining chips. Accumulation of grinding dust on clothing should be avoided. Graphite-based (G-1) or proprietary salt-based powders recommended for extinguishing magnesium fires should be conveniently located in the work area. If large amounts of fine particles, or *fines*, are produced, they should be collected in a waterwash-type dust collector designed for use with magnesium. Special precautions pertaining to the handling of wet magnesium fines must be followed.

The accumulation of magnesium dust in a water bath also can present a hazard. Dust of reactive metals like magnesium or aluminum can combine with the oxygen in the water molecule, leaving hydrogen gas trapped in a bubbly froth on top of the water. A heat source may cause this froth to explode.

Some solvents, chemical baths, and fluxes used for cleaning, welding, brazing, or finishing of magnesium alloys contain chromates, chlorides, fluorides, acids, or alkalies. Adequate ventilation, protective clothing, and eye protection must be used when working with these materials to avoid toxic effects, burns, or other injuries that they may cause.

APPLICATIONS

MAGNESIUM AND MAGNESIUM alloy welded applications, welding procedures, and joint designs (in Figure 2.1) are quite similar to those of other metals, especially aluminum and its alloys. As noted elsewhere in this chapter, attention must be directed to reactivity (primarily oxidation) during welding, cracking due to restraint, filler metal selection, welding and preheat procedures, and the welding process. Another important consideration is postweld behavior relating to galvanic and stress corrosion tendencies in service or after repair welding.

The selection of a welding process is dependent on a variety of factors. Welding of wrought material is primarily done with manual and automatic gas tungsten arc welding (GTAW) and gas metal arc welding (GMAW). The commonplace repair welding of castings involves manual GTAW, which can be carefully controlled to avoid restraint cracking and distortion problems. However, the selection of a welding process or procedure often is influenced by considerations other than purely technical factors. Among these might be:

(1) Quantity to be fabricated (short-run prototype or longer production run)
(2) Whether or not tack welding can be used in place of more costly fixturing
(3) In-house familiarity with applicable processes and procedures
(4) In-house availability of desired processes and associated equipment

The following three case studies are representative of welded magnesium applications.[4]

TACK WELDING FOR SHORT-RUN PRODUCTION

TACK WELDS WERE used instead of fixturing to position some of the component pieces to minimize tooling costs on short production runs of electronic deck assemblies. These assemblies were essentially two rectangular boxes 2 by 2 by 4 in. (51 x 51 x 102 mm) as shown in Figure 2.13. Formed sheet sections of 0.050 in. (1.27 mm) thick AZ31B-H24 were tack welded into position by gas tungsten arc welding

4. These case studies are provided courtesy of ASM International and are adapted from the *Metals Handbook*, Vol. 6, 9th Ed., 431-34. Metals Park, Ohio: American Society for Metals, 1983.

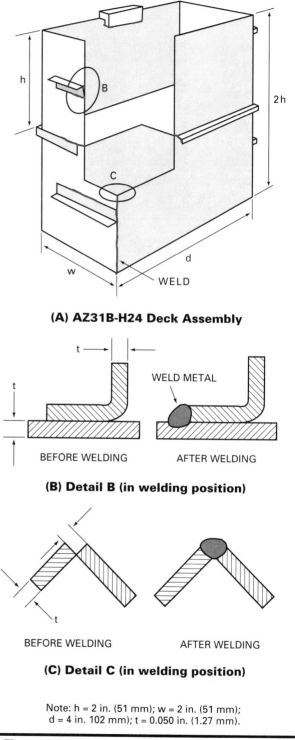

(A) AZ31B-H24 Deck Assembly

WELD METAL

BEFORE WELDING   AFTER WELDING

(B) Detail B (in welding position)

BEFORE WELDING   AFTER WELDING

(C) Detail C (in welding position)

Note: h = 2 in. (51 mm); w = 2 in. (51 mm); d = 4 in. 102 mm); t = 0.050 in. (1.27 mm).

Figure 2.13—Manual Gas Tungsten Arc Welding of Electronic Deck Assembly

(GTAW), using 1/16 in. (1.6 mm) diameter R AZ61A filler wire. The tack welds were 1/8 in. (3.2 mm) long and were spaced on 2 in. (51 mm) centers, starting at each corner. A tool plate and toggle clamp held the pieces for tack welding. Tack welds were not used to hold angle pieces.

Welding of the assembled and tack-welded components was completed by manual GTAW under the conditions shown in Table 2.32. The corner joints were welded with continuous beads about 2 in. (51 mm) long, and the flanged bottom of the top part of the assembly was joined to the sides with 1 in. (25 mm) long fillet welds. Extruded angle sections were fillet welded to the ends of the boxes with welds about 1 in. (25 mm) long (see Figure 2.13, Detail B).

The assembly was repositioned manually so that all welds could be made in either the flat or the horizontal position. A standard alternating current power supply with a high-frequency arc stabilizer was used. Helium was selected as the shielding gas because a hotter and more stable arc was produced than would have been possible with argon shielding gas. Preheating was not used, but after welding, the assemblies were stress relieved at 350 °F (177 °C) for 3 1/2 h to prevent stress-corrosion cracking. Welds were inspected visually.

MACHINE WELDING OF EXTRUDED DOOR FRAMES

AIRTIGHT DOORS FOR an aerospace application were made by welding panels of alloy AZ31B-H24 sheet to frames extruded from alloy AZ31B. The frames, which acted as stiffeners, also contained a groove for an air seal. Cross sections of similar offset butt joints in two designs of door assemblies are shown as joints A and B in Figure 2.14.

The offset lip of the extruded frames provided a single-bevel groove butt joint and supplied backing for the weld; the lap joint on the underside was not welded. The welding conditions are shown in Table 2.33.

Although production quantities were low, machine gas tungsten arc welding was used because weld quality was good and the equipment was available. Automatic travel was obtained by mounting the welding equipment on the motorized carriage of a cutting machine. Differences in welding conditions for the two joints, shown in Table 2.33, resulted from operator choice or judgment. Both procedures produced satisfactory welds, but the difference in welding speeds would have been significant had production quantities been large.

Table 2.32
Conditions for Manual Gas Tungsten Arc Welding of Electronic Deck Assembly

Joint types	Lap and corner
Weld types	Fillet and single-V-groove
Welding positions	Horizontal and flat
Preweld cleaning	Wire brushing
Preheat	None
Fixtures	Tool plate and toggle clamps
Shielding gas	Helium, at 25 ft^3/h (11.8 L/min)
Electrode	0.040-in. (1.0-mm) diameter EWP
Filler metal	1/16-in. (1.6-mm) diameter R AZ61A[a]
Torch	350 A, water cooled[b]
Power supply	300-A transformer[c]
Current, fillet welds	25 A, ac
Current, V-groove welds	40 A, ac
Postweld heat treatment	350 °F (177 °C) for 3 1/2 h

a. 36-in. (914-mm) long rod

b. Ceramic nozzle

c. Continuous duty, with high-frequency oscillator

Table 2.33
Conditions for Automatic Gas Tungsten Arc Welding of Extruded Door Frames

Joint type	Offset butt
Weld type	Single-bevel groove
Preweld cleaning	Chromic-sulfuric pickle
Welding position	Flat
Preheat	None
Shielding gas	Argon, 18 ft^3/h (8.5 L/min) for joint A; argon, 16 ft^3/h (7.5 L/min) for joint B
Electrode	1/8-in. (3.2-mm) diameter EWP
Filler metal	1/16-in. (1.6-mm) diameter ER AZ61A
Torch	Water cooled
Power supply	300 A ac (HF-stabilized)
Current (ac)	175 A for joint A; 135 A for joint B
Wire feed rate	65 in./min (27.5 mm/s)
Travel speed	20 in./min (8.4 mm/s) for joint A; 15 in./min (6.3 mm/s) for joint B
Postweld heat treatment	350 °F (177 °C) for 1 1/2 h

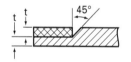

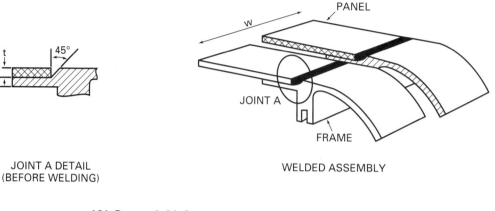

JOINT A DETAIL
(BEFORE WELDING)

WELDED ASSEMBLY

(A) Curved Airframe Door Design

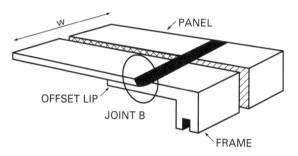

JOINT B DETAIL
(BEFORE WELDING)

WELDED ASSEMBLY

(B) Straight Airframe Door Design

Notes:
1. Frames are alloy AZ31B extrusions.
2. Panels are alloy AZ31B sheet.
3. Filler metal is magnesium alloy ER AZ61A.
4. t = 1/16 in. (1.6 mm); w = approximately 3 ft (0.9 m).

Figure 2.14—Machine Gas Tungsten Arc Welding of Extruded Door Frames

REPAIR WELDING OF A JET ENGINE CASTING

DURING AN AIRCRAFT jet engine overhaul, fluorescent-penetrant inspection revealed a 2 1/2 in. (63.5 mm) long crack near a rib in the cast AZ92A-T6 compressor housing shown in Figure 2.15. The thickness of the section containing the crack ranged from 3/16 to 5/16 in. (4.8 to 7.9 mm). Repair welding was permissible. The welding conditions are shown in Table 2.34.

The part was vapor degreased to remove surface grease and dirt and was soaked in a commercial alkaline paint remover. The crack was then marked with a felt-tip marker, and the part was stress relieved at 400 °F (204 °C) for 2 h. The crack was removed by slotting the flange through to the periphery. Each side of

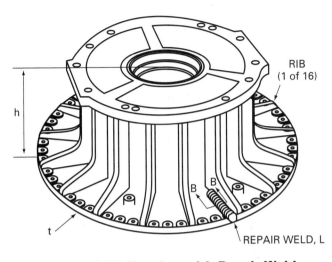

(A) AZ92A-T6 Housing with Repair Weld

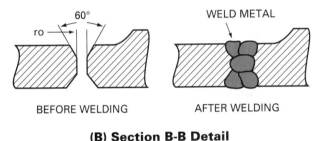

BEFORE WELDING   AFTER WELDING

(B) Section B-B Detail

Notes:
1. Housing height, h = 9.33 in. (237 mm);
   diameter, d = 21.84 in. (555 mm).
2. Repair weld length, L, is approximately 2.5 in. (64 mm).
3. Thickness, t = 3/16 to 5/16 in. (4.8 to 7.9 mm); root opening,
   ro = 3/32 in. (2.4 mm).

**Figure 2.15—Repair Welding of
Compressor Housing**

**Table 2.34
Conditions for Manual Gas Tungsten Arc Welding
of Compressor Housing**

Joint type	Butt
Weld type	60° double-V-groove repair
Shielding gas	Argon, 20 ft³/h (9.4 L/min)[a]
Electrode	1/16-in. (1.6-mm) diameter EWTh-2
Filler metal	1/16-in. (1.6-mm) diameter R AZ101A
Torch	Water cooled
Power supply	300-A transformer, with high-frequency starting
Current	Under 70 A, ac[b]
Postweld stress relief	400 °F (204 °C) for 2 h[c]
Inspection	Fluorescent penetrant

a. Also used for backing

b. Current was regulated by a foot switch

c. Also preweld

the slot was beveled to approximately 30 degrees from vertical to form a 60-degree double-V-groove. The area to be welded was cleaned with a power wire brush with stainless steel bristles. Manual welding was done using gas tungsten arc welding without preheating.

The welding technique maintained a low-amperage arc (less than 70 A) directed onto the base metal while filler metal was deposited on the sides of the groove, working from the innermost point outward. After a molten weld pool formed, the arc was weaved slightly while depositing a bead on the sides of the groove. During welding, heat input was adjusted by a foot-operated current-control rheostat to maintain a uniform weld pool. After welding was completed on one side of the slot, the casting was turned over. Excess drop-through and areas of incomplete penetration were removed by grinding. The underside was then welded by the same technique used for the first side. After welding, the casting was stress relieved at 400 °F (204 °C) for 2 h and inspected by the fluorescent-penetrant method.

SUPPLEMENTARY READING LIST

American Society for Metals. "Properties and selection: Nonferrous alloys and pure metals." *Metals Handbook*, Vol. 2, 9th Ed., 525-609. Metals Park, Ohio: American Society for Metals, 1979.

——————. "Welding, Brazing, and Soldering." *Metals Handbook*, Vol. 6, 9th Ed., 431-4. Metals Park, Ohio: American Society for Metals, 1983.

American Welding Society. *Brazing Handbook*, 4th Ed., 351-8. Miami, Florida: American Welding Society, 1991.

——————. *Recommended Practices for Resistance Welding*, C1.1-66, 45-76. Miami, Florida: American Welding Society, 1966.

——————. *Specification for Magnesium Alloy Welding Electrodes and Rods*, ANSI/AWS A5.19-92. Miami, Florida: American Welding Society, 1992.

Ayner, S. H., *Introduction to Physical Metallurgy*, 2nd Ed., 498-507. New York: McGraw-Hill Book Co., 1974.

Busk, Robert S. *Magnesium Products Design*, Chapter 5, 85-122. New York: Marcel Dekker, Inc., 1987.

Fenn, R. W., Jr., and Lockwood, L. F. "Low temperature properties of welded magnesium alloys." *Welding Journal* 39(8): 352s-6s, 1960.

Kenyon, D. M. "Arc behavior and its effect on the tungsten arc welding of magnesium alloys." *Journal of Institute of Metals* 93: 85-9, 1964-65.

Koeplinger, R. D., and Lockwood, L. F. "Gas metal arc spot welding of magnesium." *Welding Journal* 43 (3): 195-201, 1964.

Lockwood, L. F. "Automatic gas tungsten arc welding of magnesium." *Welding Journal* 44(5): 213s-20s, 1965.

——————. "Gas metal arc welding of AZ31B magnesium sheet." *Welding Journal* 42(10): 807-18, 1963.

——————. "Gas shielded stud welding of magnesium." *Welding Journal* 46(4): 168s-74s, 1967.

——————. "Now you can dip braze magnesium." *Product Engineering* 36: 113-16, March 15, 1965.

——————. "Pulse arc welding of magnesium." *Welding Journal* 49(6): 464-75, 1970.

——————. "Repair welding of thin wall magnesium sand castings." *Transactions of the American Foundrymens Society* 75: 530-40, 1967.

——————. "Spot welding of wrought HK31A, HM21A and ZE10A magnesium alloys." *Welding Journal* 39(9): 369s-78s, 1960.

Lockwood, L. F., and Klain, P. "The arc welding of wrought magnesium-thorium alloys." *Welding Journal* 37(6): 255s-64s, 1958.

Portz, A. G., and Rothgery, G. R. "Flash welded magnesium rings meet space age needs." *Welding Design and Fabrication* 36(1): 44-5, 1963.

Sibley, C. R. *Arc welding of magnesium and magnesium alloys*. Bulletin 83. New York: Welding Research Council, November 1962.

COPPER AND COPPER ALLOYS

PREPARED BY A COMMITTEE CONSISTING OF:

H. Castner, Chairman
Edison Welding Institute

C. W. Dralle
Ampco Metal, Inc.

C. E. Fuerstenau
Lucas-Milhaupt Inc.

D. B. Holliday
Westinghouse Marine Division

P. W. Holsberg
U. S. Department of the Navy

D. Medley
Wisconsin Centrifugal Casting

T. Mertes
Ampco Metal, Inc.

D. Peters
Copper Development Association

M. N. Rogers
Batesville Casket Company, Inc.

J. Turriff
Ampco Metal, Inc.

K. G. Wold
Siemens Power Corporation

WELDING HANDBOOK COMMITTEE MEMBER:
R. M. Walkosak
Westinghouse Electric Corporation

COPPER AND COPPER ALLOYS

INTRODUCTION

COPPER AND MANY of its alloys have a face-centered cubic lattice that accounts for its good formability and malleability. In pure form, copper has a density of 0.32 lb/in.^{3} (8.94 Mg/m^3), about three times that of aluminum. Electrical and thermal conductivity of copper is slightly lower than silver, but about one and one-half times that of aluminum. Copper and copper alloys are used for their electrical and thermal conductivity, corrosion resistance, metal-to-metal wear resistance, and distinctive aesthetic appearance.

The greatest single use of copper results from its high electrical conductivity. Copper is widely used for electrical conductors and for the manufacture of electrical equipment. Copper is the electrical conductivity standard of the engineering world with the rating of 100% IACS (International Annealed Copper Standard). The electrical conductivity of all materials are compared to the IACS standard. Some specially processed copper forms can reach 102% IACS.

Copper is resistant to oxidation, fresh and salt water, alkaline solutions and many organic chemicals. This good corrosion resistance makes copper alloys ideally suited for water tubing, valves, fittings, heat exchangers, chemical equipment, and bearings. Copper reacts with sulfur and ammonia compounds. Ammonium hydroxide solutions attack copper and copper alloys rapidly.

The pleasing color, relatively good strength, and good formability make copper and copper alloys highly favored for architectural applications such as decorative furnishings and roofing.

Copper and most copper alloys can be joined by welding, brazing, and soldering. These joining processes and applications are explained in this chapter. This chapter also describes the major classes of copper alloys, their metallurgy and processing, and how alloying elements affect their joining characteristics. Various sections of the chapter are identified with specific alloy groups. Readers interested in specific alloys may wish to skip to those sections.

ALLOYS

METALLURGY

MANY COMMON METALS are alloyed with copper, mostly within the limits of solid solution solubility. The principal alloying elements in copper alloys are aluminum, nickel, silicon, tin, and zinc. Small quantities of other elements also are added to improve mechanical properties, corrosion resistance, or machinability; to provide response to strengthening heat treatments; or to deoxidize the alloy.

CLASSIFICATION

COPPER AND COPPER alloys are classified into nine major groups:

(1) Coppers – 99.3% Cu minimum
(2) High-copper alloys – up to 5% alloying element
(3) Copper-zinc alloys (brass)
(4) Copper-tin alloys (phosphor bronze)
(5) Copper-aluminum alloys (aluminum bronze)
(6) Copper-silicon alloys (silicon bronze)
(7) Copper-nickel alloys
(8) Copper-nickel-zinc alloys (nickel-silvers)
(9) Special alloys

These alloys are further divided into the wrought and cast alloy categories shown in Table 3.1. The Unified Numbering System (UNS) has a five-digit number. Copper alloys C1xxxx to C7xxxx are wrought alloys, and C8xxxx to C9xxxx are cast alloys. Therefore, an alloy manufactured in both a wrought form and cast form can have two numbers depending upon method of manufacture. Copper and copper alloys have common names such as oxygen-free copper, beryllium copper, Muntz metal, phosphor bronze, and low-fuming bronze. These common or trade names are being replaced with UNS numbers.

Physical properties of copper alloys important to welding, brazing, and soldering include melting temperature range, coefficient of thermal expansion, and electrical and thermal conductivity. Physical properties for some of the most widely used copper alloys are listed in Table 3.2. The table data show that when alloying elements are added to copper, electrical and thermal conductivity decreases drastically. The electrical and thermal conductivity of an alloy will significantly affect the welding procedures used for the alloy.

Small additions of some elements (e.g., iron, silicon, tin, arsenic, and antimony) improve the corrosion and erosion resistance of copper alloys. Lead, selenium, and tellurium are added to copper alloys to improve machinability. Bismuth is beginning to be used for this purpose when lead-free alloys are desired.

Boron, phosphorus, silicon, and lithium are added to copper as deoxidizers during melting and refining. Silver and cadmium increase the softening temperature of copper. Cadmium, cobalt, zirconium, chromium, and beryllium additions to copper form precipitation hardening alloys that increase the strength of copper.

Many commercial copper alloys are single-phase solid solutions. Some copper alloys have two or more microstructural phases. These alloys can be hardened by precipitation of intermetallic compounds or by quenching from above the critical transformation temperature, which results in a martensitic transformation.

Solid-solution copper alloys are generally easily cold worked, although the force to cold work and the rate of work hardening increases with alloy content. Two-phase alloys harden more rapidly during cold working but usually have better hot-working characteristics than

do solid solutions of the same alloy system. Ductility decreases and yield strength increases as the proportion of the second phase increases.

MAJOR ALLOYING ELEMENTS

Aluminum

THE COPPER-ALUMINUM ALLOYS may contain up to 15 percent aluminum as well as additions of iron, nickel, tin, and manganese. The solubility of aluminum in copper is 7.8 percent, although this is slightly increased with the usual addition of iron. Alloys with less than 8 percent aluminum are single-phase, with or without iron additions. When the aluminum is between 9 and 15 percent, the system is two-phase and capable of either a martensitic or a eutectoid type of transformation. Increasing amounts of aluminum increase tensile strength, increase yield strength and hardness, and decrease elongation of the alloy. Aluminum forms a refractory oxide that must be removed during welding, brazing, or soldering.

Arsenic

ARSENIC IS ADDED to copper alloys to inhibit dezincification corrosion of copper-zinc alloys in water. Arsenic additions to copper alloys do not cause welding problems unless the alloy also contains nickel. Arsenic is detrimental to the welding of copper alloys that contain nickel.

Beryllium

THE SOLUBILITY OF beryllium in copper is approximately 2 percent at 1600°F (871°C) and only 0.3 percent at room temperature. Therefore, beryllium easily forms a supersaturated solution with copper that will precipitate in an age-hardening treatment. Because thermal conductivity and melting point decrease with increasing beryllium content, the higher beryllium content alloys are more easily welded. Beryllium forms a refractory oxide that must be removed for welding, brazing, or soldering.

Boron

BORON STRENGTHENS AND deoxidizes copper. Boron deoxidized copper is weldable with matching filler metals, and other coppers are weldable with boron-containing filler metals.

Table 3.1
Classification of Copper and Copper Alloys

Category	Description	Range of UNS Numbers [a]
	Wrought alloys [b]	
Copper	Copper-99.3 percent minimum	C10100-C15760
High-copper alloys	Copper-96 to 99.2 percent	C16200-C19750
Brasses	Copper-zinc alloys	C20500-C28580
Leaded brasses	Copper-zinc-lead alloys	C31200-C38590
Tin brasses	Copper-zinc-tin alloys	C40400-C49080
Phosphor bronzes	Copper-tin alloys	C50100-C52400
Leaded phosphor bronzes	Copper-tin-lead-alloys	C53200-C54800
Aluminum bronzes	Copper-aluminum alloys	C60600-C64400
Silicon bronzes	Copper-silicon alloys	C64700-C66100
Miscellaneous brasses	Copper-zinc alloys	C66400-C69950
Copper-nickels	Nickel-3 to 30 percent	C70100-C72950
Nickel-silvers	Copper-nickel-zinc alloys	C73150-C79900
	Cast alloys [c]	
Coppers	Copper-99.3 percent minimum	C80100-C81200
High-copper alloys	Copper-94 to 99.2 percent	C81300-C82800
Red brasses		C83300-C83810
Semi-red brasses }	Copper-tin-zinc and	C84200-C84800
Yellow brasses }	copper-tin-zinc-lead alloys	C85200-C85800
Manganese bronze	Copper-zinc-iron alloys	C86100-C86800
Silicon bronzes and silicon brasses	Copper-zinc-silicon alloys	C87300-C87900
Tin bronzes	Copper-tin alloys	C90200-C91700
Leaded tin bronzes	Copper-tin-lead alloys	C92200-C94500
Nickel-tin bronzes	Copper-tin-nickel alloys	C94700-C94900
Aluminum bronzes	Copper-aluminum-iron and copper-aluminum-iron-nickel alloys	C95200-C95900
Copper-nickels	Copper-nickel-iron alloys	C96200-C96900
Nickel-silvers	Copper-nickel-zinc alloys	C97300-C97800
Leaded coppers	Copper-lead alloys	C98200-C98840
Special alloys		C99300-C99750

a. Refer to ASTM/SAE Publication DS-56/HS 1086, *Metals and Alloys in the Unified Numbering System,* 6th Ed., 1993. ASTM, Philadelphia, Pa., and Society of Automotive Engineers, Warrendale, Pa.

b. For composition and properties, refer to *Standards Handbook, Part 2-Alloy Data, Wrought Copper and Copper Alloy Mill Products,* 8th Ed., New York: Copper Development Association, Inc., 1985.

c. For composition and properties, refer to *Standards Handbook, Part 7-Data/Specifications, Cast Copper and Copper Alloy Products,* New York: Copper Development Association, Inc., 1970.

Cadmium

THE SOLUBILITY OF cadmium in copper is approximately 0.5 percent at room temperature. The presence of cadmium in copper up to 1.25 percent causes no serious difficulty in fusion welding because it evaporates from copper rather easily at the welding temperature. A small amount of cadmium oxide may form in the molten metal, but it can be fluxed without difficulty. Cadmium-copper rod is Resistance Welding Manufacturers Association Class 1 alloy. The small amount of cadmium strengthens pure copper while maintaining a very high conductivity. This combination of properties makes this material ideal for electrodes used for resistance welding high-conductivity alloys such as aluminum. Cadmium-alloyed copper has been largely replaced by an overaged chromium-copper because of federal restrictions regarding the use of heavy metals in manufacturing.

Chromium

THE SOLUBILITY OF chromium in copper is approximately 0.55 percent at 1900 °F (1038 °C) and less than 0.05 percent at room temperature. The phase that forms during age hardening is almost pure chromium. Chromium coppers can develop a combination of high strength and good conductivity. Like aluminum and

Table 3.2
Physical Properties of Typical Wrought Copper Alloys

Alloy	UNS No.	Melting Range		Coefficient of Thermal Expansion at 68-572 °F (20-300 °C)		Thermal Conductivity at 68 °F (20 °C)		Electrical Conductivity, % IACS
		°F	°C	µin./(in.•°F)	µm/(m•K)	Btu/(ft•h•°F)	W/(m•K)	
Oxygen-free copper	C10200	1948-1991	1066-1088	9.8	17.6	214	370	101
Beryllium-copper	C17200	1590-1800	866-982	9.9	17.8	62-75	107-130	22
Commercial bronze	C22000	1870-1910	1021-1043	10.2	18.4	109	188	44
Red brass	C23000	1810-1880	988-1027	10.4	18.7	92	159	37
Cartridge brass	C26000	1680-1750	916-955	11.1	20.0	70	121	28
Phosphor bronze	C51000	1750-1920	955-1049	9.9	17.8	40	69	15
Phosphor bronze	C52400	1550-1830	843-999	10.2	18.4	29	50	11
Aluminum bronze	C61400	1905-1915	1041-1046	9.0	16.2	39	67	14
High-silicon bronze	C65500	1780-1880	971-1027	10.0	18.0	21	36	7
Manganese bronze	C67500	1590-1630	866-888	11.8	21.2	61	105	24
Copper-nickel, 10%	C70600	2010-2100	1099-1149	9.5	17.1	22	38	9
Copper-nickel, 30%	C71500	2140-2260	1171-1238	9.0	16.2	17	29	4.6
Nickel-silver, 65-15	C75200	1960-2030	1071-1110	9.0	16.2	19	33	6

beryllium, chromium can form a refractory oxide on the molten weld pool that makes oxyfuel gas welding difficult unless special fluxes are used. Arc welding should be done using a protective atmosphere over the molten weld pool.

Iron

THE SOLUBILITY OF iron in copper is approximately 3 percent at 1900 °F (1038 °C) and less than 0.1 percent at room temperature. Iron is added to aluminum bronze, manganese bronze, and copper-nickel alloys to increase their strength by solid solution and precipitation hardening. Iron increases the erosion and corrosion resistance of copper-nickel alloys. Iron must be kept in solid solution or in the form of an intermetallic to obtain the desired corrosion resistance benefit, particularly in copper-nickel alloys. Iron also acts as a grain refiner. Iron has little effect on weldability when used within the alloy specification limits.

Lead

LEAD IS ADDED to copper alloys to improve machinability or bearing properties and the pressure tightness of some cast copper alloys. Lead does not form a solid solution with copper and is almost completely insoluble (0.06 percent) in copper at room temperature. Lead is present as pure, discrete particles and is still liquid at 620 °F (327 °C). Leaded copper alloys are hot-short and susceptible to cracking during fusion welding. Lead is the most harmful element with respect to the weldability of copper alloys.

Manganese

MANGANESE IS HIGHLY soluble in copper. It is used in proportions of 0.05 to 3.0 percent in manganese bronze, deoxidized copper, and copper-silicon alloys. Manganese additions are not detrimental to the weldability of copper alloys. Manganese improves the hot working characteristics of multiphase copper alloys.

Nickel

COPPER AND NICKEL are completely solid soluble in all proportions. Although copper-nickel alloys are readily welded, residual elements may lead to embrittlement and hot cracking. There must be sufficient deoxidizer or desulfurizer in the welding filler metal used for copper-nickel to provide a residual amount in the solidified weld metal. Manganese is most often used for this purpose.

Phosphorus

PHOSPHORUS IS USED as a strengthener and deoxidizer in certain coppers and copper alloys. Phosphorus is soluble in copper up to 1.7 percent at the eutectic temperature of 1200 °F (649 °C), and approximately 0.4 percent at room temperature. When added to copper-zinc alloys, phosphorus inhibits dezincification. The amount of phosphorus that is usually present in copper alloys has no effect on weldability.

Silicon

THE SOLUBILITY OF silicon in copper is 5.3 percent at 1500 °F (816 °C) and 3.6 percent at room temperature. Silicon is used both as a deoxidizer and as an alloying element to improve strength, malleability, and ductility. Copper-silicon alloys have good weldability, but are hot-short at elevated temperatures. In welding, the cooling rate through this hot-short temperature range should be fast to prevent cracking.

Silicon oxide forms on copper-silicon alloys at temperatures as low as 400 °F (204 °C). This oxide will interfere with brazing and soldering operations unless a suitable flux is applied prior to heating.

Tin

THE SOLUBILITY OF tin in copper increases rapidly with temperature. At 1450 °F (788 °C), the solubility of tin is 13.5 percent; at room temperature, it is probably less than 1 percent. Alloys containing less than 2 percent tin may be single-phase when cooled rapidly.

Copper-tin alloys tend to be hot-short and to crack during fusion welding. Tin oxidizes when exposed to the atmosphere, and this oxide may reduce weld strength if trapped within the weld metal.

Zinc

ZINC IS THE most important alloying element used commercially with copper. Zinc is soluble in copper up to 32.5 percent at 1700 °F (927 °C) and 37 percent at room temperature. A characteristic of all copper-zinc alloys is the relative ease that zinc will volatilize from the molten metal with very slight superheat.

Zinc is also a residual element in aluminum bronze and copper-nickel and may cause porosity or cracking, or both.

MINOR ALLOYING ELEMENTS

CALCIUM, MAGNESIUM, LITHIUM, sodium, or combinations of these elements are added to copper alloys as deoxidizers. Very little of these oxidizing elements remain in copper alloys and are seldom a factor in welding.

Antimony, arsenic, phosphorus, bismuth, selenium, sulfur, and tellurium may cause hot cracking when alloyed in single-phase aluminum bronze and in copper-nickel alloys. The small amounts of antimony added to brasses have little influence on their weldability.

Carbon is practically insoluble in copper alloys unless large amounts of iron, manganese, or other strong carbide formers are present. Carbon embrittles copper alloys by precipitating in the grain boundaries as graphite or as an intermetallic carbide.

EFFECTS OF ALLOYING ELEMENTS ON JOINING

THE HIGH ELECTRICAL and thermal conductivity of copper and certain high-copper alloys has a marked effect on weldability. Welding heat is rapidly conducted into the base metal and may promote incomplete fusion in weldments. Preheating of copper alloys will reduce welding heat input requirements necessary for good fusion.

Copper alloys are often hardened by mechanical cold working, and any application of heat tends to soften them. The heat-affected zone (HAZ) of these weldments will be softer and weaker than the adjacent base metal. The HAZ tends to hot crack in severely cold-worked metal. In practice, there is a time-temperature reaction, so that a minimum preheat and interpass temperature control can keep the softening of the HAZ to a minimum.

Precipitation hardening in copper alloys is obtained when copper is alloyed with beryllium, chromium, boron, nickel-silicon, and zirconium. Alloys with these elements are classified as precipitation hardening.

For optimum results, components to be precipitation hardened should be welded in the annealed condition, followed by the precipitation hardening heat treatment. Welding, brazing, or soldering precipitation-hardened alloys may result in reduction of mechanical properties due to overaging.

Copper alloys with wide liquidus-to-solidus temperature ranges, such as copper-tin and copper-nickel, are susceptible to hot cracking. Because low-melting interdendritic liquid solidifies at a lower temperature than the bulk dendrite, shrinkage stresses may produce interdendritic separation during solidification. Hot cracking can be minimized by the following:

(1) Reducing restraint during welding
(2) Minimizing heat input and interpass temperature
(3) Reducing the size of the root opening and increasing the size of the root pass

Certain elements such as zinc, cadmium, and phosphorus have low boiling points. Vaporization of these elements during welding may result in porosity. Porosity can be minimized by increasing travel speed and using filler metal having low percentages of these volatile elements.

Surface oxides on aluminum bronze, beryllium copper, chromium copper, and silicon bronze are difficult to remove and can present problems when welding, brazing, or soldering. Surfaces to be joined must be clean, and special fluxing or shielding methods must be used to prevent surface oxides from reforming during the joining operation.

COPPER AND HIGH-COPPER ALLOYS

Oxygen-Free Copper

OXYGEN-FREE COPPERS (UNS Nos. C10100 to C10800) contain a maximum of 10 ppm oxygen and a minimum of 0.01 percent total of other elements. Oxygen-free copper is produced by melting and casting under a reducing atmosphere that prevents oxygen contamination. No deoxidizing agent is introduced in production of this type of copper, but oxygen can be absorbed from the atmosphere during heating at high temperatures. Absorbed oxygen can cause problems during subsequent welding or brazing of the copper.

Oxygen-free copper has mechanical properties similar to those of oxygen-bearing copper, but the microstructure is more uniform. Oxygen-free copper has excellent ductility and is readily joined by welding, brazing, and soldering. Silver may be added to oxygen-free copper to increase the elevated temperature strength without changing the electrical conductivity. The addition of silver prevents appreciable softening of cold-worked copper during short-term elevated temperature exposure. Silver increases the allowable creep stress or provides resistance to creep rupture over long time periods. The silver addition does not effect the joining characteristics.

Oxygen-Bearing Copper

OXYGEN-BEARING COPPERS include the electrolytic tough-pitch grades (UNS Nos. C11000-C11900) and fire-refined grades (UNS Nos. C12500-C13000).

Fire-refined coppers contain varying amounts of impurities including antimony, arsenic, bismuth and lead. Electrolytic tough-pitch coppers contain minimal impurities and have more uniform mechanical properties. The residual oxygen content of electrolytic tough-pitch and fire-refined copper is about the same. Impurities and residual oxygen may cause porosity and other discontinuities when these coppers are welded or brazed.

A copper-cuprous oxide eutectic is distributed as globules throughout wrought forms of oxygen-bearing copper. Though this condition does not effect mechanical properties or electrical and thermal conductivity, it makes the copper susceptible to embrittlement when heated in the presence of hydrogen. Hydrogen diffuses rapidly into the hot metal, reduces the oxides, and forms steam at the grain boundaries. The metal will rupture when stressed.

When oxygen-bearing coppers are heated to high temperatures, the copper oxide tends to concentrate in the grain boundaries causing major reduction in strength and ductility. Fusion welding of oxygen bearing copper for structural applications is not recommended. Embrittlement will be less severe with a rapid solid-state welding process such as friction welding. Appropriate silver brazing procedures and soft soldering can be successfully used to join oxygen-bearing copper.

Phosphorus-Deoxidized Copper

PHOSPHORUS-DEOXIDIZED COPPER (UNS Nos. C12000 and C12300) has 0.004 to 0.065 percent residual phosphorus. The electrical conductivity of these coppers decreases in proportion to the residual phosphorus. When the phosphorus content is 0.009, electrical conductivity is about 100% IACS. Electrical conductivity is about 85% IACS, for a phosphorus content of 0.02, and electrical conductivity is about 75% IACS for a phosphorus content of 0.04.

Free-Machining Copper

FREE-MACHINING COPPERS HAVE UNS Nos. C14500 through C14710 and contain additions of lead, tellurium, and selenium. Copper has very low solid-solution solubility for these elements. Lead disperses throughout the matrix as fine, discrete particles, while tellurium and sulfur form hard stringers in the matrix. These inclusions reduce the ductility of copper but enhance its machinability. Fusion welding is not recommended for free-machining coppers because these alloys are hot-short and very susceptible to cracking. Free-machining coppers are joined by brazing and soldering.

Precipitation-Hardenable Copper Alloys

PRECIPITATION-HARDENABLE COPPER ALLOYS have UNS Nos. C15000, C15100, C17000-C18400, and C64700-C64730. Small amounts of beryllium, chromium, or zirconium can be added to copper to form alloys that respond to precipitation hardening heat treatment to increase mechanical properties. These copper alloys are solution annealed (to a soft condition about RB 50) by heating to an elevated temperature that puts the alloying elements into solution. Rapid cooling by water quenching keeps the alloying elements in solid solution. The parts are then aged at temperatures of 600 to 900 °F (316 to 482 °C). During aging, a second phase precipitates within the matrix that inhibits plastic deformation, resulting in greatly enhanced mechanical properties. The solution annealed alloy can be cold worked prior to aging to achieve higher strength. Exposing precipitation-hardened alloys to welding or brazing temperatures will overage the exposed area. Overaging softens and results in lower mechanical properties. Mechanical property degradation is dependent upon the temperature and time at temperature. Welding may only overage the HAZ, but brazing may overage the entire part.

COPPER-ZINC ALLOYS (BRASS)

COPPER ALLOYS IN which zinc is the major alloying element are generally called *brasses* (UNS Nos. C20500-C49080, C66400-C69950, C83300-C86800). Some copper-zinc alloys have other common or trade names, such as commercial bronze, Muntz metal, manganese bronze, and low-fuming bronze. Other elements are occasionally added to brasses to enhance particular mechanical or corrosion characteristics. Additions of manganese, tin, iron, silicon, nickel, lead, or aluminum, either singly or collectively, rarely exceed 4 percent. Some of these special brasses are identified by the name of the second alloying element; two examples are aluminum brass and tin brass.

Addition of zinc to copper decreases the melting temperature, the density, the electrical and thermal conductivity, and the modulus of elasticity. Zinc additions increase the strength, hardness, ductility, and coefficient of thermal expansion. Hot-working properties of brass decrease with increasing zinc content up to about 20 percent.

The color of brass changes with increasing zinc content from reddish to gold to a light gold and finally to yellow. Selection of a welding filler metal may depend on matching the brass color when joint appearance is important.

Most brasses are single-phase, solid-solution copper-zinc alloys with good room-temperature ductility. Brasses containing about 36 percent or more zinc have two microstructural phases designated *alpha* and *beta*. The beta phase improves the hot-working characteristics of brass, but has little effect on electrical and thermal conductivity.

For joining considerations, brasses may be divided into three groups:

(**1**)   Low-zinc brasses (zinc content 20 percent maximum) have good weldability.

(**2**)   High-zinc brasses (zinc content greater than 20 percent) have only fair weldability.

(**3**)   Leaded brasses are considered unweldable, but they can be brazed and soldered satisfactorily.

The cast brasses contain from 2 percent to 41 percent zinc (UNS Nos. C83300-C85800) but often have one or more additional alloying elements, including tin, lead, nickel, and phosphorus. Cast alloys are generally not as homogeneous as the wrought products. In addition to welding complications caused by lead and other alloy elements, the variation in microstructure may cause difficulty. Cast alloys without lead are only marginally weldable, and leaded brasses are generally unweldable.

Manganese bronzes (UNS Nos. C86100-C86800) are actually high tensile brasses that contain 22 to 38 percent zinc with varying amounts of manganese, aluminum, iron, and nickel. Manganese bronze is weldable provided that lead content is low. Gas shielded arc welding methods are recommended. Manganese bronzes can be brazed and soldered with special fluxes.

COPPER-TIN ALLOYS (PHOSPHOR BRONZE)

ALLOYS OF COPPER and tin contain between 1 percent and 10 percent tin. These alloys are known as *phosphor bronzes* because 0.03 to 0.04 percent phosphorus is added during casting as a deoxidizing agent. The wrought alloys have UNS Nos. C50100-C52400. The cast copper-tin alloys (UNS Nos. C90200-C91700) are similar in nature to the wrought alloys but often have additions of zinc or nickel and contain high amounts of tin, up to 20 percent. There are also leaded copper-tin alloys (UNS Nos. C92200-C94500).

In the completely homogenized condition, these alloys are single-phase alloys with a structure similar to alpha brass. The alloys with a tin content over 5 percent are difficult to cast without dendritic segregation and the formation of a beta phase. During cooling, this beta phase gives rise to a delta phase, which can be embrittling.

In the wrought form, copper-tin alloys are tough, hard and highly fatigue-resistant, particularly in the cold-worked condition. The phase diagram predicts the precipitation of a copper-tin compound at room temperature. However, this is not observed. Some very fine precipitation may occur during cold working and this would explain the very high strengths achieved in wrought material. Electrical and thermal conductivities are low for the low-tin-content alloys and very low for those with high-tin content.

Copper-tin alloys have a narrow plastic range and must be hot worked at temperatures from 1150 to 1250 °F (621 to 677 °C). Low-tin alloys (under 4 percent) have the best hot-working properties.

All of the copper-tin alloys have good cold-working properties and high strength and hardness in the hard-rolled tempers. After cold working, these alloys can be rendered soft and malleable by annealing at temperature between 900 and 1400 °F (482 and 760 °C), depending on the properties desired. In a stressed condition, these alloys are subject to hot cracking. The use of high preheat temperatures, high-heat input, and slow cooling rates should be avoided.

Leaded copper-tin alloys (UNS Nos. C53400 and C54400) contain 2.0 to 6.0 percent lead to improve machinability. Welding of these alloys is not recommended. However, welds can be made in some alloys. Leaded copper-tin alloys are often two-phase structures, have a wide freezing range, and may be severely cored unless homogenized. Weldability decreases as lead content increases. Leaded copper-tin alloys can be welded with care by the shielded metal arc welding (SMAW) process. Inert gas welding processes are not recommended because welds will contain porosity. These alloys may be brazed and soldered if not strained while in the hot-short range.

Arc welding of cast leaded copper-tin alloys (C92200-C94500) is not recommended; but these alloys can be brazed and soldered with care.

COPPER-ALUMINUM ALLOYS (ALUMINUM BRONZE)

COPPER-ALUMINUM ALLOYS called *aluminum bronzes* (UNS Nos. C60600-C64400, C95200-C95900), contain from 3 to 15 percent aluminum, with or without varying amounts of iron, nickel, manganese, and silicon. There are two types of aluminum bronzes, and the types are based on metallurgical structure and response to heat treatment. The first type includes the alpha or single-phase alloys (less than 7 percent aluminum) that cannot be hardened by heat treatment. The second type includes the two-phase, alpha-beta alloys. Both types have low electrical and thermal conductivity that enhances weldability.

The alpha aluminum bronzes are readily weldable without preheating. Aluminum bronzes with an aluminum content below approximately 8.5 percent have a tendency to be hot-short, and cracking may occur in the HAZ of highly stressed weldments.

The single-phase alloy welded with an electrode that straddles the aluminum solubility limit results in the weldment being two-phase at elevated temperature but has a lower alpha-beta content at room temperature. The resulting weldment has better hot-working characteristics at elevated temperatures and sufficient strength at room temperature to match the single-phase (alpha) base metal. Low residual-element content in UNS No. C61300 improves most welding properties.

Generally, aluminum bronzes containing from 9.5 to 11.5 percent aluminum have both alpha and beta phases in their microstructures. These two-phase alloys can be strengthened by heat treatment to produce a martensitic-type structure and tempered to obtain desired mechanical properties. Microstructures after heat treatment are analogous in many respects to those found in steels. Hardening is accomplished by quenching in water or oil from 1550 to 1850 °F (843 to 1010 °C), followed by tempering at of 800 to 1200 °F (427 to 649 °C). The specific heat treatment depends upon the composition of the alloy and the desired mechanical properties.

Two-phase alloys have very high tensile strengths compared to most other copper alloys. As the aluminum content of these alloys increases, ductility decreases and hardness increases. Alpha-beta alloys have a plastic range wider than the alpha alloys, and this contributes to their good weldability.

Aluminum bronzes resist oxidation and scaling at elevated temperatures due to the formation of aluminum oxide on the surface. All aluminum oxide must be removed before welding, brazing, or soldering.

The single-phase aluminum bronzes (UNS Nos. C60600, C61300, C61400) are produced as wrought alloys only, although the single-phase nickel aluminum bronzes (UNS Nos. C63200-C95800) are produced both as castings and wrought products. Duplex aluminum bronzes are produced both cast and wrought and have similar characteristics. The complex aluminum bronzes are temper annealed to resist dealuminification in a sea water environment and, unless welding is of a very minor nature, the part should be retempered or annealed after welding.

Nickel-aluminum bronzes contain from 8.5 to 11 percent aluminum and from 3 to 5 percent nickel; both have alpha and kappa in the microstructure. Alloys with an aluminum content in the upper end of the range can contain the eutectoid phase, gamma 2 (γ_2), when cooled slowly from elevated temperature. A temper anneal at 1150 to 1225 °F (620 to 663 °C), followed by a rapid air cool is recommended for alloys exposed to corrosive environments.

Nickel-aluminum bronze is susceptible to cracking when welded. Therefore, procedures for welding heavy sections recommend the use of a nickel-free filler metal (ECuAl-A2, ERCuAl-A2) to fill the joint because of its greater ductility. ECuAl-A2 or ERCuAl-A2 filler metal is recommended for the root pass and for all but the last two or three cover passes. The final two or three passes are made with a nickel- aluminum bronze filler metal (ECuNiAl, ERCuNiAl). If there is any possibility of the ECuAl-A2 or ERCuAl-A2 weld metal being exposed by subsequent drilling, machining, or similar process, this procedure should not be employed.

Nickel-aluminum bronze weldments should be temper annealed for service in corrosive environments.

COPPER-SILICON ALLOYS (SILICON BRONZE)

COPPER-SILICON ALLOYS (UNS No. C64700-C66100, C87300-C87900), known as silicon bronzes, are industrially important because of their high strength, excellent corrosion resistance, and good weldability. The wrought alloys (C64700-C66100) contain from 1.5 to 4 percent silicon and 1.5 percent or less of zinc, tin, manganese or iron. With the exception of alloy C87300, the cast silicon bronze alloys have higher zinc levels (4 to 30 percent) to improve castability.

The addition of silicon to copper increases tensile strength, hardness, and work-hardening rates. The ductility of silicon bronze decreases with increasing silicon content up to about 1 percent. Ductility then increases to a maximum value at 4 percent silicon. Electrical and thermal conductivity decreases as the silicon content increases. Silicon bronzes should be stress relieved or annealed prior to welding, and should be slowly heated to the temperature desired. Silicon bronzes are hot-short at elevated temperatures and should be rapidly cooled through the critical temperature range.

Iron additions increase tensile strength and hardness. Zinc or tin additions improve the fluidity of molten bronze and improve the quality of castings and of welds made using oxyfuel gas welding.

COPPER-NICKEL ALLOYS

COMMERCIAL COPPER-NICKEL ALLOYS (UNS NOS. C70100-C72950, C96200-C96900) have nickel contents ranging from 5 to 45 percent. Copper-nickel alloys most commonly used in welded fabrication contain 10 and 30 percent nickel and minor alloying elements such as iron, manganese, or zinc. Resistance to erosion corrosion requires that any iron should be in solid solution. Thermal processing of the copper-nickel alloy must be done in a manner that does not cause precipitation of iron compounds.

The copper-nickel alloys have moderately high tensile strengths that increase with nickel content. These alloys are ductile and relatively tough, and they have a relatively low electrical and thermal conductivity.

Like nickel and some nickel alloys, the copper-nickel alloys are susceptible to lead or sulfur embrittlement. Phosphorus and sulfur levels in these alloys should be a maximum of 0.02 percent to ensure sound welds. Contamination from sulfur-bearing marking crayons or cutting lubricants are likely to cause cracking during welding.

Most copper-nickel alloys do not contain a deoxidizer, which means that fusion welding requires addition of deoxidized filler metal to avoid porosity.

Special compositions of some copper-nickel alloys that contain titanium can be obtained. These special alloys are recommended for autogenous welding of thin sheet.

Silicon is added to cast alloys for added fluidity during casting and for added strength of the cast structure.

COPPER-NICKEL-ZINC ALLOYS (NICKEL-SILVER)

NICKEL IS ADDED to copper-zinc alloys to make them silvery in appearance for decorative purposes and to increase their strength and corrosion resistance. The resulting copper-nickel-zinc alloys (UNS Nos. C73200-C79900, C97300-C97800) are called nickel-silvers. These alloys are of two general types:

(1) Single-phase alloys containing 65 percent copper plus nickel and zinc
(2) Two-phase (alpha-beta) alloys containing 55 to 60 percent copper plus nickel and zinc

The welding metallurgy of these alloys is similar to that of the brasses.

JOINING PROCESS SELECTION

COPPER AND COPPER alloys can be joined by welding, brazing, and soldering processes. Table 3.3 summarizes the applicability of the most commonly used processes for major alloy classifications.

ARC WELDING

COPPER AND MOST copper alloys can be joined by arc welding. Welding processes that use gas shielding are generally preferred, although shielded metal arc welding (SMAW) can be used for many noncritical applications.

Argon, helium, or mixtures of the two are used as shielding gases for gas tungsten arc welding (GTAW), plasma arc welding (PAW), and gas metal arc welding (GMAW). In general, argon is used when manually welding material that is either less than 0.13 in. (3.3 mm) thick or has low thermal conductivity, or both.

Table 3.3
Applicable Joining Processes for Copper and Copper Alloys

Alloy	UNS No.	Oxyfuel Gas Welding	SMAW	GMAW	GTAW	Resistance Welding	Solid-State Welding	Brazing	Soldering	Electron Beam Welding
ETP Copper	C11000-C11900	NR	NR	F	F	NR	G	E	G	NR
Oxygen-Free Copper	C102000	F	NR	G	G	NR	E	E	E	G
Deoxidized Copper	C12000 C123000	G	NR	E	E	NR	E	E	E	G
Beryllium-Copper	C17000-C17500	NR	F	G	G	F	F	G	G	F
Cadmium/Chromium Copper	C16200 C18200	NR	NR	G	G	NR	F	G	G	F
Red Brass - 85%	C23000	F	NR	G	G	F	G	E	E	—
Low Brass - 80%	C24000	F	NR	G	G	G	G	E	E	—
Cartridge Brass - 70%	C26000	F	NR	F	F	G	G	E	E	—
Leaded Brasses	C31400-C38590	NR	NR	NR	NR	NR	NR	E	G	—
Phosphor Bronzes	C50100-C52400	F	F	G	G	G	G	E	E	—
Copper-Nickel - 30%	C71500	F	F	G	G	G	G	E	E	F
Copper-Nickel - 10% Nickel-Silvers	C70600 C75200	F G	G NR	E G	E G	G F	G G	E E	E E	G —
Aluminum Bronze	C61300 C61400	NR	G	E	E	G	G	F	NR	G
Silicon Bronzes	C65100 C65500	G	F	E	E	G	G	E	G	G

E = Excellent   G = Good   F = Fair   NR = Not Recommended

Helium or a mixture of 75% helium-25% argon is recommended for machine welding of thin sections and for manual welding thicker sections or alloys having high thermal conductivity. Small additions of nitrogen or hydrogen to the argon shielding gas may be used to increase the effective heat input.

The SMAW process can be used to weld a range of thicknesses of copper alloys. Covered electrodes for SMAW of copper alloys are available in standard sizes ranging from 3/32 to 3/16 in. (2.4 mm to 4.8 mm). Other sizes are available in certain electrode classifications.[1] Submerged arc welding has been used for welding of copper alloys, although use of this process is not widespread.

Arc welding should be done in the flat position whenever practical. GTAW or SMAW is preferred for welding in positions other than flat, particularly in the overhead position. GMAW with pulsed power and small diameter electrodes is also suitable for the vertical and overhead positions with some copper alloys. Higher thermal conductivity and thermal expansion of copper and its alloys result in greater weld distortion than in comparable steel welds. The use of preheat, fixtures, proper welding sequence, and tack welds can minimize distortion or warping.

1. See ANSI/AWS A5.6, *Specification for Covered Copper and Copper Alloy Arc Welding Electrodes.*

OXYFUEL GAS WELDING

COPPER AND MANY copper alloys can be welded with the oxyfuel gas welding (OFW) process. The OFW process should only be used for small, noncritical applications, including repair welding. The relatively low heat input of the oxyacetylene flame makes welding slow compared to arc welding. Higher preheat temperatures or an auxiliary heat source may be required to counterbalance the low heat input, particularly with alloys with high thermal conductivity or with thick sections. Except for oxygen-free copper, a welding flux is required to exclude air from the weld metal at elevated temperatures.

LASER WELDING

LASER BEAM WELDING (LBW) of copper and its alloys has very limited applications. The primary difficulties with LBW of copper are the high reflectivity to the incident laser beam and the high thermal conductivity of copper and copper alloys. Copper reflects approximately 99 percent of the incident light energy of the far infrared wavelength of the CO_2 laser. This is the reason copper is commonly used for mirrors in CO_2 laser beam delivery systems. Reflectivity is temperature dependent; when a material gets hotter, the absorption of the incident light increases. However, the high thermal conductivity of copper prevents the metal from getting hotter, thereby maintaining high reflectivity.

Lasers with shorter wavelengths have successfully welded some copper alloys. Copper has slightly higher absorption of the incident light of Nd:YAG lasers with a wavelength of 1.06 µm. Plating copper with a thin layer of higher absorbing metal, such as nickel, has been demonstrated to improve coupling efficiency.

LBW is a fusion welding process so that the same considerations as other fusion processes apply. Higher cracking susceptibility may be encountered with copper alloys with wide liquidus-to-solidus temperature ranges due to the high solidification stresses resulting from the rapid cooling rates of LBW.

ELECTRON BEAM WELDING

COPPER AND ITS alloys can be readily joined by electron beam welding (EBW). The EBW process has been successfully applied for welding thin and thick gage copper alloys both in and out of vacuum. Filler metal can be added to a weld with an auxiliary wire feeder.

ULTRASONIC WELDING

COPPER-TO-COPPER ultrasonic welding is an attractive technique for microelectronic interconnections. Special care must be paid to joint preparation and cleaning or inconsistent joint quality will result.

RESISTANCE SPOT WELDING

THE EASE OF resistance spot welding (RSW) of copper and copper alloys varies inversely with their electrical and thermal conductivity. Many of the lower conductivity copper alloys are readily spot welded.

Spot welds can be made in sheet copper alloys having an electrical conductivity 30% IACS or less, including beryllium copper, many brasses and bronzes, nickel-silver, and copper-nickel alloys. Weld quality becomes less consistent as the electrical conductivity increases. Copper alloys with electrical conductivity over about 60 percent cannot be spot welded with conventional methods. Resistance spot welding of unalloyed copper is not practical.

RSW electrode forces for copper alloys are usually set to 50 to 70 percent of those used for the same thickness of steel. Welding current is higher and welding time is shorter than that used for steel. Tungsten or molybdenum-tipped RSW electrodes are preferred to minimize electrode sticking.

RESISTANCE SEAM WELDING

IT IS DIFFICULT to seam weld copper alloys because of excessive shunting of welding current, high thermal conductivity, and low electrode contact resistance. Seam welding is generally not practicable when the electrical conductivity exceeds 30 percent IACS. Copper alloys that can be spot welded can usually be seam welded.

FLASH WELDING

FLASH WELDING TECHNIQUES produce very good results on copper and copper alloys. The design of the equipment must provide accurate control of all factors, including upset pressure, platen travel, flash-off rate, current density, and flashing time.

Leaded copper alloys can be flash welded, but the integrity of the joint depends upon the alloy composition. Lead content of up to 1.0 percent is usually not detrimental.

Rapid upsetting at minimum pressure is necessary as soon as the abutting faces are molten because of the relatively low melting temperature and narrow plastic range of copper alloys. Low pressure is usually applied to the joint before the flashing current is initiated so that platen motion will begin immediately after flashing starts. Termination of flashing current is critical. Premature termination of current will result in lack of fusion at the weld interface. Excessive flashing will overheat the metal and result in improper upsetting.

FRICTION WELDING

ALTHOUGH LIMITED IN application, friction welding offers several advantages for joining copper and copper alloys. The heat-affected zone is very narrow, and the joint contains no cast metal microstructure. Joint properties are excellent.

The process can be used to join copper to itself, to copper alloys, and to other metals including aluminum, silver, carbon steel, stainless steel, and titanium.

HIGH-FREQUENCY WELDING

COPPER AND COPPER alloy tubing is frequently manufactured from strip in a tube mill using high-frequency resistance welding. The edges of the weld joint are resistance heated to welding temperature utilizing the skin effect with high-frequency current. The heated edges are forged together continuously in the tube mill to consummate a weld.

SOLID-STATE WELDING

COPPER CAN READILY be welded without melting using various combinations of temperatures, pressures, and deformations. Annealed copper can be cold welded at room temperature because of its excellent malleability. Copper tubing can be welded and pinched off using commercially available steel dies. Copper and copper alloys also can be diffusion welded and explosive welded.

BRAZING

COPPER AND ITS alloys are readily joined by brazing using an appropriate filler metal and flux or protective atmosphere. Any of the common heating methods can be used. Certain precautions are required with specific base metals to avoid embrittlement, cracking, or excessive alloying with the filler metal. Special fluxes are required with some alloys that form refractory surface oxides.

SOLDERING

COPPER AND MOST copper alloys are readily soldered with commercial solders. Most copper alloys are easily fluxed, except for those containing elements which form refractory oxides (e.g., beryllium, aluminum, silicon, or chromium). Special fluxes are required to remove refractory oxides that form on the surfaces of these alloys.

Soldering is primarily used for electrical connections, plumbing and other room-temperature applications. Joint strengths are much lower than those of brazed or welded joints.

WELDING

GENERAL CONSIDERATIONS

Filler Metals

COVERED ELECTRODES AND bare electrode wire and rods are available for welding copper and copper alloys to themselves and to other metals. These filler metals are included in the latest editions of: ANSI/AWS A5.6, *Specification for Covered Copper and Copper Alloy Arc Welding Electrodes*, and ANSI/AWS A5.7, *Specification for Copper and Copper Alloy Bare Welding Rods and Electrodes*. AWS classifications of filler metals for welding copper and copper alloys are listed in Table 3.4.

Copper Filler Metals. Bare copper electrodes and rods (ERCu) are generally produced with a minimum copper content of 98 percent. These electrodes and rods are used to weld deoxidized and electrolytic tough pitch copper with the gas metal arc welding (GMAW), gas tungsten arc welding (GTAW), plasma arc welding (PAW), and sometimes oxyfuel gas welding (OFW) processes. The electrical conductivity of ERCu electrodes is 25 to 40% IACS.

Covered electrodes (ECu) for shielded metal arc welding (SMAW) are designed for welding with direct current electrode positive (DCEP). The welding current is 30 to 40 percent higher than normally required for carbon steel electrodes of the same diameter.

Copper-Zinc (Brass) Filler Metals. Copper-zinc welding rods are available in the following classifications: RBCuZn-A (naval brass), RBCuZn-B (low-fuming brass), and RBCuZn-C (low-fuming brass). These welding rods are primarily used for OFW of brass and for braze welding copper, bronze, and nickel alloys. The RBCuZn-A welding rods contain 1 percent tin to improve corrosion resistance and strength. Electrical

Table 3.4
Filler Metals for Fusion Welding Copper Alloys

AWS Classification		Common Name	Base Metal Applications
Covered Electrode[a]	Bare Wire[b]		
ECu	ERCu	Copper	Coppers
ECuSi	ERCuSi-A	Silicon bronze	Silicon bronzes, brasses
ECuSn-A	ERCuSn-A	Phosphor bronze	Phosphor bronzes, brasses
ECuSn-C	ERCuSn-A	Phosphor bronze	Phosphor bronzes, brasses
ECuNi	ERCuNi	Copper-nickel	Copper-nickel alloys
ECuAl-A2	ERCuAl-A1	Aluminum bronze	Aluminum bronzes, brasses silicon bronzes, manganese bronzes
	ERCuAl-A2	—	
ECuAl-B	ERCuAl-A3	Aluminum bronze	Aluminum bronzes
ECuNiAl	ERCuNiAl	—	Nickel-aluminum bronzes
ECuMnNiAl	ERCuMnNiAl	—	Manganese-nickel-aluminum bronzes
	RBCuZn-A	Naval brass	Brasses, copper
	RBCuZn-B	Low-fuming brass	Brasses, manganese bronzes
	RBCuZn-C	Low-fuming brass	Brasses, manganese bronzes
	RBCuZn-D	—	Nickel-silver

a. See ANSI/AWS A5.6, *Specification for Covered Copper and Copper Alloy Arc Welding Electrodes.*

b. See ANSI/AWS A5.7, *Specification for Copper and Copper Alloy Bare Welding Rods and Electrodes,* and ANSI/AWS A5.8, *Specification for Filler Metals for Brazing and Braze Welding.*

conductivity of these rods is about 25% IACS, and the thermal conductivity is about 30 percent that of copper.

RBCuZn-B welding rods contain additions of manganese, iron, and nickel that increase hardness and strength. A small amount of silicon provides low-fuming characteristics. The RBCuZn-C welding rods are similar to RBCuZn-B rods in composition except that they do not contain nickel. The mechanical properties of as-deposited weld metal from both rods are similar to those of naval brass.

Copper-zinc filler metals cannot be used as electrodes for arc welding because of the high zinc content. The zinc vapor would volatilize from the molten weld pool, resulting in porous weld metal.

Copper-Tin (Phosphor Bronze) Filler Metals. Copper-tin, or phosphor bronze welding electrodes and rods include ECuSn-A, ERCuSn-A and ECuSn-C. The ECuSn-A composition contains about 5 percent tin, and the ECuSn-C composition has about 8 percent tin. Both electrodes are deoxidized with phosphorus. The electrodes can be used for welding bronze, brass, and also for copper if the presence of tin in the weld metal is not objectionable. These electrodes frequently are used for casting repairs. ERCuSn-A rods can be used with the GTAW process for joining phosphor bronze. The ECuSn-C electrodes provide weld metal with better strength and hardness than do ECuSn-A electrodes and are preferred for welding high-strength bronzes. Preheat and interpass temperature of about 400 °F (203 °C) is required when welding with these electrodes, especially for heavy sections.

Copper-Silicon (Silicon Bronze) Filler Metals. Copper-silicon (silicon bronze) electrodes are used in bare wire form (ERCuSi-A) for GMAW, for GTAW, and sometimes for OFW. Copper-silicon wires contain from 2.8 to 4.0 percent silicon with about 1.5 percent manganese, 1.0 percent tin, and 1.0 percent zinc. This filler wire is used for welding silicon bronzes and brasses as well as to braze weld galvanized steel. The tensile strength of copper-silicon weld metal is about twice that of ERCu weld metal. The electrical conductivity is about 6.5 percent IACS, and the thermal conductivity is about 8.4 percent of copper.

ECuSi covered electrodes are used primarily for welding copper-zinc alloys using direct current electrode positive. This electrode is occasionally used for welding silicon bronze, copper, and galvanized steel. The core wire of these covered electrodes contains about 3 percent silicon with small amounts of tin and manganese. The mechanical properties of the weld metal are usually slightly higher than those of copper-silicon base metal.

Copper-Aluminum (Aluminum Bronze). ERCuAl-A1 filler metal is an iron-free aluminum bronze. It is used as a surfacing alloy for wear-resistant surfaces having relatively light loads, for resistance to corrosive media such as salt and brackish water and for resistance to some commonly used acids. This alloy is not recommended for joining applications.

ECuAl-A2 covered electrodes for SMAW contain from 6.5 to 9 percent aluminum. ERCuAl-A2 bare wire electrodes for GTAW, GMAW and PAW contain from 8.5 to 11 percent aluminum. ERCuAl-A2 weld metal has a higher strength than the ECuAl-A2 weld metal. Both filler metals are used for joining aluminum bronze, silicon bronze, copper-nickel alloys, copper-zinc alloys, manganese bronze, and many combinations of dissimilar metals.

ECuAl-B covered electrodes contain 7.5 to 10 percent aluminum, and produce deposits with higher strength and hardness than the ERCuAl-A2 electrodes. These electrodes are used for surfacing applications and for repair welding of aluminum bronze castings of similar compositions.

ERCuAl-A3 electrodes and rods are used for repair welding of similar composition aluminum bronze castings using GMAW and GTAW processes. Their high aluminum content produces welds with less tendency to crack in highly stressed sections.

Copper-nickel-aluminum electrodes and bare-wire electrodes (ECuNiAl and ERCuNiAl) are used to join and repair both wrought and cast nickel aluminum bronze materials. These electrodes may be used for applications requiring good corrosion resistance and erosion or cavitation resistance in both salt and brackish water.

ECuMnNiAl covered electrodes and ERCuMnNiAl bare filler metal are used to join manganese-nickel-aluminum bronzes of similar compositions. These electrodes are used in applications requiring resistance to cavitation, erosion, and corrosion.

Copper-Nickel. Copper-nickel covered electrodes, (ECuNi) and bare electrode wire and rods (ERCuNi) are nominally 70 percent copper and 30 percent nickel. These filler metals contain titanium to deoxidize the weld pool and are used for welding all copper-nickel alloys.

Weld Joint Design

RECOMMENDED WELD JOINT designs for copper and copper alloys are shown in Figures 3.1 and 3.2. Figure 3.1 shows joint designs that are appropriate for GTAW and SMAW. Figure 3.2 shows joint designs for GMAW. These joint designs have larger groove angles than those used for steel. The larger groove angles are required to provide adequate fusion and penetration for copper alloys that have high thermal conductivity.

Surface Preparation

WELD JOINT FACES and adjacent surfaces should be clean and free of oil, grease, dirt, paint, and oxides prior to welding. Wire brushing is not a suitable cleaning method for copper alloys that develop a tenacious surface oxide, such as the aluminum bronzes. These alloys should be cleaned by appropriate chemical or abrasive methods. Degreasing is also recommended using a suitable solvent.

Preheating

THE RELATIVELY HIGH thermal conductivity of copper and the high-copper alloys results in the rapid conduction of heat from the weld joint into the surrounding base metal. This makes achieving fusion and weld penetration difficult. Loss of heat from the weld area can be minimized using higher energy processes or higher welding currents. Preheating the base metal prior to welding is the most common method used to counteract the effects of thermal conduction.

Selection of a preheat temperature for a given application depends upon the welding process, the alloy being welded, the base metal thickness and to some extent the overall mass of the weldment. Thin sections or high-energy welding processes, such as electron beam or laser welding, generally require less preheat than do thick sections or low-energy welding processes. The use of the GMAW process normally requires lower preheat than GTAW or OFW. When welding conditions are similar, copper requires higher preheat temperatures than copper alloys because of high thermal conductivity. Aluminum-bronze and copper-nickel alloys should not be preheated. Suggested preheat temperatures are given in later tables for the particular alloy families and welding processes.

When preheat is used, the base metal adjacent to the joint must be heated uniformly to the specified temperature. The temperature should be maintained until

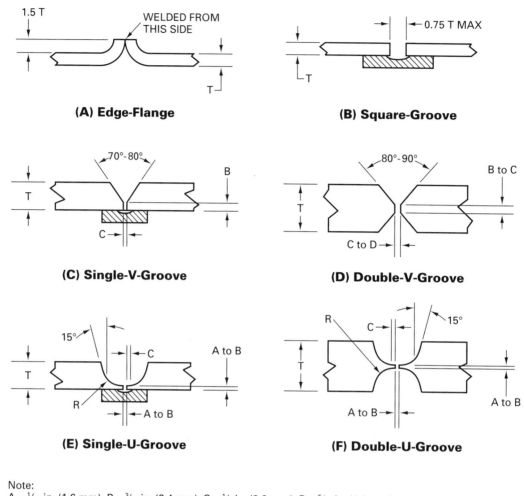

(A) Edge-Flange

(B) Square-Groove

(C) Single-V-Groove

(D) Double-V-Groove

(E) Single-U-Groove

(F) Double-U-Groove

Note:
A = $^1/_{16}$ in. (1.6 mm), B = $^3/_{32}$ in. (2.4 mm), C = $^1/_8$ in. (3.2 mm), D = $^5/_{32}$ in. (4.0 mm),
R = $^1/_8$ in. (3.2 mm), T = thickness.

Figure 3.1—Joint Designs for Gas Tungsten Arc and Shielded Metal Arc Welding of Copper

welding of the joint is completed. When welding is interrupted, the joint area should be preheated before welding is resumed.

Postweld Heat Treatment

POSTWELD HEAT TREATMENT (PWHT) of copper and copper alloys may involve annealing, stress relieving, or precipitation hardening. The need for PWHT depends upon the base metal composition and the application of the weldment. PWHT may be required if the base metal can be strengthened by a heat treatment or if the service environment can cause stress-corrosion cracking.

Copper alloys that include the high-zinc brasses, manganese bronzes, nickel-manganese bronzes, some aluminum bronzes, and nickel-silvers are susceptible to stress-corrosion cracking. Stresses induced during welding of these alloys can lead to premature failure in certain corrosive environments. These alloys may be stress relieved or annealed after welding to reduce stresses. Copper alloys that respond to precipitation hardening include some high coppers, some copper-aluminum alloys, and copper-nickel castings containing beryllium or chromium. If these alloys are not heat-treated, the hardness in the weld area will vary as a result of aging or overaging caused by the welding heat.

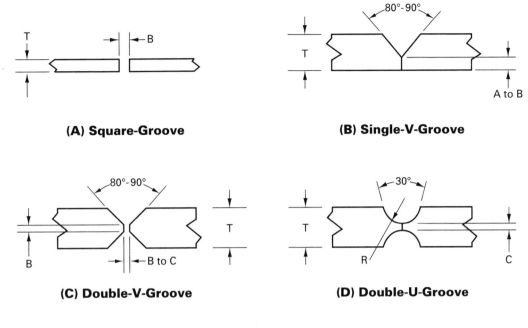

(A) Square-Groove

(B) Single-V-Groove

(C) Double-V-Groove

(D) Double-U-Groove

Note:
A = $^1/_{16}$ in. (1.6 mm), B = $^3/_{32}$ in. (2.4 mm), C = $^1/_8$ in. (3.2 mm), R = $^1/_4$ in. (6.4 mm), T = thickness.

Figure 3.2—Joint Designs for Gas Metal Arc Welding of Copper

Stress Relief

STRESS RELIEF IS intended to reduce stresses from welding to relatively low values without effectively reducing mechanical properties. Stress relief is accomplished by heating the weldment to a temperature that is below the recrystallization temperature of the base metal. Typical stress relieving temperatures for some copper alloys are given in Table 3.5. Heating time must be adequate for the entire weldment to reach temperature. The weldment is usually held for at least one hour at the stress relief temperature and then slowly cooled. Weldments thicker than 1 in. (25.4 mm) must be held for longer periods, usually for 1 hour per in. (1 h per 25.4 mm) of thickness.

Annealing

ANNEALING IS USED to reduce stresses and to homogenize weldments of hardenable copper alloys to produce a metallurgical structure that will respond to heat treatment satisfactorily. Annealing is carried out

Table 3.5
Typical Stress Relieving Temperatures for Weldments of Copper Alloys

Common Name	UNS No.	Temperature*	
		°F	°C
Red brass	C23000	550	288
Admiralty brass	C44300-C44500	550	288
Naval brass	C46400-C46700	500	260
Aluminum bronze	C61400	650	343
Silicon bronze	C65500	650	343
Copper-nickel alloys	C70600-C71500	1000	538

* Heat slowly to and hold at temperature for at least one hour.

at temperatures considerably higher than those used for stress relieving, as shown in Table 3.6. Stress relaxation proceeds rapidly at the annealing temperature. Extended annealing times or annealing at the top of the temperature range can cause excessive grain growth that may reduce tensile strength and can cause other undesirable metallurgical effects.

Heat Treatment

HEAT TREATABLE COPPER alloys of greatest commercial use are those with UNS Nos. C17000, C17200, C17300, C17500, C17510, C18200, and C15000. These materials can be supplied in any form or condition the user requires. These heat treatable copper alloys are hardened by either of the following procedures:

(1) Solution anneal, then cold work, and then age harden.
(2) Solution anneal, then age harden from the solution anneal state without cold working.

For each heat treatable copper alloy, the solution anneal temperature and age hardening temperature vary depending upon alloy chemical composition.

When a heat treatable copper alloy has been welded or brazed, and to a lesser extent soldered, the mill supplied condition has been altered. To return the base metal, the heat-affected zone and the weld to approximately the mill supplied condition, it is necessary to heat treat the welded assembly. This involves solution annealing, followed by age hardening. It usually is not feasible to perform any cold work on a welded assembly after solution anneal.

Solution Annealing. The solution anneal for beryllium-copper UNS No. C17200 alloy (1.9 percent beryllium) is to heat the part to 1450 °F (788 °C). Depending on the thickness, the part is held at temperature for 30 minutes to 3 hours. Because beryllium-copper forms a tenacious and continuous oxide surface when heated in air or an oxidizing atmosphere, a slightly reducing atmosphere should be used to produce clean and bright parts after quenching.

The quenching is critical to avoid any precipitation of the beryllium intermetallic constituent. Rapid water quenching from the solution anneal temperature is the best method to insure retention in solid solution. For parts or castings that may crack, quenching in oil or forced air may be used but may result in some precipitation of beryllium intermetallic constituent.

Age Hardening. Solution annealed products are soft, having a hardness of 45 to 85 HRB. Weldments are seldom cold worked after the solution treatment. Hardening is accomplished by aging the part or welded assembly in a furnace at a temperature of 550 to 750 °F (290 to 400 °C) for about 3 hours. Again, to prevent oxidation, a slightly reducing atmosphere is preferred. After age hardening, the parts or assemblies may be returned to room temperature in any manner, i.e.,

Table 3.6
Annealing Temperature Ranges for Copper and Copper Alloys

Common name	UNS No.	Temperature Range*	
		°F	°C
Phosphor-deoxidized copper	C12200	700-1200	371-649
Beryllium-copper	C17000, C17200	1425-1475	774-802
Beryllium-copper	C17500	1675-1725	913-941
Red brass	C23000	800-1350	427-732
Yellow brass	C27000	800-1300	427-704
Muntz metal	C28000	800-1100	427-593
Admiralty	C44300-C44500	800-1100	427-593
Naval brass	C46400-C46700	800-1100	427-593
Phosphor bronze	C50500-C52400	900-1250	482-677
Aluminum bronze	C61400	1125-1650	607-899
Aluminum bronze	C62500	1100-1200	593-649
Silicon bronze	C65100, C65500	900-1300	482-704
Aluminum brass	C68700	800-1100	427-593
Copper-nickel, 10%	C70600	1100-1500	593-816
Copper-nickel, 30%	C71500	1200-1500	649-816
Nickel-silver	C74500	1100-1400	593-760

* Time at temperature - 15 to 30 min

furnace cooled, water quenched, or air quenched. The cooling method is usually immaterial.

UNS No. C17200 alloy, solution annealed and age hardened, will have a hardness of 35 to 40 HRC, depending on thickness.

Other Heat Treatable Copper Alloys. All beryllium-copper, chromium-copper, and zirconium-copper alloys may be heat treated in the manner described above for beryllium-copper alloy UNS No. C17200. Only the temperatures are adjusted to give optimum mechanical properties.

Fixturing

THE THERMAL COEFFICIENTS of expansion of copper and its alloys are about 1.5 times that of steel (see Table 3.2), so that distortion will be greater with copper alloys. Appropriate measures to control distortion and warping include suitable clamping fixtures, to position and restrain thin components, and frequent tack welds to align the joint for welding thick sections. The ends of the tack welds should be tapered to ensure good fusion with the first weld beads.

The root pass of multiple-pass welds should be rather large to avoid cracking. Fixturing and welding procedures must be designed to limit restraint of copper alloys that are likely to hot crack when highly restrained.

Backing strips or backing rings are used to control root penetration and fusion in groove welds. Copper and copper alloys that have the same or similar chemical composition as the base metal can be used for backing. Removable ceramic backing also may be suitable for use with copper and copper alloys.

Safe Practices

COPPER AND A number of alloying elements in copper alloys (arsenic, beryllium, cadmium, chromium, lead, manganese, and nickel) have low or very low permissible exposure limits as set by the American Conference of Governmental Industrial Hygienists. Special ventilation precautions are required when brazing, welding, soldering, or grinding copper or copper alloys to assure the level of contaminants in the atmosphere is below the limit allowed for human exposure. These precautions may include local exhaust ventilation, respiratory protection, or both. Refer to the latest edition of ANSI/AWS Z49.1, *Safety in Welding and Cutting*, for proper procedures.

Welding copper alloys containing appreciable amounts of beryllium, cadmium, or chromium may present health hazards to welders and others. Where copper alloys containing these elements are welded on more than an occasional basis, the user should consult the Occupational Safety and Health (OSHA) guidelines for the specific element. Exposure to welding fumes containing these elements may cause adverse health effects. Refer to the Supplementary Reading List at the end of this chapter for detailed literature citations.

Copper and zinc fume and dust can cause irritation of the upper respiratory tract, nausea, and metal fume fever. They may also cause skin irritation and dermatitis as well as eye problems. Cadmium and beryllium fume are toxic when inhaled.

Fluxes used for welding, brazing and soldering certain copper alloys may contain fluorides and chlorides. Fume from these fluxes can be very irritating to the eyes, nose, throat, and skin. Some fluorine compounds are toxic. Furnaces or retorts that use a flammable brazing atmosphere must be purged of air prior to heating. Furnaces using controlled atmospheres must be purged with air before personnel are permitted to enter it to avoid their suffocation.

Good personal hygiene should be practiced, particularly before eating. Food and beverages should not be stored or consumed in the work area. Contaminated clothing should be changed.

For a more detailed explanation of safe practices, see Chapter 16, "Safe Practices," *Welding Handbook,* 8th Ed., Vol. 1.

WELDING COPPER

FUSION WELDING OXYGEN-BEARING copper is difficult. The high oxygen and impurity level in fire-refined copper make this material particularly difficult to weld. Electrolytic tough-pitch copper (UNS No. C11000) has somewhat better weldability but must be welded with caution. Although preheat and high heat input are necessary to counteract the high thermal conductivity of these materials, high heat input degrades weld properties. Therefore, inert-gas shielded arc processes are recommended over OFW. Solid-state processes can also be effective for these materials.

Oxygen-free (UNS No. C10200) and deoxidized copper (UNS No. C12000) should be selected for welded components when the best combination of electrical conductivity, mechanical properties, and corrosion resistance are desired.

Copper is welded with ECu and ERCu filler metals whose composition is similar to the base metal although other compatible copper filler metals may be used to obtain desired properties.

The high thermal conductivity of copper often requires preheating to achieve complete fusion and adequate joint penetration. Preheating requirements depend upon material thickness, the welding process and heat input. Figures 3.3 and 3.4 illustrate the effects of preheat temperature on penetration in copper. Typical preheat temperatures are given in Table 3.7.

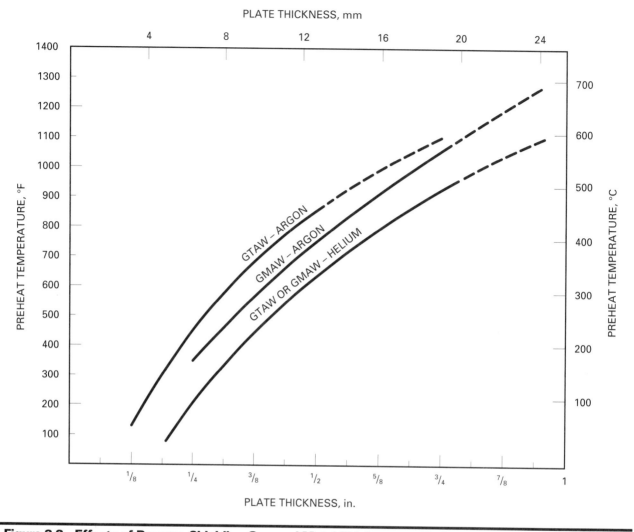

Figure 3.3—Effects of Process, Shielding Gas, and Metal Thickness on Preheat Requirements for Welding Copper

Gas Tungsten Arc Welding

GAS TUNGSTEN ARC welding (GTAW) is best suited for joining sections of copper up to 0.125 in. (3.2 mm) thick, but flat position welding of thicker sections also is performed successfully. Pulsed current is helpful for welding positions other than flat. Typical joint designs for GTAW of copper are shown in Figure 3.1.

Shielding Gases.   Argon shielding gas is preferred for GTAW of copper up to 0.06 in. (1.5 mm) thick and helium is preferred for welding sections over 0.06 in. (1.5 mm). Compared to argon, helium produces deeper penetration or permits higher travel speed, or both, at the same welding current.   Figure 3.4 illustrates the differences in penetration in copper with argon and helium shielding gases. Helium produces a more fluid weld pool that is cleaner, and the risk of oxide entrapment is considerably reduced. Mixtures of argon and helium result in intermediate welding characteristics. A mixture of 75% helium-25% argon produces a good balance between the good penetration of helium and the easier arc starting and greater arc stability of argon.

Welding Technique.   Either forehand or backhand welding may be used for welding copper.   Forehand welding is preferred for all welding positions and provides a more uniform, smaller bead than with backhand welding.

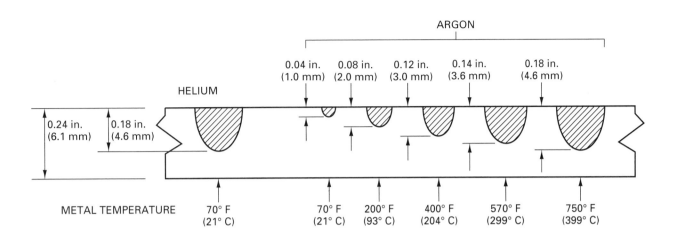

Figure 3.4—Effect of Shielding Gas and Preheat Temperature on Weld Bead Penetration in Copper when Gas Tungsten Arc Welded with 300 A dc at a Travel Speed of 8 in./min (3.4 mm/sec)

Stringer beads or narrow weave beads should be used for copper. Wide oscillation of the arc should be avoided because it exposes each edge of the bead to the atmosphere. The first bead should penetrate to the root of the joint and should be sufficiently thick to provide time for weld metal deoxidation and to avoid cracking of the weld bead.

Typical preheat temperatures and welding conditions for GTAW of copper are shown in Table 3.7. These conditions should only be used as a guide for establishing welding procedures. The high thermal conductivity of copper precludes recommending welding conditions suitable for all applications. The welding conditions should be adjusted to produce the desired weld bead shape. The limitation on travel speed is the weld bead shape. At excessive speeds, weld beads tend to be very convex in shape, causing underfill along the edges and poor fusion on subsequent weld passes.

Properties. Typical mechanical and electrical properties of copper weld metal are shown in Table 3.8. The data represents GTAW specimens tested in both the as-welded and annealed condition.

Gas Metal Arc Welding

ARGON OR A mixture of 75% helium-25% argon is recommended for gas shielding in gas metal arc welding (GMAW) of copper. Argon is normally used for 0.25 in. (6.4 mm) thickness and under. The helium-argon mixture is used for welding of thicker sections since preheat requirements are lower, joint penetration is better, and filler metal deposition rates are higher.

ERCu copper electrodes are recommended for GMAW of copper. These electrodes have the highest conductivity of any copper electrodes but contain minor alloying elements to improve weldability. The resulting weld has a lower conductivity than the base material. Copper alloy electrodes (copper-silicon and copper-aluminum) may be used to obtain desired joint mechanical properties when good electrical or thermal conductivity is not a major requirement. Electrode size will depend upon the base metal thickness and the joint design.

The filler metal should be deposited by stringer beads or narrow weave beads using spray transfer. Wide electrode weaving may result in oxidation at bead edges. Minimum conditions for spray transfer with steady current, copper electrodes and argon shielding are given in Table 3.9. Pulsed current can be used to achieve spray transfer over a wider range of welding currents.

Suggested joint designs for GMAW of copper are shown in Figure 3.2. Typical preheat temperatures and welding conditions are given in Table 3.10. These should be used as guidelines in establishing suitable welding conditions that are substantiated by appropriate tests. The forehand welding technique should be used in the flat position. In the vertical position, the progression of welding should be uphill. GMAW of copper should not be done in the overhead position. Pulsed current improves weld bead shape and operability for welding copper in positions other than flat.

Table 3.7
Typical Conditions for Manual Gas Tungsten Arc Welding of Copper

Metal Thickness	Joint Design[a]	Shielding Gas	Tungsten Electrode Diameter	Welding Rod Diameter	Preheat Temperature	Welding Current, A[b]	No. of Passes
0.01-0.03 in. (0.3-0.8 mm)	A	Ar	0.02, 0.04 in. (0.5, 1.0 mm)	— —	— —	15-60	1
0.04-0.07 in. (1.0-1.8 mm)	B	Ar	0.04, 0.062 in. (1.0, 1.6 mm)	0.062 in. (1.6 mm)	— —	40-170	1
0.09-0.19 in. (2.3-4.8 mm)	C	He	0.094 in. (2.4 mm)	0.094, 0.125 in. (2.4, 3.2 mm)	100°F (38°C)	100-300	1-2
0.25 in. (6.4 mm)	C	He	0.125 in. (3.2 mm)	0.125 in. (3.2 mm)	200°F (93°C)	250-375	2-3
0.38 in. (9.6 mm)	E	He	0.125 in. (3.2 mm)	0.125 in. (3.2 mm)	450°F (232°C)	300-375	2-3
0.5 in. (12.7 mm)	D	He	0.125, 0.156 in. (3.2, 4.0 mm)	0.125 in. (3.2 mm)	650°F (343°C)	350-420	4-6
0.63 in. & up (16 mm & up)	F	He	0.188 in. (4.8 mm)	0.125 in. (3.2 mm)	750°F min (399°C min)	400-475	As req'd

a. See Figure 3.1.

b. Direct current electrode negative.

Table 3.8
Typical Properties of Gas Tungsten Arc Weld Deposits of Copper

Test and Conditions	Tensile Strength		Yield Strength		Elongation, %	Impact Strength[a]		Electrical Conductivity, % IACS
	ksi	MPa	ksi	MPa		ft lb	J	
All weld metal test								
As-welded	27-32	186-220	15-20	103-138	20-40	20-40	27-54	—
Annealed at 1000°F (538°C)	27-32	186-220	12-18	83-124	20-40	—	—	—
Transverse tension test								
As-welded	29-32	200-220	10-13	69-159	—	—	—	—
Deposited metal conductivity[b]								
Oxygen-free copper	—	—	—	—	—	—	—	95
Phosphorous deoxidized copper	—	—	—	—	—	—	—	83
Phosphor bronze	—	—	—	—	—	—	—	37
Silicon bronze	—	—	—	—	—	—	—	26

a. Charpy keyhole specimens

b. Copper base metal welded with the given filler metal

Plasma Arc Welding

COPPER CAN BE welded with the plasma arc welding (PAW) process using ERCu filler metal. Argon, helium, or mixtures of the two are used for orifice and shielding gases depending on base metal thickness. As with GTAW, arc energy is higher with helium-rich mixtures. Hydrogen should not be added to either gas for welding copper.

Table 3.9
Approximate Gas Metal Arc Welding Conditions for Spray Transfer
with Copper and Copper Alloy Electrodes and Argon Shielding

Type	Electrode Diameter in.	Electrode Diameter mm	Minimum* Welding Current, A	Arc Voltage, V	Filler Wire Feed in./min	Filler Wire Feed mm/s	Minimum Current Density kA/in.^{2}	Minimum Current Density kA/mm^2
ERCu (copper)	0.035	0.9	180	26	345	146	191	0.30
	0.045	1.1	210	25	250	106	134	0.21
	0.062	1.6	310	26	150	63	101	0.16
ERCuAl-A2 (aluminum bronze)	0.035	0.9	160	25	295	125	170	0.26
	0.045	1.1	210	25	260	110	134	0.21
	0.062	1.6	280	26	185	78	91	0.14
ERCuSi-A (silicon bronze)	0.035	0.9	165	24	420	178	176	0.27
	0.045	1.1	205	26-27	295	125	131	0.20
	0.062	1.6	270	27-28	190	80	88	0.14
ERCuNi (copper-nickel)	0.062	1.6	280	26	175	74	91	0.14

* Direct current electrode positive

Table 3.10
Typical Conditions for Gas Metal Arc Welding of Copper

Metal Thickness	Joint Design [a]	Shielding Gas	Electrode Diameter	Electrode Feed	Preheat Temperature	Welding Current, A [b]	Travel Speed	No. of Passes
Up to 0.19 in. (up to 4.8 mm)	A	Ar	0.045 in. (1.1 mm)	180-315 in./min (76-133 mm/s)	100-200°F (38-93°C)	180-250	14-20 in./min (6-8 mm/s)	1-2
0.25 in. (6.4 mm)	B	75% He-25% Ar	0.062 in. (1.6 mm)	150-210 in./min (63-89 mm/s)	200°F (93°C)	250-325	10-18 in./min (4-8 mm/s)	1-2
0.38 in. (9.6 mm)	B	75% He-25% Ar	0.062 in. (1.6 mm)	190-230 in./min (80-97 mm/s)	425°F (218°C)	300-350	6-12 in./min (2-5 mm/s)	1-3
0.50 in. (12.7 mm)	C	75% He-25% Ar	0.062 in. (1.6 mm)	210-270 in./min (89-114 mm/s)	600°F (316°C)	330-400	8-14 in./min (3-6 mm/s)	2-4
0.63 in. and up (16 mm and up)	D	75% He-25% Ar	0.062 in. (1.6 mm)	210-270 in./min (89-114 mm/s)	800°F (427°C)	330-400	6-12 in./min (2-5 mm/s)	As req'd
0.63 in. and up (16 mm and up)	D	75% He-25% Ar	0.094 in. (2.4 mm)	150-190 in./min (63-80 mm/s)	800°F (427°C)	500-600	8-14 in./min (3-6 mm/s)	As req'd

a. Refer to Figure 3.2.

b. Constant-potential power source; direct current electrode positive using a short arc length that provides steady and quiet operation.

Shielded Metal Arc Welding

COPPER MAY BE welded with ECu covered electrodes, but weld quality is not as good as that obtained with the gas shielded welding processes. Best results with shielded metal arc welding (SMAW) are obtained when welding deoxidized copper. The electrodes may be used to weld oxygen-free and tough-pitch coppers, but the welded joints will contain porosity and oxide inclusions.

Copper may be welded with an alloy covered electrode, such as ECuSi or ECuSn-A electrodes. These electrodes are used for the following:

(1) Minor repair of relatively thin sections
(2) Fillet welded joints with limited access
(3) Welding copper to other metals

Joint designs should be similar to those shown in Figure 3.1. A grooved copper backing may be used to control the root surface contour.

Electrode size selected should be as large as practical for the base metal thickness. Welding should be done using direct current electrode positive of sufficient amperage to provide good filler metal fluidity. Either a weave or stringer bead technique may be used to fill the joint. Flat position welding using a preheat of 500 °F (260 °C) or higher is used for joints thicker than 0.13 in. (3.3 mm).

Oxyfuel Gas Welding

THE OXYGEN-FREE and deoxidized coppers can be welded using the oxyfuel gas welding (OFW) process, but welding travel speed is slower than for arc welding.

ERCu welding electrodes and appropriate flux are used for oxyacetylene welding (OAW) welding.

Preheat and auxiliary heating are recommended with thicknesses over 0.13 in. (3.3 mm) to obtain good fusion.

Type ERCu or ERCuSi filler metal can also be used for OFW of copper, depending upon the desired joint properties. When commercial flux designed for welding copper alloys is used, the welding rod and the joint surfaces should be coated with flux.

The OFW flame should be neutral when flux is used, and slightly oxidizing when welding without flux. The welding tip size should be one to two sizes larger than the tip used for the same thickness of steel. Typical welding tip sizes and joint designs are given in Table 3.11.

Backhand welding is generally preferred for the flat position. Backhand technique can give a thicker bead than forehand welding, and oxide entrapment is less. Control of the molten weld pool is greatly improved when the joint axis is tilted about 10 to 15 degrees and the direction of welding is uphill.

Long seams should not be tack welded. The initial root opening should increase along the joint length with a taper that will close gradually as welding proceeds along the joint. A rule-of-thumb is to increase the root opening 0.015 units for each unit of joint length.

Completed weld beads may be peened to relieve welding stresses and increase the weld metal strength by cold working. Peening may be done either while the weld metal is still warm or after it cools to room temperature.

Other Processes

COPPER CAN BE electron beam welded and copper tubing is high-frequency resistance welded. The solid-state processes, including friction welding and cold welding, also are effective for welding copper.

Table 3.11
Suggested Joint Designs and Welding Tip Sizes for Oxyacetylene Welding of Copper

Metal Thickness			Root Opening		Welding Tip	
in.	mm	Joint Design	in.	mm	Drill Size No.	Remarks
0.06	1.5	Edge-flange	0	0	55 to 58	—
0.06	1.5	Square-groove	0.06-0.09	1.5-2.3	55 to 58	—
0.13	3.3	Square-groove	0.09-0.13	2.3-3.3	51 to 54	—
0.19	4.8	60° to 90° single-V-groove	0.13-0.18	3.3-4.6	48 to 50	Auxiliary heating required
0.25	6.4	60° to 90° single-V-groove	0.13-0.18	3.3-4.6	43 to 46	Auxiliary heating required
0.38	9.6	60° to 90° single-V-groove	0.18	4.6	38 to 41	Auxiliary heating required
0.50-0.75	12.7-19.0	90° double-V-groove	0.18	4.6	38 to 41	Weld both sides simultaneously in vertical position

WELDING HIGH-COPPER ALLOYS

HIGH-COPPER ALLOYS include beryllium-copper (UNS Nos. C17000-17500), cadmium-copper (UNS No. C14300), chromium-copper (UNS No. C18200), chromium-zirconium-copper (UNS No. C18150), zirconium-copper (UNS No. C15000), and nickel-silicon-chromium-copper (UNS No. C18000).

Cadmium-copper has good electrical conductivity and is strengthened by cold working. Beryllium-chromium- and zirconium-copper can be strengthened by a precipitation-hardening heat treatment, either alone or in combination with cold working. Nickel-silicon-chromium alloys are strengthened by precipitation hardening, either alone or in combination with cold working.

Welding will soften the HAZ in the precipitation-hardened, high-copper alloys by annealing or overaging. The characteristics of each alloy and its condition should be considered when establishing welding procedures and manufacturing sequences. For maximum properties, beryllium- chromium- and zirconium-copper assemblies should be welded before either heat treatment or cold working.

Cadmium- and Chromium-Coppers

THE PROCEDURES RECOMMENDED for arc and OFW of copper are good bases to use for developing welding procedures for cadmium- and chromium copper. These alloys have lower electrical and thermal conductivity than copper, and they can be welded at lower preheat temperatures and heat input than those required for copper.

Cadmium-copper can be joined by the gas shielded welding processes, OFW, and flash welding. OFW welding of cadmium-copper requires a flux containing sodium fluoride and either fused borax or boric acid, or both, to dissolve cadmium oxides. Chromium-copper can be welded with gas shielded processes and flash welding, but OFW should not be used because of problems caused by chromium oxide formation on weld faces.

Beryllium-Coppers

TWO TYPES OF beryllium-copper are available. One type (low-beryllium copper) contains about 0.5 percent beryllium and 1.5 or 2.5 percent cobalt and has relatively good electrical conductivity. The other type (high-beryllium copper) contains about 2 percent beryllium and 0.2 percent cobalt or nickel. It has good strength in the precipitation-hardened condition but low electrical conductivity, about 20% IACS.

High-beryllium copper is more readily welded than low-beryllium copper. Addition of beryllium to copper lowers the melting point, increases the fluidity of the molten metal, and decreases thermal conductivity, all of which contribute to better weldability.

A difficulty common to beryllium-copper is formation of surface oxide. Beryllium forms a tenacious oxide that inhibits wetting and fusion during welding. Cleanliness of faying surfaces and surrounding surfaces before and during welding is essential for good results.

Sound welds can be made in the low-beryllium alloy, but cracking during welding or postweld heat treatment is a problem. Low-beryllium copper can be joined more readily with a filler metal of higher beryllium content.

Beryllium-copper components can be repaired by SMAW with aluminum bronze electrodes or by GTAW with silicon bronze filler metal when welds with high mechanical properties are not required. Joint designs similar to those shown in Figures 3.1 and 3.2 are suitable for GTAW and GMAW respectively.

Typical conditions used for welding beryllium copper are given in Table 3.12. These data may be used as a guide in establishing suitable welding conditions.

Stabilized AC power is preferred for manual GTAW welding of thin sections [less than 0.25 in. (6.4 mm)] to take advantage of its surface cleaning action. Direct current electrode negative (DCEN) is recommended for welding heavier sections and can be used for manual and mechanized GTAW, provided adequate gas shielding is used to prevent oxidation.

Preheat is not usually required for welding sections of 0.13 in. (3.3 mm) and less in thickness. Preheat temperatures for high thermal conductivity alloys should be those recommended for copper. For the high-strength alloys, preheat temperatures of 300 to 400 °F (136 to 204 °C) are sufficient.

After welding, optimum mechanical properties are obtained by a solution anneal heat treatment followed by cold working, if possible, and age hardening, as the data in Table 3.13 indicate. The characteristics of the weld metal must be considered in planning a postweld heat treatment when the filler metal composition is different from that of the base metal.

Components in the precipitation-hardened condition should not be welded because of the danger of cracking the heat-affected zone. Thin sections should be welded in the solution annealed condition.

In multipass welding of heavy sections, early passes are overaged by the heat of later passes. Therefore, where multiple-pass welding is required, the base metal should be in the overaged condition because it is more stable metallurgically in this condition than in the solution heat-treated condition.

Table 3.12
Typical Conditions for Arc Welding Beryllium Coppers

| Variable | GTAW | | | | GMAW |
| | Manual | | Automatic | | Manual |
	Alloy C17200	Alloy C17500	Alloy C17200	Alloy C17200	Alloy C17000
Thickness, in. (mm)	0.125 (3.2)	0.25 (6.4)	0.020 (0.5)	0.090 (2.3)	1.0 (25.4)
Joint design	C[a]	E[a]	B[a]	B[a]	D[b]
Preheat temp., °F (°C)	70 (21)	300 (149)	70 (21)	70 (21)	300 (149)
Filler metal diam., in. (mm)	0.125 (3.2)	0.125 (3.2)	— —	0.062 (1.6)	0.062 (1.6)
Filler metal feed, in./min (mm/s)	— —	— —	— —	— —	190-200 (80-85)
Shielding gas	Ar	Ar	Ar	65% Ar-35%He	Ar
Welding power	ACHF	ACHF	DCEN	DCEN	DCEP
Welding current, A	180	225-245	43	150	325-350
Arc voltage, V	—	22-24	12	11.5	29-30
Travel speed, in./min (mm/s)	— —	— —	27 (11)	20 (8)	— —

a. Refer to Figure 3.1.

b. Refer to Figure 3.2.

Table 3.13
Typical Mechanical Properties of Welded Joints in Beryllium Copper

| Alloy and Condition[*] | Tensile Strength | | Yield Strength | |
	ksi	MPa	ksi	MPa
C17200 (Cu-2Be)				
As-welded	60-70	414-484	30-33	207-228
Aged only	130-155	896-1069	125-150	861-1034
Solutioned and aged	150-175	1034-1207	145-170	1000-1172
C17500 (Cu-2.5 Co-0.5Be)				
As-welded	50-55	345-379	30-45	207-310
Aged only	80-95	552-655	65-85	448-586
Solutioned and aged	100-110	689-758	75-85	517-586

* Welded in the solution heat-treated condition

WELDING COPPER-ZINC ALLOYS (BRASSES)

COPPER-ZINC alloys can be joined by arc welding, OFW, resistance spot, flash, and friction welding processes. The electrical and thermal conductivity of brasses decreases with increasing zinc content so that high-zinc brasses require lower preheat temperatures and welding heat input than low-zinc brasses. Since zinc vaporizes from molten brass, zinc fuming is the major problem when welding brasses and is worse for the high-zinc brasses. Other alloying elements such as aluminum and nickel may slightly increase cracking tendency and increase oxide formation. For these reasons, low-zinc brasses have good weldability, high-zinc brasses only fair weldability, and leaded alpha brasses (64 - 95% Cu) are not suitable for welding. Lead makes these copper-zinc alloys very sensitive to hot cracking. The low-leaded alpha-beta brasses are weldable under conditions of low restraint, provided low weld strength can be tolerated.

Gas Tungsten Arc Welding

THE BRASSES ARE commonly joined by gas tungsten arc welding (GTAW) in sections up to 0.38 in (9.7mm) thick. Thin brass sheets can be welded together without filler metal addition, but addition of filler metal is recommended when welding sections over 0.062 in. (1.6 mm) thick. Phosphor bronze (ERCuSn-A) filler metal provides a good color match with some brasses, but silicon bronze (ERCuSi-A) filler metals reduce zinc fuming. Aluminum bronze (ERCuAl-A2) filler metal can be used to provide good joint strength for high-zinc brasses, but aluminum bronze filler metal is not effective in controlling zinc fuming so that welds tend to be porous.

V-groove weld joints having a groove angle of 75 to 90 degrees should be used to insure good joint penetration for thicknesses over 0.188 in. (4.8 mm). A preheat temperature of 200 to 600 °F (93 to 316 °C) should be used on heavier sections. The preheat temperature can be lowered for the high-zinc brasses.

Welding procedures can be designed to minimize zinc fuming by directing the welding arc onto the filler rod or the molten weld pool rather than on the base metal. The base metal is heated to fusion temperature by conduction from the molten weld pool rather than by direct impingement of the arc.

Gas Metal Arc Welding

GAS METAL ARC welding (GMAW) is used primarily to join relatively thick sections of brass and is suitable for thicknesses over 0.13 in. (3.3 mm). Zinc fuming is more severe with GMAW than with GTAW. Argon shielding is normally used, but helium-argon mixtures can provide higher heat input. Silicon bronze (ERCuSi-A), phosphor bronze (ERCuSn-A) or aluminum bronze (ERCuAl-A2) bare electrodes are recommended. The phosphor bronze electrode will produce weld metal having good color match with most brass, but the silicon bronze electrode has better fluidity. The aluminum bronze electrode is best for welding high-strength brasses to produce weldments with equivalent base metal strength. V-groove weld joints with a 60 to 70 degree groove angle or U-groove joints are recommended when using aluminum bronze filler metal.

A preheat in the range of 200 to 600 °F (93 to 316 °C) is recommended for the low-zinc brasses because of their relatively high thermal conductivity. Preheat should not be used for GMAW of high-zinc brasses, but can be used to reduce the required welding current, and thus reduce zinc fuming. During welding, the arc should be directed on the molten weld pool to minimize zinc fuming.

Shielded Metal Arc Welding

BRASSES CAN BE welded with phosphor bronze (ECuSn-A or ECuSn-C), silicon bronze (ECuSi), or aluminum bronze (ECuAl-A2) covered electrodes. The selection criteria for covered electrodes is similar to that previously described for bare electrodes used with GMAW.

The weldability of brasses with shielded metal arc welding (SMAW) is not as good as with GMAW, and relatively large groove angles are needed for good joint penetration and avoidance of slag entrapment. For best results, welding should be done in the flat position, using a backing of copper or brass.

The preheat and interpass temperature for the low brasses should be in the 400 to 500 °F (204 to 260 °C) range, and in the 500 to 700 °F (260 to 371 °C) range for the high brasses. Low preheat temperature will provide better weld joint mechanical properties when using phosphor bronze electrodes. Fast welding speed and a welding current in the high end of the recommended range for the electrode should be used to deposit stringer beads in the joint. The arc should be directed on the molten weld pool to minimize zinc fuming.

Oxyfuel Gas Welding

THE OXYFUEL GAS welding (OFW) procedures that are used for copper are also suitable for the brasses. The low brasses are readily joined by OFW, and the process is particularly suited for piping because it can be performed in all welding positions. Silicon bronze (ERCuSi-A) welding rod or one of the brass welding rods (RBCuZn-A, RBCuZn-B, or RBCuZn-C) may be used.[2] Brass welding rods containing 38 to 41 percent zinc develop a significant proportion of the hard, strong beta phase in the weld metal. This beta phase is soft and ductile at elevated temperatures, and cracking is not a problem.

Very little zinc oxide appears on the molten weld metal surfaces when OFW with a neutral or slightly oxidizing flame. When a strongly oxidizing flame is used, an oxide film forms on the molten weld metal sur face that suppresses evaporation of zinc, provided the weld metal is not overheated.

For OFW of high brasses, RBCuZn-B or RBCuZn-C welding rods are used. These low-fuming rods have compositions similar to the high brasses. A flux of AWS classification FB3-C, FB3-D, or FB3-K is required, and the torch flame should be adjusted to slightly oxidizing to control fuming.[3] Preheating and an auxiliary heat source may also be necessary.

WELDING COPPER-NICKEL-ZINC ALLOYS (NICKEL-SILVER)

NICKEL-SILVERS ARE seldom welded, although welding of nickel-silver is similar to brass having comparable zinc content. Nickel-silvers are frequently used in decorative applications where color match is important. No zinc-free filler metals are available that give a good color match for gas shielded arc welding. GTAW without filler metal addition is usually restricted to welding thicknesses of 0.094 in. (2.4 mm) or less. Square groove butt, lap, or edge joints should be used.

For a wide range of thicknesses, OFW may be performed using RBCuZn-D welding rods with a slightly oxidizing flame An AWS classification FB3-D brazing flux should be applied to both the joint area and the welding rod before and during welding.

If SMAW is to be performed, manganese bronze or copper-nickel filler metal will result in a close color match. Care must be exercised to prevent undercutting, since the melting points of these alloys are appreciably above nickel-silvers.

Annealing is recommended prior to welding severely restrained cold-worked material. Postweld stress relief is suggested for components subjected to corrosive environments. Preheat is recommended to avoid cracking of single-phase alloys, which have poor elevated-temperature ductility and susceptibility to hot cracking.

Resistance welding is practical because nickel-silvers have conductivities that are among the lowest of all.

copper alloys. Welding machines should have low inertia heads and electronic controls for best results. Power requirements are usually 125 to 150 percent of those required for comparable thicknesses of steel. Nickel-silvers can be successfully stud welded.

WELDING COPPER-TIN ALLOYS (PHOSPHOR BRONZE)

THE COPPER-TIN alloys, called *phosphor bronzes*, have rather wide freezing ranges, solidifying with large, weak dendritic grain structures. Welding procedures are designed to prevent the tendency of the weld to crack. Hot peening of each layer of multiple-pass welds will reduce welding stresses and the likelihood of cracking. Welding of leaded copper-tin alloys is not recommended. However, some leaded alloys can be welded if care is exercised. Weldability decreases with increasing lead content. Shielded metal arc welding (SMAW) generally gives better results on leaded alloys than does gas metal arc welding (GMAW).

Joint Preparation

A SINGLE-V-groove weld should be used to join copper-tin alloys in the thickness range of 0.15 to 0.50 in. (4 to 13 mm). The groove angle should be 60 to 70 degrees for GMAW and 90 degrees for SMAW. For greater thicknesses, a single- or double-U-groove weld having a 0.25-in. (6.4-mm) groove radius and a 70 degree groove angle is recommended for good access and fusion. A square-groove weld can be used for thicknesses under 0.15 in. (3.8 mm).

Preheat and Postheat

PHOSPHOR BRONZE WELD metal tends to flow sluggishly because of its wide melting range. Preheating to 350 to 400 °F (177 to 204 °C) and maintaining this interpass temperature improves metal fluidity when welding thick sections. The maximum interpass temperature should not exceed 400 °F (204 °C) to avoid hot cracking. Preheat is not essential when using GMAW spray transfer. For maximum weld ductility or stress corrosion resistance, postweld heat treatment at 900 °F (482 °C), followed by rapid cooling to room temperature, is recommended.

Gas Metal Arc Welding

GMAW IS RECOMMENDED for joining large phosphor bronze fabrications and thick sections using direct

---

2. For information on the brass welding rods, see ANSI/AWS A5.8, *Specification for Filler Metals for Brazing and Braze Welding*, and ANSI/AWS A5.27, *Specification for Copper and Copper Alloy Rods for Oxyfuel Welding*.
3. See ANSI/AWS A5.31, *Specification for Fluxes for Brazing and Braze Welding*.

current electrode positive, ERCuSn-A filler metal, and argon shielding.

Table 3.14 gives suggested GMAW welding parameters that can be used to establish welding procedures for phosphor bronzes. The molten weld pool should be kept small using stringer beads at rather high travel speed. Hot peening of each layer will reduce welding stresses and the likelihood of cracking.

Gas Tungsten Arc Welding

GAS TUNGSTEN ARC welding (GTAW) is used primarily for repair of castings and joining of phosphor bronze sheet with ERCuSn-A filler metal. As with GMAW, hot peening of each layer of weld metal is beneficial. Either stabilized ac or dc, electrode negative welding current can be used with helium or argon shielding.

Shielded Metal Arc Welding

PHOSPHOR BRONZE covered electrodes (ECuSn-A or ECuSn-C) are available for joining bronzes of similar chemical compositions. Filler metal should be deposited as stringer beads using direct current electrode positive, to obtain the best mechanical properties. Postweld annealing at 900 °F (482 °C) is not always necessary, but it is desirable for maximum ductility, particularly if the welded assembly is to be cold worked. Moisture, both on the work and in the electrode coverings, must be strictly avoided. Baking the electrodes at 250 to 300 °F (121 to 149 °C) immediately before use reduces moisture content in the covering to an acceptable level.

Oxyfuel Gas Welding

OXYFUEL GAS WELDING (OFW) is not recommended for joining the phosphor bronzes. The wide heat-affected zone and the slow cooling rate may result in hot cracking since phosphor bronzes are hot short. In an emergency, OFW with ERCuSn-A welding rods can be used if arc welding equipment is not available. If a color match is not essential, braze welding can be done with an OFW torch and RBCuZn-C welding rod. One should use a commercial brazing flux and a neutral flame.

WELDING COPPER-ALUMINUM ALLOYS (ALUMINUM BRONZE)

Weldability

SINGLE-PHASE ALUMINUM bronzes containing less than 7 percent aluminum are hot-short and difficult to weld. Weldments in these alloys may crack in the HAZ. Single-phase alloys containing more than 8 percent aluminum and two-phase alloys are considered weldable when using welding procedures designed to avoid cracking.

UNS Nos. C61300 and C61400 alloys (7 percent aluminum) are frequently fabricated for use in heat exchangers, piping, and vessels. UNS No. C61300 alloy is often preferred because of its good weldability. Both cast and wrought alloys with a higher aluminum content may be joined by arc welding. Problems with fluxing aluminum oxide from the weld metal precludes the use of oxyfuel gas welding (OFW).

Table 3.14
Parameters for Gas Metal Arc Welding of Phosphor Bronze

Metal Thickness		Joint Design						Arc Voltage, V	Welding[b] Current, A
		Groove Type	Root Opening		Electrode Diameter[a]				
in.	mm		in.	mm	in.	mm			
0.06	1.5	Square	0.05	1.3	0.030	0.8		25-26	130-140
0.13	3.3	Square	0.09	2.3	0.035	0.9		26-27	140-160
0.25	6.4	V-groove	0.06	1.5	0.045	1.1		27-28	165-185
0.50	12.7	V-groove	0.09	2.3	0.062	1.6		29-30	315-335
0.75	19.0	Note c	0-0.09	0-2.3	0.078	2.0		31-32	365-385
1.00	25.4	Note c	0-0.09	0-2.3	0.094	2.4		33-34	440-460

a. ERCuSn-A phosphor bronze electrodes and argon shielding

b. Direct current electrode positive

c. Double-V-groove or double-U-groove

Filler Metals

WELDING ELECTRODES AND filler metals recommended for joining the weldable aluminum bronzes are given in Table 3.15. Typical mechanical properties of weld metal deposited by arc welding are shown in Table 3.16. Weld metal deposited with gas metal arc welding (GMAW) is slightly stronger and harder than that deposited with covered electrodes. This is attributed to the higher welding speeds and better shielding with GMAW.

Joint Design

FOR SECTION THICKNESS up to and including 0.13 in. (3.3 mm), square-groove welds are used with a root opening of up to 75 percent of the thickness. For thicknesses of 0.15 to 0.75 in. (3.8 mm to 19 mm), a single-V-groove weld is used. The groove angle should be 60 to 70 degrees for GTAW and GMAW and 90 degrees for SMAW A double-V- or double-U-groove weld should be used for section thicknesses over 0.75 in. (19 mm) U-groove joints should have a 0.25-in. (6.4-mm) groove radius.

Preheat

PREHEAT IS OFTEN unnecessary when welding the aluminum bronzes. The preheat and interpass temperatures should not exceed 300 °F (149 °C) for alloys with less than 10 percent aluminum, including the nickel aluminum bronzes. The weldments should be air cooled to room temperature.

When the aluminum content is from 10 to 13 percent, a preheat of 300 °F (149 °C) and interpass temperature of about 500 °F (260 °C) is recommended for thick sections. Rapid air cooling of the weldment is necessary.

Gas Metal Arc Welding

GAS METAL ARC welding (GMAW) is suitable for aluminum bronze sections of 0.18 in. (4.6 mm) and

Table 3.15
Suggested Filler Metals for Arc Welding Aluminum Bronzes*

UNS No.	SMAW	GTAW or GMAW
C61300 C61400 C61800 C62300	ECuAl-A2	ERCuAl-A2
C61900 C62400	ECuAl-B	ERCuAl-A2
C62200 C62500	ECuAl-B	ERCuAl-A3
C63000 C63200	ECuNiAl	ERCuNiAl
C63300	ECuMnNiAl	ERCuMnNiAl

* Also see ANSI/AWS A5.6, *Specification for Covered Copper and Copper Alloy Arc Welding Electrodes*, and ANSI/AWS A5.7, *Specification for Copper and Copper Alloy Bare Welding Rods and Electrodes.*

Table 3.16
Typical Mechanical Properties of Aluminum Bronze Weld Metal

Electrode[a]	Tensile Strength		Yield Strength[b]		Elongation in 2 in. (51 mm), %	Brinell Hardness, HB[c]
	ksi	MPa	ksi	MPa		
			Gas Metal Arc Welding			
ERCuAl-A2	79	545	35	241	28	160
ERCuAl-A3	90	621	45	310	18	207
ERCuNiAl	104	717	59	407	22	196
ERCuMnNiAl	110	758	67	462	27	217
			Shielded Metal Arc Welding			
ECuAl-A2	77	531	35	241	27	140
ECuAl-B	89	614	47	324	15	177
ECuNiAl	99	683	58	400	25	187
ECuMnNiAl	95	655	56	386	27	185

a. Refer to specifications ANSI/AWS A5.6 and ANSI/AWS A5.7 (see Table 3.15 footnote).

b. 0.5 percent offset.

c. 3000 kg load.

thicker. Argon shielding is used for most joining and surfacing applications, while a 75% argon-25% helium mixture is helpful when welding thick sections where increased welding heat and penetration are required. To maintain proper gas coverage, the welding torch should be tilted 35 to 45 degrees in the forehand direction of travel with an electrode extension of 0.38 to 0.50 in. (9.6 to 13 mm). When welding in a position other than flat, a pulsed current power source or globular type of metal transfer is used. Table 3.17 gives suggested welding parameters for various electrode sizes. Minimum welding conditions for GMAW spray transfer with ERCuAl-A2 electrodes are given in Table 3.9.

Gas Tungsten Arc Welding

GAS TUNGSTEN ARC welding (GTAW) is recommended for critical applications, regardless of section thickness, with either stabilized ac or dc power. Alternating current with argon shielding provides arc cleaning action during welding to remove oxides from the joint faces. For better penetration or faster travel, direct current electrode negative should be used with helium but can be used with argon or a mixture of argon and helium. Preheat is used only for thick sections with this process.

Shielded Metal Arc Welding

SHIELDED METAL ARC welding (SMAW) of aluminum bronze is done with the covered electrodes that are listed in Table 3.15. Direct current electrode positive should be used with these electrodes. Representative welding current ranges for aluminum bronze covered electrodes are given in Table 3.17. Use of a short arc length and stringer or weave beads are recommended. To avoid inclusions, each bead must be thoroughly cleaned of slag before the next bead is applied. SMAW should only be used where it is inconvenient or uneconomical to use GMAW, because SMAW welding speeds are significantly lower.

WELDING COPPER-SILICON ALLOYS (SILICON BRONZE)

THE SILICON BRONZES have good weldability. Characteristics of these bronzes that contribute to this are their low thermal conductivity, good deoxidation of the weld metal by the silicon, and the protection offered by the resulting slag. Silicon bronze weld metal has good fluidity, but the molten slag is viscous. Silicon bronzes have a relatively narrow hot-short temperature range just below the solidus and must be rapidly cooled through this critical range to avoid weld cracking.

Heat loss to the surrounding base metal is low, and high welding speed can be used. Preheat is unnecessary, and interpass temperature should not exceed 200 °F (93 °C). For butt joints, the groove angle of V-groove welds should be 60 degrees or larger. Square-groove welds can be used to join sections up to 0.13 in. (3.3 mm) thick with or without filler metal. Copper backing may be used to control melt-through.

Gas Tungsten Arc Welding

THE SILICON BRONZES are readily gas tungsten arc welded (GTAW) in all positions using ERCuSi-A welding rods. Aluminum bronze welding rod ERCuAl-A2 may also be used. Welding is performed with dc power using argon or helium shielding. Welding with ac power using argon shielding takes advantage of the arc cleaning action. Representative welding conditions for

Table 3.17
Typical Operating Parameters for Arc Welding Aluminum Bronze

Electrode Size		Gas Metal Arc Welding		Shielded Metal Arc Welding Current, A[*]
in.	mm	Arc Voltage, V	Welding Current, A[*]	
0.030	0.8	25-26	130-140	—
0.035	0.9	26-27	140-160	—
0.045	1.1	27-28	165-185	—
1/16	1.6	29-30	315-335	—
5/64	2.0	31-32	365-385	50-70
3/32	2.4	33-34	440-460	60-80
1/8	3.2	—	—	100-120
5/32	4.0	—	—	130-150
3/16	4.8	—	—	170-190
1/4	6.4	—	—	235-255

* Direct current electrode positive

GTAW silicon bronzes in thicknesses of from 0.06 to 0.50 in. (1.5 mm to 12.7 mm) are given in Table 3.18.

Gas Metal Arc Welding

GAS METAL ARC welding (GMAW) may is used for joining the silicon bronzes in sections over 0.25 in. (6.4 mm) thick. ERCuSi-A electrodes, argon shielding, and relatively high travel speeds are used with this process. When making multiple-pass welds, the oxide should be removed by wire brushing between passes. Representative welding conditions for GMAW of butt joints are given in Table 3.19. The welding conditions necessary for spray transfer with ERCuSi-A electrodes are listed in Table 3.9.

Shielded Metal Arc Welding

SILICON BRONZES CAN be welded with ECuAl-A2 or ECuSi covered electrodes. Square-groove welds are suitable for thicknesses up to 0.156 in. (4 mm); single- or double-V-groove welds are used for shielded metal arc welding (SMAW) with thicker sections.

Welding in the flat position is preferred, but ECuSi electrodes can be used to weld in the vertical and overhead positions. Preheat is not needed, and the interpass temperature should not exceed 200 °F (93 °C). Stringer beads should be deposited with a welding current near the middle of the manufacturer's recommended range for the electrode size used. The arc length should be short, and the travel speed should be adjusted to give a small weld pool.

Oxyfuel Gas Welding

OXYFUEL GAS WELDING (OFW) should only be used when arc welding equipment is not available. Silicon bronzes can be oxyfuel gas welded using ERCuSi-A welding rod with a suitable flux and using a slightly oxidizing flame. Fixturing should not unduly restrict movement of the components during welding, and welding should be performed rapidly. Either forehand or backhand welding can be used, with the former preferred for thin sections.

WELDING COPPER-NICKEL ALLOYS

COPPER-NICKEL ALLOYS are readily welded with gas shielded arc welding processes, SMAW, SAW and OFW processes. Preheat is not required and the interpass temperature should not exceed 350 °F (177 °C). Surfaces to be welded shall be clean, free of oxides and other contamination, including sulfur that may cause HAZ intergranular cracking. A 70Cu-30Ni filler metal (ERCuNi and ECuNi) is recommended for welding all the grades of copper-nickel. Where a color match is required, filler metal of matching composition should be used.

Gas Tungsten Arc Welding

COPPER-NICKEL ALLOYS can be welded in all positions using the gas metal arc welding (GTAW) process. Direct current electrode negative, is recommended, although alternating current is used for automatic welding if arc length is accurately controlled. Manual welding is normally used for sheet and plate of thicknesses up to 0.25 in. (6.4 mm) and for tube and pipe. Thicker sections may be welded with GTAW, but GMAW would be more economical and reduce heat input.

The GTAW process provides high quality welds capable of meeting stringent X-ray acceptance standards. Multipass GTAW welds are best deposited using a stringer-bead technique with an arc length of 0.125 to 0.188 in. (3.2 mm to 4.7 mm).

Argon or helium may be used for shielding gas, but argon is generally used to provide improved arc control. Weld quality and soundness depend on careful attention to arc length and filler metal addition. ERCuNi filler metal contains titanium to deoxidize the weld and

Table 3.18
Typical Welding Rods and Welding Currents for Gas Tungsten Arc Welding of Silicon Bronze

Thickness		Welding Rod Diameter		Welding Current,[*] A
in.	mm	in.	mm	
0.06	1.5	0.062	1.6	100-130
0.13	3.3	0.094	2.4	130-160
0.19	4.8	0.125	3.2	150-225
0.25	6.4	0.125, 0.188	3.2 - 4.8	150-300
0.50	12.7	0.125, 0.188	3.2 - 4.8	250-325

* Direct current electrode negative with argon shielding in the flat position.

Table 3.19
Typical Conditions for Gas Metal Arc Welding of Silicon Bronze

Thickness in.	Thickness mm	Groove Type	Root Face in.	Root Face mm	Root Opening in.	Root Opening mm	Pass No.	Welding Current, A [a]	Arc Voltage, V	Electrode Feed in./min [b]	Electrode Feed mm/s	Travel Speed in./min	Travel Speed mm/s
0.25	6.4	Square	—	—	0.06	1.5	1	300	26	215	91	15	6
0.25	6.4	Square	—	—	0.13	3.3	1	305	21	305	129	15	6
0.25	6.4	60° single-V	0.06	1.5	0	0	1	300	26	215	91	13	5
0.38	9.6	60° single-V	0.06	1.5	0	0	1	300	26	215	91	10	4
0.38	9.6	60° single-V	0.13	3.3	0	0	1	300	26	215	91	21	9
							2	300	26	215	91	18	8
0.38	9.6	60° single-V	0.13	3.3	0	0	1	300	26	215	91	15	6
							2	300	26	215	91	16	7
							3	300	26	215	91	36	15
0.38	9.6	60° double-V	0.06	1.5	0	0	1	310	26	215	91	24	10
							2	310	26	215	91	16	7
0.38	9.6	60° single-V	0.06	1.5	0	0	1	310	26	215	91	18	8
							2	310	26	215	91	21	9
0.50	12.7	60° single-V	0.06	1.5	0	0	1	315	21	305	129	12	5
							2	315	21	305	129	13	5
							3	315	21	305	129	12	5
0.50	12.7	60° single-V	0.06	1.5	0	0	1	320	21	305	129	13	5
							2	320	21	305	129	7	5
0.50	12.7	60° single-V	0.13	3.3	0	0	1	310	26	215	91	18	8
							2	310	26	215	91	12	5
							3	310	26	215	91	18	8
0.50	12.7	60° double-V	0.06	1.5	0	0	1	310	26-28	215	91	12	5
							2	310	26-28	215	91	13	5

a. Direct current electrode positive and argon shielding

b. 0.062-in. (1.6-mm) diameter ERCuSi-A electrode

avoid porosity. Autogenous welds or welds without sufficient filler metal addition can contain porosity.

Autogenous welds are possible in sheet up to 0.06 in. (1.5 mm) thick, although weld porosity may be a problem in the absence of deoxidation from the filler metal. Close control of arc length is recommended to minimize porosity.

Gas Metal Arc Welding

COPPER-NICKEL ALLOYS are welded with the gas metal arc welding (GMAW) process using direct current electrode positive current with either spray or short-circuiting transfer (see Table 3.20). Best results are obtained when welding section thicknesses of 0.25 in. (6.4 mm) and greater with spray transfer using either steady or pulsed current. The approximate welding conditions for steady current GMAW spray transfer with a 0.062 in. (1.6 mm) diameter ERCuNi electrode are given in Table 3.9. Pulsed current offers advantages when joining thin sections or welding in positions other than the flat position.

While argon is the recommended shielding gas for most applications, argon-helium mixtures give increased penetration in thick sections.

Proper joint design is important to obtain good complete fusion. V-groove and U-groove joints are recommended for GMAW. A typical V-groove has a 75 degree groove angle with a 0.062 in. (1.6 mm) root face and root opening. A single-V-groove design is satisfactory for section thicknesses from 0.28 to 0.50 in. (7.1 mm to 12.7 mm). Above 0.50 in. (12.7 mm), a double-V-groove or U-groove is recommended to reduce distortion. Copper or copper alloy backing or ceramic backing tape may be used with single-V-groove welds to control root surface contour. Care must be exercised to prevent gas entrapment in the root pass if X-ray quality is required.

Shielded Metal Arc Welding

COPPER-NICKEL ALLOYS may be welded with the shielded metal arc welding (SMAW) process using ECuNi covered electrodes. The electrode diameter for a particular application should be one size smaller than a comparable steel electrode for a similar steel application.

Copper-nickel weld metal is not as fluid as carbon-steel weld metal. Careful electrode manipulation is required to produce a good bead contour, and it is essential to maintain a short arc length. A weave bead is preferred with the weave width not exceeding three times the electrode core diameter. Stringer-bead technique may be used for welding deep groove welds. Slag must be thoroughly removed from each bead before depositing the next bead.

Representative joint designs and welding conditions for SMAW of 0.25 in. (6.4 mm) thick plate are given in Table 3.21. Square-groove joints with a root opening about one half the section thickness can be used on section thicknesses of less than 0.125 in. (3.2 mm). For thicker sections, V-groove and U-groove joint designs similar to those recommended for GMAW should be used.

Oxyfuel Gas Welding

COPPER-NICKEL ALLOYS can be welded using the oxyfuel gas welding (OFW) process, but its use should be limited to applications where arc welding equipment is not available. Welding is performed using ERCuNi welding rods with a soft and slightly reducing flame. An oxidizing flame will form a cuprous oxide that will dissolve in molten metal, reduce corrosion resistance, and cause embrittlement. Preheat is not recommended. Liberal use of a flux made especially for nickel or copper-nickel alloys is necessary to protect the welding rod and base metal from oxidation.

Table 3.20
Representative Conditions for Gas Metal Arc Welding of Copper-Nickel Alloy Plate

Thickness		Electrode Feed[a]		Arc Voltage, V	Welding Current,[b] A
in.	mm	in./min	mm/s		
0.25	6.4	180-220	76-93	22-28	270-330
0.38	9.6	200-240	85-102	22-28	300-360
0.50	12.7	220-240	93-102	22-28	350-400
0.75	19.0	220-240	93-102	24-28	350-400
1.0	25.4	220-240	93-102	26-28	350-400
over 1.0	over 25.4	240-260	102-110	26-28	370-420

a. ERCuNi electrode, 0.062-in. (1.6-mm) diameter

b. Direct current electrode positive and argon shielding

Table 3.21
Representative Conditions for Shielded Metal Arc Welding
of 1/4 in. (6.4 mm) Thick, 90% Copper-10% Nickel Alloy Plate

| Welding Position | Joint Design | | Root Opening | | Weld Pass | Electrode Size | | Welding[b] Current, A |
	Groove Type	Groove Angle, Degrees	in.	mm		in.	mm	
Flat	Square	—	0.13	3.3	1, 2	1/8	3.2	115-120
Vertical[c]	Double-V	75-80	0.09-0.13	2.3-3.3	1, 2	3/32	2.4	85-90
Vertical[c]	Fillet	80	0	0	1	3/32	2.4	85
Horizontal	Single-V	75-80	0.06-0.13	1.5-3.3	1[d]	3/32	2.4	100
					2	1/8	3.2	100
Flat and overhead	Single-V	75-80	0.09-0.13	2.3-3.3	1[e]	1/8	3.2	110-115
					2	3/32	2.4	95-100

a. ECuNi covered electrodes.

b. Direct current electrode positive.

c. Direction of welding is up.

d. Backing weld pass. Back gouge before welding the other side (Pass 2).

e. Back gouge the root of the joint before completing the back weld (Pass 2).

Submerged Arc Welding

COPPER-NICKEL ALLOYS may be welded with the submerged arc welding (SAW) process. Section thicknesses greater than 0.50 in. (12.7 mm) are practical. V-groove and U-groove joint designs similar to those used in GMAW are satisfactory. Commercially available fluxes designed for welding copper-nickel should be used. Welding conditions, varying according to the flux used, are provided by the flux manufacturer. Careful attention to bead layer sequence is essential when multipass welds are deposited in a deep groove to ensure complete fusion while maintaining a flat bead contour. X-ray quality results are obtained when the technique is performed correctly.

BRAZING

BRAZING IS AN excellent process for joining copper and copper alloys. Surface oxides are easily fluxed during brazing except refractory oxides on aluminum bronzes (containing more than 8 percent aluminum), which require special techniques. When brazing is selected as the joining process, important considerations are brazing temperature, type of loading, joint strength, galvanic corrosion, and interaction between the base and filler metals at the service temperature.

All of the common brazing processes can be used except for special cases, such as resistance or induction brazing of copper and copper alloys that have high electrical conductivity.[4]

Both lap and butt joints may be used for brazements. The joint clearance must provide for capillary flow of the selected brazing filler metal throughout the joint at brazing temperature, and the thermal expansion characteristics of the alloy must be considered. A joint clearance of 0.001 to 0.005 in. (0.025 to 0.13 mm) will develop the maximum joint strength and soundness. Larger clearances may be used if reduction in joint strength is acceptable. When designing a brazed joint for a specific application, the properties and compatibility of the base metal-filler metal combination must be properly evaluated for the environment in which the brazed joint will operate.

For electrical conductivity applications, brazing filler metals generally have low electrical conductivity

---

4. Additional information may be found in the *Welding Handbook*, 8th Ed., Vol. 2, 380-422, and the *Brazing Handbook*, 279-296. American Welding Society, 1991.

Table 3.22
Guide to Brazing Copper and Copper Alloys

Material	Commonly Used Brazing Filler Metals	AWS Brazing Atmospheres[c]	AWS Brazing Flux	Remarks
Coppers	BCuP-2[a], BCuP-3, BCuP-5[a], RBCuZn, BAg-1a, BAg-1, BAg-2, BAg-5, BAg-6, BAg-18	1, 2, or 5	FB3-A, C, D, E, I, J	Oxygen-bearing coppers should not be brazed in hydrogen-containing atmospheres.
High coppers	BAg-8, BAg-1	Note b	FB3-A	—
Red brasses	BAg-1a, BAg-1, BAg-2, BCuP-5, BCup-3, BAg-5, BAg-6, RBCuZn	1, 2, or 5	FB3-A, C, D, E, I, J	—
Yellow brasses	BCuP-4, BAg-1a, BAg-1, BAg-5, BAg-6, BCuP-5, BCuP-3	3, 4, or 5	FB3-A, C, E	Keep brazing cycle short.
Leaded brasses	BAg-1a, BAg-1, BAg-2, BAg-7, BAg-18, BCuP-5	3, 4, or 5	FB3-A, C, E	Keep brazing cycle short and stress relieve before brazing.
Tin brasses	BAg-1a, BAg-1, BAg-2, BAg-5, BAg-6, BCuP-5, BCuP-3 (RBCuZn for low tin)	3, 4, or 5	FB3-A, C, E	—
Phosphor bronzes	BAg-1a, BAg-1, BAg-2, BCuP-5, BCuP-3, BAg-5, BAg-6	1, 2, or 5	FB3-A, C, E	Stress relieve before brazing.
Silicon bronzes	BAg-1a, BAg-1, BAg-2	4 or 5	FB3-A, C, E	Stress relieve before brazing. Abrasive cleaning may be helpful.
Aluminum bronzes	BAg-3, BAg-1a, BAg-1, BAg-2	4 or 5	FB4-A	—
Copper-nickel	BAg-1a, BAg-1, BAg-2, BAg-18, BAg-5, BCuP-5, BCuP-3	1, 2, or 5	FB3-A, C, E	Stress relieve before brazing.
Nickel silvers	BAg-1a, BAg-1, BAg-2, BAg-5, BAg-6, BCuP-5, BCuP-3	3, 4, or 5	FB3-A, C, E	Stress relieve before brazing and heat uniformly.

a. Protective atmosphere or flux is not required for brazing copper with BCuP fillers.

b. Furnace brazing without flux is possible if the parts are first nickel or copper plated. Braze following the procedures recommended for nickel or copper.

c. Hydrogen, inert gas, or vacuum atmospheres usually are also acceptable (AWS Type 7, 9, or 10). Brazing atmospheres are listed below:

AWS Brazing Atmosphere	Source	Maximum Dew Point of Incoming Gas	AWS Brazing Atmosphere	Source	Maximum Dew Point of Incoming Gas
1	Combusted fuel gas (low hydrogen)	Room temp.	5	Dissociated ammonia	-65°F (-54°C)
2	Combusted fuel gas (decarburizing)	Room temp.	6A	Cryogenic and purified N_2+H_2	-90°F (-68°C)
			6B	Cryogenic and purified N_2+H_2+CO	-20°F (-29°C)
			6C	Cryogenic and purified N_2	-90°F (-68°C)
3	Combusted fuel gas, dried	-40°F (-40°C)	7	Hydrogen, deoxygenated and dried	-75°F (-59°C)
4	Combusted fuel gas, dried (carburizing)	-40°F (-40°C)	9	Purified inert gas	—
			10	Vacuum	—

compared to copper. Nevertheless, a braze joint when properly designed will not add appreciable resistance to the circuit. For example, silver filler metal has little effect on the resistance in properly fitted braze joints having a joint clearance of 0.003 in. (0.08 mm).

FILLER METALS

ALL OF THE silver (BAg), copper-phosphorus (BCuP), gold (BAu), and copper-zinc (RBCuZn) filler metals are suitable for brazing copper, provided their liquidus temperature is sufficiently lower than the melting range of the base material.[5] The brazing filler metals that are commonly used for copper and copper alloys are listed in Table 3.22. Filler metals also are listed by chemical composition and brazing temperature range in Table 3.23.

All BAg filler metals may be used with any copper or copper alloy. BAu filler metals are used for electronic applications where the vapor pressure of the brazing filler metal is important.

BCuP filler metals can be used to braze most copper alloys, including some copper-nickel alloys containing less than 10 percent nickel. Any copper-nickel alloy should be evaluated by an appropriate brazing test because of the possibility of creating brittle joints. No flux is required to braze copper with BCuP.

5. All brazing filler metals are covered in ANSI/AWS A5.8, *Specification for Filler Metals for Brazing and Braze Welding.*

Beryllium-copper should not be brazed with BCuP since it will result in porous joints with low strength.

When corrosion resistance is not important, the RBCuZn filler metals may be used to join the coppers, copper-nickel, copper-silicon, and copper-tin alloys. The RBCuZn liquidus temperature is too high for brazing the brasses and nickel silvers. Torch brazing the aluminium bronzes precludes the use of high brazing temperature RBCuZn.

FLUXES AND ATMOSPHERES

RECOMMENDED BRAZING FLUXES and furnace atmospheres for brazing coppers and copper alloys are shown in Table 3.22. Flux classification FB3-A, FB3-C, and FB3-D are suitable for use with BAg and BCuP filler metals for brazing all copper alloys except the aluminum bronzes. A more reactive flux classification FB4-A is used for aluminum bronze. Flux classifications FB3-C, FB3-D, and FB3-K are required with RBCuZn filler metals because of their high brazing temperatures.

Combusted fuel gases are economical brazing atmospheres for copper and copper alloys, except for oxygen-bearing copper. Atmospheres with a high-hydrogen content cannot be used when brazing oxygen-bearing copper; hydrogen diffuses into the copper, reduces copper oxide, and forms water vapor that will rupture the copper. Inert gases that have proper dew points also are suitable atmospheres for brazing copper and copper alloys.

Table 3.23
Commonly Used Brazing Filler Materials for Copper and Copper Alloys[*]

AWS Classification	UNS No.	Composition, wt. %								Brazing Temperature Range	
		Ag	Cu	Zn	Cd	Sn	Fe	Ni	P	°F	°C
BAg-1	P07450	44-46	14-16	14-18	23-25	—	—	—	—	1145-1400	618-760
BAg-1a	P07500	49-51	14.5-16.5	14.5-18.5	17-19	—	—	—	—	1175-1400	635-760
BAg-2	P07350	34-36	25-27	19-23	17-19	—	—	—	—	1295-1550	702-843
BAg-3	P07501	49-51	14.5-16.5	13.5-17.5	15-17	—	—	2.5-3.5	—	1270-1500	688-816
BAg-5	P07453	44-46	29-31	23-27	—	—	—	—	—	1370-1550	743-843
BAg-6	P07503	49-51	33-35	14-18	—	—	—	—	—	1425-1600	774-871
BAg-7	P07563	55-57	21-23	15-19	—	4.5-5.5	—	—	—	1205-1400	652-760
BAg-8	P07720	71-73	Bal.	—	—	—	—	—	—	1435-1650	780-899
BAg-18	P07600	59-61	Bal.	—	—	9.5-10.5	—	—	—	1325-1550	718-843
BCu-1	C14180	—	99.9 min	—	—	—	—	—	0.75	2000-2100	1093-1149
RBCuZn-A	C47000	—	57-61	Bal.	—	0.25-1.0	—	—	—	1670-1750	910-955
RBCuZn-C	C68100	—	56-60	Bal.	—	0.8-1.1	0.25-1.2	—	—	1670-1750	910-955
RBCuZn-D	C77300	—	46-50	Bal.	—	—	9-11	—	0.25	1720-1800	938-982
BCuP-2	C55181	—	Bal.	—	—	—	—	—	7.0-7.5	1350-1550	732-843
BCuP-3	C55281	4.8-5.2	Bal.	—	—	—	—	—	5.8-6.2	1325-1500	718-816
BCuP-4	C55283	5.8-6.2	Bal.	—	—	—	—	—	7.0-7.5	1275-1450	691-788
BCuP-5	C55284	14.5-15.5	Bal.	—	—	—	—	—	4.8-5.2	1300-1500	704-816

[*] Refer to ANSI/AWS A5.8, *Specification for Filler Metals for Brazing and Braze Welding.*

Vacuum is a suitable brazing environment, provided neither base metal nor filler metal contains elements that have high vapor pressures at brazing temperature. Zinc, phosphorus, and cadmium are example of elements that vaporize when heated in vacuum.

SURFACE PREPARATION

GOOD WETTING AND flow of filler metal in brazed joints can only be achieved when the joint surfaces are clean and free of oxides, dirt, and other foreign substances. Standard solvent or alkaline degreasing procedures are suitable for cleaning copper base metals. Mechanical methods may be used to remove surface oxides, but care should be taken to leave the metal free of undesirable films or deposits. Chemical removal of surface oxides requires an appropriate pickling solution.

Workers must be trained in proper safety practices for handling and mixing acid solutions to prevent injury when acid cleaning is used. Workers also must be supplied with protective clothing and equipment including eye, face, and body protection. Work areas must be properly ventilated and equipped with safety showers and eyewash stations. Proper dilution and disposal of acid solutions also is required.

Typical chemical cleaning procedures are as follows:

Copper

IMMERSE IN COLD 5 to 15 percent by volume sulfuric acid for 1 to 5 minutes; rinse in cold water followed by hot water rinse, and air blast dry.

Beryllium-Copper

USE THE FOLLOWING two steps:

(1) Immerse in 20 percent by volume sulfuric acid at 160 to 180 °F (71 to 82 °C) until the dark scale is removed, then water rinse.

(2) Dip in cold 30 percent nitric acid solution for 15 to 30 seconds, then rinse in hot water, and air blast dry.

Chromium-Copper and Copper-Nickel Alloys

IMMERSE IN HOT 5 percent by volume sulfuric acid for 1 to 5 minutes. Cold water rinse followed by hot water rinse, and air blast dry.

Brass and Nickel-Silver

IMMERSE IN COLD 5 percent by volume sulfuric acid-cold rinse, hot rinse, and air blast dry.

Silicon Bronze

IMMERSE FIRST IN hot 5 percent by volume sulfuric acid, rinse in cold water, and immerse in a cold mixture of 2 percent by volume hydrofluoric acid and 5 to 15 percent by volume sulfuric acid for 1 to 10 minutes. Cold rinse followed by hot rinse, and air blast dry.

Aluminum Bronze

TOUGH ALUMINUM OXIDE can be removed or loosened using a strong alkali solution of sodium hydroxide (10 wt. percent) at 170 °F (75 °C). Immerse for 2 to 5 minutes and cold rinse followed by successive immersions in two solutions:

(1) Cold 2 percent hydrofluoric acid and 3 percent sulfuric acid mixture for 1 to 5 minutes; cold water rinse, then immerse in:

(2) A solution of 5 percent by volume sulfuric acid at 80 to 120 °F (27 to 49 °C) for 1 to 5 minutes, cold water rinse, hot water rinse, and air blast dry.

Copper Plate

IT IS OFTEN desirable to copper plate the faces of copper alloys that contain strong oxide-forming elements to simplify brazing and fluxing requirements. Copper plate about 0.001 in. (0.025 mm) thick is used on chromium-copper alloys while about 0.0005 in. (0.013 mm) thickness is sufficient on beryllium copper, aluminum bronze, and silicon bronze.

BRAZING COPPER

OXYGEN-FREE, HIGH conductivity copper and deoxidized copper are readily brazed by furnace or torch methods. Boron-deoxidized copper is sometimes preferred when brazing at high temperatures because grain growth is less pronounced than with the other coppers.

Oxygen-bearing coppers are susceptible to oxide migration and hydrogen embrittlement at elevated temperatures. These coppers should be furnace brazed in an inert atmosphere or torch brazed with a neutral or slightly oxidizing flame.

The copper-phosphorus and copper-silver-phosphorus filler metals (BCuP) are considered self-fluxing on copper. A flux is beneficial for massive copper assemblies where prolonged heating results in excessive oxidation. During brazing, the filler metal loses some phosphorus, which results in a slight increase in the remelt temperature. Joints brazed with phosphorus-containing filler metal should not be exposed to sulfurous atmospheres at elevated temperature. Exposure for long periods results in corrosive attack of the joint.

With the copper-zinc (RBCuZn) filler metals, the recommended brazing temperatures should not be exceeded to avoid volatilization of zinc and resulting porosity in the joint. When torch brazing, an oxidizing flame will reduce zinc fuming. The corrosion resistance of these filler metals is inferior to that of copper.

A lap joint will develop the full strength of annealed copper at room temperature when the overlap is at least three times the thickness of the thinner member. As the service temperature increases, the brazing filler metal strength decreases more rapidly than does the strength of the copper or copper alloy, and failure will eventually occur through the joint. The tensile strength at room, elevated, and sub-zero temperatures for single-lap brazed joints in copper is shown in Table 3.24. Typical creep properties for tough-pitch copper brazed with BAg-1A, BAg-6, and BCup-5 filler metals are shown in Figure 3.5. At 77 and 260 °F (25 and 127 °C), failures were in the base metal.

BRAZING HIGH-COPPER ALLOYS

Beryllium-Copper

THE SURFACES OF beryllium-copper components must be cleaned prior to brazing. The oxide scale can be removed by pickling.

The 2 percent beryllium-copper can be brazed by either of two methods. The more common procedure involves simultaneous brazing and solution heat treatment at 1450 °F (788 °C) in a furnace. The silver-copper eutectic filler metal, BAg-8, is generally used with an AWS flux classification FB3-A. The furnace temperature is quickly lowered to 1400 °F (760 °C) to solidify the brazing filler metal. The brazement is then quenched in cold water, and finally age hardened at 600 to 650 °F (316 to 343 °C).

A second method, used with thin sections that can be heated rapidly (preferably in one minute or less), permits brazing at a temperature below the beryllium-copper solution anneal temperature. The brazement can be precipitation hardened without having to be solution treated. Fast heating rates can be attained using an OAW torch or by resistance heating. Brazing is satisfactorily performed using BAg-1 filler metal and flux classification FB3-A, but other silver brazing filler metals may be suitable for special applications.

The 0.5 percent beryllium-copper alloys are solution heat-treated at 1700 °F (927 °C), quenched in cold water, and age hardened between 850 and 900 °F (454 and 482 °C). This alloy can be brazed rapidly with BAg-1 filler metal at about 1200 °F (649 °C) after being precipitation hardened, but the base metal is overaged during brazing resulting in hardness and strength property loses.

Chromium-Copper and Zirconium-Copper

CHROMIUM-COPPER AND ZIRCONIUM-COPPER are solution heat-treated at 1650 to 1850 °F (700 to 1010 °C), quenched, and age hardened at 900 °F (482 °C). Brazing with BAg filler metals and a fluoride-type flux should be performed after a solution heat treatment and before age hardening. The mechanical properties of the base metal after brazing and precipitation hardening will be lower than those of the solution treated and age-hardened material that has not been brazed.

Brazing followed by solution treatment and age hardening of chromium-copper results in near optimum

Table 3.24
Tensile Strength of Single-Lap Brazed Joints in Deoxidized Copper*

Brazing Filler Metal	Tensile Strength					
	-321°F (-196°C)		72°F (22 °C)		400°F (204°C)	
	ksi	MPa	ksi	MPa	ksi	MPa
BAg-1	30.1	208	19.0	131	9.7	67
BAg-6	28.0	193	17.6	121	—	—
BAg-8	24.7	170	17.6	121	—	—
BCuP-2	17.9	123	18.6	197	10.7	74
BCuP-4	21.4	150	19.1	132	—	—
BCup-5	21.9	151	17.9	123	10.8	74

* Specimens were made from 0.25 in. (6.4 mm) thick sheet; joints had an overlap of 0.15 in. (3.8 mm) and no braze fillet.

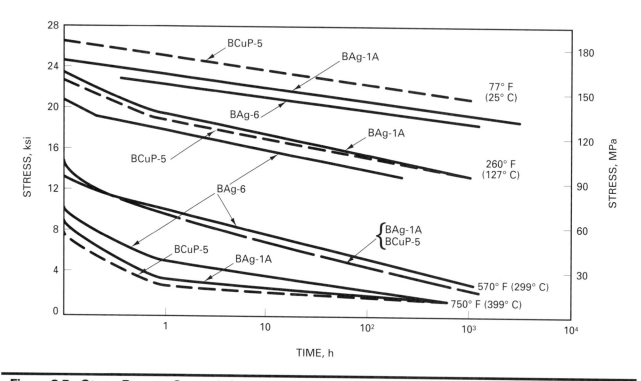

Figure 3.5—Stress-Rupture Strength Curves for Copper Brazed with Three Filler Metals Using a Plug and Ring Creep Specimen

mechanical properties. Distortion caused by quenching from the solution anneal temperature should be evaluated for each application.

Cadmium-Copper

CADMIUM-COPPERS ARE BRAZED in the same manner as deoxidized copper.

BRAZING COPPER-ZINC ALLOYS (BRASS)

ALL BRASSES CAN be brazed with BAg and BCuP, and RBCuZn filler metals. AWS classification FB3-C flux is used with BAg and BCuP filler metals, and FB3-K with RBCuZn filler metals.

Zinc fuming above 750 °F (399 °C) is reduced by fluxing the parts before furnace brazing, even when a protective atmosphere is used. Torch brazing with an oxidizing flame is also used to reduce zinc fuming. Brasses subject to cracking when heated too rapidly (i.e., leaded brasses) should be heated uniformly and

slowly to the brazing temperature. Sharp corners and other stress risers that localize thermal strain during heating should be avoided. Good practice dictates heating parts slowly to brazing temperature.

Brasses containing aluminum or silicon require cleaning and brazing procedures similar to those used for aluminum or silicon bronze. Lead in brass may alloy with the brazing filler metal and cause brittleness, especially when the lead content exceeds 3 percent. This can reduce the strength of the joint. Leaded brasses require complete flux coverage both to prevent the formation of lead oxide or dross and to maintain good flow and wetting during brazing.

Stress-relieving before brazing and slow, uniform heating minimizes the tendency of high-leaded brasses to crack. Rapid heating to the brazing temperature is to be avoided.

BRAZING COPPER-TIN ALLOYS (PHOSPHOR BRONZE)

IN A STRESSED condition, phosphor bronzes are subject to cracking during heating. Good practice dictates that the parts to be brazed are processed as follows:

(1) Stress relieve or anneal before brazing.

(2) Support with suitable fixtures for maintaining a stress-free condition during brazing.

(3) Apply flux to completely cover the joint surfaces.

(4) Use a slow heating cycle.

All of the phosphor bronzes can be brazed with BAg and BCuP filler metals, although alloys with low-tin content can be brazed with RBCuZn filler metal.

The phosphor bronze parts are sometimes made by compacting and sintering powdered metal. Before they are brazed, areas of powdered metal parts away from the braze joint must be treated to seal pores in order to restrict the penetration of the brazing filler metal. Pores can be sealed by painting the surface of a powdered metal part with a colloidal graphite suspension and then baking at a low temperature followed by a cleaning procedure.

BRAZING COPPER-ALUMINUM ALLOYS (ALUMINUM BRONZE)

ALUMINUM BRONZE is brazed with BAg brazing filler metals and AWS classification FB4-A flux. Refractory aluminum oxide forms at brazing temperature on the surfaces of bronze containing more than 8 percent aluminum. The brazing procedures must prevent aluminum oxide formation to obtain satisfactory flow and wetting of joint surfaces. In furnace brazing, flux should be used in addition to the protective atmosphere.

Copper plating on joint surfaces prevents formation of aluminum oxide during brazing. Copper plated parts are brazed using procedures suitable for brazing copper.

BRAZING COPPER-SILICON ALLOYS (SILICON BRONZE)

COPPER-SILICON ALLOYS should be cleaned and then either coated with flux or copper plated before brazing; this prevents the formation of refractory silicon oxide on joint surfaces during brazing. Copper plating is recommended as the method that produces best results. Mechanical or chemical cleaning is used to remove oxide contamination from the joint surfaces.

Silver brazing filler metals and AWS classification FB3-C flux are used to braze copper-silicon alloys.

When furnace brazing, a flux should be used in combination with a protective atmosphere. Silicon bronze is subject both to intergranular penetration by the filler metal and to hot-shortness when stressed. Components should be stress-relieved before brazing, adequately supported by fixtures during heating, and brazed below 1400 °F (760 °C).

BRAZING COPPER-NICKEL ALLOYS

COPPER-NICKEL ALLOYS are brazed with BAg and BCuP filler metals, although BCuP filler metal forms a brittle nickel phosphide with alloys containing more than 10 percent nickel. The structure and properties of joints brazed with BCuP filler metal should be thoroughly evaluated for the intended application. AWS classification FB3-C flux is suitable for most applications.

The base metal surfaces must be free of sulfur or lead that could cause cracking during the brazing cycle. Solvent or alkaline degreasing procedures are used to remove grease or oil. Surface oxides can be removed by either mechanical or chemical cleaning, or both.

Copper-nickel alloys in a cold-worked condition are susceptible to intergranular penetration by molten filler metal. They should be stress-relieved before brazing to prevent cracking.

BRAZING COPPER-NICKEL-ZINC ALLOYS (NICKEL-SILVER)

NICKEL-SILVERS CAN be brazed using the same filler metals and procedures for brazing brasses. If RBCuZn filler metal is used, precautions should be taken when brazing at relatively high temperatures. These alloys in the stressed condition are subject to intergranular penetration by the filler metal and should be stress-relieved before brazing. Nickel-silvers have poor thermal conductivity and should be heated slowly and uniformly to brazing temperature in the presence of sufficient flux.

BRAZING DISSIMILAR METALS

DISSIMILAR COPPER ALLOYS can be brazed readily. Copper alloys are brazed to steel, stainless steel, and nickel alloys. Suggested brazing filler metals for dissimilar metal combinations are given in Table 3.25.

Table 3.25
Recommended Filler Metals for Brazing Copper Alloys to Other Metals

Carbon and Low Alloy Steels	Cast Iron	Stainless Steels	Tool Steels	Nickel Alloys	Titanium Alloys	Reactive Metals	Refractory Metals
BAg	BAg	BAg	BAg	BAg	BAg	BAg	BAg
BAu	BAu	BAu	BAu	BAu	—	—	—
RBCuZn	RBCuZn	—	RBCuZn	RBCuZn	—	—	—
—	—	—	BNi	—	—	—	—

SOLDERING

COPPER AND COPPER alloys are among the most frequently soldered engineering materials. The degree of solderability, as shown in Table 3.26, ranges from excellent to difficult.[6] No serious problems arise in soldering most copper alloys, but alloys containing beryllium, silicon, or aluminum require special fluxes to remove surface oxides.

The high thermal conductivity of copper and some of its alloys require a high rate of heat input when localized heating is used.

6. Additional information on soldering, solders, and fluxes may be found in the *Welding Handbook*, Vol. 2, 8th Ed., 423-447, and the *Soldering Manual*, 2nd Ed., American Welding Society, 1977.

SOLDERS

THE MOST WIDELY used solders for joining copper and its alloys are the tin-lead solders, although in drinking water application tin-antimony or tin-silver solder is used to eliminate possible lead contamination of the water. Tin readily alloys and diffuses into copper and copper alloys. Copper alloys accept a certain amount of tin into solid solution, but one or more intermetallic phases (probably Cu_6Sn_5) will form when the solid solubility limit is exceeded. An intermetallic phase forms at the faying surfaces of solder joints, and faying surfaces tends to be brittle, so that their thickness should be minimized by proper selection of process variables

Table 3.26
Solderability of Copper and Copper Alloys

Base Metal	Solderability
Copper (Includes tough-pitch, oxygen-free, phosphorized, arsenical, silver-bearing, leaded, tellurium, and selenium copper.)	Excellent. Rosin or other noncorrosive flux is suitable when properly cleaned.
Copper-tin alloys	Good. Easily soldered with activated rosin and intermediate fluxes.
Copper-zinc alloys	Good. Easily soldered with activated rosin and intermediate flux.
Copper-nickel alloys	Good. Easily soldered with intermediate and corrosive fluxes.
Copper-chromium and copper-beryllium	Good. Require intermediate and corrosive fluxes and precleaning.
Copper-silicon alloys	Fair. Silicon produces refractory oxides that require use of corrosive fluxes. Should be properly cleaned.
Copper-aluminum alloys	Difficult. High aluminum alloys are soldered with help of very corrosive fluxes. Precoating may be necessary.
High-tensile manganese bronze	Not recommended. Should be plated to ensure consistent solderability.

and service conditions.As the thickness of the interme-tallic layer increases, the strength of the soldered joint will decrease, and service at elevated temperatures accelerates this change.

FLUXES

ORGANIC AND ROSIN types of noncorrosive fluxes are excellent for soldering coppers and may be used with some success on copper alloys containing tin and zinc, if surfaces to be soldered are precleaned. These fluxes are used for soldering electrical connections and electronic components. A light coat of flux should be applied to precleaned faying surfaces.

The inorganic corrosive fluxes can be used on all the copper alloys, but they are required only on those alloys that develop refractory surface oxides, such as silicon and aluminum bronze. Aluminum bronze is especially difficult to solder and requires special fluxes or copper plating. Inorganic chloride fluxes are useful for solder-ing the silicon bronzes and copper-nickels.

Oxide films reform quickly on cleaned copper alloys, and fluxing and soldering should be done immediately after cleaning. Copper tube systems soldered with 50% tin-50% lead or 95% tin-5% antimony solder require a mildly corrosive liquid flux or petrolatum pastes con-taining zinc and ammonium chlorides. Many liquid fluxes for plumbing application are self-cleaning, but there is always a risk of corrosive action continuing after soldering.

A highly corrosive flux can remove some oxides and dirty films, but there is always an uncertainty whether uniform cleaning has been achieved and whether corro-sive action continues after soldering. Optimum solder-ing always starts with clean surfaces and a minimum amount of flux.

SURFACE PREPARATION

SOLVENT OR ALKALINE degreasing and pickling are suitable for cleaning copper base metals, with mechani-cal methods used to remove oxides. Chemical removal of oxides requires proper choice of a pickling solution followed by thorough rinsing. Typical procedures used for chemical cleaning are the same as those described previously for brazing.

COATED COPPER ALLOYS

THE MOST COMMONLY employed coating for cop-per alloys are tin, lead, tin-lead, nickel, chromium, and silver. Soldering procedures depend upon coating char-acteristics. Except for chromium plate, none of the coating materials present any serious soldering diffi-culty. Before soldering chromium-plated copper, the chromium must be removed from the joint faces. Ther-mal conductivity of the base metal must be considered.

FLUX REMOVAL

AFTER THE JOINT is soldered, flux residues that may prove harmful to the serviceability of the joint must be removed. Removal of flux residues is especially impor-tant when the joints will be exposed to humid environ-ments. Organic and inorganic flux residues that contain salts and acids should be removed completely.

Non-activated rosin flux residues may remain on the soldered joint unless appearance is important or the joint area is to be painted or otherwise coated. Acti-vated rosin fluxes are treated in the same manner as organic fluxes for structural soldering, but they should be removed for critical electronic applications.

Zinc chloride fluxes leave a fused residue that absorbs water from the atmosphere. Removal of this residue is best accomplished by thorough washing in a hot solution of 2% concentrated hydrochloric acid [2.5 oz./gal (20 ml/L)], followed by a hot water rinse and air blast drying. The acid solution removes the white zinc oxychloride crust, which is insoluble in water. Com-plete removal sometimes requires additional rinsing in hot water that contains some washing soda (sodium carbonate), followed by a clear water rinse. Mechani-cal scrubbing may also be necessary.

The residues from the organic fluxes are quite soluble in hot water. Double rinsing is always advisable. Oily or greasy paste flux residues may be removed with an organic solvent. Soldering pastes are emulsions of petroleum jelly and a water solution of zinc-ammonium chloride. The corrosive chloride residue must be removed from the soldered joint.

When rosin residues are to be removed, alcohol or chlorinated hydrocarbons are used. Certain rosin acti-vators are soluble in water but not in organic solvents. This flux residue is removed using organic solvents, fol-lowed by water rinsing.

MECHANICAL PROPERTIES

THE MECHANICAL PROPERTIES of a soldered joint depend upon a number of process variables in addition to the solder composition. These variables include the solder thickness in the joint, base metal composition, type of flux, soldering temperature, soldering time, and cooling rate.

Soldered joints are normally designed to be loaded in shear. Shear strength and creep strength in shear are the important mechanical properties. For specialized appli-cations such as auto radiators, peel strength and frac-ture initiation strength may be important, and in a few cases, tensile strength is of interest.

Shear Strength

SHEAR STRENGTH IS determined using single- or double-lap flat specimens or sleeve-type cylindrical specimens. Testing is done with a cross-head speed of 1.0 or 0.1 in./min (25.4 or 2.54 mm/min). The shear strengths of copper joints soldered with lead-tin solders are shown in Figure 3.6. The maximum joint shear strength is obtained with eutectic composition solder (63% tin-37% lead). Shear strength may decrease up to 30 percent if the joints are aged at room temperature or at moderately elevated temperature for several weeks prior to testing.

Shear strength of soldered joints decreases with increasing temperature, as shown in Figure 3.7 for two solders commonly used in copper plumbing. Many solders remain ductile at cryogenic temperatures, and their strengths increase significantly as the temperature decreases below room temperature.

Creep Strength

THE CREEP STRENGTH of a soldered joint in shear is considerably less than its short-time shear strength, sometimes below 10 percent. The creep shear strengths of copper joints soldered with three solders are shown in Figure 3.8. The 50% tin-50% lead and 95% tin-5% antimony solders have about the same short-time shear strengths, but the tin-antimony solder has much greater long-time creep strength.

Tensile Strength

TYPICAL TENSILE STRENGTHS of soldered butt joints made with five compositions of tin-lead solder are presented in Table 3.27. Soldered joints are much stronger in tension than in shear. Tensile strength increases with increasing tin content up to the eutectic composition. Soldering is not recommended for butt joints in copper because voids in the solder layer will cause premature failure through the solder when the joint is loaded.

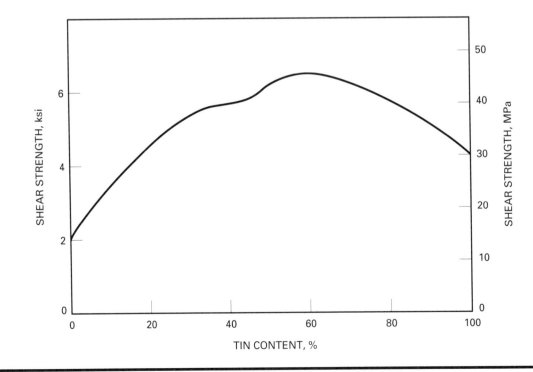

Figure 3.6—Shear Strengths of Copper Joints Soldered with Tin-Lead Solders

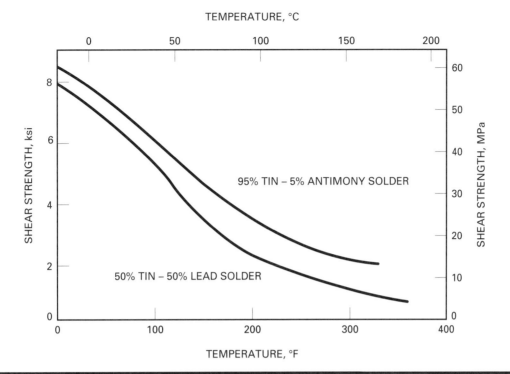

Figure 3.7—Shear Strengths at Elevated Temperatures for Soldered Copper Joints

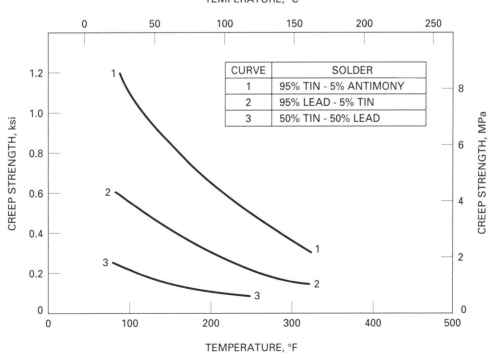

Figure 3.8—Creep Strengths at Elevated Temperatures for Soldered Copper Joints

Table 3.27
Tensile Strength of Soldered Copper Butt Joints

Solder Composition, %		Tensile Strength	
Tin	Lead	ksi	MPa
20	80	11.3	77.9
30	70	13.8	95.1
40	60	16.8	115.8
50	50	18.2	125.5
63	37	19.6	135.1

APPLICATIONS

COOLING FANS FOR ELECTRIC MOTORS

COOLING FANS FOR explosion-proof electric motors are fabricated from copper alloys because of their non-magnetic and non-sparking properties (see Figure 3.9).[7] The central hub of each fan is a centrifugally cast tin-bronze alloy that contains 1.5 percent tin and 0.3 percent phosphorus. The conical base, outer ring, and fins are 3/16 in. (4.8 mm) thick C61300 aluminum bronze. Components are assembled and welded with the gas tungsten arc welding (GTAW) process using argon shielding gas and 3/32 in. (2.4 mm) diameter ERCuAl-A2 filler metal. Direct welding current electrode negative is used at 135 amperes.

7. Information on this application was provided by Ampco Metal, Inc.

Courtesy of Ampco Metal, Inc.

Figure 3.9—Cooling Fans for Electric Motors Welded Using the GTAW Process

SEAWATER DISTILLERS FOR SHIPBOARD INSTALLATION

A SUBMERGED TUBE distilling plant that produces 3000 gallons (11 356 L) a day of fresh water aboard a ship is shown in Figure 3.10.[8] This distilling plant is fabricated from ASTM B171M Type 90/10 copper-nickel (C70600) plate and ASTM B111 tube ranging in thickness from 0.1 to 1 in. (2.5 to 25 mm).

Copper alloy C70600 is used because the excellent seawater corrosion resistance of this alloy gives this design long-term dependability. Copper alloy C70600 is also important for this particular distilling plant, used aboard a minesweeper in which the materials must have low magnetic permeability. This distiller also is economical to operate because it uses waste heat.

The distilling plant is welded with the GMAW and GTAW processes using ERCuNi filler metal and argon shielding gas. Welding parameters are shown in Table 3.28. No preheat is used but welding is restricted to a minimum material temperature of 60 °F (16 °C). Maximum interpass temperature is 300 °F (149 °C). The unit is not postweld heat treated.

8. Information on this application was provided by Aqua-Chem, Inc.

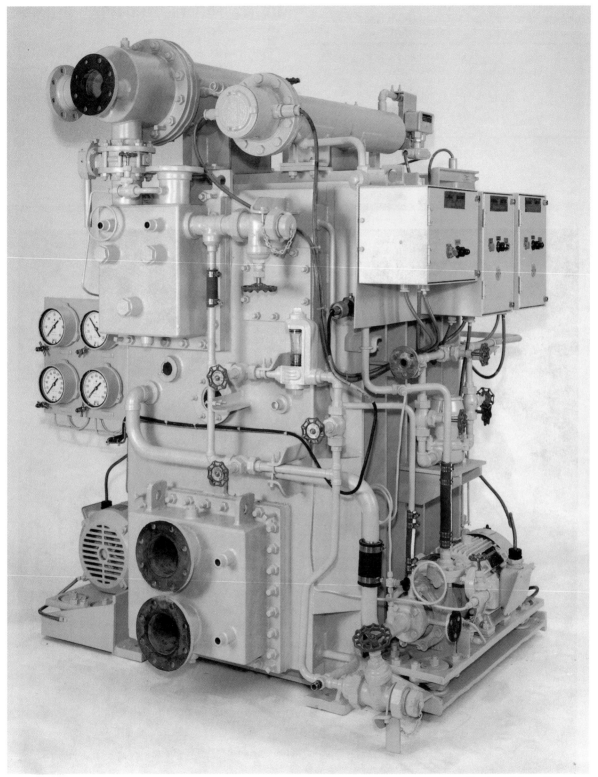

Figure 3.10—Submerged Tube Seawater Distilling Plant Fabricated with Copper-Nickel Alloy

Table 3.28
Welding Parameters for Copper-Nickel Alloy
Seawater Distilling Plant

Gas Tungsten Arc Welding	
Current	225 A, DCEN

Gas Metal Arc Welding	
Current	250 A, DCEP
Voltage	27 V

DISCHARGE ELBOW AND MOTOR MOUNT FOR CENTRIFUGAL PUMP

A DISCHARGE ELBOW for a vertical centrifugal pump, shown in Figure 3.11, is fabricated from aluminum bronze plate and castings to resist corrosive attack from seawater.[9] The majority of the elbow is C61300 plate, and the flange rings are temper annealed, centrifugal cast C95200 aluminum bronze. Most of the welds

9. Information on this application was provided by Ampco Metal, Inc.

Courtesy of Ampco Metal, Inc.

Figure 3.11—Pump Discharge Elbow Fabricated with Aluminum Bronze

on this part are GMAW, although the root passes in full penetration joints are deposited with GTAW. Welding parameters are given in Table 3.29. The part is welded at room temperature with a maximum interpass temperature of 500 °F (260 °C). No postweld heat treatment is performed.

BERYLLIUM-COPPER TOROIDAL FIELD COILS

BERYLLIUM-COPPER COMBINES high strength and good electrical conductivity. Therefore, this alloy is considered for fabrication of large conductors. An example of one potential application is for toroidal field coils for fusion reactor research.[10] A concept toroidal field coil is shown in Figure 3.12.

Fabrication of this coil would require butt welding of 1.1 in. (28 mm) thick beryllium-copper plates to form a coil that would be 20 ft (6.1 m) high by 12 ft (3.65 m) wide. Butt welds must have strengths that are equivalent to the base metal. The beryllium-copper alloy C17510, Cu-0.3Be would be used. Butt welds have been produced in double-U-joint preparations using the GMAW process. Filler metal was 0.045-in. (1.14-mm) diameter C17200, Cu-1.9Be alloy and shielding gas was 75% helium-25% argon at a flow rate of 100 ft^3/hr (47.2 L/min). Plates were preheated to 300 °F (149 °C). Welding parameters are given in Table 3.30. Completed welds can be postweld heat treated to develop ultimate tensile strengths of 99 ksi (683 MPa) and yield

10. Information on this application was provided by Princeton University Plasma Physics Laboratory and Edison Welding Institute.

Table 3.29
Welding Parameters for Aluminum Bronze Discharge Elbow

Gas Tungsten Arc Welding

Filler metal:	3/32 or 1/8 in. (2.4 or 3.2 mm) diam. ERCuAl-A2
Shielding gas:	Helium or argon/helium
Gas flow rate:	35 to 40 ft^3/hr (17 to 19 L/min)
Current:	190 to 220 A, DCEN

Gas Metal Arc Welding

Filler metal:	1/16 in. (1.6 mm) diameter ERCuAl-A2
Shielding gas:	Argon
Gas flow rate:	45 to 55 ft^3/hr (21 to 26 L/min)
Current:	320 A, DCEP
Voltage:	29 to 32 V

Table 3.30
Welding Parameters For Butt Welds in 1.1 in. (28 mm) Thick Beryllium-Copper

	Root Pass	Fill Passes
Wire Feed Speed:	525 in./min (222 mm/s)	575 in./min (243 mm/s)
Travel Speed:	18 in./min (7.6 mm/s)	12 in./min (5.1 mm/s)
Current:	390 A, DCEP	430 A, DCEP
Voltage:	30 V	30 V

strengths of 85 ksi (586 MPa). Welding, grinding, and machining of beryllium-copper require special precautions compared to other copper alloys because of greater potential health hazards. Testing of the shop environment is necessary to ensure that controls are adequate to protect workers.

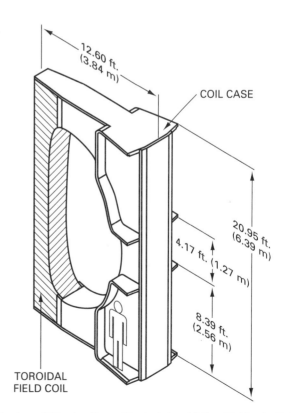

12.60 ft. (3.84 m)

COIL CASE

20.95 ft. (6.39 m)

4.17 ft. (1.27 m)

8.39 ft. (2.56 m)

TOROIDAL FIELD COIL

Courtesy of Princeton Plasma Physics Lab and Edison Welding Institute

Figure 3.12—Toroidal Field Coil and Coil Case

COPPER-NICKEL WELD SURFACING

THE AUTOMATIC GAS tungsten arc welding process with hot wire addition of filler metal is shown in Figure 3.13 being used to clad a carbon steel flange with copper-nickel.[11] The flange is ASTM A285 Grade C steel and is welded to a copper-nickel pipe. This assembly is

11. Information on this application was provided by Westinghouse Electric Corp., Marine Division.

used on a shipboard turbine condenser. The first layer of GTAW weld cladding is nickel (ERNi) and the second layer is copper-nickel. The copper-nickel is deposited with 1/16 in. (1.6 mm) diameter ERCuNi filler metal and 75% helium-25% argon shielding gas at flow rates of 85 to 95 ft³/hr (40 to 45 L/min). Preheat temperature is 60 °F (16 °C) minimum and the maximum interpass temperature is 150 °F (66 °C). GTAW current is direct current electrode negative. The welding parameters are listed in Table 3.31.

Courtesy of Westinghouse Electric Corporation, Marine Division

Figure 3.13—GTAW Cladding A Steel Flange with Copper-Nickel

Table 3.31
Welding Parameters For Hot Wire GTAW Cladding of a Carbon Steel Flange

Arc amperage	340 to 370 A
Arc voltage	14 to 16 V
Hot wire amperage	85 to 105 A (ac)
Wire feed speed	90 to 100 in./min (38 to 42 mm/s)
Travel speed	3.5 to 5.0 in./min (1.5 to 2.1 mm/s)

ALUMINUM BRONZE OVERLAY

IN THIS APPLICATION, aluminum bronze bearing surfaces are applied to a cast steel equalizer crown cushion using the GMAW process.[12] A vital part for an electric mining shovel, the equalizer crown cushion, shown in Figure 3.14, weighs 1030 lb (467 kg). Aluminum bronze bearing surfaces reduce friction and have excellent wear, impact, and galling resistance.

This application uses the manual GMAW process with ERCuAl-A2 filler metal and argon shielding gas. Welding current is 250 A, direct current electrode positive, and welding voltage is 26 V. The surfacing welds are deposited in two passes to a thickness of 0.375 in. (9.5 mm).

12. Information on this application was provided by Ampco Metal, Inc.

Courtesy of Ampco Metal, Inc.

Figure 3.14—Surfacing a Mining Shovel Casting

CONDENSER WATER BOX FABRICATION

THE FABRICATION OF a water box for a shipboard condenser that cools turbine exhaust steam is shown in Figure 3.15.[13] The formed head is ASTM B402 Type 90/10 copper-nickel (C70600), and the flanges are ASTM A285 Grade C carbon steel. The welding

process is SMAW using 1/8 and 5/32 in. (3.2 and 4.0 mm) diameter ENiCu-7 electrodes. Typical welding current for 1/8 in. (3.2 mm) diameter electrodes is 90 to 130 A, direct current electrode positive. When 5/32 in. (4.0 mm) diameter electrodes are used, welding current is 120 to 170 A. A minimum of 60 °F (16 °C) preheat is required, and maximum interpass temperature is limited to 150 °F (66 °C).

13. Information on this application was provided by Westinghouse Electric Corporation, Marine Division.

Courtesy of Westinghouse Electric Corporation, Marine Division

Figure 3.15—Shielded Metal Arc Welding of a Condenser Water Box

WELDED COPPER PLUMBING FITTINGS

LARGE-DIAMETER, WELDED PLUMBING fittings (see Figure 3.16) have inherent advantages compared to cast plumbing fittings for some applications.[14] Welded fittings can be made that are lighter weight and can be formed into more configurations than cast fittings. Large diameter, phosphorus-deoxidized copper (C12200) fittings are used for commercial and industrial plumbing systems as well as heating, ventilation, and air-conditioning systems. The fitting shown (see Figure 3.16) was welded with the GTAW process using silicon bronze (ERCuSi-A) filler metal and 50% argon-50% helium shielding gas. Direct current electrode negative was used at 225 to 275 amperes. Welds were made without preheat or postweld heat treatment.

14. Information on this application was provided by Elkhart Products Corporation and Copper Development Association.

Courtesy of Elkhart Products Corp. and Copper Development Association

Figure 3.16—Welded Plumbing Fitting

SUPPLEMENTARY READING LIST

GENERAL

American National Standards Institute. ANSI/ASC Z49.1, *Safety in welding, cutting and allied processes.* (Available from American Welding Society, Miami, Florida) New York: American National Standards Institute, 1988.

American Society For Metals. *Metals handbook*, Vol. 6, 9th Ed. Metals Park, Ohio: American Society for Metals, 1983.

——————. "Copper." *Metals handbook*, Vol. 2, 9th Ed., 239-51 and 275-439. Metals Park, Ohio: American Society for Metals, 1979.

American Welding Society. *Brazing handbook*, 4th Ed. Miami, Florida: American Welding Society, 1991.

——————. *Soldering manual*, 2nd Ed. Miami, Florida: American Welding Society, 1978.

——————. *Welding handbook*, Vol. 2, 8th Ed. Miami, Florida: American Welding Society, 1991.

International Nickel Company. *Guide to the welding of copper-nickel alloys*. New York: International Nickel Company, 1979.

WELDING

Brandon, E. "The weldability of oxygen-free boron deoxidized low phosphorus copper." *Welding Journal* 48(5): 187s-94s, 1969.

Bray, R. S., et al. "Metallographic study of weld solidification in copper." *Welding Journal* 48(5): 181s-5s, 1969.

Dawson, R. J. C. "Welding of copper and copper base alloys." Bulletin 287. New York: Welding Research Council, September 1983.

Dimbylow, C. S., and Dawson, R. J. C. "Assessing the weldability of copper alloys." *Welding and Metal Fabrication* 46(9): 461-71, 1978.

Fisher, S. M., et al. "The structural integrity of copper-nickel to steel shielded metal arc weldments." *Welding Journal* 62(3): 37-43, 1983.

Gutierrez, S. H. "Understanding GTA welding of 90/10 copper nickel." *Welding Journal* 70(5): 76-8, 1991.

Hartsell, E. W. "Joining copper and copper alloys." *Welding Journal* 52(2): 89-100, 1973.

Hashimoto, K., et al. "Laser welding copper and copper alloys." *Journal of Laser Applications* 3(1): 21-5, 1991.

Johnson, L. D. "Some observations on electron beam welding of copper." *Welding Journal* 49(2): 55s-60s, 1970.

Kelley, T. J. "Ultrasonic welding of Cu-Ni to steel. *Welding Journal* 60(4): 29-31, 1981.

Littleton, J., et al. "Nitrogen porosity in gas shielded arc welding of copper." *Welding Journal* 53(12): 561s-5s, 1974.

Mustaleski, T. M., McCaw, R. L., and Sims, J. E. "Electron beam welding of nickel-aluminum bronze." *Welding Journal* 67(7): 53-9, 1988.

Rogerson, J. H. "Significance of welds in copper and cupronickel alloy welds. Their relevance to design and quality standards." *Welding Review* 3(4): 192-4, 1984.

Ruge, J., Thomas, K., Eckel, C., and Sundaresan, S. "Joining of copper to titanium by friction welding." *Welding Journal* 66(8): 28-31, 1986.

Sandor, L. W. "Copper-nickel for ship hull constructions—welding and economics." *Welding Journal* 61(12): 23-30, 1982.

————. "Pulsed HMA spot welding of copper-nickel to steel." *Welding Journal* 63(6): 35-50, 1984.

Savage, W. F., et al. "Microsegregation in 70Cu-30Ni weld metal." *Welding Journal* 55(6): 165s-73s, 1976.

————. et al. "Microsegregation in partially heated regions of 70Cu-30Ni weldments." *Welding Journal* 55(7): 181s-7s, 1976.

Tumuluru, M., and Nippes, E. F. "Weld ductility studies of a tin-modified copper-nickel alloy." *Welding Journal* 69(5): 197s-203s, 1990.

Wold, K. "Welding of copper and copper alloys." *Metal Progress* 108(8): 43-7, December 1975.

BRAZING

Belkin, E., and Nagata, P. K. "Hydrogen embrittlement of tough-pitch copper by brazing." *Welding Journal* 54(2): 54s-62s, 1975.

Chatterjee, S. K., and Mingxi, Z. "Tin-containing brazing alloys." *Welding Journal* 70 (5): 118s-22s, 1991.

Datta, A., Rabinkin A., and Bose, D. "Rapidly solidified copper-phosphorous base brazing foils." *Welding Journal* 63 (10): 14-21, 1984.

Gibbay, S. F., Dirnfeld, F., Ramon, J., and Wagner, H. Z. "Mechanism of tough-pitch copper embrittlement by silver brazing alloys." *Welding Journal* 69 (10): 378s-81s, 1990.

Jones, T. A. and Albright, C. E. "Laser beam brazing of small diameter copper wires to laminated copper circuit boards." *Welding Journal* 63(12): 34-47, 1984.

McFayden, A. A., Kapoor, R. R., and Eagar, T. W. "Effect of second phase particles on direct brazing of alumina dispersion hardened copper." *Welding Journal* 69(11): 399s-407s, 1990.

Mottram, D., Wronski, A. S., and Chilton, A. C. "Brazing copper to mild and stainless steel using copper-phosphorus-tin pastes." *Welding Journal* 65(4): 43-6, 1986.

Munse, W. H. and Alagia, J. A. "Strength of brazed joints in copper alloys." *Welding Journal* 36(4): 177s-84s, 1957.

SOLDERING

Dirnfeld, S. F. and Ramon, J. J. "Microstructure investigation of copper-tin intermetallics and the influence of layer thickness on shear strength." *Welding Journal*, 69(10): 373s-7s, 1990.

Ohriner, E. K. "Intermetallic formation in soldered copper-based alloys at 150 to 250 C. *Welding Journal* 66(7): 191s-201s, 1987.

Ramon, J. J., and Dirnfeld, S. F. "A practical way to measure the strength of small soldered joints." *Welding Journal*, 67(10): 19-21, 1988.

Reichennecker, W. J. "Effect of long term elevated temperature aging on the electrical resistance of soldered copper joints." *Welding Journal* 62(10): 290s-4s, 1983.

————. "Effect of solder thickness and joint overlap on the electrical resistance of soldered copper joints." *Welding Journal* 60(5): 199s-201s, 1981.

————. "Electrical conductivity of copper-tin intermetallic compound Cu_3Sn in the temperature range -195 °C to +150 °C." *Welding Journal* 59(10): 308s-10s, 1980.

Saperstein, Z. P., and Howes, M. A. H. "Mechanical properties of soldered joints in copper alloys." *Welding Journal* 48 (8): 317s-27s, 1969.

SAFETY AND HEALTH

Ditschun, A. and Sahoo, M. "Production and control of ozone during welding of copper-base alloys." *Welding Journal* 62(8): 41-6, 1983.

National Institute for Occupational Safety and Health. *Occupational safety and health guidelines for beryllium and its compounds potential human carcinogen.* Washington, D.C.: National Institute for Occupational Safety and Health, 1988.

————. *Occupational health guideline for cadmium fume.* Washington, D.C.: National Institute for Occupational Safety and Health, September 1978.

————. *Occupational health guideline for chromium metal and insoluble chromium salts.* Washington, D.C.: National Institute for Occupational Safety and Health, September 1978.

CHAPTER 4

NICKEL AND COBALT ALLOYS

PREPARED BY A COMMITEE CONSISTING OF:

J. P. Hunt, Chairman
Inco Alloys International, Incorporated

H. R. Conaway
Rocketdyne Division

J. Meyer
Nooter Corporation

J. W. Tackett
Haynes International, Incorporated

WELDING HANDBOOK COMMITTEE MEMBERS:
C. W. Case
Inco Alloys International, Incorporated

P. I. Temple
Detroit Edison

CHAPTER 4

NICKEL AND COBALT ALLOYS

INTRODUCTION

NICKEL AND COBALT alloys offer unique physical and mechanical properties, and unique resistance to corrosion attack. Were it not for the need for these unusual properties, these alloys probably would not be made because of their high cost. These alloys are useful in a variety of industrial applications because of their resistance to attack in various corrosive media at temperatures from 400 °F (200 °C) to over 2000 °F (1090 °C), in combination with good low- and high-temperature mechanical strength. In their demanding industrial environments, nickel and cobalt welds must duplicate the attributes of the base metal to a very high degree. Welding, heat treating, and fabrication procedures should be established with this in mind.

High-quality weldments are readily produced in nickel and cobalt alloys by commonly used welding processes. Not all processes are applicable to every alloy; metallurgical characteristics or the unavailability of matching or suitable welding filler metals and fluxes may limit the choice of welding processes.

Welding procedures for nickel and cobalt alloys are similar to those used for stainless steel, except the molten weld metal is more sluggish, requiring more accurate weld metal placement in the joint. Thermal expansion characteristics approximating those of carbon steel are more favorable than those of stainless steel. Thus, warping and distortion during welding is not severe.

The mechanical properties of nickel and cobalt alloy base metals will vary depending on the amount of hot or cold work remaining in the finished form (sheet, plate, or tube). Some modification in the procedures may be needed if the base metal is not in the fully annealed condition.

In general, the properties of welded joints in fully annealed nickel and cobalt alloys are comparable to those of the base metals. Postweld treatment is generally not needed to maintain or restore corrosion resistance in most nickel and cobalt alloys. In most media, the corrosion resistance of the weld metal is similar to that of the base metal. Welds made on Ni-Mo alloy N10001 and Ni-Si cast alloys commonly are solution annealed after welding to restore corrosion resistance to the HAZ.[1]

Over-alloyed filler metals are often used (sometimes in lieu of postweld heat treatment) to fabricate components for very aggressive corrosive environments. The over-matching composition offsets the effects of weld metal segregation when using a matching composition. Examples are the use of filler metal NiCrMo-3 products to weld the "super" stainless alloys, containing 4 to 6 percent molybdenum, and the use of filler metal NiCrMo-10 to fabricate components of the base metal Ni-Cr-Mo alloy C-276 (UNS N10276).[2]

Postweld heat treatment may be required for precipitation hardening in specific alloys. Postweld stress relief may be necessary to avoid stress-corrosion cracking in

---

1. Nickel and nickel alloy base-metal identifications in this chapter generally will be based upon descriptions and UNS numbers (e.g., Ni-Mo alloy N10001) from *Metals and Alloys in the Unified Numbering System*, 6th Ed., Warrendale, Pa.: Society of Automotive Engineers, 1993 . In some cases, a common identifier from a commercial designation also may be included [e.g., Ni-Mo alloy B (UNS N10001)]. When this data is available in a table (e.g., Table 4.2, Page 220), a short form may be used in the text (e.g., alloy B).
2. Nickel and nickel alloy filler metal designations generally will be based upon American Welding Society specifications. ANSI/AWS A5.11 specifies welding electrodes (e.g., ENiCrMo-10) for shielded metal arc welding, and ANSI/AWS A5.14 specifies bare welding electrodes and rods (e.g., ERNiCrMo-10). In usages that may apply with either product form, the E or ER prefix may be omitted (e.g. filler metal NiCrMo-10).

applications involving hydrofluoric acid vapor or certain caustic solutions. For example, Ni-Cu alloy 400 (UNS N04400) immersed in hydrofluoric acid is not sensitive to stress-corrosion cracking, but it is when exposed to the aerated acid or the acid vapors.

The choice of welding process will be based on the following:

(1) Alloy to be welded
(2) Thickness of the base metal
(3) Design conditions of the structure (temperature, pressure, type of stresses, etc.)
(4) Welding position
(5) Need for jigs and fixtures
(6) Service conditions and environments

METAL CHARACTERISTICS

NICKEL HAS A face-centered-cubic (FCC) structure up to its melting point. Nickel can be alloyed with a number of elements without forming detrimental phases. Nickel in some aspects bears a marked similarity to iron, its close neighbor in the periodic table. Nickel is only slightly denser than iron, and it has similar magnetic and mechanical properties. The crystalline structure of pure nickel, however, is quite different from that of iron. Therefore, the metallurgy of nickel and nickel alloys differs from that of iron alloys.

Cobalt, unlike nickel, exhibits two crystallographic forms and undergoes a transition from a face-centered-cubic (FCC) structure above 750 °F (417 °C) to a hexagonal-closed-packed (HCP) structure below. On cooling, the transformation is extremely sluggish, and at room temperature the metal is typically the metastable FCC form.

At room temperature, the transformation, which involves the coalescence of stacking faults, is easily triggered by mechanical stress. The addition of certain elements such as nickel, iron, and carbon (within the solubility range) suppresses the transformation temperature, stabilizing the FCC form. On the other hand, additions of chromium, molybdenum, tungsten, and silicon, have the opposite effect. In the solution-annealed condition, wrought cobalt alloys exhibit the FCC structure. With the exception of alloy 188 (UNS R30188), which is FCC stable, the various cobalt alloys are metastable at room temperature and tend to transform to the HCP structure under the action of mechanical stress or during heat treatment (at temperatures below the transformation temperature).

Typical physical and mechanical properties for pure nickel and pure cobalt are given in Table 4.1.

A more complete listing of chemical composition, and physical and mechanical properties, by alloy, is given in Tables 4.2, 4.3, 4.4, and 4.5.

Table 4.1
Physical and Mechanical Properties of Cobalt and Nickel

Property	Units	Cobalt	Nickel
Density	lb/in^3 (gm/cm^3)	0.322 (8.91)	0.321 (8.89)
Melting point	°F (°C)	2723 (1495)	2647 (1453)
Coef. of thermal expansion [68 °F (20 °C)]	in./(in.•°F) [m/(m•°C)]	7.7 x 10^{-6} (13.9 x 10^{-6})	7.4 x 10^{-6} (13.3 x 10^{-6})
Thermal conductivity [77°F (25°C)]	Btu/(hr•ft•°F) [W/(m•K)]	40 (69)	53 (92)
Electrical resistivity	Ω/cir mil/ft (μΩ•cm)	47 (7.8)	58 (9.7)
Modulus of elasticity in tension	psi (kPa)	30.6 x 10 (211 x 10^6)	29.6 x 10^6 (204 x 10^6)
Tensile strength, annealed	ksi (MPa)	37 (255)[a]	67 (462)
Yield strength, 0.2% offset	ksi (MPa)	28-41 (193 - 283)[b]	21.5 (148)
Elongation in 2 in. (51 mm)	percent	0-8	47

a. Compressive strength is about 117 ksi (807 MPa).

b. Compressive yield strength is about 56 ksi (386 MPa).

Table 4.2
Nominal Chemical Composition of Typical Nickel Alloys

Alloy[a]	UNS Number	Ni[b]	C	Cr	Mo	Fe	Co	Cu	Al	Ti	Nb[c]	Mn	Si	W	B	Other
						Commercially Pure Nickels										
200	N02200	99.5	0.08	—	—	0.2	—	0.1	—	—	—	0.2	0.2	—	—	—
201	N02201	99.5	0.01	—	—	0.2	—	0.1	—	—	—	0.2	0.2	—	—	—
205	N02205	99.5	0.08	—	—	0.1	—	0.08	—	0.03	—	0.2	0.08	—	—	0.05Mg
						Solid-Solution Alloys										
400	N04400	66.5	0.2	—	—	1.2	—	31.5	—	—	—	1	0.2	—	—	—
404	N04404	54.5	0.08	—	—	0.2	—	44	0.03	—	—	0.05	0.05	—	—	—
R-405	N04405	66.5	0.2	—	—	1.2	—	31.5	—	—	—	0.1	0.02	—	—	—
X	N06002	47	0.10	22	9	18	1.5	—	—	—	—	1	1	0.6	—	—
NICR 80	N06003	76	0.1	20	—	1	—	—	—	—	—	2	1	—	—	—
NICR 60	N06004	57	0.1	16	—	bal.	—	—	—	—	—	1	1	—	—	—
G	N06007	44	0.1	22	6.5	20	2.5	2	—	—	2	1.5	1	1	—	—
IN 102	N06102	68	0.06	15	3	7	—	—	0.4	0.6	3	—	—	3	0.005	0.03Zr, 0.02Mg
RA 333	N06333	45	0.05	25	3	18	3	—	—	—	1	1.5	1.2	3	—	—
600	N06600	76	0.08	15.5	—	8	—	0.2	—	—	—	0.5	0.2	—	—	—
601	N06601	60.5	0.05	23	—	14	—	—	1.4	—	—	0.5	0.2	—	—	—
617	N06617	52	0.07	22	9	1.5	12.5	—	1.2	0.3	—	0.5	0.5	—	—	—
622	N06622	59	0.005	20.5	14.2	2.3	—	—	—	—	—	—	—	3.2	—	—
625	N06625	61	0.05	21.5	9	2.5	—	—	0.2	0.2	3.6	0.2	0.2	—	—	—
686	N06686	58	0.005	20.5	16.3	1.5	—	—	—	—	—	—	—	3.8	—	—
690	N06690	60	0.02	30	—	9	—	—	—	—	—	0.5[d]	0.5[d]	—	—	—
725	N07725	73	0.02	15.5	—	2.5	—	—	0.7	2.5	1.0	—	—	—	—	—
825	N08825	42	0.03	21.5	3	30	—	2.25	0.1	0.9	—	0.5	0.25	—	—	—
B	N10001	61	0.05	1	28	5	2.5	—	—	—	—	1	1	—	—	—
N	N10003	70	0.06	7	16.5	5	—	—	—	—	—	0.8	0.5	—	—	—
W	N10004	60	0.12	5	24.5	5.5	2.5	—	—	—	—	1	1	—	—	—
C-276	N10276	57	0.01[d]	15.5	16	5	2.5[d]	—	—	0.7[d]	—	1[d]	0.08[d]	4	—	0.35V[d]
C-22	N06022	56	0.010[d]	22	13	3	2.5[d]	—	—	—	—	0.5[d]	0.08[d]	3	—	0.35V[d]
B-2	N10665	69	0.01[d]	1[d]	28	2[d]	1[d]	—	—	—	—	1[d]	0.1[d]	—	—	—
C-4	N06455	65	0.01[d]	16	15.5	3[d]	2[d]	—	—	—	—	1[d]	0.08[d]	—	—	—
G-3	N06985	44	0.015[d]	22	7	19.5	5[d]	2.5	—	—	0.5[d]	1[d]	1[d]	1.5[d]	—	—
G-30	N06030	43	0.03[d]	30	5.5	15	5[d]	2	—	—	1.5[d]	1.5[d]	1[d]	2.5	—	—
S	N06635	67	0.02[d]	16	15	3[d]	2[d]	—	0.25	—	—	0.5	0.4	1[d]	0.015[d]	0.02La
230	N06230	57	0.10	22	2	3[d]	5[d]	—	0.3	—	—	0.5	0.4	14	0.015[d]	0.02La
214	N07214	75	0.10	16	—	3	—	—	4.5	—	—	0.5[d]	0.2[d]	—	0.01[d]	0.01Y, 0.1Zr[d]
						Precipitation-Hardenable Alloys										
301	N03301	96.5	0.15	—	—	0.3	—	0.13	4.4	0.6	—	0.25	0.5	—	—	—
K-500	N05500	66.5	0.10	—	—	1	—	29.5	2.7	0.6	—	0.08	0.2	—	—	—
Waspaloy	N07001	58	0.08	19.5	4	—	13.5	—	1.3	3	—	—	—	—	0.006	0.06Zr
R-41	N07041	55	0.10	19	10	1	10	—	1.5	3	—	0.05	0.1	—	0.005	—
80A	N07080	76	0.06	19.5	—	—	—	—	1.6	2.4	—	0.3	0.3	—	0.006	0.06Zr
90	N07090	59	0.07	19.5	—	—	16.5	—	1.5	2.5	—	0.3	0.3	—	0.003	0.06Zr
M 252	N07252	55	0.15	20	10	—	10	—	1	2.6	—	0.5	0.5	—	0.005	—
U-500	N07500	54	0.08	18	4	—	18.5	—	2.9	2.9	—	0.5	0.5	—	0.006	0.05Zr
713C[e]	N07713	74	0.12	12.5	4	—	—	—	6	0.8	2	—	—	—	0.012	0.10Zr
718	N07718	52.5	0.04	19	3	18.5	—	—	0.5	0.9	5.1	0.2	0.2	—	—	—
X750	N07750	73	0.04	15.5	—	7	—	—	0.7	2.5	1	0.5	0.2	—	—	—
706	N09706	41.5	0.03	16	—	40	—	—	0.2	1.8	2.9	0.2	0.2	—	—	—
901	N09901	42.5	0.05	12.5	—	36	6	—	0.2	2.8	—	0.1	0.1	—	0.015	—
C 902	N09902	42.2	0.03	5.3	—	48.5	—	—	0.6	2.6	—	0.4	0.5	—	—	—
IN 100[e]	N13100	60	0.18	10	3	—	15	—	5.5	4.7	—	—	—	—	0.014	0.06Zr,1.0V
						Dispersion-Strengthened Alloys										
TD Nickel	N03260	98	—	—	—	—	—	—	—	—	—	—	—	—	—	2 Th O$_2$
TD NICR	N07754	78	—	20	—	—	—	—	—	—	—	—	—	—	—	2 Th O$_2$

a. Several of these designations use parts of or are registered trade names. These and similar alloys may be known by other designations and trade names.
b. Includes small amount of cobalt, if cobalt content is not specified.
c. Includes tantalum (Nb+Ta).
d. Maximum value.
e. Casting alloys.

Table 4.3
Nominal Chemical Composition of Standard ASTM Nickel Casting Alloys

Alloy	UNS Number	Composition, wt. %											
		Ni	C	Cr	Mo	Fe	Th	Al	Ti	Cu	Mn*	Si*	W
					ASTM A 297-79*								
HW	N08001	60	0.5	12	—	25	—	—	—	—	2.0	2.5	—
HX	N06006	66	0.5	17	—	15	—	—	—	—	2.0	2.5	—
					ASTM A 494-79								
CY-40	N06040	72	0.4*	16	—	11*	—	—	—	—	1.5	3.0	—
CW-12M-1	N30002	55	0.12*	16.5	17	6	—	—	—	—	1.0	1.0	4.5
CZ-100	N02100	95	1.0*	—	—	3*	—	—	—	1.25*	1.5	2.0	—
M-35-1	N24135	68	0.35*	—	—	3.5*	—	—	—	30	1.5	1.25	—
N-12M-1	N30012	65	0.12*	—	28	5	—	—	—	—	1.0	1.0	—

* Maximum

Table 4.4A
Typical Physical and Mechanical Properties of Nickel Alloys
(U.S. Customary Units)

Alloy	UNS Number	Density, lb/cu. in.	Melting Range, °F	Coefficient of Thermal Expansion at 70–200 °F, μin./(in.·°F)	Thermal Conductivity at 70 °F, Btu/ (ft·h·°F)	Electrical Resistivity at 70 °F, Ω cir mil/ft	Tensile Modulus of Elasticity, 70 °F, 10^6 psi	Tensile Strength at Room Temperature, ksi	Yield Strength at Room Temperature, ksi
200	N02200	0.321	2615-2635	7.4	43	57	29.6	68	25
201	N02201	0.321	2615-2635	7.4	46	46	30.0	55	20
400	N04400	0.319	2370-2460	7.7	13	307	26.0	80	40
R-405	N04405	0.319	2370-2460	7.7	13	307	26.0	80	35
K-500	N05500	0.306	2400-2460	7.6	10	370	26.0	140[a]	90[a]
502	N05502	0.305	2400-2460	7.6	10	370	26.0	130	85
600	N06600	0.304	2470-2575	7.4	9	620	30.0	90	40
601	N06601	0.291	2374-2494	7.6	7	725	29.9	107	49
625	N06625	0.305	2350-2460	7.1	6	776	30.0	130	70
713C	N07713	0.286	2300-2350	5.9	12[b]	—	29.9	123[c]	107[c]
706	N09706	0.291	2434-2499	7.8	7	592	30.4	175[a]	145[a]
718	N07718	0.296	2300-2437	7.2	6	751	29.8	190[a]	160[a]
X-750	N07750	0.298	2540-2600	7.0	7	731	31.0	170[a]	110[a]
U-500	N07500	0.290	2375-2540	6.8	8	723	31.0	176[a]	110[a]
R-41	N07041	0.298	2400-2500	6.6	7[d]	820	31.2	160[a]	120[a]
Waspaloy	N07001	0.296	2556-2576	6.8	8	761	30.6	185[a]	115[a]
800	N08800	0.287	2475-2525	7.9	7	595	28.5	90	40
825	N08825	0.294	2500-2550	7.8	6	678	28.0	90	40
20Cb3	N08020	0.292	2498-2597	8.3	—	625	28.0	90	40
901	N09901	0.297	—	7.2	—	662	28.0	175	130
B	N10001	0.334	2375-2495	5.6	7	811	25.9	121	57
C-276	N10276	0.323	2310-2450	6.3	6	779	29.8	121	58
G	N06007	0.300	2300-2450	7.5	8	—	27.8	103	56
N	N10003	0.320	2375-2550	6.4	7	835	31.3	115	45
W	N10004	0.325	2400	6.3	—	—	—	123	53
X	N06002	0.297	2300-2470	7.7	5	712	28.6	114	52

a. Heat-treated condition

b. At 200 °F

c. As-cast

d. At 300 °F

Because nickel has wide solubility for a number of other metals, many different commercial alloys are available. Nickel and copper have complete solid solubility. Iron and cobalt are soluble in nickel to a very high degree. The limit of solubility of chromium in nickel is 35 to 40 percent and is about 20 percent for molybdenum. Additions of these major alloying elements, that is, copper, chromium, molybdenum, iron, and cobalt have no adverse effect on weldability, and in most cases they have a beneficial effect on weldability. In general, commercially pure nickel and nickel-copper alloys have similar weldability. Most of the other nickel alloys behave like the stainless steels.

Like the austenitic stainless steels, the nickel alloys have one crystalline structure up to their melting point. Since the nickel alloys do not undergo a phase change,

the grain size of the base metal or weld metal cannot be refined by heat treatment alone. The grain size can be reduced only by hot or cold work, such as rolling or forging, followed by a proper annealing treatment.

Cobalt cast and wrought alloys in fabricable forms are relatively few in number. Those cobalt alloys commonly welded generally contain two or more of the elements nickel, chromium, tungsten, and molybdenum. The weldability of these alloys is generally good.

In relatively small amounts, alloying elements such as manganese, silicon, niobium, carbon, aluminum, and titanium are not detrimental to welding of nickel or cobalt alloys. When elements such as aluminum or titanium are added (several percent) to facilitate precipitation hardening, good shielding of the weld zone is imperative in limiting oxide formation.

Table 4.4B
Typical Physical and Mechanical Properties of Nickel Alloys
(Metric Units)

Alloy	UNS Nunmber	Density, kg/m³	Melting Range, °C	Coefficient of Thermal Expansion at 21-93°C, μm/m/°C	Thermal Conductivity at 21°C, W/(m•K)	Electrical Resistivity at 21°C, μΩ•cm	Tensile Modulus of Elasticity, 21°C, GPa	Tensile Strength at Room Temperature, MPa	Yield Strength at Room Temperature, MPa
200	N02200	8885	1435-1446	13.3	70	9.5	204	469	172
201	N02201	8885	1435-1446	13.3	79	7.6	207	379	138
400	N04400	8830	1298-1348	13.9	20	51.0	179	552	276
R-405	N04405	8830	1298-1348	13.9	20	51.0	179	552	241
K-500	N05500	8470	1315-1348	13.7	16	61.5	179	965[a]	621[a]
502	N05502	8442	1315-1348	13.7	16	61.5	179	896	586
600	N06600	8415	1354-1412	13.3	14	103.0	207	621	276
601	N06601	8055	1301-1367	13.7	12	120.5	206	738	338
625	N06625	8442	1287-1348	12.8	9	129.0	207	896	483
713C	N07713	7916	1260-1287	10.6	19[b]	—	206	848[c]	738[c]
706	N09706	8055	1334-1370	14.0	12	98.4	210	1207[a]	1000[a]
718	N07718	8193	1260-1336	13.0	11	124.9	205	1310[a]	1103[a]
X-750	N07750	8248	1393-1426	12.6	11	121.5	214	1172[a]	758[a]
U-500	N07500	8027	1301-1393	12.2	12	120.2	214	1213[a]	758[a]
R-41	N07041	8249	1315-1371	11.9	11[d]	136.3	215	1103[a]	827[a]
Waspaloy	N07001	8193	1402-1413	12.2	12	126.5	211	1276[a]	793[a]
800	N08800	7944	1357-1385	14.2	11	98.9	196	621	276
825	N08825	8138	1371-1398	14.0	10	112.7	193	621	276
20Cb3	N08020	8083	1370-1425	14.9	—	103.9	193	621	276
901	N09901	8221	—	13.0	—	110.0	193	1207	896
B	N10001	9245	1301-1368	10.1	11	134.8	179	834	393
C-276	N10276	8941	1265-1343	11.3	11	129.5	205	834	400
G	N06007	8304	1260-1343	13.5	13	—	192	710	386
N	N10003	8858	1301-1398	11.5	11	138.8	216	793	310
W	N10004	8996	1315	11.3	—	—	—	848	365
X	N06002	8221	1260-1354	13.9	8	118.3	197	786	359

a. Heat-treated condition

b. 93°C

c. As-cast

d. 149°C

Table 4.5
Nominal Chemical Compositions of Typical Cobalt Alloys

Alloy[a]	UNS Number	Composition, wt. %														
		C	Ni	Cr	Co	W	Ta	Mo	Al	Ti	Fe	Mn	Si	B	Zr	Other
						Wrought Alloys										
L-605	R30605	0.1	10	20	Bal	15	—	—	—	—	3[b]	1.5	0.5	—	—	—
188	R30188	0.1	22	22	Bal	14	—	—	—	—	3[b]	1.2	0.4	—	—	0.08 La
MP 35N	R30035	0.05	35	20	35	—	—	10	—	—	—	—	—	—	—	—
S-816	R30816	0.38	20	20	Bal	4	—	4	—	—	4	1.2	0.4	—	—	4 Nb
54Co-26Cr	R31233	0.06	9	26	Bal	2	—	5	—	—	3	0.8	0.3	—	—	0.08 N
						Cast Alloys										
21	R30021	0.25	2.5	27	Bal		—	5.5	—	—	2[b]	—	—	—	—	—
X-40	R30031	0.5	10.5	25.5	Bal	7.5	—	—	—	—	—	0.75	0.75	—	—	—
FSX-414	—	0.25	10	29	Bal	7.5	—	—	—	—	1	—	—	0.01	—	—
WI 52	—	0.45	—	21	Bal	11	—	—	—	—	2	0.25	0.25	—	—	2 Nb
MAR-M 302	—	0.85	—	21.5	Bal	10	9	—	—	—	—	—	—	0.005	0.2	—
MAR-M 509	—	0.6	10	23.5	Bal	7	3.5	—	—	2	—	—	—	—	0.5	—

a. Several of these designations use parts of or are registered trade names. These and similar alloys may be known by other designations and trade names.

b. Maximum.

The weldability of both nickel and cobalt alloys is sensitive to residual elements such as sulfur, lead, zirconium, boron, phosphorous, and bismuth. These elements are practically insoluble in nickel and cobalt alloys and can undergo eutectic reactions which may cause hot cracking during solidification of welds. All commercially important nickel and cobalt alloys have specification limits covering some of those elements that are difficult to control. Boron and zirconium, in very small amounts, are added to certain alloys to improve their high-temperature performance, but weldability is diminished when such additions are made. The harmful effects of sulfur on ductility are controlled by the addition of small amounts of magnesium to wrought product forms and to filler metals.

When using coated electrodes in shielded metal arc welding, the losses of magnesium across the arc are so great that ineffectively small amounts of magnesium are recovered in the weld metal. Under these circumstances, control of sulfur is accomplished by adding manganese and niobium, which can be recovered in substantially greater amounts than can magnesium.

Fusion welds made without the addition of filler metal in nickel, nickel-copper alloys, and nickel-molybdenum alloys can develop weld-metal porosity if the welds are contaminated by oxygen, nitrogen, or carbon monoxide. Titanium and other gas-fixing elements are present in the nickel and nickel-copper alloy filler metals to combine with gas contaminants and prevent weld-metal porosity. Invariably, the weld-metal composition differs from the base-metal composition for weldability reasons. Mechanized or automatic welding is recommended for autogenous welding, and then only under closely controlled welding conditions. It is generally wiser to add filler metal, even if in small amounts, because weld quality is more predictable with the additives in the welding wire that "fix" or control gas and residual elements.

ALLOY GROUPS

IT IS CONVENIENT to classify nickel and cobalt alloys into four groups:

(1) Solid-solution-strengthened alloys
(2) Precipitation-hardened alloys
(3) Dispersion-strengthened alloys
(4) Cast alloys

Tables in the preceding section list the alloy grades and chemical compositions applicable to this chapter. A small group of iron-based alloys also are applicable; these contain large percentages of the elements nickel, cobalt, and chromium, or two of these. The weldability characteristics of the chromium-containing alloys closely resemble those of the heat-resisting grades of nickel and cobalt alloys. The names and chemical compositions of these highly alloyed iron-based alloys are given in Table 4.6.

Although iron is the dominant element in these alloys, nickel- or cobalt-based welding filler metals are recommended by the base metal manufacturers for joining them. For example, Fe-Ni-Cr alloy 800 (UNS N08800) is almost always welded using one of several nickel alloy filler metals, the specific selection of which depends on the service conditions for the welded component. Since the environment that will be encountered and the high-temperature joint strength efficiency are both critical factors in choosing the best filler metal, the base metal manufacturer usually should be consulted.

Figure 4.1 shows the stress-rupture strength of Fe-Ni-Cr alloy 800HT (UNS N08811) compared to the filler metals ERNiCrCoMo-1 and ENiCrCoMo-1. Although the compositions of the filler metals don't match the base metal, their overmatching strengths make them a good choice for welding the base metal in applications at high temperatures. Additionally, the filler metals have better oxidation resistance than the base metal at the elevated temperature. At exposure temperatures below 1400°F (760°C), other filler metals are recommended, such as ERNiCr-3 and ENiCrFe-2. This example is typical of many filler metal recommendations where service conditions determine the most suitable choice.

Table 4.6
Highly Alloyed Iron-Based Alloys

Alloy[a]	UNS Number	Nominal Composition, wt. %										
		Ni[b]	Cr	Co	Fe	Mo	Ti	W	Nb[c]	Al	C	Other
					Solid-Solution Types							
20Cb3[d]	N08020	35	20	—	36	2.5	—	—	0.5	—	0.04	3.5 Cu,1 Mn, 0.5 Si
800	N08800	32.5	21.0	—	45.7	—	0.40	—	—	0.40	0.05	—
800HT	N08811	33.0	21.0	—	45.8	—	0.50	—	—	0.50	0.08	—
801	N08801	32.0	20.5	—	46.3	—	1.13	—	—	—	0.05	—
802	N08802	32.5	21.0	—	44.8	—	0.75	—	—	0.58	0.35	—
19-9 DL	S63198	9.0	19.0	—	66.8	1.25	0.30	1.25	0.4	—	0.30	1.10 Mn, 0.60 Si
N-155	R30155	20.0	21.0	20.0	32.2	3.00	—	2.50	1.0	—	0.15	0.15 N
RA330	N08330	36.0	19.0	—	45.1	—	—	—	—	—	0.05	—
556	R30556	21.0	22.0	20.0	29.0	3.00	—	2.50	0.1	0.30	0.10	0.5 Ta, 0.02 La
					Precipitation Types							
A-286	S66286	26.0	15.0	—	55.2	1.25	2.00	—	—	0.02	0.04	—
903	N19903	38.0	—	15.0	41.0	0.10	1.40	—	3.0	0.70	0.04	—

a. Several of these designations use parts of or are registered trade names. These and similar alloys may be known by other designations and trade names.

b. Includes small amount of cobalt, if cobalt content is not specified.

c. Includes tantalum, if tantalum content is not specified.

d. While niobium (Nb) is the preferred designation for the 41st element and is used as a column heading in the table, the abbreviation for the element's alternate designation, columbium (Cb), is retained in this alloy designation until such time as the alloy designation is changed.

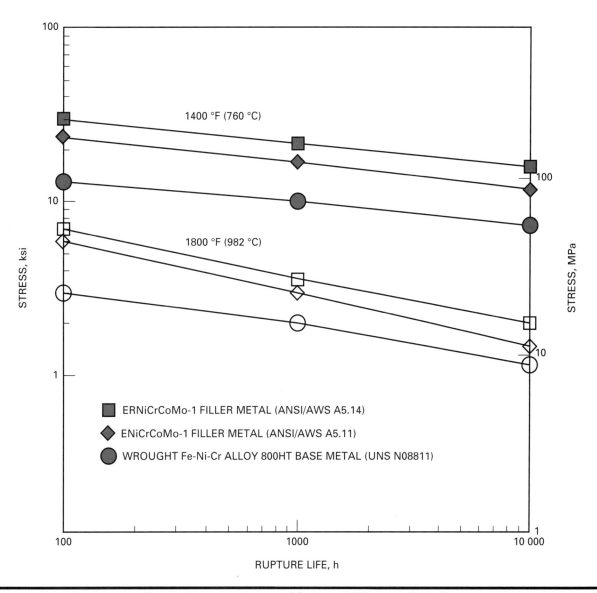

Figure 4.1—Stress-Rupture Values of Selected Alloys at High Temperatures

SOLID-SOLUTION-STRENGTHENED ALLOYS

ALL NICKEL ALLOYS are strengthened by solid solution. Additions of aluminum, chromium, cobalt, copper, iron, molybdenum, titanium, tungsten, and vanadium contribute to solid-solution strengthening. Aluminum, chromium, molybdenum, and tungsten contribute strongly to solid-solution strengthening while others have a lesser effect. Molybdenum and tungsten improve strength at elevated temperatures.

Pure Nickel

NICKEL 200 AND the low-carbon version, nickel 201, are most widely used where welding is involved. Of these, the low-carbon nickel (201) is preferred for applications involving service exposure to temperatures above 600 °F (315 °C) because of its increased resistance to graphitization at elevated temperatures. This graphitization is the result of excess carbon being precipitated intergranularly in the temperature range of 600 to 1400 °F (315 to 760 °C) when nickel 200 is held there for extended time. Major applications for the two alloys are food processing equipment, caustic handling equipment, laboratory crucibles, chemical shipping drums, and electrical and electronic parts.

Nickel-Copper Alloys

NICKEL AND COPPER form a continuous series of solid solutions with a face-centered-cubic crystal structure.

The principal alloys in this group are alloy 400 and the free-machining version of it, R-405. These alloys have high strength and toughness, and they find industrial importance primarily because of their corrosion resistance. The alloys have excellent resistance to sea or brackish water, chlorinated solvents, glass etching agents, sulfuric acids, and many other acids and alkalis.

Nickel-copper alloys are readily joined by welding, brazing, and soldering with proper precautions. Welding filler metals differ somewhat in chemical composition to improve strength and to eliminate porosity in the weld metal. Welding without the addition of filler metal is not recommended for manual gas tungsten arc welding. Some automatic or mechanized welding procedures do not require the addition of filler metal, but these are few.

Welding filler metals applicable to this alloy group also are used widely in the welding of copper alloys.

Nickel-Chromium Alloys

COMMON ALLOYS INCLUDE alloys 600, 601, 690, 214, 230, G-30, and RA-330. Alloy 600, which is the most widely used one, has good corrosion resistance at elevated temperatures along with good high-temperature strength. Because of its resistance to chloride-ion stress-corrosion cracking, it finds wide use at all temperatures and has excellent room temperature and cryogenic properties. Alloy 690, with 30% chromium, has even better resistance to stress-corrosion cracking.

Alloy 230, a Ni-Cr-W-Mo alloy, has excellent high-temperature strength and resistance to oxidizing and nitriding environments. Alloys 601 and 214 exhibit outstanding oxidation and scaling resistance at temperatures up to 2200 °F (1200 °C) and extend the temperature range achieved with the Ni-Cr alloy 600 through the addition of 1.4% aluminum. Alloy G-30, with 30% chromium, has superior resistance to commercial phosphoric acids as well as to many highly oxidizing acids.

Filler metal NiCrMo-11 (G-30) is used widely in many corrosion applications for welding molybdenum-containing alloys.

Alloy 617 is a nickel-chromium-cobalt-molybdenum alloy with an exceptional combination of metallurgical stability, strength, and oxidation resistance at high temperatures, as well as corrosion resistance in aqueous environments. The corresponding NiCrCoMo-1 filler metals for welding the base metal also are used to weld other base metals that will see service in severe applications. For example, alloy 800 base metal intended for service in the 1500 to 2100 °F (820 to 1150 °C) range is welded almost exclusively with NiCrCoMo-1 filler metals. Other filler metals may be used, but the base metal supplier should be consulted regarding the actual service conditions. In general, all of the nickel-chromium base metals possess outstanding weldability.

Nickel-Iron-Chromium Alloys

ALLOYS 800, 800HT, 825, 20Cb3, N-155, and 556 are included in the Fe-Ni-Cr alloy group. Alloys 800, 800HT, N-155, and 556 are widely used for high-temperature applications because of their high-temperature strength and their resistance to oxidation and carburization. Alloys 825 and 20Cb3 are used in corrosive environments under 1000 °F (540 °C) because of their resistance to reducing acids and to chloride-ion stress-corrosion cracking.

Nickel-Molybdenum Alloys

THE PRINCIPAL ALLOYS in this group are alloys B, B-2, N, and W. They contain 16 to 28% molybdenum and lesser amounts of chromium and iron. They are used primarily for their corrosion resistance and are not normally used for elevated temperature service. They are readily weldable.

Alloys B and B-2 have good resistance to hydrochloric acid and other acids, while alloy N was developed for resistance to molten fluoride salts. Alloy W is used in welding filler metals (NiMo-3) possessing good corrosion and oxidation resistance.

Nickel-Chromium-Molybdenum Alloys

INCLUDED IN THIS group are alloys C-22, C-276, G, S, X, 622, 625, and 686. They are designed primarily for corrosion resistance at room temperature, as well as for resistance to oxidizing and reducing atmospheres at elevated temperatures. Alloy 625 has additions of 9% molybdenum and 4% niobium, which enhance its room and high-temperature strength and corrosion resistance. Alloys X and S are used widely for high-temperature applications. All of these alloys have good weldability, and filler metals for welding them are available.

Cobalt-Chromium-Nickel-Tungsten Alloys

THE MOST NOTABLE alloys in this group are L-605, 188, S-816, and the 54Co-26Cr alloy (R31233).

The loss in ductility experienced by some cobalt alloys in the 1200 to 1800 °F (650 to 980 °C) range is characteristic of high-alloy compositions. This behavior is believed to be associated with precipitation of carbides and intermetallic compounds, which adversely affect cracking resistance. The combination of minimum joint restraint and low-energy input during welding make welding possible with predictable results.

Contamination by molten copper will cause base-metal cracking in all of these alloys. Since the more common wrought form in cobalt alloys is sheet, the use of copper backing bars must be evaluated carefully. Sometimes copper backing bars are nickel or chromium plated, or an austenitic stainless steel backing bar is substituted for copper, with a resultant loss in thermal conductivity.

CAST ALLOYS

MANY NICKEL ALLOYS can be used in cast form as well as wrought form. Several common commercial casting alloys are listed in Table 4.3; however, cast forms usually can be purchased also within typical alloy families listed in Table 4.2.

Some alloys are designed specifically for casting. Table 4.3 gives the composition of several nickel-based ASTM casting alloys. Casting alloys, like wrought alloys, can be strengthened by solid-solution or precipitation hardening. Precipitation-hardening alloys high in aluminum content, such as alloy 713C, will harden during slow cooling in the mold and are considered unweldable by fusion processes. However, surface defects and service damage are frequently repaired by welding. It should be understood that a compromise is being made between the convenience of welding and the cast strength and ductility.

Most nickel and cobalt cast alloys will contain significant amounts of silicon to improve fluidity and castability. Most of these cast alloys are weldable by conventional means, but as the silicon content increases, so does weld-cracking sensitivity. This cracking sensitivity can be avoided using welding techniques that minimize base metal dilution.

Nickel castings that are considered unweldable by arc welding methods may be welded using the oxyacetylene process and a very high preheat temperature.

Cast nickel alloys containing 30% copper are considered unweldable when the silicon exceeds 2% because of their sensitivity to cracking. However, when weldable grade castings are specified, weldability is quite good, and such welds will pass routine weld-metal inspections using methods such as radiography, liquid-penetrant testing, and pressure tests.

Cobalt cast alloys are found in gas turbine and other applications where good oxidation and sulfidation resistance, as well as strength, are needed. Table 4.5 gives the composition of some of these alloys. Note the silicon level is quite low, and weldability is considered good in these alloys.

PRECIPITATION-HARDENABLE ALLOYS

THESE ALLOYS ARE strengthened by controlled heating, which precipitates a second phase known as *gamma prime*, from a supersaturated solution. Precipitation occurs upon reheating a solution-treated and quenched alloy to an appropriate temperature for a specified time. Each alloy will have an optimum thermal cycle to achieve maximum strength in the finished aged condition. Some cast alloys will age directly as the solidified casting cools in the mold.

The most important phase from a strengthening standpoint is the ordered face-centered-cubic gamma prime that is based upon the compound Ni_3Al. This phase has a high solubility for titanium and niobium; consequently, its composition will vary with the base-metal composition and temperature of formation. Aluminum has the greatest hardening potential, but this is moderated by titanium and niobium. Niobium has the greatest effect on decreasing the aging rate and improves weldability.

These alloys are normally welded in the solution-treated condition (soft). During welding, some portion of the HAZ is heated into the aging temperature range. As the weld metal solidifies, the aging HAZ becomes subjected to welding stresses. Under certain postweld combinations of temperature and stress, the weld HAZ may crack. This is known as *strain-age cracking*. Alloys high in aluminum are the most prone to this type of cracking. The problem is less severe where niobium, which retards the aging action, has been substituted for a significant portion of the aluminum. Consequently, the weld HAZ can remain sufficiently ductile and can yield during heat treatment to relieve high welding stresses without rupture. The relative weldability of several precipitation-hardenable alloys is indicated in Figure 4.2. Where weldability is of concern, the selection of ERNiFeCr-2 filler metal is usually made.

Nickel-Copper Alloys

THE PRINCIPAL ALLOY in this group is K-500. Strict attention to heat-treating procedures must be followed to avoid strain-age cracking. Its corrosion resistance is similar to the solid-solution alloy 400. The alloy has been in commercial existence for well over 50 years and is routinely welded, using proper care, with the gas tungsten arc welding process. Weld metal properties using filler metals of matching composition seldom develop 100% joint efficiencies, thus a common consideration by the designer is to locate the weld in an area of low stress. ERNiFeCr-2 filler metal also has been used to join this alloy, but an evaluation of service environment and the differing aging temperatures between the two alloys must be made. The base metal supplier should be consulted for applicable recommendations.

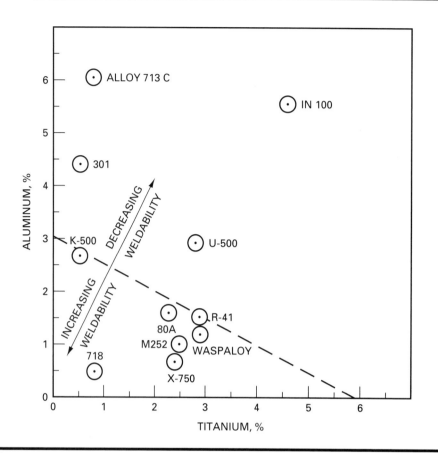

Figure 4.2—Relationship Between Estimated Weldability of Precipitation-Hardenable Nickel Alloys and Their Aluminum and Titanium Content

Nickel-Chromium Alloys

THE NICKEL-CHROMIUM age-hardenable alloys are strengthened by either an aluminum-titanium reaction or an aluminum-titanium-niobium reaction. Chromium, ranging from about 13 to 20 percent, is added to ensure good high-temperature oxidation resistance. Principal alloys in the aluminum-titanium group that are difficult to weld are alloys 713C (casting), X-750, U-500, R-41, and Waspalloy. Alloys 718 and 706 are highly weldable because of the more sluggish strengthening reaction in their Al-Ti-Nb alloying system. The principal areas of application for these alloys are gas turbine components, aircraft parts, and spacecraft, where the strength-versus-weight relationship is important.

Nickel-Iron-Chromium Alloys

THE PRINCIPAL ALLOY in this group is alloy 901. Its weldability is similar to alloy X-750; however, most applications are in forgings that require little welding. When welding is required, the same precautions are necessary to avoid strain-age cracking as with other aluminum-titanium hardened alloys.

DISPERSION-STRENGTHENED NICKEL

NICKEL AND NICKEL-CHROMIUM alloys can be strengthened to very high strength levels by the uniform dispersion of very fine refractory oxide (ThO_2) particles throughout the alloy matrix. This is done using powder metallurgy techniques during manufacture of the alloy. When these metals are fusion welded, the oxide particles agglomerate during solidification. This destroys the original strengthening afforded by dispersion within the matrix. The weld metal will be significantly weaker than the base metal. The high strength of these base metals can be retained with processes that do not involve melting the base metal. Contact the base metal supplier for current recommendations for specific conditions.

SURFACE PREPARATION

CLEANLINESS IS THE single most important requirement for successful welding of nickel and cobalt alloys. At high temperatures, these alloys are susceptible to embrittlement by many low-melting substances. Such substances are often found in materials used in normal manufacturing processes.

Both nickel and cobalt alloys are embrittled by sulfur, phosphorus, and metals with low melting points such as lead, zinc, and tin. Lead hammers, solders, and wheels or belts loaded with these materials are frequent sources of contamination. Detrimental elements are often present in oils, paint, marking crayons, cutting fluids, and shop dirt.

Figure 4.3 is an example of cracking in a sheet of nickel 200 that was improperly cleaned with a dirty rag prior to welding. When such embrittlement occurs, the alloy cannot be reclaimed by any kind of treatment. It is destroyed. The contaminated area must be removed mechanically or by thermal cutting remote from the cracked area, after thorough surface cleaning.

Many cases of contamination have been reported when parts in service must be repaired or modified by welding. It is particularly important that adequate attention be given to cleaning before applying heat from any source. Processing chemicals also may contain harmful ingredients. Figure 4.4 is an example of sulfur and lead embrittlement in Ni-Cu alloy 400 installed in a vessel that had experienced prior process service.

The depth of attack will vary with the type of embrittling element, its concentration, the base metal, heating time, and temperature. Nickel 200, for example, is easily embrittled, whereas Fe-Ni-Cr alloy 800 is less sensitive to embrittlement. Such knowledge provides little comfort, however, and the fundamental rule when welding nickel and cobalt alloys is: Unless proven safe, all foreign material must be considered harmful in the presence of heat.

The use of copper backing bars in welding cobalt alloy sheet requires special attention. Contamination by even a minute amount of molten copper leads to liquid-metal stress-cracking in cobalt-based metals. For example, a small amount of copper, inadvertently transferred onto the surface of a sheet from a backing bar, leads to severe cracking when melted by the heat of welding. This may be prevented by either plating the backing bar with nickel or chromium or using a stainless steel backing bar.

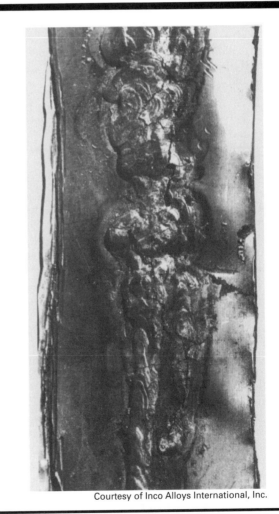

Courtesy of Inco Alloys International, Inc.

Figure 4.4—Combined Effects of Sulfur and Lead Contamination. Specimen Removed from Fatty-Acid Tank Previously Lined with Lead and Not Properly Cleaned Before Installation of Ni-Cu alloy 400 Lining

Figure 4.3—Cracking from Sulfur Contamination in the Heat-Affected Zone of a Weld in Nickel Sheet

In all these alloys, oxides formed by previous thermal operations must be thoroughly removed from the surfaces to be welded because they inhibit wetting and fusion of the base metal with the weld metal. Their presence can cause weld inclusions and poor weld bead contour since they melt at a much higher temperature than either the base metal or filler metal.

Wire brushes used for bead cleaning should be made of austenitic stainless steel. However, they will not remove tenacious oxides such as those that occur on welds in the age-hardenable alloys. Such oxides must be removed by grinding with an aluminum oxide or silicon carbide wheel. Also, carbide deburring tools are widely used for oxide removal.

If a component to be welded will not be subsequently reheated, a cleaned area extending 2 in. (51 mm) from the joint on each side will normally be sufficient to prevent damage by contamination from foreign materials. The cleaned area should include the edges of the workpiece.

The means used for removal of foreign materials prior to welding or heat treating will depend on the nature of the foreign materials. Shop dirt, marking crayons and ink, and materials having an oil or grease base can be removed by vapor degreasing or swabbing with suitable solvents. Paint and other materials not soluble in degreasing solvents may require the use of alkaline cleaners or special proprietary compounds. If alkaline cleaners containing sodium sesquisilicate or sodium carbonate are used, they must be removed prior to welding. Wire brushing will not completely remove the residue; spraying or scrubbing with hot water is required.

A process chemical such as caustic that has been in contact with the base metal for an extended period of time may be embedded and require grinding, abrasive blasting, or swabbing with a 10% (by volume) hydrochloric acid solution, followed by a thorough water wash. Manufacturers' safety precautions must be followed during the use of solvents and other cleaners.

Finally, the rough surface, or *skin* of castings may contain sand (silica) that must be removed prior to welding. Chipping, grinding, or machining are the most common forms of casting skin removal.[3]

---

3. Additional information on the cleaning of nickel and cobalt alloys may be found in the *Metals Handbook*, 8th Edition, Vol. 2, 607-10, 661-63.

ARC WELDING

NICKEL AND COBALT alloys are weldable by all the processes commonly used for steel and other base metals. Welded joints can be produced to stringent quality requirements in the precipitation-hardenable group, as well as the solid-solution group.

APPLICABLE PROCESSES

SOME ARC WELDING processes broadly applicable to nickel and cobalt alloys are identified by individual alloy in Table 4.7. Note that the shielded metal arc welding (SMAW) and gas metal arc welding (GMAW) processes are not applicable to the welding of the precipitation-hardenable alloys.

Covered electrodes for welding the age-hardenable alloys suffer from dramatically reduced mechanical properties of the weld and interbead slag adhesion, while the GMAW process results in high heat input, to which most of the age-hardenable alloys are sensitive.

Heat Input Limitations

HIGH HEAT INPUT during welding may produce undesirable changes in nickel and cobalt alloys. Some degree of annealing and grain growth will take place in the heat-affected zone (HAZ). The heat input of the welding process and the interpass preheat temperature will determine the extent of these changes.

High heat input may result in excessive constitutional liquation, carbide precipitation, or other harmful metallurgical phenomena. These, in turn, may cause cracking or loss of corrosion resistance.

The grain size of the base metal must also be considered in the proper choice of the welding process and technique. Coarser grain size increases the tendency for underbead cracking because the boundary area has a higher level of carbides and other intermetallic compounds that promote liquation cracking. Table 4.8 shows that the lower heat input processes must be used for welding most of the nickel-based alloys when the grain size is coarse.

When problems occur, either the welding technique should be modified to decrease the heat input, or another welding process of lower heat input should be substituted. Stringer beads and alteration of bead shape are examples of technique modifications that have been employed.

If inadequate inert-gas protection is encountered using the high heat input processes or a high interpass preheat temperature is allowed, a heavy oxide film may

Table 4.7
Arc Welding Processes Applicable to Some Nickel and Cobalt Alloys

Alloy[a]	UNS Number	Process[b]			
		SMAW	GTAW, PAW	GMAW	SAW
Commercially Pure Nickel					
200	N02200	X	X	X	X
201	N02201	X	X	X	X
Solid-Solution Nickel Alloys (Fine Grain)[c]					
400	N04400	X	X	X	X
404	N04404	X	X	X	X
R-405	N04405	X	X	X	—
X	N06002	X	X	X	—
NICR 80	N06003	X	X	—	—
NICR 60	N06004	X	X	—	—
G	N06007	X	X	X	—
RA 333	N06333	—	X	—	—
600	N06600	X	X	X	X
601	N06601	X	X	X	X
625	N06625	X	X	X	X
20Cb3	N08020	X	X	X	X
800	N08800	X	X	X	X
825	N08825	X	X	X	—
B	N10001	X	X	X	—
C	N10002	X	X	X	—
N	N10003	X	X		
Precipitation-Hardenable Nickel Alloys					
K-500	N05500	—	X	—	—
Waspaloy	N07001	—	X	—	—
R-41	N07041	—	X	—	—
80A	N07080	—	X	—	—
90	N07090	—	X	—	—
M 252	N07252	—	X	—	—
U-500	N07500	—	X	—	—
718	N07718	—	X	—	—
X-750	N07750	—	X	—	—
706	N09706	—	X	—	—
901	N09901	—	X	—	—
Cobalt Alloys					
188	R30188	—	X	X	—
L-605	R30605	X	X	X	—

a. Several of these designations use parts of or are registered trade names. These and similar alloys may be known by other designations and trade names.

b. SMAW - Shielded metal arc welding
   GTAW - Gas tungsten arc welding
   PAW - Plasma arc welding
   GMAW - Gas metal arc welding
   SAW - Submerged arc welding

c. Fine grain is ASTM Number 5 or finer. See Table 4.8 for recommended processes for coarse-grain base metals.

Table 4.8
Effect of Grain Size on Recommended Welding Processes[a]

Alloy	Grain Size[b]	Gas Metal Arc[c]	Electron Beam	Gas Tungsten Arc	Shielded Metal Arc
600	Fine	X	X	X	X
	Coarse	—	—	X	X
617	Fine	X	X	X	X
	Coarse	—	—	—	X
625	Fine	X	X	X	X
	Coarse	—	—	X	X
706	Fine	—	X	X	X
	Coarse	—	—	—	X
718	Fine	—	X	X	X
	Coarse	—	—	—	X
800	Fine	X	X	X	X
	Coarse	—	X	X	X
AISI Type 316 Steel	Fine	X	X	X	X
	Coarse	—	—	X	X
AISI Type 347 Steel	Fine	X	X	X	X
	Coarse	—	—	X	X

a. Processes marked X are recommended.

b. Fine grain is smaller than ASTM Number 5; coarse grain is ASTM Number 5 or larger.

c. Spray transfer.

form on the weld face. The oxide will change a normally smooth weld surface to a rough surface that is difficult to clean and inspect. The oxide makes subsequent weld beads more susceptible to oxide inclusion defects.

Corrosion Resistance

CORROSION RESISTANCE IS seldom adversely affected by welding in most alloys. Filler metals are usually selected to be close in chemical composition to the base metal, and the resulting weld metal exhibits comparable corrosion resistance in most environments.

However, the corrosion resistance of some base metals is adversely affected by welding heat in the HAZ adjacent to the weld. For example, nickel-molybdenum and nickel-silicon alloys should receive a postweld annealing treatment to restore corrosion resistance in the weld HAZ. For most alloys, postweld thermal treatments are usually not needed to restore corrosion resistance after welding.

The nickel-chromium and nickel-iron-chromium alloys, like some austenitic stainless steels, can exhibit carbide precipitation in the weld heat-affected-zone. But, in most environments, such sensitization does not impair corrosion resistance in nickel-based alloys as it does in the austenitic stainless steels. Many alloys are stabilized by additions of titanium or niobium to prevent HAZ corrosion.

While postweld heat treatments are usually not required, there are notable exceptions for service in specific environments. For example, Ni-Cr-Fe alloy 600 for fused caustic service and Ni-Cu alloy 400 for hydrofluoric acid service require a postweld stress relief to avoid stress cracking in service.

SHIELDED METAL ARC WELDING

THIS PROCESS IS used primarily for welding nickel and solid-solution-strengthened alloys. These alloys are readily welded in all positions with the same facility as steel and using welding techniques similar to those used in making high-quality welds in stainless steel. More shallow depth of fusion and relatively sluggish molten weld metal require minor variations in technique, as previously noted.

Shielded metal arc welding is seldom used to weld the precipitation-hardenable alloys. The alloying elements that contribute to precipitation hardening are difficult to transfer across the welding arc. Structures that are fabricated from these age-hardenable alloys are welded with better results by one of the gas-shielded processes. Should this process be used to weld age-hardenable alloys, interpass bead cleaning to remove oxides is critical to making a sound weld. Also, joint efficiencies will be significantly lower than those made using the gas tungsten arc welding process.

Electrode Position and Manipulation

MOLTEN WELD METAL in these alloys is less fluid than molten steel, impeding the process of spreading and wetting groove faces. Thus, the filler metal must be more accurately placed in the joint, and some electrode weaving is necessary to produce a desirable bead contour.

When possible, welding should be done in the flat position for welding ease and the economy of faster deposition rates. Table 4.16 lists typical welding currents for several different joints welded in the flat position. The electrode position is at a drag angle of 20 degrees and a work angle of 0 degrees. That position facilitates control of the molten flux and avoids slag inclusions. It is important that a short arc be consistently maintained.

When welding must be done in the vertical or overhead positions, lower current is necessary and the electrode diameter should be sufficiently small to provide proper control of the molten weld metal. For vertical welding, the electrode should be held at approximately 90 degrees to the joint and a work angle of 0 degrees. Uphill is generally recommended for best results.

Overhead welding is similar to vertical welding with the only change necessary being a slight shortening of the arc and lowering the current strength by 5 to 15 amperes.

With square- and V-groove welds, the electrode should be held normal to the joint. On the other hand, the electrode should be held at a work angle of about 30 degrees when welding U-groove joint designs (see Figure 4.14, joint designs for submerged arc welding, Page 250). This is done to obtain good fusion with the groove faces. The work angle should be 40 to 45 degrees for fillet welding.

Welding Technique and Procedure

BECAUSE NICKEL ALLOY weld metal does not spread or flow easily, it must be placed where required. Thus, it is necessary to weave the electrode. The amount of weave will depend on such factors as the following:

(1) Joint design
(2) Welding position
(3) Type of electrode

For example, pure nickel welds are relatively more sluggish in their molten state than are nickel-chromium alloy welds, thus the width of weave will differ. However, when a weave is used, it should not be wider than three times the electrode core wire diameter. Beads that are too wide can lead to slag inclusions; a large weld pool leads to a poor bead shape and possible disruption of the gas shield around the arc. Poor shielding may result in contamination of the weld metal.

There should be no pronounced spatter. When excessive spatter does occur, it usually indicates one or more of the following conditions:

(1) The arc is too long.
(2) The amperage is too high.
(3) The polarity is not DCEP.
(4) The electrode coating has absorbed moisture.

Excessive spatter also can be caused by magnetic arc blow such as when using nickel-chromium electrodes to weld 9% nickel steel. The use of high amperage to overcome poor fluidity will lead to the following:

(1) Electrode overheating
(2) Reduced arc stability
(3) Spalling of the electrode coating
(4) Porosity in the weld

When the welding arc is stopped for any reason, the arc length should be shortened slightly and the travel speed increased to reduce the weld pool size. This practice reduces the possibility of crater cracking as the weld cools, eliminates the relatively large rolled leading edge associated with large weld craters, and prepares the way for the restarting.

The manner in which the restart is made will have a significant influence on the soundness of the weld in the restart areas. A reverse or "T" start is recommended. The arc is started at the leading edge of the crater and carried back to the extreme rear of the crater at a normal drag-bead speed. The direction is then reversed, weaving started, and the weld continued. This restart method has several advantages:

(1) The correct arc length can be established in the groove away from the unwelded joint.
(2) Some preheat is applied to the relative cold (possibly cracked) weld crater.
(3) The first drops of quenched or rapidly cooled filler metal are placed such that they are remelted after the "T" is completed.
(4) Normal weld progression begins.

The opportunity for weld metal porosity is minimized using this restart technique. Many welding procedures require that weld craters be ground or deburred out prior to arc restart. In any case, the reverse or "T" restart technique still should be used.

Another commonly used restart technique involves placing the restart metal (apt to be porous) where it can readily be removed by grinding. The restart is made 1/2 to 1 in. (13 to 25 mm) behind the weld crater on top of the previous weld bead, and this arc strike is later ground level with the rest of the bead. This technique is often used when welds must meet stringent radiographic inspection standards. The technique produces high-quality welds with less welder skill than the reverse or "T" restart technique. Starting tabs may also be employed such that the used portion of the weld contains no arc strikes.

The general procedures described above are suitable for all alloys, with some slight modifications needed to suit the characteristics of individual alloys. Commercially pure nickel, for example, is less fluid than the solid-solution alloys and requires careful attention to accurate filler metal placement in the joint.

Stringer beads are recommended for the Ni-Mo alloys. If a weave technique is used, the weave is restricted to 1.5 times the electrode core wire diameter. Welding in other than flat position is not recommended for the Ni-Mo alloys. Porosity in the initial arc start can be a problem with Ni-Mo electrodes. This can be eliminated by using a starting tab for arc starting. But the best practice is to grind all arc starts and stops to expose sound metal. Testing with liquid penetrant examination of ground start and stop areas can be employed to ensure the weld soundness.

GAS TUNGSTEN ARC WELDING

GAS TUNGSTEN ARC welding (GTAW) is widely used in the welding of nickel and cobalt alloys, especially for the following applications:

(1) Thin base metal
(2) Root passes when the joint will not be back welded
(3) When the back side of the joint is inaccessible
(4) When flux residues from the use of coated electrodes would be undesirable

The GTAW and plasma arc welding processes are also the best joining processes for welding the precipitation-hardenable alloys (see Table 4.7).

Shielding Gases

THE RECOMMENDED SHIELDING gas is helium, argon, or a mixture of the two. Additions of oxygen, carbon dioxide, and nitrogen can cause porosity in the weld or accelerated erosion of the tungsten electrode and should be avoided. Small quantities of hydrogen (about 5%) may be added to argon for single-pass welds. The hydrogen addition produces a hotter arc and a smoother bead surface in single-pass welds. However, hydrogen may cause porosity in multiple-pass welds with some alloys.

The choice of shielding gas for arc characteristics and depth of fusion shape should be based on trial welding for the particular production weld. In manual GTAW welds, the shielding gas is usually argon. For mechanized welding of thin base metal without filler metal, helium has shown some definite advantages over argon, namely improved soundness in nickel and nickel-copper alloys and increased welding speeds of up to 40 percent.

However, should a problem be encountered, it is best to add filler metal. The arc voltage for a given arc length is about 40 percent greater with helium; consequently, the heat input is greater. Since welding speed is a function of heat input, the hotter arc permits greater speed. However, arc initiation in helium is more difficult below 60 amperes. Thus, for small parts and very thin base metal, argon shielding is a better choice. In some cases, a high-frequency current has been imposed to aid in initiation and maintenance of the arc in helium.

Electrodes

EITHER PURE TUNGSTEN or those alloyed with thorium may be used. A 2% thoriated electrode will give good results for most GTAW welding. The thoriated electrodes yield longer life, resulting from low vaporization of the electrode and cooler operation. It is important to avoid overheating the electrode through the use of excessive current.

Arc stability is best when the tungsten electrode is ground to a flattened point. Cone angles of 30 to 60 degrees with a small flat apex are generally used. The point geometry, however, should be designed for the particular application and can vary from sharp to flat. With higher amperages, the use of a larger diameter flat area is often desirable. The shape of the electrode has an effect on the depth of fusion and bead width, with all other welding conditions being equal. Thus, the welding procedure should spell out its configuration.

Welding Current

THE POLARITY RECOMMENDED for both manual and mechanized welding is direct current electrode negative (DCEN). Frequently incorporated in the welding machine is a high-frequency circuit to enhance arc initiation and a current-decay unit to gradually decrease the size of the weld crater when breaking the arc.

Alternating current can be used for mechanized welding if the arc length is closely controlled. Superimposed high-frequency power is required for arc stabilization. High-frequency power is also useful with dc power to initiate the arc.

Touch arc starting can cause tungsten contamination of the weld metal. Also, inadvertent dipping of the electrode into the weld pool can cause contamination of the electrode itself. If contamination occurs, the electrode should be cleaned or broken past the point of contamination and reshaped.

The use of high-frequency arc starting allows the welder to select the starting point of the weld before the welding current starts, eliminating the possibility of arc strikes on the base metal. Similarly, an abrupt break of the welding arc can lead to a porous, rough or cracked weld crater. A current-decay unit gradually lowers the current before the arc is broken to reduce the weld pool size and complete the bead smoothly. However, the same thing can be accomplished by increasing travel speed. Some welding procedures will contain a schematic illustration similar to Figure 4.13 to show the welder correct starting and stopping techniques.

Filler Metals

FILLER METALS FOR the GTAW process are generally similar to the base metals with which they are used.

However, a weld is a casting with an inherent dendritic structure, as opposed to the relative uniform grain size of the wrought base metal. Based on this knowledge, adjustments in chemical composition are frequently made to bring the base metal and weld metal properties into closer agreement. Alloying of the filler metal is also done to resist porosity formation and hot cracking caused by the high localized arc current resulting in high weld pool temperatures. Also, deliberate over-alloying is frequently done to make the filler metal tolerant to dilution by other metals, as found necessary in nickel cladding on carbon steel and in dissimilar metal welding of many different kinds.

The inherent reduction in stress-rupture ductility imposed by the dendritic structure of a weld is to some extent offset by varying the chemical composition. Likewise, the fatigue resistance of cast structures is lower than base metal with the same chemical composition. Thus, as with any alloy group when special service conditions apply, the difference between a weld and the base metal requires astute analysis by metallurgists and engineers. Figure 4.1 is a classic case in point, illustrating the need for filler metal changes as the anticipated service temperature changes, such that joint efficiency at any designated service temperature is optimum.

Welding Technique and Procedure

THE WELDING TORCH should be held with the work angle at 0 degrees and the travel angle of nearly 0 degrees. If the drag angle is more than 35 degrees, air may be drawn into the shielding gas and cause porosity in the weld metal with some nickel alloys.

The electrode extension beyond the gas nozzle should be short but appropriate for the joint design. For example, 3/16 in. (5 mm) maximum is used for butt joints in thin base metal, whereas up to 1/2 in. (13 mm) may be required for some fillet welds.

The shortest possible arc length must be maintained when filler metal is not added. The arc length should not exceed 0.050 in. (1.3 mm); 0.020 in. (0.5 mm) is preferable. Figure 4.5 illustrates the effect of arc length on the soundness of welds made in nickel-copper alloy 400 without the addition of filler metal.

An adequate quantity of shielding gas must be delivered to the weld zone at all times during welding, and it must be delivered with very little turbulence. As low as 8 ft³/h (4 L/min) has been used on thin base metal and as high as 30 ft³/h (14 L/min) for thicker base metal. To minimize turbulence, gas nozzles should be large enough to deliver the shielding gas to the weld zone at low velocity. The use of special nozzles (such as a gas lens nozzle) designed to minimize turbulence is recommended. The use of too high a gas flow rate can increase turbulence and can result in undesirable weld pool cooling.

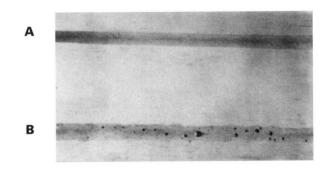

Figure 4.5—Effect of Arc Length on Soundness of Autogenous Gas Tungsten Arc Welds in a Nickel-Copper Alloy: (A) Proper Arc Length, (B) Excessive Arc Length (Porous Weld)

Since filler metals contain elements that improve cracking resistance and control porosity, optimum benefit from these elements is obtained when the weld bead contains at least 50 percent filler metal. The weld pool should be kept quiet, and agitation of the weld pool with the welding arc is to be avoided.

The hot end of the filler metal should be kept within the gas shield to avoid oxidation of the hot tip. Filler metal should be added at the leading edge of the weld pool to avoid contact with the tungsten electrode.

Shielding of the weld root surface (back side of the joint) is usually required. If a complete joint penetration weld is made with the root surface exposed to air, the weld metal on that side will be oxidized and porous. Root shielding can be provided with inert gas dams (in pipe), grooved backing bars or a backing flux. If a flux is used, it should be of a thick consistency, applied in a heavy layer and allowed to dry thoroughly. Figure 4.6 is a radiograph of a weld made using a flux insufficiently dried; note the high level of porosity in the weld. As a matter of good practice, all flux should be removed after the weld is finished. Most welding fluxes are highly corrosive at elevated temperatures.

Square-groove welds can be made in base metal up to 0.10 in. (2.5 mm) thick in a single pass. In addition to proper arc length, travel speed should be adjusted such that the weld pool is elliptically shaped; a teardrop-shaped weld pool is prone to centerline cracking during solidification. Also, travel speed has an effect on porosity in some alloys. In general, porosity will be at a minimum within some range of welding speed.

In summary, all solid-solution alloys except high silicon casting alloys are readily weldable with the GTAW process. Commercially pure nickel and nickel-copper alloys require some care, as described, to prevent porosity in welds made without filler metal. Filler metals normally contain deoxidizing elements to inhibit porosity.

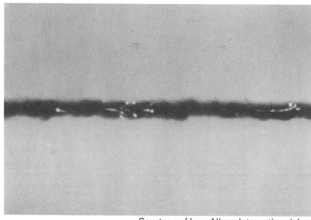

Courtesy of Inco Alloys International, Inc.

Figure 4.6—Radiograph Showing Porosity Caused by Wet Backing Flux for Nickel-Copper GTAW Weld

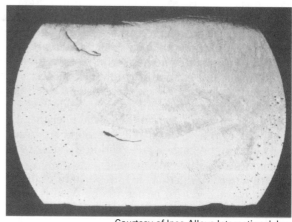

Courtesy of Inco Alloys International, Inc.

Figure 4.7—Oxide Stringers in Alloy X-750 Weld Resulting from Inadequate Cleaning of Refractory Oxides from Weld Faces Before Continuation of Welding

Therefore, filler metal additions are generally the best means of avoiding porosity when welding these alloys. Nickel alloys containing chromium are less prone to weld-metal porosity. The chromium-containing filler metals also contain other alloying additions to overcome hot-cracking tendencies.

Welding Precipitation-Hardenable Alloys

THE GTAW PROCESS is the most widely used for welding the precipitation-hardening alloys because it provides excellent protection against oxidation and loss of hardening elements. Using the same filler metal, the mechanical properties of GTAW welds will be somewhat higher than welds made using the gas metal arc welding (GMAW) process. Hot cracking tendencies in boron-containing alloys are reduced using the GTAW process as compared with the GMAW process. A special precaution must be observed when the base metal is high in aluminum and titanium. During welding, some of the aluminum and titanium form refractory oxides on the weld bead surface. In multipass welds, these oxides must be removed before the next weld bead is made. Otherwise, because of their high melting points, these refractory oxides become oxide inclusions in the weld. Figure 4.7 shows oxide stringers in alloy X-750 resulting from poor bead cleaning before subsequent passes are made.

There are two systems of precipitation-hardenable alloys. One is the nickel-aluminum-titanium system such as alloy X-750. The other is the nickel-niobium-aluminum-titanium system such as alloy 718. Both alloy systems have good weldability. The significant difference between the two systems is the time for precipitation to occur. The aluminum-titanium system responds

rapidly to precipitation-hardening temperatures. The niobium-aluminum-titanium system responds more slowly, which improves the weldability of these alloys. The delayed precipitation reaction enables the alloys to be directly postweld aged, without annealing, with less possibility of base-metal cracking.

When cracking occurs in the aluminum-titanium system, the results are catastrophic. Figure 4.8 illustrates a failure in 2 in. (51 mm) thick alloy X-750 plate welded in the age-hardened condition and re-aged after welding without intermediate stress relief. The failure could have been avoided with the following fabrication sequence:

(1) Anneal base metal
(2) Weld
(3) Stress relieve
(4) Age

Cooling rates also are critical. The rule with these alloys is that they must receive an appropriate heat treatment after they are welded and before they are precipitation hardened. Also, fast heating through the precipitation-hardening range is important to avoid staying in the precipitation-hardening temperature range too long. The base-metal supplier should be consulted before a welding procedure is accepted so that very precise instructions can be provided to the fabrication shop. In both systems, heat input during welding should be held to a moderately low level to obtain the highest joint efficiency.

Most all precipitation-hardenable alloys are subject to strain-age cracking if correct design, welding, and

Figure 4.8—Failure in 2 in. (51 mm) Thick Plate of Alloy X-750 Caused By Welding in the Aged Condition

containing niobium have a greater resistance to cracking because of the slow hardening response of the niobium precipitate compared to the aluminum-titanium precipitate.

The relative weldability of various alloys is shown in Figure 4.2. All of the alloys below the dashed line are routinely welded with proper adherence to proven heat treatment and welding schedules and have, for decades, been of use in thousands of industrial applications.

The need for management and reduction of residual stresses is related to the fact that a rapid decrease in the ductility of certain alloys at the aging temperature does not permit plastic flow to occur readily.

During the hardening thermal treatment, an overall volume contraction occurs. Welding stresses, coupled with this contraction, increase the tendency for strain-age cracking. Solution annealing after welding will relieve welding stresses and thereby decrease the potential for strain-age cracking. However, as mentioned previously, it is important to heat the weldment through the hardening temperature range rapidly to decrease precipitation reactions. Charging the welded assembly into a furnace already at or above the annealing temperature is a good first step, but the rate of heating is related to the complexity of the welded assembly, how it is supported in the furnace, and the overall mass of the assembly.

Welding stresses can be lowered using a number of techniques:

(1) Welding parts in an unrestrained condition
(2) Application of a suitable preweld heat treatment
(3) Welding the joint with appropriate sequence
(4) Annealing at some intermediate stage
(5) Changing the metallurgy of the alloy by deliberate heat treatment

For example, one metallurgical treatment is to overage the components prior to welding. Such treatment is designed to precipitate the age-hardening constituent in massive form. This treatment adversely affects the final hardness and yield strength of the alloy, but at least it can be welded because residual welding stresses are lower and further aging cannot occur. This procedure is used with alloys containing relatively large amounts of aluminum and titanium such as U-500 (UNS N07500). (See Figure 4.2.)

Welding of precipitation-hardened components should be avoided. Welding in this condition will result in re-solutioning and overaging in the HAZ, resulting in lowered properties. A postweld heat treatment is then necessary to restore the properties in this zone. In alloys with high aluminum and titanium content, cracking will occur immediately as the component is exposed to the aging temperature on its way up to its annealing temperature.

heat-treating practices are not followed completely. Strain-age cracking is seldom a problem in the niobium-aluminum-titanium system, but there have been reports of cracking of base metal welded in the aged condition, which is subsequently re-aged under highly restrained conditions. From an optimum property viewpoint, this is an unusual and undesirable procedure dictated by complexity of assemblies built in stages, which merely proves that highly weldable alloys can be made to crack in complex sequential assemblies.

Strain-age Cracking. Strain-age cracking in precipitation-hardenable nickel alloys is the result of residual stresses that exceed the yield point of the material. The high residual stresses are developed during the aging process and are increased further by forming, machining, and welding. Most precipitation-hardenable alloys are subject to strain-age cracking. However, alloys

In summary, welding of precipitation-hardenable alloys requires sound metallurgical knowledge of the base metal to be welded and procedures for fabrication, assembly, and welding that are designed to take advantage of metal and weld properties obtainable.

GAS METAL ARC WELDING

THE GAS METAL arc welding (GMAW) process can be used to weld all the solid-solution nickel alloys except high-silicon castings, but it is an inferior choice of process for welding many of the age-hardenable alloys.

The dominant mode of metal transfer is spray transfer, but short circuiting and pulsed spray welding are widely employed. Spray transfer of filler metal is more economical because it uses higher welding currents and larger diameter welding wires, but the pulsed spray welding method using smaller welding wire and lower currents is more amenable to welding positions other than flat. Both methods are widely used in the production of low-dilution weld cladding on less corrosion-resistant base metal (such as carbon and low alloy steels).

Globular transfer can also be used, but the erratic depth of fusion and uneven bead contour are conducive to defect formation, and therefore it is seldom employed.

Shielding Gases

THE PROTECTIVE ATMOSPHERE for GMAW is normally argon or argon mixed with helium. The optimum shielding gas will vary with the type of metal transfer used.

Using spray and globular transfer, good results are obtained with pure argon. The addition of helium, however, has been found to be beneficial. Increasing helium content leads to progressively wider and flatter beads and less depth of fusion.

Helium alone has been used, but it creates an unsteady arc and excessive spatter.

The use of oxygen or carbon dioxide addition, common to some base metals, should be avoided when welding nickel and cobalt alloys because even small amounts will result in heavily oxidized and irregular bead faces. Such additions also cause severe porosity in nickel and nickel-copper alloys.

GMAW gas flow rates range from 25 to 100 ft³/h (12 to 47 L/min), depending on joint design, welding position, nozzle size, and whether a trailing shield is used.

When short circuiting transfer welding is used, argon with a helium addition gives the best results. Argon alone provides a desirable "pinch" effect, but it can also produce excessively convex weld beads, which can lead

to incomplete fusion defects. With helium added, the weld bead is flatter with good wetting, reducing the possibility of incomplete fusion between weld beads.

Gas flow rates for short circuiting transfer welding range from 25 to 45 ft³/h (12 to 21 L/min). As the percentage of helium is increased, the gas flow rate must be increased to provide adequate weld protection.

The size of the gas nozzle can have important effects on welding conditions. For example, when using 50% argon - 50% helium and a gas flow rate of 40 ft³/h (19 L/min), with a 3/8 in. (9.5 mm) diameter gas nozzle, the maximum current before weld bead oxidation begins to occur is about 120 amperes. When the cup size is increased to 5/8 in. (16 mm) diameter, a current of 170 amperes can be used before weld bead oxidation begins to occur. Thus, delivering the shielding gas to the weld pool at low velocity is important.

When pulsed spray welding is used, argon with an addition of helium provides the best results. Good results have been obtained with helium contents of 15 to 20%. The flow rate should be at least 25 to 45 ft³/h (12 to 21 L/min). Excessive rates of gas flow can interfere with the arc.

Filler Metals

FILLER METALS FOR the GMAW process are identical, almost without exception, to those used with the gas tungsten arc welding process.

Welding Current

THE RECOMMENDED POLARITY is direct current electrode positive with electrode diameters of 0.035 in. (0.9 mm), 0.045 in. (7.1 mm), and 0.062 in. (1.6 mm). The specific size depends on the mode of metal transfer desired and the base metal thickness.

Constant-potential power sources are normally used, but constant-current dc units have been used in special cases. For short circuiting transfer, the equipment must have separate slope and secondary inductance controls. Pulsed spray welding requires two power sources, one for each of the two current ranges. Switching back and forth between the two sources produces a pulsed spray.

The development of power sources (thyristor, inverter, and transistor controlled) has facilitated the adjustment of welding conditions using pulsed frequencies (pulsed-arc) in which the pulse parameters are linked to the electrode feed speed control. In more highly developed equipment, all the welding conditions, pulse current, pulse duration, background current, and pulse frequency, are affected by this control process. These developments in power supply capabilities, plus the increasing use of microprocessors and computer-controlled welding in more critical applications, should increase the use of homogeneous (solid) welding wire at the expense of costly tubular welding wire.

Tubular wire has found limited acceptance in the welding of nickel and cobalt alloys. The foregoing changes also produce a welding arc with little or no spatter, and incomplete joint penetration can be avoided. Figure 4.9 is a schematic rendering of the foregoing discussion. At present, there is limited use of this equipment because the cost is at least double that of conventional equipment. Welding engineers need to stay in close contact with equipment suppliers as these improvements become less costly.

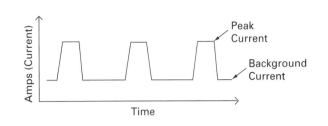

Figure 4.9—Pulsed-Arc GMAW with Automatic Control of the Pulse Current, Background Current, and Pause Time

Welding Technique and Procedure

TYPICAL CONDITIONS FOR spray, pulsed spray, and short circuiting transfer when using the GMAW process are given in Table 4.9. These data should serve as a guide in developing appropriate welding procedures.

Best results are obtained with the welding gun at both a work angle and a travel angle of 0 degrees, consistent with good visibility of the arc and good shielding. The arc length should be adjusted so that spatter is

minimal, but an excessively long arc will be difficult to control.

If manipulation of the welding gun is not correct, incomplete fusion can occur easily using short circuiting metal transfer. The gun should be advanced at a rate that will keep the arc in contact with the base metal and not the weld pool. In multipass welding, highly convex beads can increase the tendency for incomplete fusion.

With pulsed-arc welding, gun manipulation is similar to that used for the electrode in shielded metal arc

Table 4.9
Typical Conditions for Gas Metal Arc Welding of Nickel Alloys

Base Metal		Electrode					Shielding Gas	Welding Position	Arc Voltage		Welding Current, A
Alloy[a]	UNS Number	AWS Classification[b]	Diameter		Melting Rate				Average	Peak[c]	
			in.	mm	in./min	mm/s					
Spray Transfer											
200	N02200	ERNi-1	0.062	1.6	205	87	Ar	Flat	29-31	NA	375
400	N04400	ERNiCu-7	0.062	1.6	200	85	Ar	Flat	28-31	NA	290
600	N06600	ERNiCr-3	0.062	1.6	200	85	Ar	Flat	28-30	NA	265
Pulsed Spray Transfer											
200	N02200	ERNi-1	0.045	1.1	160	68	Ar or Ar-He	Vertical	21-22	46	150
400	N04400	ERNiCu-7	0.045	1.1	140	59	Ar or Ar-He	Vertical	21-22	40	110
600	N06600	ERNiCr-3	0.045	1.1	140	59	Ar or Ar-He	Vertical	20-22	44	90-120
Short Circuiting Transfer											
200	N02200	ERNi-1	0.035	0.9	360	152	Ar-He	Vertical	20-21	NA	160
400	N04400	ERNiCu-7	0.035	0.9	275-290	116-123	Ar-He	Vertical	16-18	NA	130-135
600	N06600	ERNiCr-3	0.035	0.9	270-290	114-123	Ar-He	Vertical	16-18	NA	120-130
G	N06007	ERNiCrMo-1	0.062	1.6	—	—	Ar-He	Flat	25	NA	160
C-4	N06455	ERNiCrMo-7	0.062	1.6	—	—	Ar-He	Flat	25	NA	180
B-2	N10665	ERNiMo-7	0.062	1.6	185	78	Ar-He	Flat	25	NA	175

a. Several of these designations use parts of or are registered trade names. These and similar alloys may be known by other designations and trade names.

b. See ANSI/AWS A5.14, *Specification for Nickel and Nickel Alloy Bare Welding Electrodes and Rods*

c. NA = Not Applicable.

welding. A slight pause at the limit of the weave is needed to reduce undercut.

The welding wire and contact tube must be kept clean. Dust or dirt carried into the contact tube can cause erratic feed. The contact tube should be blown out periodically and the spool of welding wire covered when not in use.

PLASMA ARC WELDING

NICKEL AND COBALT alloys can be readily joined with the plasma arc welding process. The constricted arc permits greater depth of fusion than that obtainable with the gas tungsten arc, but the welding procedures with both processes are similar. Square-groove welds can be made in base metal up to about 0.3 in. (8 mm) thick with a single pass when keyhole welding is used. Thin base metal can be welded with conventional welding, as with gas tungsten arc welding. Base metal over 0.3-in. (8-mm) thick can be welded using one of the other groove weld joint designs. The first pass can be made with keyhole welding and the succeeding passes with conventional welding. The root face should be

about 0.18 in. (5 mm) wide, compared to 0.06 in. (2 mm) for gas tungsten arc welding.

Special techniques are required for keyhole welding of thicknesses of 0.13 in. (3 mm) and greater. Upslope of the orifice gas flow and the welding current is required to initiate the keyhole; downslope of these conditions is needed to fill the keyhole cavity at the end of the weld bead.

Argon or argon-hydrogen mixtures are normally recommended for the orifice and shielding gases. Hydrogen addition to argon increases the arc energy for keyhole welding and high-speed autogenous welding. Additions up to 15 volume percent may be used, but these should be used with care because hydrogen can cause porosity in the weld metal. Therefore, the gas mixture for a specific application should be determined by appropriate tests.

Typical conditions used for automatic plasma arc welding of four nickel alloys with keyhole welding are given in Table 4.10. No filler metal was added. These conditions are not necessarily optimum. Other conditions may also produce acceptable welds, and they should be evaluated by appropriate tests prior to production to ensure product reliability.

Table 4.10
Typical Conditions for Autogenous Plasma Arc Welding of Nickel Alloys with Keyhole Welding

Alloy	Thickness		Orifice Gas Flow*		Shielding Gas Flow		Welding Current, A	Arc Voltage, V	Travel Speed	
	in.	mm	ft³/h	L/min	ft³/h	L/min			in./min	mm/s
Nickel (UNS N02200)	0.125	3.2	10	5	45	21	160	31.0	20	8
	0.235	6.0	10	5	45	21	245	31.5	14	6
	0.287	7.3	10	5	45	21	250	31.5	10	4
67Ni-32 Cu (UNS N04400)	0.250	6.4	12.5	6	45	21	210	31.0	14	6
76Ni-16 Cr-8 Fe (UNS N06600)	0.195	5.0	12.5	6	45	21	155	31.0	17	7
	0.260	6.6	12.5	6	45	21	210	31.0	17	7
46Fe-33 Ni-21Cr (UNS N08800)	0.125	3.2	10	5	45	21	115	31.0	18	8
	0.230	5.8	12.5	6	45	21	185	31.5	17	7
	0.325	8.3	14.0	7	45	21	270	31.5	11	5

* Orifice diameter: 0.136 in. (3.5 mm)
  Orifice and shielding gas: Ar-5 vol. % H_2
  Root shielding gas: Argon

SUBMERGED ARC WELDING

FILLER METALS AND fluxes are available for submerged arc welding of several solid-solution nickel alloys. Cobalt alloys and precipitation-hardenable alloys are seldom joined using this process. Nor is the process recommended for joining thick nickel-molybdenum alloys, because the high heat input and slow cooling rate of the weld results in low weld ductility and loss in corrosion resistance due to changes in chemical composition from flux reactions.

Because of its high deposition rate, the submerged arc process is an efficient method for joining thick base metal. Compared to other arc welding processes, bead surfaces are smoother, a proper flux will be self-peeling, and welding operator discomfort is less.

Suggested joint designs for submerged arc welding are shown in Figure 4.14. The double-U-groove is the preferred design for all joints that permit its use. It can be completed in less time with less filler metal and flux, and yields lower residual welding stresses.

Fluxes

SUBMERGED ARC FLUXES are available for several nickel alloys, and they are designed for use with a specific welding wire. Fluxes used to weld carbon steels and stainless steel are invariably unsuitable for welding nickel alloys. In addition to protecting the molten metal from atmospheric contamination, the fluxes provide arc stability and contribute important additions to the weld metal.

The flux cover should be only sufficient to prevent the arc from breaking through. An excessive flux cover can cause deformed weld beads. Slag is easily removed and should be discarded, but unfused flux can be reclaimed. However, in order to maintain consistency in the flux particle size, reclaimed flux should be mixed with an equal amount of unused flux.

Submerged arc fluxes are chemical mixtures and can absorb moisture. Storage in a dry area and resealing opened containers are standard practice. Flux that has absorbed moisture can be reclaimed by heating. The flux manufacturer should be consulted for the recommended procedure.

Filler Metals

SUBMERGED ARC WELDING employs the same filler metals used with the gas tungsten arc welding and gas metal arc welding processes. Weld metal chemical composition will be somewhat different as additions are made through the flux to allow the use of higher currents and larger welding wires.

Welding wire diameters are usually smaller than those used to weld carbon steels. For example, the maximum size used to weld thick base metal is 3/32 in. (2.4 mm), where 0.045 in. (1.1 mm) has been used to weld thin base metal. Table 4.11 gives deposition rates and polarity for some filler metal and flux combinations. Table 4.12 gives typical welding conditions for three alloy and flux combinations.

Welding Current

DIRECT CURRENT ELECTRODE negative (DCEN) or direct current electrode positive (DCEP) may be used. DCEP is preferred for groove joints, yielding flatter beads and greater depth of fusion at low voltage (30 to 33V). DCEN is frequently used for weld surfacing, yielding higher deposition rates, reduced depth of fusion, and reduced depth of penetration, thus reducing the amount of dilution from the base metal. However, DCEN requires a deeper flux cover and an increase in flux consumption. DCEN also increases the possibility of slag inclusions, especially in butt joints where the molten weld metal is thicker and solidification occurs from the sidewalls as well as the root of the weld.

Welding Technique and Procedure

TYPICAL CHEMICAL COMPOSITIONS from submerged arc weld metal in butt joints are given in Table 4.13. Part of the difference in chemical composition at various levels reflects dilution from the base metal and contributions by the flux. Slag inclusions are always a possibility during any welding involving flux. Slag inclusions can be controlled by appropriate joint design (Figure 4.14) and proper placement of weld beads. Figure 4.10 illustrates bead placement in a 3 in. (76 mm) thick weld in a butt joint in Ni-Cr-Fe alloy 600.

The weld face should be flat to slightly convex. Concave beads have produced centerline cracking in highly

Table 4.11
Deposition Rates for Submerged Arc Welding for Specific Filler Metal and Flux Combinations

Filler Metal and Flux	Wire Diameter			Deposition Rate	
	in.	mm	Polarity	lb/h	kg/h
ERNiCr-3 with Flux 4*	1/16	1.6	DCEN	16-18	7.3-8.2
	1/16	1.6	DCEP	14-17	6.4-7.7
	3/32	2.4	DCEN	20-21	9.1-9.5
	3/32	2.4	DCEP	16-17	7.3-7.7
ERNiCu-7 with Flux 5*	1/16	1.6	DCEN	16-17	7.3-7.7
	1/16	1.6	DCEP	14-16	6.4-7.3
	3/32	2.4	DCEN	20-21	9.1-9.5
	3/32	2.4	DCEP	16-17	7.3-7.7

* Proprietary flux from Inco Alloys International, Inc. Weight of flux consumed is approximately equal to weight of filler metal.

Table 4.12
Typical Conditions for Submerged Arc Welding for Specific Alloy and Flux Combinations

Parameter	ERNi-1 with Flux 6[a]	ERNiCu-7 with Flux 5[a]	ERNiCr-3 with Flux 4[a]
Base Metal	Nickel 200	Ni-Cu alloy 400	Ni-Cr-Fe alloy 600[b]
Filler Metal Diameter, in. (mm)	1/16 (1.6)	1/16 (1.6)	1/16 or 3/32 (1.6 or 2.4)
Electrode Extension, in. (mm)	7/8-1 (22-25)	7/8-1 (22-25)	7/8-1(22-25)
Power Source	DC, Constant Voltage	DC, Constant Voltage	DC, Constant Voltage
Polarity	DCEP	DCEP	DCEP
Current, A	250	260-280	250 with 1/16 in. (1.6 mm) wire 250-300 with 3/32 in. (2.4 mm) wire
Voltage, V	28-30	30-33	30-33
Travel Speed, in./min (mm/min)	10-12 (250-300)	8-11 (200-280)	8-11 (200-280)
Joint Restraint	Full	Full	Full

a. Proprietary flux from Inco Alloys International, Inc.

b. The conditions also apply to Fe-Ni-Cr alloy 800.

Table 4.13
Chemical Composition of Weld-Metal Sample from Welds in Butt Joints, wt. %

Filler Metal and Flux	Base Material	Ni	C	Mn	Fe	S	Si	Cu	Cr	Ti	Others
ERNiCu-7 with Flux 5*	Alloy 400	Bal.	0.06	5.0	3.5	0.013	0.90	26.0	—	0.48	—
ERNiCr-3 with Flux 4*	Alloy 600	Bal.	0.07	3.21	1.75	0.006	0.40	—	19.25	0.17	Nb + Ta = 3.39

* Proprietary flux from Inco Alloys International, Inc.

restrained joints. The number of weld beads and bead contour is most effectively controlled by voltage and travel speed.

Fully restrained welds in 6-in. (150-mm) alloy 600 plate using ERNiCr-3 filler metal and flux 4 have been made in stringently tested welds used in nuclear power applications.[4]

Chemical composition remains virtually constant throughout such joints, with no chemical element increase arising from an accumulation of the flux components. Table 4.14 shows the composition at various levels [approximately 1/2 in. (13 mm) intervals] of weld metal from the joint shown in Figure 4.10, beginning at the top of the joint progressing downward toward the weld root.

4. Flux 4 is a proprietary product (Incoflux 4) of Inco Alloys International, Inc.

Table 4.14
Chemical Composition at Various Levels*
of a 3 in. (76 mm) Thick Joint in Alloy 600
Welded with ERNiCr-3 Filler Metal and
Submerged Arc Flux 4, wt. %

Element	Level 1	Level 2	Level 3	Level 4	Level 5	Level 6
Nickel	73.6	73.5	73.6	73.5	73.7	73.6
Chromium	18.1	18.0	18.1	18.0	18.1	18.0
Niobium	3.61	3.71	3.59	3.67	3.50	3.60
Iron	0.86	0.87	0.88	0.88	0.87	1.00
Silicon	0.44	0.44	0.43	0.43	0.44	0.44
Carbon	0.05	0.05	0.05	0.05	0.05	0.05
Sulfur	0.003	0.003	0.003	0.003	0.003	0.003

* Approximately 1/2 in. (13 mm) intervals beginning at top surface. (See Figure 4.10.)

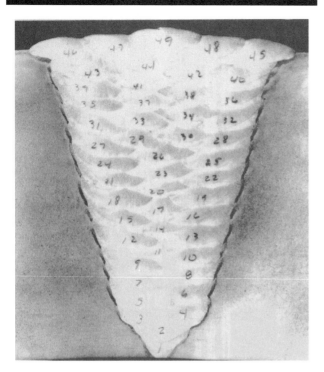

Note: Numerals indicate sequence of bead placement.

Figure 4.10—Alloy 600 Joint 3 in. (76 mm) Thick Welded with 0.062 in. (1.6 mm) Diameter ERNiCr-3 Filler Metal and Submerged Arc Flux 4

FILLER METALS AND FLUXES

Covered Electrodes

IN MOST CASES, the weld metal composition from a covered electrode resembles that of the base metal with which it is used. Invariably, its chemical composition has been adjusted to satisfy weldability requirements; usually additions are made to control porosity, enhance microcracking resistance, or improve mechanical properties.

Covered electrodes normally have additions of deoxidizing ingredients such as titanium, manganese, and niobium. ANSI/AWS A5.11, *Specification for Nickel and Nickel Alloy Welding Electrodes for Shielded Metal Arc Welding*, is used almost universally in filler metal selection. Sometimes military specifications will apply, such as the MIL-E-22200 series, but they duplicate the AWS specification in most respects.

Covering Formulations. The covering formulator uses chemistry and the physical and mechanical properties of chemicals to enhance the conditions found in the welding arc environment. Alkali metals have low ionization voltages, making ionization more predictable, and arc starting easier. A second chemical means of stabilizing a welding arc is to add compounds that increase emission. Refractory oxides make the best additives for this purpose.

Formulation for nickel electrodes must take into consideration the sluggish nature of the molten metal and the relatively high ionization potential of the elements in the electrode core wire in these alloys. The basic requirements of the covering constituents are to provide the following:

(1) Atmospheric protection from nitrogen and oxygen
(2) Arc stabilization
(3) A slag that will remove solid metal oxides by eutectic formation or solution
(4) A slag that wets and spreads vigorously and has high fluidity with good adherence to the weld face but does not result in unacceptable undercut
(5) A slag that, when cool, is easily removed

A common expression used to differentiate the welding of nickel alloys from other alloy groups is "different but not difficult." It is a correct assessment.

Covered Electrode Groups. Nickel alloy covered electrodes are divided into five alloy families, namely, Ni, Ni-Cu, Ni-Cr-Fe, Ni-Mo, and Ni-Cr-Mo alloys. Each family contains one or more electrode classifications, based upon the composition of undiluted weld metal. Table 4.15 gives the weld metal composition of the various covered electrodes available.

Covered electrodes are generally designed for use with direct current electrode positive (DCEP), i.e., electrode positive, workpiece negative. Each electrode type and diameter has an optimum current range in which it has good arcing characteristics and outside of which the arc becomes unstable, the electrode overheats, the covering spalls, or excessive spatter is encountered. Table 4.16 lists the approximate current settings for three alloy families for flat position welding. However, the current density for a given joint is influenced by base-metal thickness, welding position, type of backing, tightness of clamping, and joint design.

Recommended amperage ranges are usually printed on each carton of electrodes by the manufacturer. These ranges should be used as a guide. Actual welding conditions should be developed by making trial welds using the same electrode, base metal, thickness, and welding position to be used in production. It is important that the manufacturer's recommended current ranges not be violated.

Table 4.15
Limiting Chemical Composition Requirements for Nickel-Based Welding Filler Metal

Composition, wt.% [a,b]

AWS Classification	UNS Number[c]	C	Mn	Fe	P	S	Si	Cu	Ni[d]	Co	Al	Ti	Cr	Nb Plus Ta	Mo	V	W	Other Elements
ENi-1	W82141	0.10	0.75	0.75	0.03	0.02	1.25	0.25	92.0 min.	—	1.0	1.0 to 4.0	—	—	—	—	—	0.50
ENiCu-7	W84190	0.15	4.00	2.5	0.02	0.015	1.5	Rem	62.0 to 69.0	—	0.75	1.0	—	—	—	—	—	0.50
ENiCrFe-1	W86132	0.08	3.5	11.0	0.03	0.015	0.75	0.50	62.0 min.	—	—	—	13.0 to 17.0	1.5 to 4.0[f]	—	—	—	0.50
ENiCrFe-2	W86133	0.10	1.0 to 3.5	12.0	0.03	0.02	0.75	0.50	62.0 min.	e	—	—	13.0 to 17.0	0.5 to 3.0[f]	0.50 to 2.50	—	—	0.50
ENiCrFe-3	W86182	0.10	5.0 to 9.5	10.0	0.03	0.015	1.0	0.50	59.0 min.	e	—	1.0	13.0 to 17.0	1.0 to 2.5[f]	—	—	—	0.50
ENiCrFe-4	W86134	0.20	1.0 to 3.5	12.0	0.03	0.02	1.0	0.50	60.0 min.	—	—	—	13.0 to 17.0	1.0 to 3.5	1.0 to 3.5	—	—	0.50
ENiCrFe-7	W86152	0.05	5.0	7.0 to 12.0	0.03	0.015	0.75	0.50	Rem.	e	0.50	0.50	28.0 to 31.5	1.0 to 2.5	0.50	—	—	0.50
ENiMo-1	W80001	0.07	1.0	4.0 to 7.0	0.04	0.03	1.0	0.50	Rem.	2.5	—	—	1.0	—	26.0 to 30.0	0.60	1.0	0.50
ENiMo-3	W80004	0.12	1.0	4.0 to 7.0	0.04	0.03	1.0	0.50	Rem.	2.5	—	—	2.5 to 5.5	—	23.0 to 27.0	0.60	1.0	0.50
ENiMo-7	W80665	0.02	1.75	2.0	0.04	0.03	0.2	0.50	Rem.	1.0	—	—	1.0	—	26.0 to 30.0	—	1.0	0.50
ENiCrCoMo-1	W86117	0.05 to 0.15	0.30 to 2.5	5.0	0.03	0.015	0.75	0.50	Rem.	9.0 to 15.0	1.5	0.6	21.0 to 26.0	1.0	8.0 to 10.0	—	—	0.50
ENiCrMo-1	W86007	0.05	1.0 to 2.0	18.0 to 21.0	0.04	0.03	1.0	1.5 to 2.5	Rem.	2.5	—	—	21.0 to 23.5	1.75 to 2.50	5.5 to 7.5	—	1.0	0.50

a. The weld metal shall be analyzed for the specific elements for which values are shown in this table. If the presence of other elements is indicated in the course of the work, the amount of those elements shall be determined to ensure that their total does not exceed the limit specified for "Other Elements" in the last column of the table.

b. Single values are maximum, except where otherwise specified.

c. ASTM/SAE Unified Numbering System for Metals and Alloys.

d. Includes incidental cobalt.

e. Cobalt—0.12 maximum, when specified.

f. Tantalum—0.30 maximum, when specified.

Table 4.15 (continued)
Limiting Chemical Composition Requirements for Nickel-Based Filler Metal

Composition, wt. %[a,b]

AWS Classification Number[c]	UNS Number	C	Mn	Fe	P	S	Si	Cu	Ni[d]	Co	Al	Ti	Cr	Nb Plus Ta	Mo	V	W	Other Elements
ENiCrMo-2	W86002	0.05 to 0.15	1.0	17.0 to 20.0	0.04	0.03	1.0	0.50	Rem.	0.50 to 2.50	—	—	20.5 to 23.0	—	8.0 to 10.0	—	0.20 to 1.0	0.50
ENiCrMo-3	W86112	0.10	1.0	7.0	0.03	0.02	0.75	0.50	55.0 min.	e	—	—	20.0 to 23.0	3.15 to 4.15	8.0 to 10.0	—	—	0.50
ENiCrMo-4	W80276	0.02	1.0	4.0 to 7.0	0.04	0.03	0.2	0.50	Rem.	2.5	—	—	14.5 to 16.5	—	15.0 to 17.0	0.35	3.0 to 4.5	0.50
ENiCrMo-5	W80002	0.10	1.0	4.0 to 7.0	0.04	0.03	1.0	0.50	Rem.	2.5	—	—	14.5 to 16.5	—	15.0 to 17.0	0.35	3.0 to 4.5	0.50
ENiCrMo-6	W86620	0.10	2.0 to 4.0	10.0	0.03	0.02	1.0	0.50	55.0 min.	—	—	—	12.0 to 17.0	0.5 to 2.0	5.0 to 9.0	—	1.0 to 2.0	0.50
ENiCrMo-7	W86455	0.015	1.5	3.0	0.04	0.03	0.2	0.50	Rem.	2.0	—	0.70	14.0 to 18.0	—	14.0 to 17.0	—	0.5	0.50
ENiCrMo-9	W86985	0.02	1.0	18.0 to 21.0	0.04	0.03	1.0	1.5 to 2.5	Rem.	5.0	—	—	21.0 to 23.5	0.5	6.0 to 8.0	—	1.5	0.50
ENiCrMo-10	W86022	0.02	1.0	2.0 to 6.0	0.03	0.015	0.2	0.50	Rem.	2.5	—	—	20.0 to 22.5	—	12.5 to 14.5	0.35	2.5 to 3.5	0.50
ENiCrMo-11	W86030	0.03	1.5	13.0 to 17.0	0.04	0.02	1.0	1.0 to 2.4	Rem.	5.0	—	—	28.0 to 31.5	0.3 to 1.5	4.0 to 6.0	—	1.5 to 4.0	0.50
ENiCrMo-12	W86040	0.03	2.2	5.0	0.03	0.02	0.7	0.50	Rem.	—	—	—	20.5 to 22.5	1.0 to 2.8	8.8 to 10.0	—	—	0.50

a. The weld metal shall be analyzed for the specific elements for which values are shown in this table. If the presence of other elements is indicated in the course of the work, the amount of those elements shall be determined to ensure that their total does not exceed the limit specified for "Other Elements" in the last column of the table.

b. Single values are maximum, except where otherwise specified.

c. ASTM/SAE Unified Numbering System for Metals and Alloys.

d. Includes incidental cobalt.

e. Cobalt—0.12 maximum, when specified.

f. Tantalum—0.30 maximum, when specified.

Table 4.16
Approximate Current Settings for Nickel-Based Covered Electrodes[a]

Nickel-Copper Alloys			Nickel Alloys			Ni-Cr-Fe and Fe-Ni-Cr Alloys		
Base-Metal Thickness[b]		Current, A	Base-Metal Thickness[b]		Current, A	Base-Metal Thickness[b]		Current, A
in.	mm		in.	mm		in.	mm	
			Electrode Diameter = 3/32 in. (2.4 mm)					
1/16	1.57	50	1/16	1.57	75	≥1/16	≥1.57	60
5/64	1.98	55	5/64	1.98	80	—	—	—
3/32	2.36	60	≥3/32	≥2.36	85	—	—	—
≥7/64	≥2.77	60	—	—	—	—	—	—
			Electrode Diameter = 1/8 in. (3.2 mm)					
7/64	2.77	65	7/64	2.77	105	7/64	2.77	75
1/8	3.18	75	≥1/8	≥3.18	105	1/8	3.18	75
9/64	3.56	85	—	—	—	—	—	—
≥5/32	≥3.96	95	—	—	—	≥5/32	≥3.96	80
			Electrode Diameter = 5/32 in. (4.0 mm)					
1/8	3.18	100	1/8	3.18	110	—	—	—
9/64	3.56	110	9/64	3.56	130	—	—	—
5/32	3.96	115	5/32	3.96	135	—	—	—
—	—	—	≥3/16	≥4.75	150	≥3/16	≥4.75	105
≥1/4	≥6.35	150	—	—	—	—	—	—
			Electrode Diameter = 3/16 in. (4.8 mm)					
—	—	—	1/4	6.35	180	—	—	—
3/8	9.53	170	≥3/8	≥9.53	200	≥3/8	≥9.53	140
≥1/2	≥12.7	190	—	—	—	—	—	—

a. Selection of electrode diameter should be based on joint design. For example, smaller diameters than those listed for material 1/8 in. (3.18 mm) and over in thickness may be necessary for the first passes in the bottom of a groove joint.

b. For base-metal thicknesses less than 1/16 in. (1.57 mm), the amperage should be the minimum at which arc control can be maintained, for all three electrode groups shown.

Covered electrodes should be left in their sealed, moisture-proof containers in a dry storage area prior to use. All opened containers should be stored in a cabinet with a desiccant or held at a temperature difference 10 to 15 °F (6 to 8 °C) above the highest expected ambient temperature.

Electrodes that have absorbed moisture can be reclaimed by baking in an oven according to the manufacturer's recommendations.

Bare Rods and Electrodes

NICKEL ALLOY BARE rods and electrodes are divided into five families, as are coated electrodes.[5]

5. See ANSI/AWS A5.14, *Specification for Nickel and Nickel Alloy Bare Welding Electrodes and Rods.* Various other specifications will be found to apply, such as the Aeronautical Material Specifications (AMS) and military and federal specifications.

There are more classifications, however, in most families of bare filler metals. They are designed to weld base metals of similar composition to themselves by the gas tungsten arc (GTAW), gas metal arc (GMAW), plasma arc (PAW) and submerged arc (SAW) processes. As noted in Table 4.7, these processes are not applicable to all alloys. Some filler metals are compositionally balanced to allow them to be applied to less corrosion resistant surfaces or to join dissimilar metals, i.e., allowance is made for dilution. Also, because of high arc current and high weld pool temperatures, filler metals are alloyed to resist porosity formation and hot cracking of the weld.

Primarily, the compositional change is addition of such elements as titanium, manganese, and niobium. Frequently the major elements may be higher in the filler metal than in the base metal to diminish the effects of dilution in surfacing less corrosion-resistant base metals and to accommodate the joining of dissimilar metals.

Precipitation-hardenable weld metals will normally respond to the aging treatment used for the base metal. However, the response is usually less and the weld joint strength will be somewhat lower than that of the base metal after aging.

Precipitation-hardenable alloys may be welded with dissimilar metals to minimize processing difficulties. For example, alloy R-41 (UNS N07041) is commonly welded with ERNiCrMo-3 or ERNiMo-3 filler metals. As a result, mechanical properties are significantly lower than those of the base metal. The usual answer for this deficiency is to locate the weldment in an area of low stress.

The proper electrode diameter for gas metal arc welding will depend upon the base-metal thickness and the type of metal transfer. For spray, pulsed spray, and globular transfer, electrode diameters used are from 0.035 to 0.093 in. (0.89 to 2.36 mm). With short circuiting transfer, diameters of 0.045 in. (1.14 mm) or smaller are usually required.

Fluxes

FLUXES ARE AVAILABLE for submerged arc welding of many nickel alloys. Fluxes, in addition to protecting the molten metal from atmospheric contamination, provide arc stability and contribute important additions to the weld metal. Therefore, the filler metal and the flux must be jointly compatible with the base metal. An improper flux can cause excessive slag adherence, weld cracking, inclusions, poor bead contour, and undesirable changes in weld metal composition. Fluxes used for carbon steel and stainless steel are not suitable.

JOINT DESIGN

VARIOUS SUGGESTED JOINT designs for nickel and cobalt alloys are shown in Figure 4.11. The first consideration in designing joints for nickel and cobalt alloys is to provide proper accessibility. The root opening must be sufficient to permit the electrode, filler metal, or torch to extend to the bottom of the joint.

In addition to the basic requirements of accessibility, the characteristics of nickel and cobalt alloy weld metal require consideration. The most significant characteristic is the sluggish nature of molten weld metal. It does not spread easily, requiring accurate metal placement by the welder within the joint. Wider groove angles are used.

Secondly, the force of the arc results in less depth of fusion in nickel and cobalt alloys than in, for example, carbon steel. The lower depth of fusion makes necessary the use of a narrower root face. Increases in amperage will not significantly increase the depth of fusion. Figure 4.12 illustrates depth of fusion for materials at the same current setting.

Groove Welds

BEVELING IS NOT usually required for groove welds in base-metal thicknesses of 0.093 in. (2.36 mm) or less. Above 0.093 in. (2.36 mm) thickness, a V-, U-, or J-groove should be used, or the joint should be welded from both sides. Otherwise, incomplete joint penetration will result, leading to crevices and voids that become focal points for accelerated corrosion in the joint root. It is generally the side that is inaccessible that must withstand corrosion. Notches can also act as potential stress raisers. Complete joint penetration welds are usually required because many of the nickel and cobalt alloys are used in elevated temperature service and in severe corrosive media at all temperatures.

Root Passes

WHEN THE INSIDE is inaccessible, the gas tungsten arc process should be used for the root pass. The process produces the smoothest bead contour. To provide a good filler metal dilution of 50 percent, inserts are commonly employed. Some common starting and stopping techniques are shown in Figure 4.13.

Submerged Arc Groove Designs

SUGGESTED JOINT GROOVES for submerged arc welding are shown in Figure 4.14. The double-U-groove is the preferred design for all joints that permit its use. That design gives lower residual welding stress, and it can be completed in less time than for other grooves and with less filler metal.

Gas Metal Arc Welding

THE JOINT DESIGNS in Figure 4.14 are also used for gas metal arc welding with one important modification. For globular, spray, and pulsed-arc welding, the root radius should be decreased by about 50 percent and the groove angle increased as much as double those shown in Figure 4.14. With the gas metal arc welding process, the use of high amperage on small diameter wires produces high arc force. Such an arc is not easily deflected. Consequently, the joint design must permit the arc to be directed at all areas to be fused, i.e., U-groove joints should have a 30-degree bevel angle. This will permit proper manipulation of the arc to obtain good fusion with the groove faces.

Corner and Lap Joints

CONSISTENT WITH PREVIOUS remarks about proper access and the special characteristics of nickel and cobalt alloy (i.e., reduced spreading and depth of

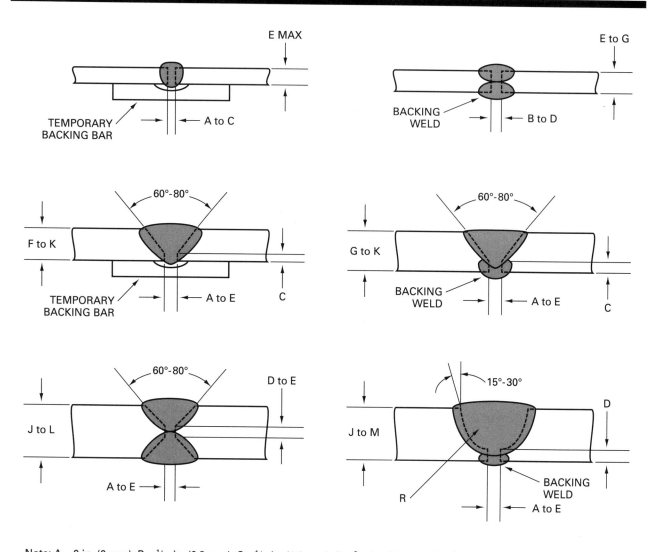

Note: A = 0 in. (0 mm); B = $^1/_{32}$ in. (0.8 mm); C = $^1/_{16}$ in. (1.6 mm); D = $^3/_{32}$ in. (2.4 mm); E = $^1/_8$ in. (3.2 mm); F = $^3/_{16}$ in. (4.8 mm); G = $^1/_4$ in. (6.4 mm); H = $^5/_{16}$ (7.9 mm); J = $^1/_2$ (12.7 mm); K = $^5/_8$ (15.9 mm); L = $1^1/_4$ in. (31.8 mm); M = 2 in. (50.8 mm); R = $^3/_{16}$ to $^5/_{16}$ in. (4.8 to 7.9 mm).

Figure 4.11—Suggested Designs for Arc Welded Butt Joints in Nickel and Cobalt Alloys

fusion), corner and lap joints may be used where high service stresses are not encountered. However, it is especially important to avoid their use at high temperatures or under thermal or mechanical cycling (fatigue) conditions.

Butt joints, where stresses act axially, are preferred to corner and lap joints, where stresses are eccentric. Sometimes they cannot be avoided. When corner joints are used, a complete joint penetration weld must be made. In most cases, a fillet weld on the inside corner will be required.

Jigs and Fixtures

IN THE WELDING of thin sheet and strip, jigs, clamps, and fixtures can reduce the cost of welding and improve the achievement of consistent high-quality welds. Proper jigging and clamping will facilitate welding by holding the joint members firmly and removing heat rapidly from the joint area, thus minimizing buckling, maintaining alignment. The weld can be subjected to compressive stress as it cools, reducing cracking tendency and increasing weld reinforcement, where it may

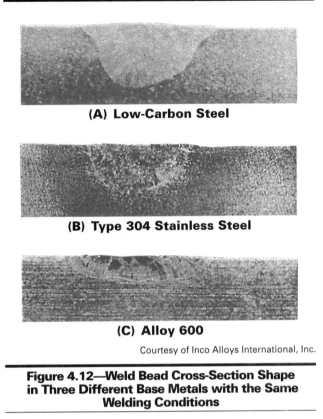

(A) Low-Carbon Steel

(B) Type 304 Stainless Steel

(C) Alloy 600

Courtesy of Inco Alloys International, Inc.

Figure 4.12—Weld Bead Cross-Section Shape in Three Different Base Metals with the Same Welding Conditions

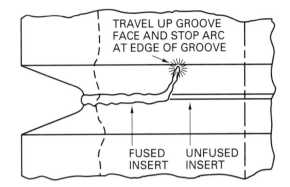

(A) Technique for Stopping Arc

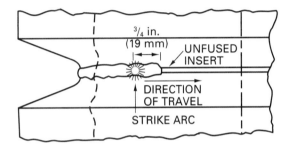

(B) Technique for Restarting Welding after Stopping Arc

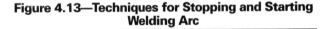

Figure 4.13—Techniques for Stopping and Starting Welding Arc

Table 4.17
Thermal Expansion Coefficients of Nickel Alloys

| Alloy | Thermal Expansion Coefficients* | |
	μin./in./°F	μm/m/°C
200	7.2	13.0
400	7.2	13.0
600	6.4	11.5
800	8.0	14.4
SAE 1020 Steel	6.7	12.1
Stainless 304	9.6	17.3
Stainless 347	9.2	16.6
90-10 Cu-Ni	9.5	17.1

* At 32 to 212 °F (0 to 100 °C)

be needed with autogeneous welds. Figure 4.15 shows a technique of increasing weld reinforcement with proper joint design and fixturing.

Since the coefficient of thermal expansion of nickel alloys does not differ greatly from that of low-carbon steel (see Table 4.17), similar amounts of clamping pressure will be needed. The clamping pressure should not be excessive, only enough to maintain alignment. Sometimes high clamping pressure is necessary, however, as when the gas tungsten arc welding process is used to make seam welds in thin base metal.

PREHEAT TEMPERATURE

PREHEAT IS NOT usually required or recommended. However, if the base metal is cold, heating to about 60 °F (16 °C) or above avoids condensed moisture that could cause weld porosity. A few specialized applications do use preheat, such as surfacing valve seats using the oxyacetylene welding process. Preheat is not necessarily detrimental, but grain growth will occur if cold-worked base metal is brought above its recrystallization temperature.

In most cases, the interpass and preheat temperatures should be kept low to avoid overheating the base metal. A maximum temperature of 200 °F (90 °C) is recommended for some corrosion-resistant alloys. Cooling methods used to reduce interpass preheat temperature should not inadvertently introduce contaminants. Examples are traces of oil from compressed air lines or mineral deposits from a water spray.

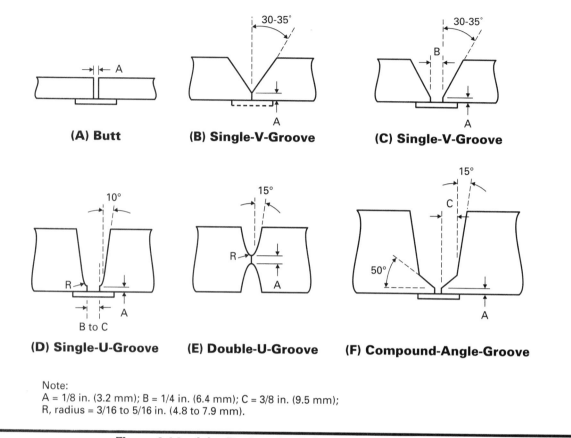

(A) Butt **(B) Single-V-Groove** **(C) Single-V-Groove**

(D) Single-U-Groove **(E) Double-U-Groove** **(F) Compound-Angle-Groove**

Note:
A = 1/8 in. (3.2 mm); B = 1/4 in. (6.4 mm); C = 3/8 in. (9.5 mm);
R, radius = 3/16 to 5/16 in. (4.8 to 7.9 mm).

Figure 4.14—Joint Designs for Submerged Arc Welding

DISSIMILAR METALS

THIS SECTION ADDRESSES exceptional procedural matters applicable to nickel and cobalt alloys that may differ somewhat from other common corrosion-resistant weld metals.[6] The procedure conditions used in obtaining certain results are given. Tabular data also are presented, especially in multi-layer surfacing where the chemical composition at the surface may be critical to service performance.

WELD CLADDING

NICKEL ALLOY WELD metals are readily applied as cladding on carbon steels, low-alloy steels, and other

base metals. All oxide and other foreign material must be removed from the surface to be clad. The procedures and precautions discussed under "Surface Preparation" in this chapter should be carefully followed.

Cracking will sometimes occur in the first layer of nickel-based cladding applied to some steels that contain high levels of sulfur, even though the base metal has been adequately cleaned. When this type of cracking occurs, all of the cracked cladding must be removed, and one thick or two thin layers of carbon-steel surfacing metal applied to the base metal. The nickel-alloy cladding can then be reapplied.

Nickel-alloy cladding can be applied to cast iron, but a trial cladding should be made to determine whether standard procedures can be used. The casting skin, or cast surface, must be removed by a mechanical means such as grinding. Cladding on cast irons with high

6. For a more general treatment of dissimilar metals welding, with an overall compatibility perspective between various metals groups, refer to Chapter 12, "Dissimilar Metals," *Welding Handbook*, 7th Ed., Vol. 4 (or to the forthcoming 8th Ed., Vol. 4).

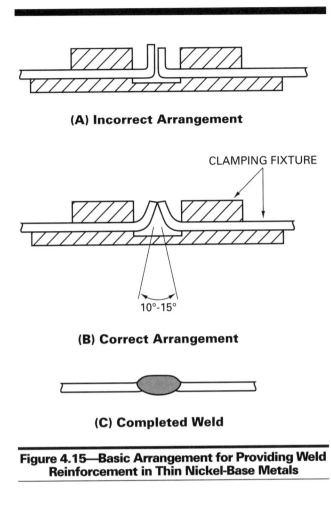

(A) Incorrect Arrangement

CLAMPING FIXTURE

10°-15°

(B) Correct Arrangement

(C) Completed Weld

Figure 4.15—Basic Arrangement for Providing Weld Reinforcement in Thin Nickel-Base Metals

contents of sulfur or phosphorus may crack because of embrittlement by those elements.

Cracking can often be eliminated by the application of a barrier layer of AWS ENiFe-CI welding electrode or AWS ENiFeT3-Ci cored wire. Those filler metals were especially developed for welding cast iron, and the weld metal is more resistant to cracking caused by phosphorus, sulfur, and carbon dilution. When cladding is applied directly to cast iron without a barrier layer, amperage should be the minimum that provides proper arc characteristics in order to hold dilution at the lowest level.

Submerged Arc Cladding

THE SUBMERGED ARC welding process produces high-quality nickel-alloy cladding on carbon steel and low-alloy steel. The process offers several advantages over gas metal arc cladding:

(1) High deposition rates, 35 to 50% increase with 0.062 in. (1.6 mm) diameter surfacing metal, and the ability to use larger electrodes.

(2) Fewer layers are required for a given cladding thickness. For example, with 0.062 in. (1.6 mm) surfacing metal, two layers applied by the submerged arc process have been found to be equivalent to three layers applied by the gas metal arc welding process.

(3) The welding arc is much less affected by minor process variations such as welding wire condition and electrical welding fluctuations.

(4) Welded surfaces of submerged arc cladding are smooth enough to be liquid-penetrant inspected with no special surface preparation other than wire brushing.

(5) Increased control provided by the submerged arc process yields fewer defects and requires fewer repairs.

Typical conditions for submerged arc cladding with various flux and filler metal combinations are shown in Table 4.18.

Chemical compositions of specific weld claddings are given in Table 4.19.

The power supply for all weld cladding applied using weaving techniques is direct current electrode negative (DCEN) with constant voltage. DCEN produces an arc with less depth of fusion, which reduces dilution. Direct current electrode positive (DCEP) results in improved arc stability and is used when stringer-bead cladding is needed to minimize the possibility of slag inclusions.

The most efficient use of the submerged arc process for cladding requires equipment with a means of weaving the electrode. Figure 4.16 shows three basic types of weaving, or oscillation. All three types provide much flatter beads with less iron dilution than is possible with stringer beads.

Pendulum weaving is characterized by a slight hesitation at both sides of the bead. It produces slightly greater depth of fusion and somewhat higher iron dilution at the edge dwell points. Straight-line weaving gives approximately the same results as pendulum weaving.

Straight-line constant velocity provides the lowest level of iron dilution. It provides for movement in a horizontal path so that the arc speed is constant throughout each weave cycle. The optimum movement is that which is programmed to have no end dwell, so that the greater depth of fusion at either side resulting from momentary hesitation is eliminated.

Iron dilution is influenced by weaving, as well as by current, voltage, and forward travel speed. Generally, iron dilution will decrease as weave width is increased, but the molten slag cannot be allowed to solidify

Table 4.18
Typical Conditions for Submerged Arc Cladding on Steel

Flux and Filler Metal Combination	Filler Metal Diameter		Current,[a] A	Voltage, V	Travel Speed		Oscillation Frequency, cycles/min	Oscillation Width		Electrode Extension	
	in.	mm			in./min	mm/min		in.	mm	in.	mm
Flux 4 and											
ERNiCr-3	0.062	1.6	240-260	32-34	3.5-5	89-130	45-70	0.9-1.5	22-38	0.9-1	22-25
	0.093	2.4	300-400	34-37	3-5	76-130	35-50	1-2	25-51	1.1-2	29-51
Flux 5 and											
ERNiCu-7	0.062	1.4	260-280	32-35	3.5-6	89-150	50-70	0.9-1.5	22-38	0.9-1	22-25
	0.093	2.4	300-400	34-37	3-5	76-130	35-50	1-2	25-51	1.1-2	29-51
	0.062	1.6	260-280	32-35	7-9	180-230	(b)	(b)	(b)	0.9-1	22-25
	0.093	2.4	300-350	35-37	8-10	200-250	(b)	(b)	(b)	1.2-1.5	32-38
Flux 6 and											
ERNi-1	0.062	1.6	250-280	30-32	3.5-5	89-130	50-70	0.9-1.5	22-38	0.9-1	22-25
ERNiCr-3	0.062	1.6	240-260	32-34	3-5	76-130	45-70	0.9-1.5	22-38	0.9-1	22-25
	0.093	2.4	300-400	34-37	3-5	76-130	35-50	1-2	25-51	1.1-2	29-51
ERNiCrMo-3	0.062	1.6	240-260	32-34	3.5-5	89-130	50-60	0.9-1.5	22-38	0.9-1	22-25

a. DCEN

b. Non-oscillating technique

Table 4.19
Chemical Composition of Submerged Arc Claddings on Steel, wt. % *

Flux and Filler Metal	Layer	Ni	Fe	Cr	Cu	C	Mn	S	Si	Ti	Nb+Ta	Mo
Flux 4 and												
ERNiCr-3	1	63.5	12.5	17.00	—	0.07`	2.95	0.008	0.40	0.15	3.4	—
	2	70.0	5.3	17.50	—	0.07	3.00	0.008	0.40	0.15	3.5	—
	3	71.5	2.6	18.75	—	0.07	3.05	0.008	0.40	0.15	3.5	—
Flux 5 and												
ERNiCu-7	1	60.6	12.0	—	21.0	0.06	5.00	0.014	0.90	0.45	—	—
	2	64.6	4.55	—	24.0	0.04	5.50	0.015	0.90	0.45	—	—
Flux 6 and												
ERNi-1	2	88.8	8.4	—	—	0.07	0.40	0.004	0.64	1.70	—	—
ERNiCr-3	2	68.6	7.2	18.50	—	0.04	3.00	0.007	0.37	—	2.2	—
Flux 7 and												
ERNiCrMo-3	1	60.2	3.6	21.59	—	0.02	0.74	0.001	0.29	0.13	3.29	8.6

* Cladding on ASTM SA 212 Grade B steel applied by oscillating technique with 0.062 in. (1.6 mm) diameter filler metal.

between weaves. As subsequent beads are made, only enough overlap of the previous bead to yield a smooth surface is required.

Techniques with no weave or oscillation (i.e., stringer beads) are sometimes used in cladding narrow areas. Figure 4.17 shows the correct position of the electrode for cladding by a non-weaving technique with submerged arc welding (SAW).

Gas Metal Arc Cladding

GAS METAL ARC welding with spray transfer is successfully used to apply nickel-alloy cladding to steel. The cladding is usually produced with mechanized equipment and with weaving of the electrode. As shown in Figure 4.18, overlaps should be sufficient to produce a level surface.

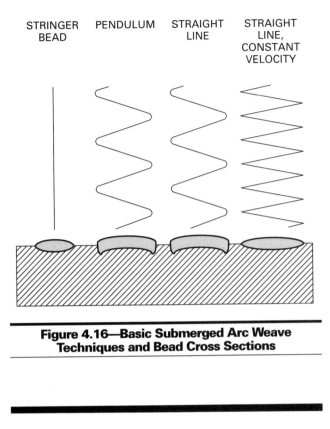

STRINGER BEAD PENDULUM STRAIGHT LINE STRAIGHT LINE, CONSTANT VELOCITY

Figure 4.16—Basic Submerged Arc Weave Techniques and Bead Cross Sections

Courtesy of Inco Alloys International, Inc.

Figure 4.18—Automatic Gas Metal Arc Cladding with Sufficient Overlap for Level Surface

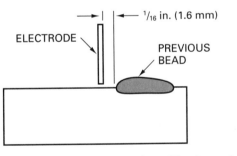

ELECTRODE $1/16$ in. (1.6 mm) PREVIOUS BEAD

Courtesy of Inco Alloys International, Inc.

Figure 4.17—Proper Electrode Position for Cladding by Non-Weaving Technique with SAW

Argon alone is often used as shielding gas. The addition of 15 to 25% helium, however, is beneficial for cladding of nickel and nickel-chromium-iron. Wider and flatter beads and reduced depth of fusion result as the helium content is increased to about 25 percent. Gas-flow rates are influenced by welding technique and will vary from 35 to 100 ft³/h (15 to 45 L/min). As welding current is increased, the weld pool will become larger and require larger gas nozzles for shielding. When weaving is used, a trailing shield may be necessary for adequate shielding. In any case, the nozzle

should be large enough to deliver an adequate quantity of gas under low velocity to the welding area.

Representative chemical compositions of automatic gas metal arc cladding are shown in Table 4.20. The cladding in this table was produced with the following welding conditions:

(1) Torch gas, 50 ft³/h (24 L/min) argon
(2) Trailing shield gas, 50 ft³/h (24 L/min) argon
(3) Electrode extension, 3/4 in. (19 mm)
(4) Power source, DCEP
(5) Oscillation frequency, 70 cycles/min
(6) Bead overlap, 1/4 to 3/8 in. (6 to 10 mm)
(7) Travel speed, 4 1/2 in./min (110 mm/min)

When nickel-copper or copper-nickel cladding is to be applied to steel, a barrier layer of nickel filler metal 61 must be applied first. Nickel weld metal will tolerate greater iron dilution without fissuring.

When cladding is applied manually, the iron content of the first bead will be considerably higher than that of subsequent beads. The first bead should be applied at a reduced travel speed to dissipate much of the penetrating force of the arc in a large weld pool and reduce the iron content of the bead. The iron content of

Table 4.20
Chemical Composition of Gas Metal Arc Cladding on Steel[a]

Surfacing Filler Metal	Current, A	Voltage, V	Layer	Ni	Fe	Cr	Cu	C	Mn	S	Si	Mg	Ti	Al	Nb+Ta
								Chemical Composition of Weld Metal, wt. %							
ERNi-1	280-290	27-28	1	71.6	25.5	—	—	0.12	0.28	0.005	0.32	—	2.08	0.06	—
			2	84.7	12.1	—	—	0.09	0.17	0.006	0.35	—	2.46	0.07	—
			3	94.9	1.7	—	—	0.06	0.09	0.003	0.37	—	2.76	0.08	—
ERNiCu-7[b]	280-300	27-29	2	66.3	7.8	—	19.9	0.06	2.81	0.003	0.84	0.008	2.19	0.05	—
			3	65.5	2.9	—	24.8	0.04	3.51	0.004	0.94	0.006	2.26	0.04	—
ERCuNi[b]	280-290	27-28	2	41.1	11.5	—	45.8	0.04	0.53	0.007	0.14	—	0.83	—	—
			3	35.6	3.1	—	60.1	0.01	0.61	0.006	0.08	—	0.43	—	—
ERNiCr-3	280-300	29-30	1	51.3	28.5	15.8	0.07	0.17	2.35	0.012	0.20	0.017	0.23	0.06	1.74
			2	68.0	8.8	18.9	0.06	0.040	2.67	0.008	0.12	0.015	0.30	0.06	2.27
			3	72.3	2.5	19.7	0.06	0.029	2.78	0.007	0.11	0.020	0.31	0.06	2.38

a. Automatic cladding with 0.062 in. (1.6 mm) diameter filler metal on SA 212 Grade B steel. See text for additional welding conditions.

b. First layer applied with ERNi-1 filler metal.

subsequent beads, as well as the surface contour of the cladding, can be controlled by elimination of weaving and maintaining the arc at the edge of the preceding bead. Such a procedure will result in a 50% overlap of beads, and the weld metal will wet the steel without excessive arc impingement. The welding gun should be inclined up to 5 degrees away from the preceding bead so that the major force of the arc does not impinge on the steel.

Shielded Metal Arc Cladding

SHIELDED METAL ARC cladding on cast and wrought steels are widely used for such applications as facings on vessel outlets and trim on valves. Typical properties are shown in Table 4.21. The electrodes shown are well suited to welding in which the weld metal will be diluted by ferrous base metal.

Table 4.21
Properties of Shielded Metal Arc Cladding on Steel

Surfacing Metal	Electrode	Electrode Diameter		Direct Current, A	Typical Cladding Properties		
		in.	mm		% Elongation in 1 in. (25.4 mm)	Hardness	
						Layer	Rockwell B
Nickel	ENi-1	3/32	2.4	70-105	45	1	88
		1/8	3.2	100-135		2	87
		5/32	4.0	120-175		3	86
		3/16	4.8	170-225			
Nickel-copper	ENiCu-7	3/32	2.4	55-75	43	1	84
		1/8	3.2	75-110		2	86
		5/32	4.0	110-150		3	83
		3/16	4.8	150-190			
Nickel-chromium-iron	ENiCrFe-3	3/32	2.4	40-65	39	1	91
		1/8	3.2	65-95		2	93
		5/32	4.0	95-125		3	92
		3/16	4.8	125-165			

The procedures outlined for shielded metal arc joining should be followed, except that special care must be taken to control dilution of the cladding. Excessive dilution can result in weld metal that is crack sensitive or has reduced corrosion resistance.

The amperage should be in the lower half of the recommended range for the electrode. The major force of the arc should be directed at the edge of the previous bead so that the weld metal will spread onto the steel with only minimum weaving of the electrode. If beads with *feather* edges are applied, more layers will be required, and the potential for excessive dilution will be greater.

The weld interface contour of the cladding should be as smooth as possible. As shown in Figure 4.19, a scalloped weld interface contour can result in excessive iron dilution, with subsequent cracking as the weld specimen is subjected to a 180-degree longitudinal bend test.

Hot-Wire Plasma Arc Cladding

HIGH-QUALITY CLADDING can be produced at high deposition rates with the hot-wire plasma arc process. The process offers precise control of dilution, and dilution rates as low as 2% have been obtained. For optimum uniformity, however, a dilution rate in the 5 to 10% range is recommended.

High deposition rates result from the use of two filler metal wires, which are resistance heated by a separate ac power source. The filler metal is in a nearly molten state before it enters the weld pool. Deposition rates for nickel-alloy weld metal are 35 to 40 lb/h (16 to 18 kg/h), approximately double those obtained with submerged arc weld cladding.

Figure 4.20 shows cross sections and the surface of a hot-wire plasma arc cladding made with ERNiCr-3 filler metal. Side-bend tests showed no cracks.

Welding conditions for hot-wire plasma arc cladding are given in Table 4.22. Chemical compositions of two-layer claddings are given in Table 4.23.

WELDING OF NICKEL ALLOY CLAD STEEL

STEELS CLAD WITH a nickel alloy are frequently joined by welding. Since the cladding is normally used for its corrosion resistance, the cladding alloy must be continuous over the entire surface of the structure, including the welded joints. This requirement influences joint design and welding procedure.

Butt joints should be used when possible. Figure 4.21 shows recommended designs for two thickness ranges [see (A) and (B)]. Both designs include a small root face of unbeveled steel above the cladding to protect the cladding during welding of the steel. The steel side should be welded first with a low-hydrogen filler metal.

(A) Cladding with Scalloped Weld Interface Contour

(B) Bend Test Specimen with Cracks Caused by Improper Weld Interface Contour

(C) Cladding with Smooth Weld Interface Contour

(D) Bend Test Specimen from Cladding Shown Directly Above

Courtesy of Inco Alloys International, Inc.

Figure 4.19—Manual Shielded Metal Arc Cladding

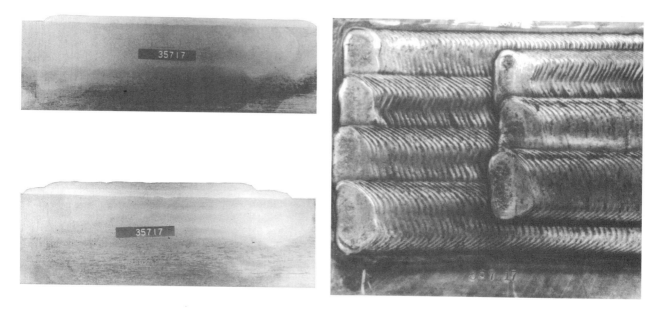

Courtesy of Inco Alloys International, Inc.

Figure 4.20—Cross Sections and Surface of Hot-Wire Plasma Arc Cladding of ERNiCr-3 Filler Metal on Steel

Table 4.22
Typical Conditions for Hot-Wire Plasma Arc Cladding

Characteristic	ERNiCu-7 Filler Metal	ERNiCr-3 Filler Metal
Filler metal diameter	0.062 in. (1.6 mm)	0.062 in. (1.6 mm)
Plasma arc power source	DCEN	DCEN
Plasma arc current	490 A	490 A
Plasma arc voltage	36 V	36 V
Hot-wire power source	AC	AC
Hot-wire current	200 A	175 A
Hot-wire voltage	17 V	24 V
Orifice gas and flow rate	75% He, 25%Ar; 55 ft^3/h (26 L/min)	75% He, 25%Ar; 55 ft^3/h (26 L/min)
Shielding gas and flow rate	Argon; 40 ft^3/h (19 L/min)	Argon; 40 ft^3/h (19 L/min)
Trailing shield gas and flow rate	Argon; 45 ft^3/h (21 L/min)	Argon; 45 ft^3/h (21 L/min)
Standoff distance	13/16 in. (21 mm)	13/16 in. (21 mm)
Travel speed	7-1/2 in./min (190 mm/min)	7-1/2 in./min (190 mm/min)
Weave width	1-1/2 in. (38 mm)	1-1/2 in. (38 mm)
Weave frequency	44 cycles/min	44 cycles/min
Bead width	2 in. (50 mm)	2-3/16 in. (56 mm)
Bead thickness	3/16 in. (5 mm)	3/16 in. (5 mm)
Deposit rate	40 lb/h (18 kg/h)	40 lb/h (18 kg/h)
Preheat temperature	250 °F (120 °C)	250 °F (120 °C)

Table 4.23
Chemical Composition of Hot-Wire Plasma Arc Cladding on Steel, wt. %*

Filler Metal	Layer	Ni	Fe	Cr	Cu	C	Mn	S	Si	Ti	Al	Nb+Ta
ERNiCu-7	1	61.1	5.5	—	27.0	0.07	3.21	0.006	0.86	2.14	0.05	—
	2	63.7	1.5	—	28.2	0.07	3.32	0.006	0.88	2.25	0.04	—
ERNiCr-3	1	68.3	8.3	18.4	0.05	0.02	2.67	0.010	0.16	0.24	—	2.16
	2	73.2	1.7	20.2	0.02	0.01	2.86	0.010	0.17	0.24	—	2.31

* Cladding on ASTM A387 Grade B steel made with 0.062 in. (1.6 mm) diameter filler metal.

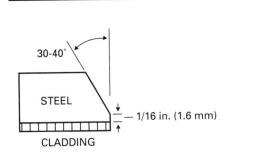

(A) Material 3/16 to 5/8 (4.8 to 16 mm) Thick

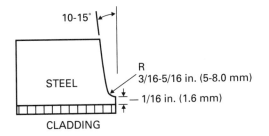

(B) Material 5.8 to 1 in. (16 to 25 mm) Thick

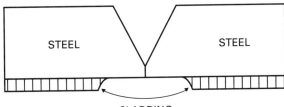

(C) Strip-Back Method of Joint Preparation

Figure 4.21—Joint Designs for Clad Steel

It is important to avoid fusion of the cladding during the first welding pass. Dilution of the steel weld with the nickel-alloy cladding can cause cracking of the weld metal. The clad side of the joint should be prepared by grinding or chipping and welded with the filler metal recommended for the alloy used for cladding. The weld metal will be diluted with steel. To maintain corrosion resistance, at least two layers, and preferably three or more, should be applied.

The strip-back method is sometimes used instead of the procedure described above. The cladding is removed from the vicinity of the joint as shown in Figure 4.21(C). The steel is then welded using a standard joint design and technique for steel, and the nickel-alloy cladding is reapplied by weld cladding. The advantage of the strip-back method is that it eliminates the possibility of cracking caused by penetration of the steel weld metal into the cladding.

Some joints, such as those in closed vessels or tubular products, are accessible only from the steel side. In such cases, a standard joint design for steel is used, and the cladding at the bottom of the joint is welded first with nickel alloy weld metal. After the cladding is welded, the joint can be completed with the appropriate nickel alloy weld metal, or a barrier layer of carbon-free iron can be applied and the joint completed with steel weld metal. If the thickness of the steel is 5/16 in. (8 mm) or less, it is usually more economical to complete the joint with nickel alloy welding filler metal. Figures 4.22A and B show fabrication sequences when both sides are accessible. Figure 4.23 shows the fabrication sequence when only one side is accessible. Figure 4.24 shows some designs other than butt joints for low-stressed welds.

NICKEL ALLOY STRIP LININGS

THE LINING OF vessels and the cladding of external surfaces with sheets or strips is accomplished by progressively fillet welding to the steel and then to

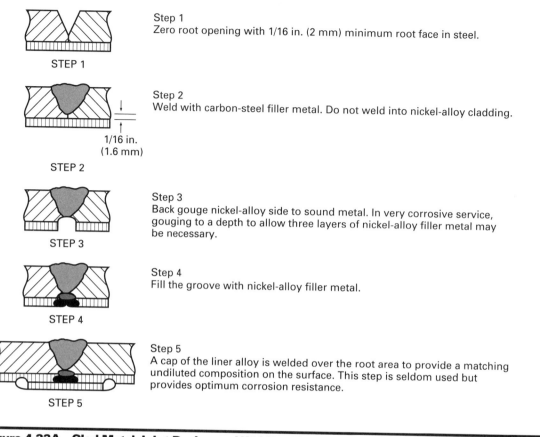

Step 1
Zero root opening with 1/16 in. (2 mm) minimum root face in steel.

STEP 1

Step 2
Weld with carbon-steel filler metal. Do not weld into nickel-alloy cladding.

1/16 in.
(1.6 mm)

STEP 2

Step 3
Back gouge nickel-alloy side to sound metal. In very corrosive service, gouging to a depth to allow three layers of nickel-alloy filler metal may be necessary.

STEP 3

Step 4
Fill the groove with nickel-alloy filler metal.

STEP 4

Step 5
A cap of the liner alloy is welded over the root area to provide a matching undiluted composition on the surface. This step is seldom used but provides optimum corrosion resistance.

STEP 5

Figure 4.22A—Clad Metal Joint Design and Welding Fabrication Sequence When Both Sides of Joint are Accessible

neighboring alloy strips or sheets. Successful service performance of these linings requires static conditions of temperature and pressure. When temperature or pressure fluctuates widely in service, fabrication by strip lining has been the cause of numerous service failures over the years. When linings are selected for fabrication, the following rules govern:

(1) The steel should be dried and free from corrosion products (rust, mill scale, etc.), preferably by grit blasting, and the surface should be ground along the line of the proposed welds.

(2) Strips should be at least 0.062 in. (2 mm) thick, cut to proper size, and formed to the required shape. The work should be planned so that long, continuous weld lengths are avoided and, as far as possible, attachment to existing joints in the steel structure is prevented. The edges and adjacent surfaces of the strips are cleaned abrasively and degreased before pressing them firmly into position and tack welding to the steel.

(3) Welding by shielded metal arc welding (SMAW) process is normal, though the gas tungsten arc welding (GTAW) and gas metal arc welding (GMAW) processes have been used. The choice of filler metal is governed by the nature of the environment to be encountered.

Some currently used joint designs are shown in Figure 4.25. These linings are used more extensively in Europe than in the United States.[7]

Figure 4.26 sets forth three procedures for welding strip linings to carbon-steel base metal. Shingle joints (A) are the least expensive, fastest to apply, and most tolerant to fit up. The second exposed weld obviously has no opportunity for iron dilution. Shingle joints have the disadvantage of the possibility of a direct leak path,

7. Refer to "Lining of vessels and equipment for chemical processes," *British Standard Practice*, CP 3003, Part 7, 1970, and to *Welding Inco Alloys*, Hereford, UK: Inco Alloys Ltd.

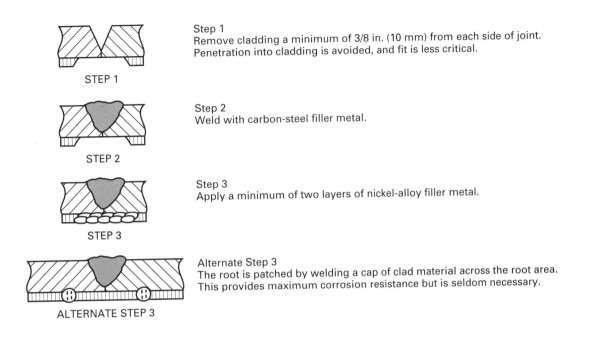

Step 1
Remove cladding a minimum of 3/8 in. (10 mm) from each side of joint. Penetration into cladding is avoided, and fit is less critical.

STEP 1

Step 2
Weld with carbon-steel filler metal.

STEP 2

Step 3
Apply a minimum of two layers of nickel-alloy filler metal.

STEP 3

Alternate Step 3
The root is patched by welding a cap of clad material across the root area. This provides maximum corrosion resistance but is seldom necessary.

ALTERNATE STEP 3

Figure 4.22B—Strip-Back Clad Metal Joint Design and Welding Fabrication Sequence When Both Sides of Joint are Accessible (Alternate Method)

which can allow contents to spread widely under the lining.

The two-bead method (B) also offers low iron dilution but requires considerable welding skill. The second bead does not penetrate the steel, which accounts for less iron pick-up. The three-bead method (C) offers the advantages of ease of fitting and multi-operator capability but requires more welding and has a greater tendency to iron dilution, especially in the first layer. Any leak path in it would involve two layers of weld, and any pin-hole in the first layer is capped off. If a pin-hole is in only the second layer, no leak to steel exists.

If necessary, attachment of nickel alloy sheets may be reinforced by spaced GMAW spot welds or by SMAW and GMAW plug welds. Caps of alloy sheet can be used to cover the plug welds to ensure full corrosion protection. Crevices are inherent in such joints; thus, mechanical cycling of such joints is worthy of close scrutiny by the equipment designer. For example, blow down of paper digester reactor vessels has caused strip linings welded as close as 6 in. (150 mm) apart to detach from the vessel wall. Today, the general rule is to use alloy clad plate if pressure or temperature fluctuates appreciably in service.

OTHER DISSIMILAR METAL APPLICATIONS

ABRASION, WEAR, AND corrosion-resistant hard surfacing alloys are routinely applied to nickel and cobalt alloys using the oxyacetylene, gas tungsten arc, and plasma arc processes, especially in valve seats and other applications where a combination of wear and corrosion resistance is demanded.[8]

Nickel-based welding filler metal also is used extensively in the welding of cast irons.[9,10]

8. For information on hardfacing, refer to Chapter 14, "Surfacing," *Welding Handbook*, 7th Ed., Vol. 2 (or to the forthcoming 8th Ed., Vol. 4).
9. For information on cast iron welding, refer to Chapter 5, "Cast Iron," *Welding Handbook*, 7th Ed., Vol. 4 (or to the forthcoming 8th Ed., Vol. 4).
10. For general information on the selection of filler metals and procedures for joining dissimilar metals, refer to Chapter 12, "Dissimilar Metals," *Welding Handbook*, 7th Ed., Vol. 4 (or to the forthcoming 8th Ed., Vol. 4).

1. Prepare joint to leave cladding protruding. Gas tungsten arc weld with appropriate nickel-base filler metal using argon backing.

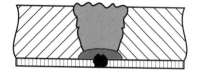

2. Deposit nickel layer and complete joint with appropriate nickel-base weld metal. If backing steel is heavy, a buffer layer of ingot-iron weld metal may allow completion of the joint using carbon-steel welding materials.

(A) Access from Steel Side

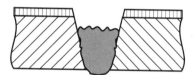

1. Partly fill joint with steel weld metal.

2. Deposit minimum of two layers of appropriate nickel-base filler metal.

(B) Access from Clad Side

Figure 4.23—Clad Metal Joint Design and Welding Fabrication Sequences When Only One Side of Joint is Accessible

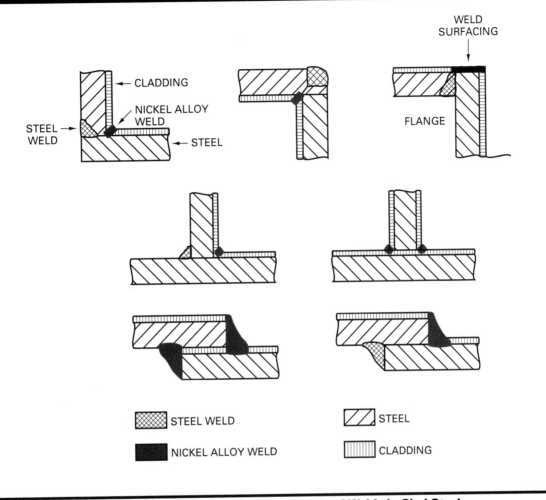

Figure 4.24—Joint Designs for Low-Stressed Welds in Clad Steel

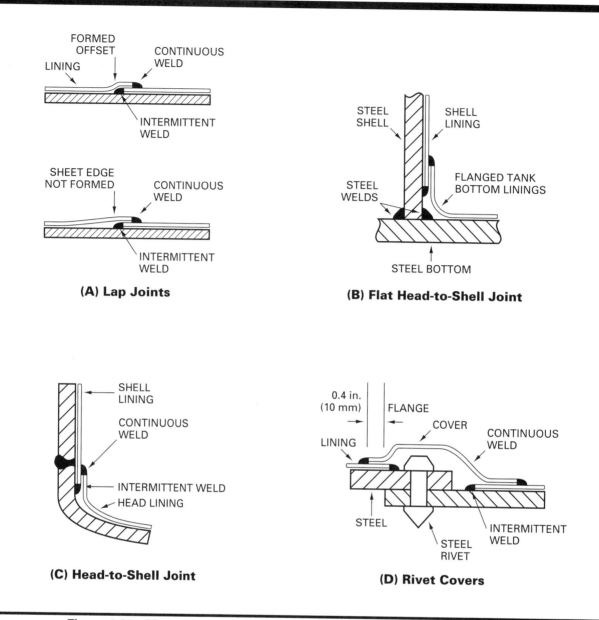

(A) Lap Joints

(B) Flat Head-to-Shell Joint

(C) Head-to-Shell Joint

(D) Rivet Covers

Figure 4.25—Methods and Designs for Welding Strip to Carbon-Steel Base Metal

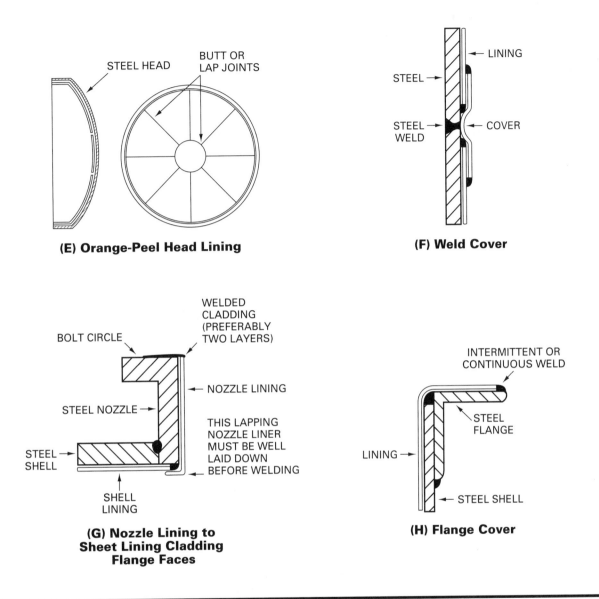

(E) Orange-Peel Head Lining

(F) Weld Cover

(G) Nozzle Lining to Sheet Lining Cladding Flange Faces

(H) Flange Cover

Figure 4.25 (continued)—Methods and Designs for Welding Strip to Carbon-Steel Base Metal

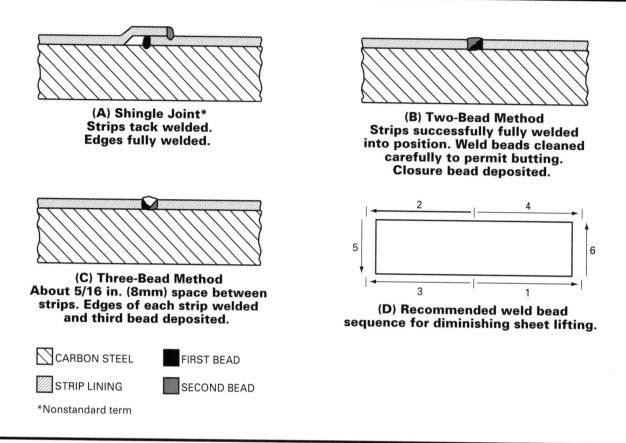

(A) Shingle Joint*
Strips tack welded.
Edges fully welded.

(B) Two-Bead Method
Strips successfully fully welded
into position. Weld beads cleaned
carefully to permit butting.
Closure bead deposited.

(C) Three-Bead Method
About 5/16 in. (8mm) space between
strips. Edges of each strip welded
and third bead deposited.

(D) Recommended weld bead
sequence for diminishing sheet lifting.

◫ CARBON STEEL   ■ FIRST BEAD

▨ STRIP LINING   ▨ SECOND BEAD

*Nonstandard term

Figure 4.26—Procedures for Welding Strip Linings to Carbon-Steel Base Metal

OXYACETYLENE WELDING

THE OXYACETYLENE FLAME produces a sufficiently high temperature for welding commercially pure nickel and some solution-strengthened nickel alloys. However, this process should only be used in those situations where suitable arc welding equipment is not available. Welds can be made in all positions.

The process is still widely used in applying abrasion and wear-resisting alloys to nickel and cobalt alloys.[11]

FLAME ADJUSTMENT

THE TORCH TIP should be large enough to provide a low-velocity, soft flame. A high-velocity, harsh flame is undesirable. Usually the tip should be the same size or one size larger than the one recommended for the same thickness of steel.

The welding torch should be adjusted with excess acetylene to produce a slightly reducing flame. When chromium-bearing alloys are welded, the flame should not be excessively reducing because the weld metal might absorb carbon.

During welding, the weld pool should be kept quiet with the cone of the flame just touching its surface. Agitation of the molten metal should be avoided. This action can result in loss of the deoxidizing elements or exposure of the molten weld metal to the surrounding atmosphere. As a result, the weld metal may be porous.

The hot end of the filler metal rod should be kept within the flame envelope to minimize oxidation.

11. See Chapter 14, "Surfacing," *Welding Handbook*, 7th Ed., Vol. 2 (or the forthcoming 8th Ed., Vol. 4).

FLUXES

FLUX IS REQUIRED for welding nickel-copper, nickel-chromium, and nickel-iron-chromium alloys. Commercially pure nickel can be welded without flux. The following mixture can be used for solution-strengthened nickel-copper alloys: barium fluoride, 60%; calcium fluoride, 16%; barium chloride, 15%; gum arabic, 5%; and sodium fluoride, 4%.

Nickel-chromium and nickel-iron-chromium alloys can be fluxed with a mixture of one part sodium fluoride and two parts calcium fluoride with 3% hematite (red iron oxide) and a suitable wetting agent added to the mixture. Such agents are Photoflo by Kodak, Dupanol*ME by DuPont, or commercial detergents.

Precipitation-hardenable nickel-copper alloys can be fluxed with a water slurry of one part lithium fluoride and two parts of either of the two previously given flux compositions.

The flux is mixed with water to produce a thin slurry. It should be applied to both sides of the joint and to the filler metal rod, and then allowed to dry before welding is started.

Borax must not be used as a flux when welding nickel alloys. It can result in the undesirable formation of a brittle, low-melting eutectic in the weld.

The flux residue must be removed from the joint for high-temperature service. Molten flux will corrode the base metal after an extended period. Unfused flux may be washed off with hot water. Fused flux is not soluble in water and must be removed mechanically by grit blasting or grinding.

NICKEL ALLOYS

THE OXYACETYLENE WELDING process is not recommended for joining the low-carbon nickel and nickel alloys, the nickel-molybdenum alloys, and the nickel-chromium-molybdenum alloys. These base metals can readily absorb carbon from the flame, and this will reduce their corrosion resistance and high-temperature properties.

Oxyacetylene welding is the only recommended joining process for the high-silicon casting alloys. The welding procedure is similar to that used for cast iron. The filler metal rod should be the same composition as the base metal. A U-groove with a 45-degree bevel angle should be used for base metal over 0.5 in. (13 mm) thick.

With the exception of the nickel-copper alloy, the precipitation-hardenable alloys should not normally be welded by the oxyacetylene process. The hardening elements are easily oxidized and fluxed away during welding.

COBALT ALLOYS

THE WROUGHT COBALT alloys are usually low in carbon. As with low-carbon nickel alloys, these alloys can absorb carbon during welding. This will alter the properties of the alloys. However, cobalt hardfacing alloys are high in carbon, and they are often deposited with the oxyacetylene process.

RESISTANCE WELDING

NICKEL AND COBALT alloys are readily welded to themselves and to other metallurgically compatible metals by spot, seam, projection, and flash welding. The electrical resistivities of the alloys range from about 57 Ω•cir mil/ft (9.5 $\mu\Omega$•cm) for commercially pure nickel to 776 Ω•cir mil/ft (129 $\mu\Omega$•cm) for a resistance-heating nickel alloy. The welding current requirements are lower with the high-resistance alloys, but the force requirements increase because of their high strengths at elevated temperatures.

For good electrical contact, the surfaces of the joint members must be clean. All oxide, oil, grease, and other foreign matter must be removed by acceptable cleaning methods. Chemical pickling is the best method of oxide removal.

SPOT WELDING

NICKEL AND COBALT alloys are spot welded in much the same manner as other metals. In many respects, these alloys are easy to spot weld. The configurations involved and the relatively short welding time tend to preclude any contamination from the atmosphere. As a result, auxiliary shielding is not normally needed during resistance spot welding.

The thermal and electrical conductivities and the mechanical properties of the alloys vary depending upon their composition and condition. Conditions for spot welding, therefore, are adjusted to account for the base metal properties. Usually, several combinations of welding conditions can produce similar and acceptable results.

EQUIPMENT

NICKEL AND COBALT alloys can be welded successfully on almost all types of conventional spot welding equipment. The equipment must provide accurate control of welding current, weld time, and electrode force. Each of these values may vary within a range without appreciably affecting weld quality. It is, however, desirable to have sufficient control over them to obtain reproducible results after the optimum values are obtained for a given application.

Upslope controls will help prevent expulsion. Dual electrode force systems are sometimes used to provide a high forging force when welding high-strength, high-temperature alloys. No significant changes in welding characteristics or static weld properties can be attributed to the use of any specific type of spot welding equipment.

For most applications, the restricted dome electrode design shown in Figure 4.27 is preferred. Truncated electrodes or ones with 5 or 8 in. (125 or 200 mm) radius faces are sometimes used for metal thicknesses in the range of 0.06 to 0.13 in. (1.5 to 3 mm). Larger nuggets and correspondingly higher shear strengths can be obtained with these electrode shapes.

Resistance Welder Manufacturers Association (RWMA) Group A, Class 1, 2, and 3 electrode alloys are recommended. Class 1 alloys are best for low-resistivity alloys and for thin sheets to minimize sticking tendencies. Class 3 electrodes are recommended for high-temperature alloys to minimize mushrooming with high electrode forces and relatively long weld times.

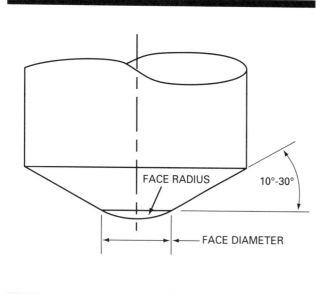

Figure 4.27—Restricted-Dome Spot Weld Electrode Design

WELDING CONDITIONS

SPOT WELDING CONDITIONS are primarily controlled by the total thickness of the joint members and, to a rather large degree, by the welding machine being used. Similar welding conditions may be suitable for making welds in the same total thickness where the number of layers differs significantly. However, for any given thickness or total accumulation, various combinations of welding current, time, and electrode force may produce similar welds.

Other conditions such as electrode size and shape are important in controlling such characteristics as metal expulsion, sheet indentation, and sheet separation. Upslope, downslope, and forging force may be used for some high-strength, high-temperature alloys to control heating rate and soundness of the weld nugget.

Alloys that have properties similar to steel behave in a like manner. The high-nickel and high-cobalt alloys generally are harder and stronger than low-carbon steel, particularly at elevated temperatures. Higher electrode forces are therefore required during spot welding. The time of current flow should be as short as possible, but sufficiently long to gradually build up the welding heat. Welding current should be set somewhat above the value that produces a weak weld, but below the setting that causes weld-metal expulsion. Current upslope is an asset in this respect.

Suitable conditions for spot welding annealed nickel alloys are given in Tables 4.24 and 4.25. A forging force is sometimes applied near the end of weld time to consolidate the weld nugget. For some alloys, forging force may be applied during a postheating impulse.

Precipitation-hardenable alloys are best welded in the solution annealed condition with settings similar to those used for similar solid-solution alloys. However, high electrode force and low welding current must be used to compensate for their high-temperature strength and high electrical resistance. The weld nuggets will be about the same size as those in the solid-solution alloys, but the shear strengths will be higher in the welded condition. Subsequent precipitation hardening will increase the shear strength by about 50 percent. Cracking will generally occur when these alloys are welded in the hardened condition. Postweld solution annealing, followed by precipitation hardening, is recommended to avoid strain-age cracking.

Some cracking may occur during spot welding of some precipitation-hardenable alloys if insufficient electrode force is used. If higher electrode force does not overcome cracking, increasing the weld time or lowering the welding current will help. Welding machines with low-inertia heads and current slope control are preferred.

Table 4.24
Typical Conditions for Spot Welding Annealed Nickel Alloys with Single-Phase Machines

Thickness[a]		Electrode Face[b] Radius		Diameter		Electrode Force		Weld Time, cycles (60Hz)	Welding Current, kA	Nugget Diameter		Minimum Shear Strength	
in.	mm	in.	mm	in.	mm	lb	N			in.	mm	lb	N
Nickel (UNS N02200)													
0.021	0.533	3	76	0.16	4.06	370	1646	4	78.8	0.12	3.05	350	1557
0.031	0.787	3	76	0.19	4.83	900	4003	4	15.4	0.18	4.57	760	3381
0.063	1.600	3	76	0.25	6.35	1720	7651	6	21.6	0.25	6.35	2400	10 676
0.094	2.387	3	76	0.31	7.87	2300	10 231	12	26.4	0.31	7.87	3600	16 014
0.125	3.175	3	76	0.38	9.65	3300	14 679	20	31.0	0.37	9.40	5600	24 910
67Ni-32Cu alloy (UNS N04400)													
0.021	0.533	3	76	0.19	4.83	300	1334	12	6.2	0.13	3.30	450	2002
0.031	0.787	3	76	0.19	4.83	700	3114	12	10.5	0.17	4.32	845	3759
0.063	1.600	3	76	0.31	7.87	2700	12 010	12	15.3	0.31	7.87	2060	9163
0.093	2.362	3	76	0.38	9.65	2760	12 277	20	22.6	0.37	9.40	3880	17 259
0.125	3.175	3	76	0.50	12.7	5000	22 241	30	30.0	0.47	11.94	5850	26 022
76Ni-16Cr-8 Fe alloy (UNS N06600)													
0.021	0.533	3	76	0.16	4.06	300	1334	12	4.0	0.12	3.05	545	2424
0.031	0.787	3	76	0.19	4.83	700	3114	12	6.7	0.18	4.57	920	4092
0.063	1.600	3	76	0.31	7.87	2070	9207	12	12.0	0.31	7.87	2750	12 233
0.093	2.362	3	76	0.38	9.65	3870	17 214	20	15.0	0.37	9.40	4400	19 572
0.125	3.175	3	76	0.44	11.18	5270	23 441	30	20.1	0.44	11.18	6400	28 468
73Ni-16Cr-7 Fe-3Ti alloy (UNS N07750)													
0.010	0.254	6	152	0.16	4.06	300	1334	2	7.3	0.11	2.79		
0.015	0.381	6	152	0.16	4.06	400	1779	4	7.4	0.11	2.79		
0.021	0.533	6	152	0.19	4.83	750	3336	6	7.5	0.14	3.56		
0.031	0.787	6	152	0.22	5.59	1750	7784	8	9.9	0.17	4.32		
0.062	1.574	10	254	0.31	7.87	4400	19 571	14	16.4	0.29	7.37		

a. Two equal thicknesses

b. Restricted dome electrode design

Table 4.25
Typical Conditions for Spot Welding Annealed Nickel Alloys with Three-Phase Frequency Converter Machine

| Thickness[a] | | Electrode Face[b] | | | | Electrode Force | | Time, cycles (60 Hz) | | | Welding Current, | Nugget Diameter | | Minimum Shear Strength | |
| | | Radius | | Diameter | | | | Weld | Pulse | Interpulse | kA | | | | |
in.	mm	in.	mm	in.	mm	lb	N					in.	mm	lb	N
						67Ni-32Cu alloy (UNS N04400)									
0.018	0.457	3	76	0.19	4.83	400	1779	13	6	1	4.3	0.17	4.32	400	1779
0.030	0.762	5	127	0.25	6.35	800	3559	13	6	1	8.5	0.18	4.57	900	4003
0.043	1.092	5	127	0.25	6.35	1600	7117	17	8	1	11.5	0.26	6.60	1750	7784
0.062	1.574	7	178	0.31	7.87	2200	9786	21	10	1	14.5	0.32	8.13	2050	9119
0.093	2.362	9	229	0.44	11.18	3800	16 903	39	9	1	22.5	0.40	10.16	5400	24 020
0.125	3.175	12	305	0.50	12.7	5000	22 241	65	10	1	31.0	0.48	12.19	7000	31 138
						73Ni-16Cr-7Fe-3Ti alloy (UNS N07750)									
0.025	0.635	3	76	0.22	5.59	2000	8896	9	8	1	6.0	0.16	4.06	900	4003
0.031	0.787	5	127	0.25	6.35	2200	9786	10	9	1	6.8	0.18	4.57	1150	5115
0.043	1.092	5	127	0.25	6.35	2700	12 010	23	5	1	8.1	0.20	5.08	1800	8007
0.062	1.575	8	203	0.31	7.87	3500	15 569	35	8	1	11.4	0.25	6.35	3300	14 679
0.093	2.362	8	203	0.44	11.18	5000	22 241	53	8	1	15.0	0.37	9.40	5700	25 355

a. Maximum thickness of multiple layers should not exceed four times this thickness. Maximum ratio for unequal thicknesses is 3 to 1.

b. Restricted dome electrode design

SEAM WELDING

THIS PROCESS IS normally used to join sheet thicknesses ranging from 0.002 to 0.125 in. (0.05 to 3.2 mm). Wheel electrodes of Resistance Welder Manufacturers Association (RWMA) Group A, Class 1 or 2 alloy are used. Individual overlapping spots are created by coordinating the welding time and wheel rotation. The wheel electrodes can be rotated continuously or intermittently. Continuous seam welding imposes limitations on the weld cycle variations that can be used. For example, a forging force cannot be applied during continuous seam welding, but it can be used with intermittent motion. High-strength alloys, such as UNS N06002 and N07750, usually are welded with forging force and intermittent drive.

Suggested conditions for seam welding two nickel alloys with continuous motion are given in Table 4.26. Table 4.27 gives conditions for welding a different nickel alloy using intermittent motion and forging force. The force must be high to consolidate the weld nugget to prevent cracking and porosity.

PROJECTION WELDING

THIS PROCESS REQUIRES die-formed projections, similar to those used for steel, in one or both parts. The parts are also usually die-formed in high-production operations. Nickel and cobalt alloys are seldom projection welded because production requirements are normally low.

Conditions for projection welding of nickel and cobalt alloys are influenced by the thickness and shape of the joint members. Generally, higher electrode forces and longer weld times than those used for steels are required because of the higher strength of nickel and cobalt alloys at elevated temperature.

FLASH WELDING

THE NICKEL AND cobalt alloys can be flash-welded to themselves and to dissimilar metals. In two respects, flash welding is well adapted to the high-strength, heat-treatable alloys. First, molten metal is not retained in the joint. Second, the hot metal at the joint is upset. This upsetting may improve the ductility of the heat-affected zone.

The machine capacity required to weld nickel and nickel alloys does not differ greatly from that required for steel. This is especially true for transformer capacity. The upset force needed for making flash welds in nickel alloys is higher than that required for steel.

Joint designs for flash welds are similar to those used for other metals. Flat, sheared, or saw-cut edges and pinch-cut rod or wire ends are satisfactory for welding. The edges of thick base metal are sometimes beveled slightly. The overall shortening of the parts due to metal lost during welding should be taken into account so that finished parts will be the proper length.

The flash welding conditions that are of greatest importance are flashing current, speed, and time, plus upset pressure and distance. With proper control of these conditions, the molten metal at the interface will be forced out of the joint, and the metal at the interface will be at the proper temperature for welding.

Generally, high flashing speeds and short flashing times are used to minimize weld contamination. Parabolic flashing is more desirable than linear flashing because maximum joint efficiency can be obtained with a minimum of metal loss.

Flash welding conditions vary with the machine size and the application. Table 4.28 gives typical conditions for flash welding 0.25 and 0.375 in. (6.3 and 9.5 mm) diameter rod of several nickel alloys.

Welding current is determined by the machine transformer tap setting. Since these alloys have higher strengths at elevated temperatures than steel, higher force is required to upset and extrude all the molten metal from the joint. The time of current flow during upsetting is critical. If the time is too long, the joint may be overheated and oxidized. If the time is too short, the plasticity of the metal will not permit sufficient upsetting to force the molten metal from the joint. A properly made flash weld will not contain any cast metal.

Table 4.26

Typical Conditions for Seam Welding Annealed Nickel Alloys with Single-Phase Machine with Continuous Motion

Thickness*		Electrode Face Width		Radius		Electrode Force		Time, cycles (60 Hz)		Welding Current, kA	Welding Speed		Width of Nugget	
in.	mm	in.	mm	in.	mm	lb	N	Heat	Cool		in./min	mm/s	in.	mm
67Ni-32Cu alloy (UNS N04400)														
0.010	0.254	0.16	4.06	3	76	200	890	1	3	5.3	75	31.7	0.09	2.29
0.015	0.396	0.16	4.06	6	152	300	1334	1	3	7.6	75	31.7	0.10	2.54
0.021	0.533	0.19	4.83	6	152	500	2224	2	6	8.7	38	16.1	0.15	3.81
0.031	0.787	0.19	4.83	6	152	700	3114	4	12	10.0	19	8.0	0.15	3.81
0.062	1.574	0.38	9.65	6	152	2500	11 120	8	12	19.0	20	8.5	0.17	4.32
73Ni-16Cr-7Fe-3Ti alloy (UNS N07750)														
0.010	0.254	0.13	3.30	3	76	400	1779	1	3	3.6	45	19.0	0.11	2.79
0.015	0.396	0.13	3.30	3	76	700	3114	2	4	3.9	36	15.2	0.12	3.05
0.021	0.533	0.16	4.06	3	76	1400	6227	3	6	8.0	30	12.7	0.14	3.56
0.031	0.787	0.19	4.83	3	76	2300	10 230	4	8	8.5	30	12.7	0.17	4.32
0.062	1.574	0.19	4.83	6	152	4000	17 792	8	16	10.3	12	5.1	0.18	4.57

* Maximum thickness of multiple layers should not exceed four times this thickness. Maximum ratio for unequal thicknesses is 3 to 1.

Table 4.27

Seam Welding of 47Ni-22Cr-18Fe-9Mo Alloy (UNS N06002) with Intermittent Motion and Forging Force

Thickness[a]		Electrode[b] Face Width		Electrode Force				Weld Times, cycles (60 Hz)			Forge time, cycles (60 Hz)	Welding Current, kA	Spots per in. (25.4 mm)
				Weld		Forge							
in.	mm	in.	mm	lb	N	lb	N	Heat	Cool	Total			
0.030	0.762	0.19	4.83	1500	6672	—	—	10	—	10	—	20.5	14
0.063	1.600	0.31	7.87	2000	8896	4000	17 792	10	2	94	15	21.5	10
0.094	2.387	0.38	9.65	4500	20 016	4500	20 016	10	2	46	25	33.0	8

a. Two equal thicknesses

b. RWMA Class III copper alloy wheel, 12 in. (305 mm) diameter, flat face, 15° double bevel

Table 4.28
Typical Conditions for Flash Welding Nickel Alloy Rods[a]

Metal	UNS Number	Diameter[b]		Upset Current Time, s	Upset		Energy Input		Joint Efficiency, %
		in.	mm		in.	mm	W·h	J	
Nickel	N02200	0.25	6.35	1.5	0.125	3.17	2.15	7740	89
		0.375	9.52	2.5	0.145	3.68	4.87	17 530	98
67Ni-32Cu	N04400	0.25	6.35	1.5	0.125	3.17	1.93	6950	97
		0.375	9.52	2.5	0.145	3.68	5.55	19 980	95
66Ni-30Cu-3Al	N05500	0.25	6.35	1.5	0.125	3.17	2.02	7270	94
		0.375	9.52	2.5	0.145	3.68	4.79	17 240	100
76Ni-16Cr-8Fe	N06600	0.25	6.35	1.5	0.125	3.17	2.15	7740	92
		0.375	9.52	2.5	0.145	3.68	5.19	18 680	96

a. Flash-off equals 0.442 in. (11.2 mm) and flashing time equals 25 seconds in all cases.

b. Ends tapered with 110° included angle

ELECTRON BEAM WELDING

ALL NICKEL AND cobalt alloys that can be successfully joined by conventional arc welding processes can also be electron beam welded. Because of its lower heat input, this process may be suitable for joining some alloys that are considered difficult to arc weld. In general, the joint efficiencies of electron beam welds will equal or exceed those of gas tungsten arc welds.

Welding in vacuum provides excellent protection against atmospheric contamination. Porosity may be a problem when welding some alloys at high welding speeds because dissolved gases do not have time to escape to the surface of the weld pool. Weaving of the beam to slightly agitate the weld pool may help the gases to escape and thus reduce porosity.

Hot cracking in the heat-affected zone may be a problem, particularly when welding thick base metal of some alloys. The use of a cosmetic (second) pass to provide a good face contour may aggravate the problem because of the high restraint imposed by the first weld pass. In this case, cracking will likely take place in the heat-affected zone of the cosmetic bead.

An example of this type of cracking is shown in Figure 4.28. The high degree of restraint imposed by the thick base metal on the narrow heat-affected zone can produce a high tensile stress during cooling. Microcracking can readily take place, particularly if the alloy tends to be hot short. An obvious solution to this problem is to use a welding procedure that does not require a cosmetic pass. Filler metal addition and subsequent mechanical finishing might be appropriate.

As with arc welding, the base metal should be in the annealed condition for welding. With precipitation-hardening alloys, the weldment should be solution treated and aged for optimum strength properties. However, distortion of the weldment must be considered when specifying a heat treatment.

Welding conditions used for electron beam welding depend upon the base metal composition and thickness, as well as the type of welding equipment. For a given thickness of base metal, various combinations of accelerating voltage, beam current, and travel speed may produce satisfactory welds.

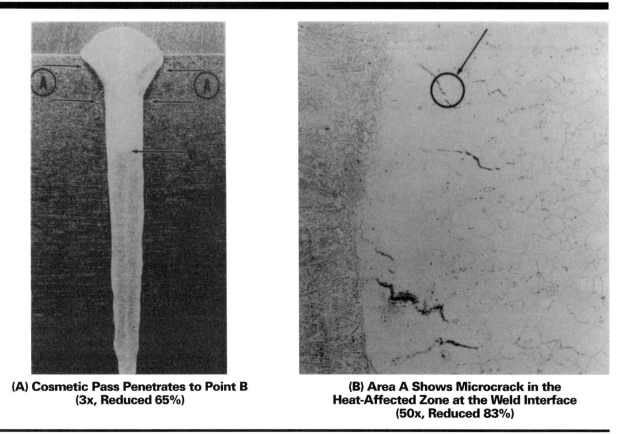

(A) Cosmetic Pass Penetrates to Point B
(3x, Reduced 65%)

(B) Area A Shows Microcrack in the
Heat-Affected Zone at the Weld Interface
(50x, Reduced 83%)

Figure 4.28—Photomacrographs of an Electron Beam Weld in 1.75 in. (45 mm) Thick
Ni-19Cr-19Fe-5Nb-3Mo Alloy (UNS N07718)

LASER BEAM WELDING

THIS PROCESS IS being evaluated for welding various nickel alloys. Welding is normally done in the open, and a gas shield is recommended to protect the weld area from oxidation. Argon or helium is suitable.

Several nickel alloys have been welded with various laser beam systems in thicknesses ranging from 0.01 to 0.38 in. (0.3 to 10 mm). The welded joint cross sections are similar to those produced by an electron beam. The metallurgical behavior of laser beam welds should be similar to that of electron beam welds. Several nickel alloys in sheet thicknesses of 0.039 and 0.079 in. (1 to 2 mm) have been successfully welded using a CO_2 laser beam with a power of up to 2.0 kW. Welding speeds of 80 and 15 in./min (2.0 and 0.38 m/min), respectively, were used.

FABRICATION FOR HIGH-TEMPERATURE SERVICE

NICKEL ALLOYS ARE used extensively in equipment for high-temperature service. The demands of this service bring out the inherent weaknesses in welds; namely, welds are castings possessing a coarse dendritic structure. Welds possess inferior hot-fatigue resistance, and long-term stress-rupture properties of welds generally are lower than for the relatively uniform grain base metals found in wrought metal forms (sheet, plate, pipe, forgings, etc.). Figure 4.1 is a classic portrayal of how such deficiencies can be handled by weld metal selection at specific temperatures, though the weld metal composition may vary radically from the base-metal composition. A knowledge of the service environment is also necessary.

DESIGN FACTORS

THE DESIGNER OF equipment intended for high-temperature service should carefully select a weld metal composition that overmatches the base metal composition in stress-rupture strength, as is illustrated in Figure 4.1. Additionally, the designer should be aware that the as-cast weld metal should be protected from the effects of high stresses by locating welds in areas of minimal stresses. For example, a horizontally positioned pipe having a longitudinal weld should be positioned to locate the weld at the top rather than at the bottom. The top location places the weld in an area of compression as the pipe sags when exposed to high temperature.

To minimize the effects of thermal and mechanical fatigue, welds should be located in areas of known low stress. Corners and areas where shape or dimensional changes occur are invariably points of stress concentration and should not contain welded joints. Butt joints are preferred because the stresses act axially rather than eccentrically, as in corner and lap joints. Figure 4.29 shows recommended designs that will allow the weld to survive the base metal. Such designs initially are costly, but often they are the only means of achieving service life equivalent to that of the base metal.

If relocation of the joint is not possible, a complete joint penetration weld should be used, and a back welding should be used if the joint root is accessible.

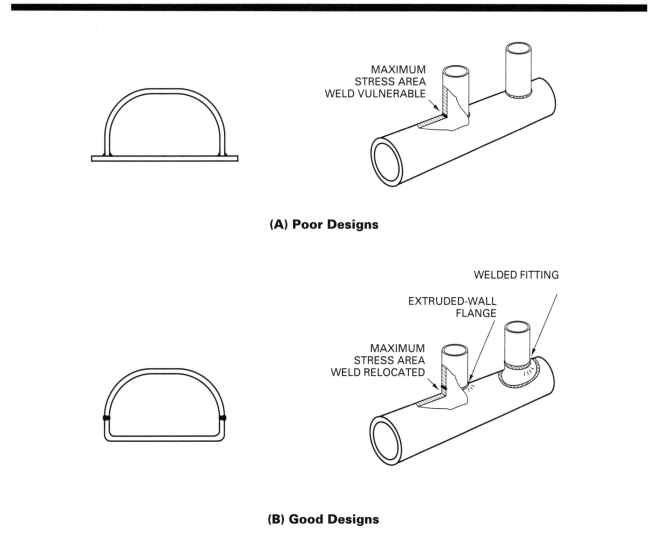

(A) Poor Designs

(B) Good Designs

Courtesy of Inco Alloys International, Inc.

Figure 4.29—Joint Designs for High-Temperature Service

WELDING PROCEDURES

AS A FUNDAMENTAL rule, welds should be complete joint penetration welds. If the design permits, no unfused areas should be left in the joint. Thermal-fatigue failures can often be traced to incomplete joint penetration welds that create stress concentrations. Figure 4.30 shows fatigue failure in a heat-treating basket emanating from incomplete joint penetration welds.

Some designs that facilitate complete joint penetration are shown in Figure 4.31. The techniques used for these designs are the following:

(1) Beveling and root opening
(2) Root opening with round forms
(3) Weld-all-around

In some applications of Ni-Cr-Fe alloy (N06600) and Ni-Cr-Fe-Al alloy (N06601), such as metal container baskets used in carburization of steel parts, failure of the weld is a foregone conclusion over a period of time.

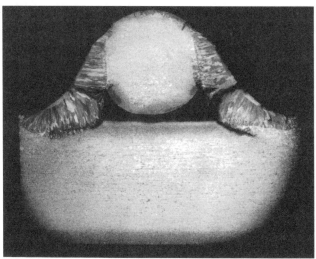

Courtesy of Inco Alloys International, Inc.

Figure 4.30—Cross Section of Joint in a Heat-Treating Basket

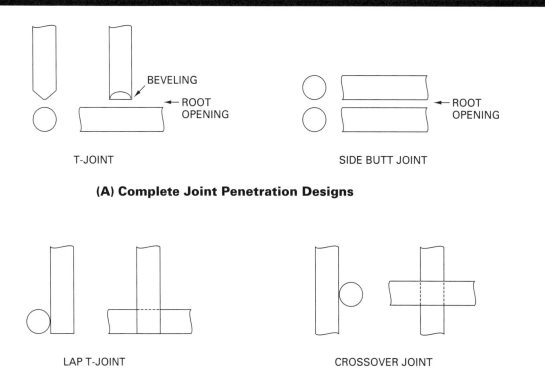

(A) Complete Joint Penetration Designs

(B) Weld-All-Around Techniques

Courtesy of Inco Alloys International, Inc.

Figure 4.31—Rod and Bar Joint Designs for High-Temperature Service

In those instances, since the base metal still has considerable life left in it, a maintenance welding schedule is introduced to repair welds at appropriate service intervals. Figure 4.30 shows a weld failure that can be repaired.

If welds must be placed in areas where changes in member size or direction occur, careful welding procedures are required to minimize the inherent higher stress concentrations. For example, it is important to avoid undercutting, incomplete joint penetration, weld craters, and excessive weld reinforcement. If complete joint penetration in rod or bar stock cannot be attained, the weld should be continuous and the joint sealed so that none of the process atmosphere (corrodent) can enter to attack the root side of the weld. [See the lap-T joint and crossover joint illustrations in Figure 4.31(B).]

Welds in heat-treating fixtures fabricated of round or flat stock should blend smoothly into the base metal without undercut. When fixtures are subject to many heating and quenching cycles, wrap-around joints or loosely fitted riveted joints are sometimes used to provide some freedom of movement.

WELDING SLAG

WELDING SLAGS ARE highly corrosive at elevated temperature, and great care must be taken to ensure their complete removal, including any weld spatter. Figure 4.32(A) shows the inside of a tubular furnace operating at red heat in which the weld slag was not removed. Corrosion through the entire weld cross section occurs rapidly. Figure 4.32(B) illustrates how catastrophic slag corrosion can be, while the surrounding base metal is in good shape.

Welding slag in oxidizing environments becomes increasingly fluid and aggressively attacks the weld and adjacent base metal. In reducing atmospheres, the welding slag acts as an accumulator of sulfur, causing failure by sulfidation in atmospheres that otherwise are adequately low in sulfur. In the Figure 4.32(A) example, the atmosphere contained 0.01% sulfur, but after only one month the weld slag contained 1.6% sulfur. Sulfur absorption by the welding slag depresses the melting point of the slag, causing the slag to become corrosive at lower temperatures than normal. Figure 4.33 is a cross section of a weld where weld slag at elevated temperature caused severe corrosive attack.

If slag cannot be removed from areas such as the tight root of a lap or crossover joint in round stock, subsequent weld beads must completely close the joint to prevent the contact of the slag with the atmosphere. The gas tungsten arc welding and gas metal arc welding processes are slag-free and often replace the use of covered electrodes.

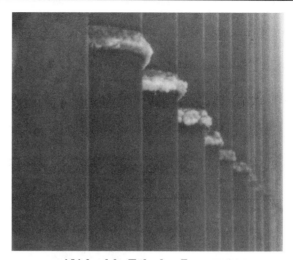

(A) Inside Tubular Furnace

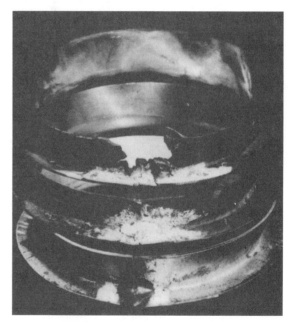

(B) Slag Corrosion Damage

Courtesy of Inco Alloys International, Inc.

Figure 4.32—Attack by Molten Welding Flux in High-Temperature Service

WELDING METALLURGY AND DESIGN

PROCESSES APPLICABLE TO nickel and cobalt alloys based on the use of fine-grain base metals (ASTM Number 5 or finer) are listed in Table 4.7. However, joining processes and procedures must reflect any special metallurgical factors of the operation. For example, some alloys are deliberately grain-coarsened to improve high-temperature service performance. Also, some

Courtesy of Inco Alloys International, Inc.

Figure 4.33—Corrosion of a Welded Joint Caused by Weld Slag During Exposure to a High-Temperature Reducing Atmosphere, with the Top Portion of the Weld Bead Showing Sulfidation Resulting from Sulfur Accumulation in the Slag

fabrication operations may cause grain-coarsening such as in the hot pressing of pressure-vessel heads. The condition of the base metal can affect the choice of welding process. For instance, the presence of a coarse grain structure (ASTM Number 5 or coarser) in the base metal restricts the use of processes having high-energy input. As seen in Table 4.8, Page 232, the gas metal arc and electron beam processes are not recommended for welding coarse-grain base metals.

BRAZING

NICKEL AND COBALT alloys can be joined readily by most of the conventional brazing processes and by diffusion brazing techniques. The severe service conditions to which these alloys are generally subjected is one reason that all phases of a brazing procedure require careful control. The chemical composition of the alloy and its physical and mechanical properties should be considered so that the assembly may be fabricated without problems. Two important factors are the effects of sulfur and low-melting-point metals and the possibility of stress cracking.

In general, cobalt alloys can be brazed with filler metals, procedures, and equipment that are similar to those used for nickel and nickel alloys.

FILLER METALS

THE SELECTION OF the brazing filler metal depends upon the service conditions of the assembly and the brazing process. It is also important to consider heat treatments that may be required for the brazement, to be certain that it will withstand the temperatures involved.

Nickel and cobalt alloys can be conventionally brazed with a number of silver, copper, nickel, cobalt, and gold filler metals. Diffusion heat treatment requires specially designed filler metals that are metallurgically compatible with the base metal. A diffusion heat treatment is used to alloy the brazing filler metal and the base metal together.

Nickel

THESE FILLER METALS generally possess excellent oxidation and corrosion resistance, as well as useful strengths at elevated temperatures. Some of them can be used to about 1800 °F (980 °C) for continuous service and to 2200 °F (1200 °C) for short-time service. They are generally designed for conventional brazing applications, but some alloys are suitable for diffusion brazing.

The filler metals are basically nickel or nickel-chromium, with additions of silicon, boron, manganese, or combinations of these, to depress the melting range below that of nickel and cobalt base metals. The brazing temperature ranges are usually between 1850 and 2200 °F (1010 and 1200 °C).

Filler metals that contain significant amounts of phosphorus to depress their melting ranges should not be used to braze nickel alloys. Brittle nickel phosphides can form at the braze interface. Boron-containing filler metals should not be used for brazing thin base metal because of their erosive action and excessive alloying with the base metal. Intergranular penetration of the base metal can also take place with some filler metals. In general, these filler metals are brittle, and this places some limitations on the design of brazed joints subject to bending or impact loading.

With diffusion brazing, a filler metal that will react with the base metal at some elevated temperature to form a molten alloy may be used. In the latter case, the alloy must be a eutectic that melts at a temperature compatible with the base metal. The brazement is heated for a sufficient period for the braze metal to diffuse with the base metal, and thus raise the joint remelt temperature to nearly that of the base metal. This treatment produces a joint that is essentially free of a distinct braze metal layer.

Filler metals are commonly applied in the form of a powder mixed with a volatile vehicle, such as an acrylic resin, for application by brush, spray, or extrusion. Precautions must be taken in using the vehicle so that no harmful residue will remain. When properly used, the vehicle should completely disappear during the heating cycle, before fusion of the filler metal. Filler metal is also available in tape and foil forms that permit the application of a controlled quantity of braze filler metal to a joint. Adjustments for density must be made when applying tape filler metals.

Cobalt

THESE FILLER METALS are generally used for their high-temperature properties and compatibility with cobalt alloys. For optimum results, brazing should be performed in a high-purity reducing or inert atmosphere. Special high-temperature fluxes are available. The brazing technique requires a degree of skill. By using diffusion brazing procedures, the brazements can be used at temperatures up to 1900 °F (1040 °C) with excursions to 2100 °F (1150 °C).

Silver

SILVER BRAZING FILLER metals may be used to join nickel and cobalt alloys to themselves and to many other metals and alloys. With proper design and brazing technique, brazed joints can develop the full strength of the base metal at room temperature. However, nickel alloys are subject to stress-corrosion cracking when exposed to molten silver filler metals. The base metal should be stress-free during brazing.

The low-melting filler metals, BAg-1, BAg-1a, and BAg-2, are commonly used, but for many corrosive environments, filler metals containing at least 50% silver are preferred. The BAg-7 filler metal is useful where stress cracking in nickel alloys might be a problem. The maximum service temperature recommended for joints brazed with silver filler metals is about 400 °F (200 °C).

Copper

MOST HIGH-NICKEL ALLOYS are capable of being brazed with BCu filler metal using the same equipment as used for steels. Minor changes in the brazing procedures may be necessary because of characteristics of some nickel alloys. Cobalt alloys should not be brazed with copper filler metals.

Copper filler metal will alloy more rapidly with nickel alloys than with steel. The molten copper will not flow far before it has dissolved enough nickel to raise its liquidus and reduce its fluidity. Therefore, the filler metal should be placed as close to the joint as possible, and the assembly should be heated rapidly to brazing temperature. A slightly rough or lightly etched surface improves the capillary flow of copper filler metal. Maximum operating temperature with copper filler metal is about 400 °F (200 °C) for continuous service and 900 °F (480 °C) for short-time exposure.

Copper phosphorus filler metals are not usually used with nickel-based alloys because of the formation of brittle nickel phosphide.

Gold

GOLD-BASED FILLER metals are used for brazing nickel and cobalt alloys where good joint ductility and resistance to oxidation or corrosion are required. Because of their low rate of interaction with the base metal, they are commonly used on thin base metal, usually with induction, furnace, or resistance heating in a reducing atmosphere or in vacuum without flux. For certain applications, a borax and boric acid flux may be used. Joints brazed with these filler metals are generally suited for continuous service up to 800 °F (425 °C).

FLUXES AND ATMOSPHERES

FLUXES, GAS ATMOSPHERES, and vacuum are used to avoid harmful reactions during brazing. Under some conditions, they may also reduce surface oxides that are present. In no case can a brazing flux or atmosphere negate the need for thorough cleaning of parts prior to brazing.

Fluxes

BRAZING FLUXES ARE usually mixtures of fluorides and borates that melt below the melting temperature of the filler metal. Standard fluxes are used on most solution-strengthened nickel and cobalt alloys. Other fluxes are available for alloys containing aluminum and titanium. Special fluxes may be needed with filler metals that require high brazing temperatures. Some fluxes are designed to have a long life during extended heating time.

The choice of flux and the technique for its use greatly influence the time required to make a braze and the quality of the joint. There is no single universal flux that is best for all brazing applications. Since there are many conditions, including base metal, filler metal, brazing process, brazing time, and joint design, there are many useful formulations of flux. Each flux is compounded somewhat differently, and each has its optimum performance region. Recommendations of the supplier should be sought, and trial runs are recommended when experience is lacking. For successful use, a flux must be chemically compatible with the base metals and filler metals involved and active throughout the brazing temperature range.

After brazing, flux removal is necessary. Flux removal is particularly important on brazements that are to operate at elevated temperatures or in corrosive environments.

Controlled Atmospheres

CONTROLLED ATMOSPHERES AND vacuum environments are employed principally to prevent the formation of oxides during brazing. Although it is possible in some nickel and nickel-copper alloys to use the atmosphere to remove the oxide films present on the metals, the best practice is to remove the oxides chemically or mechanically in a pre-braze operation so that the filler metal can wet and flow. The reactions resulting from the use of gas atmospheres and vacuum atmospheres are diverse. The general techniques of atmosphere brazing can involve the following:

(1) Gaseous atmospheres alone
(2) Gaseous atmospheres together with fluxes
(3) High vacuum
(4) Combinations of vacuum and gas atmospheres

Controlled atmospheres have the following advantages where additional fluxes are not required:

(1) The entire part is maintained in a clean, unoxidized condition throughout the brazing cycle. Parts, therefore, may sometimes be machined to finished size prior to brazing.
(2) Usually, postbraze cleaning is not necessary.
(3) Intricate sealed parts from which fluxes cannot be removed, such as electronic tubes, can be brazed satisfactorily.
(4) Large surface areas can be brazed integrally and continuously without danger of entrapped flux pockets in the joint, as in honeycomb panels.

The most common method used for brazing nickel and cobalt alloys is a controlled-atmosphere furnace. The furnace must be properly designed to maintain environment purity. There are three types of reducing atmospheres suitable for brazing, namely, combusted fuel gas, dissociated ammonia, and pure hydrogen.

An atmosphere of combusted fuel gas, containing not more than 20 grains (0.5 gm) of sulfur per 100 ft^3 (2.83 m^3), is satisfactory for nickel and nickel-copper alloys free of aluminum. Dissociated ammonia can be used with the same metals and also for nickel-chromium-iron alloys if the dew point is 80 °F (25 °C) or lower. The hydrogen in these atmospheres will reduce most metal oxides on workpieces to be brazed, provided the dew point is sufficiently low. Alloys containing chromium must be brazed at temperatures

above 1500 °F (815 °C) in dry hydrogen [-80 °F (-60 °C) dew point]. Therefore, the brazing filler metal must have a liquidus above 1500 °F (815 °C).

Oxides of aluminum and titanium cannot be reduced by hydrogen at ordinary brazing temperatures. If these elements are present in small amounts, satisfactory brazing can be done in gas atmospheres. When these elements are present in quantities exceeding 1 or 2 percent, the metal surface can be plated with a thin layer of pure metal that is easily cleaned by hydrogen, or a flux can be used in conjunction with the hydrogen. Copper plate is suitable with copper or silver filler metals. Nickel plate is generally used in conjunction with nickel or cobalt filler metals.

Brazing can also be done in a pure inert gas or a vacuum environment. The mechanism of oxide removal in vacuum is not clear. It is likely that some oxides dissociate at elevated temperatures by diffusion of oxygen into the base metal.

STRESS CRACKING

MANY NICKEL ALLOYS in contact with molten silver-based brazing filler metal have a tendency to crack when in a highly stressed condition. Alloys with high annealing temperatures are subject to this stress cracking phenomenon, particularly those that are precipitation-hardenable. The cracking occurs almost instantaneously during the brazing operation and is usually readily visible. The molten brazing filler metal will flow into the cracks and completely fill them.

The action is similar to stress-corrosion cracking, and the molten silver-based filler metal is considered as the corrosive medium. Sufficient stress to cause cracking can be produced either by cold working prior to brazing or by an applied stress from mechanical or thermal sources during brazing.

When stress cracking is encountered, its cause can usually be determined from a critical analysis of the brazing procedure. The usual remedy is to remove the source of stress. Stress cracking can be eliminated by one or more of the following actions:

(1) Use annealed base metal rather than cold-worked stock.
(2) Anneal cold-formed parts prior to brazing.
(3) Remove the source of stresses from externally applied loads, such as improper fit of parts, high fixturing forces, or unsupported parts.
(4) Redesign the parts or revise the joint design so that the fit-up does not create stresses due to improper fit or improper support.
(5) Heat at a slower rate.
(6) Select a filler metal that is less likely to cause stress cracking.

The precipitation-hardenable nickel alloys are very susceptible to stress cracking. These alloys should be brazed in the annealed or solution heat-treated condition with a filler metal of a high melting point.

POSTBRAZING HEAT TREATMENT

IT IS FREQUENTLY desirable to give brazed assemblies a postbrazing heat treatment to improve mechanical properties. The precipitation-hardenable alloys can be postbraze-hardened by dropping the temperature from the brazing temperature to the hardening temperature. Some alloys require cooling at a rate of 100 °F/h (38 °C/h) to a second, lower hardening temperature. Thus, both the brazing and hardening operations are completed in one furnace charge.

When a heat treatment is performed subsequent to brazing, the brazed joint must have sufficient strength at the heat-treating temperature to withstand the necessary handling. Postbrazing heat treatments may be a source of residual stresses in brazed joints. These stresses may cause microcracking, which may lead to premature failure in service.

SOLDERING

CONSIDERATIONS

SOLDERING CAN BE employed to join nickel and high-nickel alloys either to themselves or to any other solderable metal. Alloys containing chromium, aluminum, or titanium are more difficult to solder than are the other alloys. In designing a solder joint in a particular nickel alloy, any special characteristics of the base metal should be taken into consideration.

Many times, nickel alloys are used for a given application because of their resistance to corrosive attack. When this is a factor, the corrosion resistance of the solder also must be considered. In some cases, the joint should be located where the solder will not be exposed to the corrosive environment. The high-tin solders, such as 95% tin-5% antimony, may produce a better color match, if that is important. However, the solder may eventually oxidize in a manner different from the base metal, and the joint may then be noticeable.

If soldering is to be done on a precipitation-hardenable alloy, it should be done after heat treatment. The temperatures involved in soldering will not soften precipitation-hardened parts.

Nickel alloys are subject to embrittlement at high temperatures when in contact with lead and many other low-melting metals. This embrittlement will not occur at normal soldering temperatures. Overheating should be avoided. If welding, brazing, or other heating operations are to be done on an assembly, they must be done before soldering.

SOLDERS

ANY OF THE common types of solders may be used to join nickel alloys. A relatively high tin solder, such as the 60 tin-40 lead (wt. %) or the 50 tin-50 lead composition, is best for good wettability.

SURFACE PREPARATION

NICKEL AND NICKEL alloys heated in the presence of sulfur become embrittled. Before heating, these alloys should be clean and free from sulfur-bearing materials such as grease, paint, crayon, and lubricants.

Joints with long laps and joints that will be inaccessible for cleaning after soldering should be coated with solder prior to assembly. Coating is generally done with the same alloy to be employed for soldering. The parts may be dipped in the molten solder or the surface may be heated, fluxed, and the solder flowed on. Excess solder may be removed by wiping or brushing the joint. Nickel alloys may also be precoated by tin plating.

PROCESSES

NICKEL ALLOYS CAN be joined by any of the common soldering processes. Some minor differences in procedure may be required because of the low thermal conductivities of nickel alloys.

FLUXES

GENERALLY, ROSIN FLUXES are not active enough to be used on the nickel alloys. A chloride flux is suitable for soldering nickel or the nickel-copper alloys. Fluxes containing hydrochloric acid are required for the chromium-containing alloys.

Many of the proprietary fluxes used for soldering stainless steel are satisfactory for use on nickel alloys.

JOINT TYPES

THE STRENGTH OF soldered joints is low when compared to that of nickel alloys, which have relatively high strength. Therefore, the strength of the joint should not depend on the solder alone. Lock seaming, riveting, spot welding, bolting, or other means should be employed to carry the structural load. The solder should be used only to seal the joint.

POSTSOLDER TREATMENT

BECAUSE CORROSIVE FLUXES are required for soldering nickel alloys, the residue must be thoroughly removed after soldering. Residue from zinc chloride flux can be removed by first washing with a hot water bath containing 2 percent concentrated hydrochloric acid. Then the assembly should be rinsed with hot water containing some sodium carbonate, followed by a clean water rinse.

THERMAL CUTTING

NICKEL AND COBALT alloys cannot be cut by conventional oxyfuel gas cutting techniques. Plasma arc cutting and air carbon arc gouging are normally used with nickel alloys.

PLASMA ARC CUTTING

THIS PROCESS CUTS rapidly and can produce good cut surfaces up to 6 in. (152 mm) or greater. Generally, low travel speeds with thick base metal produce a rough surface.

A nitrogen-hydrogen mixture for the orifice gas is recommended by some equipment manufacturers for thicknesses up to 5 in. (125 mm). An argon-hydrogen mixture gives better cut surfaces for thicker base metal. Recommendations of the equipment manufacturer should be consulted concerning orifice and shielding gas selections for a particular alloy. Conditions that may be acceptable for some nickel alloys are in Table 4.29.

AIR CARBON ARC CUTTING

THIS PROCESS MAY be used for gouging of nickel alloys to form a weld groove or to remove a weld root surface or defective area. It may be used for cutting thin base metal. Type AC carbon electrodes, with alternating current or direct current electrode negative are recommended.

The cutting conditions must be adjusted to ensure complete removal of all melted material from the cut surface. The molten metal may be carburized by carbon from the electrode. Its presence during subsequent welding may cause undesirable metallurgical reactions in the weld metal.

LASER BEAM CUTTING

NICKEL AND COBALT alloys can be cut at high speeds with high-power laser beams. The power density of the focused beam permits cutting with a very narrow kerf, but there is a maximum thickness that can be cut economically with the process. The laser beam melts the metal, and a gas jet then blows the molten metal from the kerf. Clean cuts can be made with an inert gas jet.

Table 4.29
Typical Conditions for Plasma Arc Cutting Nickel Alloys

Alloy	Thickness		Gas Flow		Orifice Gas	Power, kW	Speed	
	in.	mm	ft³/h	L/min			in./min	mm/s
Commercially pure nickel	1.5	38.1	220	104	85%N$_2$-15%H$_2$	95	26	11
	3	76.2	260	123	85%N$_2$-15%H$_2$	138	6	2.5
	6	152.4	150	71	65%Ar-35%H$_2$	104	5	2.1
67Ni-32Cu (UNS N04400)	2	50.8	270	127	85%N$_2$-15%H$_2$	155	35	14.8
	3	76.2	260	123	85%N$_2$-15%H$_2$	134	5	2.1
76Ni-16Cr-8Fe (UNS N06600)	1.75	44.5	220	104	85%N$_2$-15%H$_2$	95	25	10.5
	3	76.2	260	123	85%N$_2$-15%H$_2$	135	5	2.1
	6	152.4	150	71	65%Ar-35%H$_2$	103	5	2.1
73Ni-16Cr-7Fe-3Ti (UNS N07721)	2.5	6.4	270	127	85%N$_2$-15%H$_2$	148	20	8.4
46Fe-33Ni-21Cr (UNS N08800)	1.5	38.1	220	104	85%N$_2$-15%H$_2$	92	20	8.4
	3	76.2	260	123	85%N$_2$-15%H$_2$	163	5	2.1
	6	152.4	130	61	65%Ar-35%H$_2$	98	5	2.1

SAFE PRACTICES

COMPOUNDS OF CHROMIUM, including hexavalent chromium, and of nickel may be found in fume from welding processes. The specific compounds and concentrations will vary with the welding processes and the compositions of the base and filler metals.[12] Immediate effects of overexposure to welding fumes containing chromium and nickel are similar to the effects produced by fumes of other metals. The fumes can cause symptoms such as nausea, headaches, and dizziness. Some persons may develop a sensitivity to chromium or nickel that can result in dermatitis or skin rash.

The fumes and gases should not be inhaled, and the face should be kept out of the fumes. Sufficient ventilation or exhaust at the arc, or both, should be used to keep fumes and gases from the welder's breathing zone and the general area. In some cases, natural air movement will provide enough ventilation. Where ventilation may be questionable, air sampling should be used to determine if corrective measures should be applied.

Nickel and chromium must be considered possible carcinogens under OSHA (29CFR1910.1200). Long-term exposure without proper ventilation to welding fumes, gases, and particulate may have long-term effects of skin sensitization, neurological damage, and respiratory disease such as bronchial asthma, lung fibrosis, or pneumoconiosis.[13]

Use local exhaust when cutting, grinding, or welding nickel or cobalt alloys. Exposures to fumes, gases, and dusts generated in welding should not exceed permissible exposure limits. Confined spaces require special attention. Wear correct eye, ear, body, and respiratory protection.

---

12. Consult the manufacturers Material Safety Data Sheets (MSDS) for the products being used.

13. See American Welding Society publications on safety and health, specifically publications ANSI/AWS F1.1, *Methods for Sampling Airborne Particulates Generated by Welding and Allied Processes* and ANSI/AWS F1.3, *A Sampling Strategy Guide for Evaluating Contaminants in the Welding Environment.*

APPLICATIONS

WELD REPAIR OF JET ENGINES

THE REPAIR OF high-cost jet engine parts made of nickel and cobalt alloy is essential to the efficient operation of modern aircraft. Flameholders and air seals are two such components involved in important weld-repair applications.

Flameholders

FLAMEHOLDERS OF THE military engine F100 often contain cracks, which are found during routine engine overhaul. The flameholders, made of cobalt alloy L-605, are prepared by steam cleaning, and the crack areas are wire brushed. The cracks are welded with a filler metal rod of 1/16 in. (1.6 mm) diameter alloy 188 (R30188), also a cobalt alloy, by manual gas tungsten arc welding with direct current electrode negative. Complete joint penetration welds are made in 1/16 in. (1.6 mm) thick sheet at 16 volts, 140 A, and 10 ft^3/h (4.7 L/min) argon as the torch shielding gas. The repair is completed by a postweld stress relief.

Airseals

THE WELD BUILDUP of worn airseals of a TF39 military jet engine rotor is shown in Figure 4.34. After

Courtesy of *Welding Design & Fabrication*

Figure 4.34—Gas Tungsten Arc Weld Buildup of Worn Airseals

1800 full-power engine operating cycles, the circumferential, rotating, knife-edge airseals have worn so much that they no longer seal efficiently. The seals are made of alloy 718 (UNS N07718).

The seals are prepared for repair by degreasing, machining down to a uniform height, and blasting with glass beads. The repair is made by mechanized welding with the gas tungsten arc welding process, whereby the filler metal is automatically fed with a reciprocating motion with a stroke length of 0.20 in. (5 mm) and a frequency of 4 to 6 Hz. Further welding conditions are given in Table 4.30.

After a preheat pass, the seal is built up with beads of 0.30 in. (0.8 mm) thickness, until the weld metal thickness reaches about 0.13 in. (3 mm). (See Figure 4.35.) The part is then stress relieved, and the airseals are machined to the proper operating dimensions.

Table 4.30
Welding Conditions for Repair of Airseals

Filler metal	ERNiFeCr-2
Filler metal wire diameter	0.045 in (1.14 mm)
Electrode type	W-2 ThO$_2$
Electrode diameter	0.062 in. (1.6 mm)
Electrode extension	1/2 in. (13 mm)
Stickout	1/4 in. (6 mm)
Nozzle diameter	5/8 in. (16 mm)
Shielding gas	Argon
Shielding gas flow rate	50 ft^3/h (25 L/min)
Trailing gas	Argon
Trailing gas flow rate	100 ft^3/h (50 L/min)
Wire feed speed	3 in./min (75 mm/min)
Travel speed	2-3 in./min (50-75 mm/min
Welding voltage	3.5-5.5
Welding current	10-12 A
Arc length	0.1 in. (2.5 mm)
Torch drag angle	10-15°
Torch work angle	0°
Wire push angle	45°
Wire push angle	0°

Courtesy of *Welding Design & Fabrication*

Figure 4.35—Completed Weld Buildup of Airseals

FLUE-GAS DESULFURIZATION

ELECTRIC POWER COMPANIES around the world clean the gases leaving their coal and oil-fired boilers before they go up the smokestack. Special *scrubbers* remove sulfur-bearing compounds that cause acid rain. Nickel alloys are used to fabricate flue-gas scrubbers because of their excellent corrosion resistance in these severe environments. Figure 4.36 shows four such scrubbers installed at a typical powerplant site.

Costs dictate lower nickel content of scrubber alloys where the flue-gas environments are lower in corrosiveness, at lower temperatures and weaker acidities. Typically, flue-gas scrubbers are fabricated with a number of alloys designed to meet increasingly severe environments, as shown in Table 4.31. Alloys used in the installation shown in Figure 4.36 were 904L, 625, and C-276.

Large ducts or tunnels are fabricated by cladding low-alloy structural members with thin sheets of high-alloy materials, a practice called *wallpapering*. Figure 4.37 shows typical cap welds made in this type of

Courtesy of Inco Alloys International, Inc.

Figure 4.36—Four Flue-Gas Scrubbers (foreground) at the Big Bend Power Station, Tampa, Florida

Table 4.31
Nickel Alloys Used in Corrosive Environments
of Increasing Severity

Corrosiveness of Environment	Alloys	
	Designation	UNS Number
Mild	316L	S31603
	904L	N08904
Medium	317L	S31703
	317LM	S31725
	Duplex and 6% Mo steels	—
Severe	625	N06625
	622, C-22	N06022
	C-276	N10276
	686	N06686

construction of ducting in flue-gas desulfurization installations. This steel ducting structure is wallpapered with 0.062-in. (1.6 mm) NiCrMo alloy 622 sheet.

Matching or over-matching weld filler metals are used to make all of the exposed welds. The use of one over-matching material at a fabrication site where two or more alloys are being joined prevents the inadvertent use of the wrong alloy in a critical location. For example, when the alloys 316L, 904L, 625, C-22, and C-276 are all available at one location, the use of one over-matching welding filler material such as NiCrMo-14 (alloy 686CPT, UNS N06686) prevents the inadvertent use of a lower-alloy material in the welding of a higher-alloy base metal.

For optimum corrosion resistance, fabrications must be free from the following defects (shown in Figure 4.38):

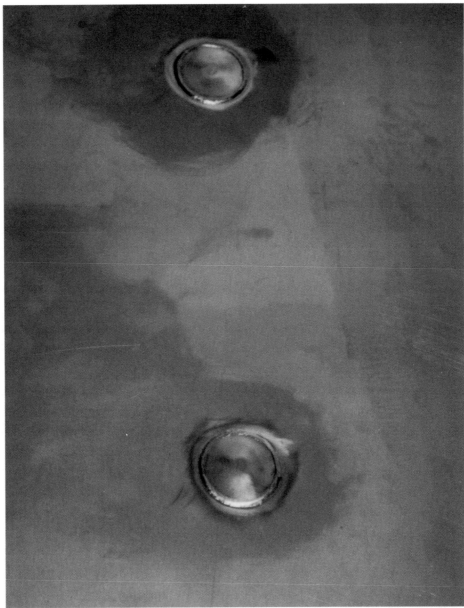

Courtesy of Inco Alloys International, Inc.

Figure 4.37—Typical Caps on Plug Welds Used in "Wallpapering" Construction of Ducting in Flue-Gas Desulfurization Installations

(1) Surface contaminants such as oil, paint, crayon marks, adhesive tape, or other debris that can act as a site for crevice corrosion

(2) Embedded iron that can initiate pitting corrosion

(3) Weld spatter that can initiate crevice corrosion

(4) Heat tint that reduces corrosion resistance of stainless steels

(5) Mechanical defects such as gouges, scratches, and arc strikes that reduce thickness and promote crevice or pitting corrosion

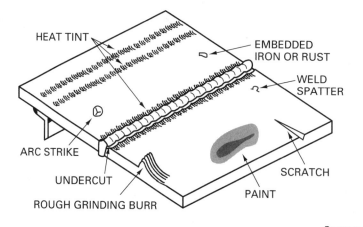

HEAT TINT

EMBEDDED
IRON OR RUST

WELD
SPATTER

ARC STRIKE

UNDERCUT

ROUGH GRINDING BURR

SCRATCH

PAINT

Courtesy of Nickel Development Institute

Figure 4.38—Typical Fabrication Defects or Surface Conditions Commonly Encountered

OFFSHORE OIL PLATFORMS

OFFSHORE OIL PLATFORMS, as well as onshore facilities, retrieve oil and natural gas from a high-pressure, high-temperature, and corrosive environment. (See Figure 4.39.) This environment calls for many of the corrosion-resistant and heat-resistant properties of nickel alloys and welding products. The downhole tubing and tools are made from alloys such as the NiCrMo alloys C-276 (N10276), G-3 (N06985), and 825 (N08825). Other downhole tools and surface hardware are made from alloys 625 (N06625) and 825 (N08825) and are welded with the NiCrMo-3 welding products. In the high-temperature areas, such as in the flare booms and tips, the NiCrMo alloy 625 and FeNiCr alloy 800HT (N08811) are used. The welding of alloy 625 is done with NiCrMo-3 filler metals, and alloy 800HT is welded with ERNiCr-3 and ENiCrFe-2 for temperature exposures below 1450 °F (790 °C). For applications at temperatures above this, the alloy 800HT is welded with NiCrCoMo-1 welding products.

Courtesy of Inco Alloys International, Inc.

Figure 4.39—Offshore Oil and Gas Production Platforms Call for the Corrosion and Heat Resistance of Nickel Alloys and Welding Products for Downhole Tubing, Tools, Surface Hardware, and Flare Booms and Tips

SUPPLEMENTARY READING LIST

Amato, I. et al. "Spreading and aggressive effects by nickel-base brazing filler metals on the alloy 718." *Welding Journal* 15(7): 341s-45s, 1972.

Chasteen, J. W. and Metzger, G. E. "Brazing of Hastelloy X with wide clearance butt joints." *Welding Journal* 58(4): 111s-17s, 1979.

Christensen, J. and Rorbo, K. "Nickel brazing below 1025 °C of untreated Inconel 718." *Welding Journal* 53(10): 460s-64s, 1974.

Coffee, D. L. et al. "A welding defect related to the aluminum content in a nickel-base alloy." *Welding Journal* 51(1): 29s-30s, 1972.

Dix, A. W. and Savage, W. F. "Short time aging characteristics of Inconel X-750." *Welding Journal* 52(3): 135s-39s, 1973.

————. "Factors influencing strain-age cracking in Inconel X-750." *Welding Journal* 50(6): 247s-52s, 1971.

Duvall, D. S. et al. "TLP bonding: a new method for joining heat resistant alloys." *Welding Journal* 53(4): 203-14, 1974.

Eng, R. D. et al. "Nickel-base brazing filler metals for aircraft gas turbine applications." *Welding Journal* 56(10): 15-21, 1977.

Jahnke, B. "High-temperature electron beam welding of the nickel-base superalloy IN-738LC." *Welding Journal* 61(11): 343s-47s, 1982.

Kamat, G. R. "Solid-state diffusion welding of nickel to stainless steel." *Welding Journal* 67(6): 44-6, 1988.

Kelly, T. J. et al. "An evaluation of the effects of filler metal composition on cast alloy 718 simulated repair welds." *Welding Journal* 68(1): 14s-18s, 1989.

Kenyon, N. et al. "Electroslag welding of high nickel alloys." *Welding Journal* 54(7): 235s-39s, 1975.

Kiser, S. "Nickel-alloy consumable selection for severe service conditions." *Welding Journal* 69(11): 30-5, 1990.

Lugscheider, E. "Metallurgical aspects of additive-aided wide-clearance brazing with nickel-based filler metal." *Welding Journal* 68(1): 9s-13s, 1989.

Matthews, S. J. "Simulated heat-affected zone studies of Hastelloy B-2." *Welding Journal* 58(3): 91s-5s, 1979.

Matthews, S. J. et al. "Weldability characteristics of a new corrosion- and wear-resistant cobalt alloy." *Welding Journal* 70(12): 331s-7s, 1991.

Mayor, R. A. "Selected mechanical properties of Inconel 718 and 706 weldments." *Welding Journal* 55(9): 269s-75s, 1976.

Moore, T. J. "Solid-state and fusion resistance spot welding of TD-NiCr sheet." *Welding Journal* 53(1): 37s- 48s, 1974.

Moore, T. J. and Glasgow, T. K. "Diffusion welding of MA 6000 and a conventional nickel-base superalloy." *Welding Journal* 64(8): 219s-26s, 1985.

Prager, M. and Shira, C. S. "Welding of precipitation-hardening nickel-base alloys." Bulletin 128, New York: Welding Research Council, February 1968.

Savage, E. I. and Kane, J. J. "Microstructural characterization of nickel braze joints as a function of thermal exposure." *Welding Journal* 63(10): 316s-23s, 1984.

Savage, W. F. et al. "Effect of minor elements on hot cracking tendencies of Inconel 600." *Welding Journal* 56(8): 245s-53s, 1977.

Savage, W. F. et al. "Effect of minor elements on fusion zone dimensions of Inconel 600." *Welding Journal* 56(4): 126s-32s, 1977.

Weiss, B. Z. et al. "Static and dynamic crack toughness of brazed joints of Inconel 718 nickel-base alloy." *Welding Journal* 58(10): 287s-95s, 1979.

Welding Research Council. "Recent studies of cracking during postweld heat treatment of nickel-base alloys." Bulletin 150, New York: Welding Research Council, May 1970.

Yoshimura, H. and Winterton, K. "Solidification mode of weld metal in Inconel 718." *Welding Journal* 51(3): 132s-8s, 1972.

CHAPTER 5

LEAD AND ZINC

PREPARED BY A COMMITTEE CONSISTING OF:

F. E. Goodwin, Chairman
International Lead Zinc Research Organization, Inc.

B. Bersch
Hoesch Stahl AG

D. Cocks
Eastern Alloys

A. Gibson
Armco, Inc.

R. M. Hobbs
BHP Steel

R. W. Jud
Chrysler Corporation

T. J. Kinstler
Metalplate Galvanizing

S. Kriner
BIEC International

A. F. Manz
A. F. Manz Associates

G. T. Spence
Cominco, Ltd.

R. G. Thilthorpe
Zinc Development Association

J. Wegria
Union Miniere

WELDING HANDBOOK COMMITTEE MEMBER:
J. M. Gerken
Consultant

CHAPTER 5

LEAD AND ZINC

WELDING AND SOLDERING OF LEAD

LEAD AND ITS PROPERTIES

LEAD CAN BE toxic if handled improperly. Anemia, damage to the nervous system, and kidney damage are possible outcomes of many years of prolonged exposure to high levels of lead fume or dust. It is suggested that the "Lead Safe Practices" section of this chapter be reviewed before undertaking any lead joining practices.

One of the first metals purified, lead has demonstrated its importance to civilization with water piping systems, in service today, constructed by ancient Romans. Indeed, the chemical symbol for lead, Pb, is from plumbum which means plumber in Latin. Also, lead roofs installed in the fifteenth century continue to provide protection from water damage. Lead continues in service after such long periods because of its ability to produce passive films, preventing significant corrosion and consequent lead release to the environment.

In spite of its shortcomings, lead has lasted as a commercial material for many reasons, among them its effectiveness as a shield from radiation and its resistance to acids such as sulfuric, chromic, and phosphoric. Lead melts at the relatively cool temperature of 621 °F (327 °C) and boils at 3092 °F (1700 °C), making fabrication by fusion processes somewhat critical if not difficult. When freshly cut, its surface is bright and silvery, but it turns dull gray when oxidized. This adds to the complications of welding. However, lead and its alloys are easily cast and mechanically worked.

Lead is a member of the cubic crystal system; the cubic nature can be observed with the unaided eye from the appearance of etched samples. Like all face-centered cubic metals, lead shows twinning lamellar microstructure upon deformation and recrystallization.

Lead is easily rolled into sheet and extruded into pipe, cable sheathing, and other products with a wide variety of cross sections. Compositions of the four grades of unalloyed lead (99.9% Pb), as specified by ASTM, are shown in Table 5.1.[1]

Two alloys commonly used in industry are tellurium lead and antimonial lead. The addition of tellurium, less than 0.1 percent, improves the work-hardening capacity of lead. The addition of antimony from 1 to 9 percent, but usually around 6 percent or less, increases the hardness and provides about twice the tensile strength of soft lead at ordinary temperatures. However, antimony lowers the melting point markedly, depending upon the percentage added. Also, antimony causes the alloy to expand upon solidification.

The thickness of sheet lead is usually specified in North America by weight in pounds per square foot; one square foot of lead 1/64 in. thick weighs 0.92 lb (0.093 m², 0.4 mm thick, weighs 0.42 kg). Because this thickness weighs approximately a pound per square foot, this most commonly used lead sheet in North America is called *one-pound sheet*. Lead pipe is usually specified by inside diameter and by wall thickness.

Lead exhibits a coefficient of thermal expansion of 16.5 x 10⁻⁶/ °F (29.7 x 10⁻⁶/ °C). Thus, a 72 °F (40 °C) temperature change, common in building applications, will cause a 0.09-in. (2.3-mm) change in a 6.6 ft (2 m) length of sheet or pipe. Lead has a volume change upon melting of 3.44 to 3.61 percent, depending upon the alloy.

LEAD WELDING PROCESSES

General

OXYFUEL GAS WELDING processes, especially oxyacetylene and oxyhydrogen, are commonly used for welding lead, also called *lead burning*. Lead and its

---

1. "Standard specification for refined lead (ASTM B29-92)," *Annual Book of ASTM Standards,* Vol. 02-04, 7-9. Philadelphia: American Society for Testing and Materials, 1995.

Table 5.1
ASTM Composition Requirements for Refined Lead, wt. %[a]

Element	Low-Bismuth, Low-Silver Pure Lead[b]	Refined Pure Lead[c]	Pure Lead[d]	Chemical-Copper Lead[e]
Sb	0.0005	0.0005	0.001	0.001
As	0.0005	0.0005	0.001	0.001
Sn	0.0005	0.0005	0.001	0.001
Sb, As, and Sn	—	—	0.002	0.002
Cu	0.0010	0.0010	0.0015	0.040-0.080
Ag	0.0010	0.0025	0.005	0.020
Bi	0.0015	0.025	0.05	0.025
Zn	0.0005	0.0005	0.001	0.001
Te	0.0001	0.0001	—	—
Ni	0.0002	0.0002	0.001	0.002
Fe	0.0002	0.001	0.001	0.002
Lead (min) by difference	99.995	99.97	99.94	99.90

a. ASTM B29-92. Values specified are maximums except for lead, which is a minimum, and copper in chemical-copper lead, which is a range.

b. This grade (UNS No. L50006), formerly termed *corroding lead*, is intended for chemical applications requiring low-silver and low-bismuth contents.

c. This grade (UNS No. L50021), formerly termed *common lead*, is intended for lead-acid battery applications.

d. This grade is UNS No. L50049.

e. This grade (UNS No. L51121), formerly termed *acid-copper lead*, is intended for applications requiring corrosion protection and formability.

alloys are easily soldered if proper care is taken not to melt the metals forming the joint. Soldering irons are usually used to heat sheet joints. Wiping, a special technique in which molten solder is poured over the joint until the base metal is wetted, uses the heat of the liquid solder for heating the base metal without additional sources of energy.

Oxyfuel Gas Welding

SEVERAL FUEL GASES can be used in oxyfuel gas welding, notably acetylene, propane, natural gas and hydrogen. Although acetylene is the preferred fuel in many countries, hydrogen is preferred in the United States because of the fine flame size that can be obtained, allowing good control of the weld-bead size. With oxyacetylene, the higher pressure needed to obtain a small flame tip disrupts control of the liquid lead pool. Welding fuel gases should be used only with equipment especially intended for them. Oxyhydrogen and oxyacetylene mixtures can be used for welding in all positions. Oxyhydrogen is especially useful in providing good weld pool control because of its localized heat input. Oxynatural gas and oxypropane are limited to the flat position because of the greater difficulty in controlling weld pool size and shape.

In all cases, the proportion of oxygen is adjusted to obtain a neutral flame. Reducing flames with organic fuels tend to deposit soot in the joint, while excessive oxygen will oxidize the lead, inhibiting wetting.

Flame intensity is controlled by varying the welding torch tip and tip size to control the flow volume of fuel gases and the flame shape. Tips from drill size 78 to 68 (0.41 to 0.79 mm diameter) are commonly used. Larger tip sizes are used for greater thicknesses and are influenced by the type of joint being welded. Butt, lap, and corner joints welded in the flat position and lap joints welded in the horizontal position require more heat than vertical welds, and will require larger tip sizes to achieve higher fuel gas flow rates. Gas pressures between 1.5 and 5 psi (10.3 and 34.5 kPa) are typically used and will produce flames between 1.5 and 4 in. (38 and 102 mm) long.

Filler metal is available in rod form between 1/8 and 3/4 in. (3.2 and 19 mm) diameter. Filler metal can also be cut from the base metal being used for the job, in widths typically of 1/4 and 1/2 in. (6 and 12 mm). The composition of the filler metal should be similar to the base metal to obtain similar melting and mechanical properties. The filler metal must be scrupulously cleaned before use, either by wire brushing or preferably by shaving off a thin layer of metal to expose fresh lead.

Neither rosin nor tallow are effective for removing surface contaminants from lead while welding. Therefore, it is essential that the mating edges, joints, and adjacent surfaces of the lead be shaved clean. If necessary, a safe and suitable solvent should be used to remove dirt, oil, and grease before shaving. Handling of the prepared surfaces should be avoided. The same care of preparation needs to be given to the lead rods and strips to be used as filler metal. More details about the methods for cleaning will be discussed in following sections of this chapter.

JOINT DESIGN

Sheet

THE THREE BASIC types of joints used for welding sheet lead are butt, corner, and lap. These are illustrated in Figure 5.1. Butt and lap welds can be used for flat work, although butt welds are preferred when lap welds give an overall thickness greater than desired. For vertical and overhead work, the lap joint is almost

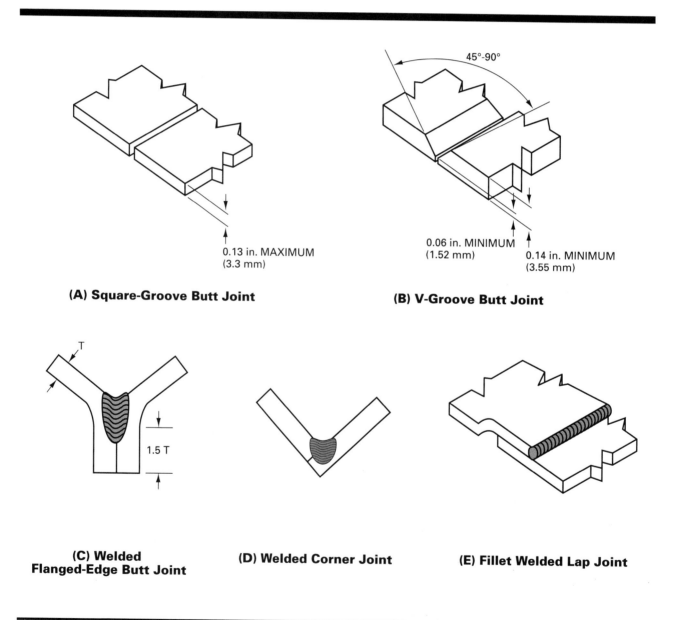

(A) Square-Groove Butt Joint

0.13 in. MAXIMUM (3.3 mm)

(B) V-Groove Butt Joint

45°-90°

0.06 in. MINIMUM (1.52 mm)

0.14 in. MINIMUM (3.55 mm)

(C) Welded Flanged-Edge Butt Joint

T

1.5 T

(D) Welded Corner Joint

(E) Fillet Welded Lap Joint

Figure 5.1—Basic Butt, Corner, and Lap Joints for Welding Lead Sheet

always used. In all cases, adequate support of the joint is necessary and can be ensured by positioning the base metal securely against supporting materials.

A square-groove, butt-joint design can be used with a 1/8-in. (3.2-mm) thick sheet lighter than 8 lb/ft^2 (39 kg/m^2). A 90-degree, V-groove butt joint is used for heavier sheet. These typical butt joints are illustrated in Figure 5.1(A) and (B). The weld grooves and sheet are shaved clean for 1.25 in. (32 mm) beyond the grooves. Butt joints are held in proper alignment for final welding by tacking or flow welding (burning-in), in which the upper meeting edges of lead are fused for a short distance as shown in Figure 5.2.

Flanged-edge butt joints and corner joints also may be used with lead weighing less than 4 lb/ft^2 (19.5 kg/m^2). Illustrations of these joints as welded are shown in Figures 5.1(C) and (D).

For the flanged-edge butt joint, the edges of the sheet to be joined are shaved clean, as are 2-in. (51-mm) widths on the top and bottom of each sheet adjacent to the edges. A flange 1.5 times the sheet thickness is then formed by bending the edge of the sheet outward. The bent edges are then fitted together and welded as a butt joint.

Corner joints are prepared by positioning lead sheets together at an angle, with one sheet always forming one full side of the angle. The other sheet is butted against it. Grooving of the joint is not necessary, regardless of thickness. The root faces of the sheets to be welded are shaved clean to a width of 2 in. (51 mm), along with the joint root of the butting member. For welding corner joints in thicker sheet, the method shown in Figure 5.3 is more suitable than the one shown in Figure 5.1(D).

Placement of the weld will depend upon whether both faces of the angle are inclined (flat position), as shown in Figure 5.3, or whether one face of the joint is vertical (horizontal position) as shown in Figure 5.4. In the case where one butting member is vertical, the weld is built up on the horizontal face and its sides fused into the vertical face. This results in a wide, horizontal weld. In the case where both faces are inclined, the weld is flat and more evenly distributed across both faces to provide more strength.

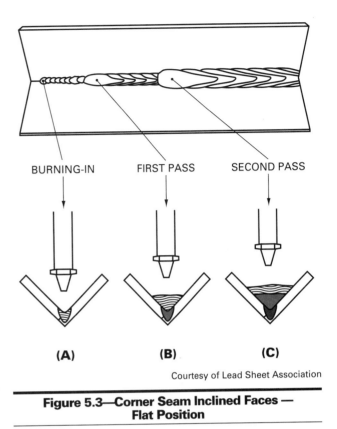

BURNING-IN FIRST PASS SECOND PASS

(A) **(B)** **(C)**

Courtesy of Lead Sheet Association

Figure 5.3—Corner Seam Inclined Faces — Flat Position

Fillet welded lap joints [Figure 5.1(E)] can be used in all positions. Typically an overlap of 1/2 to 2 in. (12 to 51 mm), is proportional to the sheet thickness. All surfaces that the filler metal will contact must be shaved clean, as well as the faying surfaces and a 1/4-in. (6-mm) border beyond the weld face.

A variation of the lap joint is the horizontal lapped seam weld on an inclined face. To allow joining of this configuration, the edge of the outer sheet is angled outward over the outer 1/4 in. (6 mm) of the inner sheet width, forming a groove between the two sheets, as shown in Figure 5.5(A). The groove angle should be approximately 45 degrees. Filler metal is fused into the groove to form the weld, as in Figure 5.5(B). Care is needed to avoid melting of the outside edge of the overlap and to fuse the weld metal to the lower sheet without cutting through.

Pipe

THE FOUR TYPES of joints used to join pipe are cup, butt, flange, and lap. The first three are shown in Figure 5.6 while the lap joint is similar to that shown in Figure 5.1(E), except that a pipe, rather than flat sheet geometry is used.

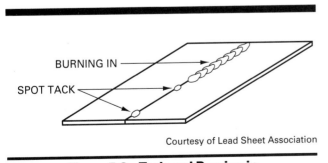

BURNING IN

SPOT TACK

Courtesy of Lead Sheet Association

Figure 5.2—Tack and Burning-in

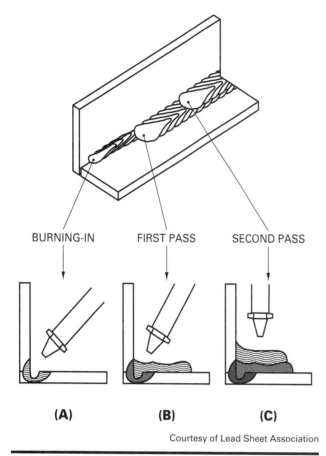

BURNING-IN   FIRST PASS   SECOND PASS

(A)   (B)   (C)

Courtesy of Lead Sheet Association

**Figure 5.4—Corner Seam One Face Vertical —
Horizontal Position**

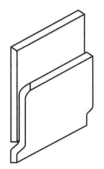 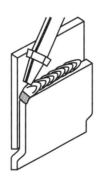

**(A) Preparation of Joint
with a 1/4-in. (6-mm)
Wide Groove**

**(B) Application of
Filler Metal in
a Single Pass**

Courtesy of Lead Sheet Association

Figure 5.5—Horizontal Lapped Seam

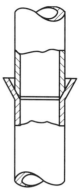

(A) Cup Joint

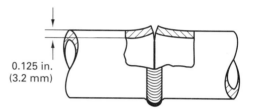

0.125 in.
(3.2 mm)

(B) Butt Joints

0.25 in.   0.063 in.
(6.3 mm)   (1.6 mm)

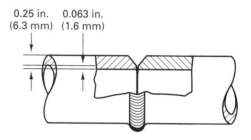

0.25 in.   0.063 in.
(6.3 mm)   (1.6 mm)

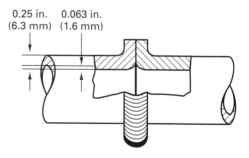

(C) Joining of Two Pipe End Flanges

Figure 5.6—Pipe Joints

To prepare the butt joint, a wooden turn pin is used to slightly flare the ends of the pipes to be joined when wall thicknesses are less than 1/8 in. (3 mm). As shown in Figure 5.6(B), this procedure yields a butt joint with a slight V-shape, allowing good filler metal penetration. It is difficult to use the turn pin with pipe thicknesses greater than 1/8 in. (3 mm), and hence the V-groove must be prepared by shaving the outer edge of the pipe wall. The V-groove should have an angle of 45 degrees and a 1/16-in. (1.6-mm) land, as shown in Figure 5.6(B). The pipe ends are butted together forming a tight joint root. The outer and inner walls of the pipe should be shaved clean for a distance of 1/2 in. (13 mm) from the joint.

To avoid the difficulties of overhead welding, a V-shaped section can be cut at the top half of the pipe and the inner wall at the bottom of the joint welded from the inside, as shown in Figure 5.7. The V-section is then repositioned and welded to the pipe from the outside. Alternatively, T-slots can be cut in the upper halves of both pipe ends. The wall sections are then bent radially outward to expose the lower half of the joint. After the lower half of the joint is welded from the inside, the wall sections are bent back to the original shape, and the T-slots in the upper part of the joint are then welded from the outside.

Lap or cup joints require that one side of the pipe joint be expanded to allow the other side to fit inside. These joints are most commonly used for vertical pipes, with the lap or cup always opening upward. In lap joints, an overlap equal to the pipe diameter is used. In cup joints, Figure 5.6(A), a flare with a length equal to half of the pipe diameter is made with a turning pin on the overlapping pipe. A 45-degree angle flare will produce a bevel-groove joint when the other pipe is inserted into the flare. Lap joints also can be used, but usually require more joint preparation.

In lap joints, an overlap equal to the pipe diameter is used. The outer pipe section used to form a lap joint should be shaved clean on the edge and both inner and outer sides of the pipe wall over the length of the joint root. The inner pipe section should be shaved clean on the outer pipe wall over the length of the lap joint in addition to a 1-in. (25-mm) length beyond the root face. Similarly, the inner pipe used in a cup joint should be shaved clean on its outer wall over the length of the cup in addition to a 1-in. (25-mm) length beyond the cup and the pipe edge. The outer pipe in a cup joint should be shaved clean on the edge and on the wall interior over the length of the cup in addition to a 1/2 in. (13 mm) length of pipe wall interior beyond the root face.

Large pipes can be joined by forming a sheet or by casting a shape in the form of a flange and sleeve. In the case of formed sheet, a seam is welded to complete the shape. The sleeve is then lapped over the pipe and welded in place as a lap joint. The flange ends are then butted together and butt welded. A section of a flange joint is shown in Figure 5.6(C).

WELDING POSITIONS

General

THE AIM OF lead welding is to make welds that penetrate fully and have a reinforcement thickness that is about one-third greater than the sheet being welded. This is controlled by the amount of heat and the amount of filler metal used. Inadequate penetration is usually the result of using too small a welding tip or flame, or progressing too quickly along the joint. Over-penetration is caused by progressing too slowly for the given torch tip and flame size. Excessive melting next to the weld, or undercutting, must be avoided because it reduces the effective joint thickness.

Figure 5.8 shows four examples of undercut seams in comparison with sound seams. Undercutting can be difficult to repair. It is caused by misdirecting the flame away from the center of the joint and the filler metal. It can happen when working outdoors in windy conditions and can also occur when the welder takes a slow and hesitant approach to the work, such as when welding is in a difficult position.

In general, a semi-circular or V-shape motion is used in welding lead. In this way the molten lead weld pool is controlled and directed by the motion of the flame, giving properly welded seams a semi-circular or herringbone appearance. Fusion of the base metal is achieved by applying the flame just long enough for the heat to penetrate to the edges of the shaved area. Then lead is melted from the filler rod into this area, with the quantity introduced determining the thickness of the seam. The cone of the flame should be just clear of the edge of the lead pool produced. The amount of lead melted from the filler for each pass should be sufficient to build up a layer approximately 0.08 in. (2 mm) thick.

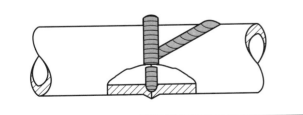

Figure 5.7—Joint Design for Welding Lead Pipe in the Horizontal Fixed Position

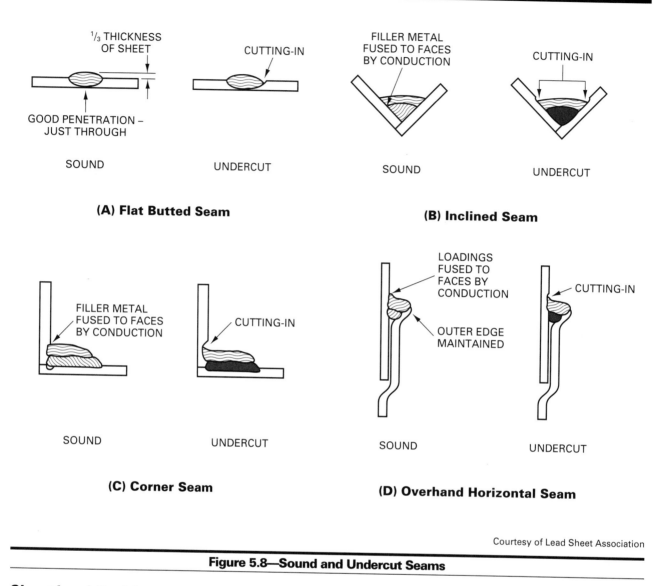

Courtesy of Lead Sheet Association

Figure 5.8—Sound and Undercut Seams

Sheet Lead Positions

Flat Position. In the flat position butt joints are most common, although flanged joints can be used for very thin sheet, and lap joints can be used if they simplify fabrication. The position of the filler strip and torch relative to the weld pool, weld bead, and joint for a flat butt joint is shown in Figure 5.9.

The torch also can be moved in a straight line rather than moving from side to side when a large flame associated with a larger welding tip is used. This technique is most appropriate when working on the bench. However, better control is obtained in field applications by progressing with a circular or V-shape motion.

The slower the progression and the more pronounced the side-to-side action, the rounder will be the pattern seam. Filler metal generally is not used on the first pass. For lap joints, the semicircular motion begins on the bottom sheet and then moves toward the lap (see Figure 5.10). The torch is then moved in a straight line back to the bottom sheet. The slower the progression and the more pronounced the side-to-side action, the rounder will be the pattern of the seam. Filler metal is added just ahead of the flame.

Procedures for producing corner joints are shown in Figures 5.3 and 5.4. The first stage in welding a corner joint is tacking or flow welding from the edges. When both sides of the angle are inclined, the first pass is brought forward, just filling the root of the weld. This is followed by a heavier second pass. The flame

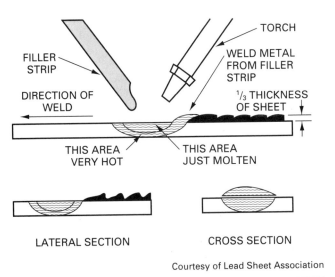

Courtesy of Lead Sheet Association

Figure 5.9—Flat Butted Seam – Position of Torch and Filler Strip

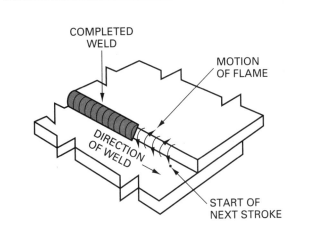

Figure 5.10—Technique For Welding Lap Joint in the Flat Position

must not be allowed to dwell on either side of the angle because this can cause undercutting, which weakens the joint. The molten weld pool also must be kept in the line of the pass by using a straight-line technique of welding and minimal weaving.

Horizontal fillet joints with one face vertical (see Figure 5.4) are first tacked and then welded with a heavy first pass. When welding in the horizontal position using the circular or V-shape motion, a soft diffused flame is most desirable. The flame for the first pass is

directed nearly vertically downward, with fusion to the vertical leg occurring only by heat conduction from the pool. The second pass is much lighter and should be pointed toward the vertical seam but without playing the torch on the vertical face. Fusion with the vertical face is again controlled by conduction.

The speed of welding butt joints in the flat position is not very different from the speed developed when oxy-fuel gas welding steel. Granjon's formula for determining ranges of typical welding speeds is:

$$\frac{0.4}{T} \leq V \leq \frac{0.8}{T} \tag{5.1A}$$

where

T = lead thickness, in.
V = welding speed, in./min

or, in metric units:

$$\frac{4.3}{T_m} \leq V_m \leq \frac{8.6}{T_m} \tag{5.1B}$$

where

T_m = lead thickness, mm
V_m = welding speed, mm/s

For 6-lb sheet lead, equivalent to 6/64 in. (2.4 mm) thick, speeds of 4.3 to 8.5 in./min (1.8 to 3.6 mm/s) are thus predicted. In practice, 6 in./min (2.5 mm/s) is typical.

Vertical Position. Lap joints are almost exclusively used when welding in the vertical position (see Figure 5.11). Welding should begin at the bottom of the joint, which in many cases is a continuation of a flat lapped seam weld, as shown in Figure 5.11(A). Otherwise, a backing should be used to support the initial filler metal. Using a pointed flame, a pool of lead is first melted at the base of the joint root on the underlapping sheet. The torch is then moved to the adjacent front sheet and the area melted into the weld pool, completely liquefying the overlap.

The welder should try to melt off the corner of the lap edge at a 45-degree angle and move the liquid lead into the weld pool by following the flame path shown by the arrow in Figure 5.11(B). The flame is then moved higher on the underlapping sheet and the process repeated, eventually carrying the weld bead to the top of the joint. Filler metal is generally not used, the lap edges being the source of weld metal. If filler metal is needed, the torch should be removed momentarily after depositing a bit of filler to allow it to solidify immediately, but not cool appreciably.

In all cases, as shown in Figure 5.11(B), the flame should be pointed downward and either perpendicular to the sheet surface or about 60 degrees into the angle of the overlap. Because this is as difficult as overhead welding, it should be avoided when possible. Instead, as shown in Figure 5.12, the edge of the lap is turned out to produce a gap less than 1/4 in. (6.4 mm) wide. The torch is held overhand and pointed into the gap. In this case, the lap is not used as filler but provides the flare in which welding is performed. A separate filler rod is used. Controlling the weld pool requires some skill.

Care needs to be taken to avoid cutting into the underlapping sheet adjacent to the weld or cutting too far into the overlap, since both faults result in weak welds. When changing from a flat lapped weld to a vertical lap weld, steps are formed on the flat weld on the bend, as shown in Figure 5.13. Welding at this bend should be done without the use of filler metal and care should be taken to avoid undercutting.

It is always easier to weld a vertical joint that is inclined to the vertical, as shown in Figure 5.14, rather than strictly vertical. This allows the edge of the outer sheet to be angled outward, forming a groove face in which filler metal can be melted. This makes it easier to maintain a relatively flat surface on which the weld pool can be maintained.

Welding of vertical sheets is even easier in the horizontal position, as shown in Figure 5.15. By making a V-groove in the horizontal section, as shown in Figure 5.15(A), welds can be made easily.

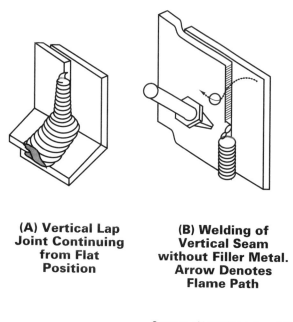

(A) Vertical Lap Joint Continuing from Flat Position

(B) Welding of Vertical Seam without Filler Metal. Arrow Denotes Flame Path

Courtesy of Lead Sheet Association

Figure 5.11—Vertical Seams

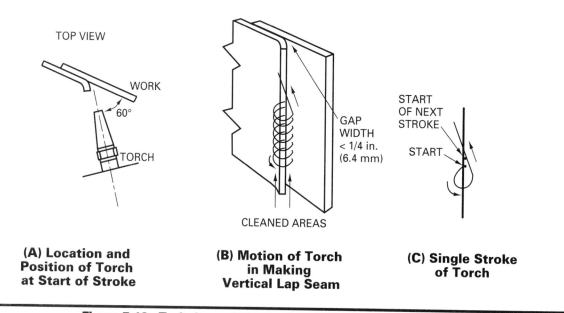

(A) Location and Position of Torch at Start of Stroke

(B) Motion of Torch in Making Vertical Lap Seam

(C) Single Stroke of Torch

Figure 5.12—Technique of Welding Lap Joint in Vertical Position

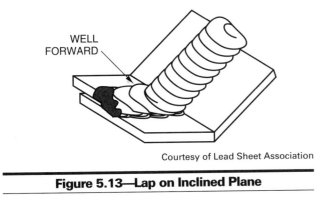

Courtesy of Lead Sheet Association

Figure 5.13—Lap on Inclined Plane

Overhead Position. Overhead welding of lead is not generally recommended because the high density of lead makes overhead pool control very difficult. If it is necessary to weld in the overhead position, a lap joint should be used. The flame should be as sharp as possible, weld beads should be small, and the operation completed as quickly as practicable. Filler metal is rarely used and only to build the bead to the required thickness. The pressure of the flame and capillary action of molten lead into the lap seam can be used by the skilled welder to control the liquid pool. An example of this position is shown in Figure 5.15(B).

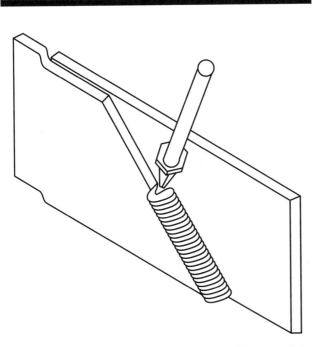

Courtesy of Lead Sheet Association

Figure 5.14—Inclined Seam on Vertical Face

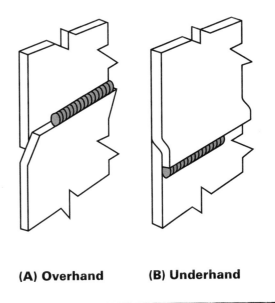

(A) Overhand   **(B) Underhand**

Figure 5.15—Joint Designs for Welding Horizontal Joints in Vertical Lead Sheet Using Overhand and Underhand Techniques

Lead Pipe Positions

FOR BEST RESULTS when welding pipe that is in the horizontal position, the work should be rotated to allow welding in the flat position. If the pipe cannot be rotated, butt joints generally are used. The V- or T-slot methods, which allow interior welding of the bottom of the pipe joint, are described in the earlier "Pipe" section on Page 295. This allows the entire joint to be welded in the flat position; otherwise, the underhand vertical position described for joining vertical and overhead sheet lap joints will be necessary. On pipes oriented vertically, lap or cup joints should be used, allowing the flat position to be used in all cases.

LEAD SOLDERING

LEAD AND LEAD alloys are easily soldered when proper care is taken not to melt the base metals, which also have relatively low melting temperatures. Solder joints in lead are generally confined to plumbing, some architectural uses, and lead-sheathed cables. The use of soldered lead joints in the chemical construction field, where highly corrosive chemicals are confined or transported, is not generally recommended. Such joints should be welded.

Solder Alloys

SOLDER ALLOYS MUST always be selected so that they can be worked without melting the base metal. Pure lead melts at 621 °F (327 °C), and one of the popular lead antimony alloys, Sb5, melts at 465 °F (240 °C). This difference includes a reasonable and practical temperature range for soldering of 135 °F (75 °C). However, to solder lead alloys that could have significantly lower melting temperatures, a lower melting solder should be used. The lowest melting temperature is obtained with the eutectic composition of 63% tin. It melts at 360 °F (180 °C) but lacks the mushy range, which can be desirable for some soldering techniques such as wiping.

A wide variety of solder compositions are covered in ASTM specification B32.[2] An alloy commonly used for joining lead sheet is SN50, which contains essentially 50% tin and 50% lead. Wiping is a special technique for soldering lead and uses alloys SN30B, SN35B, and SN40B. The numeric part of the ASTM specification indicates the percentage of tin, and these alloys also contain up to 2% antimony with the balance lead. The solders are solid up to around 360 °F (182 °C) and provide a pasty or working range of 100 °F (56 °C). Solders containing 34.5% tin, 1.25% antimony, 0.11% arsenic, and the balance lead are widely used in joining lead-sheathed power cables. The arsenic is added to promote a fine grain structure; the antimony for higher strength.

Fluxes

THE SOLDERING OF lead and its alloys can be accomplished without the use of the corrosive fluxes.

---

2. "Standard specification for solder metal (ASTM B32-95)." *Annual Book of ASTM Standards*, Vol. 02-04, 10-18. Philadelphia: American Society for Testing and Materials, 1995.

Tallow and rosin fluxes and stearic acid are generally used. Nonactivated rosin flux, which does not contain activating agents such as amines, is suitable for soldering lead. Typical compositions are shown in Table 5.2.

Surface Preparation

THE AREAS TO to be joined should be thoroughly cleaned by wire brushing or shaving. Tallow or stearic acid flux should then be applied promptly to prevent reoxidation of the cleaned areas. Only a very thin, flat film is advised to prevent the flux from spreading out beyond the area of the joint upon application of heat. Excessive use of cleaning tools should be avoided. Their overuse may cause fatigue failures due to chatter and thinning of the lead near the critical section of the joint. To confine the solder within the joint and help form and build a bead, gummed paper strips can be used. Also useful is plumber's soil, a mixture of lampblack, glue, and water.

Heating Methods

THE LOW MELTING point of lead and its alloys limits the choice of heating methods. Soldering irons are usually used for soldering sheet lead joints. In wiped joints, the heat for soldering is supplied by molten solder that is poured over parts such as pipe or cable sheathing, thereby both fusing and wetting the base metal while cooling the bulk of the solder, making it pasty and workable.

Joint Types

Lap Joints. Lap joints are more satisfactory than butt joints and should be made with a minimum lap of 3/8 in. (9.5 mm) for lead sheet up to and including 1/8 in. (3.2 mm) thick [8-lb sheet]. The joint roots of

Table 5.2
Nonactivated Rosin Fluxes

| Flux | Composition, wt. % | | | | | |
	Water White Rosin	Glutamic Acid Hydrochloride	Cetyl Pyridinium Bromide	Stearine	Hydrazine Hydrobromide	Alcohol
1	10-25	—	—	—	—	bal.*
2	40	2	—	—	—	bal.
3	40	—	4	—	—	bal.
4	40	—	—	4	—	bal.
5	40	—	—	—	2	bal.

*   Alcohol, turpentine, or petroleum

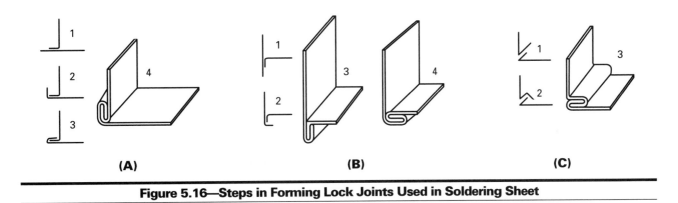

Figure 5.16—Steps in Forming Lock Joints Used in Soldering Sheet

the two sheets that form the lap should be cleaned and fluxed with tallow. The cleaned and fluxed side of the bottom sheet should extend 1/8 in. (3.2 mm) beyond the leading edge at the lap. The edge and upper side of the top sheet, to a distance of approximately 3/8 in. (9.5 mm), should also be cleaned and fluxed. The sheets are then fitted together and dressed down with a wooden or rubber mallet to fit smoothly.

Soldering is usually done with an iron and a 50% tin, 50% lead solder, first by tacking at intervals with solder. Then application of additional flux is often advisable. The flux source may be within rosin cored or stearine cored wire solder. When bar solder is used, stearine or powdered rosin should be applied to the joint.

Lock Joints. Lock joints are lap joints which have been mechanically formed before joining to prevent movement. They provide considerably more strength and are preferred whenever the joints are to be in tension. They are made in much the same way as lap joints, using locks of 1/2 in. (13 mm) or more. The solder should flow between the two lower root faces and contact with the lock. Three examples of typical lock joint geometries are shown in Figure 5.16.

Butt Joints. Butt joints are the least desirable type for joining lead sheets and are made using a process termed *solder welding*. This technique should be confined to those situations where it is impractical to use other joint designs. The abutting edges of the lead sheet are beveled with a shave hook to make an angle of 45 degrees or more with the vertical. These edges are placed firmly together and tacked at intervals of 4 to 6 in. (102 to 152 mm). Gummed paper strips pasted parallel to the seam and 1/4 to 3/8 in. (6 to 9 mm) away aid in building up the solder and reflowing it in the refinishing operation. Additional flux, as described in the lap joint section, is advisable. Solder is fed into the joint and melted by the soldering iron

as it is drawn along the joint root. Sufficient solder should be applied to build a slightly convex surface.

Pipe Joints. As important to successful results as soldering is the joint preparation. Recommended are flanged joints made in a bell and spigot manner as shown in Figure 5.17. The flared end is that into which the liquid will flow. The spigot end is beveled to fit snugly into the flared end.

The entire area to be wiped is shaved clean, lightly, as is the contacting area within the joint. A thin coat of

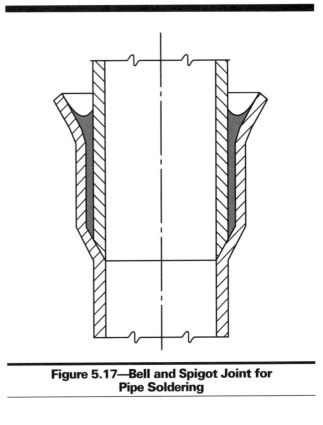

Figure 5.17—Bell and Spigot Joint for Pipe Soldering

tallow is then applied. The area beyond the joint root on both sides is then coated with plumber's soil or paper to prevent the solder from adhering at these points. The joint is assembled, the flare end is dressed down tightly by swaging with a wooden tool, and the entire assembly is braced so that it will not move during the subsequent soldering operation.

The wiping solder is heated in a ladle to a temperature of 600 °F (315 °C). To join horizontal pipe, the joint is wiped by slowly pouring the solder on top of the joint while containing the solder with a tallow-coated cloth. When the solder is in its pasty stage, the operator wipes or forms the joint. When completed, the joint should be chilled.

In vertical pipe, joints are prepared and wiped in much the same manner, except the solder is poured around the pipe at the top of the joint, and the cloth is held directly under the ladle at the bottom of the joint. The ladling or splashing-on of the solder is continued around the joint.

Branch joints in lead pipe are made by cutting a small oval-shaped hole in the main line and drawing up sufficient lead to form a collar or hub into which the beveled branch line is fitted snugly. Preparation or wiping is essentially the same as previously described.

Cup joints are similar to bell and spigot joints except that the flared end is not dressed down, and a soldering iron is used rather than wiping. These joints can be made only in the vertical position, although they can be used in any position. Preparation of the joint includes beveling of the inlet or spigot end, flaring the bell, cleaning, and fluxing with tallow only those areas to become a part of the joint. Plumber's soil or paper should be applied beyond those points. The pipes are then fitted together and spot soldered. With a sharp-pointed iron, solder is flowed around until the joint is filled about halfway. A blunted iron is used to fill the rest of the cup with solder.

Soldering of Lead Alloys

WITH THE POSSIBLE exception of alloys containing more than 2% antimony, lead alloys can be soldered in the same manner as described for lead. These alloys have a solidus temperature under 600 °F (315 °C), and so the soldering temperature factor is critical. Extreme care is necessary in soldering to avoid melting the base metal. Being heavier, ornamental castings alloyed with more than 2% antimony are more easily soldered.

Soldering Lead to Other Metals

ALL METALS THAT can be precoated with lead can be joined to lead pipe or sheet by wiping or heating with a torch or soldering iron. The precoating should be confined to those areas of the other metal that are to become an integral part of the joint. The joint should be prepared just as described for joining lead to lead.

COATING OF STEEL

LEAD IS ONE of the most common metals used to protect steel from corrosion. Because lead does not directly alloy with steel it must be alloyed with 5 to 15% tin to obtain satisfactory wetting of, and adherence to, steel sheet. Steel sheet coated with this tin alloy is called *terne* or *long terne*. Steel grades used in typical products can be commercial quality or drawing quality with compositions ranging broadly between 0.1 - 0.2% C, 0.5 - 0.9% Mn, 0.025 - 0.2% P, and 0.035 - 0.04% S. Such steels normally are continuous and aluminum-killed.

Coating thicknesses and other requirements for terne are given in ASTM specification A308.[3] Such coatings can be supplied in thicknesses up to 0.0012 in. (0.03 mm) which is equivalent to 1.12 oz/ft^2 (342 g/m^2).

A significant requirement is that most grades be bendable 180 degrees upon themselves without cracking on the outside of the bend.

Terne sheet is produced either by a continuous hot-dip coating process, in which coils of steel sheet are immersed in a molten lead-tin bath, or by the continuous electrolytic coating process, in which a fluoroborate solution of lead and tin is used to plate the steel sheet.

Welding of terne plate will be covered in the chapter "Precoated Steels" in the next volume of the *Welding Handbook*, 8th Edition.

WELDING OF DISSIMILAR METALS

THE WELDING OF lead to other metals is generally accomplished by applying a thick coating of lead to steel, copper, or other metals. The resultant product has the corrosion resistance of lead with the structural strength of other metal and is commonly referred to as *bonded lead* or *homogeneously lined lead*.

Various methods of bonding lead to other materials are available to the fabricator. These include the application of lead by hand welding or by casting lead onto the other metal. In any case, surface preparation of the metal to be covered is of paramount importance. Because pure lead does not readily wet or alloy with other metals, it is necessary to use special techniques or an alloy of lead, such as solder or antimonial lead, to coat the bare steel prior to the application of the lead

---

3. "Standard specification for steel sheet, terne (lead-tin alloy) coated by the hot-dip process (ASTM A308-94)." *Annual Book of ASTM Standards,* Vol. 01-06, 30-32. Philadelphia: American Society for Testing and Materials, 1995.

lining. Because steel is generally used as a base metal in bonded lead equipment, the procedures as applied to steel will be described in some detail. The same general techniques are applicable to other metals.

Precoating Steel with Solder

STEEL THAT IS to be solder coated should first be mechanically cleaned by grit blasting or grinding, or chemically cleaned by pickling. In handling the steel after this step, extreme care must be taken to prevent oil or grease from coming into contact with the cleaned surface.

A corrosive flux, such as zinc chloride with a composition shown in Table 5.3, is then applied, followed by the application of a 50% tin-50% lead solder. This is generally done by heating with an oxyfuel torch or by dipping the steel into a molten metal bath. While still molten, any excess solder is wiped off with a cloth. After the solder has solidified, but while still hot, the remaining flux should be removed by wiping with a cloth that has been dipped in hot water.

Precoating Steel with Antimonial Lead

THE PROCEDURE TO be followed when coating steel with antimonial lead is exactly the same as that described for coating steel with solder, but substituting 6% antimonial lead for the 50-50 solder. Either zinc chloride flux or no flux is used.

Lead Coating of Precoated Steel with an Oxyfuel Torch

IN APPLYING A lead coating to precoated steel with a torch, the flame is directed onto the steel coating until the steel coating is melted. The flame then is transferred to the lead filler rod, which is held close to the area being fused. As the lead flows, the torch is transferred to the next overlapping area. No attempt should be made to lay down the required thickness of lead in one pass. Three passes are usually required to build up a thickness of 0.25 in. (7 mm).

After the first pass, the lead should be carefully examined for pin holes and other faults. If any exist, they should be repaired by scraping down to the steel, reapplying the *tinned* coating, and filling with lead. This first layer of lead may be scraped to obtain a more uniform thickness before the second pass is applied. After each pass, the surface should be examined for faults and repaired. When the specified thickness of lead is reached, the lead may be machined or scraped to a smooth, uniform surface.

Lead Coating of Uncoated Steel with an Oxyfuel Torch

STEEL TO BE coated with lead first should be mechanically or chemically cleaned. Just before the coating operation, a flux of zinc-ammonium chloride and stannous chloride is applied to the working area. This galvanically protects the steel with a thin layer of tin.

Table 5.3
Zinc Chloride Fluxes for Soldering Coated and Uncoated Steels

Flux	Composition, wt. %						
	Zinc Chloride	Ammonium Chloride	Sodium Chloride	Stannous Chloride	Hydrochloric Acid	Petroleum Jelly	Water
1	40 oz (1130 g)	4 oz (110 g)	—	—	—	—	1 gal[a] (4 L)
2	36 oz (1020 g)	0.5 oz (15 g)	10 oz (280 g)	—	1 oz (30 g)	—	1 gal[a] (4 L)
3	25 oz (710 g)	3.5 oz (100g)	—	—	—	65 oz (1840 g)	6.5 oz (180 g)
4[b]	85 oz (2410 g)	6.5 oz (180 g)	—	9 oz (260 g)	2 oz (60 g)	—	1 gal[a] (4 L)

a. Water to make (not wt. %)

b. Also, optional wetting agent (0.1 wt. %)

The neutral flame is played on both the steel and the lead filler material. Care should be exercised to prevent overheating the flux before the lead starts to flow. Subsequent procedures are the same as those described in the preceding section. This method requires greater care to obtain a complete bond.

Lead Coating by Casting Methods

IN THIS METHOD of coating, the steel is prepared by mechanically cleaning or pickling the surface to be lined. The surface is fluxed with zinc-ammonium chloride and stannous chloride (see Table 5.3), and the underside of the sheet is heated generally. The lead filler is rubbed vigorously on the treated steel surface until a thin coating of lead adheres to the steel. Molten lead is then cast on the surface to approximately twice the desired thickness.

Any insoluble, such as dross formed in casting, will float to the surface. After the lead has cooled, this excess coating is removed by scraping to the desired thickness. The roughened surface is smoothed by flame washing, which also serves to repair any minor defects.

WELDING AND SOLDERING OF ZINC

ZINC AND ITS PROPERTIES

ZINC IS MOST commonly encountered as a protective coating on steel. Often it is alloyed with aluminum, copper, or lead to give specific qualities. Another significant use of zinc is in cast form, almost always alloyed with aluminum and lesser quantities of other metals, notably copper and magnesium. Rolled zinc also is used especially in roofing and chemical process applications where fabrication and joining are important. An alternative to galvanized zinc coatings is zinc-rich paint, containing 85 to 97% metallic zinc; these are commonly referred to as *weld-through primers*.

Zinc has a specific gravity of 7.14, which is slightly less than that of iron. It has a melting point of 787 °F (419 °C) and a relatively low boiling point of 1665 °F (907 °C). Because the boiling point of zinc is below the melting point of steels, 2700 to 2800 °F (1500 to 1570 °C), significant fuming can occur during welding of zinc-coated steel products. The electrical conductivity of commercially pure zinc is about 25 percent that of copper and increases approximately 30 percent when alloyed within the range of useful aluminum contents.

Compositions of standard zinc alloys are shown in Tables 5.4 and 5.5.

Cast zinc is easily electroplated, and it is non-sparking and non-magnetic. Zinc alloys are too brittle to be mechanically worked at room temperature but may be rolled at 300 to 500 °F (150 to 260 °C) or extruded at 480 to 650 °F (250 to 345 °C), depending upon composition. Zinc alloys have recrystallization temperatures that are at or below room temperature.

Table 5.4
Composition of Common Zinc Casting Alloys, wt. %

Element	Cast Alloys [a,b]						
	2	3	5	7	ZA-8	ZA-12	ZA-27
Al	3.5-4.3	3.5-4.3	3.5-4.3	3.5-4.3	8.0-8.8	10.5-11.5	25.0-28.0
Mg	0.02-0.05	0.02-0.05	0.03-0.08	0.005-0.020	0.015-0.030	0.015-0.030	0.010-0.020
Cu	2.5-3.0	0.25 max	0.75-1.25	0.25 max	0.8-1.3	0.5-1.2	2-2.5
Fe (max)	0.10	0.10	0.10	0.075	0.075	0.075	0.075
Pb (max)	0.005	0.005	0.005	0.003	0.006	0.006	0.006
Cd (max)	0.004	0.004	0.004	0.002	0.006	0.006	0.006
Sn (max)	0.003	0.003	0.003	0.001	0.003	0.003	0.003
Ni	—	—	—	0.005-0.020	—	—	—
Zn	Balance	Balance	Balance	Balance	Balance	Balance	Balance

a. ASTM B86-88 is the source for cast alloys 2, 3, 5, and 7; ASTM B791-91 is the source for alloys ZA-8, ZA-12, and ZA-27.

b. Zinc alloy die castings may contain nickel, chromium, silicon, and manganese in amounts of 0.2, 0.2, 0.035, and 0.05%, respectively. No harmful effects have ever been noted because of these elements in these concentrations.

Table 5.5
Composition of Common Wrought Zinc Alloys

Alloy Family	Alloying Elements							Impurities			
	Cu	Ti	Al	Mg	Cd	Pb	Fe	Pb	Fe	Cd	Cu
Pure Zinc	—	—	—	—	—	—	—	<0.003	<0.002	<0.003	<0.001
	—	—	—	—	0.004-0.006	—	—	<0.01	<0.002	—	—
Zn-Cu	0.7-0.9	—	0.002	—	0.005	—	0.008	<0.02	<0.008	<0.02	—
Zn-Cu-Ti	0.7-0.9	0.08-0.14	0.005	—	0.02	0.02	0.01	<0.020	<0.010	<0.020	—
	0.4-0.5	0.15	0.002	—	—	—	—	<0.003	<0.002	<0.003	—
	0.18	0.08	0.002	—	—	—	—	<0.003	<0.002	<0.003	—
Zn-Pb-Cd-Fe	0.002	—	—	—	0.04-0.06	0.06-0.08	0.008	—	<0.008	—	<0.002
	—	—	—	—	0.015	0.2-0.3	0.005	—	—	—	<0.001
	—	—	—	—	0.005	0.6-0.7	—	—	<0.002	—	<0.001
Zn-Al (Superplastic)	0.4-0.6	—	21-23	0.008-0.012	—	—	—	<0.01	—	<0.01	—

A zinc-copper-titanium alloy is widely used for rolled products and can be reduced to half its thickness without interstage annealing. Rolled zinc sheet is very ductile and should be fabricated at temperatures above 70 °F (20 °C).

WELDING OF ROLLED AND CAST ZINC

Joint Design

Rolled Zinc. Rolled zinc is best welded under controlled shop conditions; soldering is much preferred for joining during erection of structures. Sheets less than 0.040 in. (1 mm) thick are flanged or lapped and welded together without filler metal. Joints may be tacked at intervals to prevent movement during welding. In all cases, it is advisable to minimize heat input to the joint to limit recrystallization, which will weaken the joint in the heat-affected zone (HAZ). Resistance welding provides sound welds with less heat input than is possible with oxyfuel or arc welding.

If one of the latter processes is used to weld sheet thinner than 0.125 in. (3.2 mm), a square butt joint with a gap equal to the metal thickness is recommended. Thicker sheets should be beveled to form a 70 to 90 degree included angle with little or no land. Lap joints also may be used, with a fillet weld generally made on both sides of the sheets.

Resistance welding, using either spot welds or seam welds, is always done on lap or flanged joints. The thickness of the sheet dictates the minimum overlap of the joint.

Cast Zinc. Zinc castings can be welded using oxyfuel or arc welding processes. Joint geometries and welding procedures are described in Table 5.6.

Square butt joints with no gap can be used for casting section thicknesses up to 0.125 in. (3.2 mm). Thicknesses up to 0.25 in. (6.4 mm) can be joined using a gap of 0.0625 in. (1.6 mm). Over 0.25 in. (6.4 mm) in thickness, a V-groove with a 90-degree included angle with no root opening can be successfully used. The gas tungsten arc process is preferred because of its good heat-control capabilities. With oxyfuel welding processes, a narrower included angle, typically 45 degrees, is used in all of the above joints. Regardless of the process, it is likely that the mechanical properties of the joint will be inferior to the bulk casting. Careful testing should be carried out.

Surface Preparation

Rolled and Cast Zinc. Joint areas must be thoroughly mechanically cleaned and abraded. Cleaning methods include grinding, machining, and buffing; abrading with an emery board or sandpaper has given good results. These should be followed by degreasing if necessary. With castings, all electroplated coatings must be removed from joint areas to be fused into the weld.

Welding Processes

Arc Welding of Rolled Zinc. Although oxyfuel welding processes are more often used, arc welding,

Table 5.6
Weld Procedures for Zinc-Aluminum Alloy Castings up to 1/2 in. (13 mm) Thick
Using the Gas Tungsten Arc Process

Plate Thickness		Weld Preparation		Approximate Current, A	Passes Made	Filler Metal
in.	mm	Diagram	Description			
1/16	1.6		Square Butt, No Gap	40	One each side	None
1/8	3.2		Square Butt, No Gap	50	One each side	None
1/4	6.4		Square Butt, 1/16 in. (1.6 mm) Gap	85	One each side	1/16 in. (1.6 mm) wire
1/4	6.4		90° Groove, No Gap	50	Two passes, one side	1/16 in. (1.6 mm) wire
1/2	12.7		90° Groove, No Gap	50 root 80 2nd pass	4 to 5 passes, one side	1/16 in. (1.6 mm) wire
1/4	6.4		60° Bevel on Tee	60	Two passes, one side, no filler in root	1/16 in. (1.6 mm) wire
1/4	6.4		No Gap, No Penetration	60	One pass; fillet only	1/16 in. (1.6 mm) wire

specifically the gas tungsten arc welding (GTAW) process, can produce more uniform penetration and satisfactory welds in rolled zinc alloys. It is possible to limit heat input to the base metal more easily with arc welding because a very short arc length can be used. Argon shielding gas allows lower voltages, also minimizing heat input. Pulsing of high-frequency current over a base 60-cycle alternating-current power supply helps to overcome the arc instability caused by titanium, aluminum, and zinc oxides. Should the zinc vapor contaminate the tungsten electrode after a short period of time, the oxyfuel welding processes may prove preferable. A high shielding gas flow rate [at least 30 ft^3/h (14.2 L/min)] can partially remedy such contamination problems.

Manual arc welding requires considerable skill, especially in field erection. Automatic welding, usually performed under controlled shop conditions, produces the best results because the arc length and travel speed can be accurately controlled. Inert gas backing on the backside of the joint root is recommended to overcome the tendency for lack of fusion at the butting edges of the joint root, a condition caused by the formation of oxides when plain refractory or copper backup materials are used.

Table 5.7
Gas Tungsten Arc Welding Schedules for Zinc-Copper-Titanium Wrought Zinc Alloys

Sheet Thickness		AC Current, A	Arc Length		Tungsten Electrode				Argon Flow		Travel Speed		Filler Metal
					Diameter		Stick-Out						
in.	mm		in.	mm	in.	mm	in.	mm	ft³/h	L/min	in./min	mm/s	
Automatic Welding—Butt Joint[a]													
0.024	0.61	55	0.062	1.58	0.040	1.02	3/16	4.76	25	118	50	212	None
0.062	1.58	100	0.075	1.91	0.062	1.58	1/4	6.35	25	118	50	212	None
0.075	1.91	125-130	0.093	2.36	0.062	1.58	5/16	7.93	25	118	30	127	Base Metal
0.125	3.18	170-180	0.125	3.18	0.125	3.18	3/8	9.53	25	118	20	85	Pure Zinc
Automatic Welding—Fillet Overlap													
0.062	1.58	95	0.093	2.36	0.062	1.58	5/16	7.94	25	118	30	127	None
0.075	1.91	105	0.093[b]	2.36	0.062	1.58	5/16	7.94	18	85	30	127	None
0.125	3.18	140	0.156[b]	3.96	0.062	1.58	5/16	7.94	25	118	20	85	None
Manual Welding—Butt Joint													
0.040	1.02	30	0.062	1.58	0.062	1.58	5/16	7.94	25	118	6	25	Base Metal
0.075	1.91	150	0.125	3.18	0.062	1.58	5/16	7.94	25	118	18	76	Base Metal

a. Square butt joint preparation except 0.125 in. (3.18 mm) which is 60° V-groove.

b. Measured from bottom plate.

A typical schedule for GTAW of a wrought zinc alloy is shown in Table 5.7. For this example, a zinc-copper-titanium was selected. To obtain the best results, this system requires use of filler metal that will produce grain refinement in the deposit without appreciably reducing the ductility by precipitation hardening during cooling. The titanium in the filler metal produces grain refinement in the cast structure, but it also may cause a precipitation reaction, which decreases the ductility of the weld deposit. Titanium contents of the deposit should be no greater than 0.08 to 0.12 percent to be effective. The copper content should match the base metal. This composition will produce sound and ductile weld deposits.

Transverse bend tests indicate that improvements in the joint ductility can be obtained by hot rolling the weldments 15 percent at 300 °F (150 °C) and then annealing them at 550 °F (290 °C) for one hour. This produced successful 90-degree bends on a mandrel with a radius five times the thickness of the zinc.

The highest tensile ductility with moderate strengths was obtained after hot rolling and annealing. The tensile strength was 19 ksi (131 MPa) with an elongation of 5 percent across the weld deposit. The joint strength was approximately 90 percent of base metal strength.

A typical cross-section of the gas tungsten arc weld in wrought zinc is shown in Figure 5.18.

Gas metal arc welding is not used on zinc-based alloys due to excessive spatter and erratic transfer associated with this process.

Arc Welding of Cast Zinc. Zinc castings can be joined with the GTAW process if care is taken. The same precautions to control electrode contamination are needed as were previously described for rolled zinc. Considerable skill is required to prevent overheating of the casting. Clamping and alignment of the joint should be considered carefully before beginning the weld because they will have a great influence on the weld quality. A reliable workpiece lead must be provided.

Thin sections, less than 0.040 in. (1 mm), can be welded without filler metal. For all zinc-aluminum casting alloys, a filler metal corresponding to the ZA-8 (1% Cu, 8.5% Al) is suitable. The wider freezing ranges of other alloys that contain more copper and aluminum make their use as fillers unsuitable. Argon shielding with a flow rate of 30 ft³/h (14.2 L/min) will give satisfactory protection. A 0.375 in. (9.5 mm) GTAW torch cup is suitable. If a zinc oxide film develops in the weld pool, a higher gas

Figure 5.18—Gas Tungsten Arc Weld in 0.125 in. (3.2 mm) Thick Zinc-Copper-Titanium Alloy with Pure Zinc Filler Metal

flow rate should be used. An alternating current of 50 amperes should be used with a potential of 18 to 20 volts. A 0.0625 in. (1.6 mm) thoriated tungsten electrode is recommended.

The flat position welding technique should be used for welds to allow easy control of the liquid metal pool. Butt and V-groove joints will require that the torch be moved in a weaving manner, with a slight hesitation on each side while filler metal is added in the center. Corner joints require that more heat be added, in which case the torch should be guided in a circular motion, bringing heat to the two butting members and withdrawing to allow filler metal to be added.

Oxyfuel Welding of Rolled Zinc. A small welding torch tip should be used to minimize heat input into the base metal. For example, a tip suitable for welding 0.031 in. (0.8 mm) thick steel is suitable for welding 0.125 in. (3 mm) thick zinc. To minimize surface oxidation, the flame should be neutral or slightly reducing.

The filler metals may be either pure zinc or the same composition as the base metal. The filler rod diameter should be approximately two-thirds the thickness of the metal being welded, to a maximum of 0.16 in. (4 mm). The tendency for overheating the base metal may be minimized by using larger diameter rods or by directing the flame of the torch onto the filler rod.

Since holes may be formed if the torch is held perpendicular to the surface, the torch should be maintained at an angle of 15 to 45 degrees to the work, depending upon the sheet thickness. The thinner the material, the smaller the angle. In groove welding, the flame is directed toward the rod, which is raised out of the flame after each drop is deposited. The torch should be oscillated to prevent burning through the sheet. Backhand or forehand welding technique is satisfactory. The joint must have adequate support to prevent excessive dropthrough. Back-up materials may be graphite, fire clay, or plaster of paris. To improve strength and ductility, the weld joint may be peened. This operation, however, must be done between 200 to 300 °F (90 to 150 °C); peening at or below room temperature or above 300 °F (150 °C) may crack the weld.

Oxyfuel Welding of Cast Zinc. The oxyfuel gas welding process has been successfully used to repair zinc alloy die castings as thin as 0.035 in. (0.8 mm). Joint preparation of defects in castings consists of removing all grease, oil, dirt, oxides, and any nickel or chrome plating from the weld area. After cleaning, the defect is grooved through the full thickness of the casting to an included angle of 45 degrees.

If the casting is large and complicated, the whole structure must be preheated to approximately 250 °F (120 °C) because localized heating may result in serious distortion. The joint area should be further heated with a torch until fine droplets appear on the surface at the joint area. Extra heating must be confined to the repair area. Should these droplets appear beyond the joint area, the part has been overheated and might be destroyed. The welding torch should be adjusted to give a soft, lazy, soot-free reducing flame. The part must be properly supported during welding with either carbon paste, fire clay, or plaster.

A filler metal of the composition of AC-43A alloy (3% Cu, 4% Al) is often used because it has a low melting point; however, the ZA-8 alloy (1% Cu, 8.5% Al) offers a joint with higher strength. The preheat is maintained during welding by occasionally sweeping the flame over the casting. The filler rod or a small puddling spade may be used to break up and scrape away surface oxide as filler metal is slowly added. The weld may be shaped with a puddling spade while it is in a mushy condition.

Resistance Welding of Rolled Zinc. Wrought zinc alloys may be resistance spot or seam welded by using relatively high welding currents and low electrode forces. Compared to steel, zinc alloys have approximately twice the electrical conductivity, much lower melting ranges, approximately the same specific heats, and higher thermal conductivities. Because of the softness and generally greater thicknesses of zinc alloy stock, a low-inertia mechanical follow-up is needed for welding electrode control. Normal copper alloy electrodes are not recommended because of the tendency of the electrode to be contaminated and alloyed by zinc after repeated welds. A dispersion-strengthened alloy such as CDA Alloy 15760 should be used for long welding campaigns. A suitable spot-welding machine would have 150-kVA, single-phase, alternating current. If the zinc is reasonably clean, only degreasing is necessary. Otherwise, mechanical cleaning is required.

Internal defects, or cavities, are common in zinc-alloy spot-weld nuggets. These can be minimized by using a welding schedule employing a forging force. The forging force has little effect on spot-weld strength. A typical spot weld in wrought zinc is shown in Figure 5.19.

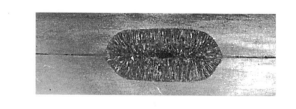

Figure 5.19—Typical Resistance Spot Weld in 0.125 in. (3.2 mm) Thick Zinc-Copper-Titanium Alloy

Electrode spot-welding and seam-welding schedules for wrought zinc alloys are given in Tables 5.8 and 5.9.

Direct tension-tensile shear ratios of spot-weld strengths usually are about 0.35 for 0.075 in. (1.9 mm) thick zinc-copper-titanium alloy. This ratio is a measure of spot-weld ductility and should be at least 0.5 for the weld to be considered ductile.

As roll speed is increased, current also must be increased while the cool time is decreased to maintain adequate penetration and proper nugget spacing. Welding speed may be more than double in thinner sections. A typical seam weld in wrought zinc is shown in Figure 5.20.

Resistance Welding of Cast Zinc. Resistance welding techniques for joining zinc alloy die castings

are very similar to those used for wrought zinc. Mechanical cleaning of surfaces prior to welding is required to remove the high-resistance films which promote excessive weld metal expulsion. Because of the tendency of aluminum and zinc aluminum casting alloys to alloy with copper spot-welding electrodes, dispersion-strengthened electrodes such as CDA Alloy 15760 aluminum-oxide-strengthened copper electrodes are desirable. A truncated cone geometry is preferable.

Spot-welding schedules given in Table 5.8 apply also to die-cast material for thicknesses greater than 0.075 in. (1.9 mm). For lesser thicknesses, the indicated welding current should be reduced by 10 percent. A typical die-cast spot weld is shown in Figure 5.21.

Limited tests indicate that seam-welding schedules given in Table 5.9 also can apply to die-cast alloys,

Table 5.8
Resistance Spot Welding Schedules for Wrought Zinc-Copper-Titanium Alloy

Thickness		Electrode Diameter*		Electrode Force		Weld Time, Cycles	Approximate Weld Current, A	Minimum Overlap		Minimum Weld Spacing		Weld Diameter		Tensile Strength	
in.	mm	in.	mm	lb	N	Cycles	A	in.	mm	in.	mm	in.	mm	lb	N
0.010	0.25	1/2	12.7	100	445	1	7000	3/8	9.5	1/4	6.4	0.075	1.91	60	267
0.020	0.51	1/2	12.7	180	801	3	9500	3/8	9.5	5/16	7.9	0.120	3.05	200	890
0.030	0.76	1/2	12.7	260	1157	5	12 000	7/16	11.1	3/8	9.5	0.160	4.06	330	1468
0.040	1.02	5/8	15.9	350	1557	6	17 000	1/2	12.7	1/2	12.7	0.195	4.95	460	2046
0.050	1.27	5/8	15.9	420	1868	8	18 000	9/16	14.3	9/16	14.3	0.220	5.59	600	2669
0.060	1.52	5/8	15.9	500	2224	9	19 000	5/8	15.9	3/4	19.1	0.250	6.35	740	3292
0.070	1.78	3/4	19.1	580	2580	11	20 500	11/16	17.5	7/8	22.2	0.270	6.86	880	3914
0.080	2.03	3/4	19.1	660	2936	13	21 500	3/4	19.1	1	25.4	0.290	7.37	1000	4448
0.090	2.29	3/4	19.1	740	3292	14	23 000	3/4	19.1	1-1/8	28.6	0.300	7.62	1150	5115
0.100	2.54	7/8	22.2	820	3648	16	24 500	7/8	22.2	1-1/4	31.8	0.315	8.00	1280	5693
0.110	2.79	7/8	22.2	900	4003	17	26 000	7/8	22.2	1-3/8	34.9	0.330	8.38	1420	6316
0.120	3.05	7/8	22.2	980	4359	18	27 000	1	25.4	1-1/2	38.1	0.340	8.64	1550	6894
0.130	3.30	7/8	22.2	1050	4671	19	28 000	1	25.4	1-5/8	41.3	0.360	9.14	1680	7473

* Electrode material — RWMA Group A, Class 2
Electrode contour — 3 in. (76.2 mm) radius dome

Table 5.9
Resistance Seam Welding Zinc-Copper-Titanium Alloy

Gauge		Electrodes*						Squeeze Roll Force		Time		Welding Speed		Weld Spacing		Tensile Strength	
		W		E		D				Heat Cycles	Cool Cycles						
in.	mm	in.	mm	in.	mm	in.	mm	lb	N	Cycles	Cycles	in./min	mm/s	in.	mm	lb	N
0.010	0.25	3/8	9.5	3/16	4.8	5	127	150	668	2	16	50	21	1/4	6.4	40	178
0.024	0.61	3/8	9.5	3/16	4.8	5	127	250	1112	4	20	50	21	5/16	7.9	127	565
0.040	1.02	3/8	9.5	3/16	4.8	5	127	400	1780	6	30	50	21	1/2	12.7	195	868
0.075	1.91	5/8	15.9	3/8	9.5	8-7/8	225	800	3560	8	70	40	17	7/8	22.2	840	3738
0.100	2.54	5/8	15.9	3/8	9.5	8-7/8	225	1000	4450	12	90	40	17	1-1/4	31.8	1230	5474

* Electrodes: Group A, Class 2 with 15-degree bevel and face radius of 3 in. (76.2 mm)

Figure 5.20—Typical Resistance Seam Weld in 0.075 in. (1.9 mm) Thick Zinc-Copper-Titanium Alloy

Figure 5.21—Typical Resistance Spot Weld in 0.075 in. (1.9 mm) Thick Zinc Die Casting Alloy

Figure 5.22—Typical Resistance Seam Weld in 0.075 in. (1.9 mm) Thick Zinc Die Casting Alloy

Table 5.10
Soldering Fluxes

Formula	1	2	3	4
		Parts by Weight		
Hydrochloric Acid	1	1	2	—
Zinc Chloride	—	—	85	25
Ammonium Chloride	—	—	6.5	10
Stannous Chloride	—	—	9	—
Alcohol	—	1	—	—
Wetting Agent	—	—	0.23	0.18
Water	1	1	128	64

although problems may be encountered with surface melting at the weld zone. A typical resistance seam weld cross-section is shown in Figure 5.22.

SOLDERING OF ROLLED AND CAST ZINC

Soldering of Rolled Zinc

WROUGHT ZINC ALLOYS can be successfully soldered. If the zinc surface has been weathered, it will be necessary to clean it to restore a bright metal finish before fluxing. The surface should be cleaned with solvent and flux sparingly applied, not poured on the surface. A short time should be allowed for the metal-flux reaction to take place. A solution of zinc chloride can be used as a flux, which also can be mixed with ammonium chloride in the proportions shown in Table 5.10. Some proprietary non-acid fluxes also are available.

The melting point of a good flux should be below that of the solder, so that melting and flow will be produced from the heat of the soldering iron. A 60/40 or 50/50 tin/lead solder should be used. Antimony-bearing solder is not satisfactory because it produces brittle joints. Strong, neat joints are not so easily made with solders containing less than 45% tin.

The joint faying surface may be abraded slightly to begin fusing action. Neither preheating nor pretinning are needed in normal practice. Flame heating is not recommended. Heat and solder may be applied with a large soldering iron. It is important that the soldering iron moves slowly in one direction so that the solder, following the iron, flows into the seam. The iron must not be heated to redness nor allowed to dwell in one spot because of the risk of melting the base metal. A back-and-forth movement of the soldering iron may cause the zinc coating to dissolve in the solder. This makes the solder gritty, sluggish, and brittle. If a non-corrosive flux is desired, as in electronic assemblies, faying surfaces must be precoated with solder.

The soldering iron should be cleaned with ammonium chloride solution from time to time. When completed, the soldered joint should be washed with water or wiped with a wet rag to remove excess acid and flux.

Best results are obtained using a solder stick with a cross section of 0.06 by 0.19 in. (1.6 by 4.8 mm). The narrow edge permits the bar to touch the surfaces close to their intersection. The wide edge is better for finished fillets.

Solder joints should be formed with a lap at least 0.06 in. (1.6 mm) wide, and preferably 0.4 to 0.6 in. (10 to 15 mm) wide. Butt joints are much weaker and should be avoided when there is the possibility of the joint being stressed during use. When soldering a joint

Table 5.11
Typical Soldered Joint Strengths

| Thickness | | Type Material | Solder Type | Solder Liquidus | | Shear Strength | |
in.	mm			°F	°C	ksi	MPa
0.075	1.9	AG40*	40 Sn-60 Pb	455	235	3.0	21
0.075	1.9	AG40*	60 Sn-40 Pb	374	190	4.4	30
0.100	2.5	Zn-Cu-Ti	5Ag-17Zn-78Cd	480	249	>8.0	>55

* Surface nickel-plated prior to soldering.

more than 0.6 in. (15 mm) long, the joint should be tacked with solder at intervals of about 2 in. (50 mm) which will prevent movement. Typical solder strengths for alloys are shown in Table 5.11.

Soldering of Cast Zinc

IT IS NEARLY impossible to solder as-cast zinc casting alloys. However, the same tin/lead alloys used for soldering rolled zinc can be used to solder zinc castings if the faying surfaces are first electroplated with copper or nickel. A thick layer of either metal should be electroplated onto the surfaces to be soldered. This also prevents contamination of the base metal by the solder, which can cause intergranular corrosion in service. Soldering should be done as quickly as possible to minimize interdiffusion. An acidified zinc chloride flux such as composition 3 in Table 5.10 is suitable. Spacer wires of 0.113 in. (2.9 mm) diameter can be used to control gap size if appropriate fixturing is not available.

Strengths of the joints are highly dependent upon the bond strength of the underlying electroplated layers. In all cases, the strengths of the joints are far below those of the parent metals.

ZINC AND ZINC ALLOY COATED STEELS

IN A WIDE variety of products, steel is galvanized with zinc coatings to improve corrosion resistance. Among the many galvanized steel products are building frames, bridge girders and beams, electric power pylons, television transmitting towers, automobile and truck chassis, railway rolling stock, piers, and deck equipment of ships. The chapter "Precoated Steels" in the next volume of the *Welding Handbook*, 8th Edition, will describe the galvanizing processes and the welding of steels coated by the various processes. This chapter also will treat the welding of steels coated with zinc-rich paint and with thermally sprayed zinc.

SAFETY AND HEALTH

EXCESSIVE EXPOSURE TO lead or zinc constitutes a potential health hazard. The specific precautions which should be followed to avoid this hazard are found in ANSI/ASC Z49.1, *Safety in Welding, Cutting and Allied Processes*.[4] This standard always should be consulted before starting to cut or weld metals that contain lead or zinc.[5]

LEAD SAFE PRACTICES

Health Effects

THE EXCESSIVE ABSORPTION of lead and lead compounds into the human body over time may result in lead poisoning, sometimes called *plumbism*. Lead taken into the body, through ingestion or inhalation, is absorbed by either the gastrointestinal tract or the lungs into the circulating bloodstream.

Excessive lead absorption may result in a variety of toxic effects. Some of these effects may be impairment of the production of hemoglobin, resulting in anemia, and damage to the nervous system, which may result in

---

4. The latest edition is available from the American Welding Society, 550 N.W. LeJeune Road, Miami, Florida 33126.
5. OSHA standards contain more general information. See U.S. Dept. of Labor, Occupational Safety and Health Administration, Final standard on Occupational Exposure to Lead, *Federal Register*, Vol. 43(220), 52952-53014, Nov. 14, 1978.

a range of neurologic disorders including mental retardation, especially in children. Kidney damage is also a possible outcome of many years of excessive lead intake. Death from lead poisoning is a rare, but a possible, outcome. For this reason, medical surveillance and health monitoring programs are recommended for all who work routinely with lead, and they are legal requirements in many countries. Workers whose blood lead levels have been found to exceed prescribed limits should be removed from further exposure in order to reduce the body burden to safe levels through excretion.

Ventilation

BECAUSE OF THE potentially toxic nature of lead, welding should be performed in well-ventilated locations to prevent inhalation of lead fumes. The American Conference of Governmental Industrial Hygenists recommended in the mid-1990s a lowering of the threshold limit value for lead from 150 to 50 micrograms per cubic meter. As an added precaution, respirators are recommended. Workers in confined areas, such as at the bottom of a deep, open tank, should be provided with a positive air supply directed into the tank by a fan or blower to a position below the breathing level. In enclosed areas such as pressure vessels or closed tanks, each worker should be required to wear an approved air-supply respirator or mask. Dust from lead fume or lead oxide also will accumulate on the clothing of those welding lead. Accumulated dusts from this clothing has been suspected of adversely affecting the health of others, such as children exposed to such clothing when it is brought home for laundering. Appropriate precautions should be taken.

ZINC SAFE PRACTICES

Health Effects

METAL FUME FEVER, more commonly called *zinc chills*, *smelter shakes*, or *brass founder's ague*, may follow exposure to zinc fumes released during welding operations on zinc-coated or zinc-containing metals, including brazing rods. The chills are caused by colloidal zinc oxide (0.3 to 0.4 micron diameter) penetrating to the lungs. Larger sizes of zinc oxide adhere to the trachea and cause no symptoms. This fever is an acute self-limiting condition without known complications, after-effects, or chronic form.

The illness begins a few hours after exposure, or more frequently during the night, and may cause a sweet taste in the mouth, dryness of the throat, coughing, fatigue, yawning, weakness, head and body aches, nausea, vomiting, chills, and fever rarely exceeding 102 °F (39 °C). A second attack seldom occurs during repeated exposure unless there has been an interval of several days between exposure.

The American Conference of Government Industrial Hygienists publishes threshold limit values for zinc oxide, which at this writing is set at 5 milligrams per cubic meter.[6] The values are reviewed annually and are subject to change.

Repeated exposure to moderate concentrations of zinc oxide in air has not proved permanently harmful; however, exposure to concentrations high enough to cause discomfort to welders or operators should be avoided.

Ventilation

ACCORDING TO THE American standard ANSI/ASC Z49.1, if welding of materials containing zinc is to be done in a confined space, ventilation must be provided. If adequate ventilation cannot be provided, personnel who may be exposed to fumes must be equipped with hose masks or be supplied respirators approved by the U.S. Mine Safety and Health Administration (MSHA) and by the National Institute for Occupational Safety and Health (NIOSH).

---

6. See *1994-1995 Threshold Limit Values for Chemical Substances and Physical Agents and Biological Exposure Indices.* Cincinnati, Ohio: American Conference of Governmental Industrial Hygienists, 1994.

APPLICATIONS

WELDING LEAD SHEET

THE WIDESPREAD USE of lead in the chemical industries is a direct result of the ease with which it can be joined by welding, as well as its exceptional resistance to corrosion by a wide variety of chemicals. Lead welding is used for lining tanks, conveyors, ducts, and other equipment used to store or mix chemical solutions. For example, it is used to produce equipment such as heat exchangers in a phosphoric acid evaporator (see Figure 5.23), electrostatic precipitators, table tops and piping for chemical laboratories, acid-resistance floors in chemical plants, an oxidation auto-clave in a cobalt refinery for manufacturing sulfuric acid by the chamber and contact processes, and numerous other equipment used in the chemical processing industries.

In the United Kingdom and to a lesser extent elsewhere, welded sheet lead is used for roof coverings, wall cladding, flashing, and gutters, as shown in Figure 5.24. Welded lead sheet also is used worldwide as a waterproof membrane to protect underground parking garages and other structures.

Lead welding has been used extensively in the preparation of lead-shielded components for gamma-ray attenuation. This is particularly true of mobile power reactor equipment such as that used on naval ships. Where large attenuations of gamma radiation intensities are desired, the use of lead instead of iron may result in a weight savings of 30 percent. It is important that the lead be bonded properly to structural supports by lead welding and that no void exists between the lead and structural members.

Lead welding has been used extensively in the installation of nuclear energy sources in power plants and submarines. Lead may be in the form of cast slabs, sheet lead, or lead-bonded steel. To install cast slabs, they are first positioned and secured to the steel by edge bonding. The areas of the steel where this edge bonding is performed are first cleaned and then tinned with solder, using a zinc chloride flux. The area between

Figure 5.24—Lead Roof

adjoining cast lead slabs or between the lead and the structural steel also are filled by overlaying lead to the required thickness.

When sheet lead is used to make up the required thickness, each sheet is edge welded to the steel structure and to the lead sheet beneath it. Gaps between adjoining sheets on the steel structure are filled by overlaying with lead. Where lead bonded steel is used, the steel backing plate is welded to the steel structure and the lead edge is bonded to the structure. The gaps are again filled by overlaying lead to the required thickness.

LEAD WELDING OF BATTERY GRID LUGS

OVER EIGHTY PERCENT of all lead consumed is used in lead-acid batteries. The economic and reliable production of batteries depends upon properly applied lead-joining techniques. A method typically used to bond battery grid lugs to internal straps is called *COS* or *cast-on-strap* fusion. In the cast-on-strap joining process, molten lead alloy (normally lead-antimony alloy) is poured into a preheated, coated mold. The mold is designed to produce a casting in the shape of the desired strap. The complete cell or group of plates is lowered so that the grid lugs are immersed in the strap metal. The strap metal solidifies about the grid lugs, forming a strap to collect current from each lug and providing a complete metallurgical bond between

Figure 5.23—Lead-Bonded Steel Tubesheet for Use in a Phosphoric-Acid Evaporator

lug and strap to give a corrosion-resistant joint with low electrical resistance.

Control of the liquid strap metal temperature is critical. The grid lugs partially melt at the surface, producing a fused bond to the molten strap metal. The grid lugs in the COS operation are normally barely melted, as opposed to being completely melted as in the hand-burning operation. The joining of the grid lug to the strap more closely resembles a soldering operation than a welding operation. This is particularly true with lead-calcium alloy or very low antimony alloy grids.

In the cast-on-strap operation, the molten metal in the mold must be sufficiently hot to partially melt the surface of the grid lug, but not too hot, or the lug will be completely melted and no bond will occur. While the lug is being heated on its surface to its melting point, the molten strap metal in the immediate vicinity of the lug is losing heat to the grid lug. Too low a metal temperature results in no melting of the grid lug but merely solidification of the metal around the lug with no bond. Because the tip of the grid lug can receive heat from a larger region, often only a joint at the very tip of the grid lug is attained.

To enhance the ability of the cast-on-strap alloy to bond to the grid lugs, the grid lugs are often prefluxed, or a thin layer of strap metal is welded on prior to the cast-on-strap fusion operation. The flux consists of an inorganic or organic coating material which reacts with any oxides present on the grid-lug surface to convert the oxides to soluble products in the flux layer. The grid lugs are often brushed to remove any adhering oxide prior to fluxing. When the cleaned, prefluxed lugs are immersed in the molten strap alloy, good bonds should occur.

If the lug is to bond effectively to the strap metal, the gases released from the flux must be allowed to escape from the molten metal. If the molten lead alloy contains too low an antimony content, there may not be sufficient eutectic liquid to permit the gases to escape from the molten metal before it solidifies. This results in trapped gases adjacent to the grid lug, giving high electrical resistance for the joint and also potential areas for penetration of corrosion product.

LEAD-SHEATH CABLE JOINTS

WHILE LOW-POWER ELECTRIC cable and most telecommunications cable have been converted from lead to polyethylene sheathing in recent years, lead sheath is still largely used for higher-voltage cable sheaths. At the spliced joint of these cables, lead welding must be used to ensure the integrity of the joint. Because of the increased bulk of the spliced conductors, jointed lead-sheath cable has a greater diameter than the cable itself and so requires the use of a lead sleeve encompassing the joint area.

A lead sleeve with the proper diameter is selected to contain the spliced conductors, and cut to a sufficient length to overlap the lead sheathing several inches on both sides of the joint. The inside of both ends of the sleeve are then scraped clean for approximately 1 in. (25 mm) and immediately fluxed with tallow or stearic acid. The outside of the sleeve also is scraped clear for approximately 2 to 3 in. (51 to 76 mm) from the ends, depending upon the diameter, and fluxed. The sleeve is placed around one end of the cable. Splicing of the conductors is then completed.

Then the areas on the cable sheathing that are to become the root faces of each joint are scraped clean and fluxed. The sleeve is centrally located over the joint, and both ends of the sleeve are dressed or drawn down with a wooden tool until they fit snugly against the clean and fluxed areas of the sheathing. An additional thin coating of flux is applied. Gummed paper or plumber's soil is applied to prevent adherence of solder at points beyond the joints. The joints are then wiped in a manner similar to that used for lead pipe.

A sealing solder that melts at approximately 200 °F (93 °C) is recommended as a precaution against porosity in the wiped joints and the cable sheathing. The sealing solder, in the form of a thin stick, is applied over the entire joint as soon as possible after the wiped solder has solidified. Its residual heat melts the sealing solder, which is then smoothed out over the joint with a wiping cloth. A typical sealing solder contains 52.5% bismuth, 32% lead, and 15.5% tin.

SUPPORTS FOR SHEET LEAD-LINED TANKS

LEAD IS WIDELY used for lining highly corrosive vessels such as sulfuric acid tanks and precipitators. Such vessels are often quite large, necessitating consideration of how the lead lining should be supported on the outer vessel walls. Where high temperatures or rapid cycles of heating and cooling are encountered, within a loose lead-lined tank or where the depth of the tank is more than 8 ft (2.3 m), reinforcement of the lead lining is necessary. This reinforcement is generally provided by vertical steel strapping, or by the so-called *button* type of fastening. All reinforcing methods are applied after the lead lining is installed.

Vertical steel strapping is attached by welding stubs on the steel lining through precut holes in the lead or with preplaced bolts in the steel or wood tank. Half-oval steel straps are placed over the threaded studs or bolts and drawn up tightly. A lead strip of the same weight and composition as the lead binding itself is then placed over the steel strap and welded on each side to the lead lining for corrosion protection of the steel strapping. Good vertical strapping practice is to place the straps about 18 in. (457 mm) apart; they should never be more than 24 in. (610 mm) apart. Studs and bolts are generally placed on 18 in. (457 mm) centers; 12 in. (305 mm) centers are occasionally used, but the centers are never more than 24 in. (610 mm) apart.

With the button type of fastening, a large steel washer and a round-head bolt are used on steel and wood tanks. The bolts are evenly spaced over the entire lead lining at about 18 in. (457 mm) centers. After a bolt is drawn up tightly, a lead cap completely encircling the steel washer and bolt head is welded to the lead lining.

An acid-brick lining on top of the lead is recommended where exceptionally severe conditions prevail, particularly when abrasion or erosion is a factor, where the vessel operates under high pressure or a vacuum, or where acids and alkalies must be handled in the same vessel.

COMMON ZINC PRODUCTS

ROLLED ZINC IS used in roofing and flashing applications, decorative trim, and for plumbing and automotive applications. It also finds wide use in dry-cell battery containers and fuses.

Die-cast and gravity-cast zinc are more accurately and economically cast than aluminum components, and they plate readily. Although most zinc castings are used for decorative trim, they also are used where strength and appearance are needed, such as in safe deposit box doors. Zinc castings frequently can replace steel forgings, steel sheet assemblies, brasses, bronzes, and plastics.

SOLDERING OF ROLLED-ZINC RAIN GUTTERS

ROLLED-ZINC RAIN gutters are widely used in both Germany and France and see specialized use in other countries, particularly where more traditional construction methods are used. These gutters are typically used with tile or metal roofs that have life expectancies in excess of 100 years, and therefore permanent joining of sections of the rain gutters is desired. In this application, zinc offers excellent corrosion resistance. Corrosion typically is less than 1 micron per year, even in the presence of decaying organic matter. Therefore, the zinc rain gutters do not require painting. Zinc can be manufactured with a stable gray patina that is common in traditional construction. Rolled zinc also will not stain surrounding materials. In humid climates, its mildicidal and fungicidal properties prove useful to surrounding materials. Rolled zinc is easy to cut, form,

stretch, and lockseam with hand-tools. Portable machinery also is available for these operations.

Materials Required

RAIN GUTTERS TYPICALLY are made from zinc-copper-titanium alloy rolled to a thickness of 0.02 in. (0.5 mm). The alloy has creep resistance superior to pure zinc and is widely used in architectural applications. The joining process uses a soldering iron that is heated with a propane torch. A torch alone is not sufficient because pressure should be applied to the joint as it is soldered in order to maintain the required gap between the two surfaces being joined. A caustic flux is used and is a mixture of water [1.1 gal (4 L)], zinc chloride [40 oz (1130 g)], and ammonium chloride [3.9 oz (110 g)].

The solder composition used is 40% tin with less than 0.5% antimony. This typically is used in the form of an extruded triangular bar, which gives a higher surface area for heating and always provides a pointed section when needed. It is important that the solder alloy be essentially free of copper, bismuth, iron, arsenic, aluminum, and cadmium because these will reduce mechanical or corrosion properties.

Soldering Procedure

THE TWO SECTIONS of the rain gutter can usually be purchased in a preformed condition. If cut to length, the cut edges are first deburred. A lap joint is used with an overlap distance of 0.4 in. (10 mm). The metal surfaces to be joined should be essentially free of grease other than the light rolling-mill oil. A bright metal finish gives best soldering results. If the surfaces are heavily oxidized, they should be cleaned with steel wool to a width of the 0.4 in. (10 mm) lap. The joint is then set up with a 0.008 to 0.02 in. (0.2 to 0.5 mm) solder gap. Solder gaps greater than 0.02 in. (0.5 mm) lead to capillarity problems and incomplete filling of the lap. Pretinning of the joint surface is not required except where the sheet thickness is greater than 0.03 in. (0.8 mm).

Figure 5.25(A) shows the application of flux to the solder joint. This flux is generously applied and allowed to penetrate into the gap. Only a single coating is necessary if the work is to be soldered immediately. Note that no backing material is used for the solder joint even though the rain gutter joint is suspended between two hangers. Hanger spacing is approximately 18 in. (455 mm).

The next step is to tack the joint with solder, as is shown in Figure 5.25(B). This keeps the gap as narrow as possible during the subsequent steps. The soldering iron is then heated and maintained at around 484 °F (250 °C). The soldering iron is applied back and forth

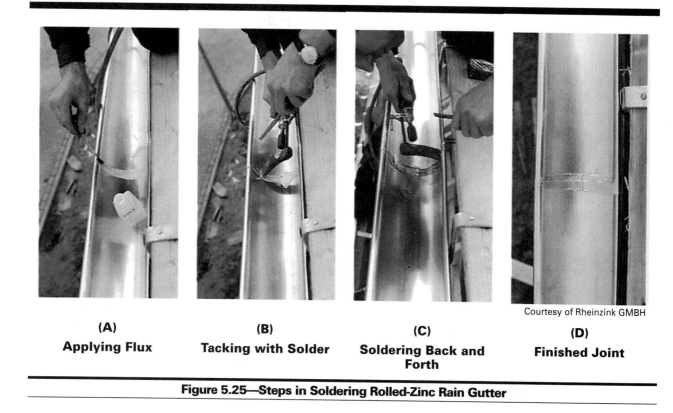

| **(A)** | **(B)** | **(C)** | **(D)** |
| Applying Flux | Tacking with Solder | Soldering Back and Forth | Finished Joint |

Courtesy of Rheinzink GMBH

Figure 5.25—Steps in Soldering Rolled-Zinc Rain Gutter

on the joint as shown in Figure 5.25(C) until the entire joint reaches the melting point of the solder. During this heating the liquid portion of the flux will boil away, leaving behind molten salts of zinc ammonium chloride. The solder is then applied and melts into the joint, filling out the gap by capillary flow. The advancing liquid solder front drives the molten flux out ahead of it until the entire gap is filled with solder. The motion of the soldering iron controls gap filling. It is drawn over the full width of the lap, constantly adding solder to the joint. Firm pressure is transmitted through the soldering iron to the joint to keep the gap as narrow as possible during soldering, encouraging capillarity. Care must be taken not to overheat the joint, as this will cause excessive oxidation. Poor heat transfer will then result.

The joint is then wiped with a wet rag to remove any residual flux. Wiping while the flux is still hot and liquid allows for the most efficient removal. The finished gutter is shown in Figure 5.25(D).

Soldering affords creation of a permanent joint in this application with small portable tools with a minimum expenditure of time. Savings are realized over other materials in which a permanent joint may be needed. Savings also result because subsequent surface treatment, such as painting, is not required.

SUPPLEMENTARY READING LIST

LEAD

Canary, W. W., and Lundin, C. D. "Properties of weld metal in a lead-calcium alloy." *Welding Journal* 52(8): 347s-354s, 1973.

Lead Development Association. *Lead sheet in building – A guide to good practice*. London: Lead Development Association.

Lead Industries Association. *Lead work for modern plumbing*. Ed., Borcina, D. M., New York: Lead Industries Association, 1952.

U. S. Department of Health and Human Services, *Preventing lead poisoning in construction workers*. NIOSH Pub. 91-116, Cincinnati, Ohio: National Institute for Occupational Safety & Health, August 1991.

U. S. Department of Labor, Occupational Safety & Health Administration, *Working with lead in the construction industry*. OSHA 3126, April 1991, Washington, D. C.: U. S. Government Printing Office, 1991.

ZINC

Ando, S., et al. "Effects of aluminum content on electron beam weldability of Zn-Al alloy castings." *Welding International* 6(1): 9-15, 1992.

Akbarkhanzadeh "A personal sampler for measuring shift exposure to welding fumes." *American Occupational Hygiene*, 28(4): 365-71, 1984.

Goodwin, F. E. "Notes on experiments on the joining of zinc castings." *Cast Metals* 1(1): 9-15, 1988.

Gregory, E. N., Herrschaft, D. C., and Cole, J. F. "Fume extraction when welding zinc-coated steels." *American Industrial Hygiene Journal* 32: 170-3, 1971.

Herrschaft, D. C. *The welding of zinc alloys*. Canadian Institute of Mining & Metallurgy, 13th Annual Conference of Metallurgists, 1974.

International Lead Zinc Research Organization, Inc. "The welding of workpieces made of Zamak." Zinc, Cadmium & Alloys, No. 37, 1964, 102-7, in French, English translation. Research Triangle Park, N.C.: International Lead Zinc Research Organization, Inc.

Wildenthaler, L. and Cary, H. B. "A progress report on fume extracting system for gas metal arc welding unit." *Welding Journal* 50(9): 623-35, 1971.

Zinc Development Association. *Joining and fixing zinc alloy die castings*. Brochure, London: Zinc Development Association, 1980.

PLASTICS

PREPARED BY A COMMITTEE CONSISTING OF:

R. A. Grimm, Chairman
Edison Welding Institute

A. Benatar
Ohio State University

D. A. Grewell
Branson Ultrasonics Corporation

P. Pottier
Laramy Products Co., Inc.

R. Rivett
Emerson Electric Company

V. K. Stokes
GE Corporate Research and Development

WELDING HANDBOOK COMMITTEE MEMBER
J. Feldstein
Foster Wheeler Energy International

CHAPTER 6

PLASTICS

INTRODUCTION

THE WIDESPREAD USE of polymeric materials (plastics) is usually considered a recent phenomenon, but some of the best polymers such as silk, skin, and wood have been around a long time. The last 50 to 60 years have seen the increasing penetration of synthetic polymers into the world economy, for the most part based on petroleum feedstocks. This penetration is likely to continue for several reasons:

(1) Manufacturing simplification—Complex parts can be made in one step, rather than in several, resulting in high manufacturing productivity.

(2) Useful properties—The chemistry of polymer synthesis offers a wide range of structures and has been exploited with a continuous improvement in polymer properties. Design of polymers is becoming increasingly understood and should lead to standardization of particular polymers. In addition to corrosion resistance, light weight and fatigue resistance, composite structures (fiber bundles held together with polymer) offer strength-to-weight ratios several times better than other materials.

(3) Adequate raw material base—While a small fraction (approximately 5 percent) of crude oil is converted to polymeric materials, polymers have a higher value than fuels. Additionally, they can be made from renewable materials, and the infrastructure for recycling is being established.

With increased use of polymers, there is a corresponding need to join them. As with any material, polymers are joined in three basic ways: mechanical fastening, fusion processes or welding, and adhesive bonding. Mechanical fastening and adhesive bonding of polymers are similar to the way these processes are used for other materials and are not the focus of this chapter. Rather, the balance is devoted to fusion bonding of polymers.

Fusion bonding has many advantages, including:

(1) High productivity—Process times can be on the order of seconds or less.

(2) Strong joints—With process optimization, processes can often produce welds as strong as the bulk polymer. As with metals, a heat-affected zone is produced.

(3) Relative insensitivity to surface preparation—Depending on joint design, surface layers are softened or melted and ejected from the bond line.

(4) Joints between substrates that are difficult to bond by other methods—Fusion processes are particularly suited to joining polymers that are hard to join with adhesives, such as polyethylene or fluoropolymers.

(5) Enhanced recyclability—Bond-line material is usually the same as the bulk polymer.

A polymer molecule consists of a chain of atoms that can be likened to a piece of cooked spaghetti in terms of shape and flexibility. Per this analogy, thermoplastic polymers consist of many strands that are intertwined but still capable of moving past each other. Thermoplastics soften or melt on heating, or both, and can be welded. Thermoplastics are usually formed by putting the heated polymer in a cooled mold.

Continuing the analogy, thermosetting polymers (unsaturated polyesters, phenolic, epoxy, most ureas and urethanes) can be compared to clumped spaghetti. The strands are not capable of much independent motion, and the mass behaves as a unit. Thermoset polymers are usually formed by polymerization of prepolymers (small polymer segments but larger than monomers) in a heated mold. Because thermosetting polymers do not melt when heated (they decompose), they are not weldable by themselves. Instead, they are joined with adhesives and mechanical fasteners.

Thermoplastic polymers are often divided into two categories: amorphous and semicrystalline. The chains in amorphous polymers are randomly coiled and intertwined with other chains. On heating, they pass through a temperature called *the glass transition temperature* (Tg) at which softening begins. They do not "melt." Rather, as the temperature is raised above Tg, the viscosity continues to decrease until it reaches a level useful for molding or welding. Because the crystal structures formed in semicrystalline polymers melt sharply at temperatures higher than Tg, they appear to have a definite melting temperature. However, even when melted, they form very viscous materials. Pure semicrystalline polymers are usually opaque while amorphous polymers are often transparent.

Examples of common amorphous polymers include acrylic, polycarbonate, polyvinyl chloride, ABS, polystyrene, polysulfone, polyetherimide. Semicrystalline polmers include polyethylene, polypropylene, polyphenylene sulfide, polyetheretherketone and polyamides.

Each thermoplastic has a processing window defined as the temperature range in which useful flow is achieved without polymer decomposition. Welding processes must be controlled to prevent decomposition.

In contrast to metals, most polymers are thermal and electrical insulators. As a result, the amount of heat required for welding thermoplastics will be much less than for a weld of equivalent size in metals. Polymers are usually formulated with additives such as fillers to lower cost, pigments for color, reinforcing agents such as fibers, and a host of other additives to achieve specific properties. These additives can affect joining, whether by fusion processes or adhesives, and best conditions must usually be determined for a particular application.

Welding of polymers results in a heat-affected zone around the weld. The pressure used during welding and the resulting flow of polymer normally causes an increase in the kinds of crystalline microstructures in the weld zone.

This can easily be seen in a semicrystalline polymer, such as polypropylene, in which squeeze flow and fast cooling rates produce a range of crystal structures. As with metals, heat-affected zones can be weaker than the bulk material. Because of residual stresses in the heat-affected zone, corrosion by aggressive agents or attack by solvents will be fastest in this zone.

Table 6.1 shows several processes used to weld thermoplastic materials that are analogous to processes for metals welding.

While metals form melt pools that flow readily, polymers form viscous liquids that must be pushed into the desired locations. The resulting pressure often results in orientation of the chains in the flow direction. This can cause bond-line properties to be anisotropic. For example, impact or tensile strengths may be lower in the plane of the bond line than perpendicular to it.

Table 6.1
Analogous Welding Processes

Plastics	Metals
Hot-gas welding	Gas welding
Spin welding	Spin welding
Resistance welding	Spot or seam welding
Ultrasonic welding	Ultrasonic welding

WELDING METHODS FOR THERMOPLASTICS

WELDING OF THERMOPLASTICS depends on having the polymer chains of the substrates achieve intimate contact and intermingle. Methods commonly used to bring this about are classified here into broad categories, depending upon the method used to impart chain mobility and intermingling. One involves application of energy from heated sources that melt the polymer, such as hot plates, hot gases, dielectric or microwave generators, and infrared heating. Another involves melting the polymer through application of energy to an implant which remains in the bond line, such as those used with resistance and induction heating. A third category describes processes such as spin, vibration, and ultrasonic welding in which polymer fusion is caused by friction or hysteresis heating of the polymer. A fourth involves nonthermal methods such as solvents to dissolve a polymer, which is then redeposited as the solvent evaporates.

THERMAL METHODS USING EXTERNAL HEATERS

Hot-Plate Welding

IN HOT-PLATE welding of thermoplastics, the polymeric substrate is brought into contact with a heated platen to soften or melt the polymer. The parts are then separated from the platen and pressed together, causing the polymer chains to intermingle and weld. Strength develops on cooling. This process usually uses a flat platen for heating (Figures 6.1 and 6.2), and clamps to hold the two pieces to be joined. Heat is applied to fuse the polymer; forging and cooling create the weld. The process can be divided into three phases: heating, withdrawal of the hot plate, and joining of the pieces.

Several factors need to be controlled to achieve successful hot plate welds:

(1) Temperature of the platen.

(2) Displacement of the polymer during heating and joining—Displacement is defined here as the amount of melted material that is squeezed out of or displaced from the joint area. Displacement occurs in both the heating and joining steps.

(3) Change-over time—This is the time between separation of the parts from the heated platen and when they are joined or pressed together.

(4) Pressure applied during the heating and joining processes—This can be controlled by limiting the displacement in each step.

Other factors requiring control depend on the polymer or configuration that is being welded. Some polymers, such as polyesters, polyamides (nylons®), and polycarbonates absorb moisture from the atmosphere and can hydrolyze and sometimes generate bubbles in the weld zone, resulting in degraded polymer and weak welds. In these materials, it is important either to dry the polymer before welding or, because dried polymers are used for molding, to conduct the welding operation immediately after molding.

During the heating process, molten polymers tend to stick to the heated platen. The most common method for preventing this is to coat the platen with Teflon, which is serviceable up to 518 °F (270 °C). For higher temperatures, the platen can be coated with a mold-release agent, or aluminum-bronze platens can be used. For small bonds, such as lap shear samples, mold release has been shown not to affect bond strength even though traces remain in the bond zone. For each application, particularly for larger bonds, trials must be conducted to ensure that the mold release agent does not affect bond strength. Suitable temperatures for hot-plate welding must ultimately be determined by testing. Potente and Natrop recommend that for semicrystalline polymers, the hot plate be held at approximately 176 °F (80 °C) above the melting point of the polymer.[1] For amorphous polymers, which do not exhibit melting points, an approximate temperature of 320 °F (160 °C) above the glass transition temperature of the polymer is recommended. Higher temperatures can be used, provided they are below the decomposition temperature of the polymer being welded.

When a polymer contacts the hot plate, the resulting temperature distribution has a maximum at the point of contact. If weld displacement is not controlled, pressure (remaining on the pneumatic cylinders used to operate the equipment) can force nearly all of the melted polymer from the weld zone. Melted or softened polymer can also be ejected from the weld zone during the joining or forging process. If nearly all of it is ejected from the weld zone, it results in a sharp transition in polymer morphology and a "dry," usually weak joint. A common practice is to apply sufficient pressure during the heating phase until the whole bonding surface is in contact with the heated platen and then to release the pressure (or stop any further displacement) during the rest of the heating step. Too low a pressure during the forging or pressing operation can result in the formation of voids in the bond line as the polymer contracts on cooling.

Strongest welds and better service lifetimes are obtained when the stressed zones, resulting from sharp transitions in polymer morphology, are minimized. The change-over time, or time between withdrawal of the heated parts from the platen until they are pressed together, must be short to prevent surface cooling and solidification. Because of the insulating properties of

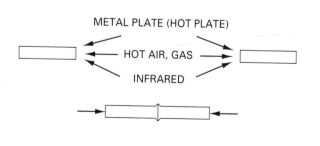

METAL PLATE (HOT PLATE)

← HOT AIR, GAS →

INFRARED

Figure 6.1—Schematic of Heated Tool (Hot Plate, Hot Gas, or Infrared)

1. Potente, H. and Natrop, J. "Computer aided optimization of the parameters of heated tool butt welding." *Polymer Engineering and Science* 29(23): 1649, 1989.

Figure 6.2—Laboratory Hot-Plate Welding Machine

most thermoplastics, heat conduction from the mass of molten polymer immediately behind the weld surface may be too slow to keep it in a molten and weldable state. Welding conducted with partially solidified surfaces will be weak (cold joint) because the polymer chains will not intermingle completely. Particularly in production situations, control of ambient conditions such as the temperature, air movement and change over time are necessary for reproducible welding. A study interrelating these factors for butt fusion welding of polyethylene pipes has been developed.

Dissimilar thermoplastics can be welded provided they are compatible and relative viscosity levels are similar. Comparable viscosities are usually obtained by using two platens that heat each polymer to the required temperature. Welding is possible only when the coefficients of thermal expansion and the polar surface tensions of the two polymers are close.

Hot-plate welding cycle times can range from seconds to minutes, although for massive parts where large amounts of polymer must be melted, times of several hours are required. Watson and Murch have examined the use of higher temperatures and shorter heating times as a way to improve the productivity of the hot-plate welding process; sound welds are produced if the displacement is controlled.[2]

Hot-plate welding is used for the construction of automobile battery cases where high reliability is required. The most common use of the process is in the joining of thermoplastic pipe for gas, water and sewer pipelines.

Evaluation of weld integrity has been a problem, and the conventional method of destructive tensile testing does not provide sufficient information, for example, to detect cold joints. Work at the Paton Institute in the Ukraine showed that, when thin sections from the bead of a joint are heated to 176 to 194 °F (80 to 90 °C),

2. Watson, M. and Murch, M. "Recent developments in hot plate welding of thermoplastics." *Polymer Engineering and Science* 29(19): 1382, 1989.

relaxation of the polymer causes reopening of the bond line in welds where polymer intermingling has not occurred.[3] A related procedure involves removing the weld bead and flexing it. A bead from a good joint will remain intact.

Infrared Welding

HEATING BY INFRARED lamps has been developed recently as a method for heating large polymeric structures. Hot-plate welding usually involves contact of polymer with the heated platen, although heating can be conducted by radiation if the pieces are held close to the platen. In infrared welding, infrared lamps scan the joining surface to melt the polymer. When a suitable temperature is reached, as controlled by sensors, the infrared lamps are withdrawn and the parts are joined or pressed together.

Hot-Gas Welding

HOT-GAS OR hot-air welding is similar to brazing or other welding processes in that a weld is created by fusion of a filler rod into a joint. Joint designs are similar to those used in metal welding and include butt, fillet, and lap joints. Joining surfaces are beveled and prepared in much the same manner as with metals.

The process involves blowing a stream of heated inert gas or air against both a filler rod (made of the same material that is being welded) and the welding surfaces of the parts to be joined. The gas stream fuses the surface of the filler rod as well as the surface of the beveled joint area. Pressure applied to the filler rod forces it into the joint causing the polymer chains to intermingle, creating a weld. It is neither necessary nor desirable to melt the filler rod completely since this will result in slower welding and an increased probability of polymer decomposition.

Some basic joint designs that can be produced by hot-gas welding are shown in Figure 6.3. This technique is also useful for tacking and plugging holes (rosette welds). In some cases, rollers are used to follow the hot-gas welding equipment providing additional pressure for consolidation of the polymers.

Key conditions that must be controlled in hot-gas welding include gas type, temperature and flow rate, angle of the filler rod to the workpiece, weld speed, and postweld pressure. Air or inert gas is heated by passing it through an electrical heating unit. Air is commonly used when oxidation of the polymer is not considered a problem. More expensive inert gases are used for many applications where higher quality welds are required.

Dry nitrogen is preferred, although carbon dioxide and other inert gases may also be suitable. For the most economical use of gas, newer welding units have valving arrangements that replace the inert gas flow with air whenever the welding unit is not actually being used for welding.

Gas flow rate and temperature are balanced to provide a set of conditions that will minimize polymer decomposition but still provide enough energy at the weld zone for efficient welding. Polymer decomposition is often, but not always, indicated by discoloration. The gas must be reasonably free of moisture and totally free of contaminants such as oil. Flow rates can be controlled with flow meters [0.53 to 2.1 cu ft/min (15 to 60 L/min)] or pressure regulators [0.15 to 7.3 psi (1 to 50 kPa)].

Gas temperatures can be measured with a thermometer or thermocouple at the distance from the tip that will be used in actual welding. Some typical temperatures used in hot-gas welding are listed in Table 6.2. These welding temperatures are only guidelines since the polymer temperature also is affected by other factors such as rod shape, flow angle, gas flow rate, and polymer filler levels

Manual Welding. Operator skill and experience are of greatest importance for manual hot-gas welding. At the present time, there is no widely recognized certification procedure for hot-gas welding although some corporations have in-house certification programs at different levels of development.

For actual welding, the parts are held in place and beveled to provide an approximate fit for the shape of the filler rod. Heated gas is directed to the end of the weld and the filler rod until the polymer faces become shiny due to melting. To maintain the proper temperature and to heat the polymer uniformly, the torch is moved in a fanning or weaving motion (approximate fanning frequency is two per second) while proceeding along the weld. For most plastics, filler rods are held at a 90 degree angle to the workpiece since this the most effective angle for applying pressure to the filler rod as it fills the cavity and minimizes stretching of the filler rod. Overheated rods, produced by slow welding rates or excessive temperature, will become too soft and rubbery to allow application of pressure. If the material being welded is elastomeric and does not transmit pressure effectively, a roller can be used to apply the pressure as the weld is being made. An alternative is to use a speed-welding tip which also has the capability of consolidating the polymers.

The composition of the filler rod polymer should be identical to that of the substrate being welded. Ideally, the shape of the filler rod should be a close match to the cavity it is to fill. While filler materials are often provided in round, triangular and oval cross sections, triangular rods usually provide welds with better

3. Esaulenko, G.B. and Kondratenko, V. Yu and Bezruk, L.I. "Study of strength characteristics of polyethylene butt welded joints and development of testing methods." IIW Doc. XVI-584-90.

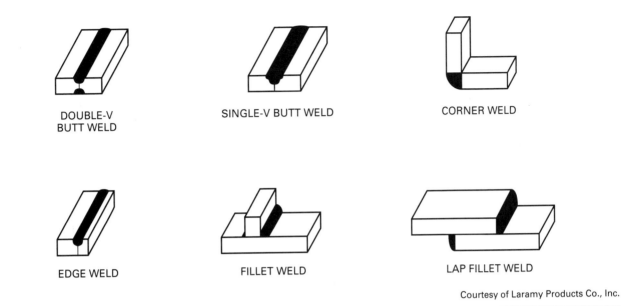

DOUBLE-V
BUTT WELD

SINGLE-V BUTT WELD

CORNER WELD

EDGE WELD

FILLET WELD

LAP FILLET WELD

Courtesy of Laramy Products Co., Inc.

Figure 6.3—Common Joint Designs for Hot-Gas Welding

Table 6.2
Typical Hot-Gas Welding Conditions[a]

	PVC Type I[b]	PVC Type II[b]	PVC Plasticized	Polyethylene Regular (LDPE)	Polyethylene Linear (HDPE)	Polypropylene	Chlorinated Polyester	FEP Fluorocarbon	Acrylic
Welding Temperature[c]	500-550 °F (260-288 °C)	475-525 °F (246-274 °C)	500-525 °F (260-274 °C)	500-550 °F (260-288 °C)	550-600 °F (288-316 °C)	550-600 °F (288-316 °C)	600-650 °F (316-343 °C)	550 - 650 °F (288-343 °C)	600-650 °F (316-343 °C)
Welding Gas	Air	Air	Air	Inert	Inert	Inert	Air	Air	Air
Butt-Weld Strength, %	75-90	75-90	75-90	80-95	50-80	65-90	65-90	80-95	75-85
Maximum Continuous Temperature	159 °F (71 °C)	145 °F (63 °C)	150 °F (66 °C)	140 °F (60 °C)	210 °F (99 °C)	230 °F (110 °C)	249 °F (121 °C)	249 °F (121 °C)	140 °F (60 °C)
Bending and Forming Temperature	250 °F (121 °C)	250 °F (121 °C)	100 °F (38 °C)	244 °F (118 °C)	270 °F (132 °C)	300 °F (149 °C)	350 °F (177 °C)	550 °F (288 °C)	280 °F (138 °C)
Cementable	yes	yes	yes	no	no	no	no	no	yes
Specific Gravity	1.35	1.35	1.35	0.91	0.95	0.90	1.40	2.15	1.19
Support Combustion	no	no	no	yes	yes	yes	no	no	yes

a. Courtesy of Laramy Products Co., Inc.

b. Type I PVC is normal, unplasticized material; Type II is rubber modified.

c. Temperatures measured 1/4 in. (6 mm) from tip.

appearance. Stronger welds are usually produced when the number of passes is minimized. If multiple passes are needed, such as when filling a large cavity, triangular rods are used because they fit together more easily than other shapes. A finished weld should overlap the beveled edge of the base material. In most cases, polymer will extend above the level of the substrate material. If desired, the surface can be smoothed by removing the excess material, but care must be taken not to leave any notches or grooves in the vicinity of the weld as they will weaken the joint.

Besides filler rods, polymer strips can also be used to make or reinforce conventional joints. In these cases, a ribbon of thermoplastic is heated and joined to a heated band on the substrate. Application of pressure creates the weld.

In general, weld quality is determined by several factors:

(1) Material weldability—Polymers vary in thermal stability. Those in which the welding temperature is near the decomposition temperature require greater care to prevent overheating.

(2) Match of filler material to substrate—The filler material should be identical to the substrate to allow weld formation and avoid stress cracking [see Figure 6.4(A)].

(3) Weld configuration—A root gap of 1/32 in. (0.8 mm) should be provided to allow the viscous, melted polymer to be pushed into the weld root to create a complete seam. [Poor penetration is shown in Figure 6.4(B).]

(4) Welding conditions (temperature, speed, pressure)—Underheating can result in poor flow and penetration, porosity, and distortion [see Figure 6.4(B), (C), and (D)]. Overheating will decompose the polymer, which is sometimes seen as scorched regions in the weld zone [see Figure 6.4(E)]. The speed must be selected to allow melting of the surfaces without melting through the filler material. Pressure must be sufficient to force the melted polymers to intermingle but not so great as to create notches at the surface of the weld or create stresses in the weld zone.

(5) Number of weld passes—Three or more passes are recommended for hermetic seals.

(6) Joint cleanliness—Stronger welds are obtained when joint surfaces are clean and the gas free of contaminants or oxygen.

(7) Notch avoidance—Excessive pressure can gouge the weld zone and create notches, weakening the weld.

(8) Welder skill level.

Weld quality can be expressed as a quality index which is the ratio between tensile strength of the welded assembly to that of the parent material. For reference

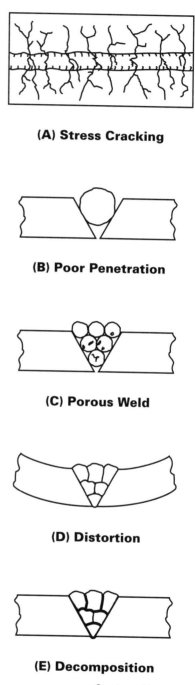

(A) Stress Cracking

(B) Poor Penetration

(C) Porous Weld

(D) Distortion

(E) Decomposition

Courtesy of Laramy Products Co., Inc.

Figure 6.4—Hot-Gas Welding Problems

purposes, tests conducted according to the German standard DIN 53455 can be used since quality index will vary with weld geometry and the type and rate of stressing.

Nondestructive testing of hot gas welds is usually conducted visually or by spark tests.

Speed Welding. Speed welding is a variant of manual hot-gas welding in which a tip is fitted to the hot-gas welding gun. (See Figure 6.5.) One purpose of this tip is to provide for feeding of the filler rod through a heated chamber where the surface melts quickly. Filler rod is drawn through the tip as the welder proceeds and the pressure foot forces the surface-melted filler rod into the joint. These features result in a substantial increase in the rate of welding, up to 40 in./min (1000 mm/min), as compared to 5 to 12 in./min (120 to 300 mm/min) for manual welding.

Because the temperatures in the heating chamber are those provided by the welding gas, it is important not to insert the filler material until the start of welding. If it is overheated by not starting the weld immediately, the filler rod may melt, char, and break. Examination of the speed-welding tip shows that part of the gas is directed toward the rod heating chamber and the contact zone while the rest of the gas is directed ahead of the contact zone where it preheats the substrate.

When starting a weld, the filler rod is inserted into the tip, and the welding gun is immediately placed in contact with the joint while holding it perpendicular to the substrate. The filler rod is immediately fed until it contacts the substrate at the start of the weld. If necessary, the torch is lifted to allow the rod to pass under the pressure foot. The torch is pulled along the weld zone to start the weld. Once the weld is started, the

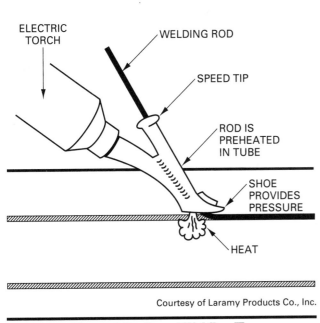

ELECTRIC TORCH

WELDING ROD

SPEED TIP

ROD IS PREHEATED IN TUBE

SHOE PROVIDES PRESSURE

HEAT

Figure 6.5—Speed-Welding Tip

angle of the torch is reduced to approximately 45 degrees to allow for preheating of the weld zone. The angle between the welder and the base material determines the welding rate. A balance must be struck between the rate of filler-rod heating and the rate of preheating. If preheating is excessive, often indicated by discoloration of the polymer in the weld zone, the angle of the torch can be raised to increase the distance between the preheating gas and the substrate. The weld should have the appearance of a smoothed, slightly rounded surface extending over the joint with a small bead on each side.

To stop a weld, the torch is stopped, again brought to a 90-degree angle, and the filler rod cut off with the pressure foot. Any filler rod remaining in the speed-welding tip must be removed immediately. Once started, a weld should be completed in a continuous, uniform process. If it is necessary to interrupt a weld, the above stop-start sequence is followed.

Automated Welding. The hot-gas welding process can be automated by incorporating the above features into a unit with feedback controls. Such units have been developed for welding joints in pond liners and other structures with long bond lines.

A variant is extrusion welding where, instead of filler rod, molten polymer is extruded into the bond area. On application of pressure, the molten polymer is squeezed into the bond area, creating the weld. This method is particularly useful when welding thicker sections. Material that is 1 in. (25.4 mm) thick can be welded in one pass to produce joints of good quality.

Hot-gas welding of thermoplastics is a low cost, versatile, and portable process for creating plastic welds. It is widely used for joining of plastic tankage and ductwork, for creating seams in pond liners and thermoplastic roofing, and for the repair of plastic pipelines and the attachment of valving to these pipelines. Hot-gas welding also has been used successfully in the repair of thermoplastic automobile bumpers.

IMPLANT WELDING

Resistive Implant

RESISTANCE OR RESISTIVE implant welding of thermoplastics differs from resistance welding of metals (spot or seam welding). With thermoplastics, a conductive material is placed in intimate contact with the substrates and heated by passage of an electric current. The heated element melts the thermoplastic which is then pressed together to create the weld. The conductive element remains implanted in the bond line (see Figure 6.6).

The resistive element can have a variety of forms including wires, braids or carbon fiber materials. The

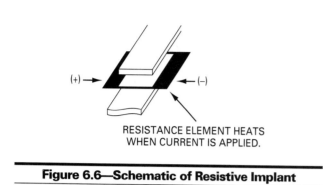

RESISTANCE ELEMENT HEATS
WHEN CURRENT IS APPLIED.

Figure 6.6—Schematic of Resistive Implant

method is capable of forming welds in large structures with complex shapes, but it has not been extensively developed for general bonding.

Power supplies can be simple and inexpensive. Since heating occurs by conventional resistive heating, both ac and dc sources can be used. Power supplies have ranged from variable transformers to arc welding supplies with capabilities up to several hundred amps. Electrical contact with the resistive element usually is achieved by connecting the terminals to ends projecting from the bond line. Joining of conductive materials, such as carbon-reinforced composites, can be complicated by electrical contact of the element with the substrate (shunting), but for thermoplastics and nonconductive composites, it is a relatively straightforward process.

In one instance, the process was used for joining thermoplastic bumpers to reinforcing members. The most common application, however, is in the electrofusion process for joining of thermoplastic pipelines used in the gas industry. Sockets are produced with resistance wires molded into the internal surface. The wires lead to terminals, also molded into the socket, which are used for establishing electrical contact. The process involves inserting the pipe ends into the socket and connecting the wire ends to a portable power supply programmed to provide a measured amount of power. The process has the advantage that it is relatively automated and reliable. It has a further advantage that the pipe is positioned before the weld is made, and no further movement is needed. For comparison, hot-plate welding of pipelines requires removal of the hot plate and movement of the pipe during the joining step. The electrofusion process has been adopted by British Gas for joining pipes up to 7 in. (180 mm) in diameter. A disadvantage of the process is that the sockets are relatively expensive, but process convenience and joint reliability are leading to its increased use.

High-Frequency Welding

HIGH-FREQUENCY WELDING OCCURS when alternating electromagnetic fields interact with materials causing them to heat. If these materials are incorporated into a bond line between thermoplastic substrates, the heat generated can cause polymer fusion and bond formation. Usually the materials that are heated by the fields (susceptible materials or susceptors) are premixed with the thermoplastic. Rates of heating can be quite rapid, and these processes are used in mass production joining applications.

Induction, dielectric, and microwave heating are combined in this overview even though not all involve implants. The processes differ mainly in the frequencies employed, the method of applying the field to the workpiece and the types of materials that are heated. Table 6.3 shows, in simplified form, some of these interactions.

This table gives a generalized comparison of the processes and represents most of the current,

Table 6.3
Interactions of Electromagnetic Fields with Substrates as a Function of Frequency

Method	Frequency Range	Application Method	Susceptible Materials
Induction	Up to 10 MHz	Noncontacting coil close to workpiece	Metals, metal powder, and magnetic oxides
Dielectric	27.16 MHz (mainly)	Contact electrodes on either side of workpiece	Polar but nonconducting polymers, e.g., PVC, polyamides
Microwave	915 MHz, 2.45 GHz	Metal-lined chamber or exit of waveguide	Polar but nonconducting polymers. Ion-containing materials, e.g., carbon black. Metals reflect energy.

commercially-used process technology. Within each category, additional detail can be included that would modify or extend application areas. In addition, development work on these processes is expanding the list of susceptible materials with the view of providing greater selectivity and control.

Induction Welding

ONE MECHANISM FOR induction heating occurs when electromagnetic fields induce eddy currents in electrically conductive materials such as metals or other conductors (susceptors) that transfer heat to a contacting thermoplastic. The mechanism of eddy current heating is illustrated in Figure 6.7(A). For efficient heating, smaller particles require higher frequencies. Welding of thermoplastics using magnetic metal powders or oxide is conducted in the 500 kHz to 8 MHz frequency range although methods are being developed at higher frequencies.

A second mechanism, hysteresis or dipole heating, occurs when the materials are magnetic, but this process switches off when certain characteristic (Curie) temperatures are reached. Both mechanisms operate simultaneously when ferromagnetic materials such as iron or nickel interact with the field but, unless the material were present as very small particles, the predominant one is usually eddy current heating. Induction heating with both mechanisms operating is illustrated in Figure 6.7(B).

Nearly all polymers are insulators and are therefore essentially transparent to electromagnetic radiation at the frequencies of induction heating.

In commercial practice, induction welding of thermoplastics uses a molded or extruded shape made from metal or metal oxide in a thermoplastic matrix. This shape is placed in the bond line, the parts are fixtured, and the field is applied. After the polymer has fused sufficiently, pressure is applied to force the mass throughout the weld zone.

Weld times can range from fractions of a second to minutes depending primarily upon coil design, coil-workpiece distance, applied power, and susceptor composition.

Coils are usually custom-designed for a particular weld configuration and production rate. They are usually made from square or round copper tubing and are water-cooled to prevent melting. A discussion of coil designs is beyond the scope of this review, but several sources of information are available.[4] Since the strength of an electromagnetic field (and therefore its heating ability) decreases with the inverse square of distance, consistent welding often requires precise control of coil-workpiece distance. Parts are usually welded using fixtures and fixed coils.

Induction welding of thermoplastics can accommodate various joint designs, some of which are shown in Figure 6.8. The flat-to-flat joint is the simplest and is suited to continuous joining with minimal modification of the bonding surfaces, while flat-to-groove joints will keep the susceptor material in a precise location. Tongue-and-groove, shear, and step joints all provide high strength and shear-resistant joints. The tongue-and-groove joint is considered to provide the best strength and hermeticity. A general guideline is that the volume of the susceptor material should be equal to or slightly greater than required to fill the anticipated cavity.

---

4. Consult the readings by Leatherman and by Zinn and Semiatin in the Supplementary Reading List for this chapter, Page 350.

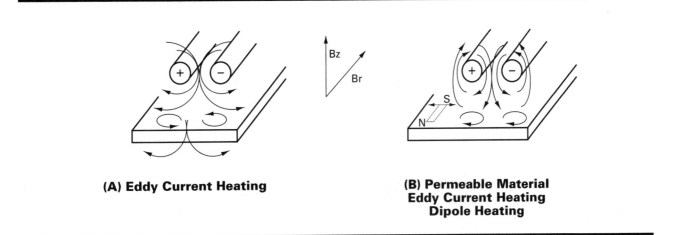

(A) Eddy Current Heating

**(B) Permeable Material
Eddy Current Heating
Dipole Heating**

Figure 6.7—Induction Heating Mechanism (Heating = I^2R)

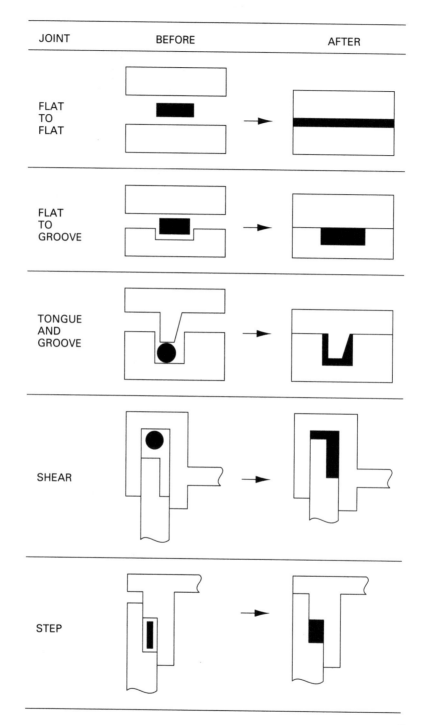

JOINT	BEFORE	AFTER
FLAT TO FLAT		
FLAT TO GROOVE		
TONGUE AND GROOVE		
SHEAR		
STEP		

Courtesy of Emabond Systems

Figure 6.8—Induction Joint Designs

Particularly on solid-state machines, the coil can be located at substantial distances from the power supply, allowing flexibility and increased automation potential. Workstations are light in weight and small enough to be mounted at the end of a robot arm for continuous joining. If the coil is fixed, the small size allows for greater freedom in moving long bond lines under the coil.

Dielectric Welding

DIELECTRIC HEATING OF thermoplastics is conducted at the FCC-assigned frequency of 27.12 MHz. Electromagnetic fields at this frequency interact with unbalanced electric charges (polar bonds) that are present in molecules such as polyvinyl chloride or polyamides. Thermoplastics with minimal charge imbalance (nonpolar bonds) are relatively unaffected by the field. Examples are polyethylene and polypropylene.

Electrodes or dies are used to transmit the field to layers of thermoplastic placed between them. One electrode must have the shape of the required seal while the other can be flat. When the field is applied, the polar thermoplastic heats until it softens or melts. The electrode applies pressure to create the seal and can conduct heat from the surface of the layers thereby preventing surface melting and sticking. If necessary, materials that are transparent to the field can be placed between the electrode and workpiece.

Dielectric welding is commonly conducted using automated sealing equipment. Key control conditions include heating time, usually ranging from 1 to 4 seconds; power, which is proportional to the area of the seal and materials; and pressure, which must be adequate for joining the parts together but not so high as to pinch off the bond area. Misalignment of dies must be controlled.

Microwave Heating

MICROWAVE HEATING IS conducted at the FCC-assigned frequencies of 915 MHz and 2.45 GHz. The 915 MHz frequency is used for higher power applications, while the 2.45 GHz frequency is used for kitchen-type microwave ovens. These fields are applied generally to substrates either in chambers or ovens which can have openings of approximately 10 ft by 10 ft (3 m by 3 m). Wave guides, which are usually rectangular metal tubes, can be used to supply more directed microwave energy.

Microwaves heat thermoplastics that have dipoles in the molecular structure, much as dielectric heating. Polyvinyl chloride, polyamides, and urethanes are examples of materials that can be heated. Polyethylene and polypropylene are examples of materials that are transparent. In addition to heating molecular dipoles, microwaves also will heat ion-containing species such

as carbon black or ion-containing polymers. Selective heating of bond lines in thermoplastics can be achieved by welding relatively transparent thermoplastics with more sensitive materials in the bond line. Sensitizers can include polyol or polyamine formulations, polar polymers that are compatible with the substrates or, in recent work, chiral particles (particles of optically active materials that have the ability to change the orientation of polarized light).

FRICTION AND ULTRASONIC METHODS

FRICTION WELDING DESCRIBES processes in which frictional heating, caused by rubbing two thermoplastic parts together under pressure, softens and melts the polymer surfaces and creates a joint. Usually, one of the parts is rubbed against the other which is held fixed. In spin welding, one of the parts is spun against the fixed part (see Figure 6.9). In vibration welding, one part is vibrated with a linear motion against the other fixed part (see Figure 6.10). Other motions capable of generating the required frictional heat have been used or proposed (see Figure 6.11).

Ultrasonic welding, also covered in this section, is made possible through hysteresis heating of joints by properly focused ultrasound waves.

Spin Welding

IN ITS SIMPLEST form, frictional spin welding is a fast and economical process for producing joints when one of the parts has a circular bonding area. Weld times for small parts range from 1 to 3 seconds. Examples of spin welded parts include tubes (welded to themselves), liquid-filled compasses, floats, and fittings.

Drill presses and lathes can be used as power sources for spin welding, but commercially designed spin welders usually provide more consistent results. Two types of equipment are available, both of which are capable

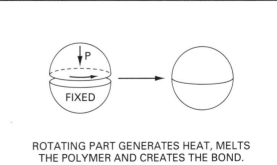

ROTATING PART GENERATES HEAT, MELTS THE POLYMER AND CREATES THE BOND.

Figure 6.9—Spin Welding

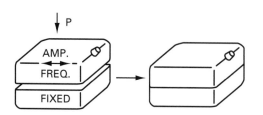

FRICTIONAL HEAT FUSES POLYMER AND CREATES THE WELD. MOTION MAY BE ORBITAL.

Figure 6.10—Vibration Welding

of producing sound joints. In one variant (a driven machine), power continues to be delivered to the rotating part as axial pressure is applied and melting occurs. At a predetermined point, power is shut off and the joint is allowed to consolidate. In the other variant (an inertial machine), the moving part is held in a rotating flywheel. When the appropriate rotational speed is achieved in the flywheel, the parts are pressed together allowing the stored energy to soften and melt the polymer through frictional heating. The inability to stop the rotation at a selected relative angular position between the fixed and rotating parts has limited application of spin welding. However, this deficiency has been eliminated with microprocessor-controlled machines.

Relative velocities can range from 6.6 to 66 ft/s (2 to 20 m/s) with weld pressures ranging from 25 to 145 psi (170 to 1000 kPa), but actual weld conditions must be determined for each particular application. The welding

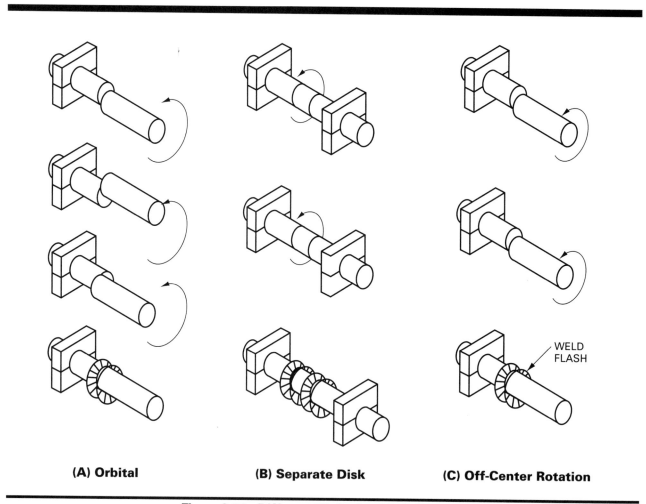

(A) Orbital   **(B) Separate Disk**   **(C) Off-Center Rotation**

Figure 6.11—Different Methods of Spin Welding

process causes some softened or melted polymer (flash) to be squeezed from the weld zone. Parts can be designed with joints to contain the flash or to minimize its formation.

Thick-walled parts will have a substantial differential rotational speed between the inner and outer walls, resulting in uneven heat generation. This can cause uneven melting or weld stresses and should be avoided either by part design (see joint designs in Figure 6.12) or by process selection (see Figure 6.11).

Once sufficient polymer has been softened or melted, rotation should be stopped quickly to prevent shearing the bond area. This can be accomplished either by using shear pins that break at a predetermined shear force or by using a clutch.

Other process variants, in which use is made of off-axis rotation, orbital motion, or a separate rotating member (which may be included in the joint or withdrawn immediately prior to application of weld pressure), can provide parts with preselected relative orientation. Off-axis rotation and orbital motion avoid differential velocities when welding thick-walled specimens or solid rods.

Most thermoplastics can be welded using spin or friction welding processes. Typical weld strengths are shown in Table 6.4.

Thermoplastics should be welded in a dried state, particularly if they are capable of hydrolysis (polyamides, polyesters, and polycarbonates, for example). As with most fusion bonding techniques, spin welding is sensitive to the amount and types of additives such as fillers. High filler levels can reduce flow of polymer in the bond line and create a joint held together by

Table 6.4
Weld Strengths for Spin Welded Joints

Parent Material	Percentage of Parent Material Strength
Polyethylene	70-95
Acrylic	75-85
Polyamide	50-70
Acetal	30-70

mechanical interlocking rather than the stronger joints created when the polymer chains flow and intermingle.

What constitutes a high filler level is not clearly defined, but levels over 20 percent should be of increasing concern. Filler level is one of the factors that must be considered when determining suitable welding conditions.

Joints should be designed to provide a maximum weld area and to reduce differential velocity. In addition, joint design can provide for self-alignment and direction or containment of flash. Some examples of joint designs are shown in Figure 6.12.

Linear Vibration Welding

VIBRATION WELDING IS a special case of friction welding where one part is rubbed against another fixed part in a linear back-and-forth motion. Pressure applied during the welding step helps to generate heat which melts the polymer and forces the chains to intermingle.

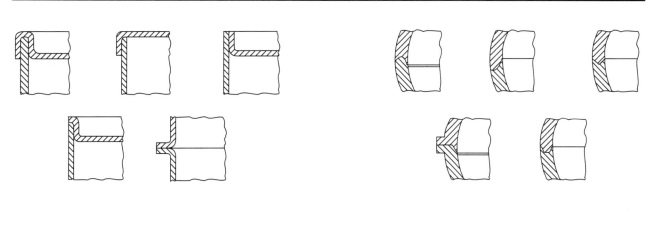

(A) For High-Duty Application   **(B) For Low-Duty Application**

Courtesy of *Welding in the World*, Journal of the International Institute of Welding

Figure 6.12—Joint Designs for Spin Welding of Circular Seams

The vibrating part is fixtured onto a spring which consists of two parallel plates (one bearing the fixture and part) connected by metal columns. The spring moves as a stiff pendulum and is driven by electromagnetic or hydraulic drivers. The spring arrangement serves several functions that are important to the success of vibration welding.

(1) The columnar arrangement allows for application of pressure to the bond line.

(2) The part, fixture, and spring form a resonant assembly thereby reducing the amount of power required to effect a weld. Since the spring stiffness and mass is fixed, care must be exercised in designing the fixture and part to maintain a resonant system. Otherwise, welding may not be efficient or machine damage can occur. The combined weight of the vibrated part and fixture must usually fall below a critical value, which is characteristic for each machine, in order to meet the requirement for resonance.

(3) The spring accurately positions the parts at the end of the weld, in the instant before the polymer has resolidified. Accuracy of a few thousands of an inch (tenths of a millimeter) are normal if the parts are held securely.

Vibration welding usually produces parts in weld times of a few seconds. In some special cases, however, times of up to several minutes are useful. Since one of the parts is vibrated against the other, the fixturing must ensure that both rubbing surfaces move relative to each other. Parts can be held in fixtures by mechanical means, vacuum or even double-sided adhesive tape, but the bond areas usually need support, particularly when under weld pressures. In some cases, knurling of the fixture under the bond area is useful to keep it from moving, provided the knurling does not damage the part.

Machine and weld parameters that are important in vibration welding include:

(1) Frequency—Vibration welding machines usually operate in the range of 100 to 300 Hz. Machines driven by electromagnetic drivers operate at low multiples of line frequency, 100 and 200 Hz in countries with 50 Hz current and 120 and 240 Hz in countries with 60 Hz current. Hydraulic machines have some capability of varying frequency, for example from 200 to 280 Hz. The latter machines allow for a more convenient fine tuning of the resonant system.

(2) Pressure—Weld pressures can range from tens to thousands of pounds per square inch (from kilopascals to megapascals) depending on the material being welded, weld area, and machine power. Increases in weld pressure will result in increased rates of heat generation and shorter weld times. In general, a given machine will have sufficient power to weld a limited weld area, 16 in.^{2} (10 000 mm^2), for example, and higher pressures can reduce the possible weld area. Identification of suitable welding conditions will often involve an examination of different weld pressures. Too low a pressure can result in insufficient melting and ejection of powdered thermoplastic from the bond line. Higher, but still inadequate pressures, can result in a filigree of molten polymer being ejected. For sound welds, the ejected flash should be a solid mass or bead.

(3) Amplitude—Vibrational amplitudes range from a few mils to several hundred mils (a few tenths of a millimeter to several millimeters). With other variables fixed, higher amplitudes result in faster heating rates and shorter weld times. Flash traps can be designed into bond lines if flash is objectionable. Weld flash results from the surface layers of the polymer being ejected from the bond area and, in many cases, even a grossly contaminated bond area will not detract from producing a sound weld.

Vibration welding is capable of producing excellent bonds, as indicated in Table 6.5. Lower values in this table are from unoptimized welds. It should be possible to achieve the higher weld factors on a consistent basis.

Vibration welding can be used for nearly all thermoplastics including polymers ranging from polyethylene and polypropylene to high melting point, engineering polymers such as acetal, and polyetherimides. It is believed that hygroscopic polymers such as polyamides do not have to be dried before welding if they have been stored at humidities less than 50 percent.

Penetration during the welding process can be described as the thickness of material that is ejected from the bond line during the welding process. Stokes found that the strongest welds are produced when the penetration equals or exceeds 0.01 in. (0.25 mm).[5]

5. Stokes, V. K. "Vibration welding of thermoplastics." *Polymer Engineering and Science* 28(15): Part III, 989, and Part IV, 998, 1988.

Table 6.5
Strengths Possible by Vibration Welding

Polymer	Percentage of Parent Material Strength
Polycarbonate	80 - 99+
Polybutylene terephthalate	70 - 95
Polyetherimide (Ultem)	30 - 99+
Polyphenylene oxide	50 - 100

The dependence of penetration on process conditions of vibration welding is shown in Figures 6.13 to Figure 6.15. Penetration is more rapid with higher amplitudes and higher pressures, as can be seen directly in Figures 6.13 and 6.14. Figure 6.15 shows that higher frequencies result in more rapid penetration than lower frequencies, even though the amplitude is higher in the latter case.

In fundamental studies of vibration welding, Stokes has shown that the process proceeds in the four stages shown in Figure 6.16.[6] Stage I involves the start of vibration and heating below temperatures where the polymer begins to melt. Stage II is the onset of melting. Stage III is the steady-state region where polymer is ejected from the bond line as rapidly as it is being generated. As a result, the bond line temperature in friction welding can be expected to be well within the processing window of the polymer. Furthermore, this temperature is self-limiting. Stage IV occurs at the end of the welding cycle when the vibration ceases and the weld is allowed to cool.

6. Stokes, V. K. "Vibration welding of thermoplastics." *Polymer Engineering and Science* 28(11): Part I, 718, and Part II, 728, 1988.

As with all friction welding processes, heat-affected zones are narrow. This results in stressed regions that, in some cases, can accent polymer limitations. For example, the poor solvent resistance of polycarbonates can be expected to become worse in the stressed bond line. Such factors should be considered when selecting the welding process, welding conditions and joint design to avoid or minimize process risk. Suitable joint designs for vibration welding generally incorporate flat joining surfaces to allow application of pressure directly to the bond line. One guideline is that flanges should be 2.5 times the wall thickness. Some typical designs are shown in Figure 6.17.

Joints must be designed to allow for the linear vibratory motion. Bond lines are usually flat and parallel. Welds have been made with curved bond lines that are plus or minus 10 degrees out of the plane of the vibration.

In vibration welding, part size is limited by machine cavity size. One of the largest parts to be vibration welded was an automotive duct that was molded in two parts and joined at a seam around the periphery. Some other applications have included automotive intake manifolds, fuel tanks, taillight assemblies, pump housings, and ductwork of various types.

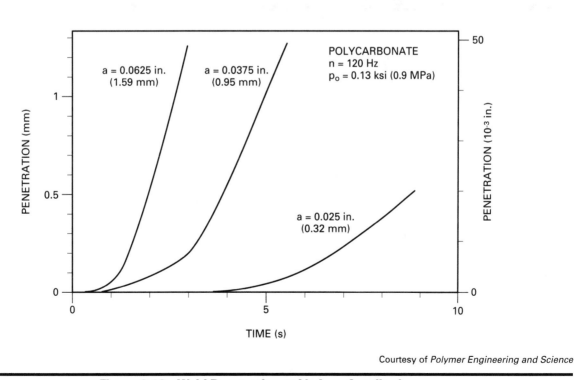

Courtesy of *Polymer Engineering and Science*

Figure 6.13—Weld Penetration at Various Amplitudes

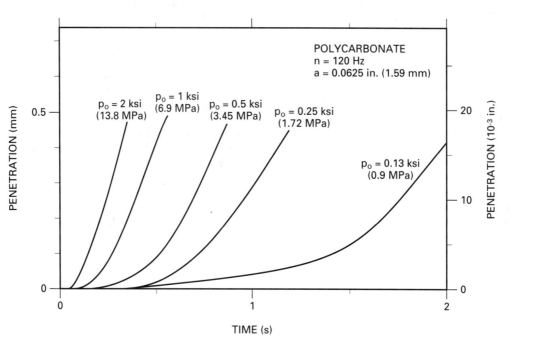

Courtesy of *Polymer Engineering and Science*

Figure 6.14—Weld Penetration at Various Pressures

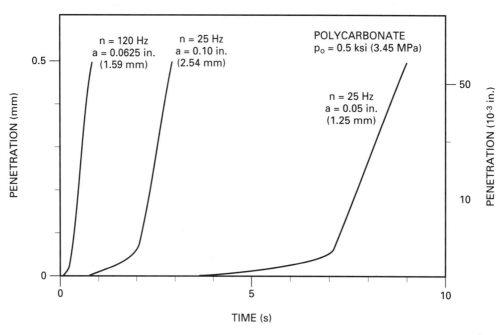

Courtesy of R. Grimm

Figure 6.15—Weld Penetration at Various Frequencies

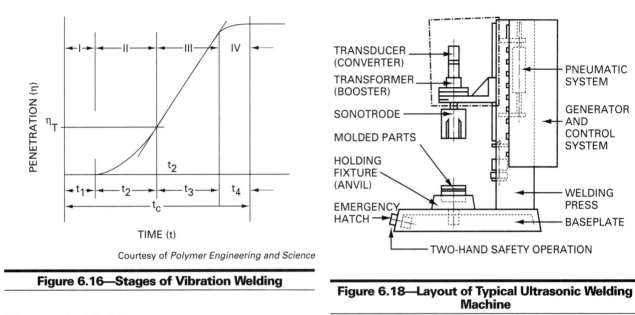

Courtesy of *Polymer Engineering and Science*

Figure 6.16—Stages of Vibration Welding

Ultrasonic Welding

IN ULTRASONIC WELDING, heating is accomplished by subjecting the parts to a fixed frequency oscillation in the range from 10 to 40 kHz. Oscillation amplitude usually ranges from 0.8 to 3 mils (20 to 80 microns). In properly designed and located joints, these oscillations are focused to produce heat and polymer fusion by hysteresis.

The elements of an ultrasonic welding machine are shown in Figure 6.18. The power supply increases both line frequency and voltage, and the output is fed to the transducer or converter. This unit contains one or more layers of a piezoelectric material sandwiched between pieces of metal. The voltage across the piezoelectric disc generates oscillations in the metal mass.

Figure 6.18—Layout of Typical Ultrasonic Welding Machine

Another less common form of an ultrasonic transducer is a magnetostrictive type. In this case, a magnetostrictive material, such as nickel, is enclosed in an electromagnet. The alternating magnetic field causes the material to expand and contract. Magnetostrictive transducers require high currents and low voltages versus the high voltages and low currents used by the more common piezoelectric transducer.

After generation in the converter, the ultrasonic waves are transmitted through a booster which can either increase or decrease their amplitude. Finally, the waves are transmitted into the parts by a horn or sonotrode which can further increase or decrease the amplitude. These three components, the converter,

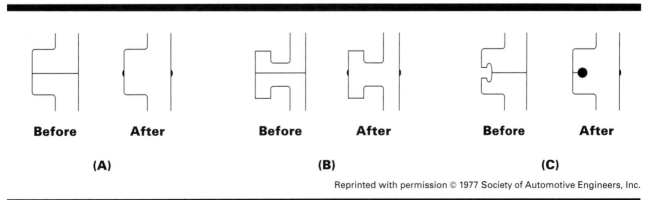

Before **After** **Before** **After** **Before** **After**

(A) **(B)** **(C)**

Figure 6.17—Joint Designs for Vibration Welding

booster and horn, are referred to as the *ultrasonic column* and are enclosed in an actuator. During the welding process, the actuator containing the column is lowered until the horn contacts one of the parts. The other part to be joined is fixtured to the base. The normal sequence of welding involves three steps:

(1) The column is lowered and the ultrasonic power activated.

(2) The joining surfaces are melted as welding occurs.

(3) The power is turned off and the parts held together to allow solidification.

The above three steps are often completed within a few seconds with actual weld times on the order of fractions of a second to a few seconds depending on the material being welded.

When ultrasonic energy is transmitted through a solid substrate, such as a rod, heating occurs in areas of alternating compressive and tensile stresses. The ultrasonic wave traveling through the part has a sinusoidal form with its amplitude defined by horn amplitude, mechanical coupling to the horn and fixture, mechanical properties of the part, and part geometry. In an ideal case, where a rod is simply attached to the horn with the free end hanging in the air (Figure 6.19), the amplitude is maximum at the horn-part interface. A quarter wavelength down the rod, the amplitude is zero. At this point there is no movement or vibration of the rod and stresses are very high. In the other extreme case, where the rod base is supported by a solid

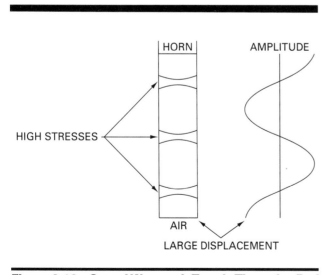

Figure 6.19—Sound Wave as it Travels Through a Rod

fixture, the areas of heating are shifted by a quarter wave length. In this case, the areas that were previously moving up and down freely, are now constricted and heat accordingly.

Studies of heating patterns in a rod are necessary to gain understanding of the process. However, it is normally impractical to design parts with joints at quarter wavelength positions. To overcome this, energy directors are used to direct the stresses and heat generation to the desired bond line. Typical joint geometries and dimensions of some energy directors are shown in Figure 6.20.

Since energy directors act as stress risers or concentrators, they make it possible to place a joint at an antinode, where there is movement but little stress. Because the lower part is held stationary by the fixture, the energy director is subjected to a large amount of alternating deformation which causes melting and welding. If an energy director is placed near a node, where there is little movement, amorphous polymers can often be welded, although with longer weld times. For semicrystalline polymers, placing the energy director near or at a node will result in no welding.

Another basic joint type used in ultrasonic welding of thermoplastics is the shear joint in which an interference fit between the two joining surfaces acts as a stress riser or concentrator. This type of joint again works best at antinodes, but is superior to the standard energy director if the joint is located at a nodal point. Figure 6.21 shows a number of different possible shear joint configurations

These joints are especially good for welding components requiring a hermetic seal. The tongue-and-groove shear joint, shown in Figure 6.21(C), increases the cross-sectional weld area, and prevents any misalignment or bending of the interference faces. Parts, when placed in the welding position, should not contain areas that bind or are otherwise in contact since this will draw weld energy away from the desired weld zone. The amount of interference in shear joints should be small, 0.005 to 0.015 in. (0.127 to 0.38 mm), to allow rapid melting and flow in the bond line. Abrasion of one or both of the bonding surfaces with a medium to coarse-grit sandpaper results in surface roughness that acts as a large number of minute energy directors to aid welding.

Early in the practice of ultrasonic welding, it was found that results varied depending on the horn-to-weld zone distance. These variations were categorized by adopting the terms *near-* and *far-field welding*. Far-field welding is defined as the condition where the horn is more than 1/4 in. (6 mm) away from the weld. Crystalline materials do not readily lend themselves to far-field welding. Near-field welding occurs when the horn is within 1/4 in. (6 mm) from the weld. Both crystalline and amorphous materials weld more readily in

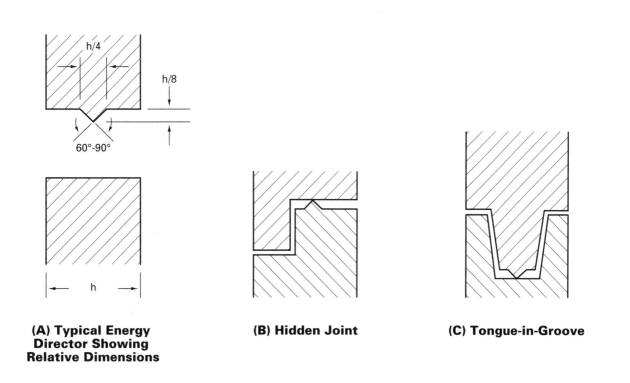

**(A) Typical Energy
Director Showing
Relative Dimensions**

(B) Hidden Joint

(C) Tongue-in-Groove

Figure 6.20—Cross Section Views of Typical Energy Director Joints

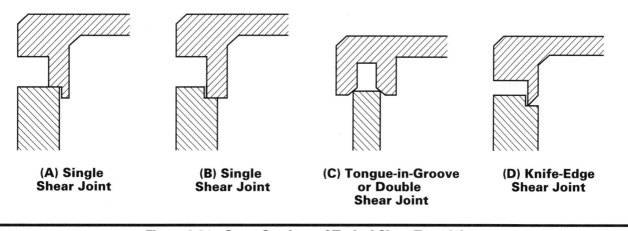

**(A) Single
Shear Joint**

**(B) Single
Shear Joint**

**(C) Tongue-in-Groove
or Double
Shear Joint**

**(D) Knife-Edge
Shear Joint**

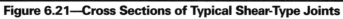

Figure 6.21—Cross Sections of Typical Shear-Type Joints

the near-field mode. The terms near- and far-field are still useful but are now a subset of the heating patterns that are produced as a result of transmission of ultrasonic energy through a material.

With ultrasonic welding, semicrystalline polymers tend to weld less readily than amorphous ones. Because semicrystalline polymers have a well-defined melt temperature, they can solidify prematurely as the melted resin flows along the cold bond line. If the fused polymer solidifies prior to spreading across the complete bond area, the mechanical coupling through the partial bond is sufficiently great that it is difficult to remelt the interface. The partial bond does not normally perform as well as a fully-formed one. Application of increased power during the welding process, through the use of higher amplitudes and energy settings, will tend to circumvent this problem.

During welding it is important that vibrations occur in desired locations. It is possible for parts to set up harmonic or resonant vibrations that can be destructive to the parts being welded. For example, one common type of ultrasonic weld involves joining a flat lid or disc onto a box. Figure 6.22 shows some of the vibrational modes that can occur in the disc-shaped part. They can create sufficient heating at nodal points, such as at the center of a disc, to melt a hole. In this instance, there are several methods for avoiding the problem. The first and most effective is to stiffen the vibrating region by either making the wall thicker or by adding rib stiffeners. Another method is to dampen the vibration by placing a stiff pad of foam rubber against the vibrating area. It

is important that the foam rubber is fixed to a rigid body, such as the base of the welding machine, so that it does not start to vibrate as well. Material property changes can also be used to reduce stray vibrations. This can be accomplished by adding fillers to the material to make it more rigid. However, filler levels over 20 percent can decrease weldability.

Additional important design criteria include:

(1)   Preparation of joints that lie on a single plane—Having the joint on a plane will tend to ensure that the wave amplitude will be uniform across the bond area and will provide uniform melting and weld strength. In addition, planar bond areas will allow the use of properly tuned horns (half-wavelength or multiples thereof). It is difficult, and in some cases impossible, to design horns with segments at different lengths that will provide uniform energy.
(2)   Design of sufficient horn contact area to prevent damage and allow efficient energy transfer.
(3)   Design parts to avoid transmission of the ultrasonic energy through bends and varying cross sections as this can result in hot spots and damage to the part.

Fixturing of the lower part is important to obtain good welding. Simple designs usually allow for better fixturing and better part insertion and removal. The fixture should hold the part as close to the weld zone as possible. If the lower part is not clamped securely, it could vibrate with the top part. In the worst case, the

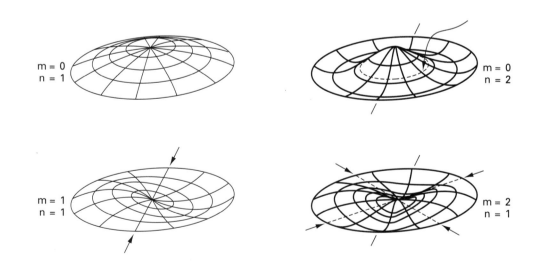

Figure 6.22—Possible Modes of Vibration in a Simple Disk

lower part will vibrate in-phase with the top, thus eliminating any stresses in the bond line and any welding.

Welding of Large Parts.

Ultrasonic welding is usually used to weld small parts. However, welding of large parts is possible by the use of techniques such as:

(1) Gang Welding—Several ultrasonic welding units are mounted on the same welding assembly with each welding a separate part or segment of the weld zone. This process is commonly used in manufacturing environments.

(2) Sequential Welding—This involves step-wise welding of a long weld zone.

(3) Scan Welding—Large parts, usually light weight, are welded using ultrasonic welding machines and a rotating anvil. Ultrasonic sewing machines are one example of this process. The vertical motion of the horn allows thin materials to be welded as they pass under the horn.

Spot Welding, Staking, and Insertion.

Ultrasonic spot welding is useful for joining panels of similar plastics. In this case, the tip of the horn is usually small [0.25 to 0.5 in. (6 to 12 mm)] and rounded with a projection extending from the center. During welding, the projection and tip produce a molten region and, on cooling, a weld at the interface.

Joining of dissimilar materials is possible by a process called *staking*. In this process, a stud is molded into one of the parts and a hole in the other. An ultrasonic horn shapes the stud into a rivet shape which holds the parts together (Figure 6.23). Ultrasonic staking is a rapid process well-suited to mass production applications.

Another practical application of ultrasonic welding is insertion. This is a method by which a foreign material, usually metal, is inserted into a thermoplastic part. Figure 6.24 shows a schematic of this process. Usually, there is an interference between the insert and the plastic part. It is also common for the insert to have knurled surfaces for better mechanical interlocking. The horn is brought into contact with the insert, and the insert is driven into the part. As the insert descends, the thermoplastic melts, fills in around the rings, and bonds to the insert resulting in a strong interlocking assembly.

Operating Conditions.

In older-style machines, there were six controlling conditions:

(1) Welding time—This is the time during which the ultrasonic energy is applied to the part. If the weld time is too long, part damage can occur. If the weld time is too short, incomplete fusion can result.

(2) Weld pressure—This is the force normal to the horn interface which promotes intimate part contact and squeeze flow of the molten polymer. Insufficient weld pressure results in incomplete part contact and poor energy transmission. Excessive pressure results in flow of molten polymer out of the joint area, unfavorable molecular orientation and weak joints.

(3) Horn amplitude—This is the vibrational amplitude at the tip of the horn. Insufficient amplitude results in poor energy transmission and the need for longer weld times. Excessive amplitude can cause part damage and overload the power supply.

(4) Trigger pressure—This is the weld pressure at which the ultrasonic energy is switched on. By using a pretrigger, it is possible to reduce the initial power requirements of the power supply and prevent overloading. However, pretriggering the ultrasonic energy can result in part damage due to vibration of the parts before application of pressure and support of sensitive areas.

(5) Down speed—This is the speed at which the ultrasonic column descends toward the parts. It controls the rate at which weld pressure increases to its maximum value. If the down speed is too high, particularly with shear joints, the parts can be pressed together in a mechanical, rather then welded, joint. With energy director joints, too rapid down speeds can result in unwelded zones under the tip of the energy director.

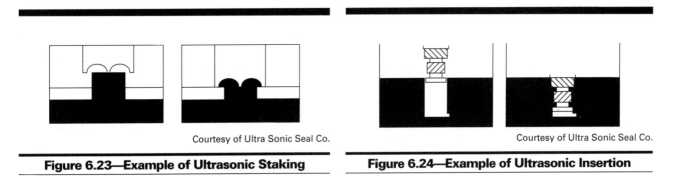

Courtesy of Ultra Sonic Seal Co.

Figure 6.23—Example of Ultrasonic Staking

Courtesy of Ultra Sonic Seal Co.

Figure 6.24—Example of Ultrasonic Insertion

(6) Hold time—This is the time during which the horn continues to apply pressure to the parts after the ultrasonic energy is turned off. Too short a hold time may result in a weak joint as stresses might be introduced before the parts have completely solidified. However, hold times are normally less than one second.

In most applications, the most important conditions are the amplitude, weld time, and pressure. To determine the proper equipment setting, it is usually necessary to make a number of welds under different conditions. During these trials, it is often possible to determine which condition needs to be varied and in which direction. For example, if the welds are incomplete, increasing the weld time will normally promote better welds. If part damage is occurring at the part/horn interface, decreasing the weld time and increasing the amplitude or weld pressure, or both, may reduce this problem.

Frequency is usually not a variable: ultrasonic welding machines are manufactured to operate at a fixed frequency within the range from 10 to 60 kHz. Higher frequencies operate with small amplitudes and are good for small, delicate parts while lower frequency machines are good for welding larger parts. Those operating at 20 kHz are a good compromise and represent the most common, general-purpose ultrasonic welding machines.

In more modern equipment, the number of possible controlling conditions has increased significantly. In the past, the amount of energy used to make a weld was controlled by varying the weld time. While this is effective, the actual amount of energy delivered to the weld zone can vary from one weld to the next, even though they are made under apparently identical conditions. With newer, microprocessor-controlled machines, the power supply monitors and often records the power during the welding cycle, integrates this curve in real time, and switches off the machine when a preset amount of energy has been delivered. In this mode, the actual amount of energy into the welds is precisely metered to the preset amount. It is also possible to control the amount of polymer melting or displacement (collapse). The welding machine monitors the position of the ultrasonic column by linear encoders, determines when the preset amount of displacement has occurred, and shuts off the power.

Weldability. The ultrasonic weldability of thermoplastics is defined by a number of mechanical and rheological properties. Crystalline materials are difficult to weld under far-field conditions unless the parts are designed properly. Far-field welding usually requires slightly more power than near-field since some of the energy is damped as the sound wave travels through the part. When welding crystalline materials, which have a well-defined melt temperature and can have a high-loss modulus, it is very difficult to get enough energy into the weld zone without melting the part at the part/horn interface.

In contrast, amorphous materials usually weld well in both near- and far-field conditions. Figure 6.25 summarizes the weldability of many common polymers and also between combinations of dissimilar polymers. Welding between dissimilar polymers is usually only possible when they are miscible in each other and have similar melting temperatures.

Fillers, reinforcing fibers and other polymer additives can have a strong effect on ultrasonic weldability. It becomes increasingly difficult to weld materials that have more than 15 to 20 percent filler level. If the percentage of filler is less than 10 percent, the weldability of the material is usually not reduced and can even be improved if material stiffness increases. The strengths of joints made with materials containing more than 20 percent filler tend to be relatively lower. When welding materials of this type, high forge pressures and high amplitudes usually work best. Welding materials with a 30 percent or greater filler level is generally very difficult. However, welding of highly filled or reinforced composites is greatly improved when pure thermoplastic resin energy directors are used.

SOLVENT WELDING

SOLVENT BONDING DIFFERS from most fusion bonding techniques for joining thermoplastics in that polymer mobility is provided by dissolution of the polymer in a solvent rather than solely by heat. Once a suitable solvent is found for the polymer, the bond surface is wetted and the parts are squeezed together. Evaporation of the solvent can provide bonds with high strengths.

Advantages of the process include:

(1) Low cost
(2) No thermal effects on the polymer
(3) Fast initial strength
(4) Tolerance of contaminants

Disadvantages include:

(1) Increasing regulation of solvent emissions, especially in manufacturing situations
(2) Difficulty in selecting the right solvent
(3) Imprecise process control
(4) Low capability for gap filling
(5) Marring the exposed surface of appearance parts
(6) Slow solvent evaporation
(7) Crazing

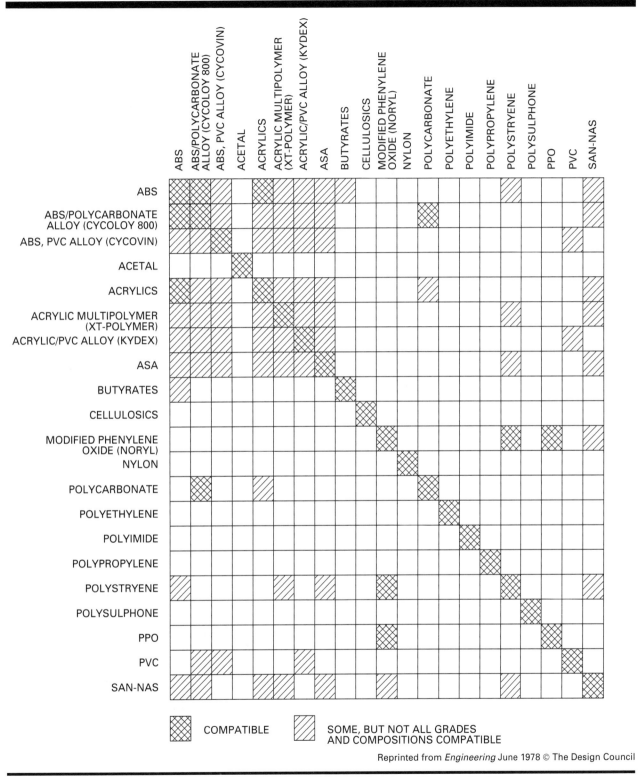

COMPATIBLE

SOME, BUT NOT ALL GRADES
AND COMPOSITIONS COMPATIBLE

Reprinted from *Engineering* June 1978 © The Design Council

Figure 6.25—Ultrasonic Weldability of Common Thermoplastics

The disadvantages, particularly the problem of solvent emissions, are causing manufacturers to seek alternative joining methods. However, the ease and low cost of the process are likely to engender its continued use for low-volume applications.

A critical aspect of solvent bonding is solvent selection. This can be speeded by determining and matching the solubility conditions for the solvent and the polymers. Appropriate solubility conditions can be achieved by blending solvents. This is frequently done, but for an additional reason. Combinations include a solvent that evaporates quickly to provide quick bond strength, and another that evaporates more slowly, to reduce the amount of stress in the bond, since it will allow polymer chains to seek less stressed configurations.

Semicrystalline polymers, as a class, are much less soluble in solvents than amorphous ones. In some cases, however, solvent bonding can be assisted by heat, or else solvents can be chosen that will interact with the polymer, chemically, in a reversible manner. Phenolic solvents can be used for bonding crystalline polyamides, for example, because the group on phenol breaks the hydrogen bonds in the polyamides and thereby reduces that barrier to solubility.

Tendencies for solvents to stress crack or craze the polymer substrate should be examined prior to welding a new application. Application of the selected solvents to a polymer part that is stressed may cause it to break at a much lower strength due to crazing and crack

initiation. Low boiling solvents (high volatility) tend to cause more crazing than high boiling ones.

Solvent bonding is capable of providing excellent bond strengths but, depending on the system, attainment of full bond strength can be a slow process. In one case, a year was required for full bond strength development.

Table 6.6 shows some typical solvents for solvent welding.[7]

---

7. For other typical solvents for various applications, see Chapter 8, "Solvent Cementing of Plastics." A.H. Landrock, *Adhesives Technology Handbook*, Park Ridge, N.J.: Noyes Publications, 1985.

Table 6.6
Representative Solvents for Solvent Welding

Polymer	Possible Solvent
Polystyrene	Toluene and ethyl acetate
Polyvinyl chloride	Tetrahydrofuran and many others
Nylon 6/6®	10-15 percent aqueous phenol
Polycarbonate	Methylene chloride and dichloroethane
Acrylic	Halogenated solvents
Polyphenylene oxide and alloys	Halogenated solvents, ketones, xylene

WELD QUALITY ASSESSMENT

ONE GOAL OF plastics welding in manufacturing is the production of 100 percent yields of high-quality welds that will have long service lifetimes, often under service conditions that are beyond design conditions.

A common practice is to identify all welding conditions and then to vary some until a suitable weld is produced. All parts are then welded using those conditions. A better practice is to examine process conditions using a series of designed experiments that permit the plotting of a weld optimization curve. Figure 6.26 shows an idealized example of such a curve where two variables are examined. All parts welded within the flat area are considered of acceptable quality. Conditions that are located at the center of the flat area will be most tolerant of process variation, such as molded part inconsistency, machine wear, or machine drift. Without knowing what the weld optimization curve looks like, one can very easily operate at the edge of the flat spot

and, as soon as there is any process variation, poor welds can result.

For manual operations and some welding processes where control is not as advanced, several other methods for assessing weld quality can be used, but reliability is not as high. Hot gas welds and welds in butt joints for thermoplastic gas distribution pipe usually are examined visually. Weld quality is largely the result of using well-trained personnel and standardized procedures.

Other nondestructive techniques are still being developed for pipe inspection; these include ultrasonic scanning and analysis by soft x-rays.

Depending on the part shape, ultrasonic techniques such as C-scan can be used for periodic inspection but usually not for routine analysis. Techniques used are heating a slice of the weld bead and observing whether or not the seam reappears as the polymer relaxes, or removing the weld bead and flexing it to generate failure.

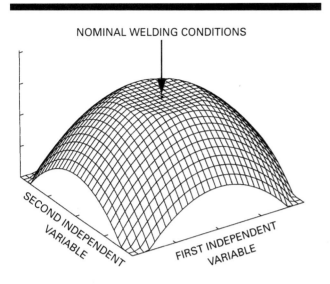

NOMINAL WELDING CONDITIONS

SECOND INDEPENDENT VARIABLE

FIRST INDEPENDENT VARIABLE

**Figure 6.26—Idealized Optimization Curve
Showing Most Reliable Operating Conditions**

As mentioned earlier, welded or fusion bonded polymers form heat-affected zones. Depending on the polymer, these zones can exhibit ranges of crystalline structure, just as in metals welding. The usual way of studying these zones is by slicing a thin section from the weld zone (microtoming) and examining the section under a microscope using polarized light. A discussion of polymer morphology is beyond the scope of this chapter, but Figure 6.27 shows some weld zones that can be produced.

The photo shows some of the structures present in an uncontrolled hot plate weld in polypropylene. At the edges are crystallites (spherulites) of the bulk polymer showing the crystals growing from nuclei at the center of each. Oriented crystals, the lamellar structures, can be seen in zones where squeeze flow occurred. Toward the center of the weld zone, small spherulites are seen because the cooling rates were fast and crystals did not have time to grow into large spherulites. The material at the weld center is most probably a region of transcrystalline growth where the crystals have grown perpendicular to the weld line. Transcrystalline growth is

Courtesy of R. Grimm

Figure 6.27—Polymer Morphology in Hot-Plate Welded Polypropylene

desirable in adhesive bonding since it increases surface energy and enhances adhesion. In welding, however, it can indicate a weak zone because it usually is formed by dirt on the coupon surface causing rapid nucleation and crystallization within the polymer. This may prevent adequate polymer chain intermingling.

Failure in the lamellar region can sometimes be detected by the formation of fibrillar fracture surfaces. Stressed zones, just as with metal structures, have different chemical potentials than bulk material and will behave differently toward environmental agents.

HEALTH AND SAFETY ISSUES

A THOROUGH DISCUSSION of health and safety issues relating to the welding of plastics is beyond the scope of this chapter. Moreover, because this area is changing rapidly, the following discussions should not be regarded as definitive. What is attempted is to point out some of the issues and indicate some references that can provide more information. The American Conference of Governmental Industrial Hygienists publishes an annually updated booklet on threshold limit values for biological exposure. It includes information on chemical hazards as well as on physical agents such as ultrasonic energy, electromagnetic radiation, noise, vibration and other agents.[8]

Health and safety issues can arise in the welding of thermoplastics in two primary areas; one results from exposure to polymer decomposition products while the second is process related.

POLMER DECOMPOSITION

UNDER SUITABLE WELDING conditions, the thermoplastic is melted or softened and the weld is created without the generation of decomposition products. These are the conditions that create the strongest and soundest welds. A useful guideline is to use welding conditions that are similar to those used in the normal processing of the polymer but below temperatures where onset of decomposition occurs.

Smoke or fumes that might be produced under abnormal conditions are a greater concern. The smoke or fumes are generally pyrolysis products (not combustion products) and the components may be toxic or otherwise harmful. Different polymers will produce different mixtures of pyrolysis products, so each case must be considered separately. Ventilation is a possibility, but the best solutions are to install suitable process controls and use conditions that avoid decomposing the polymer.

Extra care should be used during welding of chlorine- and bromine-containing polymers (either bonded to the polymer chain or as part of some, but not all, flame retardants). Overheating can result in the formation of hydrogen chloride or bromide which are toxic and can overcome welders, particularly in enclosed areas. Fluorine-containing polymers, such as Teflons, require positive ventilation when being welded using methods where the temperature of the heat source exceeds decomposition temperatures (e.g., hot-gas, infrared, or laser welding).

A booklet published by the Suppliers of Advanced Composite Materials Association describes safe handling procedures for different composite materials.[9]

PROCESS-RELATED ISSUES

VARIOUS PROCESS-RELATED safety issues can arise depending on the process. Welding machines routinely provide safety shields and features such as dual palm buttons which must read and obey safety instructions that are provided with welding equipment.

Unlike many metal welding processes, welding of plastics involves mechanically pressing the two melted substrates together because thermoplastics do not flow like molten metals. The pressures involved can be substantial and care must be exercised to avoid harm by unintended contact with these mechanisms.

Obviously, the temperatures used in most plastics welding can cause serious burns. Thus respect for hot plates and infrared lamps and their ability to burn people is required. Gases that exit from a hot-gas welding tool reach temperatures that can easily create a burned stripe when a hand is carelessly used to test the temperature. Laser beams, while much less powerful than those used for welding of metals, can cause burns and beams can be inadvertently reflected from workpieces.

8. American Conference of Governmental Industrial Hygienists (ACGIH). *Threshold limit values for chemical substances and physical agents and biological exposure indices*. This booklet can be purchased from Executive Secretary, ACGIH, 1330 Kemper Meadow Drive, Suite 600, Cincinnati, Ohio 45240-1634.

9. Suppliers of Advanced Composite Materials Association (SACMA). *Safe handling of advanced composite materials components: health information*. Available from SACMA, 1600 Wilson Road, Suite 1008, Arlington, Va. 22209, April 1989.

Electromagnetic fields are produced by induction generators, dielectric welding machines (RF welding) and microwave welding units. Electronic devices for controlling heart rhythms (pacemakers) should not be exposed to electromagnetic radiation such as those generated by induction generators. Wedding rings, wrist watches and belt buckles are common items that should not be exposed to high intensity induction fields because they can be heated and cause burns. The fields produced by induction generators, particularly at higher frequencies (1 to 10 MHz) can cause interference with communication unless they are properly grounded and shielded. Limits for both low- and high-frequency electromagnetic radiation have been established by the American Conference of Governmental Industrial Hygienists (ACGIH).[10]

Dielectric and microwave generators are normally operated at frequencies assigned by the Federal Communications Commission. Shielding of the units so that the field is contained is common practice, either through design of the equipment or by separate enclosures. Meters are available for measuring field strengths within this frequency range. The Radiation Control for Health and Safety Act governs exposure levels. A recent report on the subject of electromagnetic radiation is available and contains leading references.[11]

Hearing protection is recommended for personnel in the vicinity of sonic, ultrasonic and vibration welding machines. While ultrasonic welding is conducted at frequencies above the normal hearing range, harmonics that can be generated in the parts being welded can be audible and intense. Vibration and sonic welding machines operate at audible frequencies. Machine design is important in limiting the intensity of sound generation but the best solution is to operate equipment in sound-deadening enclosures. Ear muffs or other equipment designed for ear protection is a second option. Federal guidelines (OSHA) specify limits on exposure to high-intensity sound.[12]

---

10. American Conference of Governmental Industrial Hygienists. *Threshold limit values for chemical substances and physical agents and biological exposure indices.*

11. Nair, I., Morgan, M. G. and Florig, H. K., *Biological effects of power frequency electric and magnetic field*s. This background paper is available from NTIS as report PB89-209985, May 1989.

12. Refer to American Conference of Governmental Industrial Hygienists. *Threshold Limit Values.*

APPLICATIONS

EACH OF THE processes for welding of thermoplastics has characteristics that may make it more suitable for a particular application than others. Selection of a welding process will depend on many factors including production rate, capital cost, weld size and shape (curvature), and polymer type and formulation. Categorization, however, must be used with caution because processes can be adapted to different applications. After the following reviews of various process applications, the last subsection, "Process Selection," Page 349, will summarize some applications guidelines.

HOT-PLATE WELDING

HOT-PLATE WELDING is commonly used for joining piping from polymers such as polyethylene, polypropylene, and polyvinylchloride (PVC) because the process is economical, the equipment is relatively mobile, and the welds are reliable. The pipe shown in Figure 6.28 is similar to that used for natural gas distribution lines. Prior to welding, a tool is used to square the pipe ends. This step cleans the bonding surface and ensures they make uniform contact with the hot plate. Heating times will vary from approximately 30 seconds for 1-in.

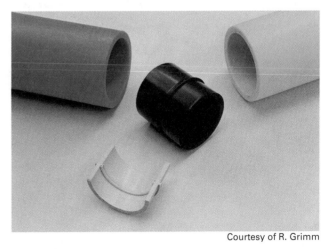

Courtesy of R. Grimm

Figure 6.28—Thermoplastic Pipe

(25.4-mm) diameter pipe to 5 to 6 minutes for 10-in. (254-mm) diameter pipe. A heating time around 2 minutes at a hot-plate temperature of 430 °F (221 °C) was used to weld the samples shown in the photograph.

Hot-plate welding has an advantage when the application requires welding parts that have complex curvature or three-dimensional bond lines. Heated platens can be machined to fit the bond lines of both parts such as for the tail-light assembly shown in Figure 6.29. Approximately 10 to 15 second heating times were used for assembly of this polycarbonate tail light. Metal platens are normally coated with a nonstick coating.

HOT-GAS WELDING

IMPORTANT ADVANTAGES OF hot-gas welding and its variant, extrusion welding, include portability, low capital cost, and the ability to make bond lines of indefinite length. These advantages have been applied to the construction of the tanker truck body shown in Figure 6.30. It is made from 3/8-in. (9.5-mm) thick polypropylene sheets that were welded using hot-gas welding. Weld zones were chamfered prior to welding. Seams over 10 ft (3 m) in length were welded using dry nitrogen gas welding and a speed tip.

RESISTIVE IMPLANT WELDING

AN ELECTROFUSION SOCKET used for resistive implant welding of polyethylene gas-distribution pipe and a section of a finished joint are shown in Figure 6.31. Prior to welding, the surface of the pipe is scarfed to clean the surface and to provide the correct dimensions for insertion into the socket and to remove oxidation. The inside surface of the socket is wound with resistance wires that remain in the finished joint. The ends of the wire element are located in the socket stubs for connection to the power supply.

Courtesy of Triple "M" Plastic Products

Figure 6.30—Polypropylene Truck Body by Hot-Gas Welding

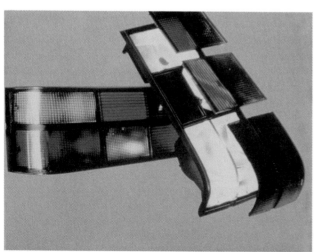

Courtesy of Branson Plastic Joining, Inc.

Figure 6.29—Automotive Tail Light Assembly

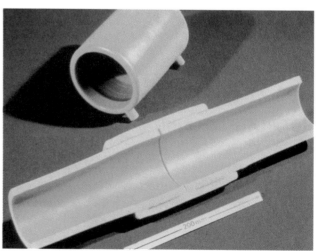

Courtesy of The Welding Institute

Figure 6.31—Electrofusion Joint in Gas Pipe

This process has several advantages. It is particularly useful for repair of buried thermoplastic pipe because the process does not require moving the pipe during the welding process, as is required with hot-plate welding, In addition, since the process is automated, less operator skill is required. A disadvantage is that the sockets are expensive relative to the cost of hot-plate welding; however, use is increasing because of process convenience and reliability.

INDUCTION IMPLANT WELDING

THE POLYETHYLENE LAWN mower fuel tank and shroud assembly shown in Figure 6.32 was welded by induction using a metal-filled polyethylene implant in the bond line. To maintain esthetics, this application used a tongue-in-groove joint with the tongue displaced to one side of the bond line causing the molten polymer to flow to the non-appearance side of the weld. The 60-in. (1.52-m) bond length was welded in approximately 20 seconds with a 2-kw power supply.

VIBRATION WELDING

THE PVC PIPE tee shown in Figure 6.33 was molded in two halves and joined using vibration welding. Molding of the tee in one part would have required a more complex and expensive mold because of the tee shape and the presence of recessed (undercut) areas on the interior face. Vibration welding of the more easily molded tee-halves was completed in 10 to 20 seconds.

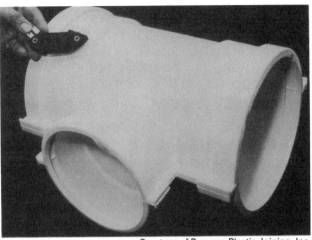

Courtesy of Branson Plastic Joining, Inc.

Figure 6.33—Vibration Welded PVC Pipe Tee and Ignition Module

Projections were molded onto the flanges to help hold the bonding surfaces during the welding process. Without them, the pipe tee might have flexed and resulted in uneven welding.

The ignition module shown at the upper left corner of the photograph was also vibration welded and shows some of the smaller parts that can be welded by vibration welding. Welding times for the module are around one second.

ULTRASONIC WELDING

THE FLOPPY DISKETTES shown in Figure 6.34 were assembled using ultrasonic welding. The magnetic recording sheet is placed between upper and lower halves of the diskette housing. Precise alignment is provided by rings molded into the halves that nest together when assembled. Triangular ridges of thermoplastic (energy directors) are located in opposition to the rings. In some configurations, bosses topped with energy directors are molded into the diskette half.

Weld times are on the order of 300 to 500 milliseconds. During this time, at least four welds are created simultaneously.

PROCESS SELECTION

THE MOST SUITABLE process for each application should be selected. Table 6.7 provides some selection conditions and observations.

Capital costs will vary depending on the size and capability of equipment. In general, solvent, hot gas, and extrusion welding are lowest in capital cost. A

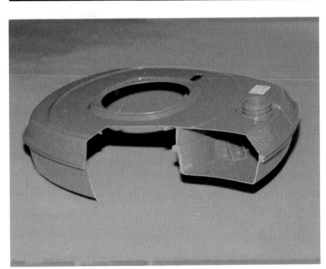

Courtesy of Emabond Systems

Figure 6.32—Lawn Mower Fuel Tank and Shroud

Courtesy of R. Grimm

Figure 6.34—Floppy Diskettes Welded Using Ultrasonic Welding

mid-range includes methods such as hot plate, infrared, ultrasonic, spin, and vibration welding. Because of the power supplies and fixturing, induction and dielectric are often more expensive than other methods, although there can be a considerable amount of overlap. While initial capital costs may be a factor in process selection, the cost per welded part is a more important factor.

Table 6.7
Process Selection Conditions

Process	Weld Time	Maximum Weld Size
Ultrasonic	Low ms to seconds	Very small to several inches (gang for larger)
Vibration	High ms to seconds	Up to several feet
Solvent	Seconds to minutes	Up to several inches
Spin	High ms to seconds	Up to several inches
Dielectric	ms to seconds	Up to several feet
Induction	Seconds	Up to several feet; indefinite is possible in continuous mode
Hot plate	Seconds to minutes	Up to 5 ft (1.5 m)
Infrared	Seconds to minutes	Up to several feet (developmental)
Resistive Implant	Minutes	Several feet
Extrusion	Continuous	Indefinite
Hot gas	Continuous	Indefinite

SUPPLEMENTARY READING LIST

American Welding Society. "Welding of plastics." *Welding Handbook*, Ch. 56, 6th Ed., Vol. 3B, Miami, Florida: American Welding Society, 1971.

Benatar, A. and Cheng, Z. "Ultrasonic welding of thermoplastics in the far-field." *Polymer Engineering and Science* 29(23): 1699-704, 1989.

Deutsche Institut Normen, *Welding of thermoplastics, Principles*. DIN 16960, February 1974.

Esaulenko, G.B., Kondratenko, V. Yu, and Bezruk, L.I. *Study of strength characteristics of polyethylene butt welded joints and development of testing methods*. International Institute of Welding Doc. No. XVI-584-90.

"Friction welding of plastics." *Welding in the World* 16(11/12): 1978.

Gabler, K. and Potente, H. "Weldability of dissimilar thermoplastics – Experiments in heated tool welding." *Journal of Adhesion* 11: 145-63, 1980.

Gumbleton, H. "Hot gas welding of thermoplastics – An introduction." *Joining and Materials* 2(5): 215-8, 1989.

Landrock, A.H. *Adhesives technology handbook*. Park Ridge, N.J.: Noyes Publications, 1985.

Leatherman, A. "Induction bonding finds a niche in an evolving plastics industry." *Plastics Engineering* 27(4), 1981.

Mengason, J. "New designs through vibration welding." Automotive Engineering Congress, Paper 770235, Society of Automotive Engineers. Detroit, February 1977.

Michaeli, W., Netze, C., and El Barbari, N. *Local changes of properties through the high-temperature welding process*. International Institute of Welding Doc. No. XVI-567-89.

Michel, P. "An analysis of the extrusion welding process." *Polymer Engineering and Science* 29(19): 1376, 1989.

Muccio, E.A. "Welding of plastics." *Plastic Part Technology*, 249-263. Materials Park, Ohio: ASM International, 1991.

Pimputkar, S.M. "The dependence of butt fusion bond strength on joining conditions for polyethylene pipe." *Polymer Engineering and Science* 29(19): 1387-1395, 1989.

Potente, H. and Natrop, J. "Computer aided optimization of the parameters of heated tool butt welding." *Polymer Engineering and Science* 29(23): 1649-1654, 1989.

Potente, H. "Ultrasonic welding – Principles and theory." *Materials Design* 5: 228-234, 1984.

Stokes, V. K."Vibration welding of thermoplastics, Part I: Phenomenology of the welding process." *Polymer Engineering and Science* 28(11): 718-727, 1988.

————. "Vibration welding of thermoplastics, Part II: Analysis of the welding process." *Polymer Engineering and Science* 28(11): 728-739, 1988.

————. "Analysis of the friction(spin)-welding process for thermoplastics." *Journal of Materials Science* 23: 2772-2785, 1988.

Swartz, H.D. and Swartz, J.L. "Focused infrared melt fusion: Another option for welding thermoplastic composites." Proceedings of the SME Joining Composites '89 Conference. Garden Grove, Calif., March 28-29, 1989.

Taylor, N.S. and Watson, M.N. "Welding techniques for plastics and composites." *Joining and Materials* 1(8): 24, 1988.

Thomas, D.W. *Making better plastic welds.* Lyndonville, Vt.: Laramy Products Company, Revised 1962.

Watson, M.N., Ed. *Joining Plastics in Production.* Abington, Cambridge, U.K.: The Welding Institute, 1988.

Watson, M.N., *Advances in joining plastics and composites.* TWI International Conference. Abington, Cambridge, U.K.: Abington Publishing, 1991.

Zinn, S. and Semiatin, S.L. "Coil design and fabrication: Basic design and modifications." *Heat Treating*: 32, June 1988.

Zinn, S. and Semiatin, S.L. *Elements of induction heating: Design, control, and automation.* Metals Park, Ohio: ASM International, 1988.

Zentralverband der Elektrotechnik- und Elektronikindustrie e. V. (ZVEI). *Ultrasonic assembly of thermoplastic mouldings and semi-finished products.* Fachverband Elektroschweissgerate, ZVEI, Stresemannalle 19, D-6000 Frankfurt (Main) 70, Germany.

COMPOSITES

PREPARED BY A COMMITTEE CONSISTING OF:

A. Benatar, Co-Chairman
Ohio State University

C. T. Lane, Co-Chairman
Duralcan USA

W. A. Baeslack
Ohio State University

S. Rokhlin
Ohio State University

WELDING HANDBOOK COMMITTEE MEMBERS:
H. R. Castner
Edison Welding Institute

M. J. Tomsic
Plastronic Inc.

CHAPTER 7

COMPOSITES

WELDING OF POLYMERIC COMPOSITES

INTRODUCTION

POLYMERIC COMPOSITE MATERIALS were developed because no single, homogeneous material could be found that had all of the desired properties for a given application.

Fiber-reinforced polymeric composites were developed initially for aerospace applications. Aluminum alloys, which typically have high strength and high stiffness at low weight, have provided good performance and have been the main materials used in aircraft structures over the years. However, corrosion and fatigue problems have been experienced with aluminum alloys. In response, fiber-reinforced polymeric composites were developed and have been used successfully in various structural applications.

Inexpensive polymeric composites, such as fiberglass, are used today not only in high-tech aerospace applications, but also in a wide variety of consumer products.

Polymeric composites can be divided into classes in various manners. One classification system separates them according to reinforcement forms such as *particulate-reinforced*, *fiber-reinforced*, or *laminate* composites. The fiber-reinforced composites can be further classified into those containing *short* or *long* fibers.

Particulate-reinforced polymeric composites can include, as a second phase, spheres, rods, flakes, and shapes of roughly equal dimensions in the three axes. These types of composites are usually polymers containing particles that extend rather than reinforce the material. Sometimes these systems are called *filled systems*, because the filler particles are added for the purpose of cost reduction rather than reinforcement (see Figure 7.1).

Fiber-reinforced composites contain reinforcements having lengths much greater than their cross-sectional dimensions. The ratio of the length to the diameter is called the *aspect ratio*. A fiber-reinforced composite is

Figure 7.1—Particulate Composite

considered to be a *short-fiber composite* when the aspect ratio is small, typically less than 100 [see Figure 7.2(A)]. When the aspect ratio of the fiber is such that further increase in length does not increase the elastic modulus of the composite, the composite is considered to be a *long-fiber composite* [see Figure 7.2(B)]. Most long-fiber composites contain fibers that are comparable in length to the overall dimensions of the composite part.

Laminated composites are composed of two or more layers, with two of their dimensions being much larger than the third (see Figure 7.3).

For long-fiber-reinforced composites, the fibers provide almost all of the load-carrying characteristics, of which strength and stiffness are the most important. A variety of fibers can be used. Table 7.1 lists some of the common fibers.

The purpose of the composite matrix is to bind the fibers together by virtue of its cohesive and adhesive characteristics and to transfer load to and between fibers. The matrix is the weakest part of the composite. Resins available today won't withstand the high stresses that the fibers can withstand. The matrix keeps the reinforcing fibers in the proper orientation and position so that they can carry the intended loads. The matrix also helps to distribute the loads more uniformly throughout the material. When the composite is under load, damage may occur in the matrix or fibers

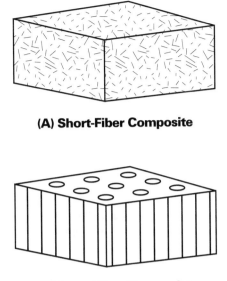

(A) Short-Fiber Composite

(B) Long-Fiber Composite

Figure 7.2—Fibrous Composites

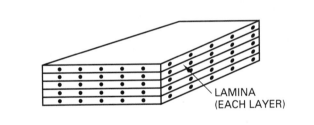

LAMINA
(EACH LAYER)

Figure 7.3—Laminated Composites

Table 7.1
Common Materials Used as Fibers

Steel	Boron
Tungsten	Graphite
E-glass	Kevlar
S-glass	Al_2O_3 whiskers
SiC whiskers	

depending on their properties. Debonding between the matrix and fiber surface can occur and reduce the composite's performance. The most common types of resins for polymeric matrices are epoxy resins, polyamide resins, polyester resins, and thermoplastic resins.

THE WELDING PROCESS

THE PROCESS OF fusion bonding or welding of polymeric composites is accomplished through diffusion and entanglement of the matrix molecules. Joining of polymeric-composite parts is critical to the manufacture of structures. Polymeric-matrix composites can be classified into two groups: *thermosetting-matrix* and *thermoplastic-matrix* composites.

Thermosetting-matrix composites cannot be welded because of the cross-linking of the polymer chains; they can be joined only by mechanical fastening or adhesive bonding, or sometimes both. For thermoplastic-matrix composites, the polymer chains are held together by secondary chemical bonds. Upon heating, these bonds weaken and break, and the polymer chains are free to move and diffuse. Therefore, in addition to mechanical fastening and adhesive bonding, fusion bonding or welding can be used to join thermoplastic-matrix composites.

Thermoplastic matrices can be further divided into two types: *amorphous* and *semicrystalline*. Amorphous thermoplastics have chains that are randomly arranged. Semicrystalline polymers have both amorphous and crystalline regions. In the amorphous regions, the chains are arranged randomly. In the crystalline regions, chain segments are closely packed and atoms are sufficiently close to form crystalline cells. For amorphous polymers, the critical temperature for welding is the glass transition temperature. For semicrystalline polymers, the critical temperature for welding is the melting temperature.

Most welding processes that are suitable for the joining of thermoplastics also can be used to fusion-bond composites. The processes for welding of plastics and composites can be classified in two groups. Processes in the first group use an *external* heat source such as a hot plate, hot gas, resistively or inductively heated implants, and infrared or laser welding. Processes in the second group use *internal* heat generation within the plastic; these processes include dielectric and microwave heating, friction heating (spin welding), vibration welding, and ultrasonic welding.

Fusion bonding or welding of thermoplastic-matrix composites offers some advantages over conventional mechanical fastening and adhesive bonding. Mechanical fastening introduces holes in the parts, which can result in stress concentrations and weakening of the parts. In adhesive bonding, the adhesive (usually a thermoset) may not be compatible with the thermoplastic composite. In fusion bonding, no new materials are introduced, and the stress distribution is more uniform than in mechanically fastened joints.

The fusion bonding process can be divided into five steps:

(1) Surface preparation to remove contaminants
(2) Heating and melting of the thermoplastic matrix on the weld surfaces
(3) Pressing to promote flow and wetting
(4) Intermolecular diffusion and entanglement of the polymer chains
(5) Cooling and resolidification of the thermoplastic

Surface Preparation

SURFACE PREPARATION FOR thermoplastic composites is important if the surfaces have been contaminated (for example, by mold release, oil, or grease). Typically, a contaminated surface may be treated mechanically or chemically, except when surface texture needs to be preserved. Dirt, oil, and grease can be removed by standard cleaning and degreasing techniques. Mold release that has transferred onto the part is more difficult to remove; therefore, it is best to use a nontransferable mold release.

Benatar and Gutowski studied the effect of using the nontransferable mold release Frekote 34 on welding of composites.[1] They found that with proper care, the small amount of mold release that transferred to the part did not affect the welding process.

Heating

HEATING OF THERMOPLASTIC composites may be accomplished in many ways. The most attractive techniques are those that heat and melt the thermoplastic only on the surface in the vicinity of the bond. These surface-heating techniques are faster and more energy efficient than other tecniques.

The presence of reinforcements can affect the conduction of heat in composite materials. In fiber-reinforced plastics, the thermal conductivity can be anisotropic, which can further complicate the heating process. For example, Figure 7.4 shows the thermal contours developed in a unidirectional graphite thermoplastic composite heated by a point source. Because of the high thermal conductivity of the fibers, the thermal contours are elliptical in shape with the major axis being in the longitudinal direction. Reinforcements with high thermal conductivities also cool the weld surfaces very quickly. Therefore, for composites with very conductive reinforcements, Benatar has shown that it is beneficial to have a resin-rich surface to provide some thermal insulation.[2] In his work, Benatar modeled the

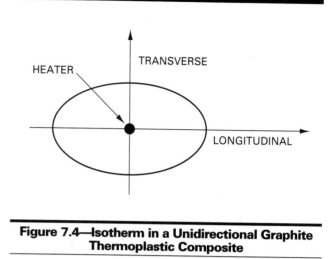

Figure 7.4—Isotherm in a Unidirectional Graphite Thermoplastic Composite

heat flow during welding using both finite element and finite difference programs.

During heating of fiber-reinforced thermoplastics, if no pressure is applied, the composite deconsolidates and forms large voids. This is due to stored elastic energy in the fibers that is released upon heating and melting of the matrix.[3,4] Surface heating offers the advantage of limiting the volume of molten polymer, thereby minimizing deconsolidation. It is generally recommended that fiber-reinforced thermoplastics be supported during heating in order to minimize deconsolidation and warping.

Pressing

PRESSING THE HEATED parts deforms the molten surface asperities and produces intimate contact at the interface. Intimate contact is necessary for intermolecular diffusion. Figure 7.5 shows a typical interface between two composite parts. A complete description of the process is quite complicated due to the irregularity of the composite surfaces, the non-uniform temperature field, and air entrapment. However, if one concentrates on one surface asperity (as in the inset diagram in Figure 7.5), its deformation and flow can be modeled by squeeze flow.

Squeezing flows are encountered in many situations involving the processing of plastics. Grimm has

1. Benatar, A. and Gutowski, T.G. "Methods for fusion bonding thermoplastic composites." *SAMPE Quarterly* 18 (1): 34-41, 1986.
2. Benatar, A. "Ultrasonic welding of advanced thermoplastic composites." Ph. D. thesis, Department of Mechanical Engineering, Massachusetts Institute of Technology, 1987.

3. Ensio, R. "The cause of deconsolidation and its effects on mechanical properties of thermoplastic composites." S.B. thesis, Department of Mechanical Engineering, Massachusetts Institute of Technology, 1984.
4. Gutowski, T. G. "Resin flow/fibre deformation model for composites." *SAMPE Quarterly* 16 (4): 58-64, 1985.

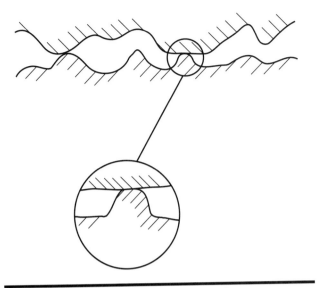

Figure 7.5—Two Surfaces in Contact

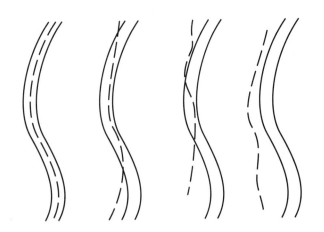

Figure 7.6—Reptation Model for Diffusion of Polymer Chains

presented a review of the work done on the squeezing flow of polymeric liquids.[5]

During welding of composites, the presence of reinforcement in the surface asperities can substantially increase the effective viscosity. Experiments have shown that, for composites with fiber volume fraction of 60 to 70 percent, the effective viscosity of the composite can be 10 to 400 times the effective viscosity of the neat resin.[6,7] Therefore, it is beneficial to have a resin-rich layer on the weld surface, thereby keeping the viscosity as small as possible.

Intermolecular Diffusion

WELD STRENGTH DERIVES from intermolecular diffusion across the bond surface and chain entanglement. The motions of individual linear polymer chains is modeled using the reptation theory developed by DeGennes.[8,9] A polymer chain is viewed as if it is confined by an imaginary tube representing the constraints imposed by adjacent chains (see Figure 7.6). The chain-like molecule moves by wriggling within a tube and only the end segments are free to move outside the tube (just like reptiles move, thereby the name reptation model). Once an end segment moves out, adjacent segments also can move out of the tube. Over time the chain can slip out of its original tube completely, and a new tube is imagined around it.

Wool and coworkers have studied the intermolecular diffusion process in detail.[10] They use the reptation model to describe chain diffusion during crack healing of polymers. They show that end segment motion is sufficient to provide strength at the interface. In the reptation model, segmental motion occurs more rapidly than diffusion of the center of mass of the molecule, thereby resulting in rapid healing of the weld interface.[11]

The reptation model was derived for amorphous thermoplastics at temperatures above the glass transition temperature. For amorphous polymers, the diffusion time depends on the welding temperature relative to the glass transition temperature. For semicrystalline polymers, intermolecular diffusion can take place only

5.   Grimm, R. J. "Squeezing flow of polymeric liquids." *American Institute of Chemical Engineers Journal* 24(3): 427, 1978.
6.   Dara, P. H. and Loos, A. C. "Thermoplastic matrix composite processing model." Interim Report No. 57, NASA - Virginia Tech Composites Program, NAG-1-343, 1985.
7.   Gutowski, T. G., Cai, Z., Soll, W. and Bonhomme, L. "The mechanics of composites deformation during manufacturing processes." *Proceedings of the First Conference on Composite Materials.* Dayton, Ohio: American Society for Composites, 1986.
8.   DeGennes, P. G. "Reptation of a polymer chain in the presence of fixed obstacles." *Journal of Chemical Physics* 55: 572, 1971.
9.   DeGennes, P. G. "Entangled polymers." *Physics Today*, June 1983.

10.   Wool, R. P. and O'Connor, K. M. "A theory of crack healing in polymers." *Journal of Applied Physics* 52(10): 2953, 1981.
11.   In addition to Wool and O'Connor, also see:
Dara, P. H. and Loos, A. C. "Thermoplastic matrix composite processing model." Interim Report No. 57, NASA - Virginia Tech Composites Program, NAG-1-343, 1985.
Jud, K., Kausch, H., and Williams, J.G. "Fracture mechanics studies of crack healing and welding of polymers." *Journal of Materials Science* 16: 204-10, 1981.
Willet, J. L., O'Connor, K. M., and Wool, R. P. "Mechanical properties of polymer-polymer welds: Time and molecular weight dependence." *ACS Polymer Reprints* 26(2): 123, 1985.
Wool, R. P. and Rockhill, A. T. "Molecular aspects of fracture and crack healing in glassy polymers." *ACS Polymer Reprints* 21(2): 223, 1980.

at temperatures above the melting temperature. Since at the melting temperature a semicrystalline polymer resembles an amorphous polymer at a temperature much higher than the glass transition temperature, then the diffusion time is very short. Estimates for the diffusion times for two semicrystalline polymers, poly-ether-etherketone (PEEK) and a DuPont™ polyamid (J Polymer), show that they are essentially instantaneous compared to the heating and pressing times.[12]

Cooling

COOLING IS THE final step in the welding process. As the thermoplastic cools and resolidifies, structural integrity is achieved in the bond and the part. The composite should be held under pressure until it has solidified sufficiently to suppress deconsolidation and warping. Also, the rate of cooling can affect the composite properties by affecting the microstructure. This is particularly important for semicrystalline matrices, because it will affect the percentage of crystallinity. Generally, a high-percentage crystallinity gives the matrix its solvent resistance. However, high-percentage crystallinity is usually associated with reduced toughness in composites.[13]

For PEEK and probably other polymers, the crystallization behavior can be modeled based on the Avrami kinetic equation.[14,15] In the Blundell and Osborn studies of crystallinity growth in PEEK, they determined that for cooling rates in excess of 1300 °F (700 °C) per minute the spherulitic growth can be greatly reduced, thereby reducing the solvent resistance. For cooling rates below 50 °F (10 °C) per minute, they found that crystallinity increases to above 35 percent, which can result in a morphology affecting mechanical properties of the part, such as the fracture toughness.[16]

REQUIREMENTS FOR JOINING

TECHNIQUES FOR FUSION bonding or welding of composites have to meet several basic requirements.

First, and most importantly, the welding process must be able to produce a strong bond repeatedly and reliably, while minimizing deconsolidation and warping. Second, the technique must be flexible enough to accommodate parts of different geometries, different matrices, and different reinforcements. This is particularly important in the aerospace industry where thermoplastic composites are used in small-lot production. Third, a means must exist for inspecting the weld either during welding or after the weld is completed. The fourth and final requirement is that the joining technique be cost effective.

Few test methods have been developed specifically for fusion-bonded or welded joints. However, fusion bonded joints are very similar in geometry and mechanical behavior to adhesively bonded joints. Therefore, one can use many of the tests that are available for adhesively bonded joints.[17] The stress analysis of welded joints also is similar to the analysis of adhesive joints, and often the same analysis techniques and equations can be used.[18] A major effort is under way in the welding community to develop standards for the testing and evaluation of fusion-bonded joints.

WELDING PROCESSES USING EXTERNAL HEATING

A VARIETY OF processes for welding thermoplastics also can be used for welding thermoplastic-matrix composites. As mentioned earlier, these welding techniques can be grouped according to whether they use external or internal heating. The processes that rely on external heat sources, which are discussed in this section, include hot-plate, hot-gas, inductively or resistively heated implant welding, infrared welding, and laser welding. Other processes using internal-heating will be discussed in a subsequent section.

12. Benatar, "Ultrasonic Welding."
13. Such an association has been reported by:
Blundell, D.J., Chalmers, J. M., Mackenzie, M.W., and Gaskin, W. F. "Crystalline morphology of the matrix of PEEK-carbon fibre aromatic polymer composites. I. Assessment of Crystallinity." *SAMPE Quarterly* 16(4): 22-30, 1985.
Blundell, D. J. and Osborne, B. N. "Crystalline morphology of the matrix of PEEK-carbon fibre aromatic polymer composites. II. Crystallisation behaviour." *SAMPE Quarterly* 17(1): 1-6, 1985.
Cogswell, F. N. "Microstructure and properties of thermoplastic aromatic polymer composites." *SAMPE Quarterly* 14(4): 33-7, 1983.
14. Blundell and Osborne, "Crystalline morphology," 1-6.
15. Brophy, J. H., Rose, R. M., and Wulff, J. "The structures and properties of materials." Vol. II, *Thermodynamics of Structure.* New York: John Wiley, 1964.
16. Cogswell, F. N. "Thermoplastic aromatic polymer composites," 33-7.

17. Refer to "Adhesives," 1994 *Annual Book of ASTM Standards,* Vol. 15.06. Philadelphia: American Society for Testing and Materials, 1994.
18. For descriptions of these techniques and applications, refer to:
Adkins, D. W. "Strength and mechanics of bonded scarf joints for repair of composite materials." Ph.D. thesis, University of Delaware, December 1982.
Delale, F., Erdogan, F., and Aydinoglu, M. N. "Stresses in adhesively bonded joints: A closed-form solution." *Journal of Composite Materials* 15: 249-71, 1981.
Goland, M. and Reissner, E. "The stresses of cemented joints." *Journal of Applied Mechanics,* Trans. ASME 1(1): A17, 1944.
Grimes, G. C. and Greimann, L. F. "Analysis of discontinuities, edge effects, and joints." *Structural Design and Analysis,* Part II, Ed. Chamis, C. C., 135-230. New York: Academic Press, 1974.
Hart-Smith, L. J. *Joining of Composite Materials,* ASTM STP 749, Ed. Kedward, K. T., 3-31. Philadelphia: American Society for Testing and Materials, 1981.

Hot-Plate Welding

HOT-PLATE WELDING IS a technique where the parts are brought into contact with a hot plate. For low-temperature polymer matrices, the plate surfaces are usually coated with polytetrafluorethylene (PTFE) to keep the parts from sticking to the hot plate. For high-temperature polymer matrices, special bronze alloys may be used to reduce sticking. In some cases, non-contact hot-plate welding is used, where the parts are brought very close to the hot plate without actually touching it. In these cases, the hot plate is elevated to very high temperatures and the composite surfaces are heated by convection and radiation.

Hot-plate welding is illustrated schematically in Figure 7.7. After the surfaces have softened, the plate is removed and the parts are pressed together. This technique is especially good for mass production of small parts. It is quite tolerant of variations in material properties and welding conditions and is very popular for welding of thermoplastics. However, it is not a flexible technique and is not often used in small production lots of parts with varying geometries. Because heating and pressing are done at different times, this technique is difficult to use for composites with highly thermally conductive reinforcements; the surfaces cool and resolidify before the parts can be aligned and pressed together.

Hot-Gas Welding

IN HOT-GAS WELDING, a heated gas is blown over a welding rod and the joint surfaces (see Figure 7.8). The molten rod is used to fill-in the joint and weld the two parts. Hot-gas welding is suitable for small weld areas, but it is quite slow even for small areas [welding rates are about 2 to 12 in./min (0.8 to 5 mm/s)]. This method is very flexible because it usually is done manually or semiautomatically. It is used primarily for low-cost composites with lower-melting-temperature matrices for small-volume production of parts with many varying geometries. Hot-gas welded joints cannot be used in high-strength applications because the joint area is small (on the order of the part thickness) and it cannot compensate for discontinuity in the reinforcement across the joint.

Resistance-Implant Welding

RESISTANCE-IMPLANT WELDING IS accomplished by passing electric current through a resistive element (see Figure 7.9). If the element is placed at the bond interface, the technique can be used to heat and melt the

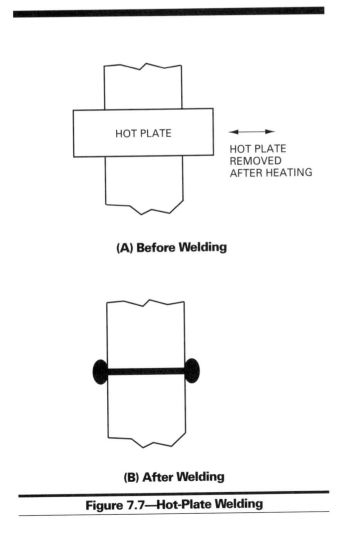

(A) Before Welding

(B) After Welding

Figure 7.7—Hot-Plate Welding

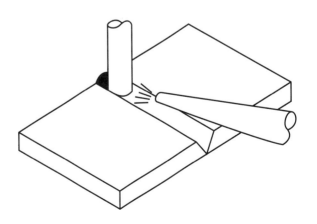

Figure 7.8—Hot-Gas Welding

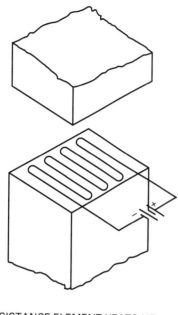

RESISTANCE ELEMENT HEATS UP
WHEN CURRENT IS APPLIED

Figure 7.9—Resistance-Implant Welding

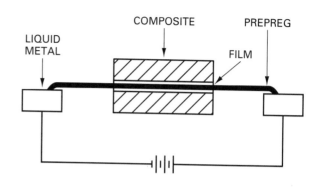

Figure 7.10—Resistance Welding of Thermoplastic Composites Using a Unidirectional Graphite Prepreg as the Resistive Element

thermoplastic and weld the parts. The heating element remains implanted at the interface, and therefore, it is important for it to be compatible with the composites. In the work of Benatar and Gutowski, a single graphite prepreg ply was used as the resistive element.[19]

Houghton found that clamping the electrical contacts onto the prepreg produced a poor connection and resulted in nonuniform heating.[20] The electrical contact was improved by placing the prepreg in baths of liquid metal, and it resulted in very uniform heating. Today, a variety of methods are available for producing uniform contact with the fibers, including the use of special solder.

Experimentally, Houghton determined that the bond strength can be improved by sandwiching the prepreg between two layers of thermoplastic films (see Figure 7.10). The films provided thermal insulation and aided in quickly producing intimate contact at the interface. By using the polymer film sandwich, Houghton was able to achieve bond strength of about 70 percent of the strength of compression-molded composites.

To speed-up the heating and fusion bonding process, *dual resin bonding* is sometimes used. This is done by using an amorphous polymer that is compatible with the semicrystalline matrix to sandwich the heating element.

Resistance-implant welding is suitable for welding of composites that have conductive reinforcements. It can be used with nonconductive composites where the presence of a single ply with conductive fibers at the weld interface is not critical to the joint.

Induction Welding

INDUCTION IMPLANT WELDING is based on the principle of a magnetic field producing eddy currents within a conductor; this results in heat generation due to the internal electrical resistance of the material (see Figure 7.11). The heating element remains embedded at the interface; therefore, it is important that it be compatible with the composites.

Induction welding of composites with nonconductive fibers is possible by placing conductive particles at the interface.[21] For composites with electrically conductive fibers, the magnetic field may be used to heat the bulk by heating the fibers. It is more desirable to heat preferentially the joint interface by placing materials there that are more conductive than the fibers. This more

19. Benatar and Gutowski, "Methods for fusion bonding thermoplastic composites," 34-41.

20. Houghton, W. "Bonding of graphite reinforced thermoplastics using resistance heating." S.B. thesis, Department of Mechanical Engineering, Massachusetts Institute of Technology, 1984.

21. For descriptions of techniques and applications, refer to:
Leatherman, A. F. "Introduction of induction bonding of reinforced thermoplastics." *Proceedings of the 41st Annual Technical Conference*, 214-6. Society of Plastics Engineers, 1983.
Schwartz, M. M. *Composite Materials Handbook*. New York: McGraw-Hill, 1984.
Wagner, B. E. "Expanded market opportunities for reinforced thermoplastics with electromagnetic welding." *Proceedings of the 39th Annual Conference of the Reinforced Plastic/Composites Institute*, 3. The Society of Plastics Industry, 1984.

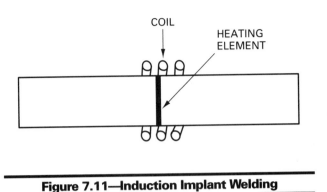

Figure 7.11—Induction Implant Welding

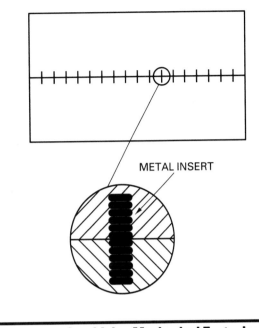

Figure 7.13—Combining Mechanical Fastening With Welding by Using Inductively Heated Metallic Inserts

conductive interface can include any conductive material like a metal mesh or a dispersion of metal particles in a thermoplastic film. Benatar and Gutowski placed a prepreg made of thermoplastic and nickel-coated graphite fibers at the joint interface between two graphite-reinforced thermoplastic composites.[22] Sandwiching the heating element between two polymer films (as shown in Figure 7.12) produced the best bond, which had a bond strength of 50 percent compared to compression-molded parts.

Heated Inserts or Screws

INSERTS OR SCREWS may be used to fasten the composites mechanically, and later the screws may be inductively heated to melt the thermoplastic around them and to weld the parts. Figure 7.13 illustrates this technique.

---

22. Benatar and Gutowski, "Methods for fusion bonding thermoplastic composites," 34-41.

The damage caused by mechanical fastening may be minimized by using miniature screws or inserts. Theoretically, this combination of mechanical fastening and welding would provide a stronger bond than just fusion bonding. Experimentally, Benatar and Gutowski found that miniature heated screws [0.02 in. (0.5 mm) diameter] can easily be inserted into a composite; however, insertion caused extensive damage to the parts by breaking many fibers.[23] This technique remains impractical until a less damaging insertion method can be developed.

Infrared and Laser Heating

INFRARED OVENS FREQUENTLY are used to heat plastics prior to a stamping or pressing operation. Heating is achieved with either infrared or laser sources by electromagnetic radiation being absorbed at the surface. The bond surfaces are exposed to lasers or infrared lamps until the thermoplastic is melted. Then, the parts are quickly aligned and pressed together until the thermoplastic cools and resolidifies (see Figure 7.14). In general, infrared and laser welding are very similar to hot-plate welding, except that the heating is done by conversion of electromagnetic radiation into heat.

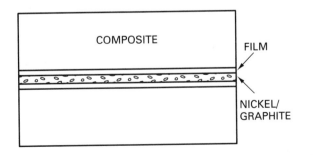

Figure 7.12—Configuration of Parts for Induction Welding of Thermoplastic Composites Using Nickel-Coated Graphite Fiber Prepreg as Heating Element

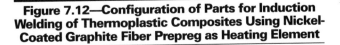

---

23. Benatar and Gutowski, "Methods for fusion bonding thermoplastic composites," 34-41.

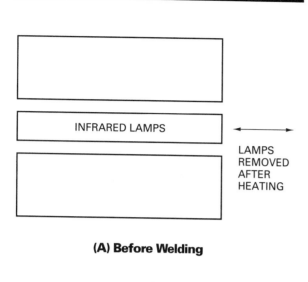

(A) Before Welding

INFRARED LAMPS

LAMPS REMOVED AFTER HEATING

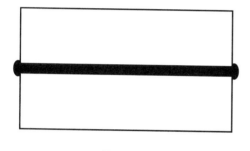

(B) After Welding

Figure 7.14—Infrared Welding

Manual attempts by Benatar and Gutowski to follow the welding sequence described above were not successful because the surfaces cooled and resolidified before the parts could be pressed together.[24] Such difficulties can be overcome by automating the process and by limiting this technique to small joint areas. Swartz and Swartz successfully used an automated, focused infrared welding system to join composites.[25] Beyeler and his associates were very successful in using lasers for continuous melting and consolidation of thermoplastic

prepregs.[26] In this case, the process was semiautomated, the joint area was small, and the flexibility of the prepreg allowed for heating until pressure was applied.

WELDING PROCESSES USING INTERNAL HEATING

IN ADDITION TO the above processes that use external heat sources, a second group of processes for joining thermoplastic composites rely upon internal heat generation within the polymer. These methods include dielectric and microwave heating, friction heating (spin welding), vibration welding, and ultrasonic welding.

Dielectric and Microwave Heating

DIELECTRIC AND MICROWAVE fields are commonly used to heat and melt thermoplastics. Depending upon the frequency of the electromagnetic field and upon the nature of the polymer, the molecules may be excited in many ways. The resulting molecular motions cause energy to be dissipated and the material to be heated. Therefore, this technique cannot be used with composites that have electrically conductive reinforcements. The addition of electrically conductive fibers to the polymer produces a very conductive composite that shields the bond interface from the electromagnetic field.

Volpe studied the shielding of graphite-epoxy composites to electromagnetic fields for a range of frequencies.[27] For frequencies around the microwave range (1 to 100 GHz), composite shielding is in excess of 60 dB. Therefore, microwave fields cannot penetrate the composite and reach the interface. Dielectric heating operates at a lower frequency range than microwave heating (at around 1 to 100 MHz). For this frequency range, the graphite composite has a lower shielding effectiveness (around 20 dB); therefore, it may be possible to penetrate the composite and heat the bond interface only. However, Benatar and Gutowski found that the electromagnetic radiation could not penetrate the conductive graphite composites and reach the joint interface, resulting in little or no heating of the bond area.[28]

24. Benatar and Gutowski, "Methods for fusion bonding thermoplastic composites," 34-41.

25. Swartz, H. D. and Swartz, J. L. "Focused infrared – A new joining technology for high performance thermoplastics and composite parts." *Proceedings of the 5th Annual North American Welding Research Conference.* Columbus, Ohio: Edison Welding Institute, 1989.

26. Beyeler, E., Phillips, W., and Guceri, S. I. "Experimental investigation of laser-assisted thermoplastic tape consolidation." *Journal of Thermoplastic Composite Materials* 1: 107, 1988.

27. Volpe, V. "Estimation of electrical conductivity and electromagnetic shielding characteristics of graphite/epoxy laminates." *Journal of Composite Materials* 14: 189-98, 1980.

28. Benatar and Gutowski, "Methods for fusion bonding thermoplastic composites," 34-41.

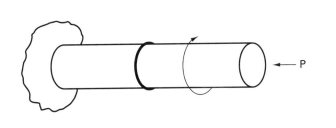

Figure 7.15—Friction Welding

Spin Welding

IN SPIN WELDING, the parts are heated through rubbing friction. Typically, one part is held stationary while the other is rotated and pressed against it (see Figure 7.15). This method is suitable for mass production of small cylindrical parts. For fiber-reinforced thermoplastics, the rotation sometimes can result in fiber misalignment.

Vibration Welding

VIBRATION WELDING IS another form of internal heat generation by friction. The friction results from the linear back-and-forth movement of one part against a second stationary part (see Figure 7.16). This technique, also known as linear friction welding, is used on small and intermediate-size parts of up to about 3.3 ft (1 m) long. With proper fixturing, vibration welding can be quite flexible and enables the welding of parts with varying geometries. The linear motion sometimes can cause fiber misalignment during welding of fiber-reinforced thermoplastics.

Ultrasonic Welding

ULTRASONIC WELDING IS a technique where vibrational energy causes cyclical deformation of surface asperities and results in heat dissipation. This melts the surface asperities, which then flow and bond the parts through fusion. Often, man-made asperities (referred to as *energy concentrators* or *directors*) are molded into one of the parts (see Figure 7.17).

Benatar and Gutowski found that in ultrasonic plunge welding of small composites, weld strengths could be achieved equivalent to the strength of compression-molded parts.[29] They also used two techniques, *sequential welding* and *scan welding*, for large parts. In sequential welding, a section of the composite parts is plunge welded, then the parts are indexed, and the adjoining section is welded. This continues until the parts are fully welded. In scan welding, the parts are moved under the ultrasonic horn at a constant velocity while they are being welded. Both methods were equally successful, producing welds with bond strengths of 80 percent of the compression-molded strength.

In later work, Benatar and Gutowski also determined experimentally and theoretically that by monitoring the dynamic mechanical impedance of composite parts as they are being welded, one can indirectly monitor the quality of the weld.[30,31] This offers the potential for closed-loop control ultrasonic welding of composites.

29. Benatar and Gutowski, "Methods for fusion bonding thermoplastic composites," 34-41.
30. Benatar, A., and Gutowski, T. G. *Proceedings of the 33rd International SAMPE Symposium and Exhibition.* "Ultrasonic welding of advanced thermoplastic composites." Covina, Calif.: Society for the Advancement of Material and Processing Engineering, 1988.
31. Benatar, A. and Gutowski, T. G. "Ultrasonic welding of PEEK graphite APC-2 composites." *Polymer Engineering & Science* 29(23): 1705-21, 1989.

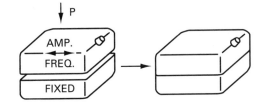

FRICTIONAL HEAT FUSES POLYMER AND CREATES
THE WELD. MOTION MAY BE ORBITAL.

Figure 7.16—Vibration Welding

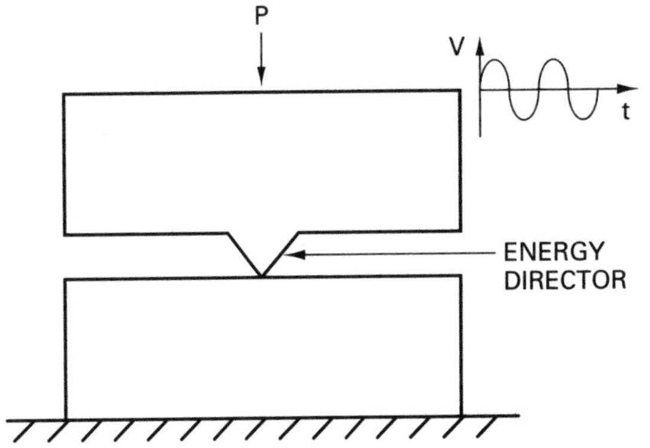

Figure 7.17—Energy Director

APPLICATIONS OF POLYMERIC COMPOSITES

THE USE OF polymeric composites is rapidly increasing due to the many advantages that they offer. Polymeric composites are used in a wide range of applications from high-tech aerospace applications to consumer products. They also are being considered for new applications as well as material replacements in existing applications. In aerospace applications, thermoplastic composites are slowly replacing thermoset polymeric composites for use in ailerons, elevators, rudders, tails, fins, stiffeners, and numerous interior components. As thermoplastic matrices replace thermosets, welding is being used to join the components together instead of adhesive bonding. In automotive applications, thermoplastic composites are used in bumper beams, load floors, batteries and battery trays, dash frames, and knee bolsters. A variety of welding processes are used to assemble these components including hot-plate, vibration, and ultrasonic welding. The electrical and appliance industries also use thermoplastic composites for many metal replacement parts such as in computer access panels, electronic packaging, and water pumps for washers. Ultrasonic and vibration welding are the common methods for joining thermoplastic composites in the appliance industry.

A variety of polymeric composites are welded using the processes described earlier. Selection of the welding process depends upon numerous factors including cost, production speed, geometry of the parts, polymer and reinforcement types, and size of the weld area.

For example, the personal alert safety system shown in Figure 7.18 is ultrasonically assembled. It is made from general-purpose polycarbonate with 10% by weight of short glass fibers. Ultrasonic welding is used to weld the semicircular caps to the base part of the device. Then, ultrasonic staking is used to join the cover to the base.

A different approach is used to weld the air intake resonator shown in Figure 7.19. In this case, the nylon 6 polymer with 33% by weight glass fibers is welded using induction implant welding. A ferromagnetic filler is used to make a gasket which is then placed along the complex joint area. This ferromagnetic material is then heated causing the thermoplastic matrix to melt and produce a hermetic seal.

Courtesy of Branson Ultrasonics Corp. and Detex Corp.

Figure 7.18—Personal Alert Safety System

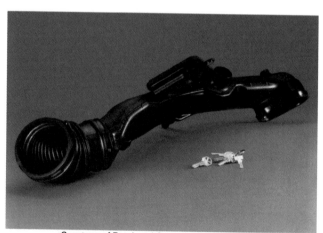

Courtesy of Emabond Systems/Ashland Chemical Company

Figure 7.19—Air Intake Resonator

WELDING OF METAL-MATRIX COMPOSITES

METAL-MATRIX COMPOSITES (MMCs) include a surprisingly wide range of materials. For present purposes, they will be defined as a metal or alloy reinforced with a second material that is chemically distinct and that has a clear interface between itself and the matrix material.

MMCs are part of a broader class of *engineered materials*, so-called because the combination of matrix and reinforcement can be custom-tailored to produce unique physical and mechanical properties. An example of a property that has been tailored is the high stiffness-to-weight ratio (specific modulus) that is important in the aerospace industry. Other properties that have been improved with engineered MMCs are wear resistance, thermal conductivity, thermal expansion, and elevated-temperature fatigue strength.

In general, the properties of most MMCs can be estimated by the *Rule of Mixtures*:

$$X_c = X_m V_m + X_r V_r \qquad (7.1)$$

where

   X = property considered
   V = volume fraction
   c = composite
   m = matrix
   r = reinforcement

For example, the specific modulus of 6061 aluminum reinforced with 20 volume percent of silicon carbide particles is estimated by this equation to be 18.0 Msi (124 GPa). Actual tensile data at 16.3 to 17.5 Msi (112 to 121 GPa) agrees quite well with this estimate.

Another attraction of MMCs is the ability to use traditional, secondary fabrication technologies from the metal-working industry. With some minor process modifications, many MMCs can be forged, extruded, rolled, sheared, machined, and welded. Some can even be shape-cast. Since welding is involved in a large part of industrial production, MMCs must be capable of being welded in order to enjoy widespread use.

COMMON MATRIX ALLOYS

BECAUSE OF THEIR origins in the aerospace industry, most metal-matrix composites (MMCs) emphasize high performance for light weight. Thus, low-density alloys are the most highly developed matrix materials. Magnesium offers the lowest density among the commonly used light metals; it is 37% lighter than aluminum. However, magnesium creeps at low temperatures, has relatively low strength, and is prone to a variety of corrosion problems.

Aluminum is easy to process, inexpensive, and has a density only 35% that of steel. It can be alloyed to fairly high strengths and, with a modulus of 10 Msi (69 GPa), is 57% stiffer than magnesium. Aluminum retains useful strength at moderate temperatures and has excellent corrosion resistance. For these reasons, it is the most widely used matrix material.

Titanium is more than twice as strong as aerospace-grade aluminum alloys. Although it is 65% heavier than aluminum, it is also 65% stiffer. Its exceptional temperature and corrosion resistance make titanium a favorite choice for demanding environments, where its high cost is acceptable.

Copper, iron, nickel, and tungsten have specialty uses as matrix alloys in electronic circuits, metal forming dies, jet engines, and military applications. Their relatively high densities are offset by excellent physical and mechanical properties that cannot be obtained by any conventional alloys.

COMMON REINFORCEMENTS

THE MOST COMMON reinforcement materials for metal-matrix composites (MMCs) are ceramics such as silicon carbide (SiC), aluminum oxide (Al_2O_3), titanium carbide (TiC), boron carbide (B_4C), and others. Other materials such as graphite (C) or titanium diboride (TiB_2) also are used. Metal filaments such as boron, steel, and tungsten are used occasionally. The choice of reinforcement material generally depends on the desired properties of the composite, its compatibility with the matrix material, and the MMC processing route.

It should be noted here that most ceramics are simply the reaction product between metals and oxygen, carbon, or other non-metallics. Many ceramics are thermodynamically unstable in the presence of liquid metals, especially the light metals, and suffer reactions that can severely degrade the properties of MMCs. The shape of the reinforcement is usually selected based on the desired performance characteristics of the MMC. For ease of discussion, we use the shapes described below to broadly classify most MMCs.

Fibers and filaments have extremely high aspect ratios (length/diameter). They are used as continuous reinforcements in products such as sheet and tubing.

The resulting mechanical properties of the MMC are highly anisotropic. However, the design of such MMC components takes advantage of these properties. Sometimes this extreme anisotropy is undesirable, and the MMC is built up from single plies of reinforced composite with different fiber orientations. Sample representations of such three-ply or five-ply MMCs, respectively, would be 0°/90°/0° or 0°/45°/90°/135°/0°.

In spite of the exceptional properties obtained, these fiber-reinforced composites suffer from the high cost of their fibers as well as from not being amenable to many traditional metal-working operations. Chopped fibers with lower aspect ratios and random orientations are used as preforms for some MMC casting processes. While the properties are more isotropic and the costs are lower, these MMCs again are limited by their secondary fabrication options.

Whiskers are small monocrystalline materials with a high aspect ratio of 50-100. Their small size [typically diameters less than 40 μin. (1 μm)] lends them to a variety of MMC production methods, although powder metallurgy is the most popular. Whisker-reinforced MMCs usually are used in the form of extrusions, forgings, and rolled sheet. The properties of the finished MMC component range from isotropic to slightly anisotropic. Costs are moderate to high, as are the mechanical and physical properties.

Because of their submicron diameters and high aspect ratios, loose mineral whiskers can be carcinogenic when ingested into the pulmonary tract. However, whiskers consolidated within MMCs appear to present no known health problems at this time.

Particles represent the low end of the reinforcement spectrum in many ways. They are relatively inexpensive and their low aspect ratio (1-5) encompasses everything from polycrystalline microspheres, to grinding grit, to monocrystalline platelets. Properties of the resulting MMCs are isotropic and significantly improved over the unreinforced matrix material. Most importantly, these MMCs can be fabricated by most standard metal-working processes.

DESIGNATIONS FOR METAL-MATRIX COMPOSITES

BECAUSE MORE ALUMINUM metal-matrix composites (MMCs) are produced than are MMCs of all other matrix alloys combined, the Aluminum Association developed a standard designation system for MMCs that has since been adopted by the American National Standards Institute. ANSI 35.5-1992 provides that aluminum-matrix MMCs be identified with designations in the following format:

matrix/reinforcement/volume%form

where

matrix = metal or alloy designation of matrix
reinforcement = chemical formula for reinforcement
volume% = volume percentage (without the % sign)
form = f, fiber or filament
      c, chopped fiber
      w, whiskers
      p, particulates

For example, 2124/SiC/25w describes an MMC of alloy 2124 reinforced with 25 volume percent of silicon carbide whiskers; 6061/Al_2O_3/10p is alloy 6061 reinforced with 10 volume percent of alumina particles; and A356/C/5c is a casting alloy with 5 volume percent of chopped graphite fibers. This designation system will be used throughout the remainder of this chapter to reference specific MMCs, both for aluminum-matrix and other MMCs. Simplified abbreviations will be used when discussing general matrix and reinforcement combinations, such as Al/B and Mg/SiC, or Fe/TiC_p and Al/SiC_w, where the subscripts may be added to indicate the reinforcement form, as in the last two examples. Some specific MMCs and their properties, preceded by those of their matrix alloys, are listed in Table 7.2.

The proliferation of MMCs and the number of specialty welding techniques creates an enormous variety of potential combinations. Obviously all welding processes are not suitable for every MMC. The following will address some of the more widely used MMCs and their welding processes as listed in Table 7.3.

PROCESSING OF METAL-MATRIX COMPOSITES

UNDERSTANDING THE MICROSTRUCTURE of any material to be welded is always important. This means that the processing route of metal-matrix composites (MMCs) also must be considered. Possible matrix and reinforcement combinations, product forms, and final costs are all determined by the initial MMC processing method. The following examples describe the most commonly used processing methods (solid-state, semisolid, and liquid), followed by some other specialty methods.[32]

32. Details of MMC processing techniques may be found in:
  Geiger, A. L. and Walker, J. A. "The processing and properties of discontinuously reinforced aluminum composites." *The Journal of the Minerals, Metals, and Materials Society* 43(8): 8-15, 1991.
  Hoover, W. R. "Recent advances in castable metal matrix composites." *Conference proceedings: Fabrication of particulate-reinforced metal composites,* Eds. Masounave, J. and Hamel, F. G., 115-23. Materials Park, Ohio: ASM International, 1990.
  Lavernia, E. J. and Wu, Y. "Spray-atomized and codeposited 6061 Al/SiC_p composites." *The Journal of the Minerals, Metals, and Materials Society* 43(8): 16-23, 1991.

Table 7.2
Typical Properties of Representative Metal-Matrix Composites[a]

Common Alloys and MMCs	Density		Specific Modulus		Yield Strength (0.2% Offset)		Tensile Strength		Elongation, %	$T_{solidus}$		Coefficient of Thermal Expansion[b]	
	lb/in.³	kg/m³	Msi	GPa	ksi	MPa	ksi	MPa		°F	°C	10^{-6}/°F	10^{-6}/°C
ZC71	0.066	1830	6.4	44	50	345	53	365	3.3	851	455	15.0	27.0
+SiC/12p	0.072	1995	9.1	63	54	372	57	393	0.6	851	455	10.3	18.5
6061	0.098	2715	10.0	69	40	276	45	310	20	1080	582	14.1	25.4
+Al₂O₃/20p	0.106	2935	14.1	97	51	352	54	372	4.0	1080	582	10.3[c]	18.6[c]
+SiC/20w	0.101	2800	17.5	121	64	441	85	586	3.6	1080	582	8.4	15.1
+B/50p	0.094[c]	2600[c]	33.4	230	—	—	200	1379	—	1080	582	7.9[c]	14.2[c]
Ti-6Al-4V	0.160	4430	16.5	114	160	1103	170	1172	10	3002	1650	5.3	9.5
+SiC/35f	0.145[c]	4015[c]	30.9	213	—	—	250	1724	—	3002	1650	4.3[c]	7.7[c]
H13[d]	0.280	7750	30.5	210	200	1379	264	1820	—	2660	1460	6.6	11.8
+TiC/45p	0.233	6450	—	—	—	—	169[e]	1165[e]	—	2660	1460	3.2	5.7

a. Longitudinal with respect to reinforcement orientation after solutionizing, quenching, and artificially aging to peak strength (T6).

b. From 68 to 572 °F (20 to 300 °C).

c. Calculated data.

d. Properties based on comparable tool steel (S7) tempered at 1000 °F (538 °C).

e. Transverse rupture strength.

Solid-State Processing Methods

SOLID-STATE METHODS ARE one of the most popular because possible reactions between the matrix and reinforcement are avoided. One of the oldest techniques is diffusion bonding of continuous fibers between metal foils to make reinforced panels for aerospace applications. Materials such as Ti/SiC and Al/B can be combined to form high-performance MMC structures.

Powder metallurgy is one of the most widely used solid-state processing techniques. Particles of the desired matrix alloy and ceramic reinforcement (including whiskers) are carefully mixed under a controlled environment. The mixture is vacuum hot-pressed at temperatures near the solidus and extruded into structural shapes or preforms for forging or rolling. Since the matrix is never fully molten, reactions between it and the reinforcement are not kinetically favored. Therefore, high performance combinations such as Al/B_4C or Al/SiC are commonly used.

Table 7.3
Representative Metal-Matrix Composites and Welding Process Combinations

Process	Fe/TiC$_p$	Ti/SiC$_f$	Al/B$_f$	Al/SiC$_w$	Al/Al₂O₃$_p$
Arc welding	—	Lim.	Lim.	Lim.	OK
Laser and electron beam welding	—	—	—	—	Lim.
Capacitor-discharge welding	—	—	—	OK	OK
Resistance welding	—	OK	OK	OK	OK
Friction welding	—	—	—	OK	OK
Brazing and diffusion bonding	OK	OK	OK	OK	OK

OK = Acceptable combination

Lim. = Limited application

— = no data, or unsuitable

Semisolid Processing Methods

SEMISOLID PROCESSING PROVIDES another method to control reactions between the matrix and reinforcement while producing MMCs. Spray deposition co-deposits reinforcement particles with an atomized spray of molten matrix alloy onto a metal chill plate. Since the solidification rate exceeds 10^3 °C/s, interfacial reactions are minimized. This also allows the use of highly alloyed metals, such as 8090 aluminum, as the matrix because no microsegregation occurs.

Rheocasting mixes a discontinuous reinforcement into the matrix at temperatures above the solidus and below the liquidus of the matrix alloy to reduce reinforcement settling and reactions with the matrix. A key advantage is the resulting composite microstructure has very low shear strength when reheated to slightly below the solidus and can be reprocessed in a variety of ways including shape casting, forging, and extrusion.

Liquid Processing Methods

INFILTRATION OF THE reinforcement by the molten matrix is one of the original fabrication methods for MMCs. The main advantage is that the reinforcement can be placed only where it is needed using squeeze casting. However, interfacial reactions can be a problem. Newer methods take advantage of certain reactions to infiltrate beds of powder and form high-volume fraction (~0.6) master alloys that can be remelted and diluted for shape casting.

Stir casting uses a variation of traditional ingot metallurgy technology. Careful selection of alloy chemistry, reinforcement composition, and processing conditions allows these MMCs to be cast into standard foundry ingots or direct-chill cast as extrusion billets or rolling slabs. Although the reinforcement and matrix combinations are limited, these MMCs offer excellent properties at costs ranging from 1 to 20% of other discontinuously reinforced MMCs. Most importantly, they can be readily processed by traditional fabrication technologies, including welding.

Other Methods

MANY OTHER SPECIALTY methods are available for fabricating MMCs. One of the more interesting is the in situ formation of reinforcements within the molten matrix. Essentially, a metastable alloy composition is chosen from the phase diagram of several metals and carbon or boron. During solidification, several volume percent of micron-size boride or carbide particles precipitate in the matrix alloy. Analogously, gas species may be bubbled through the melt to accomplish the same end. Electrodeposition of the matrix onto the reinforcement is another novel means for producing MMCs. Many other methods are available for fabricating MMCs, and describing all of them is beyond the scope of this chapter.

PROCESSING AND WELDING METALLURGY OF METAL-MATRIX COMPOSITES

MOST FUSION WELDING processes require that temperatures in the joint exceed the solidus and often the liquidus of the workpiece material. The resulting melting and resolidifying of a metal-matrix composite (MMC) matrix can dramatically alter its structure and properties. Aluminum alloys in the 2xxx and 7xxx series are particularly susceptible to this problem. In some cases, the level of the primary alloying elements may be so high as to cause grain boundary segregation in the weld metal during postweld solidification. More generally, precipitates in the heat-affected zone coarsen as they are overaged by heat from the adjacent weld. During fusion welding of an MMC that has a heat-treatable aluminum alloy matrix, the heat-affected zone (HAZ) will soften in a similar fashion as would happen in the unreinforced heat-treatable aluminum alloy. This gives these MMC weldments less strength than that of the MMC base material.

Chemical Reactions

REACTIONS BETWEEN THE matrix and the reinforcement in the molten weld pool are of even more concern than is matrix-alloy degradation during melting and resolidification. MMCs produced by solid-state and semisolid processes are often based on matrix and reinforcement combinations that are not thermodynamically stable. Some common examples of chemical reaction problems follow.

One of the most well-known reinforcement and matrix reactions during MMC welding is that of Al/SiC. When silicon carbide is exposed to molten aluminum for sufficient time and temperature, the following reaction will consume the reinforcement:

$$4Al[l] + 3SiC[s] \Rightarrow Al_4C_3[s] + 3Si[Al_l] \qquad (7.2)$$

where
     [l] indicates liquid phase
     [s] indicates solid phase
     [Al_l] indicates in solution with liquid aluminum

This irreversible reaction occurs quite readily at temperatures above 1346 °F (730 °C) in alloys with low silicon levels.[33] Not only is the silicon carbide

33. Iseki, T., Kameda, T., and Maruyama, T. "Interfacial reactions between SiC and aluminum during joining." *Journal of Materials Science* 19: 1692-98, 1984.

reinforcement partially consumed, but the resulting aluminum carbide phase is acicular. Furthermore, it is soluble in aqueous environments, including humid air.[34] Thus, fusion welding processes are not generally recommended for Al/SiC MMCs unless the weld-pool temperature is kept low, or the weld-pool chemistry is altered. At least seven weight percent of silicon is required to inhibit the reaction. It is preferable to have this composition in the matrix alloy, but use of a silicon-containing filler material (such as 4043 or 4045) also can be effective.

Another option is to use a filler containing several percent of an active metal like titanium. Since the titanium carbide reaction is thermodynamically more favorable than the aluminum carbide reaction, dissociated silicon carbide can be replaced by the equally effective titanium carbide as a reinforcement. Excess titanium will remain in solution with the aluminum or reprecipitate as a fine intermetallic phase of Ti_3Al or TiAl as in the following reaction:[35]

$$5Ti[Al_l] + 3Al[l] + SiC[s] \Rightarrow$$

$$TiC[s] + Si[Al_l] + Al[l] (+ Ti_3Al + TiAl) \qquad (7.3)$$

Of course, this also means than in the case of welding Ti/SiC composites one must contend with the degradation of the SiC by the titanium matrix. Although the SiC $\Rightarrow$ TiC conversion is not as detrimental as the aluminum carbide one, the reaction still will significantly alter the interfacial microstructure and crystallinity of the reinforcement, and consequently its load-transfer characteristics.

Since graphite is essentially pure carbon, it is directly reduced by both aluminum and titanium to their respective carbides at sufficient times and temperatures. For small amounts of superheat, this is a replacement reaction that occurs at the interface of the solid reinforcement. At high temperatures, however, the carbon goes completely into solution and reprecipitates during solidification as the carbide phase. Boron carbide also is soluble in these two metals, but titanium diboride is stable.

Aluminum oxide in dilute melts of aluminum-magnesium-silicon alloys (i.e., 6xxx series) will undergo a self-limiting reaction to form a surface layer of spinel:

$$3Mg[Al_l] + 4Al_2O_3[s] \Rightarrow$$

$$3MgAl_2O_4[s] + 2Al[l] \qquad (7.4)$$

This spinel layer does not affect the properties of the MMC.[36] However, the resulting magnesium depletion of the matrix can significantly decrease the MMC's tensile strength.

The above reaction (7.4) also occurs quite rapidly in Mg/Al_2O_3 composites, and with the same self-limiting result. However, since the magnesium carbide reaction is not thermodynamically favorable in most systems, magnesium composites reinforced with boron carbide, silicon carbide, and graphite are not degraded by fusion welding processes.

Porosity was once a major concern in welding powder metallurgy aluminum MMCs. Hydrated aluminum oxides from the surfaces of the aluminum powder dissociated in the molten weld pool to liberate large amounts of hydrogen. This required a high-temperature, vacuum degassing step prior to welding to alleviate this problem.[37] However, most manufacturers of aluminum powder-based MMCs have modified their production streams to eliminate this problem.

Weld-Pool Dynamics

HEAT AND MASS transfer in the weld pool are significantly altered by the presence of a solid phase. Intuitively, one can recognize that a weld pool containing 10-30% solids will be more viscous than a molten pool of the unreinforced matrix alloy. This reduced fluidity can be further complicated by the presence of the reaction products discussed above, which often lower fluidity even further. This results in a sluggish weld pool that is more susceptible to trapped gas porosity as well as to reduced weld penetration and lack of fusion defects. These disadvantages are somewhat offset by the lower thermal and electrical conductivity of most composites. For a given power input, welds can be made with greater penetration or higher travel speeds than in conventional alloys.

Redistribution of the reinforcement in the solidified joint can result in unanticipated defects. Segregation of the ceramic particles during solidification can result in very high local concentrations which behave as

34. Luhman, T. S., Williams, R. L., and Das, K. B. "Development of joint and joining techniques for metal-matrix composites." TR84-35. Watertown, Mass.: Army Materials and Mechanics Research Center, August 1984.

35. Meinert, K. C., Martukanitz, R. P., and Bhagat, R. B. "Laser processing of discontinously reinforced aluminum composites." *Conference proceedings*. Dayton, Ohio: American Society for Composites, 1992.

36. Holcomb, S. "Effect of spinel formation on mechanical properties of alumina particulate-reinforced aluminum composites." *Symposium Proceedings: Advances in production and fabrication of light metals and metal matrix composites*, Eds. Avedesian, M. M., Larouched, L. J., and Masounave, J., 643-649. Montreal, Quebec: Canadian Institute of Mining, Metallurgy and Petroleum, 1991.

37. Ahearn, J. S., Cooke, C., and Fishman, S. G. "Fusion welding of SiC-reinforced Al composites." *Metal Construction*, 192-7, April 1982.

inclusions in the weld. Also, excessive segregation patterns in the weld metal can contribute to inconsistent properties as the bead varies from being unreinforced to being heavily reinforced.

WELDING PROCESSES

Weld Preparation

JOINT PREPARATION OF most metal-matrix composites (MMCs) is significantly different than their unreinforced counterparts. Often the reinforcement is a hard ceramic that is quite abrasive to cutting tools. Steel band-saw blades and backgouging tools are unsuitable for joint preparation. Tools tipped with tungsten carbide used at speeds below 328 ft/min (100 m/min) are usually recommended. However, diamond-tipped or plated tools running at speeds above 1312 ft/min (400 m/min) can offer increased productivity and savings for high-volume welding. Feed rates above 0.12 in./tooth (0.30 mm/tooth) are recommended for cutting tools. Constant and medium-to-heavy pressures should be maintained for saws, and flood coolant is recommended for optimum tool life.

As with any welding, proper cleaning of MMC joints is an essential step to weld quality. This is particularly important for aluminum MMCs, which are prone to the same hydrogen absorption and porosity as conventional alloys if not cleaned properly. Cleaning procedures for MMC joints are the same as for unreinforced joints of the same matrix material. Generally, the workpiece must be chemically cleaned and degreased to remove any type of hydrocarbon-containing debris. Then, some form of mechanical abrasion is used to remove the surface film of oxide. In some cases, a light chemical etch may be required as well.

Standard nonmagnetic fixtures are suitable for positioning and clamping MMC joints. However, the application of such fixturing may need to be modified slightly. Most MMCs have significantly lower coefficients of thermal expansion than their unreinforced matrix alloys. Thus, joint movement is reduced by 10 to 30 percent during the welding cycle.

Temperature Control

CONTROL OF HEAT input is critical to preserve the original microstructure of the MMC. This is why solid-state processes such as friction welding or brazing are preferred by some fabricators. However, arc welding is still the method of choice for high productivity. With suitable adjustments to the power supply, heat input can be controlled quite carefully. Gas tungsten arc welding excels at producing low-dilution welds in systems with a reactive combination of matrix and reinforcement.

Temperature-control problems experienced by many early researchers can be linked directly to their use of very small specimens; a typical coupon might measure 0.1 x 0.5 x 2.0 in. (2.5 x 12 x 50 mm). This was a consequence of the extremely high cost of MMCs at that time. A grooved, copper backing bar provided a good heat sink, but was insufficient in many cases.

High-energy-density welds result in rapid solidification rates and in reduced sizes of the heat-affected zone (HAZ). For this reason, many people are interested in using laser and electron beam welding for MMCs with reactive combinations of matrix and reinforcement. However, these processes also can result in very high local temperatures in the weld pool and accelerated reactions between the matrix and reinforcement. Capacitor-discharge welding is an exception to this because of the low dilution, low temperature, and extremely short melting and solidification cycle.

Thus, temperature and residence time are key factors in controlling matrix and reinforcement reactions. In a practical sense, how these factors are used highlights the difference between brazing and capacitor-discharge welding. Brazing often requires thermal cycles of up to half an hour, but the temperature of the MMC does not exceed its solidus. Capacitor-discharge welding fully melts the joint interface, but the time above the liquidus is measured in microseconds. In both cases, the integrated area under the time-versus-temperature curve is sufficiently small to prevent undesirable reactions. Of course, as mentioned previously, the upper-bound curve can be shifted by the addition of a thermodynamically active species such as silicon or titanium (see Figure 7.20).

Welded Properties

JOINT STRENGTHS OF MMCs may be comparable to welds in the unreinforced matrix alloy, but joint efficiencies will be lower. Since the MMC has higher base properties than an unreinforced alloy, the ratio of welded to unwelded properties will naturally be lower. While most weld failures occur in the HAZ, the use of post-weld heat treatment, or naturally aging alloys such as 7005, can shift the failure into the weld metal. In the case of discontinuously reinforced MMCs, there is potential for strengthening the weld by using a reinforced filler metal.

Of more practical significance are the changes in modulus and ductility. Because the fusion zone has little or no reinforcement, the stiffness of the joint will be less than that of an unwelded section. This also must be accounted for when designing a welded structure. However, on the plus side, the weld joint also will have higher ductility than the base MMC. By careful selection of the joint type and location, the designer may be able to create damage-compliant zones that fail more

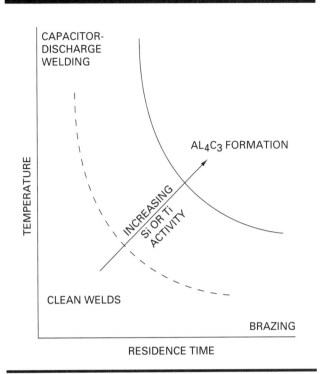

Figure 7.20—Effect of Active Species, Temperature, and Residence Time on Reactions in Weld Pool

gracefully (i.e., plastically) than the MMC base material. Since they contain no filler metal, autogenous welds do not suffer a reduction in stiffness.

As with the joining of traditional alloys, postweld heat treatment of MMCs can substantially increase the mechanical properties of the weldment. Heat treatments should be based on the standard practice for the matrix alloy and modified in accordance with the manufacturer's recommendations.

Arc Welding

ARC WELDING OF MMCs is similar to practices in crucible metallurgy. The base metal is melted in a self-constrained pool, is stirred vigorously by convection and induction forces, and then resolidifies. The resulting microstructure of this type of weld, including its defects, resembles that of a casting. Unfortunately, this *recasting* process can significantly degrade the microstructure and properties of most MMCs. The cooling rates of most welds are considerably slower than those of the original processing methods (especially spray deposition and powder metallurgy) and will not hold the excess alloying elements in solution. However, MMCs produced through ingot metallurgy are not as sensitive to these effects because their composition is

designed to be melted and resolidified repeatedly with little degradation of properties.

Because of the higher viscosity of most MMC weld pools, the geometry of the weld joint may have to be modified in some cases. To prevent lack-of-fusion defects, a root gap is recommended for most welds, and backgouging is required for two-sided welds. A minimum included joint angle of 60 degrees is recommended for arc welding, and 90 degrees is common. As a possible benefit, the lower fluidity of MMC welds may make them easier to handle in vertical or other out-of-position work.

Earlier comments about cleaning of the joint during weld preparation are particularly applicable to arc welding. In multiple-pass welds on aluminum MMCs, for example, it is recommended that the weld be vigorously scrubbed with a clean stainless steel wire brush between passes to remove any spatter, condensed magnesium oxide (a black, powdery residue from magnesium-containing fillers), or other contaminants.

Equipment and Consumables. The arc welding practice for MMCs is substantially similar to the arc welding practice for traditional metals and alloys. With 75 to 90 volume percent of the MMC still the base metal or alloy, the basic rules for equipment and filler-metal selection generally still apply. Filler metals are usually matched to the matrix alloy of the composite and the service requirements of the weld.[38]

For example, alloy 6061 is typically welded with ER4043 or ER5356. Either of these filler metals would be appropriate for an MMC with 6061 as the matrix. In the case of 6061/Al$_2$O$_3$/10p, the ER5356 is used for its assistance in keeping the reinforcement wetted. While for 6061/SiC/15w, the ER4043 would be preferred for suppressing the aluminum carbide reaction. The extra silicon in ER4045 or ER4047 may be even more effective in this regard. However, the near-eutectic composition of the resulting weld metal is unsuitable for postweld heat treatments that include solutionizing, owing to coarsening of the silicon phase.[39]

Gas Tungsten Arc Welding. This is the original inert-gas, arc-welding process developed for aluminum

38. To determine the appropriate filler metal selection for different types of MMCs, refer to the following specifications for their respective matrix materials: ANSI/AWS A5.10, *Specification for Bare Aluminum and Aluminum Alloy Welding Electrodes and Rods*; ANSI/AWS A5.16, *Specification for Titanium and Titanium Alloy Welding Electrodes and Rods*; and ANSI/AWS A5.19, *Specification for Magnesium Alloy Welding Electrodes and Rods*.
39. Lo, S. H. J., Dionne, S., Popescu, M., Gedeon, S., and Lane, C. T. "Effects of prior- and post-weld heat treatments on the properties of SiC/Al-Si composite gas metal arc-welded joints." *Symposium Proceedings: Advances in Production and Fabrication of Light Metals and Metal Matrix Composites*, Eds. Avedesian, M. M., Larouche, L. J., and Masounave, J., 575-88. Montreal, Quebec: Canadian Institute of Mining and Metallurgy and Petroleum, 1991.

and other reactive metals. Its low, controlled heat input is ideal for thin sections and complex shapes. This control of heat input is especially important for welding Al/SiC or Al/B MMCs. Atmosphere control is even more critical in welding titanium than aluminum. To prevent contamination by interstitial elements such as oxygen and nitrogen, titanium MMCs are best welded in a glovebox that can be purged and backfilled with pure argon.

For welding aluminum MMCs, gas tungsten arc welding (GTAW) is normally used with alternating current, and the arc balance is adjusted to favor either cleaning or penetration. A cleaning arc can help minimize reactions between the reinforcement and matrix by reducing heat input. However, dilution of the composite into the weld pool also is decreased. Thus, some welds do not have homogeneous microstructures and do not reflect the properties of the parent material.

An arc balanced toward cleaning also is capable of breaking up thin surface oxides present on the surface of a normal weld pool. However in the case of MMCs, convection forces can stir the reinforcement (diluted from the base material) into the oxide skin on the surface of the weld pool, where surface tension stabilizes the agglomeration as a thick crust. This layer insulates the pool and severely destabilizes the arc, causing it to skip around on the surface of the weld pool as it attempts to register on different areas. A cleaning balance only exacerbates the problem. Since the arc is unable to extract sufficient electrons to complete the electrode-positive portion of the cycle, it flares and sputters all over the pool surface.

For most MMCs, the preference is for a balanced or penetrating arc (electrode more negative). This will tighten the arc and force heat into the weld pool and joint, rather than melting the adjacent base material. This increased heat makes feeding the filler wire significantly easier and results in an improved bead profile. The degree of penetrating balance will depend on the geometry of the joint and the preferred technique of the welder.

The particle-stabilized oxide skin also alters the appearance of the weld pool. The absence of a clean pool also decreases the illumination of the work area, owing to the reduction in reflected light from the arc. Initial formation of the weld pool is often only marked by a slight "slumping" of the surface. This is an indication that subsurface melting is present. Welders should feed the filler rod into the weld pool and begin advancing the bead immediately, rather than overmelting the area waiting for a bright pool to form. Early addition of the filler will accelerate the pool's formation, increase its fluidity, and partially dissipate the oxide skin.

GTAW Procedural Examples. Some of the earliest GTAW of MMCs was done on 0.055-in. (1.40-mm)

thick sheets of fiber-reinforced Ti/W and Ti/C.[40] Manual ac welding was done at 60 A, 10 V, and 6 in./min (2.54 mm/s) under argon. The pool formation and bead appearance were good, except when the MMC sheets were not well-bonded during initial fabrication. Although reactions between the molten titanium and the tungsten or graphite fibers was severe (see Figures 7.21 and 7.22), the resulting phases did not degrade the tensile properties (see Table 7.4).

Similar work was done on 0.025-in. (0.64-mm) sheets of 6061/B/50f.[41] The conditions for manual ac welding were 20 A and 16.5 V using a travel speed of 3.9 in./min (1.67 mm/s) with a 0.063-in. (1.6-mm) diameter filler of 4043. The torch was fitted with a 0.04-in. (1.0-mm) thoriated-tungsten electrode and used 100% argon shielding gas. The low-heat input was used to minimize the weld-pool temperature; however, some formation of AlB_2 was noted. No mechanical testing was reported.

40. Kennedy, J. R. "Fusion welding of titanium-tungsten and titanium-graphite composites." *Welding Journal* 51(5): 250s-9s, 1972.
41. Kennedy, J. R. "Microstructural observations of arc welded boron-aluminum composites." *Welding Journal* 52(3): 120s-4s, 1973.

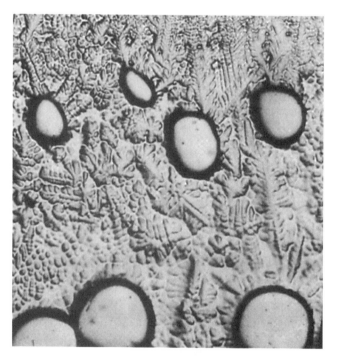

Figure 7.21—Tungsten Fiber [0.008-in. (0.2-mm) Initial Diameter] Dissolution in a Titanium Weld (x100)

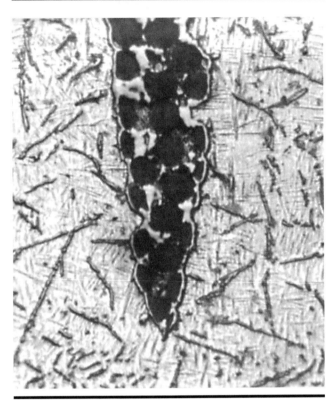

Figure 7.22—Graphite Fiber [315-μin. (8-μm) Diameter] Conversion to TiC in a Titanium Weld (x920)

Other work on welding 6061/SiC/18w is illustrative of the problems faced in arc welding discontinuously reinforced MMCs.[42] After overcoming problems with hydrogen porosity, 0.125 in. (3.18-mm) thick sheet was beveled for a 90-degree included angle with a 0.063-in. (1.6-mm) root land. The measured welding conditions were 12-14 V at 145-160 A. A filler of 4043 was fed into the weld at 6 to 8 in./min (2.5 to 3.4 mm/s) under a shield of 100% argon at a flow rate of 12 to 15 ft^3/h (5.7 to 7.1 L/min). Typical as-welded properties are reported in Table 7.5, line 2. The welds showed almost no dilution of the base material. This is a result of the low-heat input used in attempting to prevent the formation of aluminum carbide. The 0.063-in. (1.6-mm) root gap was used to overcome the viscous nature of the weld pool.

In more recent work, a series of 0.25-in. (6.4-mm) thick extrusions of 6061/Al$_2$O$_3$/20p were prepared with a 30-degree edge bevel (60-degree included angle) and a 0.063-in. (1.6-mm) root land.[43] All plates were degreased and scrubbed with a clean stainless steel brush, then fixtured to a grooved stainless steel backing bar using a 0.063-in. (1.6-mm) root gap. The ac power supply was set for 22 V and 325 A maximum with the arc balance towards penetration (a setting of 5, where 3 is balanced for a range of 0-10). A 0.094-in. (2.38-mm) diameter filler rod of ER5356 was fed into the weld at 55 in./min (23 mm/s) as the automatic, cold-wire feed traveled at 61 in./min (26 mm/s). The shielding gas of 100% argon was metered at 35 ft^3/h (16.5 L/min).

42. Ahearn, J. S., Cooke, C., and Fishman, S. G. "Fusion welding of SiC-reinforced Al composites." *Metal Construction*, 192-7, April 1982.
43. Gedeon, S. A. and Rudd, S. "Weldability of wrought DURAL-CAN aluminum matrix composites." Oakville, Ontario: Welding Institute of Canada RC461, March 1993.

Table 7.4
As-Welded GTAW Properties of Fiber-Reinforced Ti/W Composites

Tungsten Fiber, %	Specimen Type	Specific Modulus		Yield Strength (0.2% Offset)		Tensile Strength		Elongation, %
		Msi	GPa	ksi	MPa	ksi	MPa	
0	Base	14.3	99	69.2	477	88.7	612	29.0
0	Butt weld	16.7	115	73.0	503	92.8	640	17.5
0	Bead-on-sheet	15.5	107	82.4	568	101.6	701	14.0
4.5	Base	21.8	150	82.4	568	102.3	705	15.8
4.4	Butt weld	17.0	117	80.9	558	101.5	700	11.7
4.5	Bead-on-sheet	20.3	140	106.8	734	129.6	894	4.5
9.8	Base	21.2	146	95.2	656	103.5	714	3.4
9.9	Bead-on-sheet	17.2	119	105.9	730	131.3	905	4.0

Table 7.5
Welded Properties of GTAW 6061-Based Composites

Base Material (initially T6)	Yield Strength (0.2% Offset)		Tensile Strength		Elongation, %	Failure Location
	ksi	MPa	ksi	MPa		
6061-AW* (as-welded)	19.0	131	30.0	207	11	—
+SiC/18w-AW	9.3	64	26.0	179	3.1	weld
+Al₂O₃/20p-AW	19.1	132	33.1	228	6.6	HAZ

* Matrix alloy properties are typical handbook values as welded with ER5356 filler metal.

These welds passed the qualification requirements for 6061 in ANSI/AWS D1.2-90, *Structural Welding Code - Aluminum*. Typical as-welded properties are contained in Table 7.5, line 3. A cross-section of a representative weld is shown in Figure 7.23. Note the smooth bead contour, minimal porosity, and even distribution of particles in the weld.

Gas Metal Arc Welding. Owing to its high deposition rate and ease of automation, gas metal arc welding (GMAW) is preferred for most high-volume production welding. Because of the lower fluidity and reduced penetration of the composite weld pool, a root gap of 0.04 to 0.08 in. (1 to 2 mm) in conjunction with a temporary grooved backing bar is recommended for single-sided welds. However, permanent backing bars or self-backed joints are usually more practical in the field. A root gap also is preferred for double-sided welds, and the root should be thoroughly backgouged before welding the second side. Joint openings range from 60 degrees for single-V to 90 degrees for double-V butt welds with root lands of 10 to 20 percent of the plate thickness.

Similar to GTAW, the reinforcing phase interferes with arc stability during GMAW. This effect is not severe in Al/Al₂O₃ composites, and the welding proceeds normally. However, for Al/SiC composites (particularly the high-silicon types), the arc is extremely unstable. Successful welding depends on decreasing the voltage and driving the arc into the weld pool. The shorter, stiffer arc is confined by the depression in the surface of the pool, and a noisy, spray mode of droplet transfer is possible.

Building on previous work, a series of GMAW tests were conducted on 6061/SiC/20w in the mid-1980s.[44] A series of 0.25-in. (6.4-mm) plates were prepared with a 75-degree included angle and clamped to a grooved copper backing bar using a 0.094-in. (2.38-mm) root gap. The power supply was set for DCEP with a metered current of 135-140 A and 22-23 V. Two passes

were made from the same side using 0.045-in. (1.1-mm) ER5356 filler shielded with 100% argon at a travel speed of 10 to 12 in./min (4.2 to 5.1 mm/s). Similar welds also were used to join tubes with an outer diameter of 12.6 in. (320 mm) and a wall thickness of 0.75 in. (19 mm). This required 9 passes using 230-240 A and 25-27 V. There was a lack of base-material dilution in the welds and delamination adjacent to the weld in the 0.25-in. (6.4-mm) plate. (Again, the focus was on minimizing heat in the weld.) Typical as-welded properties are found in Table 7.6, lines 3-5.

A more common procedure is the multiple-pass butt weld of a 0.75-in. (19-mm) plate of 6061/Al₂O₃/20p.[45] In this example, the extrusion was prepared with a 30-degree edge bevel (60-degree included angle) and a 0.125-in. (3.2-mm) root land. After degreasing and scrubbing with a clean stainless steel brush, the plates

45. Altshuller, B., Christy, W., and Wiskel, B. "GMA welding of Al-Al₂O₃ metal matrix composite." *Symposium Proceedings: Weldability of Materials*, 305-9. Materials Park, Ohio: ASM International, 1990.

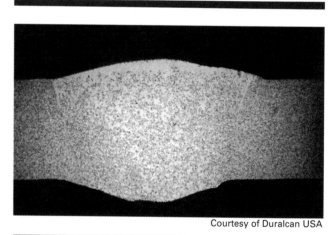

Courtesy of Duralcan USA

Figure 7.23—Cross Section of GTAW Joint in 6061/Al₂O₃/20p [0.25-in. (6.4-mm) Plate] (reduced 55%)

44. Ahearn, J. S., Cooke, D. C., and Barta, E. "Joining discontinuous SiC reinforced Al composites." Silver Spring, Maryland: Naval Surface Weapons Center TR 86-36, September 1985.

Table 7.6
Welded Properties for GMAW 6061-Based Composites using 5356 Filler Metal

Base Material (initally T6)	Form	Thickness		Yield Strength (0.2% Offset)		Tensile Strength		Elongation, %
		in.	mm	ksi	MPa	ksi	MPa	
6061-AW (as-welded)	Various	—	—	19.0	131	30.0	207	11
6061-T6*	Various	—	—	40.0	276	44.0	303	5
6061/SiC/20w-AW	Plate	0.25	6.4	21.8	150	34.4	237	4.5
6061/SiC/20w-AW	Tube	0.75	19	15.4	106	31.0	214	6.1
6061/SiC/20w-T6	Tube	0.75	19	18.8	130	36.6	252	6.6
6061/Al₂O₃/20p-AW	Plate	0.75	19	19.1	132	33.1	228	6.6
6061/Al₂O₃/20p-T6	Plate	0.75	19	27.4	189	41.0	283	3.9

* Typical handbook values for ER4043 filler metal; no data available for ER5356.

were fixtured to a grooved, stainless steel backing bar using a 0.063-in. (1.6-mm) root gap. A constant current, inverter-type power supply was set for 305 A and 26 V (DCEP). The four passes were made at different travel speeds. The root pass was made at 15.1 in./min (6.4 mm/s) to keep the arc at the leading edge of the weld pool and ensure good root penetration. The speed was slowed to 9.9 in./min (4.2 mm/s) during the second pass to ensure good sidewall fusion. Capping passes 3 and 4 were made at 13.9 in./min (5.9 mm/s). A 0.063-in. (1.6-mm) filler wire of ER5356 was used with 100% argon shielding metered at 50 ft³/h (23.6 L/min).

Typical as-welded properties are listed in Table 7.6, line 6. A representative microstructure from the fusion zone is shown in Figure 7.24. (In this figure, the base material is to the right and the weld is on the left.) There are no visible reaction products at the particle interface, and the dilution of the particles from the base material is relatively even with minimal agglomeration. Note that all of the MMCs discussed above meet the minimum tensile strength requirement of 24 ksi (165 MPa) listed in ANSI/AWS D1.2-90, *Structural Welding Code–Aluminum,* for welds in 6061.

Weld Repair.   Often MMC components are too complex or costly to scrap on the basis of cosmetic defects. This is especially true in the case of aerospace castings and forgings. GTAW is a common and effective means of repairing nonstructural defects, since low-heat input is desirable to prevent loss of temper or other changes in microstructure.

Small defects should be ground out with a diamond-plated tool. It is essential to provide good arc access to the root of the defect since the reduced fluidity of the composite weld pool may not ensure sufficient penetration. It is especially important to completely remove the defect rather than blend over it. (Liquid-penetrant inspection is one means of verifying this.) After grinding out the defect, it should be solvent-degreased and vigorously scrubbed with a clean stainless steel wire brush. For larger defects, a tungsten carbide backgouging tool may be more appropriate. (Defects of this size may require GMAW for effective repair.)

Capacitor-Discharge Welding.   The fundamentals of capacitor-discharge welding are quite simple. A stud that is to be joined to a workpiece is charged to a high capacitance and brought into sufficient proximity to

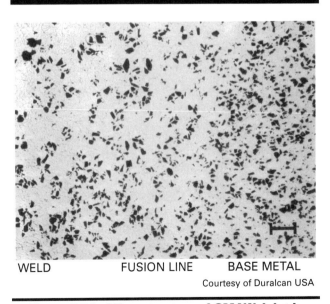

WELD   FUSION LINE   BASE METAL

Courtesy of Duralcan USA

Figure 7.24—Microstructure of GMAW Joint in 6061/Al₂O₃/20p [0.004-in (0.10-mm) Marker]

initiate an arc. Simultaneously, the physical separation and voltage potential are reduced to zero and the parts are fused. The hottest, most fluid region of the weld is expelled, carrying away any possible reaction products. The resulting microstructure and properties of the MMC joint reflect the short residence time and low heat input of the remaining material. Thus, capacitor-discharge welding is an excellent choice when it is critical to preserve the original characteristics of the MMC across the joint.

Similar and dissimilar welds have been made with the following MMCs: 6061, 2024, 6061/SiC/40p, 6061/SiC/48f, 6061/B$_4$C/30p, 2024/B$_4$C/30p, and AZ61/B$_4$C/40p.[46] Studs were 0.25 in. (6.4 mm) in diameter and sheets were 0.13 in. (3.2 mm) and 0.25 in. (6.4 mm) thick. Figure 7.25 demonstrates a low-dilution weld made between 6061/SiC/40p workpieces. Note the complete absence of porosity and the absence of lack-of-fusion defects. The formation of Al$_4$C$_3$ also was avoided (see Figure 7.26). The welding conditions used for all welds are shown in Table 7.7.

The typical arc time was 0.4 ms with a resulting solidification rate exceeding 10^6 °C/s. The minimum

---

46. Devletian, J. H. "SiC/Al metal matrix composite welding by a capacitor discharge process." *Welding Journal* 66(6): 33-39, 1987.

power input required for an acceptable weld by this process for these MMCs was 5.2 X 10^5 W/in.^{2} (8.0 x 10^8 W/m^2). Higher power inputs did not improve weld integrity significantly and resulted in larger amounts of weld metal expulsion.

Although no tensile values were reported, mechanical testing fractured the specimens in the base material rather than in the weld zone. Similar results were noted for capacitor-discharge welding using shims of 1100 aluminum that were 0.001 to 0.002 in. (20 to 40 μm) thick. The compressive force of capacitor-discharge welding, in conjunction with the low dilution and high solidification rate, can prevent other problems such as hot cracking. Although this process is perfect for attaching studs to panels or joining small parts into subassemblies, there are serious size and geometry limitations to capacitor-discharge welding.

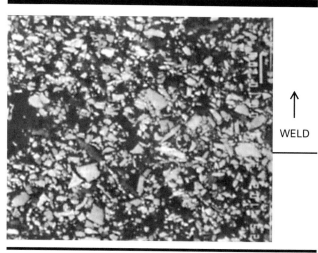

Figure 7.26—Unreacted SiC in Capacitor-Discharge Welded Joint (x800, unetched)

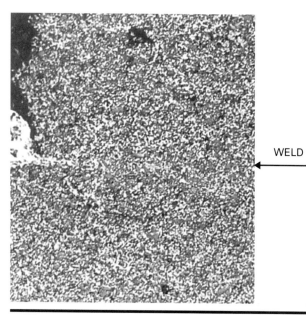

Figure 7.25—Capacitor-Discharge Welded Joint Between 6061/SiC/40p Stud and Sheet (x200, unetched)

**Table 7.7
Conditions for Capacitor-Discharge Welding of Metal-Matrix Composites**

Parameter	Setting
Drop height	3 in. (76 mm)
Drop weight	14 lb (6.4 kg)
Ignition length	0.025 in. (0.64 mm)
DC voltage	90 V
Capacitance	80 mF

Laser Beam Welding

LASER BEAM WELDING (LBW) excels at minimizing the size of the HAZ in high-speed, single-pass welding of sheet, plate, and tube. However, LBW is generally unsuitable for most MMCs because of the locally high temperatures in the weld pool that result from unequal beam coupling. These high local temperatures lead to a variety of reactions between the matrix and reinforcement, such as the aluminum carbide reaction.[47]

However, recent work on aluminum MMCs shows that by controlling the duty cycle and heat input of both continuous-wave and pulsed carbon-dioxide lasers, the amount of aluminum carbide formation in Al/SiC welds can be controlled.[48] Using the sets of parameter values in Table 7.8, maximum strength was attained at duty cycles of 67% and 74% (parameter settings C and D). Shorter duty times resulted in incomplete penetration, although the microstructure was less damaged. Longer duty times led to massive aluminum carbide formation and a decrease in mechanical properties.

However, several things should be noted concerning these results. First, the formation of aluminum carbide was reduced, not eliminated. Second, the matrix alloy was A356, an aluminum alloy with seven weight percent of silicon. Third, claims of strength improvements by such processing are relative to a nonheat-treated, slowly cooled, foundry pig with significant microstructural segregation and a coarse grain size. Thus, while LBW may have potential for joining MMCs with high-silicon matrices for nonaqueous service, there is little to suggest that it could be used to autogenously join SiC-reinforced wrought aluminum alloys. However, if filler alloys can be used, the addition of titanium or silicon-containing shims may help suppress the aluminum carbide reaction.[49]

One of the advantages of LBW for MMCs versus regular aluminum is the improved beam coupling. Because of the ceramic reinforcement, MMCs have a lower reflectivity. The downside of this is that the ceramic appears to couple more efficiently than the matrix, potentially leading to localized superheating of the particles. This exacerbates the problem of interfacial reactions.

LBW of Al/Al_2O_3 presents a quite different problem.[50] The beam does not form a stable keyhole in this material. The high-energy intensity melts the alumina particles ahead of the beam. This viscous layer of dross is transported to the rear of the keyhole, where it piles up in a ridge. As the ridge destabilizes and slumps back into the keyhole, it reacts explosively with the plasma creating a crater and temporarily decoupling the beam (see Figure 7.27).

To suppress this plasma explosion, a jet of inert gas was inclined upwards at a 5-degree angle opposite to the rotation of the tube being welded. It is theorized that the gas jet prevents contact between the trailing edge of the keyhole and the plasma inside the cavity.

47. Meinert, K. C., Martukanitz, R. P., and Bhagat, R. B. "Laser processing of discontinuously reinforced aluminum composites." *Conference Proceedings.* Dayton, Ohio: American Society for Composites, 1992.
48. Dahotre, N. B., McCay, M. H., McCay, T. D., Gopinathan, S., and Allard, L. F. "Pulse laser processing of a SiC/Al-alloy metal matrix composite." *Journal of Materials Research* 6(3): 514-29, 1991.
49. Meinert, Martukanitz, and Bhagat. "Laser processing."
50. Kawali, S. M. and Viegelahn, G. L. "Laser welding of alumina reinforced 6061 aluminum alloy composite." *Proceedings, ICALEO '91–Laser Materials Processing Symposium,* LIA Vol. 74: 156-67. Orlando, Fla.: Laser Institute of America, 1991.

Table 7.8
Laser Processing Parameters*

	A	B	C	D	E	F
On time, ms	20	20	20	20	20	20
Off time, ms	20	15	10	7	5	2
Duty cycle, %	50	57	67	74	80	91
Average power, W	1600	1830	2130	2370	2560	2900

* The following parameters were the same for all sets, A-F.

Traverse speed:	1.0 in./s (25 mm/s)
Beam mode:	TEM_{10}
Beam polarization:	Circular
Focal point:	0.02 in (0.5 mm) below surface
Shielding gas:	Argon, 8.5 ft^3/h (4.0 L/min), coaxial

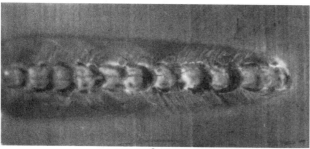

Courtesy of Laser Institute of America

Figure 7.27—Cratering from Laser Beam Welding of 6061/Al₂O₃/20p

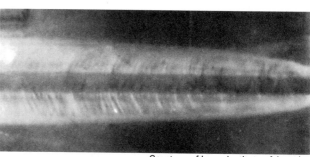

Courtesy of Laser Institute of America

Figure 7.28—Laser Beam Welding Bead Stabilized by Gas Jet

This specific jet angle is critical to stabilizing the keyhole shape without disrupting the plasma jet. Although a relatively smooth bead results (see Figure 7.28), the welds still exhibited poor fluidity and poor mechanical properties.

Electron Beam Welding

A HIGH-ENERGY PROCESS often regarded as similar to LBW, electron beam welding (EBW) is fundamentally different. While both have a narrow fusion zone and HAZ, the heat of these processes comes from quite different energy conversions. LBW thermal energy comes from light energy (photonic excitation of the target atoms), whereas EBW thermal energy comes from the kinetic energy in an accelerated electron stream. Furthermore, LBW is usually an atmospheric process, while most EBW is conducted in a vacuum. Differences in interactions involving these components (beam, atmosphere, and MMC) result in fundamentally different welding modes.

Duplication of selected LBW trials on A356/SiC/20p using EBW with similar parameters [e.g., 3.0 kW at 3.3 in./s (85 mm/s) and 0.02-in. (0.4-mm) beam diameter] yielded substantially less Al₄C₃ formation.[51] Under certain conditions of EBW, the formation of Al₄C₃ was almost eliminated (see Figure 7.29). However, similar work on 2014/SiC/15p did not show the same benefits; on the contrary, these trials resulted in a cutting rather than a welding action. A cross section of a similar area in a plate of 2014/Al₂O₃/15p reveals this lack of fusion as well as large gas pores (see Figure 7.30). This appears to be similar to the cratering problem associated with laser beam welding.

Resistance Welding

ALTHOUGH IT IS most widely used for spot welding automotive body sheet, resistance welding also includes processes such as flash butt welding of bicycle rims. The easily controlled, relatively low-heat input helps prevent reactions between the matrix and reinforcement in the molten weld nugget. However, the geometry of such joints is rather limited.

Early work on resistance welding of Al/B concentrated on generating sound spot and seam welds without crushing the boron fibers, as shown in Figures 7.31(A) and (B).[52,53] Typical seam joints in 0.02-in. (0.5-mm) thick sheets of 1100/B/50f generated lap shear strengths of 69.2 ksi (477 MPa) yielding joint efficiencies above 40%. Spot joints 0.2 in. (5.0 mm) in diameter were calculated to have greater than 75% efficiency based on longitudinal lap shear tests. Typical welding parameters used are shown in Table 7.9.

Similar work on Al-7Zn/C was less successful.[54] Spot welds between composite sheets and between composite and 2219 sheets (using an 0.003-in. [76-μm] interlayer of BAlSi-4 brazing foil) exhibited weld-metal expulsion and poor bonding. Increasing the electrode force above 400 lb (1779 N) suppressed this, but resulted in fiber packing and damage in the weld nugget (see Figure 7.32). The only other parameters noted were a current of 5000 A and weld time of 5 cycles. No mechanical data was reported.

51. Lienert, T. J., Brandon, E. D., and Lippold, J. C. "Laser and electron beam welding of SiCp reinforced aluminum A-356 metal matrix composite." *Scripta Metallurgica et Materialia* 28(11): 1341-6, May 1993.

52. Hersh, M. S. "Resistance welding of metal matrix composite." *Welding Journal* 47(9): 404s-9s, 1968.
53. Hersh, M. S. "The versatility of resistance welding machines for joining boron/aluminum composites." *Welding Journal* 51(9): 626-32, 1972.
54. Goddard, D. M., Pepper, R. T., Upp, J. W., and Kendall, E. G. "Feasibility of brazing and welding aluminum-graphite composites." *Welding Journal* 51(4): 178s-82s, 1972.

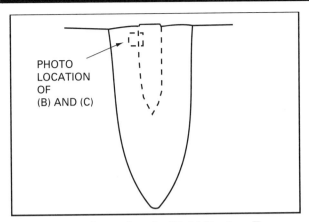

(A) Photo Location in Weld Fusion Zone

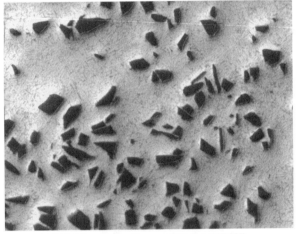

(B) Electron Beam Weld (EBW)

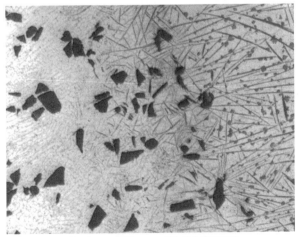

(C) Laser Beam Weld (LBW)

Courtesy of The Ohio State University

Figure 7.29—Microstructure of Fusion Zone for EBW and LBW using Identical Parameters on A359/SiC/20p (x400, reduced 74%)

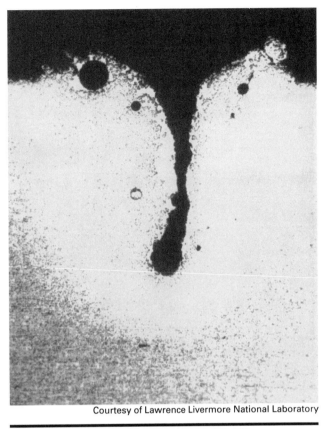

Courtesy of Lawrence Livermore National Laboratory

Figure 7.30—Electron Beam Welding on 2014/Al$_2$O$_3$/15p (x12.5)

One of the problems with resistance welding for discontinuously reinforced MMCs is reinforcement segregation in the weld nugget, as shown in Figure 7.33.[55] Little is known about this phenomenon at present. This 0.31-in. (8-mm) weld nugget required 30 000 A for 4 cycles using a 900 lb (4003 N) electrode force. The oversize fusion zone illustrates the lower electrical and thermal conductivity, and consequently weld power requirement, of the MMC as compared to the base alloy. Limited mechanical tests on such welds made with identical parameters in 0.04-in. (1.0-mm) thick sheets of 6061/Al$_2$O$_3$/20p yielded a 76% higher peel strength [668 versus 380 lb (2971 versus 1690 N)] than the minimum required for 6061.

55. Brenner, A., Budapesti Muszaki Egyetem, Mechanikai Technologia Tanszek, Hungary. Unpublished Research. February 1993.

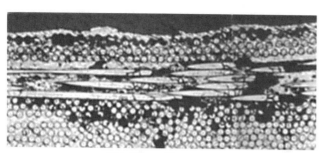

(A) Fiber Damage

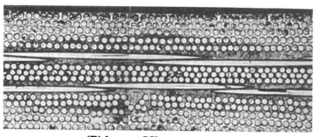

(B) Intact Microstructure

Figure 7.31—Resistance Spot Welding in 1100/B/50f Showing Fiber Damage as Compared to Intact Microstructure (x25, reduced ~30%)

Table 7.9
Typical Resistance Welding Parameters Used on Multiple-Impulse Machine (Power Rating of 150 kva)

Parameter	Value
Electrode	Class I [8 in. (200 mm) radius]
Weld Force	1000 lb (4480 N)
Current on	6 cycles
Phase shift	50%
Current decay	4 cycles
Phase shift	25%
Forge delay	2.4 cycles
Forge force	1780 lb (8006 N)

Friction Welding

AN ONGOING DEBATE about friction welding is whether it is a true solid-state process or only a semi-solid-state process. With respect to reactions between the matrix and reinforcement in MMCs, friction welding offers the advantages of solid-state joining and will be addressed as such here. Like other nonmolten processes, friction welding is excellent for joining dissimilar materials while maintaining the original properties of

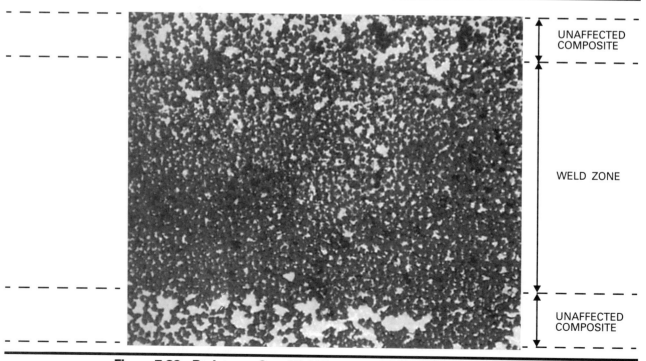

UNAFFECTED COMPOSITE

WELD ZONE

UNAFFECTED COMPOSITE

Figure 7.32—Resistance Spot Weld Between Sheets of Al-7Zn/C (x250)

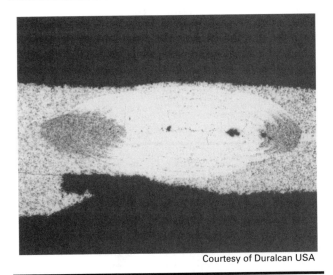

Courtesy of Duralcan USA

Figure 7.33—Resistance Spot Welding Between 0.04-in. (1.0-mm) Sheets of 6061/Al₂O₃/20p (x15, reduced 74%)

both across the joint. The rotational version is favored for joining tubing, while the linear mode is useful for bar stock and forgings.

An example of joining dissimilar materials is the rotary friction welding of 6061-T6511 to 6061/Al₂O₃/10p-T6.[56] Tubing with an outer diameter of 1.0 in. (25 mm) and a wall thickness of 0.070 in. (1.78 mm) was machined and cleaned to provide the required joint surface. Maximum joint strengths of 40.6 to 42.6 ksi (280 to 294 MPa) were obtained with fly-wheel speeds of 2625 to 3280 rpm and axial pressures of 550 psi (3.8 MPa). All failures were in the HAZ of the composite, within 0.08 to 0.20 in. (2 to 5 mm) of the bond line. Given the base material strength of 51.5 ksi (355 MPa), this yielded joint efficiencies of 79 to 83%. Although the efficiency of friction-welded joints is excellent, the need to remove the extensive flash resulting from the upset of the joint imposes some limitations on the use of this process.

Another example of rotary friction welding is the joining of 1.8-in. (45-mm) diameter bar stock of 2618/SiC/14p-T6 to itself.[57] The bars were rotated at 950 rpm before applying a friction force of 20 tons (18.8 MN) and a forge force of 30 tons (29.4 MN). The

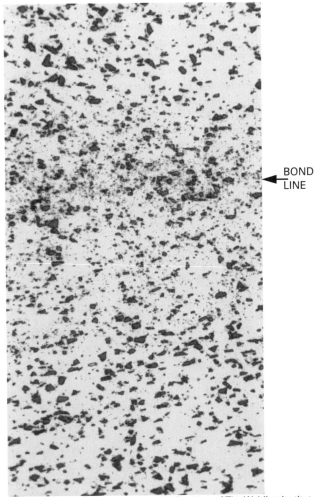

← BOND LINE

Courtesy of The Welding Institute

Figure 7.34—Unreacted SiC Particles at Bond Line of Rotational Friction Weld in 2618/SiC/14p-T6 (x200)

total joint displacement was 0.31 in. (8 mm) with a weld time of 4.6 seconds. In the as-welded condition, the specimens fractured adjacent to the bond line at an average tensile strength of 55 ksi (382 MPa) with 2.0% elongation, resulting in a joint efficiency of 84%. Figure 7.34 shows some break-up of the SiC particles at the weld interface, but no formation of Al₄C₃.

Other Joining Processes

THIS SECTION WILL describe diffusion welding, brazing, and diffusion brazing processes. The common feature in all cases is that the base material to be joined is not melted. Thus, these processes are particularly

56. Cola, M. J., Baeslack III, W. A., Altshuller, B., and Sjostrom, T. "Inertia-friction welding of a particulate-reinforced aluminum-matrix composite." *Conference Proceedings: SAMPE Metals Conference,* M424-38. Covina, Calif.: Society for the Advancement of Material and Process Engineering, October 1992.

57. Threadgill, P. L. "Joining of particulate reinforced aluminum alloy metal matrix composites." TWI 5617/3A/93. Cambridge, U.K.: The Welding Institute, March 1993.

well-suited to reactive combinations such as Ti/SiC and Al/B. In all cases, surface cleanliness is critical. Specimens are usually cleaned with acetone to remove any contaminants and mechanically abraded to disrupt the oxide skin. Using the thinnest possible interlayer of filler metal will minimize the required bonding time at temperature. These processes are able to join highly complex geometries and very thin sections.

Although used for a variety of applications, diffusion welding of MMCs is most commonly used to produce sheet and tubes of continuous fiber-reinforced MMCs. Good diffusion-welded joints are indistinguishable from the base composite, and the efficiency is typically 100% after heat treating.[58] If post-joining heat treatment is not possible, however, diffusion welding is not always desirable. This is because the entire assembly is subjected to high temperatures (and pressures) for long times, which usually degrades the MMC properties significantly.

In a comprehensive brazing study, no statistical difference was found in the bond strengths for the two fillers, BAlSi-2 and BAlSi-4, present as 0.003-in. (76-μm) foils.[58] The foils were clamped between coupons of 6061/B*/54f at 35 ksi (241 MPa) under argon at temperatures of 1076 to 1094 °F (580 to 590 °C), which is the solidus temperature for 6061.[59] Bond strengths were measured for the joint types listed in Table 7.10.

A more recent work has focused on brazing 6061/Al$_2$O$_3$/15p.[60] Although a variety of fillers were tried, the best choices were silver and BAlSi-4 (see Table 7.11). Specimens 0.75 in. (19 mm) in diameter and 3.0 in. (76 mm) long were clamped end-to-end in a spring-loaded jig at ~10 psi (~69 kPa) and heated under a vacuum of 10^{-3} torr (0.133 Pa). The brazed interface

shown in Figure 7.35 is typical. Diffusion of the braze layer away from the bond line appears to concentrate the reinforcement at the interface.

Brazing also was attempted on small coupons of Al-7Zn/C.[61] Filler foils of BAlSi-4 were melted under flowing argon at 1094 °F (590 °C) for 5 to 10 minutes without forming a good joint. By experimenting with dissimilar joints using 6061, and later with combination layers of 6061 and BAlSi-4 foils, sound joints could consistently be made. The proposed mechanism describes silicon diffusing from the molten BAlSi-4 foil into the 6061 and lowering its melting temperature. Magnesium from the 6061 then diffuses to the MMC interface and provides the necessary wetting action to form a sound joint (see Figure 7.36). This somewhat resembles the mechanism of diffusion brazing.

Further similarities between brazing and diffusion brazing were noted in a study joining 6061/Al$_2$O$_3$*/10c to 6061.[62,63] Small cylinders [0.8 in. (20 mm) long and 0.4 in. (10 mm) in diameter] were assembled with a layer of filler metal in a steel jig; however, the only compression applied was from the difference in thermal expansion during the braze cycle. The first filler tried was a 0.006-in. (150-μm) foil of 4045. Using a series of

Table 7.10
Properties of Brazed Joints for 6061/B*/54f

Joint Type	Tensile Strength		Efficiency, %
	ksi	MPa	
Single lap	74	510	48
Single-strap butt	45	310	29
Double-strap butt	119	821	78
2-degree scarf	93	641	61

58. Breinan, E. M. and Kreider, K. G. "Braze bonding and joining of aluminum boron composites." *Metal Engineering Quarterly* 9(4): 5-15, November 1989.

59. B* refers to SiC-coated boron fibers. Although expensive, the SiC coating was used as a diffusion barrier to prevent the molten aluminum from reacting with the boron fibers during the relatively long cycle times.

60. Klehn, R. "Joining of 6061 aluminum matrix-ceramic particle reinforced composites." M.S. thesis, Massachusetts Institute of Technology, Department of Materials Science and Engineering, September 1991.

61. Goddard, D. M., Pepper, R. T., Upp, J. W., and Kendall, E. G. "Feasibility of brazing and welding aluminum-graphite composites." *Welding Journal* 51(4): 178s-82s, 1972.

62. Al$_2$O$_3$* indicates a reinforcement that also contains SiO$_2$ and is actually a form of mullite.

63. Refer to Suganuma, K. and Okamoto, T. "Joining of alumina short-fibre reinforced AA6061 alloy to AA6061 alloy and to itself." *Journal of Materials Science* 22: 1580-4, 1987.

Table 7.11
Brazing Parameters and Properties for 6061/Al$_2$O$_3$/15p-T6

Filler	Thickness		Temperature		Time, min	Yield Strength (0.2% Offset)		Tensile Strength	
	in.	μm	°F	°C		ksi	MPa	ksi	MPa
Silver	0.001	25.4	1076	580	120	46.8	323	49.5	341
BAlSi-4	0.005	127	1085	585	20	46.6	321	48.7	336

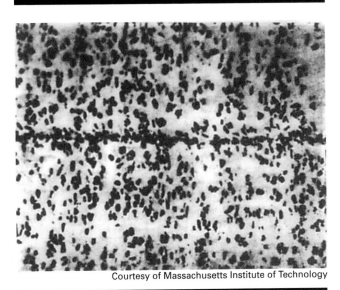

Courtesy of Massachusetts Institute of Technology

**Figure 7.35—Brazed Interface for Specimens of
6061/Al₂O₃/15p with Silver Filler Foil
(x140, reduced 74%)**

temperatures from 1076 to 1130 °F (580 to 610 °C), each specimen was held for 10 minutes under a vacuum of 5 X 10^{-5} torr (6.65 mPa). Although there was considerable scatter in the test data, tensile strengths were about 14.5 ksi (100 MPa). Other tests used similar conditions, but substituted BA03 for the braze foil [a 0.006-in. (150-μm) foil of 3003 clad with 400 μin. (10 μm) of 4045 on each side]. These joints had strengths scattered around 29.0 ksi (200 MPa).

The interfaces remaining from these two filler materials were markedly different. The 4045 foil left a thin layer of original material and some small micropores at the bond line. Since the brazing temperature was below the solidus for 3003, the BA03 interfaces exhibited a distinct layer between the MMC and the 6061. A greater propensity for pores was noted on the unreinforced side of the bond line. Additional brazing of 6061/Al₂O₃*/5c to itself, and to 6061 using BA03, demonstrated an approximately 25% stronger bond between the 3003 interlayer and the MMC than between the interlayer and the 6061.

Joining coupons of 1100/B/45f by using a 0.04-in. (1.0-μm) copper interlayer is an example of true diffusion brazing.[64] The assembly was held in a vacuum furnace at 1030 °F (554 °C) for 15 minutes or at 1060 °F (571 °C) for 7 minutes. As the temperature passed 1018 °F (548 °C), a eutectic liquid of Al-33.2Cu formed. At the brazing temperature, copper diffused into the aluminum matrix adjacent to the joint (see Figure 7.37). As the copper concentration decreased below 5.65%, the joint isothermally solidified. The concentration gradient across the joint was reduced further by homogenizing at 939 °F (504 °C) for 2 hours. Using

---

64. Niemann, J. T. and Garrett, R. A. "Eutectic bonding of boron-aluminum structural components: Part 1 - Evaluation of critical processing parameters." *Welding Journal* 53(4): 175s-84s, 1974.

6061 SHEET

BRAZE ALLOY

COMPOSITE

INTERMETALLIC PARTICLES (GRAY)

ORIGINAL BOND LINE

GRAPHITE FIBERS (DARK)

Figure 7.36—Brazed Interface Between 6061 Sheet and Al-7Zn/C with 718 as Braze Alloy (x500)

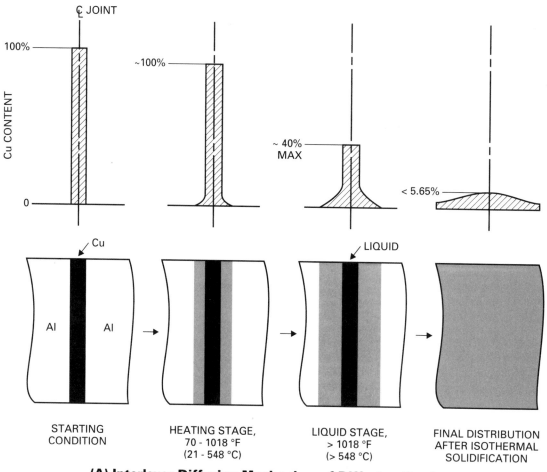

(A) Interlayer Diffusion Mechanism of Diffusion Brazing

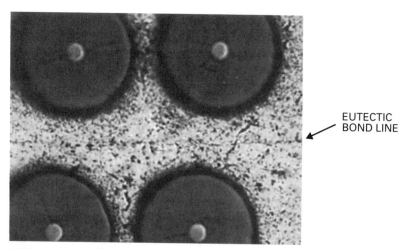

(B) Resulting Microstructure (x400, as polished)

Figure 7.37—Copper Diffusion in Brazing of Aluminum-Matrix Composite (1100/B/45f)

this method, as-bonded joint strengths of 160 ksi (1103 MPa) were achieved with resulting efficiencies of 86%. Diffusion brazing also was successfully used to incorporate titanium foil interlayers in a hybrid laminate of 1100/B/45p.

APPLICATIONS OF METAL-MATRIX COMPOSITES

THIS SECTION DESCRIBES applications involving welding of metal-matrix composites. Joining applications of polymeric composites are described in an earlier section (see Page 364).

It has been noted that the mechanical and physical properties of metal-matrix composites (MMCs) vary widely as a function of their composition and processing. Likewise, joint efficiencies vary as widely as the reinforcement geometries and the joining processes. The combination of base material properties, as-welded microstructures, and joint efficiencies determine whether a given MMC is appropriate for a desired application. The following examples illustrate this principle as well as the evolution of MMC technology over the past decades.

Various satellites and the U.S. space shuttles were some of the first structures to incorporate MMCs in their design. Tubular struts of 6061/B/50f provided exceptional stiffness without the weight penalty of titanium or steel. However, these hand-crafted, diffusion-bonded struts cost over 500 times more than an unreinforced strut.

MMCs entered markets other than aerospace when an automaker squeeze-cast aluminum into alumina chopped-fiber preforms for diesel pistons. Subsequently, automakers have used both squeeze-cast and gravity-cast techniques to make MMC automotive parts. (See Figure 7.39, Page 386.) The combination of low mass with excellent stiffness and thermal fatigue resistance resulted in the desired performance at an acceptable price. Another automaker adapted a similar MMC to cylinder liners to reduce weight while maintaining excellent wear resistance.

An aircraft manufacturer now uses extruded 6061/SiC/25p in portions of the avionics racks on military aircraft. The MMC extrusions provide the required stiffness and substantial weight savings. Although these MMCs are commercially extruded, their cost is still high because of the relatively expensive whisker reinforcement.

A specialty mountain bike is another MMC product now available to the general consumer market (see Figure 7.38). Extruded $6061/Al_2O_3/10p$ tubing is used to make about 2000 frames per month with the gas tungsten arc welding process. The 3.1-lb (1.4-kg) frames have higher specific strength and stiffness than standard aluminum or chrome-moly steel frames.

Extrusion dies for MMC components often are made from a Fe/TiC composite. These dies have the required combination of high-temperature strength, wear resistance, and fracture toughness for this demanding application.

To achieve higher pumping speeds, natural gas companies have switched from cast iron pistons to aluminum pistons with lower inertial mass. In order to reduce maintenance downtime, these companies are beginning to use a cast A359/SiC/20p-T6 insert in the ring area. These ring inserts can weigh over 66 lb (30 kg) and weld repair of occasional cosmetic defects is a welcome alternative to scrapping the casting.

Courtesy of Duralcan USA

Figure 7.38—Mountain Bike Made Using Alumina MMC

Courtesy oF Duralcan USA

Figure 7.39—Piston Insert Made Using Alumina Chopped-Fiber MMC

HEALTH AND SAFETY ISSUES

A THOROUGH DISCUSSION of health and safety issues relating to the welding of composites is beyond the scope of this chapter. In addition, because this area is changing rapidly, the following discussion should not be regarded as definitive. Below is an attempt to point out some of the relevant safety issues along with references to get additional information.

Regardless of the material being welded, standard welding safety practices should be followed at all times. Always weld in a well-ventilated area using appropriate eye protection and safety equipment. For general welding safety procedures, refer to ANSI/ASC Z49.1, *Safety in Welding, Cutting, and Allied Processes*, and to Chapter 16 of the *Welding Handbook*, Vol. 1, 8th Ed.

During the welding of polymeric composites, health and safety issues result from three primary areas; exposure to polymer decomposition products, potential exposure to small fibers and dust, and process-related safety issues.

During welding of thermoplastics composites, generation of decomposition products is possible when inappropriate excessive heating is used. Smoke or fumes that might be produced under such abnormal conditions may be toxic or otherwise harmful. Therefore, it is important to have sufficient control of the welding process so as to avoid overheating and decomposition of the material. In addition, during machining or grinding of composites, it is possible for fiber or filler dust to be released resulting in a breathing hazard. Under these circumstances, protective masks must be used even when liquid coolants are utilized. Additional information about handling of composites is available from material suppliers' safety sheets, from the American Conference of Governmental Industrial Hygienists, which publishes annually an updated book on chemical and biological hazards, and from the Suppliers of Advanced Composite Materials Association.[65,66]

Whiskers consolidated within metal-matrix composites appear to present no known health problems at this time. However, loose mineral whiskers can be carcinogenic when ingested into the pulmonary tract.

Welding process-related health and safety issues can arise independently of the material being welded. Welding machines usually incorporate safety shields, dual palm buttons, or both, which must never be disabled. In addition, one should follow the safety information that is provided with the welding equipment.

During welding, high pressures and high temperatures are used, and care must be exercised to avoid contact with the parts of mechanical pressing mechanisms. In addition, contact should be avoided with heated tools or heated gases. Exposure to infrared lamps and laser beams and to reflections from these sources should be avoided. Shielding should be used to avoid exposure to electromagnetic fields, which are produced during induction, radio frequency, and microwave welding. Meters for measuring the field strengths are available and should be used to ensure that the electromagnetic radiation levels are within accepted safety standards (see ACGIH booklet).[65] Hearing protection is required when using ultrasonic and vibration welding equipment. Sound meters could be used to ensure that sound levels are within limits specified by federal guidelines.

65. American Conference of Governmental Industrial Hygienists (ACGIH). *Threshold limit values for chemical substances and physical agents and biological exposure indices.* Cincinnati, Ohio: ACGIH.
66. Suppliers of Advanced Composite Materials Association (SACMA). *Safe handling of advanced composite materials components: health information.* Arlington, Va.: SACMA, April 1989.

SUPPLEMENTARY READING LIST

POLYMERIC COMPOSITES

ASM International. "Properties and selection: non-ferrous alloys and special-purpose materials." *Metals Handbook*, Vol. 2, 10th Ed. Metals Park, Ohio: ASM International, 1990.

Benatar, A. and Gutowski, T. G. "Methods for fusion bonding thermoplastic composites." *SAMPE Quarterly* 18(1): 34-41, 1986.

_____. "Ultrasonic welding of advanced thermoplastic composites." *Proceedings of the 33rd International SAMPE Symposium and Exhibition.* Covina, Calif.: Society for the Advancement of Material and Process Engineering, 1988.

Beyeler, E., Phillips, W., and Guceri, S. I. "Experimental investigation of laser-assisted thermoplastic tape consolidation." *Journal of Thermoplastic Composite Materials* 1: 107, 1988.

Blundell, D.J., Chalmers, J. M., Mackenzie, M.W., and Gaskin, W. F. "Crystalline morphology of the matrix of PEEK-carbon fibre aromatic polymer composites. I. Assessment of Crystallinity." *SAMPE Quarterly* 16 (4): 22-30, 1985.

Blundell, D. J., and Osborne, B.N. "Crystalline morphology of the matrix of PEEK-carbon fibre aromatic polymer composites. II. Crystallisation behaviour." *SAMPE Quarterly* 17 (1): 1-6, 1985.

Brophy, J. H., Rose, R. M., and Wulff, J. "The structures and properties of materials." Vol. II, *Thermodynamics of Structure.* New York: John Wiley, 1964.

Cogswell, F. N. "Microstructure and properties of thermoplastic aromatic polymer composites." *SAMPE Quarterly* 14(4): 33-7, 1983.

DeGennes, P. G. "Reptation of a polymer chain in the presence of fixed obstacles." *Journal of Chemical Physics* 55: 572, 1971.

DeGennes, P. G. "Entangled polymers." *Physics Today*, June 1983.

Delale, F., Erdogan, F., and Aydinoglu, M. N. "Stresses in adhesively bonded joints: A closed-form solution." *Journal of Composite Materials* 15: 249-71, 1981.

Goland, M., and Reissner, E. "The stresses of cemented joints." *Journal of Applied Mechanics*, Trans. ASME, 1(1): A17, 1944.

Grimes, G. C., and Greimann, L. F. "Analysis of discontinuities, edge effects, and joints." *Structural Design and Analysis*, Part II, Ed. Chamis, C. C., 135-230. New York: Academic Press, 1974.

Gutowski, T. G. "Resin flow/fibre deformation model for composites." *SAMPE Quarterly* 16 (4): 58-64, 1985.

Gutowski, T. G., Cai, Z., Soll, W., and Bonhomme, L. "The mechanics of composites deformation during manufacturing processes." *Proceedings of the First Conference on Composite Materials*. Dayton, Ohio: American Society for Composites, 1986.

Hart-Smith, L. J. *Joining of Composite Materials*, ASTM STP 749, Ed. Kedward, K. T., 3-31. Philadelphia: American Society for Testing and Materials, 1981.

Jud, K., Kausch, H., and Williams, J. G. "Fracture mechanics studies of crack healing and welding of polymers." *Journal of Materials Science* 16: 204-10, 1981.

Schwartz, M. M. *Composite Materials Handbook*. New York: McGraw-Hill, 1984.

Wagner, B. E. "Expanded market opportunities for reinforced thermoplastics with electromagnetic welding." *Proceedings of the 39th Annual Conference of the Reinforced Plastic/Composites Institute*, 3. The Society of Plastics Industry, 1984.

Willet, J. L. and O'Connor, K. M. *ACS Polymer Reprints* 26(2): 123, 1985.

Wool, R. D., and O'Connor, K. M. *Journal of Applied Physics* 52(10): 2953, 1981.

Wool, R. P. and Rockhill, A. T. *ACS Polymer Reprints* 21(2): 223, 1980.

METAL-MATRIX COMPOSITES

Ahearn, J. S., Cooke, D. C., and Barta, E. "Joining discontinuous SiC reinforced Al composites." *Naval Surface Weapons Center TR 86-36*, Silver Spring, Md., September 1985.

Altshuller, B., Christy, W., and Wiskel, B. "GMA welding of $Al-Al_2O_3$ metal matrix composite." *Symposium Proceedings: Weldability of Materials*, 305-9. Materials Park, Ohio: ASM International, 1990.

Breinan, E. M. and Kreider, K. G. "Braze bonding and joining of aluminum boron composites." *Metals Engineering Quarterly* 9(4): 5-15, November 1989.

Cola, M. J., Baeslack III, W. A., Altshuller, B., and Sjostrom, T. "Inertia-friction welding of a particulate-reinforced aluminum-matrix composite." *Proceedings of SAMPE Metals Conference*, M424-38. Covina, Calif.: Society for the Advancement of Material and Process Engineering, October 1992.

Goddard, D. M., Pepper, R. T., Upp, J. W., and Kendall, E. G. "Feasibility of brazing and welding aluminum-graphite composites." *Welding Journal* 51(4): 178s-82s, 1972.

Hersh, M. S. "Resistance welding of metal matrix composite." *Welding Journal* 47(9): 404s-9s, 1968.

Iseki, T., Kameda, T., and Maruyama, T. "Interfacial reactions between SiC and aluminum during joining." *Journal of Materials Science* 19: 1692-8, 1984.

Kawali, S. M. and Viegelahn, G. L. "Laser welding of alumina reinforced 6061 aluminum alloy composite." *Proceedings, ICALEO '91–Laser Materials Processing Symposium*, LIA Vol. 74: 156-67. Orlando, Fla.: Laser Institute of America, 1991.

Klehn, R. "Joining of 6061 aluminum matrix-ceramic particle reinforced composites." M.S. thesis, Massachusetts Institute of Technology, Department of Materials Science and Engineering, September 1991.

Lienert, T. J., Brandon, E. D., and Lippold, J. C. "Laser and electron beam welding of SiCp reinforced aluminum A-356 metal matrix composite." *Scripta Metallurgica et Materialia* 28(11): 1341-6, May 1993.

Lo, S. H. J., Dionne, S., Popescu, M., Gedeon, S., and Lane, C. T. "Effects of prior- and post-weld heat treatments on the properties of SiC/Al-Si composite gas metal arc-welded joints." *Symposium Proceedings: Advances in Production and Fabrication of Light Metals and Metal Matrix Composites*, Eds. Avedesian, M. M., Larouche, L. J., and Masounave, J., 575-88. Montreal, Quebec: Canadian Institute of Mining, Metallurgy and Petroleum, 1991.

Meinert, K. C., Martukanitz, R. P., and Bhagat, R. B. "Laser processing of discontinuously reinforced aluminum composites." *Conference Proceedings*. Dayton, Ohio: American Society for Composites, 1992.

Niemann, J. T. and Garrett, R. A. "Eutectic bonding of boron-aluminum structural components." *Welding Journal* 53(4): 175s-84s, 1974.

Suganuma, K. and Okamoto, T. "Joining of alumina short-fibre reinforced AA6061 alloy to AA6061 alloy and to itself." *Journal of Materials Science* 22: 1580-4, 1987.

Threadgill, P. L. "Joining of particulate reinforced aluminum alloy metal matrix composites." *TWI 5617/3A/93*. Cambridge, U.K.: The Welding Institute, March 1993.

CHAPTER 8

CERAMICS

PREPARED BY A COMMITEE CONSISTING OF:

M. L. Santella, Chairman
Oak Ridge National Laboratory

S. Kang
Seoul National University

H. Mizuhara
WESGO, Inc.

A. J. Moorhead
Oak Ridge National Laboratory

G. A. Rossi
Selee Corporation

T. N. Tiegs
Oak Ridge National Laboratory

WELDING HANDBOOK COMMITTEE MEMBER:
J. G. Feldstein
Foster Wheeler Energy International

CERAMICS

INTRODUCTION

CERAMICS ARE INORGANIC nonmetallic materials separable into two broad categories: traditional ceramics and advanced ceramics.[1] A common characteristic of ceramic materials is that they are manufactured from powders which are formed to a desired shape, and then heated to high temperature with or without the application of external pressure to achieve a final densified part.

Traditional ceramics include clay products, refractories, silicate glasses, and cements. They are most commonly made from inexpensive, readily available, naturally occurring minerals. These materials typically have low densities (or relatively high porosity contents) and normally are not used in applications where joining by techniques other than cementing is practical.

Advanced ceramics, by comparison, are made from powders that are chemically processed or synthesized and in which properties such as particle size distribution and chemical purity are closely controlled. Within the family of advanced ceramics are materials developed for their exceptional mechanical properties. This subset of advanced ceramics is often referred to as *structural ceramics*, and it includes monolithic materials such as aluminum oxide (Al_2O_3), zirconium oxide (ZrO_2), silicon carbide (SiC), silicon nitride (Si_3N_4), and silicon-aluminum oxynitrides (sialons), as well as ceramic composites like Al_2O_3 containing SiC whiskers or SiC containing titanium diboride (TiB_2) particles. Care is taken during the manufacture of structural ceramics to ensure that chemical composition is controlled and that high densities (or relatively low porosity contents) are achieved.

The technological interest in structural ceramics is directly related to their unique properties when they are compared to metals. Many ceramics are characterized by high strength, not only at room temperature but at elevated temperatures as well. Silicon carbide, for example, can maintain a tensile strength in excess of 29 ksi (200 MPa) at 2800 °F (1530 °C), the melting point of iron. Other ceramics, like Si_3N_4 and certain ceramic composites, also maintain high strength at high temperatures. Besides high strength, other properties that make ceramics attractive candidates for applications usually reserved for metallic alloys include excellent wear resistance, high hardness, excellent corrosion and oxidation resistance, low thermal expansion, high electrical resistivity, and high strength-to-weight ratio. Structural ceramics are being used or considered for use as cutting tools, bearings, machine-tool parts, dies, pump seals, high-temperature heat exchangers, and a variety of internal combustion and turbine engine parts. Typical properties for some metals and structural ceramic materials are given in Table 8.1.

Ceramic joining, especially ceramic-to-metal joining, has been the subject of much developmental research over the years. However, with high interest in using ceramics as structural components in demanding applications, such as internal combustion engines, turbine engines, and heat exchangers, has come a heightened interest in ceramic joining technologies. Industry interest in joining ceramics is the same as for joining metals. However, the development of more effective joining techniques for ceramics could have a much greater impact on their use in mass-produced components.

One of the most important functions of joining techniques is to provide the means for economic fabrication of complex, multicomponent structures. Development of effective ceramic joining techniques will be especially significant because of limitations on component manufacturing due to ceramic processing techniques and to the materials themselves. For example, deformation of densified ceramics to form complex shapes is practically impossible owing to the fact that most ceramic materials are brittle even at elevated temperatures. Also, in some development programs like those for advanced

1. The preparation of this chapter has been sponsored by the U. S. Department of Energy, Assistant Secretary for Conservation and Renewable Energy, Office of Transportation Technologies, as part of the Ceramic Technology for Advanced Heat Engines Project of the Advanced Materials Development Program, under contract DE-AC05-84OR21400 with Martin Marietta Energy Systems, Inc.

Table 8.1
Typical Properties of Some Pure Metals and Structural Ceramics[a]

Material	Strength[b]		Modulus of Elasticity		Coefficient of Linear Thermal Expansion		Electrical Resistivity,	Thermal Conductivity,
	ksi	MPa	ksi	GPa	in./in./°F	μm/m/°C	μΩ•cm	W/(m•K)
Al	4.9	34	8992	62	13.10	23.6	2.6548	221.75
Cu	10.0	69	15 954	110	9.17	16.5	1.6730	393.71
Fe	18.9	130	28 427	196	6.50	11.7	9.71	75.31
Mo	50.0	345	46 992	324	2.72	4.9	5.20	142.26
Ni	22.0	152	30 023	207	7.39	13.3	6.84	92.05
Ti	30.0	207	16 824	116	4.67	8.4	42.00	21.90
Al_2O_3	43.5	300	55 114	380	3.78	6.8	$>10^{20}$	27.20
SiC	72.5	500	69 618	480	2.33	4.2	10^7	62.80
Si_3N_4	145.0	1000	44 091	304	1.78	3.2	$>10^{20}$	10.00
ZrO_2	101.5	700	29 733	205	5.39	9.7	$>10^{17}$	2.00

a. Values given are typical values for each material at or near room temperature. Property values of both metals and ceramics can vary significantly with composition.

b. Yield strengths are given for metals; modulus of rupture strengths are given for ceramics.

heat engines, some complex parts are being made as monoliths by difficult processing schemes or by extensive machining of densified billets.

While this approach to component manufacturing is acceptable for development purposes, it is undesirable for mass production because of high costs. The difficulty of machining ceramics also makes it costly. By reducing the complexity of individual parts, significant savings in machining cost can be expected. Effective methods of joining ceramics may eliminate machining altogether in some cases.

Effective ceramic joining techniques also can play an important role in improving the reliability of ceramic structures. Because ceramics are brittle materials, they are very sensitive to flaws resulting from the quality of raw materials used in their production and to the characteristics of various processing techniques, including machining. A single flaw can cause the rejection or, if undetected, the failure of a ceramic part. Rather than dealing with complicated monolithic parts, it is easier to inspect and detect flaws in simple-shaped components before they are joined to form complex structures.

Even though there is keen interest in the development of structural ceramics and their use in new and unusual engineering applications, the electronics industry has the largest fraction of advanced ceramics actually in use. Also, while the development of materials like zirconium oxide, silicon nitride, and silicon carbide has been vigorously pursued in recent years, aluminum oxide still is the most widely used structural ceramic with a sizeable electronics commercial market.

CERAMIC MATERIALS

ALUMINUM OXIDE

ALUMINUM OXIDE (or alumina or Al_2O_3) is generally extracted from the mineral ore bauxite, which has a world production of around 35 million metric tons annually. About 90 percent of this alumina is reduced to produce aluminum metal, and about 3 percent is used for alumina specialty products including structural ceramic grades of Al_2O_3. The ceramics market for Al_2O_3 encompasses a wide range of applications from glass and chinaware to spark plugs, biomedical ceramics, and integrated circuitry.

In the United States, applications for Al_2O_3 are dominated by the electronics field. These applications generally require an insulating material with some or all of the following properties: high mechanical strength (e.g., for rectifier housings), high electrical resistivity, low dielectric losses (e.g., for transmitter tubes), high density, and translucency (e.g., for sodium vapor discharge lamps). Most of these requirements can be met only by high-purity structural grades of Al_2O_3. Numerous specialty electronic applications exist for Al_2O_3, including spark plugs, high-voltage insulators, vacuum tube envelopes, rf windows, rectifier housings, integrated circuit packages, and thick- and thin-film substrates. Some examples of electronic components that contain Al_2O_3 braze joints are shown in Figure 8.1.

The compositions of structural aluminas can range from 90 to 99.99% Al_2O_3 with the balance being impurities which are intentionally added to aid in the processing and densification of the materials. For example, alumina made especially for metallization may be doped with fluxing additives to form a microstructure that is optimum for ceramic-to-metal bonding.

ZIRCONIUM OXIDE

ZIRCONIUM OXIDE (or zirconia or ZrO_2) occurs in nature most commonly as the mineral zircon ($ZrSiO_4$) and less frequently as free zirconia. Annual production of zirconia ore reportedly has been as high as 650 000 tons, with most of it being used as refractory material and only a relatively small portion in structural ceramics.

Much of the recent technological interest in ZrO_2-based ceramics stems from the fact that ZrO_2 can exist in three different crystal structures: cubic, tetragonal, and monoclinic. The structure or combination of structures that exist in ZrO_2 ceramics can be controlled by alloying and heat treatment to produce materials with high strength and relatively high toughness. The good mechanical properties are a result of a stress-induced transformation of the metastable tetragonal crystal structure to the equilibrium monoclinic crystal structure. Two distinct types of toughened ZrO_2 have been developed. One is partially stabilized zirconia (PSZ), which consists of a cubic matrix containing particles with mainly the tetragonal structure. The other is tetragonal zirconia polycrystalline (TZP), which is processed to retain nearly 100 percent of the metastable tetragonal structure at room temperature.

Excellent resistance to thermal shock and mechanical impact damage has led to the use of PSZ as extrusion die materials. Superior wear properties and low susceptibility to stress corrosion also has led to the use of PSZ as a structural biomedical material. Wear properties, high strength, low thermal conductivity, and relatively high thermal expansion have produced strong interest in the application of PSZ in a variety of internal combustion engine components.

(A) Insulated Electrical Leads

(B) Vacuum Tube Containing Alumina Insulator

Figure 8.1—Examples of Ceramic-to-Metal Braze Joints

SILICON CARBIDE

SILICON CARBIDE (SiC) occurs naturally, but not in quantities large enough to be of significant commercial value. In the western countries, about 500 000 tons per year of SiC are produced by carbothermic reduction, reacting silica (SiO_2) with carbon at temperatures above 4000 °F (2200 °C). SiC powder can be made also by the pyrolysis of polymer precursor materials, such as polycarbosilanes and polyborosilanes, and by gas-phase synthesis techniques. One such method, for instance, is by plasma decomposition of mixtures of methane with either silicon tetrachloride, dimethyl dichlorosilane, or methyl trichlorosilane.

Major uses of SiC are for abrasives, where it is commonly used as loose or bonded grits; for a siliconizing agent in steel making; and for structural ceramic materials. Structural grades of SiC are most commonly made by two processes.

In one process, porous bodies of SiC and carbon are infiltrated with liquid silicon metal and then heat treated (reaction sintered) to produce further reaction of the carbon and silicon to form more SiC. Silicon carbide parts produced by this technique are known as *reaction-bonded SiC*, and typically have very high densities and contain about 10 to 15 percent of free, unreacted silicon.

The other process commonly used to produce SiC parts is pressureless sintering. In this case, small amounts (about 1 to 3 wt. %) of boron and carbon are mixed with fine SiC powder to promote densification, and then the parts are sintered in an inert atmosphere at temperatures above 3630 °F (2000 °C). Material produced by this method is known as *sintered SiC*, and it is routinely densified to near its theoretical value.

One of the first important engineering applications of SiC was for rotating mechanical seals where reaction-bonded SiC was found to outperform hard metals and Al_2O_3 by a wide margin. Sintered SiC has become an important material for the fabrication of sliding seals in hermetically sealed pumps, particularly where hazardous, corrosive, or abrasive media must be pumped. The corrosion resistance and mechanical properties of sintered SiC make it an attractive candidate material for many high-temperature applications. There is intense interest in SiC for automotive applications, where this material is being used to make critical components of advanced-design turbine engines.

SILICON NITRIDE

SILICON NITRIDE (Si_3N_4) does not occur in nature, and the high quality Si_3N_4 powders required for making structural material must be synthesized. Common processes for synthesizing Si_3N_4 powders include the reaction of silicon metal with nitrogen gas, the decomposition of silicon diamide, the reaction of silicon chloride and ammonia, and by the carbothermic reaction of silica, carbon, and nitrogen gas. Silicon nitride parts also can be made by the nitridation of green, or unsintered, silicon bodies at temperatures up to 2550 °F (1400 °C). This technique is referred to as *reaction*

bonding. Reaction-bonded Si_3N_4 is relatively porous and of low strength.

The production of high-density, high-strength Si_3N_4 requires high-temperature processing as well as the use of small quantities of liquid-phase sintering aids such as yttrium oxide, magnesium oxide, or aluminum oxide. These are commonly added to Si_3N_4 singly or in combination at concentrations of 4 to 15 wt. % to achieve consolidation.

Widespread use of Si_3N_4 parts has been hampered by the difficulties of producing high-quality, low-cost Si_3N_4 powders and densifying them into parts or items with closely controlled flaw populations. Nevertheless, Si_3N_4 figures prominently in all of the advanced engine concepts currently under development.

Silicon nitride is considered to be one of the toughest and strongest of ceramics at temperatures above 1830 °F (1000 °C). It also is a prime candidate for lightweight engine components requiring good wear resistance. Some automotive turbocharger rotors are currently made from Si_3N_4 to take advantage of its superior mechanical behavior and low density compared to metals. Other potential applications for Si_3N_4 include power turbine rotors, shafts for gas turbine engines, and valves and valve train parts for internal combustion engines.

SIALONS

SILICON-ALUMINUM OXYNITRIDES (or sialons) also do not occur in nature, but because the elements used to produce sialons are among the most abundant on earth, there is no possibility of a raw materials shortage for their production. Sialon powders can be produced by a number of techniques including the heating of clay and coal mixtures or sand and aluminum mixtures in nitrogen. The most common form of sialon used in industry is ß'-sialon, which is a solid solution described by the chemical formula, $Si_{6-X}Al_XO_XN_{8-X}$, where X can vary from 0 to 4.

Of the two major variations of ß'-sialons, one contains a small amount of glassy phase used to promote densification during sintering. In the other variation, a heat treatment causes the glassy phase to crystallize. The material containing the glassy phase has very high strength at low temperatures, but its strength quickly decreases with increasing temperature. The heat-treated version has lower strength but retains it to higher temperatures, and it has better high-temperature creep resistance than the glass-containing materials.

One of the most successful applications of sialons has been as cutting tools for machining cast irons, hardened steels, and nickel-based alloys. In these applications, they outperform both tungsten carbide and alumina tools because of superior high-temperature strength, thermal-shock resistance, and wear resistance. Sialons also are being used as extrusion die inserts in the production of ferrous and nonferrous metals and as die inserts and mandrels in the production of stainless and high-alloy steel tubing. The excellent strength and thermal-shock resistance of sialons also make them preferred materials for gas shrouds for automatic welding operations. Also, sialons are well suited for many seal and bearing applications.

CERAMIC-MATRIX COMPOSITES

A WAY TO improve further upon the properties of monolithic ceramics is to create ceramic-matrix composites (CMCs) by intentionally incorporating ceramic fibers, whiskers, or particles into a ceramic-matrix material. Ideally, when this is done, the best properties of both the matrix and the additive materials are retained, and the CMC has properties that can considerably exceed those of the individual monolithic materials. The two major categories of CMCs are fiber-reinforced ceramic composites and whisker- or particulate-containing ceramic composites.

Fibers are continuous or nearly-continuous filaments that are added to ceramic matrices primarily to increase toughness and creep resistance while maintaining or improving strength. Materials that can be made into fibers include Al_2O_3, mullite (a mixture of Al_2O_3 and SiO_2), SiC, and Si_3N_4.

Incorporating fibers into a ceramic matrix complicates material processing considerably. Improvements in properties achieved in many fiber-reinforced CMCs have been limited by difficulties resulting from the nonuniform distribution of fibers, degradation of fiber properties during processing, and incomplete densification of the matrix.

Many of the problems associated with processing fiber-reinforced CMCs can be overcome through the use of particles or whiskers as additives to ceramic-matrix materials. Whiskers are short, single-crystal fibers, either rod-shaped or needle-shaped, typically with aspect ratios under 100 and with diameters under 120 μin. (3 μm). Unlike fiber-reinforced CMCs, whisker- and particle-reinforced CMCs can be fabricated using conventional ceramic processing techniques.

One CMC that has received considerable attention is SiC-whisker-reinforced Al_2O_3. Compared to monolithic materials, both strength and fracture toughness are significantly improved as SiC whiskers are added to either Al_2O_3 or mullite matrices. In addition, the SiC-whisker-reinforced CMCs exhibit excellent resistance to thermal shock, wear, and oxidation. An important use for SiC-whisker-reinforced Al_2O_3 is for metal-cutting tool bits, where dramatic improvements in bit life have been achieved in certain applications.

Another promising group of CMCs are the ZrO_2-toughened Al_2O_3 group (ZTAs). These are Al_2O_3 matrix materials that derive improved strength and toughness from dispersions of partially-stabilized ZrO_2 particles.

JOINING PROCESSES

PUBLISHED TECHNICAL LITERATURE is perhaps the best source of information on ceramic joining techniques and provides the basis for the discussion presented in this chapter. Patent literature also is a good source of information. Many useful sources on this subject are listed in the Supplementary Reading List at the end of the chapter.

Ceramics can be bonded to either ceramics or metals both by direct methods, such as diffusion bonding, and by indirect methods, such as brazing. Many process variations exist in both of these categories of bonding methods. However, ceramic joining is an area with much developmental work in progress, and few examples of ceramic joining techniques have achieved widespread commercial application. Consequently, there are no standards for describing ceramic joining techniques or their use for specific materials or applications.

Another area where standards are lacking is in the mechanical property testing of ceramic and ceramic-to-metal joints, and care should be taken when comparing the test results of one study to those of another. Owing to the developmental nature of much of the work on this subject, a wide variety of techniques have been used to measure strength values of ceramic-containing joints. For example, joint strength has been measured using techniques from tensile testing, flexure testing, torsion testing, and the compression of unnotched disks.

Similarly, a variety of approaches has been used for measuring the shear strength and fracture toughness of ceramic-containing joints. Furthermore, the size of test specimens can have a significant influence on the results because of the statistical nature of the fracture process in brittle materials. Generally, as test specimen size increases, measured properties such as strength decrease because the probability of encountering a critical-sized flaw, i.e., one that will initiate failure, also increases with specimen size. For this reason, direct comparison of test data from different studies should be done only when each has used the same testing techniques and the same identically prepared specimen sizes. Standards for the testing and the interpretation of test results will undoubtedly be developed as interest increases in the commercial use of joined ceramics in structural components.

CERAMIC BRAZING

STRUCTURAL CERAMICS ARE among the most stable compounds known, and a result of their chemical stability is that they are wetted only with difficulty by liquid metals. Wetting is defined by the conditions where the contact angle, θ, of a liquid on a solid surface is less than 90 degrees. Some contact angle values for a variety of pure metals and some alloys on Al_2O_3 and Si_3N_4 surfaces are given in Table 8.2.

Because wetting ($\theta < 90°$) is necessary for producing usable braze joints, an essential consideration in ceramic brazing is the need to promote wetting of the ceramic surfaces by the braze filler metals. Wetting of ceramic surfaces can be obtained by two general methods:

(1) Applying coatings that promote wetting to the ceramic surfaces prior to brazing

(2) Alloying braze filler metals with elements that activate wetting

Sintered-Metal-Powder Process

THE MOST WIDELY used approach for brazing Al_2O_3 involves the sintered-metal-powder or moly-manganese (Mo-Mn) process. The Mo-Mn process actually refers to a method of metallizing Al_2O_3 surfaces by coating them with a mixture of metal and oxide powders followed by a sintering treatment at high temperature.

The basic features of the process actually date back to the 1930s, and were refined in the 1950s into the metallizing treatment used today. In the Mo-Mn process, a slurry consisting of powders of Mo and MoO_3, Mn and MnO_2, and various glass-forming compounds is applied to an Al_2O_3 surface in the form of a paint. The coated ceramic is then fired in wet hydrogen at a temperature near 2730 °F (1500 °C). This causes the glassy material to densify the metallic layer and to bond it to the ceramic surface. Successful completion of the process depends on factors such as the amount and viscosity of the glassy phases in both the Mo-Mn layer and the Al_2O_3, the ratio of grain size in the ceramic to the pore size in the Mo-Mn layer, and the processing temperature and atmosphere.

The Mo-Mn process has been widely accepted by industry as a standard method for metallizing Al_2O_3 surfaces, and numerous variations have been developed to extend the usefulness of the process. For instance, the metallizing mixture can be applied by brush for small production runs of unique or oddly shaped parts, or it can be applied by mass-production techniques such as spraying or silk screening when necessary.

Alumina surfaces metallized by the Mo-Mn process can be brazed with a number of standard braze filler metals such as the following: BAg-8 (72Ag-28Cu), BAu-1 (64Cu-36Au), and BAu-4 (82Au-18Ni).[2] Prior to actual brazing, nearly all metallized parts are nickel coated either by painting with a nickel-oxide paint followed by sintering in hydrogen, or they are nickel plated. The nickel plating greatly improves wetting of the Mo-Mn coating by alloys such as those mentioned

Table 8.2
Selected Data on Contact Angles for Various Liquid Metals and Alloys on Al_2O_3 or Si_3N_4 Substrates

Metal	Temperature		Contact Angle, Degrees
	°F	°C	
Al_2O_3 Substrate			
Sn	1922	1050	145
Ag	1832	1000	159
Au	2012	1100	138
Cu	2012	1100	148
Ni	2732	1500	120
Fe	2822	1550	128
Cr	3452	1900	85
Cu-2Ti	2012	1100	142
Cu-25Ti	2012	1100	15
Si_3N_4 Substrate			
Sn	2012	1100	144
Ag	2012	1100	155
Au	2012	1100	157
Cu	2012	1100	131
Cu-3Ti	2192	1200	64
Ag-28Cu	1652	900	167
Ag-28Cu-2Ti	1652	900	50
Ag-28Cu-5Si	1652	900	129

2. AWS-ASTM standard designations are used for braze filler metals. Compositions (in parentheses) are in weight percent.

above. One version of the Mo-Mn process is shown schematically in Figure 8.2, where it is compared to the active braze alloy process described next.

Even though the Mo-Mn process requires many more steps than other ceramic brazing techniques, it still is widely used and is often preferred for the following reasons:

(1) The process is well known.
(2) It is easily automated.
(3) Minor deviations in the process variables do not significantly affect the quality of the braze joints.
(4) Brazing can be done in wet hydrogen, vacuum, or inert atmospheres, giving the manufacturer a variety of processing options.

A drawback to the Mo-Mn process is that its application is limited to Al_2O_3 and possibly some other oxides.

Active Braze Alloy Process

EARLY EXPERIMENTAL WORK on reactions between liquid metals and oxides showed that when a liquid metal contains an element that forms a more stable oxide than the solid oxide on which the liquid metal is held, then wetting or spreading of the liquid occurs. Titanium and zirconium, for example, are particularly effective at wetting oxides, and it has been known since the 1950s that small additions of Ti to other metals promotes their wetting of oxides. Today, the ability and effectiveness of Ti additions to promote wetting of Al_2O_3 and other ceramics like SiC, Si_3N_4, and sialons is well established. A filler metal specially alloyed to promote wetting on ceramics is often referred to as an active braze alloy (ABA).[3]

3. When used in conjunction with an alloy name (e.g., Cusil-ABA), ABA is a registered trademark of WESGO, Inc.

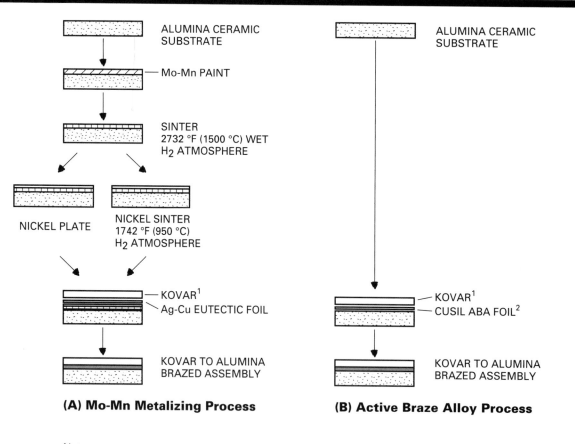

(A) Mo-Mn Metalizing Process

ALUMINA CERAMIC SUBSTRATE

Mo-Mn PAINT

SINTER 2732 °F (1500 °C) WET H_2 ATMOSPHERE

NICKEL PLATE

NICKEL SINTER 1742 °F (950 °C) H_2 ATMOSPHERE

KOVAR[1]
Ag-Cu EUTECTIC FOIL

KOVAR TO ALUMINA BRAZED ASSEMBLY

(B) Active Braze Alloy Process

ALUMINA CERAMIC SUBSTRATE

KOVAR[1]
CUSIL ABA FOIL[2]

KOVAR TO ALUMINA BRAZED ASSEMBLY

Notes:
1. Kovar is a trademark of Westinghouse Corporation.
2. Cusil ABA is a trademark of WESGO, Inc.

Figure 8.2—Schematic Representations of Two Processes for Brazing Ceramics

Commercial Active Braze Alloys. The selection of commercially available active braze filler metal is limited, particularly for high-temperature applications. Some of the active braze alloys available today are listed in Table 8.3. The majority are silver-based, and the numbers are the percentage by weight of the other constituents. The Cusil-ABA and Incusil-ABA alloys have been used to braze a wide variety of Al_2O_3 and Si_3N_4 components and test specimens.

Experimental Active Braze Alloys. Because of low industrial demand, many experimental alloys developed to wet and adhere to ceramics have not been commercialized. Nevertheless, the usefulness of some of these alloys in special conditions has been demonstrated. Some alloys of particular note include the following (compositions in wt. %):

(**1**) Alloys of Ti-48Zr-4Be and Ti-49Cu-2Be with brazing temperatures of 1922 °F (1050 °C) and 1830 °F (1000 °C), respectively, were used for joining Al_2O_3.[4] The Al_2O_3 joints made with the Ti-Cu-Be alloy withstood repeated exposure to high-temperature steam and severe thermal transients. The Ti-Cu-Be alloy also has been used for ceramic-to-metal braze joints in sensors for high-temperature steam instrumentation systems.[5]

(**2**) An alloy with the composition of Ti-25Cr-21V and a brazing temperature of 2912 °F (1600 °C) is reported to work well for brazing Al_2O_3.[6]

(**3**) A Cu-26Ag-29Ti alloy was found to produce excellent wetting on a variety of oxide ceramics.[7,8] Alumina joints made with this alloy had strengths above 165 MPa (24 ksi) at 752 °F (400 °C); ZrO_2 joints had strengths above 130 MPa (19 ksi).

(**4**) A Cu-20Au-18Ti alloy was reported to give excellent braze joints of both Al_2O_3 and ZrO_2.[7,8] With this alloy, Al_2O_3 joint strengths up to 32 ksi (220 MPa) were achieved at 750 °F (400 °C). Room-temperature strengths of 38 ksi (260 MPa) were obtained for ZrO_2 joints.

(**5**) Amorphous alloys with compositions of Cu-43Ti and Ti-30Ni have been reported to produce strong braze joints of both SiC and Si_3N_4.[9]

Vapor Coating Process

A VARIATION OF the metallizing approach is the use of vapor-deposited coatings of metals to promote the wetting of ceramics. This metallization approach has not been widely exploited for practical purposes even though the ability of metallic vapor coatings to enhance the brazing characteristics of ceramics has been known for some time.[10]

The key element of the process is the deposition of a thin coating, generally of a reactive metal such as titanium, onto ceramic surfaces prior to brazing. These coatings either isolate the ceramic from directly contacting the liquid filler metals, or they react with the ceramic during the brazing process. The coatings typically are made by the sputter coating process or the electron beam vapor coating process. Coating thicknesses of a few microns or less appear to be satisfactory

---

4. Fox, C. W. and Slaughter, G. M. "Brazing of ceramics." *Welding Journal* 43(7): 591-7, 1964.

Table 8.3
Commercial Active Braze Alloys

Alloy Composition, wt. %	Trade Name	Manufacturer or Supplier
Ag-19.5Cu-5In-3Ti	CB1	Degussa
Ag-4Ti	CB2	Degussa
Ag-26.5Cu-3Ti	CB4	Degussa
Ag-34.5Cu-1.5Ti	CB5	Degussa
Ag-1In-1Ti	CB6	Degussa
Sn-10Ag-4Ti	CS1	Degussa
Pb-4In-4Ti	CS2	Degussa
Ag-24Cu-14.5In-Ti	Lucanex 616	Lucas-Milhaupt
Ag-28Cu-0.5Ni-Ti	Lucanex 715	Lucas-Milhaupt
Ag-42Cu-2Ni-Ti	Lucanex 559	Lucas-Milhaupt
Ag-27Cu-4.5Ti	Ticusil	WESGO
Ag-27.5Cu-2.0 Ti	Cusil-ABA	WESGO
Ag-23.5Cu-14.5In-1.25Ti	Incusil-ABA	WESGO
Ti-33Ni	TiNi	WESGO
Ti-15Cu-15Ni	TiCuNi	WESGO
Au-3Ni-0.6Ti	Gold-ABA	WESGO

---

5. Moorhead, A. J., Morgan, C.S., Woodhouse, J. J., and Reed, R. W. "Brazing of sensors for high-temperature steam instrumentation systems." *Welding Journal* 60(4): 17-28, 1981.
6. Canonico, D. A., Cole, N. C., and Slaughter, G. M. "Direct brazing of ceramics, graphite and refractory metals." *Welding Journal* 56(8): 31-8, 1977.
7. Moorhead, A. J. and Keating, H. "Direct brazing of ceramics for advanced heavy-duty diesels." *Welding Journal* 65(10): 17-31, 1986.
8. Moorhead, A. J. "Direct brazing of alumina ceramics." *Advanced Ceramic Materials* 2(4): 159-66, 1987.
9. Naka, M., Tanaka, T., Okamoto, I., and Arata, Y. "Non-oxide ceramic joint made with amorphous Cu50Ti50 and Ni24.5Ti75.5 filler metals." *Transactions of the Japanese Welding Research Institute*, 12: 337-40, 1983.
10. Early work on vapor-deposited coatings was by:
  Weiss, S. and Adams, C. M., Jr. "The promotion of wetting and brazing." *Welding Journal* 46(2): 49s-57s, 1967.
  Brush, E. F., Jr. and Adams, C. M., Jr. "Vapor-coated surfaces for brazing ceramics." *Welding Journal* 47(3):106s-14s, 1968.

for promoting wetting. The following are some examples of the use of this technique:

(**1**) Sputter-deposited coatings of titanium were used to enhance the wetting of partially stabilized zirconia and alumina by Ag-30Cu-10Sn filler metal (BAg-18).[11] For brazing partially stabilized zirconia (PSZ) to cast iron or titanium, joint shear strengths in the range of 30 ksi (200 MPa) were obtained.

(**2**) Vapor-deposited coatings of hafnium, tantalum, titanium, and zirconium have been used to promote the wetting of SiC and Si_3N_4 by a wide variety of filler metals including Ag-28Cu (BAg-8), Au-18Ni (BAu-4), and Au-25Ni-25Pd (BVAu-7).[12] The joint region of a shaft consisting of Si_3N_4 (PY6) brazed to Incoloy 909 is shown in Figure 8.3.[13] The Si_3N_4 was vapor coated with titanium prior to brazing.

11. Hammond, J. P., David, S. A., and Santella, M. L. "Brazing ceramic oxides to metals at low temperatures." *Welding Journal* 67(10): 227s-32s, 1988.
12. Contributors to this work have been:
Santella, M. L. "Brazing of titanium-vapor-coated silicon nitride." *Advanced Ceramic Materials* 3(5): 457-462, 1988.
Kang, S., Dunn, E. M., Selverian, J. H., and Kim, H. J. "Issues in ceramic-to-metal joining: An investigation of brazing a silicon nitride-based ceramic to a low-expansion superalloy." *American Ceramic Society Bulletin* 68(9): 1608-17, 1989.
13. PY6 is a trademark of WESGO, Inc. Incoloy 909 is a trademark of Inco Alloys International.

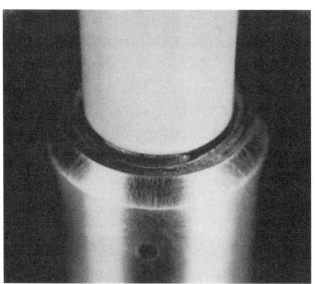

Courtesy of GTE Laboratories, Inc.

Figure 8.3—Shaft of Si_3N_4 (PY6) Brazed to Incoloy 909 with Au-5Pd-2Ni Filler Metal

Brazing Process Considerations

Brazing Joint Design.   Brazing a ceramic material to itself represents no special problems in joint design. When brazing ceramics to metals, however, a major concern is the accommodation of residual stresses that result from the differences in thermal expansion coefficients between the two materials. Ceramics have low coefficients of thermal expansion, generally lower than those of common metals to which they might usually be joined. The strains produced by the differences in thermal expansion coefficients can be large enough to cause spontaneous fracture of ceramic components. The average mismatch strain in a ceramic-to-metal joint can be roughly estimated by the expression:

$$e_m = \Delta\alpha\Delta T \qquad (8.1)$$

where

e_m = mismatch strain, unit length/unit length

$\Delta\alpha$ = the difference in the linear thermal expansion coefficient between the ceramic and the metal used in the joint, unit length/unit length/degree of temperature

ΔT = the difference between the temperature at which the braze filler metal solidifies and room temperature, degrees

This expression considerably oversimplifies the residual stress state of ceramic-to-metal joints, but it is instructive because it illustrates the effect of two important variables. Low strains, and therefore low residual stresses, are promoted by minimizing differences in thermal expansion coefficients and by brazing at the lowest temperature possible based on service conditions of the joint and good brazing techniques.

In reality there is a distribution of residual stresses in ceramic-to-metal joints, and the stresses will vary from compressive to tensile depending upon factors such as the geometry of the component and the materials used for the joint. For example, Figure 8.4 shows the results of subjecting a Si_3N_4-to-metal shaft joint like that shown in Figure 8.3 to a rigorous stress analysis. A complex distribution of varying residual stress is produced in the brazed component, as may be seen in this contour plot of maximum principal stress (in Pa) at 68 °F (20 °C) in the vicinity of the joint.

In practice there are several ways to reduce residual stresses in ceramic-to-metal joints to acceptable levels. The first is by selecting a ceramic and metal with the same or similar thermal expansion coefficients. This approach is effective, but is often not practical because

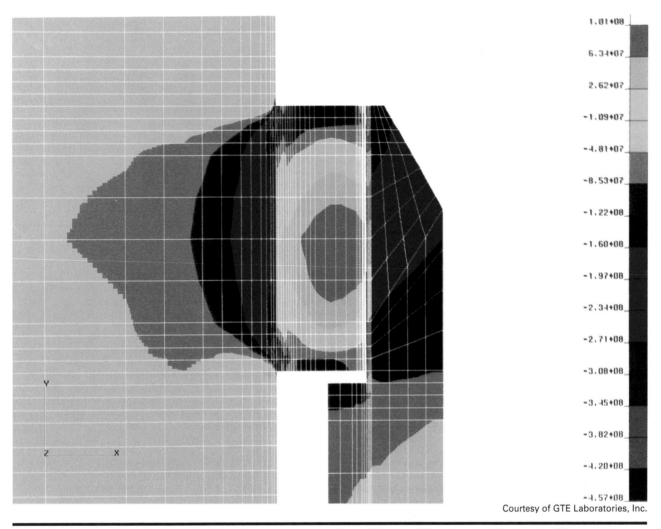

1.01+08
6.34+07
2.62+07
-1.09+07
-4.81+07
-8.53+07
-1.22+08
-1.60+08
-1.97+08
-2.34+08
-2.71+08
-3.08+08
-3.45+08
-3.82+08
-4.20+08
-4.57+08

Courtesy of GTE Laboratories, Inc.

Figure 8.4—Results of a Residual Stress Analysis for a Si₃N₄-to-Metal Joint Such as in Figure 8.3

the components of a joint are usually selected for other reasons like strength, corrosion resistance, or oxidation resistance.

Another method for reducing residual stresses is by using compliant interlayers in making the joints. In this method, for example, a thin piece of a highly ductile material, or one with a thermal expansion coefficient near that of the ceramic, is inserted between the ceramic and metal parts making the desired joint. The interlayer material accommodates the residual stresses by deforming or by isolating the ceramic from the most intense thermal expansion mismatch stresses. This approach is illustrated in Figure 8.5, which shows the use of a titanium interlayer for reducing residual stress in a joint of

ZrO_2 brazed to cast iron. The thermal expansion coefficients of ZrO_2 and titanium are nearly the same, but titanium is better able to withstand the strain, due to thermal expansion mismatch, without failing.

Graded interlayers are yet another method for reducing residual stresses. In this approach, a composite material with varying composition is placed between the ceramic and metal parts. The thermal expansion characteristics of the composite vary with its composition, and within wide limits the mismatch strains in ceramic-to-metal joints can be reasonably controlled. Joint design also can be used to minimize residual stress, and some recommendations for several types of axisymmetric joints are given in Figure 8.6.

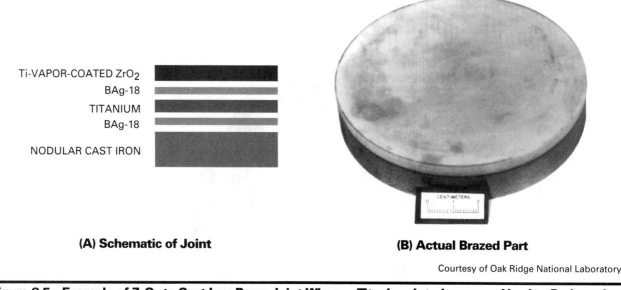

Ti-VAPOR-COATED ZrO$_2$
BAg-18
TITANIUM
BAg-18
NODULAR CAST IRON

(A) Schematic of Joint

(B) Actual Brazed Part

Courtesy of Oak Ridge National Laboratory

Figure 8.5—Example of ZrO$_2$-to-Cast Iron Braze Joint Where a Titanium Interlayer was Used to Reduce the Effects of Residual Stresses on the ZrO$_2$

Brazing Processes and Equipment. Many ceramics, particularly oxides like Al$_2$O$_3$ and ZrO$_2$, have low thermal conductivities compared to metals. As a result, ceramics take longer to heat to a set temperature than do most metals or alloys, and they are difficult to heat evenly by localized heating techniques. Also, most ceramics are electrical insulators, especially at low temperatures [less than 1800 °F (1000 °C)], and therefore cannot be directly heated by induction. Induction heating of ceramics can be done by using a metal or graphite susceptor, but this approach is difficult to control and not recommended. Because of their unique electrical and thermal properties, ceramics are best suited to furnace brazing. For most applications, vacuum or controlled-atmosphere brazing in inert gas are the processes most often used for brazing ceramics.

Since ceramics do not heat as rapidly as metals, care should be taken to control heating rates during brazing so that both materials are maintained at the same temperature. Once melted, braze filler metals will be drawn to regions where the temperature is highest. If a metal component of a ceramic-to-metal joint reaches the brazing temperature faster than the ceramic component, then the filler metal may be drawn away from the joint, and a defective joint or no joint at all may be produced. One technique for minimizing this problem is to equilibrate the temperature of assemblies just below the solidus temperature of the filler metal, followed by slow controlled heating to the final brazing temperature. An example of a controlled braze cycle for use with a Ag-35Cu-2Ti filler metal is shown in Figure 8.7.

The rapid cooling of ceramic-to-metal joints also should be avoided. If the cooling rate is too fast, the metal component will cool more rapidly than the ceramic, and any residual stress problem will be magnified. In addition, slow controlled cooling will permit some plastic deformation of ductile braze filler metals, under the stresses due to thermal expansion coefficient mismatches, and will thereby reduce the residual stress level retained in the joint at room temperature.

Surface Cleaning for Brazing. As for most brazing operations, cleanliness is an essential step in obtaining acceptable ceramic and ceramic-to-metal joints. Ceramic surfaces may contain the same kind of contaminants normally found on metal surfaces, such as traces of oil or grease, fingerprints, and metallic or abrasive particles. During either the Mo-Mn process or the active braze alloy process, most oily substances will be pyrolized to carbonaceous residues, and metallic particles may be oxidized. These residues should be removed, as they will likely result in inconsistent sintering or brazing. Surface contaminants also can adversely affect the adhesion of vapor-deposited coatings as well as the brazing characteristics of vapor-coated ceramics.

Ultrasonic cleaning or scrubbing with a mild detergent is effective for removal of loose particles and some organic oils. Metallic contaminants can be removed by immersion in dilute acids followed by rinsing in a neutralizing solution. Other possible treatments include rinsing in acetone and alcohol followed by air drying at

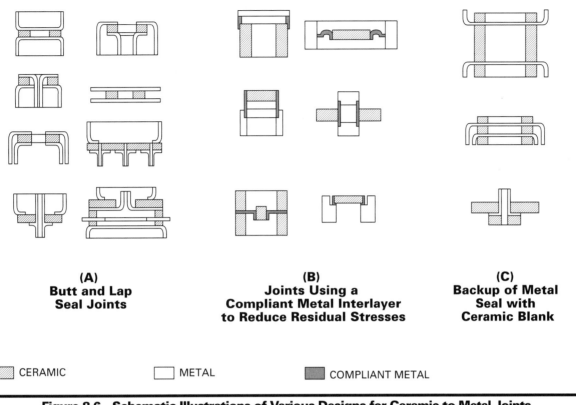

(A)
Butt and Lap
Seal Joints

(B)
Joints Using a
Compliant Metal Interlayer
to Reduce Residual Stresses

(C)
Backup of Metal
Seal with
Ceramic Blank

CERAMIC   METAL   COMPLIANT METAL

Figure 8.6—Schematic Illustrations of Various Designs for Ceramic-to-Metal Joints

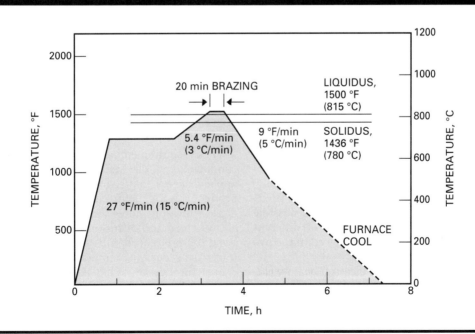

Figure 8.7—Example of a Temperature Cycle with Equilibration Below the Solidus of the Filler Metal Before
Final Brazing of the Ceramic-to-Metal Joint

212 to 400 °F (100 to 200 °C), or heating in air to the range of 1500 to 1800 °F (800 to 1000 °C).

For brazing processes other than the Mo-Mn process, it may be desirable to polish previously ground ceramic surfaces, or to heat treat them at a very high temperature (resintering), prior to the brazing operation. The reason for this is that the grinding operations often used to shape ceramic components typically produce a high degree of near-surface microcracking. This surface damage can increase the probability that residual stresses will produce catastrophic cracks during the cooling of ceramic-to-metal joints. It also can reduce the load carrying capability of both ceramic and ceramic-to-metal joints. To obtain the highest reliability in a ceramic-containing joint, surface damage on the ceramic surfaces should be removed to whatever extent possible.

OTHER CERAMIC JOINING PROCESSES

Fusion Welding of Ceramics

FUSION WELDING PROCESSES like arc welding, electron beam welding, or laser welding are not generally applied to the joining of ceramics. The reasons involve material characteristics, as well as practicality. Only a few published reports exist on this subject.

A number of mechanical and physical properties make fusion welding of ceramics difficult. Ceramics are brittle materials, in many instances exhibiting limited or no ductility, even at very high temperatures. Many studies on the welding of metals have shown that having ductility in both the base material and fusion zone are important characteristics for producing welds that are free of cracks. With limited ductilities, ceramics have little resistance to cracking as a result of thermal stresses produced by fusion welding operations.

Melting is an inherent requirement for fusion welding processes. However, many important structural ceramics like SiC and Si_3N_4 sublime rather than melt. Others, like Al_2O_3, have relatively high vapor pressures at their melting temperatures, which contribute to excessive fusion-zone porosity.

Another important reason for the dearth of information on the fusion welding of ceramics is that several other techniques exist for more easily and economically accomplishing the desired result. These and other problems notwithstanding, welding of ceramics is possible and has been demonstrated on a number of different materials. A few of these studies are reviewed in the following paragraphs.

Rice has described a study of the electron beam welding of Al_2O_3.[14] Welding of coarse-grained Al_2O_3 [grain size greater than 2400 μin.(60 μm)] resulted in extensive cracking, which was attributed to thermal expansion anisotropy of the Al_2O_3 grains. However, welds made in fine-grained Al_2O_3 [grain size = 40 to 400 μin. (1 to 10 μm)] apparently were free of cracks, and some of these joints had strengths in excess of 29 ksi (200 MPa). To achieve this level of success, the Al_2O_3 was preheated to a high, but unreported, temperature by means of a supplemental heating furnace.

Rice further described the arc welding of tantalum carbide and zirconium diboride to themselves and of zirconium diboride to tantalum metal. While details of the welding procedures were not reported, these materials apparently were preheated to high temperatures prior to welding and were slowly cooled afterwards. The quality of the welds was not discussed, but joint strengths in the range of 14.5 to 58 ksi (100 to 400 MPa) were reported.

Laser welding of Al_2O_3 also is possible, as has been shown by Maruo and coworkers.[15] They made autogenous welds in aluminas of a wide range of purity (48 to 99.5%), and in another case used an Al_2O_3 rod as a joint filler material. Use of the Al_2O_3 filler rod was found to improve joint strengths, which ranged up to 12 ksi (82 MPa). Cracking of the welds was a major problem; however, it was overcome by a combination of preheating to temperatures above 1830 °F (1000 °C) and closely controlling welding speeds.

Diffusion Welding of Ceramics

DIFFUSION WELDING IS referred to also as *solid-state welding* or *bonding* because the aim of the process is to join parts without melting any of their component materials. Two key requirements of the diffusion welding process are the following:

(1) Joint surfaces must be in the most intimate contact possible.
(2) Diffusion between the material or materials being welded must be sufficient to produce a weld in a reasonable time.

These requirements dictate that diffusion welding be conducted at elevated temperatures with pressure applied across joints. Ideally, these process conditions produce plastic deformation locally at the joint surfaces, which promotes surface contact and subsequently allows creep and diffusion to seal voids and produce a bond. A pre-placed filler metal also may be used to lower the processing temperature, pressure, or time.

Structural ceramics like Al_2O_3, SiC, Si_3N_4, and ZrO_2 do not deform readily except at very high temperatures. Compared to metals, they also are difficult to densify as

14. Rice, R. W. "Joining of ceramics." *Advances in Joining Technology*, Eds., Burke, J. J., Gorum, A. E., and Tarpinian, A., 69-111. Chestnut Hill, Mass.: Brook Hill Publishing Co., 1976.

15. Maruo, H., Miyamoto, I., and Arata, Y. "CO_2 laser welding of ceramics." *Proceedings of International Laser Processing Conference*, Anaheim, Calif., 1981.

pure materials because of the intrinsically low diffusivities of either or both of their constituent elements. Because of this, they are difficult to diffusion weld, especially without the aid of filler materials.

An example of this difficulty is provided by the work of Elssner and coworkers, who found that coarse-grained alumina [grain size about 700 μin. (18 μm)] could not be satisfactorily diffusion welded to itself because its resistance to creep prevented the formation of intimate contact at joint surfaces.[16] On the other hand, diffusion welding of the coarse-grained alumina to a fine-grained alumina [grain size about 40 μin. (1 μm)] was achieved when the fine-grained material deformed enough to produce good contact at the joint surfaces. Temperatures near 3100 °F (1700 °C) were required for producing joints in this case.

In another study, diffusion welding was used for joining dense SiC parts.[17] The technique consisted of grinding and polishing the SiC surfaces with diamond pastes of increasing fineness to a mirror-like finish, cleaning debris from the surfaces, and drying and friction fitting the mating surfaces by hand. This method was successful when applied to disks of about 0.8 in. (2 cm) in diameter. However, when billets with 0.4 in.^{2} (250 mm^2) surfaces were used, they did not bond strongly after firing. The failure to produce joints of consistently high strength was attributed to insufficient flatness and smoothness of the polished joint surfaces and the limited deformation and diffusivity of SiC, even at very high temperatures.

Diffusion welding of ceramics to metal components does not appear to be commonly done. Presumably this is related to the difficulties of maintaining the shape of metal parts that are subject to the high temperatures and pressures required by the process. Temperatures are in the range of 0.5 to 0.98 T_m, where T_m is the melting temperature of the metal in kelvin and pressures go up to the 14.5 ksi (100 MPa) range. The use of metal foils as filler metals for diffusion welding of ceramics has been more widely practiced, however. In this case, deformation of the metal foil has the advantage of improving its contact with the surfaces of the ceramic parts being joined.

Klomp has bonded tubes of polycrystalline and single-crystal alumina using foils of aluminum, copper, iron, nickel, lead, and platinum.[18] Foil thicknesses were 3940 μin. (100 μm), and the alumina tubes were 0.4 in.

O.D. by 0.24 in. I.D. (10 mm O.D. by 6 mm I.D.). The bonding was done in a hydrogen atmosphere at temperatures of about 0.9 T_m for each metal foil. The bonding times ranged up to 20 minutes. Vacuum-tight joints that could withstand thermal cycling from room temperature to the bonding temperature were produced, and bond strength had a linear relation to the melting point of the metal, as is shown in Figure 8.8. Bonding with foils of aluminum or lead required the highest pressures, and this was attributed to the stability of their surface oxides at the bonding temperature. These surface oxides were destroyed by the deformation accompanying the higher bonding pressures.

Alumina also has been diffusion welded to niobium and gold.[19,20] In the study by Bailey and Black, rods [0.37 in. (9.5 mm) in diameter] and tubes [0.4 in. O.D. by 0.24 in. I.D. (10 mm O.D. by 6 mm I.D.)] were diffusion welded end-to-end with gold foils of thickness 1770 to 2560 μin. (45 to 65 μm). The Al_2O_3 surfaces were ground flat and parallel using optical finishing techniques. The bonding was done in air at temperatures ranging from 1620 to 1980 °F (880 to 1080 °C) [the T_m of gold is 1945 °F (1063 °C)], at times between 0.5 and 100 h, and pressures up to 2.6 ksi (18 MPa).

For these materials, it was found that room-temperature bond strength increased with bonding temperature up to T_m and with bonding time up to 100 h, as illustrated in Figures 8.9 and 8.10. Optimum bond strength could be obtained with pressures in the range of 150 psi (1 MPa) as shown in Figure 8.11. Bonding temperatures above the T_m of gold resulted in very weak joints. Coating the Al_2O_3 surfaces with gold by evaporation prior to diffusion welding with gold foil also was found to increase bond strength. This was attributed to an improvement in the surface contact area produced by the coating.

There are few published reports describing the diffusion welding of Si_3N_4 and SiC with metal fillers. However, the information available indicates that many pure metals (e.g., Hf, Mo, Nb, Ni, Ta, and Zr) and alloys (e.g., iron- and nickel-based superalloys) react with these ceramics to form interface layers with complex microstructures and undesirable mechanical properties.

Microwave Joining of Ceramics

AS THE NAME implies, microwave joining relies on the use of microwaves to heat ceramic materials to temperatures high enough to accomplish the joining of parts. Microwaves can be used to heat materials rapidly and efficiently, and this technique appears to be well

16. Elssner, G., Diem, W., and Wallace, J. S. "Microstructure and mechanical properties of metal-to-ceramic and ceramic-to-ceramic joints," in *Surfaces and Interfaces in Ceramic and Ceramic-Metal Systems*, Eds., Pask, J. and Evans, A., 629-39. New York: Plenum Publishing Co., 1981.

17. Bates, C. H., Foley, M. R., Rossi, G. A., Sundberg, G. J., and Wu, F. J. "Joining of non-oxide ceramics for high-temperature applications." *American Ceramic Society Bulletin* 69(3): 350-6, 1990.

18. Klomp, J. T. "Bonding of metals to ceramics and glasses." *American Ceramic Society Bulletin* 51(9): 683-8, 1972.

19. Morozumi, S., Kikuchi, M., and Nishino, T. "Bonding mechanism between alumina and niobium." *Journal of Materials Science* 16: 2137-44, 1981.

20. Bailey, F. P. and Black, K. J. T. "Gold-to-alumina solid state reaction bonding." *Journal of Materials Science,* 13: 1045-52, 1978.

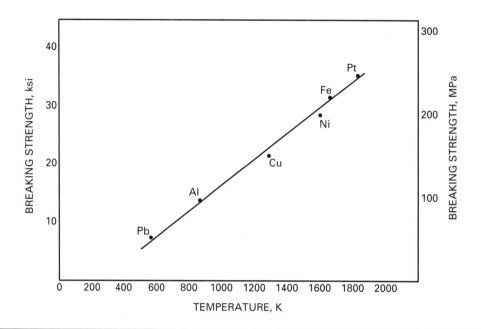

Figure 8.8—Breaking Strength of Ceramic-to-Metal Bonds as a Function of the Melting Point of the Metal Foil Used as a Filler Metal

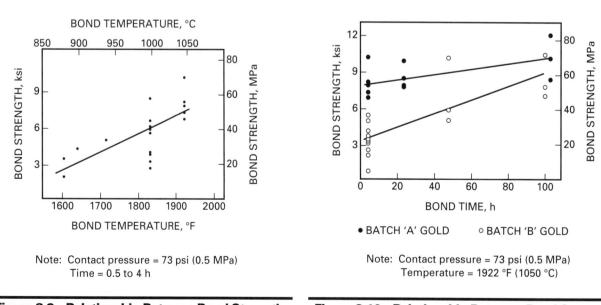

Note: Contact pressure = 73 psi (0.5 MPa)
Time = 0.5 to 4 h

Note: Contact pressure = 73 psi (0.5 MPa)
Temperature = 1922 °F (1050 °C)

Figure 8.9—Relationship Between Bond Strength and Bonding Temperature for Alumina Diffusion Welded with Gold Foil

Figure 8.10—Relationship Between Bond Strength and Bonding Time for Alumina Diffusion Welded with Gold Foil

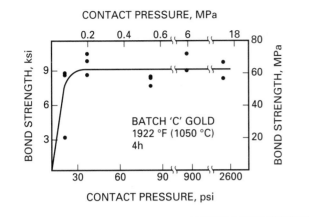

Figure 8.11—Relationship Between Bond Strength and Bonding Contact Pressure for Alumina Diffusion Welded with Gold Foil

suited to various aspects of processing ceramics, including joining. Although interest in microwave processing has grown dramatically since the mid-1980s, this technology is still largely in a developmental stage.

Microwave heating is fundamentally different from conventional heating techniques in that heat is generated within a material rather than from an external source. A consequence is that microwave-heated materials heat from the inside out, rather than the reverse as with conventionally heated materials. Advantages of this internal, volumetric heating is that both small and large parts can be heated rapidly and uniformly, a characteristic which is especially attractive for ceramic materials.

For ceramic joining applications, microwave processing offers the following possibilities:

(1)   Rapid heating with modest power levels compared to conventional techniques
(2)   Localized heating of joint regions through sample configuration and joint design
(3)   Selective heating of materials
(4)   On-line process control

Background. Microwaves are electromagnetic waves having frequencies in the range of roughly one to several hundred gigahertz, with corresponding wavelengths in the range of about 39 to 0.04 in. (1000 to 1 mm). Depending on the type of material with which they interact, microwaves can be transmitted, absorbed, or reflected in accordance with the laws of optics. Metals reflect microwaves, but many ceramics either transmit or absorb them and thus can be heated by microwaves.

Most ceramics are electrical insulators, and it is their dielectric (or nonconducting) properties that are important for describing how they interact with microwaves. Discussions of the dielectric properties used to describe the interaction of microwaves with ceramics often begin with the complex dielectric constant (or permittivity):

$$\varepsilon^* = \varepsilon_0 \, (\varepsilon'_r - j\varepsilon''_{eff}) \tag{8.2}$$

where

$$
\begin{aligned}
\varepsilon^* &= \text{complex dielectric constant (F/m)} \\
\varepsilon_o &= \text{permittivity of free space} \\
\varepsilon'_r &= \text{relative dielectric constant} \\
\varepsilon''_{eff} &= \text{effective relative electric loss factor} \\
j &= (\text{-}1)^{1/2}
\end{aligned}
$$

When microwaves pass through a ceramic, they generate electric fields within the ceramic that are attenuated by a number of mechanisms. The loss mechanisms causing attenuation of the electric fields are collectively accounted for in the loss factor, ε''_{eff}. The electric field losses result in volumetric heating, and are commonly described in terms of the loss tangent:

$$\tan\delta = \frac{\varepsilon''_{eff}}{\varepsilon'_r} = \frac{\sigma}{2\pi f \varepsilon_0 \varepsilon'_r} \tag{8.3}$$

where

σ = total conductivity of the material being heated
f = the microwave frequency
ε = one of three forms of this variable defined in Equation 16.2

The definitions and relationships just described form the basis for determining the power, P, absorbed during microwave heating:

$$P = \sigma|E|^2 = 2\pi f \varepsilon_0 \varepsilon'_r \tan\delta |E|^2 \tag{8.4}$$

where
E = magnitude of the internal electric field

This expression for P shows that the power absorbed varies linearly with the microwave frequency, the relative dielectric constant, and the loss tangent, and varies with the square of the electric field. The assumptions implicit in the power equation are rarely completely fulfilled in practice, but this expression, nevertheless, provides a useful approximation of the power absorbed and its dependence on important material characteristics.

Joining Applications. The application of microwave processing techniques to joining problems appears to be a viable and versatile approach, and several published reports on microwave joining of ceramics are available.

Meek and Blake used microwave processing to produce joints of Al_2O_3 and of Al_2O_3 to Kovar.[21] The Al_2O_3-to-Kovar joints were made using a glass as a bonding material. The glass was powdered and mixed with a material that promoted coupling with the microwaves. This approach allowed the bonding material to be heated selectively and rapidly. Reportedly, it also resulted in a bond different from that obtained with conventional processing. For this application, 16 times less energy was required for making the joint compared to conventional heating, while the time required was 14 times less. A similar approach was used for making Al_2O_3 joints. Both types of joints were reported to be hermetically sealed.

Butt joints of Al_2O_3 and mullite (a mixture of Al_2O_3 and SiO_2) were made by Palaith and coworkers.[22] The joining was accomplished without special surface preparation of the mating surfaces and without a bonding material. A load corresponding to 1300 psi (9 MPa) was applied to the joints during heating to a joining temperature in the range of 2280 to 2460 °F (1250 to 1350 °C). Joints having strengths in excess of that of the base material were produced in 4 minutes using a power level of 150 W. A novel feature of the experimental set-up was the use of acoustic waves to monitor the process and to determine when bonding was complete.

Fukushima and coworkers also have reported on the application of microwave processing to the joining of ceramics.[23] In their study, butt joints were made of Al_2O_3 of 92 to 99% purity and of Si_3N_4. Joints of 92 to 96% pure Al_2O_3 were obtained without the use of separate bonding materials, and the strengths of these joints were equal to the respective base material strengths. On the other hand, joining of 99% pure Al_2O_3 was more difficult, and the best joints required the use of a bonding material composed of 0.024 in. (0.6 mm) sheets of 92 to 96% Al_2O_3. It was concluded from these results that joining was accomplished by the preferential melting of sintering aids, which were located along the Al_2O_3 grain boundaries and contained most of the impurities. Consequently, the joints were formed without melting the Al_2O_3.

Similar behavior was observed for Si_3N_4, i.e., the ease of joining increased with the impurity content of the material. For higher purity Si_3N_4, acceptable joints required the use of a lower purity Si_3N_4 bonding material.

CERAMIC PROCESSING TECHNIQUES

FOR THE PURPOSE of this discussion, joining by ceramic processing techniques is taken to mean the joining of ceramics by techniques normally used for producing monolithic ceramic bodies. This definition is somewhat arbitrary, but is meant to include processing techniques like the use of slip interlayers as filler materials and the hot pressing of green, or unsintered, bodies in order to simultaneously densify and join ceramic parts. The application of these techniques can be complicated relative to more conventional joining techniques such as brazing, and can require considerable expertise. Ceramic processing techniques, however, can produce excellent joints in parts of complex shapes.

Goodyear and Ezis developed two techniques for joining Si_3N_4 that rely heavily on ceramic processing technologies.[24] In the first, slip-cast bonding was used to join silicon components into desired shapes. Afterward, the joined silicon bodies were nitrided to produce homogenous reaction-bonded Si_3N_4 components. Key elements of the technique began with producing the green silicon parts that were later joined. Next, the silicon parts were presintered in argon at temperatures up to 2000 °F (1100 °C) for times up to 8 h (depending on the green density of the silicon parts). These components were then joined by introducing a low-viscosity, chemically-suspended silicon slurry into a joint gap defined by the surfaces of the presintered silicon parts. The slurry solidified by extraction of the slurry medium through the porous silicon parts by capillary action, producing a coherent mass of silicon in the joint gap. Finally, the entire joined silicon assembly was converted to Si_3N_4 by reaction with nitrogen gas at temperatures up to 2660 °F (1460 °C).

Successful application of the slip-cast bonding technique requires an intimate knowledge of slip casting techniques including powder characterization, powder suspension mechanisms, and slurry viscosity control. For bonding, the configuration of the joint gap and the

21. Patents awarded for this work are:
   Meek, T. T. and Blake, R. D. "Ceramic-glass-metal seal by microwave heating." U.S. Patent No. 4,529,856; 1985.
   Meek, T. T. and Blake, R. D. "Ceramic-glass-ceramic seal by microwave heating." U.S. Patent No. 4,529,857; 1985.
   Meek, T. T. and Blake, R. D. "Method for producing ceramic-glass-ceramic seals by microwave heating." U.S. Patent No. 4,606,748; 1986.
22. Palaith, D. and Silberglitt, R. "Microwave joining of ceramics." *American Ceramic Society Bulletin* 68(9): 1601-6, 1989.
23. Fukushima, H., Yamanaka, T., and Matsui, M. "Microwave heating of ceramics and its application to joining." *Proceedings of Microwave Processing of Materials, Symposium*, Vol. 124, Eds. Sutton, W. H., Brooks, M. H., and Chabinsky, I. J., 267-72. Pittsburgh, Pa.: Materials Research Society, 1988.

24. Goodyear, M. U. and Ezis, A. "Joining of turbine engine ceramics." *Advances in Joining Technologies*, Eds., Burke, J. J., Gorum, A.E., and Tarpinian, A., 113-53. Chestnut Hill, Mass.: Brook Hill Publishing, 1976.

method of introduction of the bond slurry into it are especially important considerations.

The second technique studied by Goodyear and Ezis involved hot-press bonding of Si_3N_4 parts that were already densified by reaction sintering or hot pressing. In the initial experiments, Si_3N_4 disks were hot-press bonded with and without a filler material under the following conditions: temperature of 2910 and 3090 °F (1600 and 1750 °C), times of 1 and 3 h, and pressures of 1015 and 2465 psi (7 and 17 MPa). When used, the bonding filler material consisted of a slip of Si_3N_4 containing 2 percent by weight of MgO. Subsequent testing indicated that the highest-strength bonds were obtained at the higher conditions of temperature, time, and pressure and when the slip filler material was omitted. Actual turbine rotors were fabricated by hot-press bonding reaction-sintered blade rings to hot-pressed hubs at 3180 °F (1750 °C) under a pressure of 2465 psi (17 MPa) for 2 h. This process was later improved by simultaneously densifying the hot-pressed hub and bonding it to the reaction-sintered blade ring.

Joining Si_3N_4 and SiC with ceramic processing techniques also was the subject of a study by Bates and colleagues.[25] In this case, the parts to be joined were green, or unsintered, billets obtained by cold isostatic pressing of Si_3N_4 powder. Of three joining conditions evaluated, one had no filler material and two used a filler material consisting of Si_3N_4 powder containing 4 percent by weight of Y_2O_3, suspended in either a slip-casting or injection-molding media.

The flatness of the surfaces to be joined was limited by the capability of machining techniques for green billets. This problem was overcome by using a filler material. Both the slip-casting and the injection-molding slurries provided a plastic filler-material layer which helped fill voids and irregularities between the mating surfaces of the green Si_3N_4 billets. Final densification and bonding of the billets was accomplished by a glass-encapsulation hot isostatic pressing (HIP) technique. Theoretically dense bodies were obtained for joints made without a filler material and for joints made with the slip-casting filler material. After processing, the joint regions could not be discerned visually and were difficult to detect optically. Joints made by this technique had excellent strength at room temperature [128 ksi (880 MPa)] and at 2372 °F (1300 °C) [greater than 87 ksi (600 MPa)].

Bates and colleagues used a similar technique for joining SiC. A slurry containing SiC powder and sintering aids in a suitable media was applied between two flat surfaces of green SiC billets. The billets were then hand-pressed together and then cold isostatic pressed. Afterward, the joined green billets were densified by pressureless sintering. This processing scheme resulted in a bond-zone thickness of about 0.0024 in. (60-μm), which was clearly visible after sintering. The billets had densities near 97 percent of theoretical. Subsequent testing showed that joint specimens had average strength of 50.9 ksi (351 MPa) at room temperature and 49 ksi (338 MPa) at 2790 °F (1530 °C). The average strength of unjoined control specimens was 57.6 ksi (397 MPa) and 50.3 ksi (347 MPa), respectively, at the same two test temperatures.

The hot isostatic pressing technique also has been used to bond Si_3N_4 to steels.[26] In one case, Si_3N_4 was bonded to a carbon steel with molybdenum powder by the HIP process. The joint assembly was vacuum sealed in a capsule and HIP bonded using conditions of 2000 °F (1100 °C), 17.4 ksi (120 MPa), and 1 h. Metallographic analysis of the joints indicated that no reaction occurred between the Mo and the Si_3N_4, and bonding was attributed not to diffusion but to mechanical interlocking at the Mo/Si_3N_4 interface. Similar results were reported for joining Si_3N_4 to a stainless steel with titanium powder using the HIP process. No strength data were reported for either type of joints.

GLASS-SEALING TECHNIQUES

SMALL QUANTITIES OF sintering aids are used in the densification of most structural ceramic materials. Typically, these sintering aids react with the parent ceramic but still remain in the final microstructure as glassy grain-boundary phases.

It has been recognized for some time that glasses, particularly those used to assist densification during sintering, also have the potential to be used for joining ceramic parts. One advantage of using these glass materials as fillers for joining is that chemical compatibility with the parent ceramic is generally assured. Other advantages are that the viscosity, flow properties, and melting characteristics of glasses can be controlled over wide ranges, and that adherence of the glasses to the ceramics is usually quite good. Another desirable feature of glasses is that many compositions can be crystallized to improve their mechanical and corrosion properties. When glasses are crystallized under controlled conditions, the resulting materials are referred to as *glass-ceramics*. Both glasses and glass-ceramics have been used as filler materials for joining ceramics.

Various mixtures in the Al_2O_3-MnO-SiO_2 system were found useful for bonding alumina to itself and to a variety of metals that include tungsten, molybdenum, nickel, platinum, and Kovar.[27] These particular

25. Bates, Foley, Rossi, Sundberg, and Wu, "Joining of non-oxide ceramics," 350-6.

26. Tanaka, T., Morimoto, H., and Homma, H. "Joining of ceramics to metals." Nippon Steel Technical Report No. 37: 31-8, April 1988.
27. Klomp, J. T. and Bolden, P. J. "Sealing pure alumina ceramics to metals." *American Ceramic Society Bulletin* 49(2): 204-11, 1970.

compositions were chosen because they resist crystallization, i.e., they remain in glassy phases after cooling, and they were known to wet Al_2O_3 and metals. Surfaces to be bonded were coated with a paste (or frit) consisting of the oxide mixtures and binder, and then heated to a temperature 120 °F (50 °C) above the melting point of the glass frit. The bonding temperatures ranged from 2200 to 2550 °F (1200 to 1400 °C), and the joints were held at the bonding temperature for 2 minutes. The atmosphere during the bonding cycle was not critical, but a reducing atmosphere was used to avoid oxidation of metal components when making ceramic-to-metal joints. The Al_2O_3 joints made by this technique had good strength, were vacuum-tight, and were resistant to thermal shock and high pressure.

In another study of bonding Al_2O_3 with glasses, crystallization of the glass bonding material was specifically desired.[28] The glass chosen for a filler material had the following composition (wt. %): $64.7SiO_2$-$18.5Al_2O_3$-$9.3MgO$-$6.5ZrO_2$-$1.0CaF_2$. This composition was expected to be compatible with Al_2O_3 ceramics because MgO and SiO_2 are commonly used as sintering aids and SiO_2, Al_2O_3, and MgO readily react together. Zirconia was added as a nucleating agent to promote crystallization of the bond zone to a fine-grained microstructure, and CaF_2 was added as a fluxing agent.

It was expected that this particular glass composition could easily be crystallized to 95 percent by volume in the temperature range 1780 to 2280 °F (970 to 1250 °C). The glass was prepared by melting the individual components together at 2900 °F (1600 °C) for 16 h, followed by quenching and then grinding to a fine powder. The surfaces of the Al_2O_3 parts joined with the powdered glass filler material were prepared by grinding with 600-grit diamond abrasive followed by washing with chromic acid. The joints were formed by coating the Al_2O_3 surfaces with the powdered glass, placing the surfaces in contact, and then heating the tiles to a joining temperature in the range of 1830 to 2550 °F (1000 to 1400 °C). Joints made at 2550 °F (1400 °C) had bond-zone thicknesses of 2000 to 4000 μin. (50 to 100 μm). The degree of crystallization of the bond layers was varied by controlled heat treatments after the joining operation. Subsequent testing showed that the fracture toughness of the bond layers increased with its degree of crystallization. Similar results also were obtained from Al_2O_3 joints made with filler materials of a zinc borosilicate glass-ceramic and of mixtures of talc (hydrated magnesium silicate) and clay (hydrated aluminum silicate).

Zirconia ceramics also have been joined with glass filler materials. In one case, a $35CaO$-$50TiO_2$-$15SiO_2$ filler material was used for bonding partially-stabilized

ZrO_2 at 2552 °F (1400 °C).[29] A mixture of the powdered filler material was made into a slurry with polyvinyl butyral and methanol. This was spread onto ZrO_2 surfaces that had been roughened with 240-grit SiC paper. The joint assembly was pressed together by hand and then placed into a hot-forging apparatus. A slight pressure [29 psi (0.2 MPa)] was applied to the joint, which was then heated to 2550 °F (1400 °C) in 3 h. When the temperature reached 2550 °F (1400 °C), a compressive pressure of 650 psi (4.5 MPa) was applied and maintained as the temperature was increased to 2590 °F (1420 °C) and held for 2 h. The sample was cooled to room temperature under the 650 psi (4.5 MPa) pressure. This technique resulted in joints with bond-layer thicknesses of about 400 μin. (10 μm).

In a following study, ZrO_2 was joined with a glass having the composition (wt. %): $14.9MgO$-$19.9Al_2O_3$-$56.1SiO_2$-$9.1TiO_2$.[30] Joints were made with filler materials consisting of the powdered glass containing 33, 50, 67, and 80% pure ZrO_2 powder. This glass was chosen because of its higher strength compared to the CaO-TiO_2-SiO_2 used in initial experiments. The ZrO_2 was added primarily to control the thermal expansion behavior of the bond layer, but it also was found to improve joint strength.

Two types of glass powders were used as filler materials. In one case, the magnesia-alumina-silica (MAS) powder was produced by crushing and sieving Corning 9606 glass. In the other, the glass composition was made from its constituent oxide powders. The method of joining was similar to that described in the preceding paragraph except that the joining temperature range was 2370 to 2490 °F (1300 to 1365 °C), and much lower pressures [20 to 80 psi (0.14 to 0.55 MPa)] were applied across the joints made with these filler materials. Joints made with filler materials containing 67 and 80% ZrO_2 powder had average bend strengths of 17.7 ksi (122 MPa) and 22.9 ksi (158 MPa), respectively, at room temperature.

Glasses used for joining Si_3N_4 have been specifically formulated to approximate the composition of the grain boundary phases that result from the reaction of sintering aids with Si_3N_4 in hot-pressed or sintered material.[31] For example, glasses with compositions of $35MgO$-$55SiO_2$-$10Al_2O_3$ and of $34SiO_2$-$20Al_2O_3$-$46Y_2O_3$ have been used for joining.

The glass compositions were prepared by melting the appropriate oxide mixtures at 3000 °F (1650 °C) in air,

28. Zdaniewski, W. A., Shah, P. M., and Kirchner, H. P. "Crystallization toughening of ceramic adhesives for joining alumina." *Advanced Ceramic Materials* 2(3A): 204- 8, 1987.

29. Swartz, S. L., Majumdar, B. S., Skidmore, A., and Mutsuddy, B. C. "Joining of zirconia ceramics with a CaO-TiO₂-SiO₂ interlayer." *Materials Letters* 7(11): 407- 10, 1989.

30. Ahmad, J. et al. "Analytical and experimental evaluation of joining ceramic oxides and ceramic oxides to metals for advanced heat engine applications." Battelle Columbus Final Report to Oak Ridge National Laboratory on subcontract 86X-SB046C. Report No. ORNL/Sub/87-SB046/1, 1992.

followed by grinding and sieving to -100 mesh (particle size ≤ 150 μm). An oxynitride glass with a composition of $32SiO_2$-$19Al_2O_3$-$43Y_2O_3$-$6N_2$ also was prepared by adding Si_3N_4 to the SiO_2-Al_2O_3-Y_2O_3 glass and melting under nitrogen at about 3090 °F (1700 °C).

The joining method proceeded by first applying a slurry of glass powder to the surface of a Si_3N_4 specimen. This was then heated to near 2960 °F (1625 °C) in

31. Contributors to this silicon nitride work include:
Loehman, R. E. "Transient liquid phase bonding of silicon nitride ceramics." *Surfaces and Interfaces in Ceramic and Ceramic-Metal Systems*, Eds., Pask, J. and Evans, A., 701-11, New York: Plenum Press, 1980.
Brittain, R. D., Johnson, S. M., Lamoreaux, R. H., and Rowcliffe, D. J. "High-temperature chemical phenomena affecting silicon nitride joints." *Journal of the American Ceramic Society* 67(8): 522-6, 1984.
Johnson, S. M. and Rowcliffe, D. J. "Mechanical properties of joined silicon nitride." *Journal of the American Ceramic Society* 68(9): 468-72, 1985.

nitrogen at two atmospheres of pressure [29 psi (0.2 MPa)] to produce a glaze on its surface. The glazed pieces were then placed above unglazed pieces and heated near 2870 to 3000 °F (1575 to 1650 °C) for 30 to 60 minutes to produce a bonded Si_3N_4 specimen.

The room-temperature strengths of joints prepared by this approach were reasonably high and independent of joining time and temperature within the specified ranges. Joint strength, however, was dependent on joint thickness, with the maximum strength being obtained for bond layers of thickness 790 to 1180 μin. (20 to 30 μm). Strengths also were measured at temperatures up to 2370 °F (1300 °C) and found to be quite low. Some joined specimens also were heat treated at 2190 to 2370 °F (1200 to 1300 °C) in nitrogen at 29 psi (0.2 MPa) to promote crystallization of the bond layers. This treatment was found to cause a decrease of room-temperature strength, even though crystallization should have resulted in stronger joints.

SAFETY AND HEALTH CONSIDERATIONS

BECAUSE STRUCTURAL CERAMICS are among the most stable compounds known, they do not represent significant safety or health concerns in the monolithic state. However, ceramics, glasses, metals, and most materials, when finely divided as powder or dust, represent a potential health hazard.

Ceramic powders may be encountered when the sintered-metal-powder process is used to prepare ceramic surfaces for brazing, or when a glass-sealing technique is used for the ceramic-bonding operation. Exposure to metal powders also may be associated with the use of the sintered-metal-powder process or may occur if powdered filler metals are used for brazing.

The term *dust* has no precise definition, but it is usually taken to mean finely powdered material produced during the breakdown of solids by machining, grinding, or crushing operations. The most likely way for potentially harmful dust to be encountered during joining operations is when ceramic or metal parts are machined or when surfaces are ground in preparation for the actual joining process.

Exposure to ceramic, glass, or metal powder or dust is most likely to occur through inhalation, ingestion, absorption through the skin, or contact with the skin or eyes. The degree of hazard for any particular powdered material will depend upon its chemical composition, particle size, and on the amount of exposure. The result of exposure may range from mild irritation to, in extreme cases, death. Consequently, all powdered materials or dusts should be handled in such a way that

their introduction into the body is minimized or prevented.

The following recommendations are made when the handling of powdered materials is necessary:

(1) A suitably ventilated and equipped room always should be used. This would include efficient dust-extraction equipment, readily cleaned working surfaces and floors, and personal washing facilities nearby. Workbenches should have tough, impermeable surfaces.

(2) Eating, smoking, and drinking in work areas should be prohibited.

(3) The generation of airborne dust should be minimized, because it is clearly preferable to prevent dust rather than to control it. For instance, wet rather than dry grinding is preferred for surface preparation and machining.

(4) When dust creation cannot be avoided, all protective measures should be taken to prevent excessive exposures, including suitable protection for eye and skin contact. Respirators should be used when a process cannot be contained within a dust-extraction unit.

(5) Personal hygiene should be maintained at all times.

Specific concerns for selected ceramic materials are described below.

Aluminum oxide (alumina, Al_2O_3) powder or dust may irritate the upper respiratory tract and can cause lung damage if inhaled in large quantities over an extended period of time. Alumina is not toxic when swallowed. It is not absorbed through the skin, but may cause mild skin irritation.

Silicon carbide (SiC) may cause irritation of the respiratory tract when inhaled, and excessive exposure may result in shortness of breath or excessive mucous production. Silicon carbide is not toxic when swallowed. It is not absorbed through the skin, but may cause irritation or abrasion to the skin and eyes.

Silicon dioxide (silica, SiO_2) is moderately toxic as an acute irritating dust, and silica-related dust disease (silicosis) has been known for centuries. Silicosis is known to be associated with crystalline silica as an occupational disease. Its effects are compounded with long-term exposure and inhalation of dust particles of less than 400 μin. (10 μm) in size. It was generally believed that silicosis took years to develop or was the result of very high doses of fine dust. Workers show signs of exposure with shortness of breath. Although it has been known that high dosages could result in death, doubt still exists within the scientific community that crystalline silica causes lung cancer in humans. Thus, the Occupational Safety and Health Administration (OSHA) has proposed exposure limits, shown in Table 8.4, based on eight-hour, time-weighted averages. Silica also may be an irritant to the skin and eye.

Silicon nitride (Si_3N_4) dust may irritate the respiratory tract and eyes. Specific toxicological information is not currently available on the effects of ingestion or skin contact. Consequently, Si_3N_4 should be treated with caution, and direct contact should be avoided.

Zirconium oxide (zirconia, ZrO_2) dust may irritate the respiratory tract. Because it generally is insoluble and poorly absorbed, toxic effects from ingestion are not expected. Zirconia may cause irritation or abrasion to the skin and eyes.

Issues relevant to the use of metals, fluxes (which generally are not used for brazing ceramics), various gases, and processing equipment are outlined in Chapter 6, "Safety and Health," *Brazing Handbook*, Fourth Edition, published by the American Welding Society. Information about hazards associated with various ceramics, glasses, metals, gases, and fluxes also should be available at the work site in Material Safety Data Sheets (MSDS).

Table 8.4
Proposed OSHA Permissible Exposure Limits for Silica*

Substance	Exposure limit, mg/m³
Quartz	0.1
Cristobalite	0.05
Tridymite	0.05
Tripoli	0.1

* Proposed limits as of 1995.

INDUSTRY NEEDS

CERAMIC JOINING HAS been the subject of a large number of studies over the last four decades, and the interest in this technology has been heightened by the recent development of advanced ceramics. Effective ceramic joining techniques can provide the flexibility needed to design ceramic-containing structures. Such joining techniques can have significant favorable impact on both the economics of building ceramic structures and on the reliability of ceramic structures.

Although techniques like brazing, originally developed for metal joining, are being successfully applied to ceramic joining, there are critical needs for the development of other joining techniques and for the refinement of existing techniques.

The technology of joining ceramics for structural applications not related to the electronics industry is largely in a developmental stage. One consequence is that no industry-wide standards exist, as this chapter is written, for ceramic joining practices. Other important areas where further developments, refinements, and standardization are needed include testing, interpretation of test results, and nondestructive examination. For these reasons, the case studies described in the following section on applications are necessarily limited.

APPLICATIONS

NUCLEAR REACTOR RF FEEDTHROUGH

ONE APPLICATION THAT illustrates some of the concerns in making brazed ceramic-to-metal joints is that of a high-voltage, vacuum feedthrough for a power-injecting antenna in an experimental nuclear fusion reactor.[32] A typical feedthrough consists of an Al_2O_3 cylinder, which provides high-voltage electrical

32. Information on this application was provided by Oak Ridge National Laboratory.

insulation, to which specially designed copper rings are brazed at each end. The copper rings provide the means for making the metal-to-metal mechanical seals necessary for connecting the feedthrough to a high-vacuum system. A drawing of the feedthrough is shown in Figure 8.12; some details of the copper ring on the right end of the Al_2O_3 cylinder in the drawing are given in Figure 8.13. A photomacrograph of one section of a completed joint is shown in Figure 8.14.

The following design requirements for the feedthrough combine to make this a difficult brazing operation:

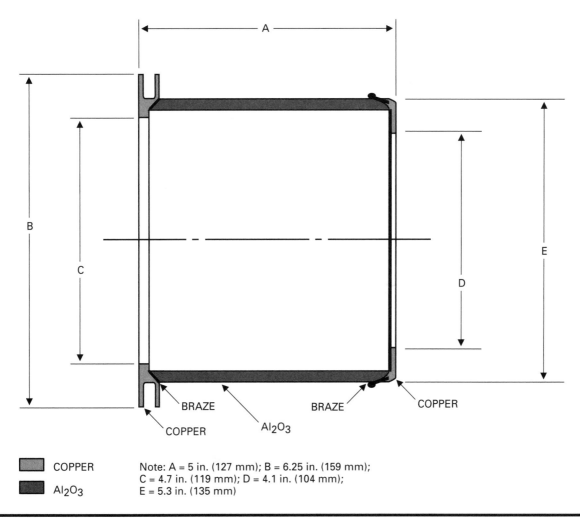

COPPER

Al_2O_3

Note: A = 5 in. (127 mm); B = 6.25 in. (159 mm);
C = 4.7 in. (119 mm); D = 4.1 in. (104 mm);
E = 5.3 in. (135 mm)

Figure 8.12—Ceramic Feedthrough Brazement of Copper to Al_2O_3

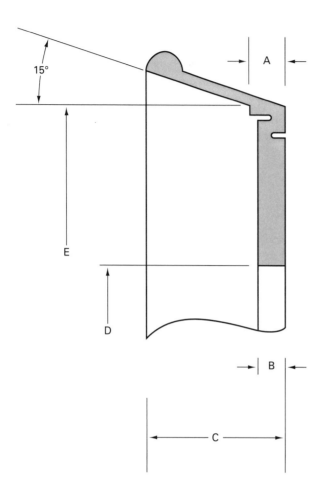

Note:
A = 0.125 in. (3.2 mm); B = 0.93 in. (23.6 mm);
C = 0.5 in. (12.7 mm); D, diameter = 2.50 in. (63.5 mm);
E, diameter = 3.62 in. (91.9 mm).

Figure 8.13—Copper Ring on the Right Side of Al$_2$O$_3$ Cylinder in Figure 8.12

(1)   leak-tight under high vacuum
(2)   high strength to permit mechanical joining into the vacuum system
(3)   electrically insulating to prevent arcing in high-voltage RF fields

Typically, the outer edges of both ends of the Al$_2$O$_3$ cylinders were beveled 15 degrees relative to their axes to help in positioning of the copper end rings and for help in controlling the distribution of residual stresses in the finished feedthroughs. The copper rings were machined to close tolerances which were calculated to provide a clearance of 0.002 in. (0.051 mm) between

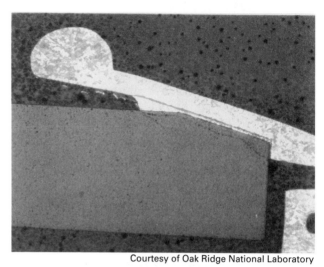

Courtesy of Oak Ridge National Laboratory

Figure 8.14—Photomacrograph of One Section of a Completed Joint (reduced 74%)

them and the Al$_2$O$_3$ cylinder for filler metal flow at the brazing temperature. Due to the high thermal expansion coefficient of copper relative to Al$_2$O$_3$ (16.5 μm/m/°C vs. 6.8 μm/m/°C, respectively), the copper rings were undersized at room temperature. Fixtures were designed to maintain alignment of the rings relative to the Al$_2$O$_3$ during heating to the brazing temperature.

As heating of the assembly progresses, the copper rings expand more than the Al$_2$O$_3$ cylinder, and they eventually slide over the bevels to position themselves in the proper location as the braze filler metal melts. On cooling, the copper rings yield and leave the ends of the Al$_2$O$_3$ cylinder, including the joints, in compression. This situation is desirable for the Al$_2$O$_3$. Figure 8.15 shows an assembly of the individual parts and fixture being loaded into a vacuum furnace.

The braze filler metal used for this application had a nominal composition of Ag-27Cu-4.5Ti, and it was chosen so that brazing could be done in a one-step process without metallization of the ceramic. The brazing thermal cycle used for the joints is shown in Table 8.5.

The slow heating rate was chosen to promote uniform heating of the entire assembly. The intermediate hold at 1292 °F (700 °C) was inserted into the heating cycle to equilibrate the temperature of the assembly in the event that its temperature was not uniform, even though a slow heating rate was used. The slow cooling rate and intermediate hold at 752 °F (400 °C) on cooling were used to promote uniform cooling of the brazed feedthrough and to promote mechanical yielding and relaxation of the copper rings. A vacuum of 5 x 10^{-5}

Courtesy of Oak Ridge National Laboratory

Figure 8.15—An Assembly of Individual Copper and Al₂O₃ Parts and Fixture Being Loaded into a Vacuum Furnace for Brazing

Table 8.5
Brazing Thermal Cycle for Joints in Nuclear Reactor RF Feedthrough

| Step | Action | Temperature Rate and Duration | |
		Fahrenheit	Celcius
1	Heating	9°/min to 1290°	5°/min to 700°
2	Holding	1290° for 30 min	700° for 30 min
3	Heating	9°/min to 1600°[a]	5°/min to 870°[a]
4	Holding	1600° for 10 min[b]	870° for 10 min[b]
5	Cooling	9°/min to 750°	5°/min to 400°
6	Holding	750° for 30 min	400° for 30 min
7	Cooling	9°/min to 390°	5°/min to 200°
8	Furnace cooling to room temperature		

a. Brazing temperature

b. To melt the filler metal and produce the brazing joint

torr or better was maintained throughout the thermal cycle. This processing produced leak-tight joints with no detectable cracking of the Al₂O₃. Feedthroughs made with this process have functioned properly in service.

CERAMIC-TO-CERAMIC JOINING

SOME OF THE issues associated with ceramic joining can be illustrated by considering the joining of Si₃N₄ with oxide glasses.[33] Desirable methods of joining ceramics require that little or no pressure be needed across joint interfaces, that the joining technique not be restricted to simple shapes, and that the processing required be reasonably compatible with common practice. Brazing fulfills these requirements, but braze filler metals generally do not have the mechanical properties at elevated temperatures, or the resistance to environmental effects, that ceramics do. In some cases of ceramic joining, the use of nonmetallic filler materials may overcome these difficulties.

---

33. Information on this application was provided by SRI International.

The method of using oxide glasses to join Si₃N₄ is outlined here as a two-step process consisting of first producing a glaze of the joining glass on the Si₃N₄ joint surfaces, and then making the actual braze joint. This approach provides for better control over joint thickness and, consequently, better control of joint strength.

The nominal composition (wt. %) of one glass used for joining Si₃N₄ is 35MgO-55SiO₂-10Al₂O₃. One technique for making a glass with this composition consists of heating a mixture of the oxide powders in platinum crucibles to 2912 to 3002 °F (1600 to 1650 °C) in air for 4 to 16 h. The molten glass is then made into a relatively coarse powder, or fritted, by pouring it directly into water. A finer, more uniform powder is then made by ball-milling the frit with Al₂O₃ balls in a plastic container. Afterwards, the powder is sieved to the desired size for joining. Slurries of the glass are made by mixing the powder in a suitable vehicle such as isopropanol or methanol.

Prior to joining, the joint surfaces of the Si₃N₄ pieces are ground, then cleaned in trichlorethylene, acetone, and methanol. A slurry of the powdered [-100 mesh (≤ 149 μm)] glass is then applied to one of the joint surfaces. Glazing is done by heating the glass-coated Si₃N₄ pieces to 2696 °F (1480 °C) for 15 min followed by 10 min at 2948 °F (1620 °C) under nitrogen gas at a pressure of 2 atm (~200 kPa). This heat treatment produces a relatively uniform layer of glass, which is then ground down until only a thin layer of it remains on the Si₃N₄.

Joints are made by placing this piece on top of an unglazed piece in a boron nitride jig, as shown in Figure 8.16. These assemblies are then heated to 2876 °F (1580 °C) for 45 min under 2 atm (~200 kPa) of nitrogen to make the joints. The joint thickness depends upon the amount of glass remaining on the glazed piece before joining. The resulting joint thickness directly influences joint strength, as is shown in Figure 8.17.

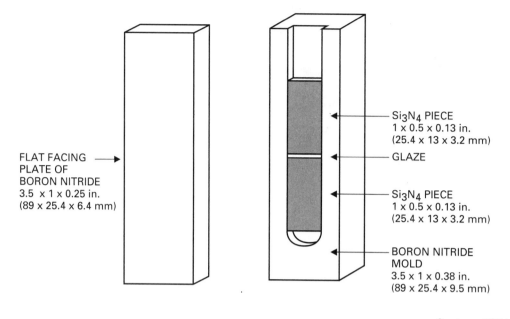

Courtesy of SRI International

Figure 8.16—Boron Nitride Jig Used in Joining Two Pieces of Si₃N₄. The Upper Piece is Glazed Before Being Joined to the Lower Piece

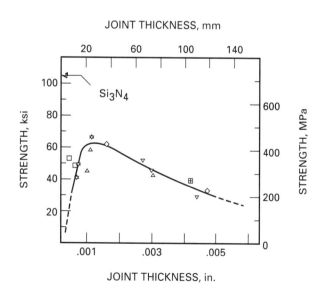

Figure 8.17—Strength versus Joint Thickness of Si₃N₄-to-Si₃N₄ Joint

SUPPLEMENTARY READING LIST

Bailey, F. P. and Black, K. J. T. "Gold-to-alumina solid state reaction bonding." *Journal of Materials Science*, 13: 1045-52, 1978.

Bates, C. H., Foley, M. R., Rossi, G. A., Sundberg, G. J., and Wu, F. J. "Joining of non-oxide ceramics for high-temperature applications." *American Ceramic Society Bulletin* 69(3): 350-6, 1990.

Brittain, R. D., Johnson, S. M., Lamoreaux, R. H., and Rowcliffe, D. J. "High-temperature chemical phenomena affecting silicon nitride joints." *Journal of the American Ceramic Society* 67(8): 522-6, 1984.

Brush, E. F., Jr. and Adams, C. M., Jr. "Vapor-coated surfaces for brazing ceramics." *Welding Journal* 47 (3): 106s-14s, 1968.

Buljan, S., Pasto, A. E., and Kim, H. J. "Ceramic whisker- and particulate- composites: Properties, reliability, and applications." *American Ceramic Society Bulletin* 68(2): 387-94, 1989.

Canonico, D. A., Cole, N. C., and Slaughter, G. M. "Direct brazing of ceramics, graphite and refractory metals." *Welding Journal* 56(8): 31-8, 1977.

Carbone, T. J. "Production processes, properties, and applications for calcined and high-purity aluminas." *Alumina Chemicals*, Ed., Hart, L. D., 99-108. Westerville, Ohio: The American Ceramic Society, 1990.

Davis, Richard C. "Hazardous materials minimization program in government electronics ceramics." *American Ceramic Society Bulletin* 70(3): 436-8, 1991.

Elssner, G., Suga, T., and Turwitt, M. "Fracture of ceramic-to-metal interfaces." *Journal de Physique*, 46, Colloque C4, 597-612, 1985.

Fox, C. W. and Slaughter, G. M. "Brazing of ceramics." *Welding Journal* 43(7): 591-7, 1964.

Fukushima, H., Yamanaka, T., and Matsui, M. "Microwave heating of ceramics and its application to joining." *Symposium Proceedings, Microwave Processing of Materials*, Vol. 124, Eds., Sutton, W. H., Brooks, M. H., and Chabinsky, I. J., 267-72. Pittsburgh, Pa.: Materials Research Society, 1988.

Goodyear, M. U. and Ezis, A. "Joining of turbine engine ceramics." *Advances in Joining Technologies*, Eds., Burke, J. J., Gorum, A. E., and Tarpinian, A., 113-53. Chestnut Hill, Mass.: Brook Hill Publishing, 1976.

Hammond, J. P., David, S. A., and Santella, M. L., "Brazing ceramic oxides to metals at low temperatures." *Welding Journal* 67(10): 227s-32s, 1988.

Hommel, R. O. and Marvin, C. G. "Silica – The next environmental hazard." *American Ceramic Society Bulletin* 71(12): 1805, 1992.

Institute of Ceramics. *Health & safety in ceramics*. Elmsford, New York: Pergamon Press, 1986.

Jack, K. H. "Nitride ceramics – The systems." First European Symposium on Engineering Ceramics: Applications, availability and advances in a high performance technology, 5-21. London: Oyez Scientific & Technical Services Ltd., 1985.

Johnson, S. M. and Rowcliffe, D. J. "Mechanical properties of joined silicon nitride." *Journal of the American Ceramic Society* 68(9): 468-72, 1985.

Kang, S., Dunn, E. M., Selverian, J. H., and Kim, H. J. "Issues in ceramic-to-metal joining: An investigation of brazing a silicon nitride-based ceramic to a low-expansion superalloy." *American Ceramic Society Bulletin* 68(9): 1608-17, 1989.

Klein, A. J. "Processing with microwaves." *Metals Progress* 1(4): 36-9, 1985.

Klomp, J. T. and Bolden, P. J. "Sealing pure alumina ceramics to metals." *American Ceramic Society Bulletin* 49(2): 204-11, 1970.

Klomp, J. T. "Bonding of metals to ceramics and glasses." *American Ceramic Society Bulletin* 51(9): 683-8, 1972.

Knoch, H. "Carbide and boride ceramics – Availability and commercialisation." First European Symposium on Engineering Ceramics: Applications, availability and advances in a high performance technology, 77-89. London: Oyez Scientific & Technical Services Ltd., 1985.

Loehman, R. E. "Transient liquid phase bonding of silicon nitride ceramics." *Surfaces and Interfaces in Ceramic and Ceramic-Metal Systems*, Eds., Pask, J. and Evans, A., 701-11. New York: Plenum Press, 1980.

Loehman, R. E. and Tomsia, A. P. "Joining of ceramics." *American Ceramic Society Bulletin* 67(2): 375-80, 1988.

Maruo, H., Miyamoto, I., and Arata, Y. "CO_2 laser welding of ceramics." *Proceedings of International Laser Processing Conference*, Anaheim, Calif., 1981.

Meek, T. T. and Blake, R. D. "Ceramic-glass-metal seal by microwave heating." U.S. Patent No. 4,529,856; 1985.

Meek, T. T. and Blake, R. D. "Ceramic-glass-ceramic seal by microwave heating." U.S. Patent No. 4,529,857; 1985.

Meek, T. T. and Blake, R. D. "Method for producing ceramic-glass-ceramic seals by microwave heating." U.S. Patent No. 4,606,748; 1986.

Mizuhara, H. and Mally, K. "Ceramic-to-metal joining with active brazing metal." *Welding Journal* 64(10): 27-32, 1985.

Mizuhara, H., Huebel, E., and Oyama, T., "High-reliability joining of ceramic to metal." *American Ceramic Society Bulletin* 68(9): 1591-9, 1989.

Mizuhara, H. and Huebel, E. "Joining of ceramic to metal with ductile active filler metal." *Welding Journal* 65(10): 43-51, 1986.

Moorhead, A. J., Morgan, C.S., Woodhouse, J. J., and Reed, R. W. "Brazing of sensors for high-temperature steam instrumentation systems." *Welding Journal* 60 (4): 17-28, 1981.

Moorhead, A. J. and Keating, H. "Direct brazing of ceramics for advanced heavy-duty diesels." *Welding Journal* 65(10): 17-31, 1986.

Moorhead, A. J., "Direct brazing of alumina ceramics." *Advanced Ceramic Materials* 2(4): 159-66, 1987.

Morozumi, S., Kikuchi, M., and Nishino, T. "Bonding mechanism between alumina and niobium." *Journal of Material Science* 16: 2137-44, 1981.

Naka, M., Tanaka, T., Okamoto, I., and Arata, Y., "Non-oxide ceramic joint made with amorphous Cu50Ti50 and Ni24.5Ti75.5 filler metals." Transactions of the Japanese Welding Research Institute, 12: 337-40, 1983.

Nicholas, M. G. and Mortimer, D. A. "Ceramic/metal joining for structural applications." *Material Science and Technology* 1(9): 657-65, 1985.

Palaith, D. and Silberglitt, R. "Microwave joining of ceramics." *American Ceramic Society Bulletin* 68(9): 1601-6, 1989.

Pattee, H. E. "Joining ceramics to metals and other materials." Bulletin 178. New York: Welding Research Council, November, 1972.

Rae, A. W. J. M., Cother, N. E., and Hodgson, P. "Nitride ceramics – Availability and commercialisation." First European Symposium on Engineering Ceramics: Applications, availability and advances in a high performance technology, 27-37. London: Oyez Scientific & Technical Services Ltd., 1985.

Rice, R. W. "Ceramic composites." *Fiber-Reinforced Ceramic Composites*, 451-95. New York: Noyes Publications, 1990.

Rice, R. W. "Joining of ceramics." *Advances in Joining Technology*, Eds., Burke, J. J., Gorum, A. E., and Tarpinian, A., 69-111. Chestnut Hill, Mass.: Brook Hill Publishing Co., 1976.

Santella, M. L., "Brazing of titanium-vapor-coated silicon nitride." *Advanced Ceramic Materials* 3(5): 457-62, 1988.

Sax, N. I. and Lewis, R. J., Sr. *Hazardous chemicals desk reference*. New York: Van Nostrand Reinhold, 1987.

Schwartz, M. M., *Ceramic Joining*, Metals Park, Ohio: ASM International, 1990.

Srinivasan, M. "The silicon carbide family of structural ceramics." *Structural Ceramics*, Ed., Wachtman, J. B., Jr., 99-159. San Diego, Calif.: Academic Press, Inc., 1989.

Stevens, R. "Oxide ceramics - new oxide systems." First European Symposium on Engineering Ceramics: Applications, availability and advances in a high performance technology, 97-110. London: Oyez Scientific & Technical Services Ltd., 1985.

Subbarao, E. C. "Zirconia – An overview." "Science and Technology of Zirconia." *Advances in Ceramics*, Vol. 3, Eds., Heuer, A. H. and Hobbs, L. W., 1-13. Columbus, Ohio: The American Ceramic Society, 1981.

Sutton, W. H. "Microwave processing of ceramic materials." *American Ceramic Society Bulletin* 68(2): 376-86, 1989.

Swartz, S. L., Majumdar, B. S., Skidmore, A., and Mutsuddy, B. C. "Joining of zirconia ceramics with a $CaO-TiO_2-SiO_2$ interlayer." *Materials Letters* 7(11): 407- 10, 1989.

Tanaka, T., Morimoto, H., and Homma, H. "Joining of ceramics to metals." Nippon Steel Technical Report No. 37, 31-8, April 1988.

Torti, M. L. "The silicon nitride and sialon families of structural ceramics." *Structural Ceramics*, Ed., Wachtman, J. B., Jr., 161-94. San Diego, Calif.: Academic Press, Inc., 1989.

Twentyman, M. E. "High-temperature metallizing: Part 1. The mechanism of glass migration in the production of metal-ceramic seals," 765-76; with Popper, P. "Part 2. The effect of experimental variables on the structure of seals to debased aluminas," 777-90; with Popper, P. "Part 3. The use of metallizing paints containing glass or other inorganic bonding agents," 791-8. *Journal of Materials Science* 10: 765-98, 1975.

Twentyman, M. E. and Popper P., "High-temperature metallizing, Part 3. The use of metallizing paints containing glass or other inorganic bonding agents." *Journal of Materials Science* 10: 791-8, 1975.

Weiss, S. and Adams, C. M., Jr. "The promotion of wetting and brazing," *Welding Journal* 46(2): 49s-57s, 1967.

Zdaniewski, W. A., Shah, P. M., and Kirchner, H. P. "Crystallization toughening of ceramic adhesives for joining alumina." *Advanced Ceramic Materials* 2(3A): 204-8, 1987.

MAINTEN- ANCE AND REPAIR WELDING

PREPARED BY A COMMITEE CONSISTING OF:

N. Sharpe, Chairman
T.C.A Corporation

L. E. Anderson
Consultant

D. A. Neuner
Dave's Aluminum Welding

D. L. Lynn
Barckoff Welding Management

L. C. Heckendorn
Intech R&D USA, Inc.

J. R. Goyert
Goyert Brothers Inc.

J. I. Danis
Exxon Company USA

P. I. Temple
Detroit Edison Company

WELDING HANDBOOK COMMITTEE MEMBER:
J. C. Papritan
Ohio State University

MAINTENANCE AND REPAIR WELDING

INTRODUCTION

MAINTENANCE AND REPAIR welding began when early man heated a broken tool in a charcoal fire and hammered the pieces into one shape. Forge welding was born with this early process. As early as the 16th century, some welding repairs were made by fusion, or melting, processes. These early techniques were inexact, however, and the success of early welding repairs was never certain.

After centuries of process improvements and technique refinements, maintenance and repair welding remains an inexact science and technology. Success is not to be taken for granted nor guaranteed. Unlike production welding, repair welding usually is complicated by more unknowns and more difficult situational factors.

The primary focus of this chapter is to chart a course through these welding factors using a logical decision model for evaluating maintenance and repair situations. This systematic approach and methodology will help ensure that all variables that can affect the welding repair are taken into consideration. This approach is helpful because those confronting welding repair situations come from varied backgrounds and technical specializations. With growing needs and advancements in the maintenance and repair of nuclear power plants, undersea structures, space stations and satellites, as well as more traditional structures, a systematic approach to welding maintenance and repair is extremely important.

A quick overview of maintenance and repair welding will introduce the kind of information needed to determine if a weld repair is appropriate or possible. Then a logical thought process, or decision model, for evaluating a repair will be described. While the suggested decision model may seem simple, those in charge of making weld repairs for a living will confirm that it embodies a difficult and complex process. The variables in the decision model represent the expert core developed through research and work experience in many repair situations. Case histories of such repairs are presented later in the chapter in the Applications section. These case studies present and analyze the important variables to consider to increase the probability of successful repairs in real-life settings.

MAINTENANCE AND REPAIR SITUATIONS

MAINTENANCE AND REPAIR are distinct and separate operations, but they have many common attributes. Both require individuals possessing a broad knowledge of welding and materials, for example. They may be distinguished, however, by a number of situational factors. Maintenance is accomplished during scheduled outages or downtime. Good preventive maintenance performance requires an ability to keep equipment in service without a major shutdown, or reconditioned within a specified time frame. Repair work, or corrective maintenance, on the other hand, involves the salvage of parts that are broken or worn and typically inoperable for continued use.

Equipment sometimes may be welded without being dismantled. However, defective parts may be removed from the equipment for subsequent welding if replacement parts are not readily available or if the repair is significant. Based upon an evaluation of the defective parts, they may be repaired in the field or at a qualified service center. Frequently, costs and time are prime considerations influencing maintenance and repair decisions, along with performance and safety criteria.

Maintenance or repair welding often is required to get equipment back on-line in a timely manner; long lead times are required for purchasing castings or forgings, for heat treatment, and for finish machining replacement parts. Old and obsolete equipment may even require the time-consuming process of reverse engineering of a replacement part, but effective replacement is almost always possible. In some instances, where old equipment is involved, the owner may feel an advantage can be gained by following the dual path of fabricating a new replacement part from specifications used in making the original part. Both the building of the replacement part and the salvaging of the old original part are carried out at the same time. This is done so that if the repaired original part does not meet expectations or again fails in service a replacement part is available. Welding may be involved in both the repair and the replacement part if the part in question required welding in its original fabrication.

MAINTENANCE AND REPAIR WELDING

MOST PLANS FOR preventive maintenance work provide for a detailed welding analysis. One the other hand, much corrective maintenance repair work requires an immediate fix and is relatively simple, and welding may be performed without a detailed welding analysis. Typical examples of the latter include the hardsurfacing of farm implements and welding repairs to car door and fender panels typically performed in automotive body shops. In general, the composition and physical properties of the base material are known. This simplifies the selection of a joining or thermal spray process to be used in the repair sequence.

On the other hand, welding repairs to more complex machinery, especially equipment that includes large castings or forgings with no prior welding fabrication requirements, often require a detailed analysis of the defect and sound engineering decisions. Such equipment typically is found in industrial settings involving power generation, diesel engines, extrusion plants, paper mills, rolling mills, galvanizing lines, and foundries. Owners of large or old equipment may not have the necessary operation and maintenance manuals, therefore a thorough study of the materials and structure design is mandatory. Some industries, on the other hand, may have historical data on equipment fabrication practices that are appropriate to the required maintenance or repair, which should expedite planning for the repair.

A SYSTEMATIC APPROACH TO MAINTENANCE AND REPAIR WELDING

PLANNING IS ESSENTIAL before performing any welding repair or preventive maintenance operation. Alternative courses of action must be analyzed and decisions made. In many cases, a large number of factors must be considered in deciding and planning the most appropriate welding repair. Having a systematic approach to help ensure that all factors are considered can be a substantial advantage. Two such approaches are presented in this chapter. The first, a welding repair planning checklist, is discussed briefly. The second, a welding repair decision model, extends the checklist items and also considers interactions among the various factors in a decision-tree format. Once this welding repair decision model has been presented, the chapter subsections that follow will expand on the issues highlighted by the various components of the model.

A systematic approach to evaluating and planning welding repairs can be extremely important, whether the defect to be repaired is in a bridge, a building, a pressure vessel, or some other equipment. Such an approach ensures that all of the variables that can affect the repair are taken into consideration. Once this information has been collected, an appropriate repair may be planned with an increased probability of success.

WELDING REPAIR PLANNING CHECKLIST

A SYSTEMATIC OUTLINE or checklist of the considerations in planning a welding repair is provided in Table 9.1. Such a step-by-step evaluation, with the user adopting or discarding each listed item as it is considered, should help anyone in deciding the best course of action. This checklist may be used by welding repair personnel in a qualified service center, the smallest job shop, or the largest chemical plant or powerplant.

Whoever is making the welding repair decision, whether a welder, foreman, technician, engineer,

Table 9.1
Systematic Checklist for Planning Welding Maintenance and Repairs

Equipment or Component Analysis
Is the equipment or component a welded fabrication?
Defect location and type
    Visual examination
    Nondestructive examination
    Destructive examination
Defect repair and service conditions
    Should it be repaired?
        Type and nature of defect
        Service history
    Should it be replaced?
        Potential for future failure
        Cost vs. Replacement
        Design flaws
    Used as is
What are the base materials?
    Document sources for this information
    Operating and maintenance manuals
    Physical methods to determine the base materials
    Heat-treated condition of the base materials
Are the base materials weldable?
    When they are? When they are not?
    Dissimilar material welds
Distortion and shrinkage
    Structural restraints
    Process restraints
Standards and codes
    Industry standards
    Customer or OEM specifications
    Applicable welding codes
    Military specifications
Preparation of the defect or worn parts
    Is the area accessible to the welder?
    Methods of material removal
    Clean and safe working environment
    Ensuring defect removal

Development of Repair Procedure
If an OEM fabrication, is procedure or assistance available
If not, then selection of the welding process
    Size and position of repair
    Type of filler metal
        Service conditions
        Physical and metallurgical nature of base metal
        Welding parameters
        Dissimilar filler metal applications
Cleaning before and during the weld repair
Peening (when and when not to use)
Material categories
Tests to determine weldability
Techniques to deal with distortion
How to minimize shrinkage
Material removal
Welding process vs. repair application
Filler materials
    Corrosion resistance
    Match base-material chemistry
    Match mechanical properties
Preheating (if required)
Postheating and postweld heat treating
Reinspection after completion of the repair
Types and size of defects
Commonly used NDE methods
Service life of equipment
Design-to-fail analysis

superintendent, or owner, should consider their personal base of experience while following the checklist. The checklist should not be considered complete or inclusive of every situation; instances will occur which will require the user to add additional specialties to fit a particular situation. In many repairs, however, the checklist may be the primary and sometimes the only means of ensuring a correct and successful welding repair.

A WELDING REPAIR DECISION MODEL

THE DECISION MODEL suggested for planning welding maintenance or repair activities is represented as a flow chart using the logic common to computer programming, with feedback loops showing when one aspect of the process may affect another. The graphical model consists of symbols that are connected to show sequences of decisions and alternative action.

When these symbols are used to describe an interactive decision process, the graphic representation is often called a *decision tree*. A welding repair decision tree is shown in Figures 9.1(A) and (B). This decision tree represents a useful thought process when one is making decisions and planning the steps necessary to achieve a successful welding repair. This should give the foreman, engineer, or welder a quick overview of the information that will be needed. One then can determine if a welding repair is feasible and the information and steps that may be related. For some, the decision tree may even provide an introduction to maintenance and repair welding. Further discussion of the main steps in the welding repair decision tree is provided in the sections that follow. After that, specific maintenance and repair applications will be presented in the Applications section.

Identify Welding Repair Need

FOR A SUCCESSFUL repair, the first step is to collect specific and accurate information. It is important to know the exact size and location of the defect, the site where the repair will be performed, and the availability of welding equipment. Information on the component to be repaired, such as model number, serial number, and part name and identification number are helpful. Photographs and sketches are valuable in formulating a good repair procedure. It is important that a worn or fracture area be carefully studied to determine how to best accomplish the repair. Experienced welders can contribute valuable information to a successful repair. Their experience often allows them to foresee problems that otherwise might not be anticipated by the plan.

A defect requiring repair will be either a fabrication defect or a service failure. The natures of these two types of defects differ and influence the repair plan.

Fabrication defects occur during the making of a weld and include porosity, slag inclusions or undercut, lack of fusion, incomplete penetration, and solidification cracking. Repair by welding involves removal of defective areas and replacement with sound metal. The repair procedure may be very simple and merely require the deposition of additional weld metal to rectify undercut.

The repair of deep-seated defects such as lamellar tearing can entail extensive excavating and rewelding or a simple seal weld. If cracking is present, conditions may have to be changed to avoid similar defects in the future.

Service failures consist of cracks caused mainly by overstressing (sometimes initiating from an existing manufacturing discontinuity), fatigue, brittle fracture, or stress corrosion. An example of a service failure would be a crack in a containment vessel that propagates through the vessel wall from the inside. Wear of all types, such as abrasion, cavitation, and flow-assisted corrosion, also must be considered in the overall repair job.

Fatigue cracking is a form of stress relief in highly stressed areas. It will not propagate because the residual stress is less than the yield strength. Fatigue cracking frequently can be left without repair. In other cases, fatigue crack growth can be monitored by periodic inspection and repaired at a convenient time.

Brittle fracture is a relatively rare occurrence compared with fatigue cracking. It is far more spectacular, however, and may cause disasters such as the breaking of a ship in half or the fragmentation of a pressure vessel. The repair can range from the removal of the cracked area followed by rewelding, to the welding by

fabrication of a new subsection which is fitted and welded into place.

Determine the Nature of the Defect

A COMPLETE HISTORY of the worn or failed part is important in determining the root cause of the defect. The starting point is to determine when, where, and how the failure occurred. Interview operators with questions about the part's service history, including length of service and other failures. Prepare sketches with dimensions and take photographs to provide a complete record.

Nondestructive and destructive examinations may be helpful in determining the nature of the failure.

Nondestructive Examination (NDE). Inspection of the worn or fractured surfaces first by visual inspection and other NDE methods helps in determining the nature and extent of the failure. Visual inspection may be supplemented with remote visual videos using computer enhancement and with stereo macrographic examinations. Whatever examination method is used, be sure to follow the instructions of the manufacturer.

NDE methods include liquid-penetrant testing (PT), magnetic-particle testing (MT), radiographic testing (RT), ultrasonic testing (UT), and eddy current (EC) examination.[1] The PT and MT methods are easy and commonly used for defect examinations in many repair situations.

Liquid-penetrant testing first requires that the surface to be examined is cleaned of all rust, paint, or other contaminators. A visible red dye or a fluorescent dye is then applied to the area in question. The red or fluorescent dye will penetrate into cracks, porosity, or other defects too small to be visible to the unaided eye. After a prescribed soak time, the excess dye is removed and the surface is wiped clean. A developer or a black light is used on the suspect area. If cracks or other surface defects are present, the dye will reveal the location of the discontinuity. This is the least expensive method of defect identification.

Magnetic-particle testing requires a trained technician with only iron powder and an alternating current with a suitable yoke. The yoke is placed across the suspect area and activated with a surge of alternating current, creating a magnetic field. If a defect is present, the magnetic field will be interrupted. The iron powder is then sprinkled over the area and will align with any defect. This technique also can be done using two prods

1. Publications of the American Society for Nondestructive Testing provide detailed information on the selection of the most appropriate method to be used for a given application. American Society for Nondestructive Testing, Inc., 1711 Arlingate Lane, P.O. Box 28518, Columbus, Ohio 43228-0518.

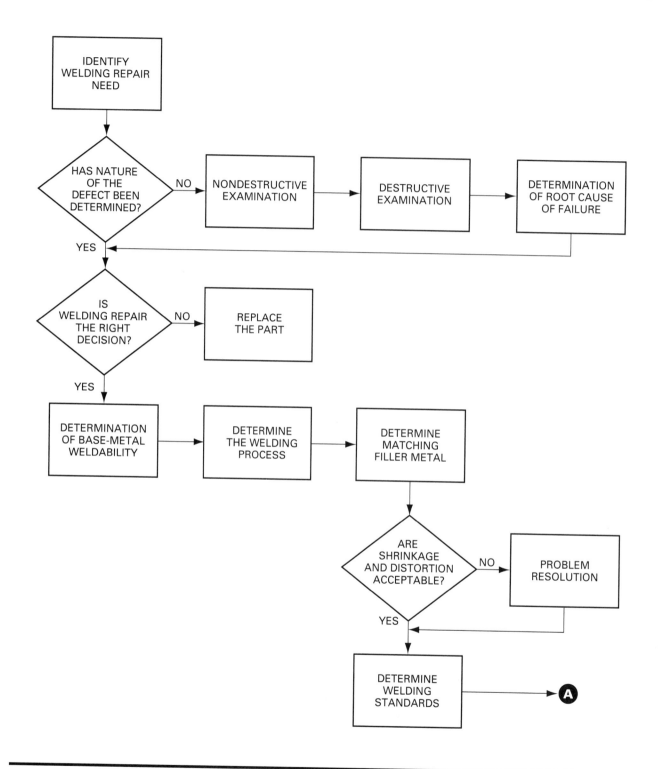

Figure 9.1—Weld Repair Decision Tree

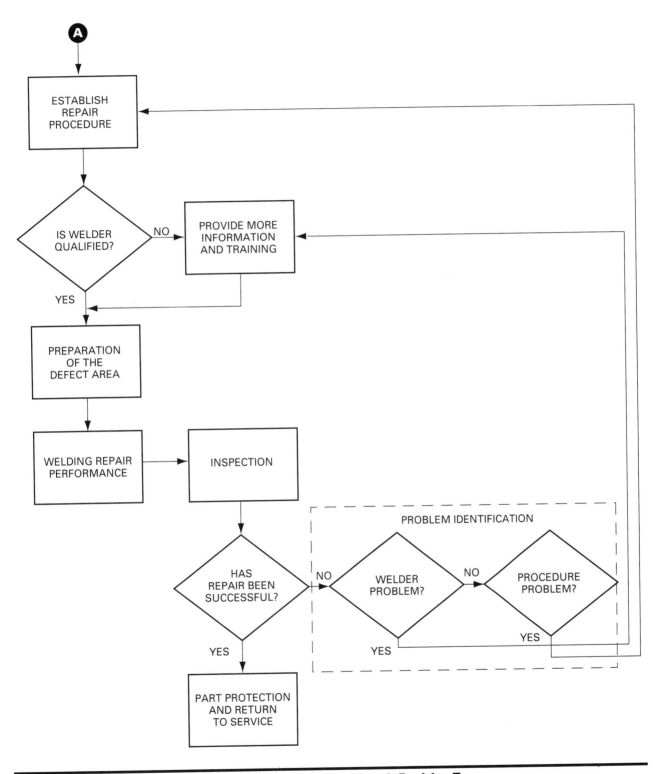

Figure 9.1(continued)—Weld Repair Decision Tree

and direct current to not only reveal surface discontinuities but also imperfections down to 1/4 in. (6.4 mm) below the surface.

Destructive Examination. Destructive examination can be used for several purposes.

(1) To examine the area of a defect such as a crack by removing a portion of the material on either side by means of a "boat" sample.
(2) Drill shavings are removed so that a material analysis can be made.

When weldability is a question, material samples from the structure requiring repair, if available, are welded. Then these welds may be destructively examined using weldability testing methods.[2]

Determination of Root Cause of Failure. The root cause of the failure must be determined in order to judge if a planned weld repair is correct. Repairing by welding an existing design using identical thicknesses of the same materials simply may be setting the components up for another identical failure. Nondestructive and destructive examinations may be helpful in determining the root cause of failure. Hardness tests may assist in determining the approximate tensile strengths of the base and weld material.

Making the Welding Repair Decision

DECIDING THAT WELDING repair is the right solution, as opposed to replacement with new parts, relies on judgment factors that follow the earlier need for identification of the material in question and any defect examinations performed. It may be found that repair or rebuild is too costly, or that a good repair is not feasible. The relative strengths of the welded joint and the original base metal may be an important consideration.

A special consideration may be the effect of repeated repairs. Depending upon the nature of the service environment, a noncritical welded structure may be repaired several times with no detrimental effects. Sometimes a lack of knowledge of the effects of repeated repairs on material properties means that the number of repairs should be restricted. Under codes and specifications published by standards-writing bodies, the number of repairs permissible may be limited. The American Petroleum Institute, for example, limits the number of repairs to two in the same area on pipe fabrication and installation, and on critical offshore structures.[3] The owner must make the final decision as to whether a future failure is best avoided through welding repair or replacement.

Determination of Base-Metal Weldability

MOST MATERIALS ARE readily weldable. However, fully retaining the service characteristics of materials after welding may require a more difficult or sophisticated procedure. It is important to know the material chemistry and hardness in order to prepare a successful welding repair procedure. This information sometimes may be found in parts manuals or by checking with the manufacturer's service department. If necessary, have the suspect material analyzed for its chemical composition.

Color is sometimes used to distinguish materials. For example, copper is reddish-brown. Brasses high in zinc are yellow. Brasses lower in zinc are more reddish; they also are harder than copper. Aluminum, magnesium, zinc, tin, and lead alloys are silvery colored. Gray cast iron is dark gray on the fracture surface. White cast iron is silvery white on a scratch surface. High-carbon steel chips have edges that are lighter in color than low-carbon steels, and they also are harder to cut. Manganese steel chips are blue and nonmagnetic.

If risks are minimal and the above methods of base-metal identification are not feasible, an analysis of the equipment can be made by categorizing the parts. The categories could be structural components such as plates, beams, bars, castings, and forgings. The structural components can be divided into six categories: mild steels, medium-carbon steels, high-carbon steels, low-alloy steels, wear-resistant steels, and corrosion- and heat-resistant steels. After categorizing, a check using a portable hardness tester, grinder, or metal file should then be used to gauge the relative hardness of the material. The harder the material, the higher the strength.[4] High-strength materials require higher strength or higher alloy rods and more sophisticated welding procedures such as additional preheat and heat-input control. Castings and forgings can be categorized in a similar manner.

One should not weld any lifting, aligning, or hold-down lugs on any parts unless the material chemistry is known. If the correct procedure is not used, the lugs may pull out at the weld, resulting in catastrophic failure. Additional damage may be caused by overwelding when adding attachments. A good rule is that the combined fillet weld should be no larger than the throat dimensions of the lug or attachments.

2. Weldability tests are described in Chapter 4, *Welding Handbook*, Vol. 1, 8th Ed., 119-21.

3. See Paragraph 9.5, API 1104, American Petroleum Institute, 1220 L Street N.W., Washington, D.C. 20005.
4. Tables relating hardness to tensile strength may be found in "Standard Test Methods and Definitions for Mechanical Testing of Steel Products (ASTM A370-94)," *Annual Book of ASTM Standards*, Vol. 01-01, 221-66. Philadelphia: American Society for Testing and Materials, 1995.

Determining the Welding Process

THE WELDING PROCESS chosen for the repair depends on several considerations, such as the extent of the needed repair. Process options for maintenance and repair welding are shown in Figure 9.2.

Whether the welding repair would be made in a shop or at a field site was once a greater consideration than it is today. With the advent of portable power sources and auxiliary arc welding equipment, most processes can be used in the field as well as in the shop. Still, the practical availability and compatibility of process equipment

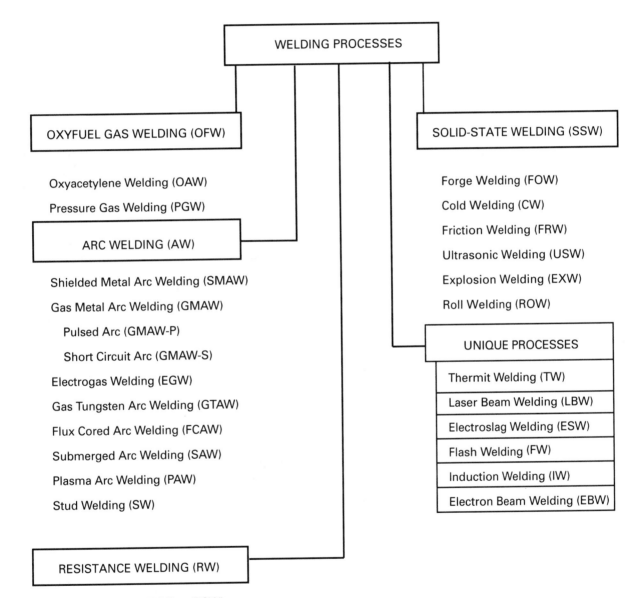

Figure 9.2—Welding Repair Process Options

both in the repair shop and in the field must be considered. Other factors related to process selection for shop versus field sites include preweld and postweld heat treatments, cost effectiveness and safety considerations such as ventilation and exhaust systems, toxic materials handling, bottled gases, fuel tanks, oil lines, and electrical cables. A maintenance check of all the equipment to be used is mandatory.

Determine Matching Filler Metal

ELECTRODE CHEMISTRY AND mechanical properties must be determined next in the planning of sound maintenance and repair procedures. The cause of a fracture can have an influence on the type of electrode selected for the repair. A good rule is to use an electrode with the same corrosion resistance and tensile strength as the parent material when full-strength welded joints are required. Fractures usually occur in parts that are highly stressed and rigid because of their function and configuration. Therefore, the weld metal must be ductile to compensate for weld metal shrinkage. Increasing the cross-sectional area at the defect will result in lower stress per unit area.

Maintenance and repair of iron castings require special filler metal considerations. Shielded metal arc welding is the most commonly used process for repairing iron castings. A variety of electrodes are available, many formulated for specific conditions. Users will find electrodes specifically formulated for gray, malleable, and nodular cast irons. Some are important for use with castings that require machining prior to being placed in service. Others are for thin-section castings. Some can produce porosity-free welds, while others have special cleaning agents added to the electrode coating.

Shrinkage and Distortion

THE NEXT QUESTION in the planning or repair process concerns the acceptability of any shrinkage or distortion. This can be calculated in the planning phase or experienced during actual repair welding. Excessive shrinkage or distortion must be analyzed and resolved in order to continue with the repair process in the decision model. Preventive measures need to be investigated and perhaps incorporated into the welding process.

Welding shrinkage commonly means the decrease in volume that occurs when molten weld metal solidifies. Other types of shrinkage describe a solid-state phenomenon. Although these conditions are unavoidable, they can be minimized and, in many cases, used to counter each other. A basic understanding of metal behavior during heating and cooling while under restraint and some applied engineering skill can help keep shrinkage and distortion in check and contribute to a successful welding repair.

Clamping a part in a fixture and tack welding tend to reduce the effects of shrinkage. Within limits, increasing the number of layers of weld metal for manual welding repair will result in restraints in the joint during welding and will result in less shrinkage. The first beads placed in a joint will help hold the position of the parts against the shrinkage effects of the inter-welded beads. However, this idea must be balanced against the potentially added distortion that may result from increasing the number of weld beads.

Welding in the flat or horizontal position, with resultant higher travel speeds, tends to reduce the net shrinkage. Preheating to reduce the heating and cooling temperature gradient also minimizes shrinkage. A suitable level of preheating depends upon the thickness of the part and the material's carbon equivalent.[5]

Distortion is a permanent or temporary deviation from a desired form or shape. Welding causes distortion because stresses develop in materials from localized unrestrained thermal expansions and contractions. It is possible under some circumstances to contain these internal stresses within an undistorted weldment. Whether distortion occurs will depend on the magnitude of the welding stresses, the distribution of the stresses in the weldment, and the strength of the members acted on by these stresses. Three types of distortion are angular distortion, longitudinal bowing, and buckling.

Angular distortion is the change in relative position of members extending from a weld zone. Five measures have been identified for decreasing angular distortion:

(1) Using the minimum amount of weld metal required to achieve the desired joint strength
(2) Depositing the weld metal with the fewest possible number of layers
(3) Avoiding a weld profile with a narrow root and a wide face
(4) Balancing the amount of weld metal about the neutral axis of the joint
(5) Presetting the members at an angle opposite to that expected to develop as a result of welding

Longitudinal distortion, or bowing of long members, is caused by shrinkage stress which develops at some distance from the neutral axis of the member. When welds of unequal size must be made at unequal distances from the neutral axis, the opposing stress should be balanced by depositing welds closest to the neutral axis and by making the welds that are farther away from the neutral axis smaller in size. Bending or

---

5. A method for determining preheat temperatures based on carbon equivalents is explained in Appendix XI of ANSI/AWS D1.1-94, *Structural Welding Code-Steel*, 285-94.

cambering the member in a bow opposite to that which will develop from welding is a practical recourse. The hot side will finally become the short side. Back-step techniques also will aid in reducing distortion.

Sheet metal often buckles during welding, whereas heavy plates do not. Buckling is caused by the inability of lateral unsupported sheet metal to resist compressive stress. Intermittent welding sequences minimize buckling by keeping these stresses at a minimum in regions near the welds.

Determine Welding Standards

WELDING REPAIR PROCEDURES should be in accordance with applicable published standards, and identifying and incorporating these into the repair plan is an essential step in the decision process. Using the guidance of established welding standards helps to ensure that welded products are safe and reliable and that welding personnel are adequately protected from hazards to health and safety.[6]

Standards include codes, specifications, recommended practices, classifications, methods, and guides. They describe the technical requirements for a material, process, product, system, or service. They also indicate the procedures, methods, equipment, or tests to determine that the requirements have been met.

Published standards become mandatory when specified by a governmental jurisdiction or when referenced by contractual or other procurement documents. Many standards or agreements also contain nonmandatory sections or appendices that may be used as reference guides or recommended practices at the discretion of users.

Establishing the Repair Procedure

THE PURPOSE OF a welding repair procedure is to identify all of the variables necessary to complete the repair. The repair procedure consists of the technical details usually specified in welding procedures plus other ancillary details involving set-up, fit-up, and other related matters affecting the total repair. Included are welding data such as welding position, size of fillet welds, thickness of metal for groove welds, and electrode size and classifications. Requirements for preheating, postheating, and postweld heat treatments and cleaning must be considered. Amperage, voltage, and sequence of weld passes have to be designated. If an automatic process is used, travel speed must be determined.

Many procedural matters will depend upon the welding process selected.[7] For example, a small power source requirement for shielded metal arc welding (SMAW) or gas tungsten arc welding (GTAW) would be to set up a 250-ampere ac/dc rectifier-type welder power source. This kind of arc welding system is used in most light fabricating shops and repair facilities. The system includes a power source, electrode cable, ground cable, and clamps. For GTAW process repairs, a high-frequency unit, GTAW torch, shielding gas, regulator, and flow meter are necessary. Tungsten electrode preparation is extremely important for arc stability to preclude tungsten pickup and give sound weld deposits. A water recirculating system is needed for water-cooled torches. This GTAW system produces sound, high-quality weld deposits. Postcleaning is not always needed. GTAW equipment can be highly specialized and costly.

The equipment setup must be in accordance with the manufacturer's instructions. Power sources and attachments must be in accordance with Occupational Health and Safety Administration rules and regulations established in the welding industry. All high-pressure gas cylinders must be safely secured in the upright position. All connections must be tight. Loose electrical connections can cause arcing at the connector, which produces poor-quality welds and is a potential fire hazard. Loose gas and water connections can result in porosity or safety hazards.

Gas metal arc welding (GMAW) and an important variation, pulsed welding (GMAW-P), make other capabilities available to the repair welder. The GMAW process uses a dc power supply, semiautomatic or automatic wire feed system, welding gun, and shielding gas supply. The shielding gas will vary according to the base material being welded. Many proprietary gases are available on the market, so selection must be made carefully.

GMAW produces strong, high-quality welds at high speeds using bare electrodes. Since cleaning agents and fluxes are not used, the amount of postweld cleaning is minimal. Flux cored arc welding (FCAW) is similar to the GMAW process except flux cored wired provides the vehicle for the weld metal.

Setting up for a field repair poses some different considerations under conditions tougher than in the shop. With the use of an engine-driven power source, however, the SMAW, GTAW, and GMAW processes can be readily set up for field operation.

Fit-Up Requirements. An important but sometimes overlooked part of the repair procedure is parts fit-up.

6. A more complete discussion of welding standards is provided in Chapter 13, "Codes and Other Standards." *Welding Handbook*, Vol. 1, 8th Ed., 411-36.

7. Refer to AWS *Welding Handbook*, Vol. 2, 8th Ed., and vendor literature for more detailed descriptions of processes and procedures.

Proper fit-up, fixturing, and tack welding, when appropriate, contribute to a successful repair.

Fit-up can be secured with the use of jigs and fixtures, whose function is to facilitate assembly and alignment of parts. Fixtures are used for three major purposes: assembly, precision fitting, and robotic applications. Light and heavy-duty clamping devices can be incorporated into dedicated fixtures for large production runs, or into adjustable fixtures that can be easily modified for several short production runs. The parts may be partially or completely welded in the fixture. If the part is tack welded and then removed from the fixture prior to welding, the fixture is called a *fitting jig*.

Tack welds should be of sufficient number and suitable proportion to hold the assembly in place during normal handling. They should be of equal quality as specified by the welding procedure, so that removal is unnecessary during subsequent welding of the joint. Auxiliary welding equipment such as turning rolls, positioners, and head and tail stocks aid the welder in positioning the work for maximum operating advantage.

When the GTAW process is used for tacking in pipe welding with consumable inserts, it is preferred to use shallow tack welds that do not fuse through the land to the pipe ID. Tack welding of plate should penetrate and fuse through the full thickness of the land when using the SMAW or GTAW process. In GTAW process tacking, the tacks should be carefully examined; crater cracks, slag, and heavy oxides should be removed before completing the production weldment. Providing a reasonable taper (feather edge) toward both ends of the tack is important to facilitate a sound tie-in with subsequent weld passes.

Thermal Requirements. High-strength, low-alloy, and heat-treated steels present special thermal requirements. Preheating is needed to preclude cracking and to maintain desired mechanical properties. In addition, preheat is desirable when welding heavy sections of low or mild carbon steels to ensure good fusion. Postweld heat treating is used for stress relief.

Preheating involves raising the temperature of the base metal above ambient temperature before the start of welding. A preheat furnace is the most satisfactory and uniform method of heating. More commonly, manual torch heating is used with natural gas mixed with compressed air. This method produces a hot flame and is clean. Electric blankets or temporary furnaces are used more extensively in field repairs. An entire part may be preheated or, if the part is large or welding is limited, a local area of the weld joint may be heated to the required temperature.

The required preheat temperature depends upon the composition of the base material, the rigidity of the members to be joined, and the welding process. Preheating may be applied to a weldment to prevent cracks, to reduce hardness in the heat-affected zone, to reduce residual stress, and to minimize distortion. Preheating softens hardened zones, reduces shrinkage stresses, and improves the base material notch toughness.

Postheating is used to control the base-metal cooling rate and facilitate the escape of hydrogen gas. The following postweld thermal treatments have been developed to achieve desired material properties:

(1) Stress-relief heat treatment – Heating to a suitable temperature below the lower critical temperature of the metal, holding long enough to reduce residual stresses, and then cooling slowly.

(2) Annealing – A generic term denoting a treatment, consisting of heating to and holding at a suitable temperature followed by furnace or slow cooling at a suitable rate, used primarily to soften metallic materials but also to simultaneously produce desired changes in other properties or in microstructure.

(3) Normalizing – Heating a ferrous alloy to a suitable temperature above the transformation range and then cooling in air to a temperature substantially below the transformation range. Materials may be put into service in either the normalized or the normalized-and-tempered condition.

(4) Hardening – Increasing hardness by suitable treatment, usually involving heating and cooling. Examples used to restore parts to service include laser and induction heat treatments or a carburizing heat treatment.

(5) Quench and temper (Austenitize, quench, and temper) – Rapid cooling from the austenitizing temperature, followed by reheating the hardened steel to some temperature below the eutectoid temperature for the purpose of decreasing hardness and increasing toughness.

(6) Normalize and temper – Reheating a normalized metal for the same purpose of decreasing hardness and increasing toughness.

(7) Austempering – A heat treatment for ferrous alloys in which a part is quenched from the austenitizing temperature at a rate fast enough to avoid formation of ferrite or pearlite and then held at a temperature just above the martensitic transformation temperature until all of the material is transformed into bainite.

(8) Martempering – A hardening procedure in which an austenitized ferrous workpiece is quenched in an appropriate medium at a temperature that completes the austenite-to-martensite transformation until the temperature of the part is uniform throughout, and then cooled in air. The treatment is frequently, but not always, followed by tempering.

(9) Solution annealing – Heating an alloy to a suitable temperature, holding at that temperature until one or more constituents enter into solid solution, then

cooling rapidly to hold these constituents in solution. Solution-annealing-and-aging treatments are used to harden precipitation-hardening materials such as 17-4 stainless steel alloys (e.g., A564 Type 630) and some copper and aluminum alloys.

Welder Qualification and Preparation

THE TECHNICAL AND job-specific training of the welder are critical to a successful welding repair. To assess technical skills, welder qualification tests frequently are prescribed by governing codes, specifications, or other work rules.[8] These tests determine the ability of the persons tested to produce acceptably sound welds with the process, materials, and procedure called for in the test. Welder qualification tests cannot foretell how an individual will perform on a particular job-specific weld. Supervisors must determine if the available welders have the knowledge and skill experience required for a particular job and make any appropriate adjustments.

Sometimes additional training may be required, for example in limited-access situations. At other times, additional information about the welding repair may need to be obtained and shared with the welder. Sometimes a mock-up of the defect can be prepared to determine the ability of the welder and the feasibility of the proposed repair.

Supervisors may wish to consider whether short instructions may be better than detailed explicit instructions, which may give some welders a heightened sense of pressure. Stressful physical conditions such as uncomfortable positions and frequent preheat requirements also can affect the quality of the welder's repair work. Decisions on human factors such as these need to be considered, and needed training, work redesign, or information be provided, before moving ahead in the welding repair planning sequence.

Preparation of the Defect Area

ONE OF THE most important considerations of a welding repair procedure is to clean the fracture area or worn part of all grease, paint, moisture, dirt, rust, spalled material, or any other condition that may be detrimental. Hydrogen adversely affects the properties of weld deposits. As the molten weld metal cools and solidifies, hydrogen is rejected from the solution and becomes entrapped in the solidifying weld metal. It will collect at grain boundaries or at any discontinuity, where it will create high pressures. These pressures

cause high internal weld stresses. These pressures and stresses can create minute cracks in the weld metal with the potential to develop into larger cracks. Hydrogen will gradually escape from the solid over a period of time.

Spalled buildup may not allow the welding arc to penetrate to solid material. All spalled material must be removed by air carbon arc gouging or by grinding to preclude trapped contaminates. After air carbon arc gouging, it is recommended to remove a 1/16 in. (1.6 mm) layer by mechanical means to remove the carburized layer before rewelding the joint. Some methods of cleaning a part are steam cleaning, grit blasting, and burning off oil and grease with an oxyfuel torch or in a burnout furnace.

A flame must not be concentrated in one area for too long; the flame should be swept back and forth across the part. Residue can then be brushed off with a clean wire brush. The area cleaned around the fracture or worn part should be large enough so that contaminants cannot reach the repair area. Proper preparation of the defect area for welding repair depends upon the defect size, whether it goes completely through the part thickness, and other factors that may be unique to the particular part.

Welding Repair Performance

THE WELDING REPAIR should be performed in accordance with all of the previous decisions regarding materials, standards, human factors, processes, and procedures.

Inspection

THE PROCESS OF visual inspection begins virtually when welding begins, to inspect during multiple-pass welding and as soon as the welding repair is completed. Checked first are visual items such as weld appearance, size, and the extent of the welding. Measurements can determine dimensional accuracy, degree of distortion, and the size of surface discontinuities. The following surface discontinuities can be readily identified by visual examination and rejected when beyond limits specified by the applicable standard: cracks, undercut, overlap, exposed surface porosity and slag, weld profile deviations, and weld surface roughness.

For detection and accurate evaluation of surface discontinuities, the weld surface must be thoroughly cleaned of all slag and oxides. The cleaning operation must be carried out carefully to avoid masking any discontinuities from view. For example, if a chipping hammer is used to remove slag, the hammer marks could mask fine cracks. Shot blasting, similarly, may peen the surface of soft weld deposits and hide discontinuities.

---

8. Welder performance qualification and certification standards are described in Chapter 14, *Welding Handbook*, Vol. 1, 8th Ed., 438-64.

Dimensional accuracy of the weldment is determined by suitable measuring methods. The conformity of weld size and contour is determined by a suitable weld gauge. The size of the fillet weld in joints whose members are at a right angle, or nearly so, is defined in terms of the length of the fillet legs. The gauge determines whether the leg size is within allowable limits and whether the weld deposit is concave or convex. Special gauges are required when the members are at an acute or obtuse angle.

For groove welds, the height of reinforcement should be consistent with the specified requirements. The weld surface appearance must meet the requirements of the applicable standard. A fabrication standard may permit a limited amount of undercut, undersize, and piping porosity, while cracks, incomplete fusion, and unfilled craters are not acceptable.

Care should be exercised when judging the quality of weldments by visual appearance alone. Acceptable surface appearance does not guarantee quality workmanship and is not a reliable indication of subsurface weld integrity. However, proper visual inspection procedures during fabrication can increase product reliability over that based only on final inspection.

Welding Repair Success or Correction

SUCCESS OF THE weld repair will be determined by visual and any other inspection procedures that may be appropriate. A successful welding repair will be followed by final cleaning and protective services before returning the part to service.

Corrective action must be taken whenever the welding repair is not successful, and this starts with the determination of the problem. Steps in problem determination are shown within the portion of Figure 9.1B surrounded by dotted lines. This involves the re-evaluation of the preparedness of the welder, and then of the appropriateness of the repair procedure, including the fit-up procedure, thermal treatments, and other procedure specifications. This also could involve the re-evaluation of the welding process selected for the welding repair, although this is not shown in the diagram.

Once the specific problem has been identified, one simply returns to the appropriate point in the welding-repair decision tree and continues again from that point in the welding-repair process sequence. The experience of the first effort and its failure analysis adds more knowledge into the decision-making process the next time through, leading to a greater likelihood that the next welding repair will be a success.

Part Protection and Return to Service

THE EXTENT OF cleaning and protective services necessary will depend largely on the process used. For example, the SMAW process will require more postweld cleaning than the GTAW or GMAW process.

Thorough postweld cleaning is required whenever slag producing welding processes have been used. The slag could be highly corrosive, as with coated stainless steel electrodes. Stainless steels are poor heat conductors and work-harden readily. These two characteristics must be considered when machining and grinding.

Rust or heat-treat scale can be removed by steam cleaning or blasting prior to painting. Cleaning between passes of a multiple-pass procedure can also reduce postweld cleaning.

When the required maintenance or repair is accomplished, final inspection is performed and is followed by metal removal operations. This may be as simple as the grinding of the weld surface to blend with the surrounding structure. Sometimes, more elaborate machining operations may be necessary to achieve original dimensions with required fits and tolerances.

Finally, preservation may take the form of painting or other protection. The last step in preservation is wrapping the part or mounting it on a skid for return to the owner for installation or storage as necessary.

APPLICATIONS

INTRODUCTION

MAINTENANCE AND REPAIR using welding is an ongoing function that occurs in every industry. This function needs to be performed in a cost-effective and quality manner. The use of a method to develop these maintenance and repair procedures has proven to be the most successful way of achieving the desired results.

CASE STUDIES

THE RANGE OF possible maintenance and repair situations that could be described in this chapter is endless. The following cases have been chosen for their general application as successful solutions. While some of these may have been field expedients at the time, these successful solutions are structured here as examples of how to use the method described in this chapter. Where certain steps in the repair decision process (Figure 9.1) may have been omitted in a particular case study, this step is simply omitted from the presentation of that case. Ideally, however, all steps are considered in deciding on a maintenance or repair welding plan. These cases can serve as illustrations of the systematic method presented in this chapter being applied in real-life, practical solutions.

Repairing Cold-Storage Tank Leaks

CORROSION IN ALL of its many forms is playing a larger and more significant role in the need for maintenance and repairs.[9] The more hazardous materials being handled in industrial applications coupled with the higher levels of pollution in our environment has lead to the use of more sophisticated alloy metals. The repair and maintenance of these metals to extend their service life is both necessary and cost effective. However, the amount of knowledge and skill has also increased and the need to follow a method to help ensure success more important than ever.

Microbially influenced corrosion (MIC) is one form of corrosion that has increased significantly with the increase pollution that has found its way into the ground water of many geographic areas of the United States. The effects of this microbial attack affects a wide range of industries from power plants, both fossil fuel and nuclear, and beverage and food manufacturing facilities to finally chemical and petroleum plants. The following example will show how the method can be used to deal with a corrosion problem.

Eight cold-storage tanks were found to have developed a leaking problem after being in service for several months (see Figure 9.3). Since the tanks were an essential part of the production cycle, the decision was made to enact a corrective action as soon as possible. The tanks were cylindrical in shape, and oriented vertically. They were constructed with 1/4 in. (6.4 mm) thick bottoms and 3/16 in. (4.8 mm) sides and tops. The steps in the weld repair decision tree (Figure 9.1) are reviewed below for these tanks.

Identify Welding Repair Need. The value and essential nature of the tanks and the seriousness of the leaking required that corrective action be initiated immediately.

Has Nature of the Defect Been Determined? Following the discovery of the leaking, the actual nature of the leak was unknown. The two leading possibilities were some sort of mistake made during

9. This case study is based on information provided by Don Lynn of Barckoff, Inc.

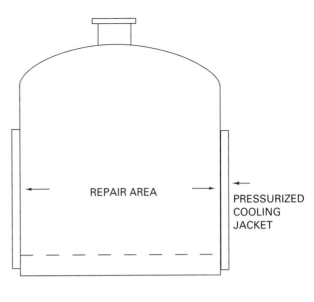

REPAIR AREA

PRESSURIZED COOLING JACKET

BOTTOM OF TANK REPLACED

Figure 9.3—Welding Repair Areas for Cold-Storage Tanks

erection of the tanks or some form of corrosion currently unknown had damaged the tanks in the course of only a few months of service. Corrosion was found through follow-up investigation to have caused the problem. Corrective actions were taken to ensure that the problem would not reappear after the repairs were made.

To determine the extent of the damage that resulted in the leakage, a liquid-penetrant inspection was made of the tank bottoms and the side walls of the tank. The liquid-penetrant test showed that both the bottoms and the walls of the tanks had corrosion nucleation sites, however, due to the greater concentration of sediment and microbes the bottoms was more severely damaged than the side walls. This was the nondestructive examination portion of determining the nature of the defect.

A destructive examination was performed in order to determine the nature of the leaking. Small sections of the tank bottom of the unit that was leaking the worst were sent to a laboratory for analysis. The results of the tests from the laboratory showed that the cause of the perforations responsible for the leaking were from microbially influenced corrosion that had resulted from using torpid water from nearby wells for the hydrotesting of the tanks. The water had contained a large amount of sediment as well as aerobic and anaerobic microbes. When the tanks were pumped out after the hydrotest, a heavy layer of sediment was allowed to remain on the bottom and a film covered the walls of the tanks.

Subsequent cleaning of the tanks at a later date before going into production did not remove the microbes that already had become attached to the tank walls and bottoms. The side walls of the tank exhibited a microbial polyp with the corrosion cell producing a cavity in the metal, as shown in Figure 9.4. The bottoms of the tank also were showing the same microbial polyps, except that the cavity had penetrated through the bottom, as shown in Figure 9.5.

Is Welding Repair the Right Decision? The decision was made that the damage was too extensive in the bottoms to repair, and so they would require replacement. The sides were determined repairable, and preparations were begun for the making of these repairs while the fabrication of the replacement bottoms was carried out (see Figure 9.3).

Determination of Base-Metal Weldability. The tanks were all constructed of ASME SA-240 Type 304L stainless steel plate material and fittings, so there was no question of weldability.

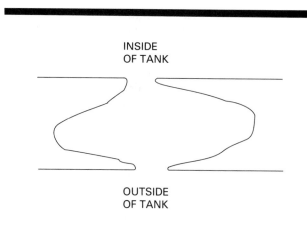

INSIDE
OF TANK

OUTSIDE
OF TANK

Figure 9.5— Penetration of Bottom of Stainless Steel Cold-Storage Tank by Microbially Influenced Corrosion Cell

Determine the Welding Process. The defects were so small the determination was made to use the gas tungsten arc welding (GTAW) process to make all of the repairs.

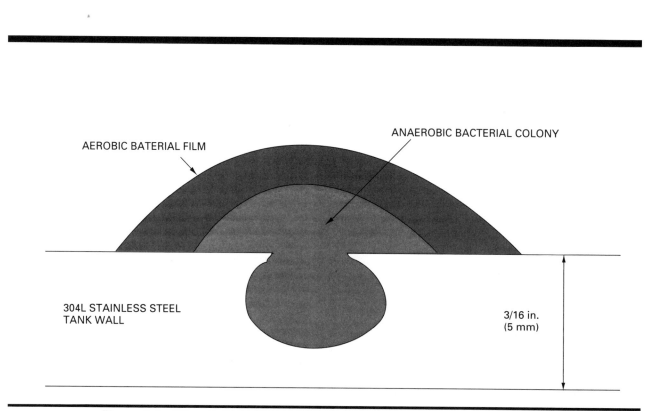

AEROBIC BATERIAL FILM

ANAEROBIC BACTERIAL COLONY

304L STAINLESS STEEL
TANK WALL

3/16 in.
(5 mm)

Figure 9.4—Microbially Influenced Corrosion Cell in Cold-Storage Tank Wall

Determine Matching Filler Metal. During the original construction, ASME SFA5.9, ER308L stainless steel filler metal was used. The nature of the defects from corrosion and the corrective action to prevent future corrosion of this type was so straightforward that no need was found to require a change in the filler metal.

Are Shrinkage and Distortion Acceptable? Although the number of repairs was very large even after the decision to replace the bottoms, the nature of the repairs required very little actual welding to restore the defective area to an acceptable condition. Therefore, shrinkage and distortion were determined to be of no consequence.

Determine Welding Standards. The tanks were atmospheric types with cooling jackets around the outside of them. The jackets were pressurized for maximum cooling efficiency, and this condition required that they be repaired in accordance with ASME Section VIII, unfired pressure vessels with qualified welding procedures and certified welders.

Establish Repair Procedure. The type of welding procedure used was required to meet several objectives. The first was that since all of the corrosion sites were not the same size, those that were well developed would require grinding to expose the defect sufficiently for repairing to begin. The second concern was the remaining immature sites which were no larger than a small pinhole and much more numerous than the well-developed sites. The object of the repair of these smaller sites was to fuse them out using just a GTAW torch and to skip the grinding process. However, before this method of repair was performed on the production defects, some assurance was required to show that it would work consistently.

The tests to demonstrate the fusing-type repair was done on pieces of the now scrap tank bottoms. Different welding parameters were tried on selected defects, and then nondestructive and destructive tests were performed on these repairs to ascertain their success. A method using standard GTAW parameters was selected.

Is Welder Qualified? The testing of different repair procedures, described subsequently, revealed that some training in the use of the proper method of weld repair was necessary before a welder could be released to work on the tanks, even though all of the welders were previously qualified for making stainless steel welds using the GTAW process as prescribed by ASME.

Provide More Information and Training. For this training of the welders, the scrap tank bottoms were employed. The welders performed the repairs in accordance with the procedure, and the results of their practice welds were evaluated using liquid-penetrant inspection.

Preparation of the Defect Area. The preparation of the defect areas required that all corrosion activity be stopped and the by-products be removed as much as possible. These two objectives were met by a combination of drying and chemical cleaning.

Welding Repair Performance. During the actual performance of the repairs, the tank walls were laid out in three-foot square grids so that the work could be carried out in a systematic fashion. The areas to be repaired were first liquid-penetrant inspected to identify all of the defects. Those identified as needing grinding before welding were taken care of first. The welding repairs were performed using standard GTAW parameters and puddling the heat around the defect until the area became molten, which tended to float out any defects. This step was followed by the addition of some filler metal to bring the weld to a complete fill. After the welding was completed in a grid, all required grinding was performed to bring the surface back to acceptable contours.

Inspection. The repair areas were liquid-penetrant inspected again following welding. Any defects that showed up at this time were ground before being repaired a second time. Very seldom was a third cycle of repairs required in any grid. The final result was an acceptable repair as determined by visual and liquid penetrant inspection.

Has Repair Been Successful? The same nondestructive method used during the original construction of the tanks was employed for the replacement bottoms to ensure that the repairs had been successful. By the time the new tank bottoms began to arrive at the plant site, the first of the tanks requiring sidewall repairs had been completed, allowing the old bottoms to be removed and the new bottoms installed. This replacement was accomplished by disconnecting the tanks from the piping and raising them up enough to cut the bottom free and into pieces for removal. The new bottoms were placed under the tank in pieces and welded together.

Part Protection and Return to Service. A hydrotest at the completion of the replacement and repairs to the tanks was performed with filtered, treated water, preventing a repeat of the original corrosion problem. Following the successful hydrotest, the tanks were immediately caustic cleaned to remove any impurities that may still remain in the tanks. The tanks were then returned to service.

Repair Welding of a Pressure Vessel

PRESSURE VESSEL TOWERS used in petrochemical processing are subjected to aggressive environments and on occasion may incur localized metal loss due to corrosion or erosion.[10] One process plant, consisting of two stripping towers, each with two reboilers, incurred metal loss around the nozzles where the reboiler return fed back into the pressure vessel.

The vessel, 80 ft (24 m) tall by 11 ft (3.5 m) in diameter, was fabricated from 3/8 in. (10 mm) thick ASME SA516-60 plate. The repair involved on-site replacement of a 7-ft (2.1-m) high, 360-degree shell course, which was done in six equal sections. (See Figure 9.6) The weld-surfaced inlet nozzles and solid stainless steel couplings also had to be welded into the new shell sections. The steps in the weld repair decision tree (Figure 9.1) are reviewed below for these components.

Identify Welding Repair Need. The metal loss around the nozzles resulted in leaks. The pressure vessel clearly indicated the need for welding repairs.

Has Nature of the Defect Been Determined? Identifying the location of metal loss in the vicinity of the inlet nozzle from the reboiler was accomplished through a preliminary ultrasonic examination performed on the stripping vessel while it was being brought down. Visual inspection was performed after the unit was brought to ambient conditions and rated safe for entry. This investigation revealed features resembling a "river delta" pattern, as shown in Figure 9.7. The metal loss extended from 4 to 8 ft (1.2 to 2.4 m) on either side of a reboiler inlet nozzle. The pattern of metal loss is shown in Figure 9.8. Subsequent examination indicated that the pattern of erosion and corrosion initiated downstream of the flange-to-nozzle neck weld on the reboiler. Apparently, the root reinforcement of the weld created turbulence, which contributed to the metal loss.

Is Welding Repair the Right Decision? Time and safety were major considerations in developing a repair strategy. Based on consultations with corrosion engineers, the use of austenitic stainless steel was identified as a viable means of preventing future shell corrosion. While it was clear that weld repairs were needed, several options were developed and considered before the final welding repair decision was made:

(**1**) Restore the carbon steel shell with matching weld metal and attach stainless steel wear plates.

10. This case study is based on information from the following source: Danis, J. I., "Repair Welding of a Gas Stripping Tower." *Welding Journal* 65(1): 18-23, 1986.

Figure 9.6—Stripper Towers with One of Six Equal Shell Sections Removed

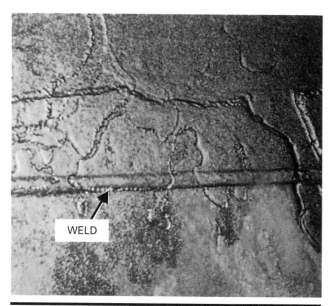

Figure 9.7—"River Delta" Pattern of Corrosion. Weld Metal Appears Less Affected than Base Metal.

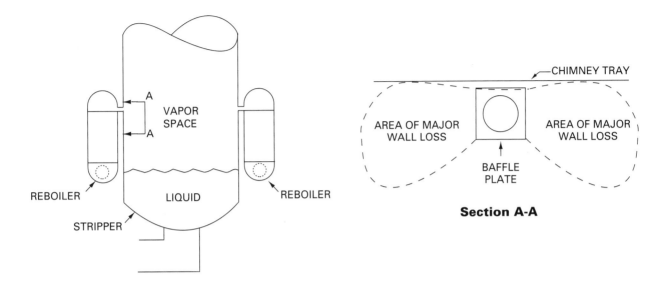

Figure 9.8—Location and Pattern of Metal Loss Around Inlet Nozzles in Gas Stripping Tower

(2) Restore the carbon steel shell with matching weld metal and surface the shell with stainless steel.

(3) Replace the corroded sections with carbon steel plates and weld surface with stainless steel.

(4) Replace the entire shell course with carbon steel and weld-surface with stainless steel.

(5) Replace the entire shell course with stainless steel clad plate.

Option 5 was selected as best satisfying the objectives of minimizing downtime while ensuring vessel life.

Determination of Base-Metal Weldability. ASME SA516-70 plate, 1/2 in. (12 mm) thick, clad with 3/16 in. (4.8 mm) nominal thickness of stainless steel, was purchased to SA-264 and SA-578 specifications. Explosion-welded cladding was selected in lieu of roll cladding due to a more favorable delivery schedule. While 304L stainless steel was desired, some plates were clad with 304 stainless steel due to its availability. Some additional requirements were determined to be necessary, since the 8 x 20 ft (2.4 x 6 m) clad plates were significantly larger than explosively clad plate previously used by this petrochemical company. These additional requirements were as follows:

(1) Ultrasonic examination from the clad side to ASME SA-578, Level 1, acceptance standards

(2) Plate flatness to meet the requirements of ASME SA-20

(3) Shear strength of cladding to be reported for information only

The ultrasonic inspection identified areas of separation in some of the clad plates. The loose areas were removed by a combination of air carbon arc gouging and grinding. Following a liquid-penetrant inspection to verify that the steel substrate was not damaged during removal of the loose cladding, the plates were rolled to the radius of the stripping vessel and weld repaired.

Determine the Welding Process. All welding for this field repair was performed using manual shielded metal arc welding (SMAW).

Determine Matching Filler Metal. Welding the affected areas of the rolled plates involved the deposition of a barrier layer of E309L filler material followed by two passes using E308L. The 309L filler metal was chosen as the barrier layer since it can accommodate dilution from the carbon steel base metal and still produce an austenitic stainless steel deposit chemistry. Type E308L was chosen for subsequent filler passes to produce a chemistry which closely matched that of the

Type 304L stainless steel cladding. Other filler metal details of the repair of various components of the shell course are included in the procedure steps shown in Table 9.2.

Are Shrinkage and Distortion Acceptable?

The clad plates were extensively deformed after explosion welding. However, since the majority of deformation was in a concave direction, it was not considered to be detrimental to subsequent rolling operations.

The reboiler nozzles, which were located in the top head of the reboilers, also experienced metal loss. Replacing them with new nozzles was evaluated. However, it was considered that distortion of the reboiler heads was likely. Consequently, it was decided to remove the heads, which bolted to the shell of the reboiler, and manually weld-surface the nozzles with a combination of E309L and E308L. Following welding and inspection, the heads were postweld heat treated (PWHT). A slight amount of nozzle distortion was measured after PWHT. This was corrected by re-machining of the affected flange faces.

Determine Welding Standards. All welding was performed in accordance with the requirements of ASME IX. The welding procedures and welder qualifications were reviewed for compliance to the code and to internal company standards.

Establish Repair Procedure. The steps in the repair procedures for various tower components are shown in Table 9.2.

Preparation of the Defect Area. The stripping tower repair plan included removing and replacing the shell course in six equal sections. It was originally desired to replace shell sections in diametrically opposite positions, to minimize the adverse effect that weld shrinkage could have on tower straightness. However, to minimize the amount of crane repositioning required to achieve this balanced welding, the welding sequence

---

Table 9.2
Procedures for Shielded Metal Arc Welding Repair of Gas Stripper Tower Components

Procedure Step (for components marked with "X")	Shell Section Girth Seams	Shell Section Vertical Seams	Inlet Nozzles from Reboiler	Stainless Steel Couplings
Align or fit up stripper nozzle with reboiler nozzle	—	—	X	—
Preheat to drive off surface moisture	X	X	X	X
Deposit root pass from interior	X[a]	X[a]	X[a]	X[b]
Fill with E7018 (E309L for stainless steel couplings)	X	X[c]	X[c]	X[c]
Liquid-penetrant inspect	—	X	X	—
Backgouge from exterior	X	X	X	—
Liquid-penetrant inspect	X	X	X	X
Fill with E7018 (E309L for stainless steel couplings)	X[d]	X	X	X
Liquid-penetrant inspect	—	—	X	X
Radiograph	—	X	—	—
Position reinforcing pad, and weld with E7018	—	—	X	—
Deposit E309L	X[e]	X[f]	X[f]	X[f]
Liquid-penetrant inspect 25% of first layer	—	X	X	X
Deposit one pass minimum of E308L	—	X	—	X
Radiograph	X	—	—	—

a. E6011 filler metal.

b. E309L filler metal.

c. Fill to stainless steel interface.

d. Develop taper between shell courses.

e. Make "seal weld" at edge of clad plate.

f. Butter layer.

shown in Figure 9.9 was used. Also shown are the presence of stiffening rings and vertical braces designed to ensure that the vessel would be self-supporting and not suffer mechanical distortion or buckling once sections were removed.

Welding Repair Performance. The weld repairs were performed using the sequence shown in Figure 9.9 and in accordance with the procedures in Table 9.2. Inspections, which were done at various stages, are described below.

Inspection. Proper quality of the back cladding was considered essential to the repair. To achieve this, the following inspection steps were imposed:

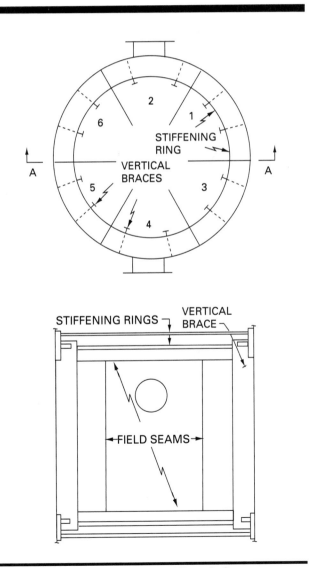

Figure 9.9—Schematic of Repair Sequence

(1) Liquid-penetrant inspection of 25 percent of the first layer of back cladding to verify that excessive dilution had not occurred.

(2) Chemical analysis of all back-cladded seams. Two areas from each seam were analyzed for chromium and nickel with a field-portable nuclear fluorescence analyzer.

(3) Ferrite-content verification in all back-cladded seams. Ferrite was measured at two locations per seam with a Severn gauge.

(4) Liquid-penetrant inspection of all back-cladded seams following hydrotest to ensure freedom from cracking.

Following replacement of the shell course, it was necessary to reinstall two major beams that supported the lower tray in the tower. Since these were to be welded directly to the stainless steel cladding rather than to the carbon steel backing material, it was necessary to verify the integrity of the cladding at the attachment area. This was accomplished by ultrasonic inspection of all material within 6 in. (150 mm) of the attachment weld from the base metal side. A similar ultrasonic examination was performed at all repaired areas in the clad plate.

Has Repair Been Successful? After postweld heat treatment of the upper and lower horizontal girth seams, to minimize the potential for environmental cracking, the vessel was hydrostatically tested. This was followed by a final liquid-penetrant inspection conducted on back-cladded seams, which verified freedom from cracking.

Part Protection and Return to Service. Immediately following the repair, the pressure vessel was returned to service. Periodic ultrasonic thickness monitoring was recommended to help ensure the long-term reliability of the repair.

Repairing Wear on Coal Crushers

WEAR ON ROLL crusher surfaces in coal crushers and pulverizers is a common problem for coal-fired power plants in the utility industry.[11] Hundreds of tons of coal must be crushed and then pulverized into powder every hour to keep combustion furnaces running. Crushers reduce coal chunks of up to 4 in. (101 mm) diameter into pieces of 1 in. (25 mm) diameter or smaller. A ring-type rotor is the crusher component that takes much of

11. This case study is based on information from the following source, which is gratefully acknowledged: Kuvin, B.F., "Weld surfacing cures coal crushers." *Welding Design & Fabrication*, 62(7): 39-41, 1989.

the beating from this highly abrasive process. (See Figure 9.10).

A pulverizer, which then grinds the smaller coal pieces into fine powder, consists of a bowl measuring nearly 9 ft (2.7 m) across at the top and three rolls, in the form of truncated cones, that rotate in the bowl. These components are shown in Figure 9.11 waiting for reclamation. The steps in the weld repair are reviewed below for selected components.

Identify Welding Repair Need. Usually, the repair and reclamation of worn coal crusher and pulverizer components occurs as regularly scheduled maintenance. Shielded metal arc welding was used to fill in uneven areas, to build up worn-down parts to original dimensions, and to apply hardfacing alloys to wear surfaces.

Has Nature of the Defect Been Determined? While the defects or needs associated with regular

maintenance welding are well known, it is important in a case such as this to measure the parts to document the amount of wear that must be repaired. During this visual inspection, loose and broken hardfaced buildup also may be removed.

Is Welding Repair the Right Decision? With regular maintenance welding, this decision is routine and is based upon previous experience as to the most economical methods of keeping the components in useful service. However, new welding processes and new filler metal alloys are evaluated to ensure that the most economical methods are being used.

Determination of Base-Metal Weldability. In this case, the base metal is known to be cast iron with a cast high-nickel coating that is 2 in. (51 mm) thick.

Determine the Welding Process. Three welding processes were planned for use on this project. For the

Courtesy of *Welding Design & Fabrication*

Figure 9.10—Suspension Arm Surfaces of Ring-Type Rotor are Restored to a Hardness of 100 R$_b$ by Manual Open-Arc Welding

pulverizer rolls, a specially designed automatic submerged arc welding (SAW) machine was used (see Figure 9.12). Also, a 16 in. (406 mm) diameter hub cover for the pulverizer bowl center was resurfaced using submerged arc welding.

To restore dimensions to the inner surfaces of the pulverizer bowls, a manual open-arc welding process was used. Following this, a semiautomatic flux cored arc welding (FCAW) process was used in mechanized cap passes to hardface the bowl inner surfaces.

Determine Matching Filler Metal. The surfacing of the pulverizer rolls with the submerged arc welding process makes use of a high-alloy iron-based filler wire, 1/8 in. (3.2 mm) in diameter, containing up to 36% total chromium, manganese, silicon, zirconium, and carbon. The pulverizer bowl is first built up with a proprietary wire, 1/16 in. (1.6 mm) in diameter, containing up to 4% total chromium, manganese, silicon, molybdenum, and carbon. Then the hardsurfacing mechanized passes use 100 HC hardfacing wire, 7/16 in. (11.1 mm) diameter, containing 37% total chromium, manganese, silicon, molybdenum, and carbon.

Are Shrinkage and Distortion Acceptable? Shrinkage and distortion are not a problem in this weld surfacing job.

Establish Repair Procedure. The repair procedures were established and are shown in Table 9.3 for selected component repairs.

Is Welder Qualified? The welders for this project were certified as having appropriate qualifications. Also, they had sufficient experience in scheduled maintenance-welding work of this magnitude.

Welding Repair Performance. The welding repairs of the pulverizer rolls and the bowl inner surfaces, including the FCAW cap passes, were performed in accordance with the procedures in Table 9.3.

Part Protection and Return to Service. All components were returned to service for another year of rugged abrasion and wear from the crushing and pulverizing of coal rocks. Each pulverizer grinds 76 tons of coal per hour.

Courtesy of *Welding Design & Fabrication*

Figure 9.11—Coal Pulverizer Bowl (background) and Three Rolls Attached to Their Journal Assemblies (foreground)

Courtesy of *Welding Design & Fabrication*

Figure 9.12—SAW Head is Moved Back and Forth to Fill in Low Areas in Resurfacing of Pulverizer Roll

Fertilizer Processing Equipment

PIPING AND STORAGE tanks in fertilizer processing equipment develop leaks that are repairable through welding.[12] The manufacture of common fertilizers uses large quantities of corrosive chemicals such as sulfuric and phosphoric acids and ammonia. Tanks and piping that carry acids have been made of A36 carbon steel and are protected by rubber inner linings. Nevertheless, they frequently corrode and leak. Repair welding is used to seal these leaks as well as to repair other equipment in this severely corrosive environment.

Pipe segments used in piping repairs are shown in Figure 9.13. These are 38 in. (965 mm) in diameter by 15 ft (4.6 m) long. They are fabricated from two 5/8 in. (16 mm) plates rolled into semicircles and welded longitudinally. The chemical storage tank uses 3/8 in.

(9.5 mm) ASTM A36 plate at the bottom and 1/8 in. (3.2 mm) plate at the top.

Other fertilizer processing equipment suffer both corrosive and abrasive wear. Mixing blades on agitators used to stir corrosive chemicals must be replaced at least annually (see Figures 9.14 and 9.15). The mixing blades are of AISI Type 317 stainless steel and are 1/2 in. (12.7 mm) thick. Screw conveyors transport both phosphate dust and fertilizer pellets, and wear surfaces such as the conveyor flights require periodic hardsurfacing. The steps in the weld-repair decision tree (Figure 9.1) are reviewed below for these components.

Identify Welding Repair Need. When carbon steel storage tanks and rubber-lined piping develop leaks from corrosive material, immediate welding repair is required. Unscheduled repairs may account for up to 75 percent of the repair and maintenance work at a fertilizer production plant. For a tank-wall leak, a temporary steel patch is welded onto the tank after the contents

12. This case study is based on information from the following source, which is gratefully acknowledged: Kuvin, B.F., "Hardfacing battles corrosion." *Welding Design & Fabrication*, 62(7): 43-45, 1989.

Table 9.3
Welding Process Parameters for Selected Repair of Coal Crusher and Pulverizer Components

SAW of Pulverizer Rolls	Manual Open-Arc Welding of Pulverizer Bowl Inner Surfaces	FCAW Cap Passes on Pulverizer Bowl Inner Surfaces

Current: 350 A
Voltage: 28 - 30 V
Travel speed:
  First pass: 40 in./min (17 mm/s)
  Second pass: 45 in./min (19 mm/s)
Wire feed speed: 59 in./min (25 mm/s)
Hardness achieved: 57 R_c
	Current: 300 A (DCEP)
	Voltage: 28 V
	Electrode stickout: 1/2 - 3/4 in. (13 - 19 mm)
	Weave: Slight weave to make beads
	    1/2 - 3/4 in. (13 - 19 mm) wide
	Hardness achieved: 40 R_c (two-layer buildup)
		Current: 300 - 375 A
		Voltage: 28 V
		Travel speed: 55 - 65 in./min (23 - 28 mm/s)
		Hardness achieved: 62 R_c

are drained to below the level of the leak. This patch and surrounding material are cut out during scheduled maintenance, after the tank has been drained and cleaned thoroughly, and a permanent patch is welded into place. The need for welding repairs usually is just as obvious with the other equipment components.

Has Nature of the Defect Been Determined?
Corrosive leaks represent a never-ending war in the maintenance and repair of fertilizer processing equipment and storage tanks. Scheduled maintenance of this equipment involves periodic inspections to check for places where chemicals have eaten through pipes, walls, or other components. This problem is so pervasive that any other type of defect may be rare. For this reason, maintenance inspectors need to be especially alert to identify correctly any other type of defect that also may occur.

Courtesy of *Welding Design & Fabrication*

Figure 9.13—Acid-Transmission Pipe Segment Will Have Flanges Welded to Ends for Bolting in Field Repairs

Is Welding Repair the Right Decision? In the case of the defects described here, welding typically has been established to be a repair method that is both efficient and cost-effective. Welding or hardfacing usually are involved in about one-fifth of all emergency repairs made at a fertilizer production plant.

Determination of Base-Metal Weldability. The base metal for the chemical storage tanks and piping is ASTM A36 carbon steel plate. The agitator mixing blades are AISI type 317 stainless steel.

Determine the Welding Process. The objectives of this job could be met by using the manual shielded metal welding (SMAW) process and semiautomatic flux cored arc welding (FCAW).

Determine Matching Filler Metal. Filler metals selected for each of the equipment components are shown along with process parameters in Table 9.3.

Establish Repair Procedure. The repair procedures were established and are shown in Table 9.3 for selected components.

Is Welder Qualified? The welders for this project were certified as having appropriate qualifications. Also, they had sufficient experience in welding repairs of this magnitude.

Preparation of the Defect Area. Chemicals were removed from the tank and piping to permit cleaning. The tank was then removed to the weld area for positioning and repair.

Welding Repair Performance. The welding repairs of the fertilizer tanks and piping system components were performed in accordance with the procedures in Table 9.3.

Inspection. Visual and liquid penetrant inspections were conducted to ensure that the system was free from cracks.

Courtesy of *Welding Design & Fabrication*

Figure 9.14—Agitator Shaft and Mixing Blades Used to Stir Phosphate Dust and Sulfuric Acid in the Production of Fertilizer

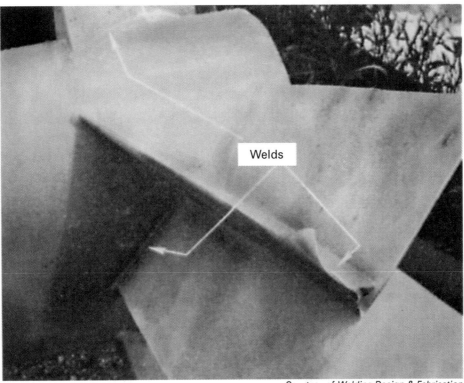

Welds

Figure 9.15—Welds Join Replacement Mixing Blades to Shaft of Agitator

Table 9.4
Welding Process and Parameters for Repair of Fertilizer Tank and Piping System Components

Process and Parameters	Fertilizer Tanks	Piping	Flanges to Piping*	Mixing Blades (Type 317 Stainless)	Conveyor Flights (Hardfacing to 60 R_c)
Process	SMAW	SMAW	FCAW	SMAW	SMAW
Current	120 -160 A DCEP	75-125 A (DCEP)	(DCEP)	140-160 A (DCEP)	180-240 A (DCEP)
Electrode diameter	1/8 in. (3.2 mm)	1/8 in. (3.2 mm)	5/64 in. (2.0 mm)	3/16 in. (4.8 mm)	1/4 in. (6 mm)
Electrode class	E7018 (AWS A5.1)	Root pass: E6010 Fill passes: E7018 (AWS A5.1)	E70T-1 (AWS A5.20)	E320 coated (AWS A5.4)	EFeCr-A1 (AWS A5.13)

* Gas and flow rate, CO_2 at 50 ft³/h (24 L/min) minimum. Wire feed speed, 275 in./min (116 mm/s). Electrode stickout, 1 in. (25 mm). Voltage, 31 V. Interpass temperature, 324 °F (163 °C).

SUPPLEMENTARY READING LIST

American Society for Metals. "Welding, brazing and soldering." *Metals Handbook*, Vol. 6, 9th Ed. Metals Park, Ohio: American Society for Metals, 1983.

Danis, J. I., "Repair welding of a gas stripping tower." *Welding Journal* 65(1): 18-23, 1986.

Easterling, K. *Introduction to Physical Metallurgy of Welding*, Chapter 1. London: Butterworth & Company Ltd., 1983.

Hellier, C. J. "What's new in nondestructive testing of welds and heat affected zones." *Welding Journal* 72(3) 39-44, 1993.

Irving, B. "Preheat: The main defense against hydrogen cracking." *Welding Journal* 71(7): 25-31, 1992.

—————. "Stress corrosion cracking: Welding's No. 1 nemesis." *Welding Journal* 71(12): 37-9, 1992.

Kjeld, F. "Gas metal arc welding for the collision repair industry." *Welding Journal* 70(4): 39-46, 1991.

Kotecki, D. "Hardfacing benefits maintenance and repair welding." *Welding Journal* 71(11): 51-3, 1992.

Kou, S. *Welding Metallurgy*, 109-25. New York: John Wiley & Sons, 1987.

Kuvin, B.F. "Hardfacing battles corrosion." *Welding Design & Fabrication* 62(7): 43-5, 1989.

—————. "Weld surfacing cures coal crushers." *Welding Design & Fabrication* 62(7): 39-41, 1989.

Masubuchi, K. "Recent research activities at MIT on residual stresses and distortion in welded structures." *Welding Journal* 790(12): 41-7, 1991.

Mechley, J. A., and Petroski, A. "Developing welding skills for today's technology." *Welding Journal* 69(9): 71-3, 1990.

Menhart, J. M. "Inplant quality standards for maintenance welding." *Plant Engineering*: 52-5, March 23, 1989.

Scruggs, D. M. "Advanced materials move into hardfacing." *Welding Journal* 70(8): 27-9, 1991.

Siewert, T. A. "Development of a weld procedure to repair joints in a railroad-type track." *Welding Journal* 67(8): 17-23, 1988.

Yosh, D. and Caploon, A. "Successful welding repair of turbine casing cracks." *Welding Journal* 71(2): 29-32, 1992.

UNDER-WATER WELDING AND CUTTING

PREPARED BY A COMMITTEE CONSISTING OF:

C.E. Grubbs, Chairman
*Global Divers &
Contractors, Inc.*

A. E. Bertelmann
*Global Divers &
Contractors, Inc.*

S. Ibarra
Amoco Research Center

S. Liu
Colorado School of Mines

D. J. Marshall
*Global Divers &
Contractors, Inc.*

D. L. Olson
Colorado School of Mines

WELDING HANDBOOK COMMITTEE MEMBER:
P. I. Temple
Detroit Edison

CHAPTER 10

UNDERWATER WELDING AND CUTTING

INTRODUCTION

UNTIL THE ADVENT of exploration and production of offshore energy resources, underwater welding was used only infrequently and with highly unpredictable results. This early *wet welding* was done at ambient pressure with the welder/diver in the water and with no mechanical barrier around the arc. Dry hyperbaric (pressure conditions in excess of surface pressure) welding was unheard of. Cutting underwater was generally limited to salvage and for the removal of piling and inland waterway obstructions.

As the number of offshore structures increased, and the older ones began to experience fatigue and corrosion damage, plus accidental damage during and after installation, the need for underwater welded repair methods, with predictably satisfactory results, increased dramatically. As the materials used in offshore structures and subsea pipelines escalated from readily weldable low-carbon steel to higher strength steels that were subject to hydrogen-induced cracking, the development of new and better welding materials and procedures had to keep pace.

This chapter addresses underwater (hyperbaric) wet and dry welding and the processes and methods that are most commonly used. Properties of weldments and specifications for qualifying welding procedures and welders are referenced, and inspection methods and procedures are discussed. The following underwater cutting processes are covered: oxygen arc, arc water, oxyfuel gas (three processes), plasma arc, shielded metal arc, and thermal cables and lances. Safety precautions and procedures are recommended to cope with the hazards associated with underwater welding and cutting, especially those that are unique to welding and cutting in the underwater environment.

BRIEF HISTORY

ALTHOUGH SOUND STRUCTURAL wet welds were not made before 1970, wet welding is not new. In 1917, underwater wet welding was used to stop leaks in seams and rivets in ship hulls. As World War II drew to a close, underwater welding was an important tool used in raising sunken vessels. This tool, however, was in the hands of divers who seldom, if ever, were skilled welders. As a result, underwater wet welding was relegated to salvage and emergency repairs and reliability of the results was unpredictable.

The first documented wet welded structural repair was made in inland waters in 1970.[1] The first major wet welded repair on an offshore structure was made in 1971.[2] Damage and repair are shown in Figures 10.1A and 10.1B. Since then hundreds of wet welded repairs have been made to offshore structures, subsea pipelines, ship hulls, and dock and harbor facilities.

In 1985, wet welding procedures and five welders were qualified at depths down to 325 ft (100 m) and subsequently used to make structural repairs at depths of 165 ft (50 m) and 320 ft (100 m) on an offshore

1. Chicago Bridge and Iron Company used wet welding in the repair of a dock retaining structure near Memphis, Tenn., in 1970.
2. This repair, supervised by C. E. Grubbs, was made on a drilling and production platform of the Humble Oil Co. (Exxon) in the Gulf of Mexico.

Courtesy of Chicago Bridge & Iron Company

Figure 10.1A—Before Underwater Wet Welded Repair, Offshore Structure K-Brace is Damaged

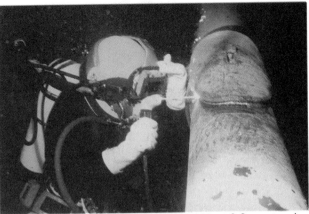

Courtesy of Global Divers & Contractors, Inc.

Figure 10.2—Welder/Diver Repairing Offshore Drilling and Production Platform at Depth of 320 ft (100 m)

Courtesy of Chicago Bridge & Iron Company

Figure 10.1B—After Underwater Wet Welded Repair, Offshore Structure K-Brace is Sound

production platform in the Gulf of Mexico. (See Figure 10.2.) As of 1995, no wet welding procedure had been qualified deeper than 325 ft (100 m). Wet welds qualified to ANSI/AWS D3.6-93, *Specification for Underwater Welding*, were first used on stainless steel at nuclear power plants in late 1987.[3]

The first dry hyperbaric weld was used to *hot tap* a 6-in. (152-mm) diameter branch pipeline to a 16-in. (406-mm) diameter trunk line in the Gulf of Mexico in

3. Refer to ANSI/AWS D3.6-93, *Specification for Underwater Welding*. Miami: American Welding Society, 1993.

1965. Water depth was 80 ft (24 m). The process was next used for a subsea pipeline tie-in about 1972 and shortly after that came into use for repairs to offshore structures. Some of the early dry welding chambers were as large as small machine shops, weighed up to 128 000 lb (58 100 kg), and were equipped with powerful hydraulic rams and clamps for forcing pipe ends into alignment for welding. Their sizes and elaborate life support and environmental control systems probably led to their being called *habitats* instead of *dry welding chambers*. Welding chambers with line-up frames and hydraulic equipment are still required for joining pipe underwater. For pipeline hot taps and structural repairs, however, weld chambers now are no larger than is required to encompass the weld area and accommodate one or two welders.

In 1978, a United States diving contractor qualified a shielded metal arc (SMAW) dry welding procedure on API 5LX65 pipe [30 in. diameter by 1 in. wall (762 mm by 25.4 mm)] at an offshore depth of 1000 ft (305 m). The welding procedure was later used for a subsea pipeline tie-in at 1012 ft (308 m). Subsequently the same contractor made many deep-water pipe welds, but 1012 ft (308 m) is believed to be the record depth for practical use of underwater welding. A North Sea contractor has made several dry welded production pipeline tie-ins offshore at depths down to 722 ft (220 m). A mechanized gas tungsten arc welding (GTAW) procedure was used for the tie-ins. The same contractor has, in hyperbaric facilities onshore, qualified pipe welds twice (1984 and 1991) manually with a combination of GTAW and flux cored arc welding (FCAW) at a pressure equivalent to 1476 ft (450 m) of seawater. Experimental work at two different European research centers have resulted in pipe-welding procedures being qualified under a pressure equivalent to 1968 ft (600 m) of seawater.

Underwater oxyfuel gas cutting was tried as early as 1908, but with little success. In 1926, the oxyhydrogen torch was developed and used for salvage of a submarine that had sunk in 130 ft (40 m) of water. Until then, oxyacetylene torches were used, but not deeper than about 25 ft (8 m). The first of several underwater cutting processes that use oxygen, with no fuel gas, was developed in 1942. The arc water gouge prototype dates to about 1980, but was not widely used until 1985.

GENERAL DESCRIPTIONS

UNDERWATER WET WELDING may be described as welding at ambient pressure with the welder/diver in the water with no physical barrier between the water and the welding arc. Metallurgically, the wet welding process is comparatively complex. However, wet welding underwater closely resembles welding in air, in that the welding arc and molten metal are shielded from the environment — water or air — by gas and slag produced by decomposition of the flux coated electrodes or flux cored wire, and the welder welds in the same "clothes" that were worn getting to the work site.

Underwater dry welding is done at ambient pressure in a chamber from which water has been displaced. Depending on the size and configuration of the chamber, the welder/diver may be completely in the chamber, or only partially in the chamber, and may work in conventional welder's attire, dive gear, or a combination of both. The various modes of dry underwater welding are discussed later in this section.

GENERAL APPLICATIONS

UNDERWATER WELDING METHODS have been used during the installation of new offshore structures, subsea pipelines and hot taps, docks and harbor facilities, and have been used for modifications and additions to underwater structures. However, underwater welding is most often required for repairs to existing structures. Maintenance and repair applications include:

(1) Offshore structures - Repair of damage caused by corrosion and fatigue. Repair or replacement of structural members damaged during installation, by objects falling overboard, boat collisions, or other accidental damage.

(2) Subsea pipelines - Repair or replacement of damaged pipeline sections and pipeline manifolds.

(3) Harbor facilities - Repair of corrosion and collision damage to sheet and H piling and to cover openings where sheet pile have been driven out of interlock during construction. Also repair and replacement of tubular braces and supports of docks and tanker mooring dolphins.

(4) Floating vessels - Permanent and temporary repairs to holes in ship and barge hulls and to hulls and pontoons of semisubmersible drill ships.

(5) Nuclear power plants - Repair of steam dryer damage caused by fatigue and mishandling, feedwater spargers, other reactor internals, and leaks in pool liners. (See Figure 10.3.)

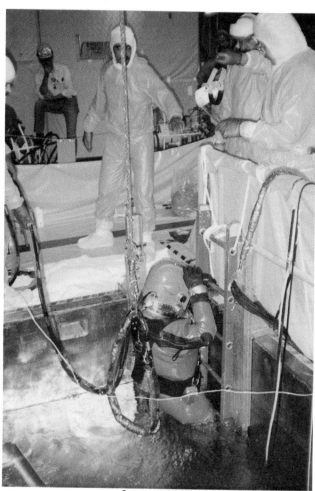

Courtesy of Global Divers & Contractors, Inc.

Figure 10.3—Welder/Diver and Support Personnel at Nuclear Power Plant During Wet Welded Repairs to a Stainless Steel Steam Dryer

Materials encountered in the above mentioned repairs included structural carbon steel with carbon equivalents (CE) from less than 0.40 to high-strength steels with 0.47 CE.[4] The carbon equivalent of some sheet pile was 0.695. Nuclear power plant material was austenitic stainless steel.

Three specific application case studies are described in the Applications section beginning on Page 495. One, the underwater wet welding repair of an offshore

_____

4. Carbon equivalent (CE) is an empirical guide relating the chemical composition of steel with hydrogen-induced cracking tendencies. Carbon equivalent formulas for underwater welding are CE = C + Mn/6 + (Cr + Mo + V)/5 + (Ni + Cu)/15. Or, if only carbon and manganese are known, CE = C + Mn/6 + 0.05.

structure, provides a good example of why the installation of weld chambers (habitats) for dry hyperbaric welding may not be a viable alternative in some cases. Another, the repair of a pipeline river crossing, used small weld chambers large enough to accommodate the welder/diver down to the waist for dry hyperbaric welding at depths of 72 and 90 ft (22 and 27 m). In the third application, the repair of a dock, 1800 ft (550 m) of fillet welds were laid down at working depths from the waterline to the mudline at more than 30 ft (9 m) to put repair plates over 83 holes in cylindrical sheet pile cells of a wharf-retaining structure. All of the underwater welding repairs described in these case studies were performed in the early 1970s and still were in excellent condition two decades later.

SAFETY IN UNDERWATER WELDING AND CUTTING

UNDERWATER WELDING AND cutting exposes the diver to the hazards he or she would encounter while doing the same tasks above water, plus all of the risks associated with commercial diving. A third risk category includes hazards that are unique to welding and cutting in an aqueous environment capable of conducting lethal electrical currents and at pressures that, in dry welding chambers, greatly increase the flammability of materials and the explosiveness of gases. Also, the arcs generated by electrical welding and cutting currents dissociate the surrounding water into gaseous bubbles of hydrogen and oxygen mixtures that, when confined in even small volumes, can be violently explosive.

GENERAL

THIS SECTION INCLUDES safety precautions and procedures that apply to all underwater welding and cutting processes. Safety procedures that are not generic are discussed separately in the sections on specific underwater welding and cutting processes. Special precautions also must be followed for the use of any cutting procedure during repair or salvage of vessels or structures with voids or compartments that could trap explosive gases or could contain hazardous materials.

It is important that safety be practiced above water as well as below; accidents above the water can endanger personnel in the water. No work that could result in falling objects should be permitted over the area in which the diver may be working, entering, or leaving the water.

Preparations

PRE-PROJECT AND PRE-DIVE discussions should clearly define the following:

(1) Work to be done
(2) Project plan
(3) Responsibilities of each individual involved in the project
(4) Contingency plans in case of emergencies
(5) All known obstructions and potential hazards that may be unique to the project.

All equipment, rigging gear, power tools, gas cylinders, manifolds, regulators, gauges and hoses, and communication systems must be checked to ensure that everything that is to be used is in good working order.

Welder/Diver Support

DIVE EQUIPMENT AND personnel dive gear, supervision and support personnel, general and emergency diving procedures plus safe practices shall, as a minimum, meet the requirements of the Association of Diving Contractors.[5] The welder/diver should be provided a stable underwater work stage (a suspended platform that supports a diver in the water) except when the welder/diver will be working on a firm bottom or seabed.

_____

5. Consensus Standard For Commercial Diving Operations

For optimum safety and efficiency, three individuals may need to be designated to provide direct full-time support to the welder/diver during welding or cutting. One person (the tender) must maintain communications with the welder/diver, transmit his instructions to others, and operate the welding or cutting current knife switch. A second individual responds to instructions from the tender to adjust amperage control, send electrodes down to the diver, or other tasks. In some settings, especially offshore, a third crew member will tend the diver's hose, maintaining the required tension or slack, and assist when the diver enters and exits the water.

In nuclear applications, additional personnel are required for diver decontamination and radiation monitoring activities. Also, remote underwater video camera technicians are generally needed to assist the health-physics personnel in ensuring that the divers avoid high radiation areas or *hot spots*.

Explosive and Hazardous Materials

IN ADDITION TO the explosive mixtures of hydrogen and oxygen that are disassociated from the water by wet welding and by arc cutting processes, the diver must be aware of other possible sources of explosive gases and of hazardous materials. During the repair or salvage of vessels or other structures with compartments and other voids that could retain explosive gases, or contain hazardous materials, the following precautionary measures are recommended:

(1)   Prior to starting work, as-built drawings should be studied to determine all areas that could contain or might entrap explosive gases that could be ignited by underwater welding or cutting operations.

(2)   Cargo manifests should be reviewed to see what hazardous materials a vessel (ship, tanker, or barge) is carrying and where it may be located.

(3)   Procedures must be developed to vent areas that could contain or entrap explosives and to remove or prevent diver exposure to chemicals or other hazardous materials.

DRY HYPERBARIC WELDING

UNDERWATER DRY WELDING is performed in an open-bottom enclosure (habitat or chamber) that not only excludes the water but, because of its physical dimensions, inhibits the welder/diver's freedom of movement and keeps the welder/diver in close proximity to the weldment and the heat treatment equipment. Safety precautions for dry underwater welding include all of those required for welding in a wet, constrictive space above water, plus additional precautions to cope with hazards associated with diving, increased pressure, complicated logistics and restricted diver ingress and egress.

In preparation for and during the underwater work, the welder/diver must have topside support as described earlier, and a designated stand-by diver to assist in an emergency. Voice communications between the welder/diver and the tender must be maintained at all times; manual signals via the life support umbilical are usually not possible when the welder is in the weld chamber. Loss of communications must be treated as an emergency and the dive aborted until communications are restored. When weld chambers are to be used in mid-water, they should be designed and installed so as to permit easiest possible ingress and egress. When weld chambers are used for joining, repairing, or hot tapping buried pipelines, with the bottom of the chamber below the sea bed, precautions must be taken to prevent the possibility of the sides of the excavation caving in and entrapping the welder.

The partial pressure (percentage) of oxygen (PPO_2) in the background gas that maintains water displacement in the weld chamber is of utmost importance to safety of the welder. Numerous underwater dry welding scenarios indicate that air can be used safely at water depths down to 90 ft (27 m). Below 90 ft (27 m), the background gas should be suitable for life support (breathing gas) and must not be capable of supporting rapid combustion of flammable materials. Helium-oxygen (HeO_2) mixtures are most commonly used with monitoring of PPO_2.

The 90-ft (27-m) depth restriction described above applies to production work in open water in an open-bottom weld chamber where the welder has quick and easy access to the water below in case of a smoldering spot on the welder's clothing. The 90-ft (27-m) depth guideline does not apply to dry welding in a closed hyperbaric chamber for qualification of welding procedures or for other purposes. In such cases, the diving contractor's safety director must dictate the background gas mixture and whether or not the chamber is to be equipped with a high-pressure water-spray system. Regardless of the gas mixture in the hyperbaric chamber, the welder's clothing and gloves should be made of fire retardant materials.

Fire in the underwater weld chamber requires the following, two of which may be controlled:

(1)   Fuel - Keep flammable materials to an absolute minimum.

(2)   Oxygen - The chamber must be equipped with a top-side continuous readout oxygen analyzer.

(3)   A source of ignition (welding sparks), which cannot be avoided.

Monitoring and controlling oxygen levels in the chamber is the primary consideration for fire prevention in the chamber.

Paints, solvents, hydrocarbons, or any other materials that could release toxic or irritating fumes should never be permitted in the chamber. If depth dictates the use of hydraulic, rather than pneumatic, powered tools, the tools with hoses and fittings must be pressure tested at 150 percent of maximum working pressure. For maximum safety, a water-glycol mixture may be used in lieu of conventional hydraulic fluid. Similarly, if pneumatic tools are to be used, the tools and hoses must be flushed clean of any excess lubricants, and in-line lubricators must not be used.

When air is used as the background gas, the weld chamber can be continuously or intermittently vented to avoid accumulation of fumes and smoke. The high cost of mixed gas precludes venting, so smoke and fume scrubbers must be used in the chamber. Similarly if the chamber background gas is other than air, and if exhaust from the welder's diving mask would raise the PPO_2 of the background gas, then the welder's exhaust gas must be discharged outside the chamber by means of an overboard dump system. Similarly, if pneumatic tools are used, an overboard dump system must be used to avoid PPO_2 buildup.

A TV camera for surveillance of activities inside the weld chamber is an important safety factor, especially if voice communications are interrupted. However, use of AC powered equipment should be kept to an absolute minimum. If AC is supplied to the chamber, all circuits must have ground fault interrupters that disconnect the current when only a few milliamperes of leakage is detected.

The welder/diver is equipped with a conventional diver's hat or mask for passage to and from the underwater work site. Once inside the chamber, the welder/diver usually switches to a lightweight life support mask with communications and some type of welder's shield or hood. Located inside the weld chamber is an emergency reserve breathing system (bail-out). The welder wears a protective jacket and gloves to prevent burns from weld splatter and for insulation against heat generated by welding and heat treatment of the weldment. The life support umbilical with breathing gas hose and communication lines is fitted with a flexible heat-shielding sleeve for protection.

WET WELDING AND ARC CUTTING

THIS SECTION ADDRESSES hazards and safe practices associated with underwater wet welding and cutting processes that use an electrical current with an electrode, cable, or lance.

Alternating current (ac) power sources should never be used for welding or cutting underwater. (The use of ac current underwater for any purpose should be avoided.) For maximum safety, diesel driven dc welding machines are preferred. (The use of gasoline driven equipment offshore is prohibited.) If welding equipment is used that requires ac primary current, it must be installed by experienced personnel using approved wiring diagrams to avoid any possibility of the ac current being short circuited into the dc welding circuit.

All power sources should be treated as potential hazards. They should be in good mechanical and electrical condition, and they should be grounded to the support vessel by a welded connection or, if on land, in accordance with the manufacturers' instructions. Welding cables and cable splices must be properly insulated and the ground clamp or lug must be securely attached to the workpiece. A welded ground connection adjacent to the area that is to be welded or cut is preferred.

A 400-amp, single-pole knife switch is installed in the circuit leading to the electrode holder or cutting torch to interrupt the current when required. Electrode holders and cutting torches must be designed specifically for use underwater.

Never use ac current for welding or cutting underwater; it can be fatal. When using dc current underwater, the following precautions and procedures are recommended to minimize the potential of electrical shock, which can cause painful discomfort and possibly adverse health effects.

(1) Diver's attire should include a wet suit in good condition, rubber or rubberized canvas gloves, foot gear, and a suitable dive hat or mask. Gloves must be free of tears and holes.

(2) Nuclear wet welding applications require use of dry suits with integral boots and gloves in order to prevent radiological contamination of the diver. Prior to donning before each dive, the dry suits must be confirmed water-tight by testing with compressed air.

(3) Diver's headgear should be equipped with a welder's lens holder assembly. The tinted lens should be dark enough to provide eye protection from the intense glare of the arc. The assembly should be hinged so as to provide unobstructed vision when not being used for welding or cutting.

(3) The knife switch, operated by the tender topside, must be left open at all times except when the diver calls for current on ("make it hot"). When the diver is at the underwater work site, the responsibility for protection against electrical shocks is shared by the diver and the tender. Normal operating procedure is for the diver to position the tip of the electrode and command, "make it hot." The tender quickly and firmly closes the knife switch and then tells the diver, "it's hot." When the electrode is consumed, or at any other time the diver wants the current interrupted, the diver commands, "make it cold." The tender opens the knife

switch and tells the diver, "it's cold." It is important that the tender verbally confirms the diver's instruction, and that the tender does so only after the appropriate action has been taken.

(4) The knife switch should be located adjacent to the tender's radio to the diver and positioned so that the possibility of accidental closure is minimized. It should be bolted or otherwise firmly attached to a stable base. Motions to open and close the switch should be quick and firm to minimize electrical arcing which can cause eye damage and erode contact areas on the switch.

(5) When the current is on, the diver should avoid shocks by not letting the electrode, electrode holder, or cutting torch touch him or her and the diver should not get between same and the grounded workpiece.

Electrical welding and cutting processes produce explosive mixtures of hydrogen and oxygen. It is important to note the difference in the mixture of gases that result from wet welding processes versus cutting processes that use oxygen.

The hydrogen and oxygen gas mixture produced by wet welding processes is flammable. Bubbles rising to the water surface will burn with a yellow-orange flame when ignited. However, there have been no reported incidents that would indicate that gases disassociated from the water during underwater welding represent any hazard to the welder/diver. Nevertheless, it must be assumed that under enough pressure the flammable bubbles, when accumulated in sufficient quantities, may be explosive. The fuel (hydrogen) and oxygen are present, and for ignition, sparks (perhaps small flakes of semi-molten slag) can be carried by bubbles up several feet into gas-filled voids above the diver. With this in mind, there are two scenarios that require procedures to protect the welder/diver:

(1) If wet welding is to be performed in a confined space, or under structural members of shapes that would hold rising gas bubbles, vent holes should be made to preclude gas entrapment, especially if the water depth is greater than about 66 ft (20 m) [3 atms].

(2) When wet welding in a hyperbaric facility (wet pot), the rising gas bubbles must constantly be vented to avoid significant accumulation. This can be done by designing the wet pot with a comparatively small vented dome at the top of the chamber, or welding can be done under a hood that funnels the gas outside of the chamber (overboard dump).

There is absolutely no doubt about the explosiveness of gases produced by underwater oxygen arc cutting processes. Several diver fatalities and injuries have been attributed to this hazard. In addition to the hydrogen and oxygen disassociated from the water, there is a significant amount of oxygen that is not consumed during the cutting process. This explosive gas mixture can be trapped, and will explode violently, in any unvented area behind or above the material being cut. Mini explosions even occur in the cupped palm of the gloved hand holding the electrode. Risks of this very real hazard can be reduced by anticipating areas where gases can accumulate and venting those areas with holes drilled or made with the arc water torch. (Material up to 3/4 in. (19.050 mm), and more if necessary, can be cut with this torch.) More ingenuity will be required to avoid accumulation of gases when making cuts on mud- and concrete-backed materials.

THERMAL CABLE AND THERMIC LANCE CUTTING

THE THERMAL CABLE and thermic lance cutting processes share the following hazardous characteristic: Both generate large quantities of hydrogen and oxygen which, upon ignition, are consumed rapidly at very high temperatures and will continue to burn until the high pressure oxygen is consumed or cut-off. Also, both processes are apt to be used during salvage operations, which are inherently hazardous.

General Precautions

THE FOLLOWING PRECAUTIONS and safety procedures apply to the use of both the thermal cable and thermic lance:

(1) Never try to speed up cutting by creating an inferno deep inside thick sections of plate or shafts. When cutting thick sections, the cable or lance should be withdrawn every three or four seconds so water can enter the cut and partially cool the superheated metal.

(2) Because of the very rapid burn-off rates [20 to 24 in./min (8 to 10 mm/s)] of cables and lances, the diver must be ever alert to the danger of letting the burning tip of the cable or lance get too close to the diver's hands.

(3) If oxygen pressure drops too low at the tip of the cable, and possibly the lance, oxidation can switch from the tip to the internals and race back toward and behind the diver. A vigilant topside crew should keep this from occurring by maintaining proper oxygen pressure.

(4) In case of emergencies, the high pressure oxygen must be cut off immediately. It is recommended that the oxygen system have both a cut-off valve and a vent valve. Venting the oxygen to the atmosphere stops flow of oxygen at the tip of the cable or lance instantly. When the oxygen is merely cut off, the cable or lance will continue to burn until pressure equalizes at the tip underwater.

Thermal Cable

THE THERMAL CABLE will readily cut ferrous metals, but attempts to use normal motions to cut nonferrous metals may result in violent explosions. Nonferrous metal removal is accomplished by melting rather than oxidation, and removal of molten metal should be done with light punching or sawing motions. Thermal cables will not cut concrete, coral, or rock; attempts to do so may create explosions. Use of thermal cables deeper than about 300 ft (91 m) is not recommended.

Thermic Lance

THE THERMIC LANCE cutting process generates such large volumes of hydrogen and oxygen that explosions sometime occur even though there may be no apparent cause of entrapment other than the surrounding water. These intermittent explosions are most apt to occur when the diver tries to remove concrete and thick nonferrous metals too quickly. These materials should be removed by slow, deliberate punching and stroking motions.

PLASMA ARC CUTTING

BECAUSE SOME PLASMA arc cutting equipment delivers higher amperage and voltage than is used by other underwater cutting processes, all plasma arc systems must be treated with utmost caution above and below water. Installation, operations, and maintenance of the equipment must be performed by qualified personnel and in conformance with the manufacturer's instructions. Cable and hose sizes and gas pressures must be determined by the manufacturer's recommendations.

Based on comparatively limited experience in using plasma arc manually underwater, and the fact that the high-frequency pilot arc initiation of a 1000-amp power source involved 6000 to 9000 volts, utmost protection must be provided to the underwater operator of the plasma arc torch. Until the results of 1992 safety studies are known, recommendations are that the diver be fully insulated by a dry suit with integral, or securely attached, gloves and boots. The diver's dress must be confirmed water-tight by an air pressure test prior to each dive.

Other underwater safety precautions shall be no less than those specified for wet welding and arc cutting.

OXYFUEL GAS CUTTING

OXYFUEL GAS CUTTING processes used underwater include oxyacetylene, oxyhydrogen, and oxygen MAPP.[6] Follow safety procedures for handling gas cylinders and hooking up manifolds, plus those for installing, testing, and maintaining regulators, gauges, and hoses.[7] Safety precautions for the diver using oxyfuel gas torches underwater are the same as for using the equipment above water, with one exception. Because of the instability of acetylene under pressure, oxyacetylene torches are hazardous when used in water deeper than about 25 ft (8 m). Attempts to use oxyacetylene below that depth may result in explosions in or behind the mixing chamber of the torch.

---

6. MAPP is a registered trademark of the BOC Group. MAPP gas is a stabilized mixture of methylacetylene and propadiene.
7. See Safety section in Supplementary Reading List and ANSI/ASC Z49.1, *Safety in Welding, Cutting and Allied Processes,* Miami: American Welding Society (latest edition).

UNDERWATER WELDING

THIS SECTION ADDRESSES underwater wet and dry welding methods, processes, and essential variables. The influence of pressure, hydrogen-enriched environment, rapid cooling rates, and how these unique variables affect the chemistry and mechanical properties of underwater weldments are discussed. Weld properties produced by dry and wet welding methods are compared. Metallurgical considerations in underwater wet welding are discussed. Also, solutions to the metallurgical deficiencies of hyperbaric dry and wet welds are proposed.

Six methods have been used for welding underwater:

(1) Welding at ambient water pressure with the welder/diver in the water without any physical barrier between the water and the welding arc (Wet Welding).

(2) Welding at ambient water pressure in a large chamber from which water has been displaced, in an atmosphere such that the welder/diver does not work in diving gear (Dry Welding in a Habitat).

(3) Welding at ambient water pressure in a simple open-bottom dry chamber that accommodates, as a minimum, the head and shoulders of the welder/diver in full diving gear (Dry Chamber Welding).

(4) Welding at ambient water pressure in a small transparent, gas-filled enclosure with the welder/diver in the water and no more than the welder/diver's arms in the enclosure (Dry Spot Welding). Variations of this method used in nuclear applications include use of automatic or semi-automatic remote dry welding processes (e.g., gas tungsten arc welding) within inverted inert gas-filled enclosures, with the welder/operator situated topside.

(5) Welding in a pressure vessel in which the pressure is maintained at approximately one atmosphere regardless of outside ambient water pressure (Dry Welding At One Atmosphere.)

(6) Welding inside of a closed-bottom, open-top enclosure at one atmosphere (Cofferdam Welding).

In the discussion that follows, Methods 2 and 3 above (Dry Welding In A Habitat and Dry Chamber Welding) will be addressed as "Dry Hyperbaric Welding." Method 4 is outmoded and will not be discussed. Methods 5 and 6 (Dry Welding At One Atmosphere and Cofferdam Welding) are covered by the latest edition of ANSI/AWS D1.1, *Structural Welding Code – Steel*.

DRY HYPERBARIC WELDING

Dry Weld Chamber Environments

WELDING UNDERWATER IN a dry environment is made possible by encompassing the area to be welded with a physical barrier (weld chamber) that excludes the water. The weld chamber is designed and custom-built to accommodate braces and other structural members whose centerlines may intersect at or near the area that is to be welded (Figure 10.4).

The chamber is usually built of steel, but plywood, rubberized canvas, or any other suitable material can be used. Size and configuration of the chamber is determined by dimensions and geometry of the area that must be encompassed and the number of welders that will be working in the chamber at the same time (Figure 10.5).

Water is displaced from within the chamber by air or a suitable gas mixture, depending upon water depth and pressure at the work site. Buoyancy of the chamber is offset by ballast, by mechanical connections of the chamber to the structure, or by a combination of both.

A dry weld chamber used for pipeline hot-taps is shown in Figure 10.6; the side panel is removed to show key items, which are listed in the accompanying notes. The same chamber is shown in Figure 10.7 ready for offshore installation on a 36 in. (0.91 m) diameter pipeline at a water depth of 90 ft (27 m). In Figure 10.8, the chamber has been *blown down* (water has been displaced), and a welder/diver is welding the vertical branchline valve assembly to the trunkline prior to installing the full-encirclement reinforcement saddle.

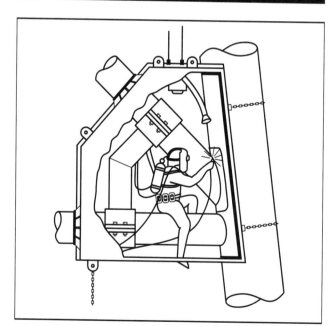

Courtesy of Global Divers & Contractors, Inc.

Figure 10.4—Weld Chamber Designed for Structural Repairs Accommodates Intersecting Structural Members

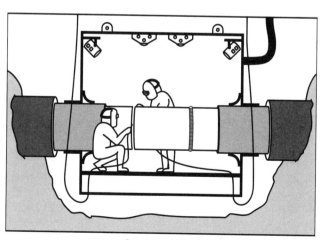

Courtesy of Global Divers & Contractors, Inc.

Figure 10.5—Weld Chamber for Pipeline Tie-In Accommodates More Than One Welder

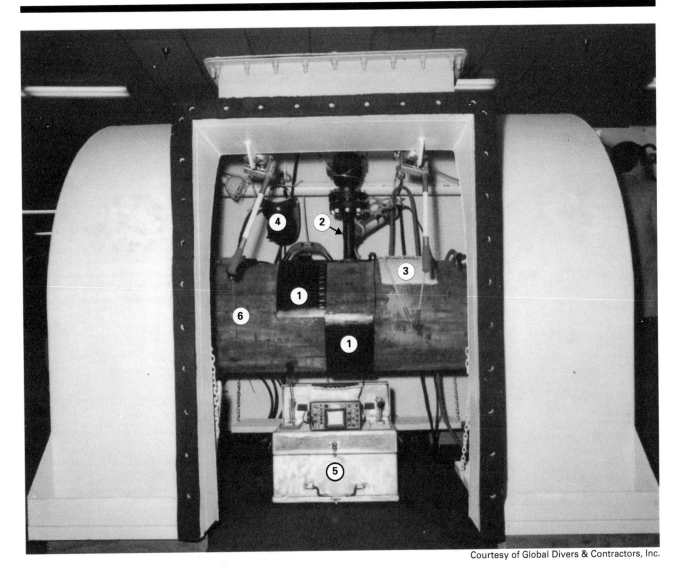

Courtesy of Global Divers & Contractors, Inc.

Note–Weld chamber items:
1. Full encirclement reinforcing saddle
2. Valve assembly for connecting pipeline
3. Preheat blanket
4. Welder's hood
5. Topside dive station box with knife switch, volt and amp meters, and radio for communications
6. Trunk line (main pipeline)

Figure 10.6—Dry Hyperbaric Weld Chamber and Key Items

Essential variables that are unique to dry hyperbaric underwater welding result from the environment, depth, and pressure. The working environment within the weld chamber restricts the welder's movements and sometimes keeps the welder from getting into the best position to see the weld joint and perform the welding. Furthermore, the welder is always in the immediate vicinity of the high-temperature weldment and heat-treating equipment. The combination of environment and depth adversely affects logistics and communications and increases safety hazards.

The diver's environmental problems can be minimized with good equipment, planning, and diver support. Hyperbaric pressure remains the predominate essential variable and adversely affects underwater welding in at least two ways, one affecting arc stability and metal transfer and the other affecting as-deposited weld metal composition.

Courtesy of Global Divers & Contractors, Inc.

Figure 10.7—Dry Weld Chamber Ready for Installation on Pipeline at Water Depth of 90 ft (27 m)

Courtesy of Global Divers & Contractors, Inc.

Figure 10.8—Welder/Diver Inside Installed Chamber Welding Branchline Valve Assembly to the Trunkline

metal, oxygen increased from an acceptable 300 ppm to a questionable level of 750 ppm. A similar trend is seen in Figure 10.9.[8]

Another report, based on examinations of dry hyperbaric weldments made from the surface to 980 ft (300 m), shows similar changes in weld metal chemistry plus reduced V-notch toughness based on Charpy tests made at 50 °F (10 °C). (See Figure 10.10.) High weld-metal oxygen content has been related to a decrease in weld metal toughness. Manganese and carbon variations can cause significant change in hardenability of weld metal.

Heat Treatment

BECAUSE OF EXTREMELY high levels of humidity in underwater welding chambers, all welding procedures should include preheating and maintenance of appropriate interpass temperatures. Temperature ranges are determined by base metal composition and thickness, arc energy (heat input), and hydrogen potential of the welding process. Heat treatment power sources, insulation, and instrumentation used for underwater dry welding are essentially the same as those used above water.

Electrical resistance heating mats typically are powered by standard characteristic power sources with open-circuit voltage of 80V. Figure 10.11 shows heating mats, partially covered with insulating blankets, installed on a 24-in. (610-mm) diameter pipe assembly in preparation for welding procedure qualification at a

Under hyperbaric conditions, the increased thermal conductivity of gases causes constriction of the welding arc and elevation of the potential drop across the arc column. Constriction of the welding arc and consequent increase in energy density promotes substantial changes in cathode and anode behavior which accentuates arc instability.

Under increased pressure, it has been reported that deposited weld-metal manganese and silicon are substantially reduced. When comparing welds made at 1000 ft (305 m) with welds made above water, manganese content decreased by 30 percent. From the surface to 250 ft (76 m), silicon content decreased by 10 percent. Weld-metal carbon content more than tripled when comparing welds made at 1000 ft (305 m) to welds made at one atmosphere. Also, for the same weld

8. Gaudiano, A. V. and Groves, D. "Underwater Dry Environment Habitat Welding." Report for Taylor Diving & Salvage, Inc.

UNDERWATER DEPTH, m

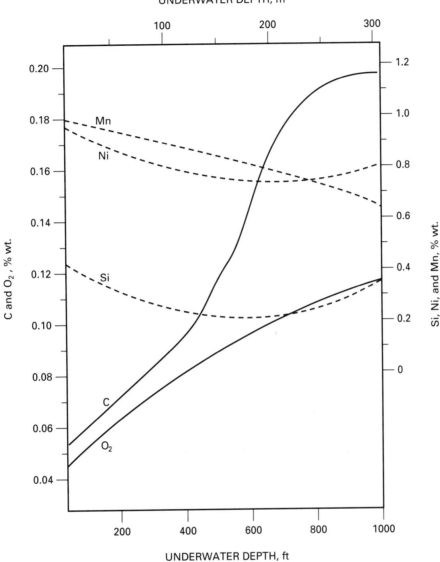

UNDERWATER DEPTH, ft

Figure 10.9—Weld Chemical Analysis versus Water Depth for Typical Low-Hydrogen Electrode

depth of 680 ft (207 m). This follows the requirements of ANSI/AWS D3.6-93 Class O/API 1104.[9]

Temperatures of weldments are monitored by topside readout thermocouples. The inspector can check positioning of heating mats and insulation via a topside television monitor, and the welder can check

temperatures with hand-held temperature probes. In addition to preheating to remove moisture and promote hydrogen release, it is possible to postheat weldments at stress relieving temperatures.

DRY WELDING PROCESSES

THE WELDING PROCESSES most commonly used for dry underwater welding are gas tungsten arc welding (GTAW), shielded metal arc welding (SMAW) and flux cored arc welding (FCAW). GTAW is most often

9. Refer to ANSI/AWS D3.6-93, *Specification for Underwater Welding*. Miami: American Welding Society, 1993. Class O underwater welds are intended to meet the requirements of some other designated code, in this case API Standard 1104, *Welding of Pipelines and Related Facilities*. Washington, D.C.: American Petroleum Institute.

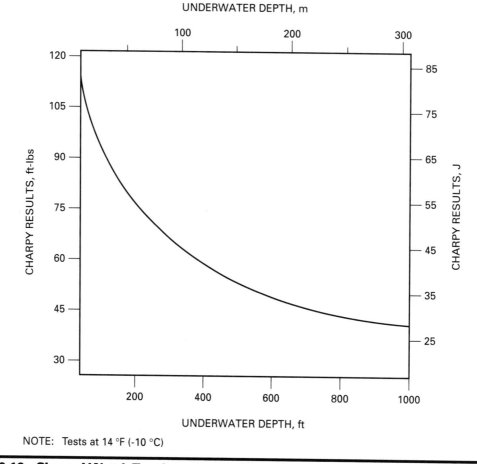

UNDERWATER DEPTH, m

NOTE: Tests at 14 °F (-10 °C)

Figure 10.10—Charpy V-Notch Toughness versus Water Depth for Typical Low-Hydrogen Electrode

used for root and hot passes due to its satisfactory arc stability under hyperbaric conditions and ease of handling. Additionally, the independent control over heat source and rate of wire feeding makes the process ideal for filling varying root openings. The other processes are usually used for filling and capping groove welds and for fillet welds due to their higher rate of deposition.

Shielded Metal Arc Welding

SMAW IS THE most commonly used process for structural repairs and is often used to fill and cap pipeline welds after GTAW has been used for the root pass. Commercially available electrodes are commonly used at depths down to about 300 ft (91 m); electrodes with specially formulated flux coatings have been used at greater depths. This process has the advantage of comparatively simple equipment, most welders are experienced in the use of the process, and it is very cost effective.

Disadvantages of the SMAW process include precautions required to keep the electrodes dry enroute to the weld chamber and the fact that the chamber should be equipped with an electrode oven. Arc stability and weld metal fluidity are adversely affected by increased pressure, and excessive fluidity makes it difficult to deposit root passes in open-root joints. For this reason, GTAW is often used to close open roots, with SMAW employed to fill and cap.

Gas Tungsten Arc Welding

COMPARED TO SMAW and FCAW, GTAW is the least sensitive to depth and pressure. The equipment requirements are a little more than for SMAW, with shielding gas composition and flow rates playing an important part in arc stability. GTAW is especially suitable for remote application of (dry spot) tack welds made without the addition of filler metal. By carefully controlling the essential variables, tack welds with highly predictable break-away strength can be

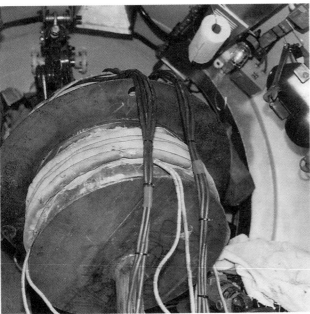

Courtesy of Global Divers & Contractors, Inc.

Figure 10.11—Heated and Insulated Pipe Sections in Hyperbaric Welding Facility for Qualification of Welding Procedure at Depth of 680 ft (207 m)

produced on underwater threaded fasteners in nuclear plants. Disadvantages for general applications include slow deposition rates and the need for greater welder skill.

Flux Cored Arc Welding

FCAW REQUIRES CONSIDERABLY more equipment than SMAW or GTAW. Commercially available flux cored wires have been used with very good results on structural repairs down to 200 ft (61 m). Specifically formulated flux cored wires have extended the depth range to 1000 ft (305 m). One of the advantages of the process, particularly for structural repairs on thick high-strength steels, is its high rate of deposition and high-heat input. The wire is contained in a pressure compensated wire feed unit which prevents moisture pick-up from the weld chamber environment. Welding is continuous with typical wire feed units holding 15-lb (6.8-kg) spools of wire.

WET WELDING

UNDERWATER WET WELDING is done by a welder/diver at ambient hydrostatic pressure with no physical barrier between the water and the weld area.

(See Figure 10.12.) The arc and the molten weld metal are shielded from the aquatic environment by a gaseous envelope or bubble composed of gases produced from decomposition of the flux coated welding electrode or flux cored wire, plus oxygen and hydrogen disassociated from the water.

The welder/diver is dressed in dive gear appropriate to the diving mode, which may be surface supplied air, mixed gas, or saturation. Depending on water temperature and the environment, the welder/diver may wear a wet suit, dry suit, hot-water heated suit, or, when in warm waters encountered at nuclear power plants, a specially cooled dry suit. Additionally, the welder/diver's head gear is equipped with a welder's lens holder assembly, and the welder/diver wears electrical shock-resistant rubber gloves.

Essential Variables

UNDERWATER WET WELDING variables must be considered in addition to essential variables associated with welding with shielded metal arc welding and flux cored arc welding above water. Factors that adversely affect wet weldments include:

Courtesy of Chicago Bridge & Iron Company

Figure 10.12—Welder/Divers Repairing Offshore Structure Using Underwater Wet Welding

(1) Hyperbaric pressure accentuates arc instability by constricting the arc plasma and increasing energy density

(2) Possibility of magnetic arc blow is increased

(3) Increased pressure causes loss of manganese and silicon and increased amounts of carbon and oxygen in the weld metal

(4) Disassociation of the water promotes hydrogen pick-up in the weldment

(5) The infinite heat-sink of the surrounding water which causes an unprecedented cooling rate

Another, and often unrecognized, wet welding variable is related to the gaseous bubble that displaces the water from the welding arc and weld pool area. The volume of the protective bubble and density of the gases vary substantially with depth and pressure. The bubble is somewhat transient in that the volume, regardless of hydrostatic pressure, constantly fluctuates as it forms and grows to the largest volume its surface tension permits, then partially deflates, regenerates, deflates again, and so on.

The continuous cyclic fluctuation in the size of the bubble is generally of no concern to the welder/diver. However, with commonly used wet welding electrodes, fluctuation of the bubble volume becomes very rapid in shallow water [< 10 ft (<3 m)] and, especially in overhead welding, adversely affects the welding process. Two phenomenon contribute to the changes in the volume of gases generated by the process of welding at different depths. For instance, from a depth of 33 ft (10 m) to zero, the volume of gas produced by decomposition of the flux coated electrode doubles in accordance with Boyle's law. Furthermore, the volume of water vapor in the bubble increases because of the lower boiling temperature of water near the surface [212 °F (100 °C)] versus 250 °F (121 °C) at a depth of 33 ft (10 m). Consequently, the excessive amount of gases accelerates fluctuation of the bubble size, and vigorously rising bubbles of excess gas create turbulence in the weld area. This, plus water vapor dilution of the beneficial protective gases, makes it difficult to make an acceptable wet overhead weld in very shallow water [<10 ft (<3 m)].

The solutions to this shallow water wet welding phenomena involves reformulation of wet welding electrode coatings and innovative design of weld joint details and the configuration of repair and construction sections. Conversely, as depth and pressure increase, the bubbles become too small to provide optimum protection to the welding arc and weld pool. The ideal solution to this problem may be the development of underwater wet welding electrodes with flux coatings formulated for different depth ranges. Based on experience with conventional wet welding electrodes, the depth ranges for carbon and low-alloy steels might be, as a minimum, 0 to 10 ft (3 m), 10 to 165 ft (3 to 50 m), and 165 to 330 ft (50 to 100 m).

Table 10.1
Boiling Point of Water at Hydrostatic Pressures of Various Depths of Sea Water

Depth		Boiling Temperature	
ft	m	°F	°C
0	0	212	100
33	10	250	121
165	50	320	160
330	100	370	188
1000	305	460	238

While increased hydrostatic pressure may be condemned for its adverse affects on wet welding, there is a commonly unrecognized trade-off. The boiling temperature of water increases with hydrostatic pressure, as shown in Table 10.1. If that superheated water is physically confined to the weld area, it can be used to help maintain interpass temperatures.

Of the various essential variables associated with wet welding of carbon and low-alloy steels, the predominate factor is the rapid rate of cooling. Wet welds made with mild steel (ferritic) welding electrodes on material with a carbon equivalent (CE) greater than 0.40 are subject to hydrogen-induced underbead cracking in the quenched and hardened heat-affected zone (HAZ), as shown in Figure 10.13.[10] Restraint during welding also appears to be a determining factor for hydrogen cracking. (See test results in Table 10.2).

10. Figures 10.13 and 10.14 and Tables 10.2, 10.3, and 10.4 are copyrighted material from Grubbs, C.E., and Seth, O.W., "Multipass all position 'wet' welding, a new underwater tool." Proceedings of Offshore Technology Conference, Paper No. OTC-1620. Houston, Texas, May 1-3, 1972, and are used courtesy of the Offshore Technology Conference.

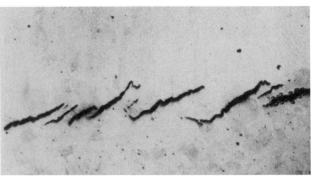

Courtesy of Offshore Technology Conference

Figure 10.13—Hydrogen-Induced Crack in the HAZ. Weld made with Mild Steel Electrodes on Base Metal with CE of 0.564 (1% Nital Etch, 40X) (Reduced 68%)

Table 10.2
Summary of Mild Steel Electrode Tests

Test Number[a]	Material Specification	Base Metal Carbon Equivalent	HAZ Max Vickers Hardness	HAZ Max Rockwell C Hardness	Ultimate Tensile Strength	Failure Location[b]	Location of Any Cracks[c]	Restraint	Porosity Level	Remarks[d]
101	A283C	0.385	401	27.8	70.5	WM	—	Yes	Fair	—
102	A283C	0.385	401	34.0	69.0	PL	—	Yes	Fair	—
103	A283C	0.306	369	26.5	64.7	WM	—	No	Excessive	—
104	A283C	0.306	313	23.0	59.3	WM	—	No	Excessive	102 ft (32 m)
105	A283C	0.355	408	<20	—	—	—	Yes	Fair	—
106	A36	0.455	446	—	56.1	WM	HAZ	Yes	Fair	—
107	A36	0.480	427	43.0	58.6	WM	—	No	Fair	63 ft (19 m)
108	A283C	0.306	374	<20	65.6	WM	—	No	Fair	—
109	A283C	0.306	336	22.5	65.8	PL	—	No	Fair	—
110	A442-60	0.382	343	25.5	62.3	WM	—	Yes	Fair	—
111	A442-60	0.382	390	32.0	70:7	PL	—	Yes	Good	—
112	A283C	0.313	389	29.5	63.0	PL	—	Yes	Good	—
113	A442-60	0.392	374	30.0	70.7	PL	—	Yes	Fair	—
114	A515-70	0.445	445	44.9	65.4	WM	HAZ	Yes	Fair	—
115	A283C	0.303	349	<20	63.6	PL	—	Yes	Fair	—
116	A283C	0.288	355	25.0	58.0	WM	—	Yes	Excessive	166 ft (51 m)
117	A515-70	0.445	472	43.5	66.8	WM	HAZ	Yes	Fair	—
118	A283C	0.303	370	22.5	—	—	—	No	Fair	—
119	A36	0.309	356	23.0	—	—	—	No	Fair	—
120	A283C	0.318	354	32.0	—	—	—	Yes	Excessive	—
121	A36	0.309	400	27.0	69.9	PL	—	No	Fair	Vertical
122	A36	0.309	379	29.5	67.2	WM	—	No	Fair	Horizontal
123	A36	0.309	405	21.0	68.0	WM	—	No	Fair	Overhead
124	A36	0.309	405	34.0	66.4	WM	—	No	Fair	Vertical
125	A36	0.309	404	28.0	71.3	WM	—	No	Fair	Horizontal
126	A36	0.309	385	26.0	68.3	WM	—	No	Fair	Overhead
127	A36	0.309	—	—	—	—	—	—	—	—
128	A36	0.309	—	—	65.3	WM	—	No	Excessive	Yield = 62.4 ksi (430 MPa)
129	A36	0.379	419	36.0	—	—	—	No	—	Cruciform - Vertical[e]
130	A36	0.379	455	29.5	—	—	—	No	—	Cruciform - Vertical[e]
131	A36	0.379	432	36.0	—	—	—	No	—	Cruciform-Horizontal[e]
132	A36	0.379	430	36.6	—	—	—	No	—	Cruciform-Overhead[e]
133	A36	0.395	425	34.5	68.5	—	—	No	—	Cruciform - Vertical and Overhead[e]
134	MS-Pipe	0.188	208	<20	—	—	—	Yes	—	—
135	Pipe	0.438	469	38.5	—	—	HAZ	Yes	—	—
136	A515-60	0.329	300	23.0	67.1	WM	—	No	Fair	—
137	A36	0.364	379	24.0	—	—	—	Yes	—	Hot Tap
138	A36	0.363	452	38.0	—	—	—	Yes	—	Lap Weld - Pressurized Pipe
139	X-60	0.393	407	37.5	74.5	WM	—	Yes	Fair	All Position - Pipe Butt Weld
140	Corten A	0.435	381	31.3	76.6	WM	—	No	Fair	—
141	A283C	0.290	344	26.0	—	—	—	Yes	Excessive	—
142	A36	0.428	466	42.2	53.0	WM	HAZ	No	—	—
143	A283C	—	380	27.9	63.8	PL	—	No	Good	—
144	A537A	0.428	394	35.0	70.0	WM	HAZ	Yes	Fair	100 ft (30 m)
145	A537A	0.428	416	36.0	66.5	WM	HAZ	Yes	Fair	150 ft (46 m)

a. Courtesy of Offshore Technology Conference. (Test numbers shown are arbitrary.)
b. WM = weld metal; PL = plate.
c. Where no cracks were found, the location was left blank.
d. Unless specified otherwise, water depth was ≤ 33 ft (10 m).
e. The cruciform cracking test is described in the *Welding Handbook*, 8th Ed., Vol. 1, Pages 120-121.

The use of austenitic stainless steel filler metal for wet welding carbon and low-alloy steels with a CE greater than 0.40 circumvents the HAZ cracking problem. However, under certain degrees of restraint, most stainless steel weldments on high CE material are subject to gross diffusion zone cracking. (See Figure 10.14 and Table 10.3.) The cracking is located adjacent to the fusion line on the base metal side where alloying elements such as chromium and nickel from the weld metal are diffused into the base metal. This forms a hard martensitic structure that is subject to hydrogen cracking. The crack shown in Figure 10.14 is in the unetched (comparatively unetched due to high hardness) zone adjacent to the weld deposit. Although the crack may appear to be in the weld metal, it is a few thousandths of an inch (a few hundredths of a millimeter) into the base metal diffusion zone.

The use of austenitic stainless steel electrodes for wet welding of austenitic stainless base metals has resulted in weldments meeting the standards expected of in-air welds. Austenitic stainless, because of its natural aversion to the formation of martensite, seems to have an affinity for wet welding. In fact, the austenitic wet welds at shallow depths [less than 40 ft (13 m)] are in some respects metallurgically superior to similar dry welds, since the joint is essentially rapid-quench-annealed during the welding process. This minimizes heat-affected zone sensitization, which is a prime contributor to intergranular stress-corrosion cracking in nuclear plant components. Similar good results have been achieved using nickel-based electrodes on austenitic stainless steel base metals at shallow depths.[11]

Wet welds made with high-nickel filler metal on materials with CE values up to 0.696 have neither HAZ nor diffusion zone cracking (see Table 10.4) and have better ductility and impact properties than wet welds made on carbon steel with ferritic or stainless steel filler metal. Unfortunately, and so far inexplicably, nickel welding electrodes are highly depth sensitive. Wet welding procedures with ferritic electrodes have been qualified down to 325 ft (100 m).[12] Wet all-position fillet and groove class A welding procedures have been qualified with nickel welding electrodes no deeper than 33 ft (10 m). It is doubtful that currently available nickel electrodes can produce acceptable ANSI/AWS D3.6-93, *Specification for Underwater Welding*, Class A all-position groove welds much deeper than about 45 ft (14 m). This is true in spite of the fact that two welding electrode manufacturers, under contract to a major construction company, spent a total of more than five years trying to mitigate the depth sensitivity of nickel welding electrodes. Three research programs were studying the problem in 1992, with hopes of formulating nickel welding electrodes that will be less depth sensitive. In 1974, employees of a major construction company improvised flux formulations and manually coated 625 nickel core wire. The "shade tree" metallurgists took the hand-made electrodes offshore and made a down flat groove weld on A537 Gr A material at a depth of 96 ft (29 m). (This is Test No. 244 in Table 10.4.) There were no cracks, porosity level was acceptable, and reduced section tensiles failed at 72.7 ksi (501 MPa). In 1994, the weld had not been duplicated at depths deeper than 33 ft (10 m).

Wet Welding Processes

Shielded Metal Arc Welding (SMAW). The versatility and effectiveness of the SMAW process generally makes it preferable to other processes used for underwater wet welding. The simplicity of the SMAW process makes it possible for the welder/diver to catch a plane to a remote area for an emergency repair and take nothing but welding electrodes (often proprietary), an electrode holder, and a welding lens holder assembly. At the job site, the welder/diver can use locally available conventional dive gear and virtually any DC welding power source. Also, successes with procedure and welder qualification with this versatile process have contributed to finished-product quality, the key criteria in process selection.

Commercially available underwater wet welding electrodes are used by most diving contractors that provide underwater welding services. Since 1970, however, one or two contractors have used proprietary welding

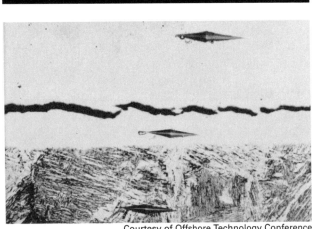

Courtesy of Offshore Technology Conference

Figure 10.14—Diffusion Zone Crack in the Base Metal. Restrained Weld Made with Stainless Steel Electrodes on Base Metal with CE of 0.479 (1% Nital Etch, 400X) (Reduced 68%)

11. O'Sullivan, J. E. "Underwater welding in nuclear plant applications." *Maintenance and Repair Welding in Power Plants*, 75-83. Miami: American Welding Society, 1992.

12. See "Wet Welding State-of-the-Art," Page 481.

Table 10.3
Summary of Austenitic Stainless Steel Electrode Tests

Test Number[a]	Material Specification	Base Metal Carbon Equivalent	HAZ Max Vickers Hardness	HAZ Max Rockwell C Hardness	Ultimate Tensile Strength	Failure Location[b]	Location of Any Cracks[c]	Restraint	Porosity Level	Remarks[d]
151	A283C	0.385	—	—	—	—	Diffusion	Yes	—	—
152	A283C	0.385	—	—	—	—	Diffusion	Yes	—	Vertical
153	A283C	0.385	—	—	—	—	Diffusion	No	—	—
154	A36	0.479	430	41.9	74.4	WM	Diffusion	Yes	Good	—
155	A442-60	0.382	376	23.1	64.7	PL	—	No	Good	—
156	A537A	0.420	430	32.0	106.0	WM	—	Yes	Good	—
157	A36	0.357	393	37.1	72.8	PL	—	Yes	Good	—
158	A283C	0.303	—	—	—	—	Diffusion	Yes	—	—
159	A517D	0.597	—	—	—	—	Diffusion	Yes	—	—
160	A517D	0.597	—	—	—	—	Diffusion	Yes	—	—
161	A283C	0.303	390	23.0	62.9	PL	—	Yes	Fair	—
162	A537A	0.468	—	—	—	—	—	Yes	Excessive	—
163	A537A	—	456	41.0	87.5	PL	—	Yes	Good	—
164	A537A	0.468	420	38.0	—	—	—	Yes	Fair	—
165	Sheet Pile	0.454	513	38.5	53.0	WM	—	No	Fair	Vertical
166	Sheet Pile	0.464	435	36.0	—	—	—	No	—	Vertical
167	Sheet Pile	0.464	461	26.5	—	—	—	No	—	Vertical
168	Sheet Pile	0.506	421	31.0	—	—	—	No	—	Vertical
169	A516-70	0.413	—	—	—	—	Diffusion	Yes	—	—
170	A516-70	0.413	426	29.5	84.6	WM	Diffusion	No	Excessive	Yield = 77.3 ksi (533 MPa)
171	A516-70	0.413	454	33.0	77.6	PL	—	Yes	Fair	—
172	A516-70	0.413	464	41.0	78.9	WM	—	No	Excessive	—
173	A516-70	0.413	441	39.0	79.2	WM	—	Yes	Excessive	—
174	A537A	0.452	446	35.0	—	—	Diffusion	Yes	—	—
175	A537A	0.452	426	30.0	—	—	Diffusion	Yes	—	—
176	S.S.	—	492	43.2	87.3	PL	—	Yes	Fair	—
177	A516-70	0.486	492	39.0	85.0	PL	—	Yes	Fair	—
178	A283C	0.290	—	—	64.3	PL	—	No	Fair	—
179	A516-70	0.510	494	44.3	64.9	WM	—	Yes	Good	—
180	A516-70	0.483	460	40.9	77.8	PL	—	Yes	Fair	—
181	A283C	0.381	372	30.7	66.6	PL	—	Yes	Fair	50 ft (15 m)
182	A283C	0.307	385	29.7	81.4	WM	—	Yes	Fair	50 ft (15 m)
183	SSS-100	0.597	440	39.7	83.1	WM	—	Yes	Fair	50 ft (15 m)
184	A537A	0.499	420	41.0	80.5	WM	—	—	Fair	—
185	A537A	0.428	436	32.2	76.7	WM	—	No	Fair	96 ft (29 m)
186	A537A	0.428	425	29.2	43.0	WM	—	No	Fair	148 ft (45 m)
187	A537A	0.499	435	27.2	—	—	Diffusion	No	Good	50 ft (15 m)
188	A537A	0.499	450	44.0	—	—	Diffusion	No	Fair	50 ft (15 m)
189	A537A	0.499	425	36.0	—	—	Diffusion	No	Fair	50 ft (15 m)
190	A537A	0.499	500	—	—	—	Diffusion	No	Good	—

a. Courtesy of Offshore Technology Conference. (Test numbers shown are arbitrary.)
b. WM = weld metal; PL = plate.
c. Where no cracks were found, the location was left blank.
d. Unless specified otherwise, water depth was ≤ 33 ft (10 m).

Table 10.4
Summary of Austenitic High-Nickel Electrode Tests

Test Number[a]	Material Specification	Base Metal Carbon Equivalent	HAZ Max Vickers Hardness	HAZ Max Rockwell C Hardness	Ultimate Tensile Strength	Failure Location[b]	Location of Any Cracks[c]	Restraint	Porosity Level	Remarks[d]
201	A283C	0.385	—	—	—	—	—	Yes	—	Slag, nonfusion
202	A517D	0.597	435	42.4	93.3	WM	—	Yes	Fair	—
203	A517D	0.597	435	41.2	84.4	WM	—	Yes	Fair	—
204	A517D	0.595	435	42.8	57.9	WM	—	Yes	Excessive	—
205	A515-70	0.445	504	39.9	78.8	WM	—	Yes	Fair	—
206	A283C	0.303	287	23.2	57.9	WM	—	Yes	Good	—
207	A515-70	0.445	484	43.0	48.6	WM	—	Yes	Fair	—
208	A537A	0.468	435	37.0	53.8	WM	—	Yes	Good	—
209	A537A	0.468	433	38.0	67.0	WM	—	Yes	Fair	—
210	Sheet Pile	0.454	504	36.0	40.0	WM	—	No	Fair	Vertical
211	A537A	0.468	452	37.0	83.8	PL	—	Yes	Fair	—
212	A537A	0.468	437	40.0	81.7	WM	—	Yes	Excessive	—
213	Sheet Pile	0.464	396	26.0	—	—	—	No	—	Vertical
214	Sheet Pile	0.464	465	34.0	—	—	—	No	—	Vertical
215	Sheet Pile	0.464	484	25.5	—	—	—	No	—	Vertical
216	A537A	0.468	419	37.0	76.6	WM	—	No	Fair	—
217	Sheet Pile	0.506	495	46.1	—	—	—	No	—	Vertical
218	Sheet Pile	0.506	505	34.5	—	—	—	No	—	Vertical
219	A283C	0.291	227	14.5	62.7	WM	—	No	Good	Vertical
220	A283C	0.291	336	16.0	62.8	WM	—	No	Fair	—
221	A283C	0.357	346	17.0	75.8	WM	—	Yes	Good	Yield = 63.5 ksi (438 MPa)
222	A283C	0.357	360	23.0	70.1	WM	—	Yes	Fair	Yield=61.8 ksi (426 MPa)
223	HY-80	0.696	426	42.0	78.8	WM	—	No	Fair	Vertical
224	HY-80	0.696	426	42.8	85.6	WM	—	No	Fair	Horizontal
225	HY-80	0.696	411	41.3	83.9	WM	—	No	Fair	Overhead
226	HY-80	0.696	431	39.0	75.4	WM	—	No	Good	—
227	A283C	0.358	281	24.0	82.0	WM	—	No	Fair	Yield = 66.4 ksi (458 MPa)
228	A283C	0.358	286	20.0	68.1	WM	—	No	Good	Yield = 63.6 ksi (439 MPa)
229	A516-70	0.413	461	38.5	76.7	WM	—	No	Good	Vertical
230	A516-70	0.413	426	38.3	73.1	WM	—	No	Good	Overhead
231	A516-70	0.413	439	39.0	72.6	WM	—	No	Fair	—
232	HY-80	0.696	437	41.2	66.9	WM	—	Yes	—	—
233	HY-80	0.696	437	41.2	66.9	WM	—	Yes	—	—
234	A517A	0.541	451	41.5	95.5	WM	—	Yes	Fair	—
235	A516-70	0.484	480	39.0	83.5	WM	—	Yes	Good	—
236	A516-70	0.483	501	45.2	54.8	WM	—	Yes	Fair	—
237	Monel	—	503	48.0	60.4	WM	—	Yes	Good	—
238	Inconel	—	495	41.0	87.0	PL	—	Yes	Good	—
239	A283C	0.108	215	<20	92.7	WM	—	Yes	Good	Yield = 73.7 ksi (508 MPa)
240	A516-70	0.527	520	44.0	67.9	WM	—	Yes	Excellent	—
241	A516-70	0.483	461	30.0	85.5	PL	—	Yes	Good	—
243	A36	0.428	470	42.7	62.7	WM	—	No	—	—
244	A537A	0.428	378	35.4	72.7	WM	—	No	Fair	96 ft (29.3 m)
245	A537A	0.428	400	33.0	72.5	WM	—	No	Fair	50 ft (15.2 m)
246	A537A	0.499	435	36.5	66.4	WM	—	No	Fair	50 ft (15.2 m)

a. Courtesy of Offshore Technology Conference. (Test numbers shown are arbitrary.)
b. WM = weld metal; PL = plate.
c. Where no cracks were found, the location was left blank.
d. Unless specified otherwise, water depth was ≤ 33 ft (10 m).

electrodes. Regardless of the basic electrode to be used underwater, it must be made waterproof. How electrodes are transported from above water to the underwater work site, and how long they can remain there before being used are essential variables of wet welding procedures. Properly processed and transported, electrodes used to qualify wet welding procedures at depths down to 325 ft (100 m) remained moisture free for more than 24 hours.

During early development of wet welding methods, innovative welder/divers experimented with virtually every available welding electrode and every conceivable method of making them waterproof. Electrodes were dipped, sprayed, taped, and even encased in heat-shrink tubing. An old U.S. Navy diving manual even recommended that, when no other waterproofing liquid was available, the diver and the diver shipmates dissolve their celluloid pocket combs and toothbrush handles in acetone.

The electrodes used most often for wet welding, regardless of how they may be waterproofed, are classified by AWS as E6013 and E7014.[13] The E6013 has a high-titania potassium flux covering and the E7014 covering is iron powder, titania. E6013 electrodes tend to have better weldability and bead appearance. The iron powder E7014 electrodes have a higher rate of deposition.

Welding underwater with covered electrodes (SMAW) is done very much as it is done above water. Amperage, rate of travel, and electrode angle depend on position of weldment and whether the weld bead is being deposited as root, fill, or cap. Direct current is used and the polarity is usually straight (electrode negative). However, there may be times or geographical locations where reversing the polarity will produce better results (less porosity). There are geographical areas such as the North Sea where reverse (electrode positive) polarity is significantly better than straight polarity.

In addition to the large amount of ferritic welding electrodes consumed during offshore wet welded repairs, a significant amount of underwater wet welding has been performed on stainless steel at nuclear power plants. Since January 1980, records show as many as 40 different occasions when stainless steel electrodes were used for repairing nuclear power plant pool liners and reactor vessel components. Repair applications ranged from simple fillet weld patch plates to rather complex repairs and modifications to steam dryers and sparger assemblies. Welding electrodes used included E308, E308L, E309, and E316.[14] Wet welds made with stainless electrodes on stainless steel material have typically been qualified in accordance with the requirement of ANSI/AWS D3.6-93 Class O/ASME Section IX.[15]

Flux Cored Wire (FCAW). Flux cored wire and other semiautomatic welding processes have been investigated in the United States and North Sea areas as potential underwater wet welding processes. However, the process has not competed with SMAW because of reported excessive porosity, failure to meet visual acceptance standards, and problems with underwater wire feeding devices.

Recent developments of nickel-based flux cored filler materials have provided improved wet weldability and halogen-free flux formulations specifically designed for wet welding applications.[16] (The presence of halogen ions on wetted stainless steel will adversely affect corrosion resistance of the steel, thereby promoting corrosion cracking.) Excellent results have been reported with cladding, fillet welds, and multipass groove welding applications. Alloy 625 has been evaluated at depths of 3 to 20 ft (1 to 6 m) on 304 stainless steel, Alloy 600 and 690 base materials, and Alloy 82 cladding materials. These underwater welding developments have resulted in code-quality mechanical properties.

Similarly, improved underwater wet welding capabilities and halogen-free flux formulations have been developed with stainless steel flux-cored wires. Excellent results have been achieved on cladding, fillet welds, and multipass groove weld applications at depths from 3 to 20 ft (1 to 6 m). Mechanical tests of type 308L stainless steel flux-cored wire on 304 stainless steel base metal have demonstrated the ability to exceed the acceptance criteria for ASME Section IX.

DRY AND WET WELDING COMPARED

Comparison of Weld Test Requirements

ONE WAY TO compare the properties of dry weldments with wet weldments is to compare the requirements for welding procedure qualifications as defined by ANSI/AWS D3.6-93, *Specification for Underwater Welding*. The following comparisons are based on the requirements for AWS D3.6-93 Class A, which originally was intended for underwater ferritic dry welds, and Class B, which originally was intended to reflect the

---

13. Refer to ANSI/AWS A5.1, *Specification for Carbon Steel Electrodes for Shielded Metal Arc Welding*. Miami: American Welding Society (latest edition).

14. Refer to ANSI/AWS A5.4, *Specification for Covered Corrosion-Resisting Chromium and Chromium-Nickel Steel Welding Electrodes*. Miami: American Welding Society (latest edition).

15. Refer to ANSI/AWS D3.6-93, *Specification for Underwater Welding*. Miami: American Welding Society. Class O underwater welds are intended to meet the requirements of some other designated code, in this case Section IX, *ASME Boiler & Pressure Vessel Code*. New York: The American Society of Mechanical Engineers, 1992.

16. Findlan, S.J., and Frederick, G.J. "Underwater wet flux-cored arc welding development of stainless steel and nickel-based materials - interim report," Electric Power Research Institute NDE Center, Charlotte, N.C. (June 23, 1992).

state-of-the-art for underwater ferritic wet welds. The differences in acceptance criteria between Class A and Class B generally reflect the expected differences in finished underwater dry welds and wet welds, respectively.[17]

Visual. The main differences in the expected surface appearance of dry and wet welds have to do with the following:

(1) Undercut which shall not exceed 1/32 in. (0.8 mm) for Class A (dry quality) welds, and with exceptions, 1/16 in. (1.6 mm) for Class B (wet quality) welds. (A proficient welder/diver has no more problem avoiding unacceptable undercut on a wet weld than he does on a dry weld.)

(2) Root underfill and melt-through. Wet welds do not meet the requirements for dry welds on open- root single-V-groove weldments without backing. Wet welds are normally used for fillet welds and groove welds with backing bars. When it is necessary that wet welds be made on single-V-groove joints without backing, such as the repair of through-thickness cracks, procedures are developed and qualified to determine if the degree of root penetration meets fitness-for-purpose requirements.

Radiographic. Proficient welder/divers produce wet welds that meet the radiographic requirements for dry welds, except for the amount of porosity. Virtually all wet welds made with ferritic welding electrodes have a comparatively high number of gas entrapments which form during the rapid solidification of the weld metal. Individual pores are usually less than 1/16 in. (1.6 mm); they are evenly dispersed throughout the weld metal and have a tendency to increase with depth.

Reduced Section Tension and Fillet Weld Shear Tests. The combined cross section of porosity and changes in weld metal chemistry do not normally begin to adversely affect the tensile strength of wet weld metal until a depth of about 200 ft (61 m). Even at 325 ft (100 m), reduced section tension tests on wet welds typically fail at loads in excess of the minimum design strength of the base metal. With the depth limitations mentioned in the previous paragraph, the tensile strength of qualified wet welds equals the strength of dry welds made above or below water. The often repeated statement that the tensile strength of wet welds

is about 80 percent of dry welds is not based on welds made by qualified welder/divers.

All-Weld-Metal Tensile Strength. ANSI/AWS D3.6-93 does not require all-weld-metal tension tests for Class B (wet quality) welds. However, when they are made for wet welds, ultimate tensile strength values are, with water-depth limitations, usually equal to those of dry welds, but elongation may be reduced to as low as 8 to 10 percent.

Root and Face, or Side Bends. For dry welds made on carbon and low-alloy steels with a design yield strength of less than 50 ksi (345 MPa), the bending radius specified by ANSI/AWS D3.6-93 is two times the thickness of the bend specimen (2T).[18] For Class B (wet quality) welds, the bend radius is 6T. This requirement is sufficiently stringent for these welds made at 325 ft (300 m). However, when made at ≤ 33 ft (10 m), they are usually capable of passing a 4T bend test. There is not enough historical data to determine what bend radii would be most appropriate for these welds made at different depth ranges.

Macroetch. Macrosections transverse to the weld are required for both welding methods and are examined at 5X magnification. Requirements for both dry-quality and wet-quality welds are the same, except wet welds may have slag plus porosity up to 5 percent of the surface area of the weld metal. Wet welds rarely fail this test.

Hardness. Hardness surveys are specified for Class A welds but not for Class B welds. For Class A welds, microhardness measurements shall not exceed Vickers 325 at 10 kg. When wet welds are tested for hardness, the maximum hardness measurement in the heat-affected zone (HAZ) usually exceeds the allowable for Class A welds.[19] However, this problem may be mitigated by use of the multiple-temper-bead welding technique described on Page 481.

Charpy Impacts. ANSI/AWS D3.6-93 specifies that Class A weld impact tests made at the minimum design service temperature (or temperature specified by the customer) shall average not less than 15 ft-lb (20 J) and the minimum measurement shall not be less than 10 ft-lb (14 J). As a result of recent developmental work, wet welds are being made with notch toughness that significantly exceed the requirements for Class A welds.

---

17. See Tables 7, 9 and 11 regarding Weld Procedure Qualification - Number and Type of Test Specimens, ANSI/AWS D3.6-93, *Specification for Underwater Welding*, pp. 64, 80, 96. Miami: American Welding Society, 1993.

18. Wet stainless steel weldments are usually qualified in accordance with the requirements of ANSI/AWS D3.6-93 for Class A or Class O/ASME Section IX welds with 2T bend radii.

19. See "Wet Welding State-of-the-Art," Page 481.

Fracture Mechanics. For a more valid method of comparing dry-weld and wet-weld properties, an underwater welding literature search was made on data available since 1980. The search disclosed only one well-documented comparison of surface welds, dry hyperbaric welds, and wet welds. The research department of a major oil company funded and directed the study. The testing and evaluation was carried out at the joining and welding research center of a leading university. The objective of the program was to compare systematically the mechanical properties of welds made under the three different conditions. Single V-groove welds with backing bars were made on 3/4 in. (19 mm) ASTM A-36 plate. Underwater dry and wet welds were made at a depth of 33 ft (10 m).

The program report provides useful information on fatigue crack propagation and fracture mechanics of welds made under the three dissimilar conditions.[20] The report states that the behavior of the wet weld was similar to the behavior of the corresponding surface and dry hyperbaric welds. Also, the absolute crack growth rates were similar to wrought ferrite-pearlite steel. Figure 10.15 provides a comparison of the fatigue crack growth characteristics of weld metal deposited by the three welding methods.

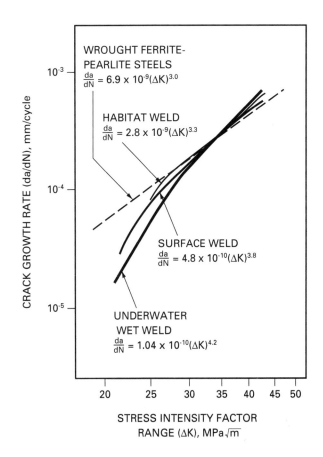

Figure 10.15—Fatigue-Crack Growth Characteristics of Surface, Dry Hyperbaric, and Underwater Wet Welds

Dry Welding Advantages

ADVANTAGES OF DRY welding are as follows:

(1) With appropriately qualified welding procedures and welder/divers, dry hyperbaric welding can be used to join any material with SMAW, GMAW, FCAW or GTAW processes, resulting in weldments with quality equivalent to above-water welds.

(2) Heat treatment processes and equipment can be used to preheat, maintain interpass temperatures, and to postheat.

(3) Water is excluded from the weld area so the cooling rate is no greater than for welds made above water, except when HeO_2 is used as background gas in the weld chamber.

(4) Welder/diver and gas shielding of weldment are comparatively unaffected by water movement such as current and surge.

(5) There are no limitations to the protective coatings that can be applied to completed weldments.

20. Matlock, K. D., Edwards, G. R., Olson, D. L., and Ibarra, S. "An evaluation of the fatigue behavior in surface, habitat, and underwater wet welds," *Second International Conference on Offshore Welded Structures*, Ed. H. C. Cotton, P15-1– P15-10. Cambridge, England: The Welding Institute, 1983. An abbreviated version was later published in *Underwater Welding*, 303-10. New York: Pergamon Press, 1983.

Dry Welding Disadvantages

DISADVANTAGES OF DRY welding are as follows:

(1) Two situations preclude the installation of a habitat or chamber for dry welding. One is the number, configuration, and size of structural members or other obstructions adjacent to the area to be welded, which make it impossible or impractical to install a habitat or weld chamber for dry welding. The other is when the installation and presence of a dry weld chamber on an offshore structure could endanger the integrity of the structure due to forces exerted on the weld chamber, and transmitted to the structure, by movement of the water caused by storms, currents, and, in shallow water, ground swells.

(2) The time and cost to complete an underwater repair with dry welding is usually significantly more than with wet welding. Time required to procure and

transport material, and to fabricate the weld chamber, delays mobilization of the job. Installation and removal of the weld chamber delays completion of the job. Costs of a job may be further increased by the need for additional lifting and rigging equipment, background gas for the weld chamber, and the need for additional support personnel.

(3) With dry welding, safety hazards that are not associated with wet welding exist because the welder/diver works within the weld chamber at hyperbaric pressures. These are discussed in the safety section of this chapter.

Wet Welding Advantages

ADVANTAGES OF WET welding are as follows:

(1) Wet welds can be made where it would be physically impossible to evacuate the water from the area to be welded.

(2) For emergency and other underwater welding requirements, the personnel, equipment, tools, and consumables are *in stock* and can be mobilized with no delay.

(3) No equipment larger or heavier than a welding power source is required, so logistics, on-site work space, and lifting equipment are minimal. Table 10.5 lists equipment requirements for underwater wet and dry welding.

(4) Wet welding projects are usually completed in significantly less time, and at lower cost, than would have been incurred if dry welding had been used. Cost comparisons are discussed later in this section.

Wet Welding Disadvantages

DISADVANTAGES OF WET welding are as follows:

(1) As discussed earlier, some of the properties of wet weldments are inferior to those of dry weldments. These areas of inferiority include increased porosity, reduced ductility, and greater hardness in the HAZ. These anomalies are caused by the extraordinarily fast cooling rate induced by the surrounding water.

(2) Until development of innovative wet welding techniques in 1994 (see Wet Welding State-of-the-Art, Page 481), the use of wet welding with ferritic welding electrodes was limited to steels with carbon equivalents not greater than 0.40 wt. %. Steels with higher carbon equivalents are subject to hydrogen-induced underbead cracking and excessive hardness in the HAZ.

(3) Sea conditions, such as ocean currents and ground swells in shallow water, make the welder/diver's

work more difficult and have adverse affects on the gaseous bubble that shields the welding arc and pool.

(4) Lack of visibility in rivers and other inland waters can seriously impair the welder/diver's capabilities unless he uses special underwater viewing devices.

Cost Comparisons

TWO RELATED CASE histories are presented to help compare underwater welding costs with dry welding and with wet welding. The costs incurred to perform any underwater welding task is usually significantly greater if dry welding is used rather than wet. Because each project may be different in many ways, the only way to compare cost, wet versus dry, is on a specific project-to-project basis.

Offshore Platform Repair – Case A. A vertical-diagonal brace on an offshore drilling and production platform was rammed by a supply boat and badly deformed. The damaged section, which extended from about 14 ft (4 m) above water to about 30 ft (9 m) below the waterline, had to be replaced. Neither the upper nor lower 3 or 4 ft (0.92 or 1.2 m) of the brace was damaged, i.e., there were good *stub ends* where the brace connected to the legs of the platform, above and below water.

The damaged section was cut and removed. During fabrication of the replacement section, a dry weld chamber was temporarily attached to the lower end, the new brace section was installed, water was displaced from the pre-installed weld chamber, and the underwater connection was made with dry welding. The weld chamber was no larger than required to accommodate the brace and one welder. The simplicity of the chamber and the fact it did not have to be installed underwater, with the advantage of the deposition rate for dry welding being better than for wet, resulted in this being a project where dry welding was more economical than wet.

Offshore Platform Repair - Case B. On the same platform involved in the preceding example, an identical brace on the opposite side of the platform had suffered similar damage. However, a major difference was that the deformation extended down to the leg, and there were some cracks in the toe of the weld that had propagated through the wall of the leg.

For this repair, the crack was repaired and the leg reinforced with a doubler plate that was welded to the end of the brace before the brace was installed. If dry welding had been used, the weld chamber would have had to accommodate six structural members (the leg, three horizontal braces, and two vertical-diagonal braces). In this case, underwater installation of the weld

Table 10.5
Basic Equipment for Underwater Arc Welding and Cutting

Category	Equipment, Tools, and Consumables
Wet Welding, Oxygen Arc Cutting, and Dry Welding	Welding power source, 300 to 600 amp, dc[a]. Knife switch, single pole, 400 amp. Voltage and amperage meters. Welding cables, 2/0 to 4/0 with ground clamp. Hand and power tools required for underwater work. Tools and parts for maintenance and repair of equipment, cables, etc.
Wet Welding	Electrode holder with 1/0 whip. Welding electrode transfer system. Welder's lens holder assembly. Welding electrodes, waterproof.
Oxygen Arc Cutting	Cutting torch with 1/0 whip. Oxygen cylinders with manifold. Oxygen regulator and hose. Cutting Electrodes. Oxygen.
Dry Welding	Dry weld chamber (habitat) with seals, ballast and/or tie-downs. Lifting equipment and tools for installing chamber. Welding electrode transfer system. Dry storage compartment. Storage oven for welding electrodes. Welding electrodes. Gas/fume scrubber with replacement cartridges[b]. Gas analyzers[b]. Preheat equipment with thermocouples, topside readout. Heat shields, flexible for life support and communications umbilicals. Out-of-chamber exhaust[b]. Welder's hood, or lens holder assembly attached to breathing apparatus. Protective clothing and gloves, fire retardant. Grinders, buffers, and chippers. Television camera with lights and topside monitor/recorder. Lighting for welder/diver visibility. Ground fault interrupters in all ac electrical circuits. Welding electrodes. Weld chamber background gas, fire retardant.[b] Nitrogen, if hot-tap pressure tests required.

a. For exothermic cutting, a 12-volt battery may be used.

b. Required if depth at bottom of weld chamber is more than 90 ft (27 m).

chamber, reattaching the brace, and removing the chamber might have taken eight to ten times as long as it took to make the repair with wet welding. The additional time, plus costs to design, procure material, fabricate and transport the weld chamber, is an example of how much more time and cost are sometimes required if dry welding is used. In this case, wet welding was preferred.

METALLURGICAL CONSIDERATIONS IN UNDERWATER WET WELDING

GUIDANCE FOR PRODUCING wet underwater welds that meet the performance standards set forth in ANSI/AWS D3.6-93, *Specification for Underwater Welding*, is provided in this section. However, a good understanding of underwater wet welding metallurgy is necessary to improve the mechanical properties as they apply at greater depths. The mechanical properties of wet weldments depend both on the nature of the resulting weld deposit and the heat-affected zone (HAZ). These microstructures in turn are greatly affected by the cooling rate after welding as well as the gases that enter the weld puddle during welding. Macrographs of surface, dry hyperbaric, and underwater wet welds are shown in Figure 10.16.[21]

The water environment acts as a fast quenching medium which hardens the HAZ and makes it susceptible to hydrogen cracking. It has been general practice that the wet underwater welding technique using mild steel electrodes is limited to steels of low-carbon equivalent (CE < 0.40 wt. %).[22] The carbon equivalent is typically given by the expression:

$$CE = \%C + \% \ Mn/6 + \%(Cr + Mo + V)/5 + \%(Ni + Cu)/15 \qquad (10.1)$$

The wet underwater welding of higher strength steels must satisfy more stringent requirements for acceptable HAZ properties, including concerns about hydrogen damage. It has been shown that high carbon equivalent steels which have been wet welded with ferritic electrodes are subject to hydrogen cracking of the HAZ.

There is a major difference between wet welding and dry hyperbaric welding. Both experience increased pressure with depth (one bar for each 10 meters of water depth), but as mentioned previously, the wet environment also influences the cooling rate during welding which affects the nature of the weld-metal phase transformation. The notation $\Delta t_{8/5}$ refers to the time for cooling from 800 °C to 500 °C. While shielded metal arc welding has $\Delta t_{8/5}$ values in the range of 8 to 16 seconds, typical wet welding procedures have been reported to have a $\Delta t_{8/5}$ between 1 and 6 seconds depending on heat input of 20 to 90 kJ/in. (0.8 to 3.5 kJ/mm) and on plate thickness.[23,24,25] Such a fast cooling rate produces significant amounts of HAZ martensite in nearly all low carbon steels. As the carbon equivalent of these steels approaches a value of 0.40

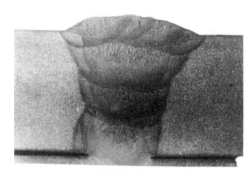

(A) Surface

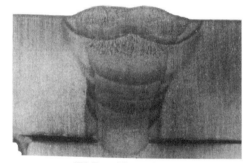

(B) Dry Hyperbaric

(C) Underwater Wet

Courtesy of The Welding Institute

Figure 10.16—Representative Macrographs of Surface, Dry Hyperbaric, and Underwater Wet Welds

21. Matlock, K. D., Edwards, G. R., Olson, D. L., and Ibarra, S. "An evaluation of the fatigue behavior in surface, habitat, and underwater wet welds," *Second International Conference on Offshore Welded Structures*, Ed. H. C. Cotton, P15-1– P15-10. Cambridge, England: The Welding Institute, 1983.
22. Grubbs, C.E., and Seth, O.W., "Multipass all position 'wet' welding, a new underwater tool." Proceedings of Offshore Technology Conference, Paper No. OTC-1620. Houston, Texas, May 1-3, 1972.
23. Christensen, N. "The metallurgy of underwater welding." *Proceedings of IIW Conference on Underwater Welding*, 71-9. New York: Pergamon Press, 1983.
24. Hausi, A., and Suga, Y., "`Cooling of underwater welds", *Transactions of the Japan Welding Society*, 2 (1), 1980.
25. Tsai, C.L., and Masubuchi, K. "Mechanisms of rapid cooling in underwater welding." *Applied Ocean Research* 1 (2): 99-110, 1979.

percent by weight, the fusion-line Vickers hardness usually exceeds 400 HV10. As the martensite content increases in the near fusion line region of the HAZ, the susceptibility to hydrogen cracking becomes a measurable concern.

Pyrometallurgy of Underwater Welds

WET SHIELDED METAL arc welds have been chemically evaluated as a function of depth (pressure) down to 656 ft (200 m).[26, 27] The variation of weld-metal manganese content as a function of depth is given in Figure 10.17.

Note the sharp decrease in manganese content from 0.6 percent by weight for the surface weld to 0.25 percent by weight for the weld made at 90 ft (27 m). This decrease in weld-metal manganese content can be directly correlated to the rapid increase in weld-metal oxygen content for the same range of depth as seen in Figure 10.18. This observation suggests the manganese content is controlled by oxidation. Similar decreases in weld metal content were found for the weld-metal silicon, as shown in Figure 10.17.

It is expected that these oxide-forming elements would respond in this manner to the increase in oxygen. The weld-metal carbon content, on the other hand, was found to increase significantly from the weld made at the surface to the welds made at 152 ft (46 m), as shown in Figure 10.19.

26. Ibarra, S. and Olson, D. L. "Wet underwater steel welding." *Ferrous Alloy Weldments*, Eds., Olson, D. L. and North, T. H., 329-78. Zurich, Switzerland: Trans Tech Publications, 1992.
27. Ibarra, S., Grubbs, C.E., and Olson, D.L. "Fundamental approaches to underwater welding metallurgy," *J. Metals.* 40(12):8-10, 1988.

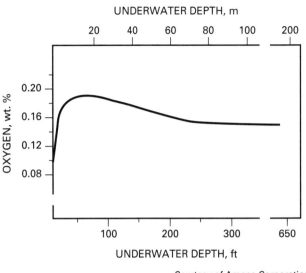

Courtesy of Amoco Corporation

Figure 10.18—Wet Underwater Weld-Metal Oxygen as a Function of Depth

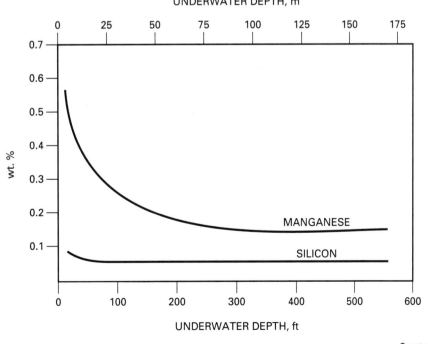

Courtesy of Amoco Corporation

Figure 10.17—Wet Underwater Weld-Metal Manganese and Silicon as a Function of Depth

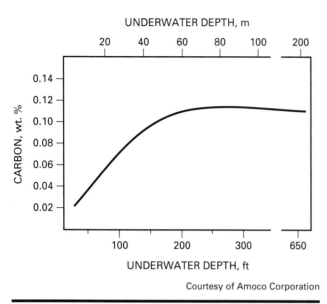

UNDERWATER DEPTH, m

Courtesy of Amoco Corporation

**Figure 10.19—Wet Underwater Weld-Metal Carbon
as a Function of Depth**

A linear relationship exists between the [C] [O] product and the total pressure, P, or depth for the shielded metal arc hyperbaric welding.[28] Using the assumption that the weld metal composition has some information of a quenched-in high-temperature reaction, but also recognizing that the welding is neither isothermal or at equilibrium, the plot of the product [C] [O] would be expected to be a linear function of the total pressure if this CO reaction is controlling. This behavior was observed for hyperbaric shielded metal arc welding (SMAW) down to the equivalent of 330 ft (100 m) in depth. The carbon monoxide is a product of the decomposition of the calcium or magnesium carbonate which is in the SMAW electrode coating to make a working and protective welding plasma and cover. The resulting increases in the weld-metal oxygen and carbon contents suggest concern about using flux coating containing carbonates for hyperbaric welding. Ferrosilicon additions to the flux reduced weld-metal oxygen but did not alleviate significantly this problem.

The carbon monoxide reaction was also found to be controlling in wet underwater welding.[29] Plotting the [C] [O] product for wet underwater welds as a function of depth (pressure) demonstrates an excellent linear

correlation down to 165 ft (50 m), as is seen in Figure 10.20. This observation suggests that the carbon monoxide reaction controls the oxygen content down to 165 ft (50 m), and the oxygen content in turn controls the oxidation of manganese and silicon, and thus, their weld-metal contents.

Between 165 and 330 ft (50 and 100 m), the weld-metal oxygen and carbon become fairly constant. Notice also that the weld-metal manganese and silicon content are fairly constant between 165 ft (50 m) and 330 ft (100 m). This observation suggests that the oxygen content controls the weld-metal manganese and silicon at depths greater than 165 ft (50 m).

From the law of mass action, it can be seen that the weld-metal oxygen content will remain constant and is not a function of pressure or water depth. The [C] [O] product will also remain constant due to a constant oxidation rate.[30] At depths greater than 165 ft (50 m), water in the plasma becomes the primary chemical factor and the [C] [O] product becomes less dominant and insensitive to depth, as is also seen in Figure 10.20.

The compositional results for depths greater than 165 ft (50 m) suggest that the H_2O decomposition reaction may be controlling. At temperatures above 1830 °F (1000 °C), water vapor begins to disassociate into hydrogen and oxygen, and dynamic equilibrium takes place.

Another indication that hydrogen is more abundant and is involved in controlling the weld-pool chemical composition is that the arc stability also decreases in extreme depth. It can be seen in Figure 10.21 that the process parameter ranges for underwater welding decrease with depth. This behavior can be explained by the high-ionization potential for hydrogen which makes it more difficult to sustain the welding arc.

Microstructural Development of Underwater Welds

LOW-CARBON STEEL weld metal microstructure consists of various fractions of ferrite, bainite (aligned carbides), and martensite. The weld microstructure shows characteristics of the solidification process with long dendritic or cellular dendritic grain growth from the edges of the weld pool toward the center of the weld bead. There are three types of ferrite associated with low-carbon steel weld metal: grain boundary ferrite, sideplate ferrite and acicular ferrite. Grain boundaries can promote the formation of large grain boundary ferrite, but are also the sites for the nucleation and growth of Widmanstatten or sideplate ferrite, which protrudes

28. Grong, O., Olson, D.L., and Christensen, N. "On the carbon oxidation in hyperbaric MMA welding." *Metal Construction* 17 (12): 810R-814R, 1985.

29. Ibarra, S. Grubbs, C.E., and Olson, D.L., "Fundamental approaches to underwater welding metallurgy," *J. Metals.* 40(12):8-10, 1988.

30. Tsai, C.L., and Masubuchi, K., "Interpretive report on underwater welding", WRC Bulletin, 224, pp. 1-37. New York.: Welding Research Council, 1977.

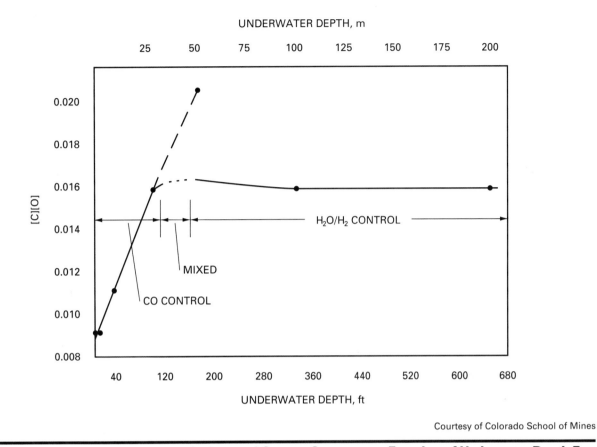

UNDERWATER DEPTH, m

Courtesy of Colorado School of Mines

Figure 10.20—Product of Weld-Metal Carbon and Oxygen Content as a Function of Underwater Depth For Weld Metal Produced With Treated E6013 Shielded Metal Arc Electrode

from the grain boundaries into the prior austenite grains. Acicular ferrite forms inside the grain and has a much finer basket-weave structure. Coarse intragranular ferrite, also known as *blocky ferrite*, may also be present in the weld metal. Other microconstituents, such as pearlite, cementite and martensite, also may result. At faster cooling rates, it is possible to form bainitic or aligned carbide and martensitic structure. Figure 10.22 illustrates the nature of these various microstructures.

Increasing weld-metal oxygen promotes ferrite nucleation due to the greater number of inclusions available to be sites for heterogeneous nucleation. Increasing manganese causes an increase in hardenability. These changes can potentially alter the weld metal microstructure and mechanical properties.

The relative amounts of each ferrite morphology, which were measured for wet welds down to 330 ft (100 m), are reported in Figure 10.23. At depths near the surface, the weld metal is mainly primary grain

boundary ferrite with 10 to 20 percent upper bainite. With increasing depths, the relative amount of primary grain boundary ferrite decreases to about 50 percent, with increasing amounts of upper bainite and sideplate ferrite. The major changes in microstructure occur in the first 165 ft (50 m) of depth, just as with the weld metal composition. At depths greater than 165 ft (50 m), the weld metal composition and microstructure remain fairly constant. This behavior is consistent with the observation of a constant weld-metal oxygen content which suggests a constant oxidation rate.

It is highly desirable to have some acicular ferrite in the microstructure instead of the significant sideplate ferrite content thereby improving toughness. The sideplate ferrite has a lath formation mechanism similar to that of acicular ferrite, but it nucleates on the existing grain boundary ferrite and grows into the remaining austenite, whereas acicular ferrite nucleates on intragranular inclusions under specific conditions. The AWS E6013 grade electrode, commonly used in underwater

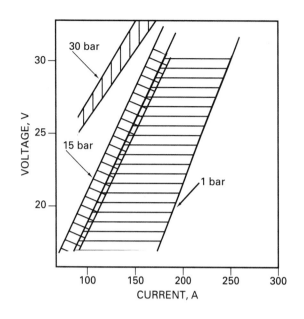

NOTE: ACCEPTABLE WELDING PARAMETER RANGE REDUCES WITH PRESSURE OR DEPTH.

Figure 10.21—The Regions in Welding Parameter Space for Successful High-Pressure Welding as a Function of Pressure or Depth

welding, does not produce the necessary weld metal composition to form acicular ferrite and sideplate ferrite results.

Acicular ferrite formation may require specific weld metal alloying additions, such as titanium and boron with the proper weld-metal oxygen and manganese contents. Shielded metal arc underwater welding electrodes have been modified to introduce titanium and boron additions to the weld pool.[31]

Research and development aimed at improving the microstructure of wet welds and reducing porosity has been continuing through a joint industry-university program.[32] During Phase I of the program, welds were made with 58 batches of experimental welding electrodes. Results of analysis of the test weldments show substantial improvement in the microstructure of the weld metal.

31. Sanchez-Osio, A., Grubbs, C.E., Liu, S., Olson, D.L., and Ibarra, S. "The effects of titanium and boron on underwater weld metal microstructure and properties"
32. Amoco Corp. Research Center, Naperville, Illinois, has provided funding for the program. The welding research center at the Colorado School of Mines, Golden, Colorado, has formulated experimental flux coatings and manufactured welding electrodes for the program. From January 1991 to February 1992, wet welds were made for the project by Global Diver and Contractors, Inc., Lafayette, Louisiana.

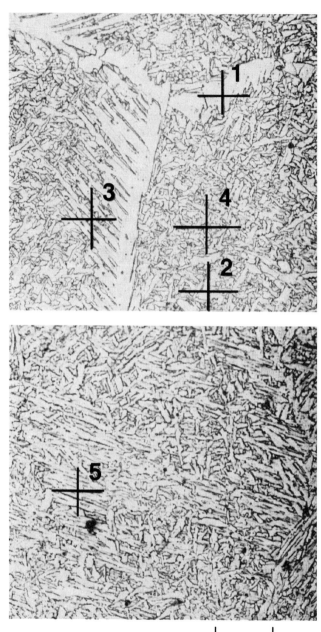

800 µin. (20 µm)

Notes:
1. Grain boundary ferrite
2. Blocky ferrite
3. Sideplate ferrite
4. Acicular ferrite
5. Upper bainite

Courtesy of Colorado School of Mines

Figure 10.22—As-Deposited Weld Microstructures Showing Various Microstructural Constituents

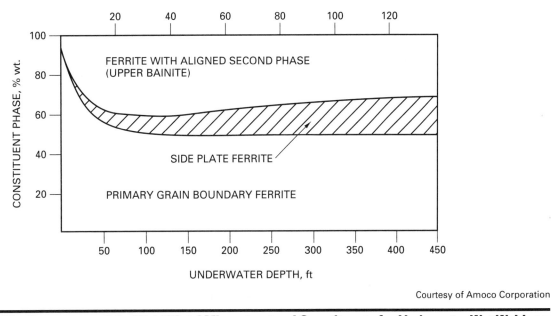

Courtesy of Amoco Corporation

Figure 10.23—The Percentage of the Weld Metal Microstructural Constituents for Underwater Wet Welds as a Function of Water Depth

A weld-metal oxygen content versus effective weld-metal manganese content (Mn+6C) diagram has been suggested to assist in selecting compositional modifications for underwater wet welding consumables.[33,34] Figure 10.24 was constructed based on microstructural and compositional analyses and represents a cooling rate found typically in wet welding. Notice that this diagram suggests that to attain the desired acicular ferrite both weld-metal oxygen and manganese content are required to increase. This graphical understanding assists the designing of new underwater welding electrode formulations to improve wet weld quality at greater depths.

Hydrogen Mitigation

THE SUSCEPTIBILITY TO hydrogen cracking is a concern for surface welding as the material strength increases above 50 ksi (350 MPa), but it can be controlled by the proper selection of welding procedures and consumables.[35,36] Since moisture drastically increases the availability of hydrogen, wet underwater welding of higher strength steels is very susceptible to hydrogen cracking.[37]

The problem of HAZ cracking resulting from wet welding of higher strength steels can be explained as a result of three factors:

(1) Hydrogen from the weld pool
(2) Microstructures that develop in the HAZ
(3) Stress levels that develop in the weld joint

To eliminate the cracking, the effect of these factors on the weld joint must be reduced.

Three possible methods to reduce or manage the hydrogen content are the following:

33. Ibarra, S. and Olson, D. L. "Wet underwater steel welding." *Ferrous Alloy Weldments*, Eds. Olson, D. L. and North, T. H., 329-378. Zurich, Switzerland: Trans Tech Publications, 1992.
34. Ibarra, S., Grubbs, C.E., and Olson, D.L. "Fundamental approaches to underwater welding metallurgy," *J. Metals*. 40(12):8-10, 1988.
35. Ibarra, S. and Olson, D. L. "Wet underwater steel welding." *Ferrous Alloy Weldments*, Eds. Olson, D. L. and North, T. H., 329-378. Zurich, Switzerland: Trans Tech Publications, 1992.
36. Kuester, K., Hoffmeister, H., and Schafstall, H.G. "Hydrogen pick-up during welding using an advanced wet welding technique (local dry spot)," Proc. of 2nd Int. GUSI-Symposium - Underwater Tech., Geesthacht, Fed. Rep. of Germany, 1987.
37. Ozaki., O., Naiman, J., and Masubuchi, K. "A study of hydrogen cracking in underwater steel welds." *Welding Journal*. 56(8): 231s-237, 1977.

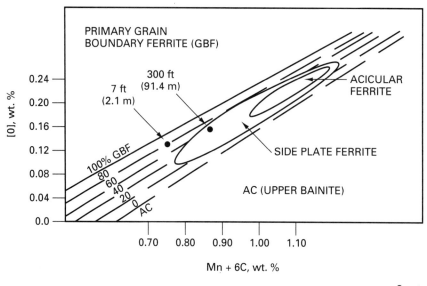

Courtesy of Amoco Corporation

Figure 10.24—Suggested Compositional Diagram for the Prediction of Weld Metal Microstructure for Underwater Wet Welds

(1)   To use consumables which can hold a high concentration of hydrogen in molten weld pools as well as in the solidified weld bead.

(2)   To alter consumable fluxes to introduce alternate gases in the welding plasma, thereby reducing the hydrogen content.

(3)   To select welding parameters which minimize weld pool hydrogen pickup.

Influence of Welding Consumables. The use of ferritic-based weld deposits will cause sufficient hydrogen reject from the weld deposit into the fully martensitic HAZ promoting underbead cracking. A method to alleviate some of the hydrogen susceptibility is to use an austenitic weld deposit which has a much larger solubility for hydrogen and thus less tendency to transport hydrogen to the HAZ.[38]

The use of austenitic stainless steel consumables results in a deposit with high thermal expansion coefficient relative to ferrite based weld metal. Thus, higher residual stresses and increased tendency for cracking also occur. The compromise solution between high solubility for hydrogen and thermal expansion mismatch is the use of high-nickel-based weld deposits since nickel based alloys have approximately the same thermal expansion as ferritic steel and are capable of managing a large hydrogen content.

Although the nickel weld metal deposit is virtually immune to the restraint cracking noted in the austenitic stainless steel welds, the use of nickel-based electrodes is restricted by their depth sensitivity. The welds produced using high-nickel electrodes can be very porous, and in some cases susceptible to embrittlement. Nickel-based welds at ten meters depths have been shown to have excessive porosity resulting from a lack of heat input at these depths. A higher heat input is required to make nickel-based consumables behave properly and obtain an acceptable weld deposit. These nickel-based wet weld deposits have been successful only for fillet welds down to ten meters depth when welded with the SMAW process.[39]

To reduce the influence of the plasma hydrogen content, other gases need to be introduced in the arc. It is already known that increasing the carbonate content of welding flux increases the CO content in the arc and thus relatively reduces the hydrogen content.[40,41]

38. Stalker, W.A., Hart, P.H.M., and Salter, G.R. "An assessment of shielded metal arc electrodes for the underwater welding of carbon manganese structural steels." Proceedings of Offshore Technology Conference, Paper No. OTC-2101, Houston, Texas, 1975.

39. See "Recent Developments-Nickel Welding," Page 484.
40. Chew, B., "Prediction of weld-metal hydrogen levels obtained under test conditions." *Welding Journal*, 52: 386s-91s, 1973.
41. Sorokin, L.I., and Sidlin, Z.A. "The effect of alloying elements and of marble in an electrode coating on the susceptibility of a deposited nickel chrome metal to pore formation." *Svar. Proiz.*, No. 11, 7-9, 1974.

Another factor affecting the increase use of carbonates is that the resulting increasing weld-metal oxygen content causes a significant reduction in weld-metal hydrogen.[42] Thus, increasing the carbonate contents of the wet underwater electrode coatings should allow for arc stability at greater depths and for the reduction of weld-metal hydrogen content and porosity. The influence of increased carbonate content in the electrode coating is expected to increase the depth at which the CO reaction is controlling.

The use of carbonates to reduce the plasma hydrogen content does promote the problem of weld-metal carbon and oxygen pickup with increasing pressure. Carbon has been reported to reduce the hydrogen solubility in ferrite and increase the hydrogen solubility in austenite. Thus, increasing carbon content will cause greater potential hydrogen supersaturation with the decomposition of austenite, reducing the favorable hydrogen-reduction effects in the arc. This problem could limit the use of increased carbonate content to shallow depths.

It was also reported that alloying the weld deposit with chromium, molybdenum, and tungsten can also reduce the hydrogen-content. Niobium, as a strong deoxidant, increases the hydrogen content. Electrodes containing ten percent $CaCO_3$ in the coating showed that, on the basis of their capacity to inhibit the process of pore formation, the alloying elements can be arranged in the following sequence: titanium, niobium, manganese, rhenium, molybdenum and tungsten.[43] Reducing the hydrogen content should also reduce the tendency for porosity.

Hydrogen pickup was reported to increase with coating thickness of rutile electrodes used in wet underwater SMAW but decreased with heat input.[44,45] The HAZ hardness also was reported to increase with coating thickness. Weld metal Charpy V-notch toughness of 26 to 30 ft-lb (35 to 40 J) at 32 °F (0 °C) has been produced by applying multilayer stringer beads using a low-heat input.

Influence of Welding Parameters. The influence of welding parameters on the underwater welding process and on the metallurgical reactions has been reported.[46] Strong relationship between weld-metal hydrogen content and welding parameters (welding potential and current) has been found.[47] Weld-metal diffusible hydrogen content as a function of welding voltage was found to increase with increasing voltage and decreasing welding current. The diffusible hydrogen content in wet underwater welds was found to decrease with increasing heat input for SMAW and FCAW.[48,49]

The third factor influencing hydrogen cracking is the applied and residual stress state. The applied stress can be reduced by improving the fit-up on the weld joint to be used in the repair. In the underwater wet welding repair of platforms, scalloped sleeves are frequently fillet welded over the damaged area. Careful fabrication of the scalloped sleeve will result in improved fit-up and thus reduced stresses. The applied stress on a weldment can also be improved by using sufficient weld metal deposit to support the required load.

The residual stresses are more difficult to manage in an environment which is not easily accessible to postweld heat treatment. There are welding practices that can reduce residual stresses which include:

(1) The use of small weld deposits

(2) The use of consumables with compatible coefficients of thermal expansion with the base materials

(3) The selection of edge preparations which reduce the size of the total weld deposit. The cross-sectional area of the total weld deposit is directly related to tendency for shrinkage.

Temper Bead Practice. Further methods to reduce stress and thus hydrogen cracking must consider the thermal experience or experiences of the wet welding practice.[50] It is very difficult to perform any significant preheat treatment procedures to the base material in wet welding of heavy sections. A promising approach

42. Sorokin and Sidlin. "The effect of alloying elements," 7-9.
43. Sorokin and Sidlin, "The effect of alloying elements," 7-9.
44. Hoffmeister, H., and Kuster, K., "Process variables and properties of underwater wet shielded metal arc laboratory welds." *Underwater Welding*, IIW Conference, Trondheim, Norway, 115-28. New York: Pergamon Press, 1983.
45. Hoffmeister, H., and Kuster, K. "Process variables and properties of wet underwater gas metal arc laboratory and sea welds of medium strength steels." *Underwater Welding*, IIW Conference, Trondheim, Norway, 121-8. New York: Pergamon Press, 1983.
46. Madator, N.M. "Influence of the parameters of the underwater welding process on the intensity of metallurgical reactions." *Welding Research Abroad*, (3): 63, 1972.
47. Kononenko, V.Y. "Effect of water salinity and mechanized underwater welding parameters on hydrogen and oxygen content of weld metal." *Welding Under Extreme Conditions*, IIW Conference, Helsinki. 113-8. New York: Pergamon Press, 1989.
48. Hoffmeister, H., and Kuster, K. "Process variables and properties of underwater wet shielded metal arc laboratory welds." *Underwater Welding*, IIW Conference, Trondheim, Norway, 115-28. New York: Pergamon Press, 1983.
49. Hoffmeister, H., and Kuster, K. "Process variables and properties of wet underwater gas metal arc laboratory and sea welds of medium strength steels." *Underwater Welding*, IIW Conference, Trondheim, Norway, 121-8. New York: Pergamon Press, 1983.
50. Matsunawa, A., Takemata, H., and Okamoto, I., "Analysis of temperature field with equiradial cooling boundary around moving heat sources - heat conduction analysis for estimation of thermal hysteresis during underwater welding." *Trans. JWRI* 9(1): 11-8, 1980.

that is being investigated is the use of temper bead practice to reduce the near fusion-line-cracking tendency. Temper bead practice is typically a weld deposition that is laid down over a previous weld deposit which has a less susceptible cracking tendency than the base metal.[51] An illustration of temper bead practice is seen in Figure 10.25. This second deposition (or temper bead) must be carefully located relative to the previous bead fusion line such that its thermal experience tempers the near-fusion-line heat-affected zone of the base material which has a cracking susceptible carbon equivalent. Temper weld beads have been found to influence the properties of wet underwater welds. Weld metal Charpy V-notch toughness have been seen to improve with increasing number of weld passes.[52,53] Using the above suggestions to reduce residual stresses as well as hardness in conjunction with a temper bead practice most likely represents the optimum practice to wet weld the higher carbon-equivalent steels. Another advantage

of using the temper bead practice is that while the temper bead is in the high-temperature austenite condition, it is a favorable reservoir for hydrogen and thus extracts some of the hydrogen from the more crack-susceptible base metal HAZ.

A temper-bead procedure has been performed with wet underwater welding with a measurable reduction of fusion-line and near-fusion-line hardness. When a platform steel with a CE of 0.40 wt. % was welded in a 10-meter tank, microscopic examination of the weldments revealed extensive hydrogen cracking and HAZ Vickers hardness values in excess of 450 HV10. After the temper bead technique was used, the HAZ Vickers hardness values decreased to approximately 300 HV10. Side bend tests were conducted on these samples, and the results showed that the samples were of acceptable quality. This reduction in hardness thus represents a HAZ microstructure that is less susceptible to hydrogen cracking.

Chemical heat-generating welding consumables have recently been investigated as an additional heat source during underwater welding.[54] Samples from a platform steel with a CE greater than 0.40 wt. % were welded in a 10-meter tank. A heat-generating thermit electrode was used after each pass to postweld heat treat the

51. Olsen, K., Olson, D.L., and Christensen, N. "Weld bead tempering of heat-affected zone." *Scandinavian Journal of Metallurgy*, 11, 163-8, 1982.

52. Hoffmeister, H., and Kuster, K., "Process variables and properties of underwater wet shielded metal arc laboratory welds." Underwater Welding, IIW Conference, Trondheim Norway, p. 115-25 New York: 1983.

53. Hoffmeister, K., and Kuster, K. "Process variables and properties of underwater wet gas metal arc laboratory and sea welds of medium strength steels." Underwater Welding, IIW Conference, Trondheim, Norway, pp. 239-46, New York: Pergamon Press, 1983

54. Ibarra, S. and Olson, D. L. "Wet underwater steel welding." *Ferrous Alloy Weldments*, Eds., Olson, D. L. and North, T. H., 329-78. Zurich, Switzerland: Trans Tech Publications, 1992.

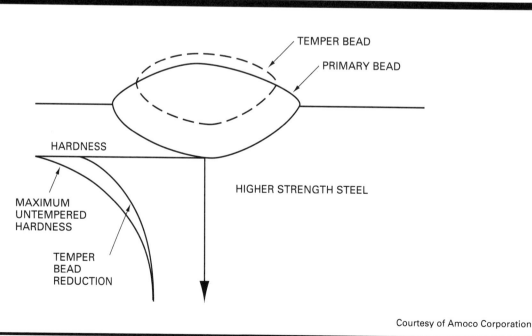

Figure 10.25—A Schematic Illustration of the Use of Temper Bead Practice to Reduce HAZ Hardness and the Susceptibility to Underbead Cracking

HAZ and reduce hydrogen cracking. Thermit-type reactions during arc welding have been used to improve the underwater temper bead practice to reduce the hydrogen-cracking susceptibility of the HAZ.

Weld Metal Porosity

WELD METAL POROSITY is the most common defect for all welding processes but is a major concern in welds that are made underwater. The amount of porosity is the most obvious difference between a wet weld and a dry weld. Pore formation results from entrapment, supersaturation of dissolved gas, or gas producing chemical reactions. The nature and amount of weld metal porosity involves at least four competing time dependent processes.[55] These processes are the nucleation, growth, transport, and coalescence of pores. The physical requirement for pore formation is that the sum of the partial pressure, P_g, of soluble gases must exceed the sum of the following pressure terms:

$$P_g > P_a + P_h + P_b \qquad (10.2)$$

where

P_a is the atmosphere pressure
P_h is the hydrostatic pressure
P_b is the pressure increase due to the curvature of the pore.

In the case of underwater welding P_h is the controlling term since it is directly related to depth.

Besides pressure considerations, pore formation must also encounter and overcome the kinetic activation barriers, such as nucleation. There is a critical pore radius which, if exceeded, will allow the pore to nucleate and grow spontaneously to unlimited dimensions.

To illustrate the effect of water pressure (depth) on porosity formation in underwater gravity welding, results from welding with three different coated electrodes are shown in Figure 10.26.[56] The ilmenite type, high titanium-oxide type, and iron-powder- iron-oxide- type electrodes of 5/32 in. (4 mm) diameter and steel base metals of 0.25 and 0.35 in. (6 and 9 mm) thicknesses were used. Increasing water depth led to higher porosity in the welds. The pore shape was observed to change from spherical to a long but narrow cylindrical

shape with increasing pressure. This observation suggests a change of porosity formation mechanism with depth. The concentration of hydrogen in the molten metal increases considerably with water pressure.

Bubble compositions of 62 to 82% H_2, 11 to 24% CO and 4 to 6% CO_2 have been reported.[57,58,59] Others determined the composition to be approximately 96 vol. % H_2, 0.4 vol. % CO and 0.06 vol. % CO_2. More specific data on a rutile iron powder (E7014) electrodes are 45% H_2, 43% CO, 8% CO_2, and 4% others. The variation in bubble composition indicates the broad nature of the chemical processes associated with wet underwater welding and is most likely due to variations in electrode covering compositions, energy input, and water depth. In the future, it may be possible to use bubble compositional analysis with proven chemical models to develop and optimize wet underwater welding electrodes.

With increasing travel speed, the pore concentration increases and goes through a maximum.[60] Porosity was also reported to increase with increasing welding current since the welding current raised the average surface temperature of the weld pool and enlarged the area of the hot granular zone over which the hydrogen absorption took place.[61] In underwater welding the relationship between porosity and welding current is strongly influenced by the type of moisture resistance coating on the electrode. Gas pickup also increased as the arc was lengthened. In general terms, hydrogen absorption and therefore porosity levels in welding should be minimized by using a low current with DCEP, high current with DCEN, a short arc, and fast travel speed.

Fatigue as a Function of Porosity

THE DESIGNERS OF offshore structures are usually concerned with the initiation and growth of fatigue cracks in fracture critical members. The cyclic fatigue stresses will develop with any natural sea movement and will increase during storms. It is important that the underwater repair of weldments be designed so that the

55. Trevisan, R.E., Schwemmer, D.D., and Olson, D.L. "The fundamentals of weld metal pore formation." *Welding: Theory and Practice,*" Eds. Olson, D. L., Dixon, R. D., and Liby, A. L., 79-115. Amsterdam: Elsevier Science Publishers, B.V., 1990.

56. Suga, Y. and Hasui, A. "On formation of porosity in underwater weld metal (The 1st Report) – Effect of water pressure on formation of porosity." *Transactions of the Japan Welding Society,* 17(1):58-64, April 1986. Also reprinted as IIW Doc. IX-1388-86. Cambridge, England: International Institute of Welding, 1986.

57. Masubuchi, K. "Underwater factors affecting welding metallurgy." AWS Conf. Proc. Underwater Welding of Offshore Platforms and Pipelines, pp. 81-98, Nov. 5-6 (1980), American Welding Society, Miami, Florida, 1980.

58. Malator, N.M. "The properties of the bubbles of steam and gas around the arc in underwater welding." *Automatic Welding,* 18(12), 25-9 1965.

59. Silva, E.A. "Gas production and turbidity during underwater shielded metal-arc welding with iron powder electrodes." *Naval Engineer's Journal* (12) 1971.

60. Trevisan, R.E., Schwemmer, D.D., and Olson, D.L. "The fundamentals of weld metal pore formation." *Welding: Theory and Practice.* Eds. Olson, D. L., Dixon, R. D., and Liby, A. L., 79-115. Amsterdam: Elsevier Science Publishers, B.V., 1990.

61. Silva, E.A. "Gas production and turbidity during underwater shielded metal-arc welding with iron powder electrodes." *Naval Engineer's Journal* (12) 1971.

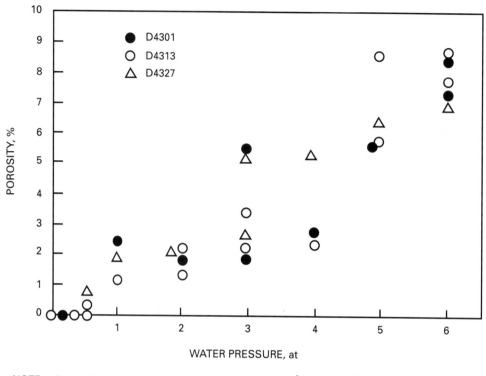

NOTE: A technical atmosphere, at, is equal to 14.22 lb/in.² or 1 kgf/cm²

Figure 10.26—The Effect of Water Pressure on Porosity

possibility of the initiation and growth of such cracks is minimized. There is concern that the growth of fatigue cracks in underwater welds is higher because of noted decreases in fracture toughness and because of the presence of varying amounts of porosity.

Drilling the tip of a fatigue crack is a common field technique used to prevent further crack growth. In the same manner, the presence of porosity in weldments can be beneficial because the pores act as pinning sites for fatigue cracks to retard or stop their growth. Multiple-pass steel weldments fabricated with E6013 electrode having three different waterproof coatings showed that the presence of some porosity was beneficial at low stresses because it retarded fatigue-crack growth in surface, habitat and underwater wet welds.[62] The fatigue crack growth data for the low- and high-porosity

underwater wet welds can be determined from Figure 10.27.

The low-porosity, dry-habitat weld was show to contain approximately 3% porosity on the fatigue-fracture surface, whereas the high-porosity underwater wet weld contained approximately 12% porosity on the fatigue-fracture surface. In comparison to normal wrought ferrite-pearlite steels, all experimental underwater welds exhibited lower growth rates for low values of stress intensity factor range ($\Delta K = K_{max} - K_{min}$, the range stress intensity factor values during a stress fluctuation). However, the wet welds tended to have higher fatigue-crack growth rate than the wrought alloys at high values of ΔK, particularly in wet welds.

Because the microstructure, hardness, and chemical compositions (except for manganese and oxygen) were similar for all of the welds, the fatigue properties primarily reflect the differences in porosity. The surface and hyperbaric habitat welds generally do not have the porosity shown in underwater wet welds. At low ΔK, the wet underwater welds have the most fatigue-crack growth resistance. The large number of pores act as pinning sites for the advancing fatigue-crack front. Increased porosity in a weldment yielded lower fatigue-crack growth at low ΔK.

62. Matlock, K. D., Edwards, G. R., Olson, D. L., and Ibarra, S. "An evaluation of the fatigue behavior in surface, habitat, and underwater wet welds." *Second International Conference on Offshore Welded Structures*, Ed. Cotton, H. C., P15-1– P15-10. Cambridge, England: The Welding Institute, 1983. An abbreviated version was later published in *Underwater Welding*, 303-310. New York: Pergamon Press, 1983.

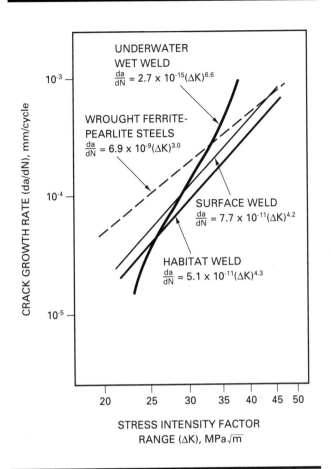

Figure 10.27—Comparison of the Effects of Porosity and Frequency on the Fatigue-Crack Growth Behavior of Surface, Dry Hyperbaric, and Underwater Wet Welds

At high ΔK, the crack growth rate is much greater for the higher porosity content. The implication is that at higher stresses, mechanical behavior of the sample acts more like that of a tensile test. The high number of pores act to reduce the cross-sectional area and thus allow the cracks to propagate at a high rate. It can be concluded that the presence of porosity can be beneficial in cases where the weldments are subjected to low stresses (such as the sleeve repair of underwater structure) and where other properties such as fracture toughness do not deteriorate.

WET WELDING STATE-OF-THE-ART

TO APPRECIATE THE significance of the results of recent advancements in wet welding, one must consider the differences in previously accepted mechanical properties of wet weldments as compared to the requirements for dry weldments. This can best be done by examining the main differences in the requirements for the qualification of Class B and Class A welding procedures as specified by AWS D3.6-93.[63]

Depth Capabilities and Limitations

THE MAXIMUM KNOWN depth at which underwater wet welding procedures have been qualified in accordance with the requirements of AWS D3.6-93 for Class B welds is 325 ft (100 m).[64] Base material was ASTM A36 with a carbon equivalent of 0.347. Welding electrodes were E6013 with a proprietary supplemental overcoat.

For the repair of an offshore structure in the Gulf of Mexico, welding procedures and five welders were qualified onshore in fresh water in a hyperbaric simulation facility, or *wet pot*, at pressures equivalent to 165 ft (50 m) and 325 ft (99 m). Prior to starting production welding offshore, additional test weldments were made at the 160 ft (49 m) and 320 ft (98 m) underwater work sites. In addition to the weldments required by AWS D3.6-93 for Class B welds, welds also were made and tested as specified by AWS D3.6-93 for Class A welds. Mechanical properties of these weldments are summarized in Table 10.6.

All aspects of the welding were significantly better in open seawater than in the confines of the freshwater filled wet pot, including arc stability, general weldability, surface appearance, ease of slag removal, and mechanical properties.

There was no difficulty in qualifying welding procedures at a depth of 165 ft (50 m). The only unusually difficult part of the qualifications was meeting the bend requirements (four of four specimens on 6T radius) at 325 ft (99 m). For the welder/diver, welding at a depth of 325 ft (99 m) was not significantly different than welding at 165 ft (50 m). The difficulty in passing the 325 ft (99 m) bend test is attributed to the pressure-induced increase in porosity and adverse affects on weld metal composition and microstructure. The success of future efforts to qualify wet welds at greater depths will depend on using welding electrodes that are specifically formulated to counteract the adverse affects of pressure and to minimize weld metal porosity.

Multiple-Temper-Bead Technique

WHEN CONVENTIONAL WET welding techniques are used to join carbon and low-alloy high-strength (carbon equivalent $\geq$ 0.40) steels with ferritic filler

---

63. See "Dry and Wet Welding Compared," Page 465.
64. Grubbs, C. E. "Qualification of welding procedures down to 325 fsw." International Diving Symposium, Houston, Texas, February, 1986.

Table 10.6
Mechanical Properties of Wet Weldments[a]

Test	Specified Minimum (AWS D3.6-93)	Test Results at Various Depths and Two Environments[b]				
		33 ft (10 m)	160 ft (49 m)[b]	165 ft (50 m)	320 ft (98 m)[b]	325 ft (99 m)
Fillet Weld Shear[c] Strength, ksi (MPa)	34.8 (240)	47.9 (330)	— —	42.6 (294)	— —	43.9 (303)
Reduced-Section[c] Tension, ksi (MPa)	58.0 (400)	77.4 (534)	68.1 (470)	65.9 (454)	64.5 (448)	61.6 (425)
All-Weld-Metal[d] Tension Yield Strength, ksi (MPa)	— —	71.6 (494)	— —	60.5 (417)	— —	56.3 (388)
Tensile Strength, ksi (MPa)	58.0 (400)	75.9 (523)	— —	63.0 (434)	— —	58.8 (405)
Elongation, %	14.0	12.5	—	9.5	—	6.5
Charpy Impacts[d] Minimum Value, ft-lb (J)	10.0 (13.6)	29[e] (39.3)	— —	23[f] (31)	— —	16[f] (22)
Average Value, ft-lb (J)	15.0 (20.3)	29[e] (39.3)	— —	25[f] (34)	— —	18[f] (24)

a. Courtesy of Global Divers & Contractors, Inc.

b. Welds made at 160 ft (49 m) and 320 ft (98 m) were made offshore in sea water. Other welds were made in fresh water in a hyperbaric chamber for qualification of welding procedures.

c. The minimum requirements shown for fillet weld shear strength tests and reduced-section tensile tests are specified by AWS D3.6-93 for both Class B (wet) welds and Class A (dry) welds.

d. The minimum requirements shown for the all-weld-metal and Charpy V-notch impact test are specified by AWS D3.6-93 for Class A (dry) welds. Those tests are not required for Class B (wet) welds.

e. Charpy V-notch impact tests conducted at 28 °F (-2 °C).

f. Charpy V-notch impact tests conducted at 32 °F (0 °C).

metal, the heat-affected zone (HAZ) of the base metal is subject to hydrogen-induced underbead cracking, excessive hardness, and insufficient V-notch toughness. Until development of the multiple-temper-bead (MTB) wet welding technique, prudent use of wet welding with ferritic filler metal was limited to materials with a carbon equivalent (CE) less than 0.40.

The multiple-temper-bead welding technique has been used for wet welded repairs on several offshore structures in the Gulf of Mexico.[65] However, it was

originally developed for use in the replacement of a vertical diagonal brace on a North Sea offshore production platform.[66] The material to be welded was BS4360 Gr50D, and proprietary E6013-type welding electrodes were used. Water depth was 33 ft (10 m). During attempts to qualify welding procedures with conventional wet welding techniques, the HAZ of all weldments had gross hydrogen-induced cracking. Ultimately, the multiple-temper-bead technique was developed and used to qualify the welding procedure and fourteen welder/divers. The test results met all the requirements of ANSI/AWS D3.6-93, *Specification for Underwater Welding*, for Class B welds. Supplemental

65. Welding procedures were qualified and repairs made by Global Divers & Contractors, Lafayette, Louisiana.

weldments were made that met the requirements of ANSI/AWS D3.6-93 for Class A welds for HAZ hardness and Charpy V-notch impacts at 14 °F (-10 °C).

During 1994, comprehensive use and testing of multiple-temper-bead welds confirmed the wet welding techniques's effectiveness in reducing hardness and preventing hydrogen cracking in the HAZ of base metals with carbon equivalents up to 0.462, including 0.20 carbon.

Nickel Electrodes and Carbon Steel Base Metal

THE OPTION OF using nickel welding electrodes to prevent hydrogen-induced underbead cracking in the HAZ of crack-susceptible, high-strength carbon steels is a highly desirable one. However until recently, the successful qualification of nickel wet welding procedures was limited to fillet welds qualified per ANSI/AWS D3.6-93 requirements for Class B welds at ≤ 33 ft (10 m).

In 1992, underwater wet groove welding procedures were qualified in accordance with the requirements of ANSI/AWS D3.6-93 Class O/ASME Section IX for above-water welds. High-nickel welding electrodes with proprietary waterproof processing were used. The base metal was ASTM A572 Gr 50. Water depth was 33 ft (10 m). Procedures were qualified for torus repairs, water service systems, and water intakes and outfalls at nuclear power plants.

Developmental work was under way in 1994-95 to increase the wet-welding depth capabilities of nickel electrodes from 33 ft (10 m) to greater than 50 ft (15 m).

Stainless Steel Electrodes and Base Metal

SINCE 1980 THERE have been more than fifty applications of underwater wet welding for repairs and modifications to nuclear-power generating facilities. Repairs have ranged from simple fillet-welded doubler plates for repairing leaks in pool liners to fairly complex repairs of boiling-water-reactor steam dryers.[67] In most cases,

austenitic stainless steel (316L) welding electrodes were used to weld austenitic stainless Type 304 base metal. Groove and fillet welds have been made in all positions down to 40 ft (12.2 m).

The use of austenitic stainless steel electrodes for wet welding of austenitic stainless base metals has produced weldments that meet all the requirements specified by ANSI/AWS D3.6-93 Class A and Class O/ASME Section IX for above-water welds. Austenitic stainless, because of its natural aversion to the formation of martensite, seems to have an affinity for wet welding. In fact, austenitic wet weldments made down to 40 ft (12 m) are in some respects metallurgically superior to similar welds made above water. This is because the weldment is essentially rapid-quench-annealed during the welding process, thereby minimizing HAZ sensitization, a prime contributor to intergranular stress-corrosion cracking in nuclear plant components.[68]

Ferritic Welding Electrodes and Carbon Steel Base Metal

THE MECHANICAL PROPERTIES of wet weldments reported in Table 10.6 accurately reflect the state-of-the-art of wet welding with ferritic electrodes as of March 1995, when the table was prepared. The values shown in the column for 33 ft (10 m) are based on the results of tests made on weldments produced in 1994 during an ongoing Joint Industry Underwater Welding Development Program. Values shown for welds made at depths greater than 33 ft (10 m) are based on tests of welds made during the qualification of welding procedures in May 1985 for underwater structural repairs on an offshore drilling and production platform.

Summary of Results. Except for bend tests at 2T radius and all-weld-metal elongation, the wet welds made at 33 ft (10 m) met or significantly exceeded the ANSI/AWS D3.6 requirements for Class A (dry) welds in all test categories shown in Table 10.6, i.e., fillet weld shear strength, reduced-section tensile strength, all-weld-metal tensile strength, and Charpy V-notch impact values at 28 °F (-2 °C).

Except for bend tests and all-weld-metal elongation, wet welds made from 160 ft (49 m) to 325 ft (99 m) met the ANSI/AWS D3.6 requirements for Class A welds in all tests that were made, i.e., fillet weld shear strength, reduced-section tensile strength, all-weld-metal tensile strength, and Charpy V-notch impact values at 32 °F (0 °C).

66. See Ibarra, S., Reed, R. L., Smith, J. K., Pachniuk, I., and Grubbs, C. E. "The structural repair of a north sea platform using underwater wet welding techniques." *Proceedings of the Offshore Technology Conference*, 57, Paper No. OTC-6652, Houston, Texas, May 1991. Also see Ibarra, S., Olson, D. L., and Grubbs. "Underwater wet welding techniques for repair of higher-carbon-equivalent steels." *Proceedings of the Offshore Technology Conference*, 135, Paper No. OTC-6214, Houston, Texas, May 1990, and Ibarra, S., Olson, D. L., and Grubbs. "Underwater wet welding of higher strength offshore steels." *Proceedings of the Offshore Technology Conference*, 67, Paper No. OTC-5859, Houston, Texas, May 1989.

67. O'Sullivan, J. E. "Wet underwater weld repair of susquehanna unit 1 steam dryer." *Welding Journal* 67(6):19-23, 1988.

68. O'Sullivan, J. E. "Underwater welding in nuclear plant applications." *Maintenance and Repair Welding in Power Plants*, 75-83. Miami: American Welding Society, 1992.

From 33 ft (10 m) to 165 ft (50 m), groove-weld tensile values declined about 15 percent with an additional loss of 5 percent at 325 ft (99 m). This is attributed to pressure-induced changes in the weld metal, including effects on microstructure, increased porosity, reduced manganese and silicon, plus increased carbon and oxygen.

It may be interesting to note that the welds made in open seawater had greater tensile strengths than the welds made in fresh water within the confines of the hyperbaric test chamber at comparable depths.

Supplemental Information. Until late 1994, even the best wet welds failed to meet radiographic requirements of ANSI/AWS D3.6 for Class A welds because of fine, evenly dispersed, but excessive porosity. During the aforementioned Joint Industry Underwater Welding Development Program, an experimental wet welding electrode was developed that, when used at a depth of 33 ft (10 m), produced welds with porosity-free Class A radiographs.

This electrode from the Joint Industry Program was designed, in addition to reduce porosity, to increase all-weld-metal elongation, which during the program had averaged 12.5 percent. All-weld-metal elongation from two welds made with this electrode averaged 14.3 percent.

During the Joint Industry Program, welds were made on ASTM A537 Class 1.75-in. (44.5-mm) thick base metal with a carbon equivalent of 0.462, including 0.20 carbon. Using techniques developed during the program, wet welds were produced that not only were free of hydrogen-induced HAZ cracks (HAZ was examined at 400X magnification), but also met Vickers 10 kg hardness test requirements according to ANSI/AWS D3.6 for Class A welds.

As indicated earlier in this chapter, wet welds have not yet been produced that have enough ductility and elongation to meet the 2T bending radii specified by ANSI/AWS D3.6 for Class A welds.

QUALIFICATIONS OF WELDING PERSONNEL

THE IDEAL UNDERWATER welder would not only have formal qualifications in welding and diving, but would also have substantial skill and experience in fitting and rigging, inspection, sketching, and photography. Specific qualifications and individual needs in these diverse areas are described in the following subsections.

Welding

THE WELDER/DIVER SHOULD be capable of qualifying for wet and dry hyperbaric welding on all-position fillet and groove welds, on pipe and plate, in accordance with the requirements of ANSI/AWS D3.6-93 and other governing codes or specifications.

Prerequisites for these welder qualifications include the training and experience to become a highly proficient top-side welder. This should be followed by in-house underwater welding training and practice, and underwater welding experience, to extend these top-side capabilities to underwater dry and wet welding. Before going to work for a diving company, the aspiring welder/diver should be sure that the company has an established practice of providing underwater welding training, that their hyperbaric facilities are adequate, and that their instructors are experienced underwater welders.

Diving

SKILLED *TOP-SIDE* WELDERS, through in-house and on-the-job diver training, could once make the transition to underwater welding without formal diver training. Now, for employee safety and for liability reasons, no reputable diving company condones this practice.

To be employed as a potential welder/diver, one must have a certificate of graduation from a recognized commercial diving school. Before becoming a full-fledged diver, an apprenticeship of about two years is required as a tender or diver/tender. During that time, the potential welder/diver participates in underwater welding and cutting classes. As capabilities develop, the diver/tender will have opportunities to work as a welder/diver.

Inspection

WELDER/DIVERS WHO also are certified welding inspectors are in great demand. The Inspection section of this chapter provides details on inspection requirements, methods and the training, qualification, and certification of NDT inspectors.

Fitting and Rigging

BUDGETS AND ESTIMATES for underwater welding projects are not based on separate individuals performing each of the many tasks that are essential to the successful and economical completion of the job. Good underwater welders, through training and experience, can rig, install, and fit up the sections they will weld. They can also operate, maintain, and make minor repairs on basic job equipment.

Photography and Drafting

UNDERWATER INSPECTION REPORTS are essential to pre-job planning, especially for the repair of underwater damage to offshore structures. The reports are used by engineers to determine the nature and extent of the damage and for planning and designing the repair sections and methods. Competing contractors use the reports for estimating time and costs.

Projects are not complete until the customer has received a clear comprehensive report on the work that was done. Nothing does more to provide the customer with the information he needs than a clear, concise written report with good photographs and three dimensional sketches. Every welder/diver should develop these skills.

UNDERWATER WELDING SPECIFICATIONS

UNTIL THE FIRST underwater welding specification was published in 1983, hundreds of underwater welds, some of them crucial to the performance of an offshore structure or subsea pipeline, were made without benefit of codes, specifications, or sanctioned guidelines. Under those circumstances, users and providers of underwater welding improvised inconsistent proprietary methods to determine if a welding procedure was suitable for its intended use, and if the welder/divers were capable of performing the underwater work. Furthermore, there were no inspection guidelines. Underwater inspection equipment and techniques were, as was the welding, in the developmental stages.

Industry needed guidelines or a specification with comprehensive information that would enable and promote the following:

(1) The engineer could select the welding method (wet or dry) that met his or her fitness for purpose requirements and could make informed critical design decisions.

(2) The purchaser could conveniently specify and obtain underwater welding that met the engineer's requirements.

(3) Contractors could prepare estimates and quote each project with a clear understanding of what was required.

(4) The work could be executed in accordance with the purchaser/contractor agreement and a comprehensive underwater welding specification.

Agencies that develop underwater welding standards include the International Institute of Welding, American Welding Society, American Society of Mechanical Engineers, Det Norske Veritas and Bureau Veritas. Documents of all agencies are referenced in this section. The American Welding Society specification is given more attention because it is the most comprehensive, and it has benefitted from users' input for a longer period of time.

AMERICAN WELDING SOCIETY

THE AMERICAN WELDING Society (AWS) began developing the ANSI/AWS D3.6, *Specification for Underwater Welding* at a meeting of D3b, Subcommittee on Underwater Welding, in the Spring of 1974. The subcommittee was established by the D3 Committee on Welding in Marine Construction and has a balanced membership of users of underwater welding, providers of underwater welding services, and governing agencies. The ANSI/AWS D3.6 specification was first published in 1983, revised and reissued in 1989, and the third edition was published in 1993.[69]

Because weld metal transfer characteristics, solidification behavior, and other properties vary with pressure and cooling rates, and each method of underwater welding may produce properties that differ from what is usual with welds made above water, the ANSI/AWS D3.6-93 specification currently defines four weld classes. Each weld class has a specific set of quality requirements that must be verified during procedure qualification. The selection of the class of weld to use for a specific application is determined by the user.

Class O welds meet the requirements of a governing "in-air" code, specification, or other standard, as well as additional requirements defined in the ANSI/AWS D3.6-93 specification.

Class A underwater welds are intended to be suitable for applications and design stresses comparable to their conventional surface welding counterparts by virtue of specifying comparable properties and testing requirements. Originally intended to apply only to dry hyperbaric underwater welding, Class A is currently also applicable to wet welding of austenitic and some nickel-based materials. Also, as a result of process evolution and process enhancements, there have been some recent

---

69. The revised D3.6 specification has been reorganized to be more consistent with other ANSI/AWS standards. To make it easier to use, the general requirements for all classes of underwater welding are in one section, with separate sections addressing the specific requirements for each of the four classes of welds. New additions include requirements for the use of stainless steel and nickel filler metals, a new depth limitation table, and clearer definitions of the four weld classes.

reports of achieving Class A criteria with underwater wet welding of carbon and low-alloy materials using flux cored arc welding (FCAW) and shielded metal arc welding (SMAW), including the specified maximum hardness and minimum V-notch impact properties.

Class B welds are defined by an intermediate set of mechanical and examination requirements. They are intended for less critical application where reduced ductility and increased porosity can be tolerated. The suitability of Class B welds for a particular application should be evaluated on a fitness-for-purpose basis. Class B requirements were originally established to reflect the state-of-the-art capabilities of underwater wet welding of carbon and low-alloy base metals using mild steel electrodes.

Class C welds satisfy lesser requirements than the other types and are intended for applications where the load-bearing function is not a primary consideration. The selection of Class C welds shall include a determination that their use will not impair the integrity of the primary structure by creating fracture initiation sites.

Procedure qualifications for Class O welds include visual and radiographic examinations, all-weld-metal tension tests, macroetch, hardness tests, and any other tests required by the referenced in-air code or specification (e.g., ANSI/AWS D1.1, Structural Welding Code–Steel or ASME Section IX).

The matrix for qualifying Class A dry groove and fillet welds includes visual and radiographic examinations, reduced-section tension tests, all-weld-metal tension tests, fillet weld shear tests, side bends or root and face bends, macroetch, hardness tests, impact tests, and fillet weld break tests.

For Class B wet welds, the matrix is the same as for Class A, except hardness, impact, and all-weld-metal tension tests are not currently specified.

Class C wet welds are not used for groove welds, and the fillet welds are subject to visual examination and bead-on-plate bridge-bend tests only. The 20-degree bend test is intended to reveal any subsurface HAZ crack.

Procedure qualification for the four classes of welds must be made under conditions that simulate the conditions under which production welds will be made, including hydrostatic pressure. In addition, for the four classes of welds, ANSI/AWS D3.6-93 specifies that a confirmation weld be made at the underwater work site and approved before production welding is started.

The ANSI/AWS D3.6-93 specification provides information required for qualifying welding procedures and welders for each of the four classes of welds, plus procedures for inspecting and testing the weldments. In addition, the specification addresses nondestructive underwater inspection methods including visual, radiographic, ultrasonic, magnetic particle, and electromagnetic (eddy current). Also included are the variables associated with dry welding and wet welding plus the number and types of test specimens required for qualifying welding procedures and welders.

BRITISH STANDARDS INSTITUTE

A BRITISH STANDARD also addresses the underwater welding issues of safety, welding procedures, welder approval, welding consumables, weld joint preparation, and inspection for hyperbaric welding.[70]

INTERNATIONAL INSTITUTE OF WELDING

UNDERWATER WELDING SPECIFICATIONS have received attention for several years from the International Institute of Welding (IIW), an organization with worldwide membership. In 1990, the IIW Select Committee on Underwater Welding established standard guidelines covering essential requirements for an underwater welding specification.[71]

AMERICAN SOCIETY OF MECHANICAL ENGINEERS

PRESSURE VESSELS AND other components in nuclear power plants are built using welding procedures and welders qualified in accordance with the requirements of the ASME *Boiler & Pressure Vessel Code*, Section IX, of the American Society of Mechanical Engineers (ASME).[72] Repairs that may be required after the vessels are put in service must meet the requirements of ASME Section XI. Because Section XI did not anticipate underwater wet welded repairs, the main Section XI Committee has established a Task Group that is developing code requirements for underwater welding that will be added to ASME Section XI. Until such an addition to ASME Section XI code becomes effective, underwater wet welding at nuclear power plants will continue to be governed by ANSI/AWS D3.6-93 Class A and O/ASME IX welds.

DET NORSKE VERITAS

IN JANUARY 1987, the Norwegian agency augmented its long list of Veritas Offshore Standards by publishing

70. BS 1415, *Specification for Process Welding Steel Pipelines on Land and Offshore*, Appendix J, "Recommendations for hyperbaric welding," London: British Standards Institute, 1984.
71.   IIW DOC SCUW 124-90, *Standard Guidelines for Specifications for Underwater Fusion Welding*, Cambridge, England: International Institute of Welding, 1990.
72.   "Welding and brazing qualifications," *ASME Boiler & Pressure Vessel Code*, Section IX. New York: The American Society of Mechanical Engineers (latest edition).

its *Recommended Practices for Underwater Welding, RP3604.* The document provides recommended practices for qualification of welding procedures and welders, handling of welding consumables, making and testing confirmation welds, and inspection of underwater production welds.

BUREAU VERITAS

THE MARINE BRANCH of Bureau Veritas developed and published in 1986 their document on underwater welding, *Underwater Welding – General Information and Recommendations.* For qualification of personnel for underwater inspection, they have published their *Procedure to Obtain the Diver-Inspector Aptitude Certificate.*

UNDERWATER INSPECTION

UNDERWATER INSPECTION OF weldments and structures is an integral adjunct to underwater welding. Indeed, inspection of underwater components often indicates the need for welded repairs, and the repairs themselves should be inspected.

Underwater structures and components are subject to forces similar to those found above water, including tension, compression, torsion, and bending forces. These forces may be compounded by wave, tide, and storm forces as well as corrosion phenomenon associated with the aquatic environment.

The methods and techniques of underwater inspection have been adopted from conventional surface methods to observe, test, and document the integrity, condition, and function of these structures and components.

INSPECTION REQUIREMENTS

THE SPECIFIC REQUIREMENTS for underwater inspections are usually defined by the client. However, the inspections usually are conducted in accordance with applicable codes and regulations such as those published by the American Welding Society (AWS), the American Petroleum Institute (API), and the American Society of Mechanical Engineers (ASME). In some cases the codes and regulations have been specifically designed to address the conditions of the underwater environment, such as ANSI/AWS D3.6-93, *Specification for Underwater Welding*, which provides detailed information for testing underwater welds.

Underwater inspection may be voluntary, as in the case of an internal operator requirement or monitoring program, or mandatory in order to meet government regulations. The Minerals Management Service (MMS) of the United States Department of the Interior, for example, has established a regulation guiding the performance of in-service inspections of offshore structures.[73] Similar state and federal regulations exist for many other types of underwater inspections including

bridges, dams, intake and outfall ducts, and nuclear facilities.

INSPECTION METHODS

UNDERWATER INSPECTION METHODS and techniques have been adopted from conventional surface methods. The unique environment usually has required some equipment or technique modification and also may favor one method over another. The American Society for Nondestructive Testing (ASNT) has recognized the following nondestructive test (NDT) methods:

(1) Radiographic Testing (RT)
(2) Magnetic Particle Testing (MT)
(3) Ultrasonic Testing (UT)
(4) Liquid Penetrant Testing (PT)
(5) Electromagnetic (Eddy-Current) Testing (ET)
(6) Neutron Radiographic Testing (NRT)
(7) Leak Testing (LT)
(8) Acoustic Emission Testing (AET)
(9) Visual Testing (VT)

Figures 10.28A, B, and C show the use of NDT methods 2, 3, and 5.

Underwater inspections are usually divided into two broad categories: fabrication and in-service.

Fabrication Inspection

COMPONENTS TO BE USED underwater are usually fabricated on land and the assemblies then transferred to the aquatic work site. Inspection of these components is typically performed at the assembly site prior to the unit being placed into service, and conventional top-side inspection techniques are employed. However, fabrication is sometimes performed in the marine

---

73. Code of Federal Regulations 30 CFR 250.142.

Courtesy of Global Divers & Contractors, Inc.

Figure 10.28A—Underwater NDT - Magnetic Particle Testing for Weld Defects

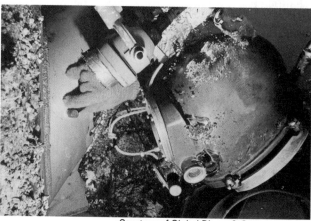

Courtesy of Global Divers & Contractors, Inc.

Figure 10.28B—Underwater NDT - Ultrasonic Testing for Measuring Wall Thickness and for Detection of Flooded Members

Courtesy of Global Divers & Contractors, Inc.

Figure 10.28B—Underwater NDT - Electromagnetic (Eddy Current) Testing for Weld Defects

environment, and may involve bolted, clamped, or welded assemblies. Inspection of these assemblies requires similar NDT methods as used top-side, modified where necessary to accommodate the underwater environment. Inspection of underwater welded components, for example, can be tested for included weld defects (porosity, cracks, lack of fusion, incomplete penetration, inclusions, etc.) as well as for external weld defects (cracks, cold lap, and undercutting).

In-Service Inspection

IN-SERVICE INSPECTIONS TEST a structure or component after it has been placed into operation. These tests particularly seek to locate the detrimental effects of stress, corrosion, and other damage that may have occurred. Most in-service defects, particularly stress-related cracking, initiate at a component's outer surface and are generally tested with NDT methods appropriate to this aspect (VT, MT, and ET). While fabrication inspections usually are one-time occurrences, in-service inspections are normally conducted at scheduled intervals over the projected life span of the unit. Properly monitored, these inspections can give in-sights into design factors, corrosion prevention, and enhanced safety.

Several examples of NDT methods as applied to underwater inspections, both in-service and fabrication inspections, are given below:

(1) Visual Testing (VT) – Visual, video, and photographic inspection and documentation of the completed assemblies

(2) Ultrasonic Testing (UT) – Flaw detection of weldments, thickness measurements, and testing the watertight integrity of submerged tubular members (flooded member detection)

(3) Magnetic Particle Testing (MT) – Testing for the presence or absence of near-surface or surface flaws (cracks) by using a magnetizing device (permanent or electromagnetic) and observing the distribution of ferro-magnetic particles applied to the surface

(4) Electromagnetic (Eddy-Current) Testing (ET) – Testing for the presence or absence of surface flaws (cracks) by using an electronic probe and a remote display device (e.g. oscilloscope) that displays magnetic eddy currents that are induced in the test piece and interpreted by the operator

(5) Radiographic Testing (RT) – Testing for internal and back-wall weld flaws (particularly useful for testing of welded pipeline joints)

PERSONNEL AND QUALIFICATIONS

UNDERWATER INSPECTIONS SHOULD be performed by trained and certified NDT inspectors. Usually these inspectors are professional divers who have had specific NDT training. The American Society of Nondestructive Testing (ASNT) has published a recommended guide for the training, qualification, and certification of NDT inspectors.[74] This document describes three basic levels of qualification. These levels may be further subdivided by the employer for specific situations where additional levels of skills and responsibilities are deemed necessary. While in the process of being trained and qualified, and prior to being certified to NDT Level 1, an individual should be considered a trainee. A trainee should work with a certified individual and shall not independently conduct any NDT, interpret nor evaluate the results of any NDT, nor report NDT results. The three basic levels of qualification are as follows:

(1) NDT Level 1 – An NDT Level 1 individual shall be qualified to properly perform specific calibrations, specific NDT, and specific evaluations for acceptance or rejection determinations according to written instructions and to record results. The NDT Level 1 individual shall receive the necessary instruction or supervision from a certified NDT Level II or III individual.

74. ASNT Recommended Practice SNT-TC-1A.

(2) NDT Level II – An NDT Level II individual shall be qualified to set up and calibrate equipment and to interpret and evaluate results with respect to applicable codes, standards, and specifications. The NDT Level II individual shall be thoroughly familiar with the scope and limitations of the methods for which qualified and should exercise assigned responsibility for on-the-job training and guidance of trainees and NDT Level I personnel. The NDT Level II individual shall be able to organize and report the results of NDT.

(3) NDT Level III – An NDT Level III individual shall be capable of establishing and designating the particular NDT method, techniques, and procedures to be used. The NDT Level III individual should be responsible for the NDT operations and shall be capable of interpreting and evaluating results in terms of existing codes, standards, and specifications. The NDT Level III individual should have sufficient practical background in applicable materials, fabrication, and product technology to establish techniques and to assist in establishing acceptance criteria when none are otherwise available. The NDT Level III individual shall have general familiarity with other appropriate NDT methods, as demonstrated by the ASNT Level III Basic examination or other means. The NDT Level III, individual in the methods in which certified, shall be capable of training and examining NDT Level I and II personnel for certification in those methods.

REMOTE UNDERWATER INSPECTION

IN ADDITION TO "hands-on" inspection of underwater structures and components, these inspections are increasingly being performed remotely by robotic or similar devices. A large industry segment has developed using Remotely Operated Vehicles (ROV's) to perform inspections and other tasks unsuitable for human performance due to cost, depth, or safety. The NDT methods remain essentially the same as when conducted manually, however certain modifications in equipment and techniques are often necessary to accommodate the particular task or machine. Although the NDT task is performed remotely, the results of the NDT still must be interpreted and evaluated by a qualified and certified NDT inspector.

UNDERWATER THERMAL CUTTING

UNDERWATER CUTTING IS an essential part of many underwater projects. Inland and coastal divers, for example, must cut away cables from fouled propellers, salvage barges, and small boats, cut sheet and H-pile for removal during repair of damaged harbor facilities, and remove piling from cofferdams and bridge foundations. These and a mydrid of other

cutting tasks constitute a large part of a construction diver's work. For the offshore oilfield diver, underwater cutting is frequently required during the installation of new offshore structures as well as the removal of obsolete drilling and production platforms. During the productive life of offshore structures and subsea pipelines, underwater cutting is essential for the removal of impaired sections in preparation for the repair of damage resulting from fatigue, corrosion, excessive loads, collisions, and other accidental damage.

COMPARISON OF PROCESSES

BASIC THERMAL PROCESSES used for cutting and otherwise removing ferrous and nonferrous materials underwater are the following:

(1) Electrical arc with oxygen (oxygen arc)
(2) Gases that support combustion with oxygen (oxyfuel gas)
(3) Shielded metal arc (flux coated electrode)
(4) Plasma arc
(5) Electrical arc with water jet (arc water)

Cutting methods based on the oxygen arc principle include the use of the following:

(1) Waterproofed tubular steel electrodes
(2) Flux coated and waterproofed tubular steel electrodes
(3) Two types of exothermic tubular steel insulated electrodes
(4) The thermal arc cable
(5) The thermit lance

Exothermic electrodes, thermal cables, and the thermit lance can be ignited by a heat source other than a welding machine.

The most commonly used underwater cutting process is oxygen arc with insulated tubular steel or exothermic electrodes, with the exothermic electrode being the most efficient. However, for rapid rough cutting in near zero visibility during salvage operations, the thermal cable or lance may be preferred. Plasma arc is used for cutting stainless steel in nuclear power plants. For removal of metal during repairs of welds and cracks in the walls of base metal, the arc water process is unparalleled and has largely replaced earlier methods using grinding, chipping, and gouging. Each underwater cutting process will be discussed further in the following sections.

OXYGEN ARC CUTTING

THE OXYGEN ARC underwater cutting process is based on maintaining an electric arc between the metal being cut and one of four different tubular steel electrodes (two tubular steel and two exothermic). When the heat of the arc raises the temperature of the base metal to the molten stage or ignition temperature, a high-velocity stream of oxygen is directed at the heated spot through the center of the electrode. The metal is then removed by being blown away or oxidized and blown through the gap (kerf) that is initiated by the cutting action. The tip of the electrode is exposed to the heat and oxidation and is consumed in the process, so electrodes must be replaced frequently. A list of required equipment for oxygen arc cutting is included in Table 10.5. Figure 10.29 shows a diver using the oxygen arc cutting process.

Tubular Steel Electrodes

THE TWO TUBULAR steel electrodes are 5/16 in. (7.9 mm) in diameter and 14 in. (356 mm) long with 0.113 in. (2.9 mm) concentric bore holes that extend through the length of the electrodes. Both types of electrodes are waterproofed and one type is flux coated.

Courtesy of Global Divers & Contractors, Inc.

Figure 10.29—Underwater Oxygen Arc Cutting with Tubular Steel Electrode

While the electrode without flux coating is still used, mostly on clean low-carbon steel, the flux coated electrode is widely preferred. The flux coating facilitates initiation and maintenance of the arc and generates a gaseous bubble that helps stabilize the arc. The waterproof coating maintains the integrity of the flux and provides electrical insulation. These electrodes make clean cuts on carbon steels but are ineffective on corrosion resistance metals, copper-based alloys, and nonconductive materials such as paint and marine growth.

Basic equipment required for underwater cutting with tubular steel electrodes include a 400 amp dc welding power source with 2/0 welding cables and a pressure-regulated supply of oxygen with 1/4 in. (6.35 mm) inside diameter hose attached to the oxygen arc cutting torch. A list of equipment, tools, and consumables are shown on Table 10.5.

The diver will require appropriate dive gear and life support equipment plus tools for removal of marine growth, heavy coats of paint, and scale. The dive helmet or mask must be equipped with a welder's lens holder assembly.

Support personnel should have tools and parts for maintenance and repair of equipment and for splicing cables and hoses. The supply of cutting electrodes and oxygen must be adequate. Oxygen purity should be not less than 99.5 percent.

Preparation for cutting with tubular steel electrodes includes venting all air from the oxygen hose, setting the regulator for proper pressure and flow, securely attaching the ground clamp or tab plate to the base metal adjacent to the area to be cut, and regulating the welding machine for correct amperage and straight polarity (electrode negative).

Cutting is started by the diver calling for current on ("make it hot"). The electrode is brought into contact with the base plate to initiate the arc, and the diver releases the oxygen to blow away or oxidize the metal. Cutting amperage will vary from 300 to 400 and oxygen pressure at the torch should be not less than 90 psi (0.62 MPa). This should be increased to 110 psi (0.76 MPa) for 1 in. (25.4 mm) thick material, and 150 psi (1.03 MPa) for 3 in. (76 mm). Electrodes are consumed at the rate of about one per minute. Electrode consumption and rate of travel will vary widely depending on the diver's skill and other factors such as thickness and chemistry of the material being cut, marine growth, underwater visibility, and sea conditions. However, for estimating purposes, one may assume one electrode and one minute for cutting 18 in. (457 mm) of 1/4 in. (6.350 mm) thick material, 12 in. (304 mm) of 1 in. (25.4 mm) thick material and up to one minute and one electrode for cutting 3 in. (76 mm) of 3 in. (76 mm) thick material. Oxygen consumption of 70 ft^3/min (2 m^3/min) will of course increase with depth and pressure in accordance with Boyle's Law.

Exothermic Electrodes

THE EXOTHERMIC ELECTRODES are available in two configurations. Both have a thin-wall steel outer shell (tube) 3/8 in. (9.53 mm) in diameter and 18 in. (457 mm) long. Both are insulated, one with a spiral tape wrap and the other with a plastic coating. Inside the tape-wrapped electrode there are seven circumferential located small diameter mild steel rods. Instead of longitudinally positioned rods, the plastic coated electrode uses a mild steel wire, spiral wrapped around a mild steel tube. With both electrodes, the mild steel interparts and the outer tube will burn independently (without electrical current) after the arc is initiated and oxygen is flowing through the electrode. The exothermic electrodes will cut cleanly through ferrous materials and will melt and blow away most other materials such as noncorrosive metals, concrete, barnacles, and other calcareous marine growth with temperatures that exceed 10 000 °F (5537 °C).

Basic equipment requirements are the same as when tubular steel electrodes are used, with two exceptions. Oxidation or burning of the electrode can be started with less welding amperage, or even a 12 volt battery. The electrode will continue to burn as long as oxygen pressure is maintained. However, the use of 150 to 300 amps during cutting will produce the best results. Also, due to increased consumption of oxygen, the inside diameter of the hose should be 3/8 in. (9.53 mm) instead of 1/4 in. (6.35 mm) used with plain tubular electrodes.

Preparation for cutting with exothermic electrodes includes venting all air from the oxygen hose and adjusting the regulator for the required pressure. If a welding power source is to be used, the ground clamp or tab plate should be securely attached to the metal that is to be cut and the welding machine set for desired amperage and for straight polarity.

If a battery is to be used for ignition of the cutting electrode, the cable terminals are connected the same as for a welding machine: positive to the conductive metal to be cut and negative to the electrode holder (torch).

The procedure followed by the diver for cutting oxidizable metals is very similar to that used with tubular steel electrodes: the arc is stuck, the base metal is preheated to oxidation temperature, and the oxygen turned on. For cutting nonconductive metals, the exothermic electrode is ignited by making contact with a tab plate (striking plate) attached to the ground cable. The nonconductive material is then removed by the intense heat generated by the oxidation of the electrode and by the impingement of the high-pressure oxygen on the melting or spalling material.

When cutting with exothermic electrodes, proper current at the electrode will result in fast effective cutting; too much results in increased electrode consumption.

Based on using 2/0 cables, the minimum required amperage will vary from 150 with 100 ft (30 m) of cable, to 170 amps with 500 ft (152 m) of cable. Oxygen consumption is greater than for tubular steel electrodes. The oxygen regulator must be capable of delivering a minimum of 70 ft^3/min (2 m^3/min) at 90 psi (0.62 MPa) over ambient pressure at the underwater work site. Proper pressure will result in the stream of oxygen extending approximately 6 in. (152 mm) into the water from the tip of the electrode.

Time required to cut and remove conductive metals and nonconductive materials, plus consumption of oxygen and electrodes, varies widely depending on the diver's proficiency, chemistry of conductive metals, type of nonconductive materials, and underwater working conditions. For estimating purposes, if oxygen pressure and flow and amperage are correct, electrodes will last 45 to 55 seconds. One electrode will cut 20 in. (508 mm) of 1/2 in. (12.7 mm) thick, 14 in. (356 mm) of 1 in. (25.4 mm) thick, 12 in. (304 mm) of 1-1/2 in. (38 mm) thick and 10 in. (254 mm) of 2 in. (51 mm) thick carbon steel if a welding power source, is used. Without a welding power source cutting efficiency is reduced by about 30 percent on 2 in. (51 mm) thick material to 40 percent on 1/2 in. (12.7 mm) material.

ARC WATER METAL REMOVAL AND CUTTING

THE ARC WATER process of removing metal underwater is based on maintaining an arc between the metal to be removed and a carbon-graphite copper coated waterproofed electrode. The resulting molten metal is removed by a jet of high-pressure water that exits the torch through an orifice located under the electrode collect.

As a cutting tool, the arc water process is limited for practical purposes to metals with a maximum thickness of about 3/4 in. (19.1 mm). For 3/4 in. (19.1 mm) thick material, the diver will usually make two passes to complete the cut. Although the arc water torch, or more accurately the arc water gouge, does not match the performance of the commonly used oxygen arc torches in cutting carbon-manganese steel, it has some unique and important uses.

As opposed to oxygen arc processes, the arc water gouge has the following capabilities:

(**1**) It can, with the aforementioned material thickness limitations, be used to cut or gouge any metal on which an electrical arc can be maintained.

(**2**) It is effective in the removal of metal for the repair of defective welds and cracks in base metal, plus the beveling of tubular members and plate edges during fit-up and preparation of joints for welding underwater.

The process leaves a clean smooth surface free of hardened scale and ready to weld.

(**3**) Oxygen arc cutting processes generate an explosive mixture of hydrogen and oxygen. The arc water process creates no hazardous gas mixture and can be used to cut vent holes to prevent accumulation of explosive mixtures during subsequent cutting with oxygen arc torches.

Basic equipment requirements include a 400 to 600 amp dc, constant current welding power source, 400 amp knife switch, one 2/0 to two 4/0 welding cables, water supply source and 1/4 in. to 3/8 in. (6.35 mm to 9.53 mm) inside diameter hose, arc water torch with 1/0 whip, and carbon-graphite electrodes.[75] The water supply source or pump should be able to produce 3.5 gallons (13 liters) per minute at 90 to 110 psi (0.62 to 0.76 MPa) greater than the hydrostatic pressure at the working depth.

The arc water electrodes are 5/16 by 9 in. (7.94 by 229 mm) long. Amperage will vary from 200 to 550 depending on length of cables and whether the process is used for shallow gouging or for cutting. Depth of groove (metal removed) will vary with rate of travel, angle of electrode, amperage, and water pressure. On carbon steel, 375 to 400 amps with 90 to 110 psi (0.62 to 0.76 MPa) water pressure and 25 to 35 degree electrode angle will produce a 7/16 in. (11 mm) wide groove 3/8 in. (9.53 mm) deep and 18 in. (457 mm) long in one minute while consuming less than one-half [7.12 in. (181 mm)] of an electrode. For light gouging, one electrode may be good for 48 in. (1219 mm) of metal removal. Results are similar on stainless steel. Reverse polarity (electrode positive) is used, except the manufacturer recommends straight polarity for copper alloyed metals if the copper content is more than 80 percent.

Preparation for use of the arc water process includes adjusting the power source for proper amperage and polarity, evacuating air from the water hose, adjusting the water supply pressure and volume, and securely attaching the ground clamp. The diver calls for current on and starts metal removal by bringing the electrode in contact with the base metal. During gouging the electrode tip is pointed toward the unremoved metal at an angle of approximately 30 degrees. For cutting material 3/8 in. (9.53 mm) thick and under, the angle should be about 60 degrees. The tip of the electrode is advanced steadily and as quickly as the desired depth of metal is removed.

---

75. Four hundred (400) amps with 2/0 cables have been used down to 330 ft (100 m). Heavy-duty gouging at 680 ft (207 m) required 600 amps and two 4/0 power cables.

OXYFUEL GAS CUTTING

UNDERWATER CUTTING TECHNOLOGY has greatly improved since divers first took a topside open-flame torch underwater in 1908. The results were less than satisfactory. The density of the aqueous environment inhibited and adversely affected stability of the flame, and the water acted as an infinite heat-sink which adversely affected the preheat process.

The process of cutting underwater with a gas torch is essentially the same as above water. A small spot of the steel to be cut is heated to the molten state, then oxygen is directed at that spot. The oxygen combines chemically with the molten metal, instantly converting it into various gases and chemical compounds, literally burning it. In addition, the pressure of the oxygen stream drives the converted material, or slag, out of the kerf.

Underwater torches differ from topside torches in that they are equipped with a perforated spacer sleeve around the torch tip which keeps the tip the proper distance from the material being cut. Oxyacetylene and oxyhydrogen torches are equipped with a third hose and valve so compressed air can be used to evacuate the water from the spacer sleeve to help stabilize the flame.

Underwater cutting with gas torches has largely been replaced by oxygen arc cutting processes. Among several reasons for this, the dominate ones are the higher degree of skill required to properly cut with gas torches, greater speed and efficiency of oxyarc cutting, and especially offshore, the logistic difficulties of supplying the numerous and bulky cylinders of hazardous gases. Because use of the oxyfuel gas cutting process is so limited, the following descriptions of the methods (oxyacetylene, oxyhydrogen and oxygen MAPP) will provide only the most basic information.

Oxyacetylene Cutting

OXYACETYLENE IS INEFFECTIVE in cutting materials other than carbon steel. The method is not used deeper than 25 ft (8 m) because of the extreme instability of acetylene at pressures over 15 psi (103 kPa).

Oxygen Hydrogen Cutting

OXYHYDROGEN WAS THE first oxyfuel gas cutting process to be successfully used in water deeper than 25 ft (8 m). Its first reported use was in 1926 during salvage of a U.S. Navy submarine which had sunk in 130 ft (40 m) of water. Although the oxyhydrogen flame is not as hot as the oxyacetylene, and it will only cut carbon steel, it was the commercial diver's choice until the advent of the oxygen arc process in the early 1960s.

Oxygen MAPP Cutting

THE OXYGEN MAPP procedure for cutting metal underwater is essentially the same as that used above water.[76] A standard oxyacetylene cutting torch equipped with a MAPP gas tip and spacer sleeve (no compressed air is required) is used underwater. With oxygen MAPP cutting, the combustion temperature is higher than other gas cutting processes: it makes clean slag-free cuts, but only on carbon steel. It is safer than oxygen arc (explosive gas mixtures are consumed in the cutting process), requires no power source, and is highly portable. However, it requires more diver skill, and the process is not as fast as oxygen arc.

PLASMA ARC CUTTING

THE UNDERWATER PLASMA arc cutting process is based on forcing an electrically charged ionized gas through a constricted orifice creating an electric arc that preheats the metal to be cut. The molten metal is then removed by the force of the high-velocity plasma jet stream issuing from the constricted orifice. Initiation of the plasma arc takes place in the chamber of the torch nozzle where an electric arc is generated between the electrode (negative) and the nozzle (positive). This non-transferred arc is referred to as the *pilot arc* and is used to transfer the negative electrical current from the electrode to the positive charged work piece. When this transfer takes place, the main plasma cutting arc is established, and the pilot arc is shut off by control circuitry in the power supply. The high temperature of the main cutting arc and the velocity of the plasma jet proceeds to melt and displace the molten metal in its path.

The plasma arc process uses gases such as nitrogen, oxygen, an argon/hydrogen mixture, and compressed air for the plasma gas. However, for underwater plasma arc cutting, nitrogen, argon/hydrogen, and compressed air are the preferred gases. Secondary gases are sometimes used to help shield the plasma arc underwater. Compressed air is used most often, but carbon dioxide and argon can also be used. The plasma gas temperature during cutting reaches temperatures between 20 000 and 50 000 °F (11 093 and 27 760 °C).

Reports on developmental work and practical use, both remote and manually controlled, indicate that for cutting metal plate underwater, the plasma arc process outperforms oxygen arc by a wide margin and at an operational cost savings of 80 to 90 percent. Scientific and developmental reports on use of the process above and underwater, have been published intermittently for almost thirty years. However, practical underwater use of the process started less than four years ago and has

---

76. MAPP is a registered trademark of the BOC Group. MAPP gas is a stabilized mixture of methylacetylene and propadiene.

been limited to cutting stainless steel at nuclear power plants.

The plasma arc process is used in nuclear power plants because it cuts stainless steel faster and cleaner (less dross) than any other process, thereby minimizing costly plant outage time and adverse alterations to the critical chemistry of plant water. Diver-operated plasma arc cutting torches have been used at several nuclear power plants and for qualifying cutting procedures and operators. On some projects where excessive underwater radiation levels prohibit the use of divers, remotely operated manipulators have been used for removal of stainless steel components.

During the use of plasma arc underwater, available amperage has ranged from 100 to 1000 depending on the power source. Both pure nitrogen and air have been used as plasma gas. The following two reports on underwater use of the process are anecdotal:

(1) Remotely controlled manipulators have been used to cut 2 to 2-1/2 in. (51 to 63.5 mm) stainless steel at the rate of 7 in. to 8 in. (178 to 203 mm) per min at 180 arc volts. Approximately one minute was required to cut 12 in. (305 mm) of 3/4 to 1-7/8 in. (19.05 to 47.6 mm) thick material with arc voltage ranging from 90 to 100 for the 3/4 in. (19.05 m) material and up to 140 for the thicker material. Cutting current ranged from 450 to 860 amps depending on geometry of the cut. Plasma gas was pure nitrogen at 40 to 70 psi (276 to 483 kPa) at main arc transfer depending on thickness of material to be cut. Maximum water depth was 35 ft (11 m).

(2) Divers have used 100 amp plasma torches with air to cut stainless steel plate at rates of 78 in. (1981 mm) per minute on 1/4 in. (6.35 mm) thick material, 32 in. (813 mm) per minute on 1/2 in. (12.7 mm) and 7 in. (178 mm) per minute on 1 in. (25.4 mm) thick material. Rate of travel and plate thicknesses would of course increase with higher amperage power sources.

A prime advantage of the process is its ability to make rapid, clean, and inexpensive cuts on virtually all electrically conductive materials.

Disadvantages are the higher cost of equipment and the perceived, or very real, safety hazards involved in manual use of the equipment underwater.

A major supplier, before marketing the equipment, has initiated a study to help define the best scientific and practical procedures available to determine to what degree the plasma arc process puts the diver at risk. The fact that the high-frequency arc initiation of a 1000-amp unit involves 6000 to 9000 volts is of real concern, even though the process takes place within the confines of the arc chamber between the electrode and the nozzle. However, even if it is determined that the

closed circuit nature of the arc initiation procedure represents no risk to the diver, the available amperage and open circuit voltage used during plasma arc cutting is greater than the maximum used by any other underwater cutting process (600 amps and 80 to 90 volts).

With one exception, where divers used conventional wet suits while operating 100 amp plasma arc torches, divers have dressed in dry suits which insulate them from the surrounding water and electrical currents. During 300 to 400 hours of underwater use of 100 amp and 200 amp plasma arc cutting equipment, no diver has reported any incidents of shock discomfort.

OTHER CUTTING PROCESSES

Thermal Cable

THE THERMAL CABLE cutting process works on the same principle as the exothermic electrode process in that the cable acts as the exothermic electrode and high-pressure oxygen flows through the coreless cable. The flexible spiral cable from which the center strand [approximately 1/8 in. (3.18 mm)] has been removed for oxygen passage, consists of six bundles of high strength steel wire encased in a plastic sheath for insulation. The cable is fitted with a threaded connector for attachment to the oxygen supply and a lug for connection to the electrical ignition source.

Advantages of the thermal cable over other oxygen arc underwater cutting methods include:

(1) Takes less skill to operate
(2) Can be used in hard to reach places and near zero visibility
(3) Will cut metals of unlimited thickness
(4) No time lost changing electrodes [cables come in lengths of 50 ft and 100 ft (15 m and 30 m)]

However, cuts are rough, it is ineffective on nonconductive material, rate of oxygen is comparatively high, and its use is not recommended deeper than 300 ft (91 m).

One must do the following to prepare for cutting underwater with the thermal cable:

(1) Set up the manifolded bank of oxygen cylinders and connect the oxygen regulator to the manifold and the oxygen hose to the thermal cable.
(2) Attach one electrical cable to the upper end of the thermal cable and another cable to a ground lug (strike plate), then lower thermal cable and ground cable to the underwater work site.

(3) Displace all water from the hose and thermal cable with oxygen.

Oxygen pressure should exceed hydrostatic pressure at the working depth by 250 to 350 psi (1.7 to 2.4 MPa) depending on cable diameters of 1/4 to 1/2 in. (6 to 12 mm). One 24 volt or two 12 volt batteries capable of supplying 400 amps/80 volts, or a welding machine, can be used for ignition of the thermal cable. Straight polarity (cable negative) is preferred.

The diver initiates the cutting procedure by insuring that oxygen pressure is adequate, calling for current on and starting oxidation (burning) of the cable by bringing it into contact with the strike plate. Cutting of oxidizable metal will begin as soon as the burning end of the cable is brought into contact with it. For cutting metal less than 2.5 in. (64 mm) thick, the thermal cable is held about 90 degrees to the base metal and moved steadily forward while maintaining contact with the material being cut. For material 2.5 in. (64 mm) and thicker, a brushing motion is used to prevent metal on either side of the intended cut line from becoming excessively hot and melting away. Thermal cables burn away at a rate of about 1.5 to 2 ft (0.5 to 0.6 m) per minute. Rate of travel (length of material cut per minute) is significantly greater than with exothermic electrodes and of course depends on the thickness of the material being cut.

The following data, combined from three sources, illustrates the factor interplay among cable burn-off rate, oxygen pressure and consumption, and time to cut 2-in. (51-mm) thick steel. Each 3 to 6 ft (0.91 to 1.82 m) of cable should cut 12 in. (305 mm) of 2-in. (51-mm) thick steel in 3 minutes and 20 seconds, while consuming 1440 ft^3 (40.8 m^3) of oxygen at 280 psi (1.93 MPa) over ambient pressure.

Thermic Lance

THE THERMIC LANCE is a 3/8-in. (9.5-mm) diameter, 10.5-ft (3.2-m) long steel pipe packed with metal alloys including aluminum, magnesium, thermit, and low-carbon steel. High-pressure oxygen is forced through the alloy filler and once ignited, the lance burns with great force generating cutting temperatures up to 10 000 °F (5537 °C).

Reported advantages of the thermic lance versus the thermal cable are that it will burn or melt through almost anything including steel and nonferrous metals, rock and concrete, and has no depth limitation. The disadvantage is that a 10.5 ft (3.2 m) lance is consumed in about six minutes. Two or more lances can be connected together, but the added length makes it awkward to handle.

Primary use of the lance is to cut very thick steel plate and propeller or other high-strength steel shafts. It is said to be capable of *punching* a hole through 12 in. (305 mm) thick steel in one minute while consuming 6 to 12 in. (152 to 305 mm) of the lance.

Basic equipment, preparation for use, source of ignition, and cutting procedure for the thermic lance are essentially the same as for the thermal cable.

Shielded Metal Arc

THE SHIELDED METAL arc process is the most basic of all cutting processes used underwater. It can be used when oxygen is not available, when no other cutting equipment is on site, or when available equipment is ineffective on nonferrous metals. It can be used for the removal of steel, brass, copper, and other copper-based alloys.

The shielded metal arc process is comparatively slow. It does not cut through or oxidize metal as the oxygen arc process does; it merely melts the metal. The molten metal does not freely run out of the kerf but must be pushed out by manipulation of the electrode. When cutting material thicker than 1/2 in. (12.7 mm), a slow, short-stroke, sawing motion is used.

The only equipment required is a dc welding power source with cables and a waterproofed electrode holder. As little as 300 amps may be used but more, up to 500 amps, is preferred. Straight polarity (electrode negative) is used. Shielded metal arc underwater cutting electrodes are commercially available in 3/16 in. (4.76 mm) and 1/4 in. (6.35 mm) diameters. However, virtually any welding electrode, waterproofed with any water repellent liquid or even wrapped in electrical tape, can be used.

APPLICATIONS

OFFSHORE STRUCTURE REPAIR

MANY EXAMPLES EXIST of the viability of underwater wet welding for the repair of offshore structures.[77] The particular repair project described next is especially noteworthy for the following reasons:

---

77. Two additional application case studies also are included in the "Cost Comparisons" section, Page 468, comparing dry and wet underwater welding costs.

(1) The engineering design of the repairs, made at five locations on the structure, project planning and execution, plus fitness-for-purpose of the repairs are all well documented.

(2) Repairs were made on highly stressed principal members and K-joints that sixteen years later were in perfect repair after continued exposure to the same cyclic loads that caused the original fatigue failures.

(3) The project provides a good example of why the installation of weld chambers (habitats) for dry hyperbaric welding sometimes may not be a viable alternative.

Damage to the structure included three 16 in. (406 mm) diameter horizontal braces torn loose from two 24 in. (610 mm) diameter struts and a 17.8 ft (5.4 m) diameter leg, leaving "cookie-cutter" holes in them. Three to four cracks, up to 8 in. (203 mm) long, radiated from each of four holes. Also, two 12 to 14 in. (305 to 356 mm) long, through-thickness cracks were at the toes of welds in the K-joints.

The damage was caused by prevailing ground swells that had induced out-of-plane bending in tubular members located only five feet below the waterline. Also, cross sections (outside diameters) of members were increased by 16 to 20 in. (406 to 508 mm) by marine growth.

Because of the high stresses that would be imposed on the repaired areas, the platform operator and the repair contractor would have preferred using dry hyperbaric welding. This would have required the installation of dry weld chambers at six locations on the platform. One of the chambers would need to be large enough to encompass or be penetrated by seven struts and braces with diameters from 16 to 24 in. (406 to 610 mm). The tops of all chambers would be only 24 in. (610 mm) below the mean waterline, so the chambers and platform would be subjected to loads imposed by the prevailing ground swells.

Analyses of material samples, which were taken from each location where underwater welding would be required, disclosed one area where the high-carbon equivalent (0.43) material would be susceptible to hydrogen-induced underbead cracking in the heat-affected zone.

Wet welding was used to avoid the possibility of further damage to the platform that might result from the installation of dry weld chambers. Welds on the high-carbon equivalent material were made with nickel electrodes to prevent hydrogen-induced cracking. Other areas were welded with mild steel (ferritic) electrodes. Welding procedures and welders were qualified at the offshore job site. Figure 10.30 shows the tubular members that were replaced and areas A through E that were repaired.

Details of how the repairs were executed, with areas identified, are described in the following steps.

(1) Holes in the 24 in. (610 mm) diameter struts were enlarged to remove some of the shorter [≤ 3 in. (≤ 76 mm)] cracks. Other cracks that radiated from the holes were beveled for V-groove repairs.

(2) Peripheral edges of the holes were beveled; these and the cracks were fit with backing bars rolled to appropriate radii.

(3) Rolled plates were fabricated and fit against the backing bars as insert plates. The resulting peripheral grooves, and the beveled cracks, were prepared for proper root gap, and single-V-groove welds were made. All welds were then ground flush with the base metal.

(4) A new horizontal brace [52 ft (15.85 m) long by 16 in. (406 mm) diameter] was fabricated, and split sleeves were tack welded to each end of the brace. One of three sleeves at one end of the brace, A, would act as a doubler plate over the butt welded insert plate in the wall of a 24 in. (610 mm) diameter strut, and the other two sleeves were welded to the adjacent horizontal brace and a vertical-diagonal brace, each 20 in. (508 mm) in diameter. The single sleeve on the other end of the new brace was fit and welded over the butt welded insert plate in the other 24 in. (610 mm) diameter strut, C.

(5) The new brace was then lowered into place and the sleeves were fit and fillet welded over the insert plates, struts, and braces.

(6) Part of a second 16 in. (406 mm) diameter brace was replaced by fabricating a section about 8 ft (2.4 m) long. One end was equipped with a split sleeve, and the other end was prepared for butt welding to the undamaged end of the existing brace. The sleeve, acting as a doubler plate, was fillet welded over the insert plate in the 24 in. (610 mm) diameter strut, D.

(7) The third brace, which had torn loose from the 17.8 ft (5.4 m) diameter leg, was determined to be redundant and was not replaced. However, the hole in the leg, B, and a buckled diaphragm plate inside the leg had to be repaired. The first step was to cut a 36 in. (914 mm) diameter hole in the leg. This removed the 16 in. (406 mm) diameter hole and surrounding deformation in the wall of the leg and provided access for material that would be required to repair the buckled diaphragm, and to close the 36 in. (914 mm) diameter hole.

To complete the underwater work at this location, B, a 3/8 in. (9.5 mm) thick by 40 in. (1016 mm) diameter rolled plate was fillet welded over the 36 in. (914 mm) diameter hole in the leg. The water level inside the leg was then pumped down below the damaged area and internal repairs were made using conventional dry

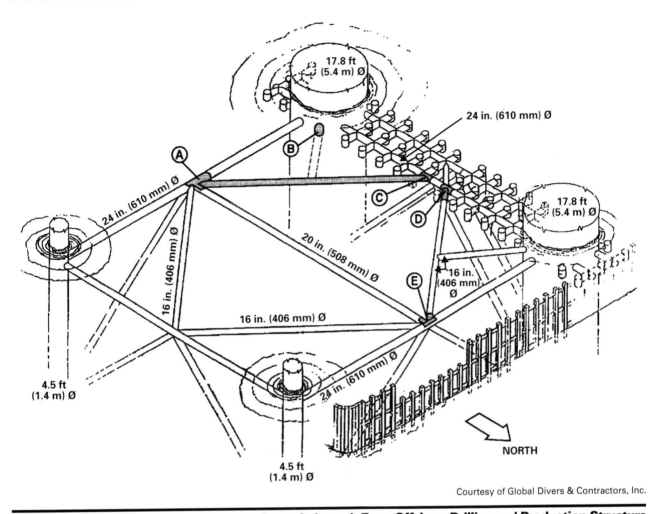

Courtesy of Global Divers & Contractors, Inc.

Figure 10.30—Repair of Fatigue Damage, Areas A through E, on Offshore Drilling and Production Structure

welding. Internal work included enlarging the 36 in. (914 mm) diameter hole to remove irregularities and hardened surfaces resulting from the underwater cutting, beveling the peripheral edge of the hole, and fitting and welding a 1/2 in. (12.7 mm) thick by about 38 in. (965 mm) diameter insert plate. The previously wet welded 40 in. (1016 mm) diameter plate acted as a backing plate for the groove welding on the insert plate.

The remaining damage consisted of two 12 to 14 in. (305 to 356 mm) long, through-thickness cracks in the heat-affected zone of K-joint weldments, E.

Pneumatic chipping tools and grinders were used to remove the cracks to within 1/8 in. (3.2 mm) of the inside wall of the tubular members and to bevel the edges for welding. Holes were drilled at ends of cracks to prevent further propagation, and groove welds were made. Crack repairs were reinforced with split sleeves

fillet welded over the cracks and to the adjacent tubular members.

Repairs were made in October 1974. Sixteen years later, all repairs were in perfect condition although numerous other fatigue failures had occurred at other locations on the platform.

PIPELINE REPAIR

THIS CASE INVOLVES the repair of an 8 in. (203 mm) diameter pipeline river crossing at working depths of 72 and 90 ft (21.95 and 27.43 m). The problem was that the pipeline buckled and ruptured, and part of the ruptured line was buried under 20 ft (6.1 m) of rock fill. (See Figure 10.31.) Strong river currents, heavy river traffic, and zero visibility exacerbated the problem of making repairs.

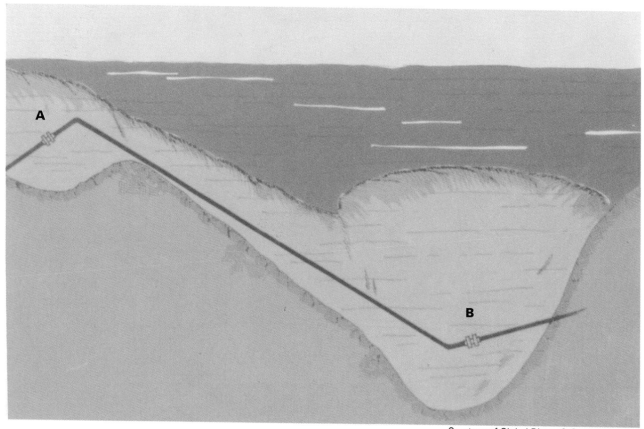

Courtesy of Global Divers & Contractors, Inc.

Figure 10.31—Repair of Pipeline River Crossing. Damaged Section Replacement was 8 in. (203 mm) Diameter x 42.5 ft (13 m) Long. Water Depth at Flange A, 72 ft (22 m). Water Depth at Flange B, 90 ft (27 m). Zero Underwater Visibility. 3.5 Knot Current.

The solution started with an airlift to uncover the damaged line, to excavate a trench for the repair section, and to provide work space at the two locations where repair connections would be made. The shallow connection would be made 6 ft (1.83 m) below the mudline at a depth of 72 ft (21.95 m), and the other would be made 30 ft (9.14 m) below the mudline at 90 ft (27.43 m).

The damaged section, approximately 50 ft (15.24 m) long, was cut at each end and brought to the surface. Slip-on pipe flanges were tack-welded to the ends of the undamaged sections underwater. Small weld chambers, large enough to accommodate the welder/diver down to his waist, were then installed, and the tack-welded flanges were dry hyperbarically welded to the pipe ends.

Mating flanges were temporarily bolted to the permanent flanges. A section of pipe was lowered and positioned so that the ends could be wet welded to the temporary flanges. The repair section template with flanges was unbolted and brought to the surface.

Using the template as a pattern, the permanent repair section was fabricated, pressure tested, and given a coating for corrosion protection.

The work barge crane was used to lower the repair section into its approximate final position. The load was then transferred to davits installed at both underwater flanged connection locations so divers could control final precise positioning of the repair section. To complete the repair, stainless steel ring gaskets were positioned between the flanges, bolts were installed and tightened, and the pipeline was pressure tested for twelve hours at 1875 psi (12.93 MPa).

This repair was made in October 1973, and the pipeline was still in service 20 years later.

DOCK REPAIR

A WHARF-RETAINING STRUCTURE comprised of nineteen cylindrical sheet pile cells and 600 ft (183 m) in length is the subject of this case. The working depth was from the waterline to the mudline, at a depth of more than 30 ft (9.14 m).

The problem was that fill material in the cells had spilled out through holes corroded in webs of the sheet piling, leaving the dock area unsupported. Underwater inspection disclosed eighty-three holes ranging up to 17 ft (5.18 m) long and 8 to 10 in. (203 to 254 mm) wide. The dock was closed to traffic, and the wharf was out of service. Heavy ship-channel traffic and zero visibility would challenge the underwater welders.

The solution was obtained through collaboration between the customer and the contractor. Since pile interlocks had a maximum design stress of 8000 lb per linear in. (143 kg per linear mm), the customer specified that the minimum tensile strength of the welds on the repair plates be not less than the design stress of the interlocks. To qualify the welding procedure and the welders, test welds were made in the ship channel. Transverse shear specimens from the simulated repairs, welded with 1/2 in. (12.7 mm) vertical fillet welds, resulted in shear strength ranging from 18 300 to 24 200 lb per linear in. (327 to 433 kg per linear mm). Austenitic welding electrodes were used to circumvent

hydrogen-induced underbead cracking in the heat-affected zone of the high-carbon equivalent (0.69) sheet pile.

Each of the 484 sheet piles that were exposed to the ship channel water was inspected from the waterline to the mudline. Areas around holes and corroded sections were cleaned to sound metal to determine the length and width of the damage that would be repaired. Repair plates were lowered into place to be fitted and welded over the holes and severely corroded sections in webs of the sheet piles. Over 700 ft (213 m) of 1/2 in. (12.7 mm) thick, 11 in. (279 mm) wide, repair plate was welded to the sheet piles with 1400 ft (426.7 m) of 1/2 in. (12.7 mm) vertical fillet welds and about 400 ft (121.9 m) of horizontal fillet welds. All welds were visually inspected with a specially designed *clear-water-flow* underwater video camera.

After completion of the underwater repairs, the port authorities replaced fill material (sand) in each of the nineteen cells, and high-pressure concrete grout was used to top-off the fill and relevel the concrete dock slab. Underwater repairs were completed in March 1972, and the dock was still in service 20 years later.

SUPPLEMENTARY READING LIST

American Welding Society. *Specification for underwater welding*, ANSI/AWS D3.6-93. Miami: American Welding Society.[78]

Christensen, N. "The metallurgy of underwater welding." *Proceedings of IIW Conference on Underwater Welding*, 71-79. New York: Pergamon Press, 1983.

Grong, O., Olson, D. L., and Christensen, N. "Carbon oxidation in hyperbaric MMA welding." *Metal Construction* 17 (12): 810R-14R, 1985.

Grubbs, C. E. and Seth, O.W. "Multipass all position 'wet' welding: a new underwater tool." Proceedings of Offshore Technology Conference, Paper No. OTC-1620, Houston, Texas, May 1-3, 1972.

Hausi, A., and Suga, Y. "On cooling of underwater welds." Transactions of the Japan Welding Society, 2(1), 1980.

Hoffmeister, K., and Kuster, K. "Process variables and properties of wet underwater gas metal arc laboratory and sea welds of medium strength steels." *Proceedings of IIW Conference on Underwater Welding*, 239-246. New York: Pergamon Press, 1983.

Ibarra, S., et al. "The structural repair of a North Sea platform using wet welding techniques." Proceedings of Offshore Technology Conference, Paper No. OTC-6652, Houston, Texas, May 1991.

Ibarra, S., et al. "Underwater wet welding techniques for repair of higher carbon-equivalent steels." Proceedings of Offshore Technology Conference, Paper No. OTC-6214, 1990.

Ibarra, S., Grubbs, C. E., and Olson, D. L. "Fundamental approaches to underwater welding metallurgy." *Journal of Metals* 40(12): 8-10, 1988.

————. "The nature of metallurgical reactions in underwater welding." Proceedings of Offshore Technology Conference, Paper No. OTC-5388, Houston, Texas, May 1987.

Ibarra, S., and Olson, D. L. "Wet underwater steel welding." *Ferrous Alloy Weldments*, Eds., Olson, D. L. and North, T. H., 329-78. Zurich, Switzerland: Trans Tech Publications, 1992.

Ibarra, S., Olson, D. L., and Grubbs, C. E. "Underwater wet welding of higher strength offshore steels." Paper No. OTC-5889, Houston, Texas, May 1989.

Inglis, M. R., and North T. H. "Underwater welding: a realistic assessment." *Welding and Metal Fabrication* 11(4): 165, 1979.

78. Fifty-two additional references concerning diving, construction standards, welding, and testing are listed on Pages 101-102 of this version, ANSI/AWS D3.6-93.

Joos, T. "Evolution of underwater cutting paces topside technology." *Welding Journal* 70(4): 63-5, 1991.

Matlock, K. D., Edwards, G. R., Olson, D. L., and Ibarra, S. "An evaluation of the fatigue behavior in surface, habitat, and underwater wet welds," *Proceedings of the Second International Conference on Offshore Welded Structures*, Ed. Cotton, H. C., P15-1 – P15-10. Cambridge, England: The Welding Institute, 1983. An abbreviated version was later published in *Underwater Welding*, 303-10. New York: Pergamon Press, 1983.

Nagarajan, V., and Loper, Jr., C. R. "Underwater welding of mild steel: a metallurgical investigation of critical factors." Proceedings of Offshore Technology Conference, Paper No. OTC-2668, Houston, Texas: May 1976.

Olsen, K., Olson, D. L., Christensen, N. "Weld bead tempering of the heat-affected zone." *Scandinavian Journal of Metallurgy* 11: 163-8, 1982.

Olson, D. L. "Metallurgical investigation of 650-foot underwater wet weld." Colorado School of Mines Report MT-CWR-088-001 (for Amoco Corporation Research), December 31, 1987.

O'Sullivan, J. E. "Wet underwater weld repair of Susquehanna Unit 1 steam dryer." *Welding Journal* 67(6): 19-23, 1988.

Perlman, M., Pense, A. W., and Stout, R. D. "Ambient pressure effect of gas metal arc." *Welding Journal* 50 (6): 231s-8s, 1971.

Suga, Y., and Hasui, A. "On formation of porosity in underwater weld metal (The 1st Report) – Effect of water pressure on formation of porosity." Transactions of the Japan Welding Society, 17(1): 58-64, April 1986. Also reprinted as IIW Doc. IX-1388-86, Cambridge, England: International Institute of Welding, 1986.

Trevisan, R. E., Schwemmer, D. D., and Olson, D. L. "The fundamentals of weld metal pore formation," *Welding: Theory and Practice*, Eds. Olson, D. L., Dixon, R. D., and Liby, A. L., 79-115. Amsterdam: Elsevier Science Publishers, B. V., 1990.

Tsai, C.L., and Masubuchi, K. "Mechanisms of rapid cooling in underwater welding." *Applied Ocean Research*, 1(2): 99-110, 1979.

Wood, B. J. and Bruce, W. A. "Underwater wet repair welding at depths approaching -600 ft," *Proceedings of AWS International Conference on Underwater Welding*, 5-13. Miami: American Welding Society, 1991.

Yemington, C. R. "Underwater NDE beyond diver depths." *Welding Journal* 69(8): 63-5, 1990.

WELDING HANDBOOK
INDEX OF MAJOR SUBJECTS

INDEX